Effective masses for electrons in units of m_0, the rest mass of the free electron

	$m^*_{(K=0)}$	$m^*_{\|\|}$	$m^*_{\perp}$	m^*_{ce}	m^*_{dse}
Si		0.92	0.197	0.26	1.09
GaAs	0.067			0.067	0.067
Ge		1.64	0.082	0.12	0.56
InP	0.077			0.077	0.077

Effective masses for holes in the valence bands of several semiconductors

	m^*_{lh}	m^*_{hh}	m^*_{sh}	Δ (eV)	m^*_{ch}	m^*_{dsh}
Si	0.16	0.48	0.24	0.044	0.36	1.150
GaAs	0.082	0.45	0.15	0.34	0.34	0.48
Ge	0.044	0.28	0.08	0.29	0.21	0.292
InP	0.08	0.4	0.15	0.11	0.3	0.42

Conversion of units

1 eV	1.6×10^{-19} J
1 cm	10^{-2} m 10^{8} Å 10^{4} μm
1 Å	10^{-10} m 10^{-8} cm 0.1 nm
0°C	273.18 K
1 tesla	1 Wb/m^2 10^4 gauss

Fundamentals of Semiconductor Devices

Betty Lise Anderson
The Ohio State University

Richard L. Anderson

Boston Burr Ridge, IL Dubuque, IA Madison, WI New York San Francisco St. Louis
Bangkok Bogotá Caracas Kuala Lumpur Lisbon London Madrid Mexico City
Milan Montreal New Delhi Santiago Seoul Singapore Sydney Taipei Toronto

The McGraw-Hill Companies

FUNDAMENTALS OF SEMICONDUCTOR DEVICES

Published by McGraw-Hill, a business unit of The McGraw-Hill Companies, Inc., 1221 Avenue of the Americas, New York, NY 10020.

Some ancillaries, including electronic and print components, may not be available to customers outside the United States.

This book is printed on acid-free paper.

1 2 3 4 5 6 7 8 9 0 DOC / DOC 0 9 8 7 6 5 4

ISBN 0–07–236977–9

Publisher: *Elizabeth A. Jones*
Senior sponsoring editor: *Carlise Paulson*
Marketing manager: *Dawn R. Bercier*
Senior project manager: *Sheila M. Frank*
Senior production supervisor: *Laura Fuller*
Lead media project manager: *Audrey A. Reiter*
Senior media technology producer: *Eric A. Weber*
Designer: *Rick D. Noel*
Cover design: *Scan Communications*
Cover image: *Courtesy IBM, IBM Microelectronics Photo Catalog, #567-CMOS 7S 6 Levels of Copper*
Senior photo research coordinator: *Lori Hancock*
Compositor: *Interactive Composition Corporation*
Typeface: *10.5/12 Times Roman*
Printer: *R. R. Donnelley Crawfordsville, IN*

Library of Congress Cataloging-in-Publication Data

Anderson, Richard L.
Fundamentals of semiconductor devices / Richard L. Anderson, Betty Lise Anderson.—1st ed.
p. cm.
Includes index.
ISBN 0–07–236977–9
1. Semiconductors. 2. Transistors. I. Anderson, Betty Lise. II. Title.

TK7871.85A495 2005
621.3815'2—dc22

2003067537
CIP

www.mhhe.com

BRIEF CONTENTS

PART 4
Bipolar Junction Transistors 551

PART 5
Optoelectronic Devices 673

CONTENTS

PREFACE

This is a textbook on the operating principles of semiconductor devices. It is appropriate for undergraduate (junior or senior) or beginning graduate students in electrical engineering as well as students of computer engineering, physics, and materials science. It is also useful as a reference for practicing engineers and scientists who are involved with modern semiconductor devices.

Prerequisites are courses in chemistry and physics and in basic electric circuits, which are normally taken in the freshman and sophomore years.

This text is appropriate for a two-semester course on semiconductor devices. However, it can be used for a one-semester course by eliminating some of the more advanced material and assigning some of the sections as "read only." The authors have attempted to organize the material so that some of the detailed derivation sections can be skipped without affecting the comprehension of other sections. Some of these sections are marked with an asterisk.

This book is divided into five parts:

1. Materials
2. Diodes
3. Field-Effect Transistors
4. Bipolar Transistors
5. Optoelectronic Devices

The first four parts are followed by "Supplements" that, while not required for an understanding of the basic principles of device operation, contain related material that may be assigned at the discretion of the instructor. For example, the use of SPICE for device and circuit analysis is briefly discussed for diodes, field-effect transistors, and bipolar transistors. While SPICE is normally taught in courses on electric circuits, it is useful to know the origin of the various parameters used to characterize devices. This material on SPICE is relegated to supplements, since not all schools cover SPICE in courses on electron circuit analysis and such courses may be taught before, concurrently with, or after the course on semiconductor devices.

Part 1, "Materials," contains four chapters and two supplements. The first two chapters contain considerable review material from the prerequisite courses. This material is included since it is used extensively in later chapters to explain the principles of device operations. Depending on the detailed content of the prerequisite courses, much of these chapters can be relegated to reading assignments.

The level of quantum mechanics to be covered in a course like this varies widely. In this book some basic concepts are included in the main chapters of Part 1; those wishing to cover quantum mechanics in more detail will find more extensive material in Supplement A to Part 1.

The basic operating principles of large and small devices of a particular type (e.g., diodes, field-effect transistors, bipolar junction transistors, photodetectors) are the same. However, the relative importance of many of the parameters involved in device operation depends on the device dimensions. In this book the general behavior of devices of large dimensions is treated first. We treat, in each case, "prototype" devices (such as step junctions and long-channel field-effect transistors), from which the fundamental physics can be learned, and then develop more realistic models considering "second-order effects." These second-order effects can have significant influence on the electrical characteristics of modern, small-geometry devices. The instructor can go into as much depth as desired or as time permits.

Topics treated that are typically omitted in undergraduate texts are:

- The differences between the electron and hole effective masses as used in density-of-state calculations and conductivity calculations.
- The differences in electron and hole mobilities (and thus diffusion coefficients) if they are majority carriers or minority carriers.
- The effects of doping gradients in the base of bipolar junction transistors (and/or the composition in heterojunction BJTs) on the current gain and switching speed.
- Band-gap reduction in degenerate semiconductors. While this has little effect on the electrical characteristics of diodes or field-effect transistors, its effect in the emitter of bipolar junction transistors reduces the current gain by an order of magnitude.
- The velocity saturation effects due to the longitudinal field in the channel of modern field-effect transistors with submicrometer channel lengths reduces the current by an order of magnitude compared with that calculated if this effect is neglected.

While the major emphasis is on silicon and silicon devices, the operation of compound semiconductor devices, alloyed devices (e.g., SiGe, AlGaAs) and heterojunction devices (junctions between semiconductors of different composition) are also considered because of the increased performance that is possible with such *band-gap engineering*.

Many of the seminal publications on semiconductor devices cited in the references at the end of each chapter through 1990 are reprinted in *Semiconductor Devices: Pioneering Papers,* edited by S. M. Sze, World Scientific Publishing Co., Singapore, 1991.

Fabrication, while an important part of semiconductor engineering, is often skipped in the interest of time. This material is introduced in Appendix C, and can be assigned as read-only material if desired.

ACKNOWLEDGEMENTS

We would like to thank, first and foremost, our spouses Bill and Claire for their love, support, patience, and help. We are also grateful to the anonymous manuscript reviewers for their comments and suggestions, as well as the staff at McGraw-Hill for all their help. We thank our students for valuable feedback on the manuscript. Finally, we would like to thank all the companies and individuals that provided photographs and data for this book.

Anderson & Anderson

PART 1

Materials

Semiconductors form the basis of most modern electronic systems (e.g., computers, communication networks, control systems). While there are applications for other materials in electronics (e.g., magnetic materials in hard drives), this book concentrates on electronic devices that are based on semiconductors.

Understanding the operation and design of semiconductor devices begins with an understanding of the materials involved. In Part 1 of this book, we investigate the behavior of electrons in materials, starting with the atoms themselves. Then we progress to electrons in crystalline semiconductors.

We will see that classical mechanics does not provide a complete picture of electron activity in solids. In principle one should instead use quantum mechanics to predict the electrons' behavior, but the application of quantum mechanics is not as simple as the more familiar classical or Newtonian mechanics. We will therefore introduce *pseudo-classical mechanics,* which modifies familiar classical equations to account for some quantum mechanical effects.

Some basic quantum mechanical concepts important for the understanding of device operation are covered in Chapter 1. (A more thorough discussion is contained in Supplement A to Part 1, found after Chapter 4.) In Chapter 2, we cover *pseudo-classical mechanics,* which allows us to predict the reaction of electrons to complicated fields, while using simple and intuitive pseudo-classical equations.

The use of pseudo-classical mechanics will also allow us to draw and use energy band diagrams. These diagrams are indispensable for understanding and predicting the motion of the electrons and holes, and thus the current in semiconductors.

In Chapter 3, we will see that conductivity of semiconductors is controlled by the number of charge carriers available to carry current. The charge carriers in semiconductors are electrons and holes. Their numbers are controlled by the concentrations of impurity elements that are intentionally added to the material. The carrier concentrations also depend on temperature, and if light is shining on the sample, the concentrations can also vary.

It will emerge that there are two major forms of current in semiconductors, drift current and diffusion current. Drift current is caused by the presence of an electric field, whereas diffusion current arises when the carrier concentrations vary with position.

Chapter 4 covers nonhomogeneous semiconductor materials, in which the doping or the material composition itself may vary. These variations can lead to internal electric fields that can enhance device performance. Most modern semiconductor devices have regions of such nonhomogeneous materials.

The Supplements to Part 1 contain additional topics relevant to semiconductor materials, including a more detailed discussion of quantum mechanics, the statistics of electrons in bound states, and phonons.

We will start with electrons in atoms. ■

CHAPTER 1

Electron Energy and States in Semiconductors

1.1 INTRODUCTION AND PREVIEW

We begin our study of semiconductors with some fundamental physics of how electrons behave in matter. The ability to control the movement of electrons in solids is the basis of semiconductor device engineering. In order to understand the electronic properties of these devices, it is necessary to understand the electronic properties of the materials from which they are made, and how those properties are affected by impurities (intentional and unintentional), temperature, applied voltages, device structures, and optical radiation.

Since solids are composed of atoms, we start by examining the electronic properties of atoms, and then extending those results to simple molecules and solids. In particular, the results for silicon (Si) and gallium arsenide (GaAs) are emphasized, currently the two most commonly used semiconductors in integrated circuits and semiconductor devices. Several other semiconductors and semiconductor alloys important in modern devices are also discussed.

As we investigate the atom, we'll be using quantum mechanics, a branch of science that is needed to accurately describe the behavior of very small objects such as atoms and electrons. We will see as we go along that quantum mechanics is based on the idea that energy can exist only in discrete packets, or quanta. The size of a quantum is so small that it doesn't affect one's results when one is computing the momentum or velocity of large objects such as automobiles or dust particles, but the quantum description is extremely important for electrons and atoms.

An understanding of quantum mechanics is not simple to obtain, and its use to calculate properties of more than a few systems in closed form is difficult. Fortunately, however, in semiconductors the behavior of electrons of interest can be determined by *pseudo-classical mechanics,* in which classical formulas such as Newton's laws and the Lorentz equation can be used, with the true electron mass replaced by an *effective mass*. As a result, in this section, a minimal discussion of quantum mechanics is presented. A somewhat greater discussion of quantum mechanics appropriate to some of the electronic processes in semiconductor devices is presented in Supplement A to Part 1, after Chapter 4.

The key to understanding semiconductors is to appreciate the physical interpretation of the mathematical results. Physical understanding is emphasized in this book.

1.2 A BRIEF HISTORY

In the early twentieth century, scientists were trying to develop models that would explain the results observed from such experiments as the scattering of X-rays, the photoelectric effect, and the emission and absorption spectra of atoms. In 1910, J. J. Thompson proposed a model of the atom in which a sphere of continuous positive charge is embedded with electrons, as shown in Figure 1.1a. Ernest Rutherford, in 1911, offered an improvement to the Thompson model: in the Rutherford model of the atomic structure, all of the positive charge and virtually all of the atom's mass were assumed to be concentrated in a small region in the center of the atom. This nucleus is often treated as a sphere with a radius on the order of 10^{-14} meters. The negatively charged electrons were assumed to orbit about the positively charged nucleus, much as planets orbit the sun or satellites orbit the earth.

In 1913, Neils Bohr assumed that the electrons in the Rutherford model of the atom orbited the nucleus in circles, as shown in Figure 1.1b. From this, he predicted that for the atom to be stable, the electrons could have only certain energies, or that the energies would be *quantized.* Energy and many other observables (properties that can be directly measured) are expressed in terms of Planck's constant. Planck's constant, h, has the value 6.63×10^{-34} joule–seconds. The energies Bohr predicted for electrons in atoms were in excellent agreement with the experimental results obtained from spectroscopic data.

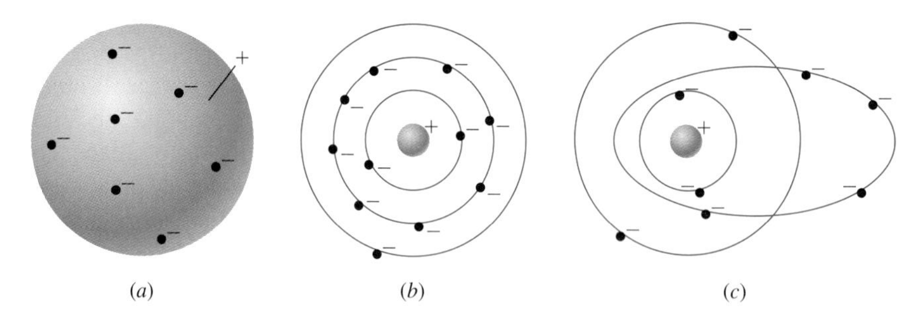

Figure 1.1 (a) The Thompson model of an atom, in which the positive charge is uniformly distributed in a sphere and the electrons are considered to be negative point charges embedded in it; (b) the Bohr model, in which the positive charge is concentrated in a small nucleus and the electrons orbit in circles; (c) the Wilson-Sommerfeld model, which allows for elliptical orbits.

In 1916, Wilson and Sommerfeld generalized the Bohr model to apply it to any physical system in which a particle's motion is periodic with time. This modification allows for the possibility of elliptical orbits, as shown in Figure 1.1c.

1.3 APPLICATION TO THE HYDROGEN ATOM

In this section, we briefly review the Bohr model of the hydrogen atom. The hydrogen atom is emphasized because *hydrogen-like impurities* are important in semiconductor devices, and these impurities can be treated in a manner analogous to the Bohr model. In Supplement A to Part 1, we will compare these results to those obtained using quantum mechanics as represented by Schroedinger's equation.

1.3.1 THE BOHR MODEL FOR THE HYDROGEN ATOM

We start with the Bohr model, in which the electrons revolve around the nucleus in circular paths. Because the mass of the nucleus is 1.67×10^{-27} kg, 1830 times that of the electron, the nucleus is considered to be fixed in space.

We consider as an example the neutral hydrogen atom, which has one orbiting electron, and we treat the electron and nucleus both as point charges. The Coulomb force between two particles with charges Q_1 and Q_2 is

$$F = \frac{Q_1 Q_2}{4\pi\varepsilon_0 r^2} = \frac{-q^2}{4\pi\varepsilon_0 r^2} \tag{1.1}$$

where r is the distance between the two charges and $\varepsilon_0 = 8.85 \times 10^{-12}$ farads/meter is the permittivity of free space (because there is only free space between the nucleus and the electron). The expression at the far right-hand side of Equation (1.1) is obtained by recognizing that the hydrogen nucleus has only one proton, so Q_1 is equal to $+q = 1.602 \times 10^{-19}$ coulombs, the elemental charge, and the charge of the electron Q_2 is equal to $-q$. The resulting negative sign in Equation (1.1) indicates that the force is attractive.

We now have an expression for the attractive (centripetal) force between the two particles, and we recall from classical mechanics that the force F on a particle is equal to minus the gradient of the potential energy, or

$$F = -\nabla E_P = -\frac{dE_P}{dr} \tag{1.2}$$

In the last expression, the gradient is taken in the r direction, and E_P is the potential energy of the electron at position r. Equation (1.2) with the aid of (1.1) can be rewritten as

$$dE_P = dE_P(r) = -F dr = \frac{q^2\, dr}{4\pi\varepsilon_0 r^2} \tag{1.3}$$

One can integrate both sides to obtain E_P, but there will be a constant of integration. The actual value of the potential energy is arbitrary (as is the choice of the constant), since the value of the potential energy depends entirely on one's choice of reference. We can choose a convenient reference by noting that the Coulomb force at infinite distance is zero. It makes sense for this case, then, to choose $r = \infty$ as a reference point, so we define the potential energy at $r = \infty$ as the *vacuum level,* E_{vac}:

$$E_P(r = \infty) = E_{\text{vac}} \tag{1.4}$$

This is the energy required to free the electron from the influence of the nucleus, essentially by moving the electron infinitely far away from it. If the electron is infinitely far from the nucleus, it cannot really be considered part of the atom—it is now a free electron in vacuum.

Now we can solve Equation (1.3) for a given value of r:

$$\int_{E_P}^{E_{\text{vac}}} dE_P = \int_r^{\infty} \frac{q^2\,dr}{4\pi\varepsilon_0 r^2} \tag{1.5}$$

where E_P is the electron potential energy at some distance r from the nucleus. Integrating both sides and rearranging, we obtain

$$E_P = E_{\text{vac}} - \frac{q^2}{4\pi\varepsilon_0 r} \tag{1.6}$$

Figure 1.2 shows a plot of the r dependence of E_P. From Equation (1.1), and since the force is equal to minus the gradient (slope) of the potential energy, we see that the force on the electron is directed toward the nucleus, or the Coulomb force is centripetal.

Since the electron is revolving in a circle of radius r around the nucleus, we know from Newtonian mechanics that its centrifugal force is equal to

$$F = \frac{mv^2}{r} \tag{1.7}$$

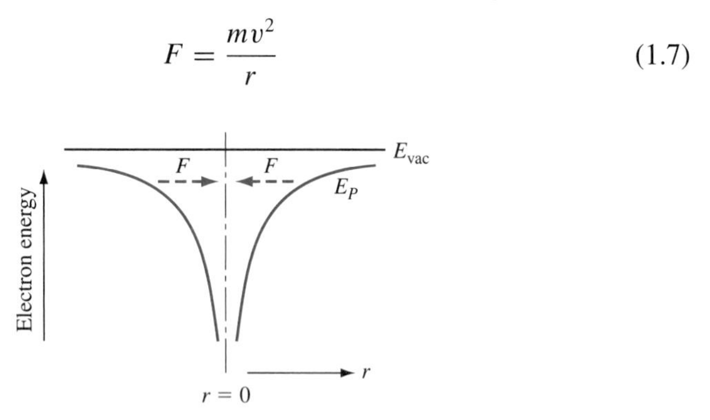

Figure 1.2 Potential energy diagram for an electron in the vicinity of a single positive point charge. The electron is considered to be a point charge.

For the atom to be stable, the net force on the electron must be zero. Equating our previous expression for the centripetal force due to the Coulomb attraction [Equation (1.1)] to the centrifugal force [Equation (1.7)], we can write

$$\frac{mv^2}{r} - \frac{q^2}{4\pi\varepsilon_0 r^2} = 0 \tag{1.8}$$

Bohr also postulated that the integral of the angular momentum around one complete orbit is an integer multiple of Planck's constant h:

$$\oint P_\theta \, d\theta = \int_0^{2\pi} mvr \, d\theta = nh \tag{1.9}$$

where n is an integer. Since the orbit is assumed circular in the Bohr model, r is a constant, and so are the potential energy E_P and the speed v. Therefore, the integral becomes

$$2\pi mrv = nh \tag{1.10}$$

There is a solution for each integer value of n, so we write

$$mv_n r_n = n\frac{h}{2\pi} = n\hbar \tag{1.11}$$

Here we have introduced a new symbol; it turns out that engineers and physicists (and now you) use the quantity $h/2\pi$ so much that there is a special character for it, $\hbar$, pronounced "h-bar." The subscripts n in Equation (1.11) indicate the particular orbital radius or speed associated with a specific quantum number n.

If we simultaneously solve Equations (1.8) and (1.11), we can derive an expression for the *Bohr radius of the nth state,* where by "state" we mean the properties associated with a particular value of n:

$$r_n = \frac{4\pi\varepsilon_0 n^2\hbar^2}{mq^2} \tag{1.12}$$

and the speed of the electron in that particular state is

$$v_n = \frac{q^2}{4\pi\varepsilon_0 n\hbar} \tag{1.13}$$

Our primary goal, however, is to find the energies associated with these states. We know that the total energy of a system is equal to the kinetic energy plus the potential energy. The kinetic energy of the nth energy level is

$$E_{K_n} = \frac{1}{2}mv_n^2 = \frac{mq^4}{2(4\pi\varepsilon_0)^2 n^2\hbar^2} \tag{1.14}$$

For the nth energy level, we can find r_n from Equation (1.12) and use that in Equation (1.6) to write for the potential energy

$$E_{Pn} = E_{\text{vac}} - \frac{mq^4}{2(4\pi\varepsilon_0)^2 n^2 \hbar^2} E_{Pn} = E_{\text{vac}} - \frac{mq^4}{(4\pi\varepsilon_0)^2 n^2 \hbar^2} \tag{1.15}$$

Thus, the total energy E_n is

$$E_n = E_{Kn} + E_{Pn} = E_{\text{vac}} - \frac{mq^4}{2(4\pi\varepsilon_0)^2 n^2 \hbar^2} \tag{1.16}$$

We say that the energy is *quantized*. It can have only discrete values associated with the quantum number n.

EXAMPLE 1.1

Find the energies and radii for the first four orbits in the hydrogen atom.

■ Solution

$$\begin{aligned} E_n &= E_{\text{vac}} - \frac{mq^4}{2(4\pi\varepsilon_0)^2 n^2 \hbar^2} \\ &= E_{\text{vac}} - \frac{(9.11 \times 10^{-31}\,\text{kg})(1.60 \times 10^{-19}\,\text{C})^4}{(2)(4)^2(3.1416)^2(9.85 \times 10^{-12}\,\text{F/m})^2(1.05 \times 10^{-34}\,\text{J}\cdot\text{s})^2}\left(\frac{1}{n}\right)^2 \\ E_n &= E_{\text{vac}} - \left(\frac{1}{n}\right)^2 (2.18 \times 10^{-17}\,\text{J}) \\ &= E_{\text{vac}} - \left(\frac{1}{n}\right)^2 (13.6\,\text{eV}) \end{aligned} \tag{1.17}$$

Here a new unit of energy is introduced, the *electron volt* (eV). The electron volt is defined as the amount of energy acquired by an electron when it is accelerated through 1 volt of electric potential. To convert between S.I. (International System) units (joules) and electron volts, use

$$\boxed{1\,\text{eV} = 1.60 \times 10^{-19}\,\text{joules}}$$

Electron volts are *not* S.I. units, and therefore must be used with care in calculations.

The Bohr radii can be calculated from Equation (1.12):

$$r_n = \frac{4\pi\varepsilon_0 n^2 \hbar^2}{mq^2} = \frac{(4)(3.1416)(8.85 \times 10^{-12}\,\text{F/m})(1.05 \times 10^{-34}\,\text{J}\cdot\text{s})^2}{(9.11 \times 10^{-31}\,\text{kg})(1.60 \times 10^{-19}\,\text{C})^2} \times n^2 \tag{1.18}$$

$$r_n = 0.053 n^2\,\text{nm}$$

The energies and Bohr radii of the first four energy levels are given in Table 1.1. These energies and radii are plotted in Figures 1.3 and 1.4, respectively.

Table 1.1 The first four Bohr energies and orbital radii for the hydrogen atom

E_n	r_n
$E_1 = E_{vac} - 13.6$ eV	$r_1 = 0.053$ nm
$E_2 = E_{vac} - 3.4$ eV	$r_2 = 0.212$ nm
$E_3 = E_{vac} - 1.51$ eV	$r_3 = 0.477$ nm
$E_4 = E_{vac} - 0.85$ eV	$r_4 = 0.848$ nm

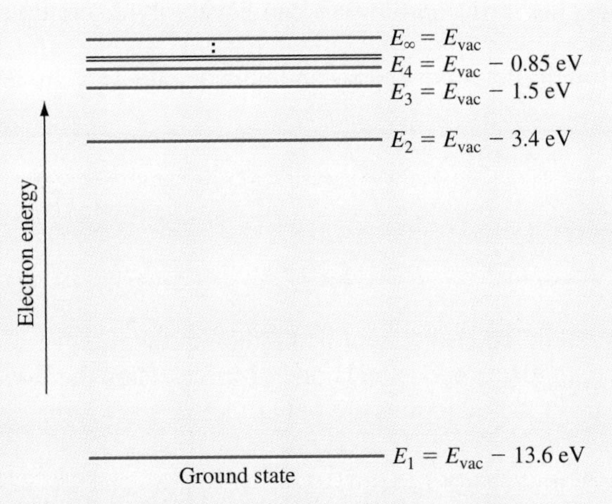

Figure 1.3 Allowed energies in the hydrogen atom. Higher energies occur increasingly close to each other, approaching the vacuum level.

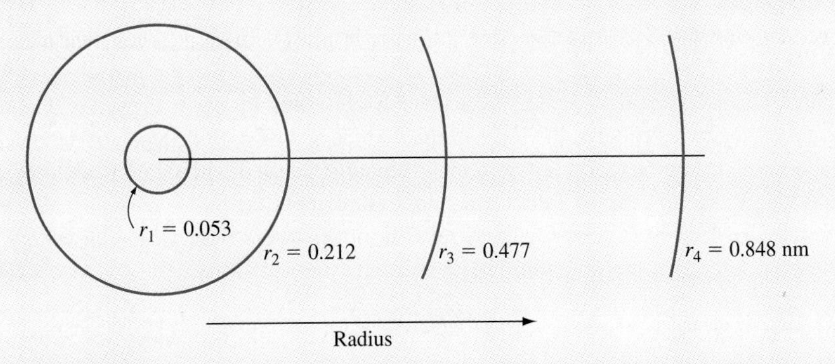

Figure 1.4 Radii of the first four atomic orbits of the hydrogen atom, according to the Bohr model.

There are several things to notice in this example. First, the differences between the vacuum energy level and the allowed energies vary as $1/n^2$. Thus the higher the quantum number, and therefore the energy, the closer together (in energy) the energy levels are. Second, the Bohr radius varies as n^2. This means that the higher the energy level, the farther the electron is from the atomic nucleus. If the electron has energy greater than E_{vac}, the Coulombic force is not enough to keep the electron bound to the atom. In this case, a hydrogen ion (H^+) is created as the electron leaves the atom.

Also, notice that we do not give a number for the energy of a particular state, but rather we express the energies as so many electron volts from some reference level (in this case E_{vac}). It is pointless to say, "This level is at 10 eV," since 10 eV could be anywhere, depending on your choice of reference. This point cannot be emphasized enough. Potential energy and thus total energy are arbitrary.

> The energy of a state must always be expressed as an energy difference—the difference between the energy of the state and some known reference, for example, $E_{vac} - E$.

Finally, it should be pointed out that although the number of possible states is infinite, once an electron occupies one of these states the hydrogen atom becomes neutral and other electrons are not attracted.

EXAMPLE 1.2

Consider a particle in a one-dimensional universe, oscillating in the parabolic potential energy shown in Figure 1.5. This represents an approximation to an electron in a modern quantum well laser.[1] The potential energy function is a parabola, and the particle is a simple harmonic oscillator. Explain the motion of the particle using the energy diagram, paying attention to where the kinetic energy is largest, smallest, and the directions of the forces.

■ Solution

Conservation of energy dictates that the particle must remain at a constant energy. Thus, it oscillates back and forth at this particular energy. When the particle is in the center, it has the smallest potential energy and the largest kinetic energy, and thus the largest velocity. Recall that the force is equal to the negative of the slope of the potential energy [Equation (1.2)]. Therefore, as the particle travels through the center, e.g., to the right, the force is to the left (the slope is positive to right of center). Thus, the particle decelerates. It continues to slow as it moves to the right. The total energy is constant, but since the potential energy is increasing, the kinetic energy decreases. When the particle gets to the edge, it will have zero kinetic energy. The particle stops for an instant, but the force is still accelerating it to the left. It picks up speed until it passes through the center, at which point the force is to the right, and the particle begins to slow once again.

[1]It is called a *quantum well* because the potential energy forms a "well" with quantized energy states.

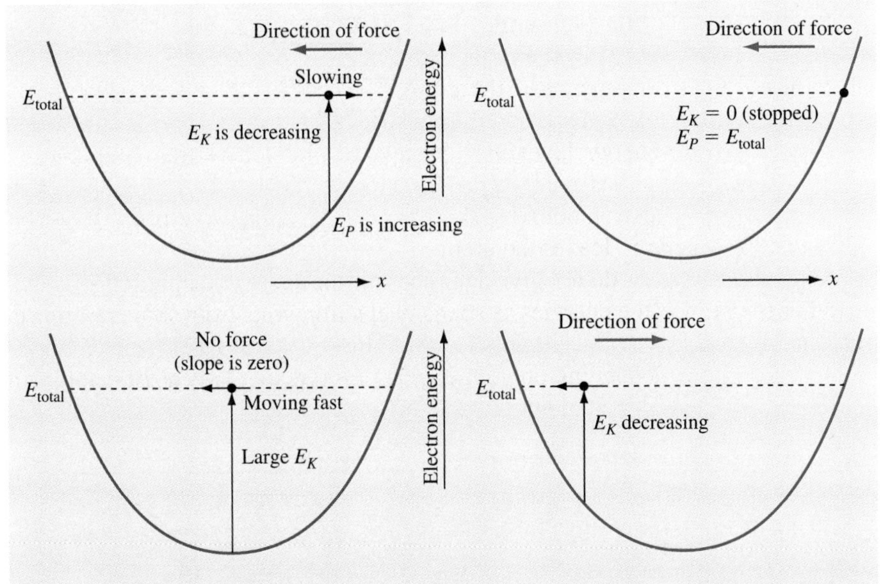

Figure 1.5 Motion of an electron in a parabolic potential. The electron oscillates back and forth at a constant total energy.

1.3.2 APPLICATION TO MOLECULES: COVALENT BONDING

We now extend the Bohr model of the hydrogen atom to the hydrogen ion, H_2^+, and the hydrogen molecule, H_2. Figure 1.6a indicates the electron energy diagrams for two isolated hydrogen nuclei. By *isolated,* we mean that the nuclei are sufficiently far apart that they do not influence each other. The energy levels for an electron are the same for each nucleus, those calculated in the previous section. They are quantized as we saw.

When the nuclei are allowed to approach each other, an electron would be influenced by both nuclei according to Coulomb's law. The electron potential energy of a single electron (H_2^+ ion) at any point is now

$$E_P = E_{\text{vac}} - \frac{q^2}{4\pi\varepsilon_0 r_1} - \frac{q^2}{4\pi\varepsilon_0 r_2} \tag{1.19}$$

where r_1 and r_2 are the distances between the electron and each of the two nuclei, respectively. Figure 1.6b shows the allowed electron energy states, which are still quantized. Notice that an electron in the ground state (lowest energy state) would be bound to one of the nuclei, but an electron in an excited state could travel back and forth between the nuclei, in effect shared by the two atoms. Since electrons tend to seek their lowest allowed energy, this condition of the electron being in one of the upper levels would not last long—the electron would quickly revert to the ground state.

Figure 1.6c shows the energy band diagram for the case where the separation is small enough that the potential energy maximum between the nuclei is below

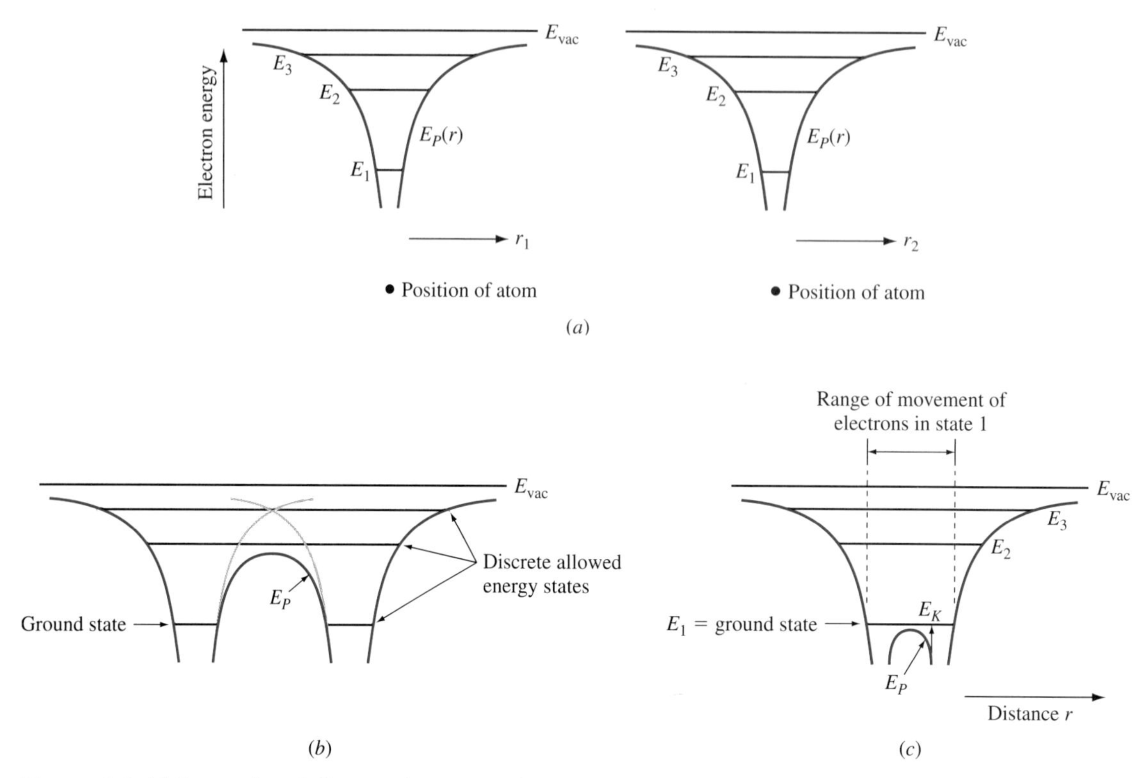

Figure 1.6 (a) Energy band diagram for two noninteracting hydrogen nuclei; (b) as the nuclei are brought together, the upper energy levels merge and electrons in those levels are shared between the atoms; (c) the nuclei are sufficiently close together that all energy levels are shared. Since the lowest level is usually the only occupied level for hydrogen, if it is occupied by two electrons the H_2 molecule is stable.

the ground state energy (E_1). In this situation, an electron in the ground state would be shared by the two nuclei, oscillating between the two positions at which $E = E_P$.

An interesting thing happens in this case, however. Since *each* nucleus has a ground state associated with it, it turns out that two electrons can occupy these ground states for a neutral H_2 molecule. Although the presence of the second electron will alter the electrostatic forces and therefore the energy band diagram slightly, the result is that both electrons will oscillate in the region indicated.

In the region between the nuclei, the kinetic energy ($E_K = E_1 - E_P$), and thus the velocity, is small. The electrons travel more slowly in this region, or on the average, the electrons spend most of their time between the two nuclei. The electrons therefore create a negatively charged "electron cloud" in this region that tends to attract the two nuclei together. If the internuclear spacing is too small, however, the potential energy E_P decreases, which increases the kinetic energy E_K since total energy E is conserved. As the kinetic energy and therefore the electron speed increases, the electron cloud effect is reduced, lessening the attractive force. At a particular spacing, the electron-cloud–induced nuclear bonding is stable, and a stable H_2 molecule results. This mechanism is referred to as *covalent bonding.*

1.3.3 QUANTUM NUMBERS AND THE PAULI EXCLUSION PRINCIPLE

A basic premise of quantum mechanics is that the energies in an atom are quantized, or exist only at certain discrete values. We earlier introduced the quantum number n, called the *principal quantum number.* It describes the energy of an electron in an allowed state. As indicated in Supplement A to Part 1, from quantum mechanical considerations (e.g., Schroedinger's equation) there are three other quantum numbers, including the azimuthal quantum number n_θ, which describes the ellipticity of an orbit. The last two describe the tilt of the orbits, and *spin*[2] of the electron. The energy is determined primarily by the principal quantum number. The other three quantum numbers can affect the energy, but only slightly.

The physical meanings of these quantum numbers are not essential to the understanding of transistors, but the *Pauli exclusion principle* is essential, because it governs which states may be occupied:

PAULI EXCLUSION PRINCIPLE

No two electrons in an interacting system can have the exact same set of quantum numbers.

For example, in the lowest energy orbit of an atom, $n = 1$. We know from freshman chemistry that this state can hold two electrons; those two electrons must have different spin quantum numbers, either $+\frac{1}{2}$ or $-\frac{1}{2}$.

In the $n = 2$ state, there are two possible orbital shapes. One orbit is spherically symmetric and holds two electrons of opposite spin (the "*s*" state). There are three elliptical orbits with the same shape but different orientations. Each of these can hold two electrons of opposite spin, bringing the maximum number of electrons in the second "shell" to eight. The periodic table is built on these quantum numbers.

In the following section, we will see how the Pauli exclusion principle affects the electron energies of solid materials such as silicon crystals.

[2]Spin is a purely quantum-mechanical parameter and has no analog in classical mechanics. It can be crudely considered to result from considering the electron to be a spinning sphere. The charge in the spinning electron then produces a magnetic field. The magnitude of this spin magnetic field is quantized and it can have two possible orientations with respect to an applied magnetic field.

1.3.4 COVALENT BONDING IN CRYSTALLINE SOLIDS

In this section we investigate the mechanisms by which silicon atoms bond to each other to form crystals. Silicon, by far the most important semiconductor material, has atomic number 14. It has two electrons in the first shell (which is full), eight in the second (also full), and four electrons in the half-full third shell. The third shell can hold up to eight electrons. There exist all of the higher shells, too, as for any atom, but they are empty except when the atom is in an excited state.

Energy Bands Silicon crystallizes into the so-called diamond structure in which each atom has four "nearest neighbors." The arrangement of the atoms is discussed briefly in Section 1.8. For now, we merely note that within the crystal, each Si atom shares one of its four outer electrons with each of its (four) nearest neighbors. Neighboring atoms are bound together by two electrons as shown in the bonding description of Figure 1.7a, where the crystal is shown in two dimensions for clarity. In this bonding model, the crystal is held together by the electrostatic forces between the bonding electrons and the positive ion cores (covalent bonding).

Since the interatomic spacing is small (on the order of a quarter of a nanometer), an electron is influenced by all nearby nuclei as well as nearby electrons. Just as for the hydrogen molecule, the potential energy of an electron is the sum of the potential energies due to each neighboring atom, as shown in Figure 1.7b. Here we are plotting the potential energy E_P as a function of position, in this case in the x direction. Note that except near the surface, the potential energy E_P is a periodic function with the periodicity of the crystal lattice.

An important feature of Figure 1.7b is that the discrete states associated with the isolated atoms (Figure 1.6a) are now broadened into energy bands in the crystal. For a crystal with N atoms, each state originating from a single atom splits into a band of N discrete states. In a crystal containing billions of atoms, these states are infinitesimally close to each other in energy, and each band is often considered to be continuous in energy. We think of each shell of the discrete Si atoms as combining into a band of allowed states in a silicon crystal.

In Figure 1.7b, the valence and conduction bands, shown in the figure, are common to all the atoms in the crystal, while the lower energy states, corresponding to the inner shells, are highly localized near the atomic nuclei. As would be expected, these lower states are only slightly influenced by the presence of neighboring atoms and thus are spread only slightly in energy compared with the higher energy states. Electrons in these two lower energy shells are tightly bound to their respective atoms and play no role in the operation of devices. We will therefore not discuss them further.

The third shell of atomic silicon splits into two bands in crystalline Si. The lower of these bands, the *valence band,* contains four electrons. These valence electrons are responsible for covalent bonding in crystalline Si.[3] The four vacant states in the third shell of atoms in Si form a band in crystalline Si called the

[3]Recall that for hydrogen, it was the electrons in the first shell that took part in covalent bonding.

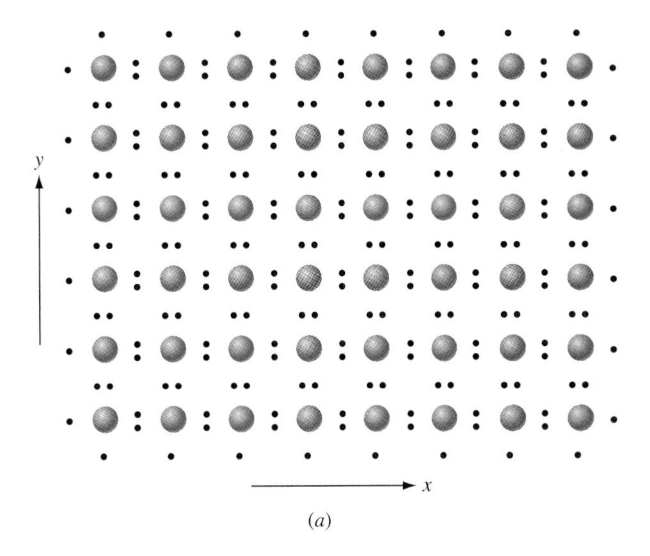

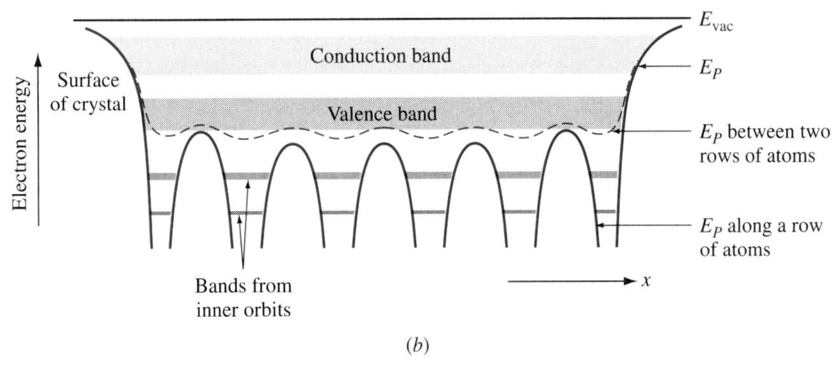

Figure 1.7 (a) Two-dimensional bonding representation of a crystalline solid; (b) potential energy for an electron in that crystal along a row of atoms (solid line) and between rows (dashed line). In this representation the electron is considered a point charge.

conduction band. This conduction band is split off from the valence band by an energy of 1.12 eV at room temperature.

Note that in the regions of space between the atoms, the kinetic energies and thus the velocities of the electrons in the valence band are relatively small. The electrons, since they are moving slowly, spend much of their time in these regions between the atoms. Thus, an electron cloud exists between atoms. The attraction between these negative clouds and adjacent positive nuclei (covalent bonding) is what holds the crystal together.

Ionic Bonding We have seen that silicon bonds covalently. Other elements in column IV of the periodic table can be expected to do the same. Apart from silicon, however, there are many other technologically interesting semiconductor materials, called *compound semiconductors.* Instead of every atom being of the same element, the crystal may consist of regular arrangements of different elements. One example is gallium arsenide (GaAs). This has the same general crystal structure as silicon, except that alternate atoms are gallium and the ones between them are arsenic. Gallium is in column III of the periodic table, and thus has three electrons in its outer shell. Arsenic, from column V, has five. Thus, when these two atoms are neighbors, they can each "fill" their outer shells with eight electrons by sharing electrons.

The electrons still spend most of their time in the region between the nuclei, but now there is a slight difference. Since the arsenic nucleus has greater positive charge, the electrons tend to be attracted toward the As side of the bond. This has, in essence, the effect of charging the As atom slightly negatively and ionizing the Ga atom slightly positively. The resulting Coulombic force helps hold the crystal together. Indium phosphide (InP) is another example of a III-V semiconductor crystal. The III-V's have a bonding character that is mostly covalent but partially ionic.

Cadmium telluride is an example of a II-VI semiconductor, and its bonding is even more ionic and less covalent. Table salt (NaCl) is an example of a I-VII crystal, although it is not a semiconductor. NaCl is considered to be primarily ionically bonded.

Electron Affinity, Ionization Potential, and Band Gap With the aid of Figure 1.8, we define some quantities useful in semiconductor electronics. The lowest energy in the conduction band is designated as E_C, the conduction band edge, while the highest energy in the valence band is denoted E_V, the valence band edge. We have not drawn the lower bands since they are not interesting to semiconductor device physics.

The *ionization energy* γ is defined as the minimum energy required to excite an electron from the top of the valence band in the crystal to the vacuum level. Most of the electrons will be in the valence band, because those states are at lower energies than those in the conduction band. The electrons most likely to be

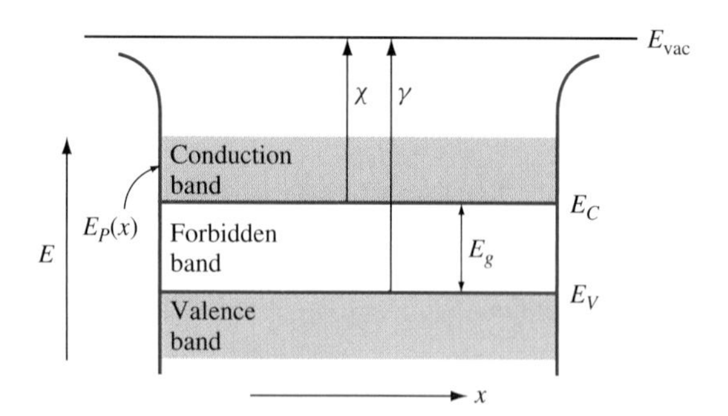

Figure 1.8 Definitions of vacuum energy E_{vac}, electron affinity χ, ionization energy γ and the energy gap E_g.

ionized are those requiring the least amount of extra energy to leave the atom, and these are at the top of the valence band:

$$\gamma = E_{\text{vac}} - E_V \tag{1.20}$$

The *electron affinity* χ is defined as the energy difference between the vacuum level and the vacant state of lowest energy, in this case E_C:

$$\chi = E_{\text{vac}} - E_C \tag{1.21}$$

The energy gap E_g is the minimum energy required to excite an electron from the valence band to the conduction band

$$E_g = E_C - E_V \tag{1.22}$$

As can be seen from Figure 1.8,

$$\gamma = E_g + \chi \tag{1.23}$$

The electron affinity and forbidden band gap (normally referred to simply as *band gap*) are fundamental properties of a semiconductor. Values for some semiconductors of interest are shown in Table 1.2. These quantities will be used extensively in the description of device operation. Also included in Table 1.2 is an insulator, amorphous SiO_2, or silicon dioxide. It is included because its properties are important in some semiconductor devices.

It should be pointed out that while the band gap of a semiconductor can be measured accurately, this is not the case for the electron affinity. The band gap can be determined by the energy required to excite electrons from the valence band to the conduction band. This can be done by measuring the photoconductivity as a function of photon energy, as will be discussed in Chapter 3. The photons penetrate into the bulk material where the band properties are well defined. To measure electron affinity, however, excited electrons must pass through the semiconductor surface to a collector in a vacuum. The band properties at the surface can differ from those in the bulk because of surface contamination or mechanical strain. As might be expected, there is some scatter in the published values for electron affinity. Those listed in Table 1.2 are considered reasonably reliable.

Table 1.2 The electron affinity and band gap for some common materials at 300 K

Semiconductor	Electron affinity χ (eV)	Band gap E_g (eV)
Si	4.05	1.12
GaAs	4.07	1.43
Ge	4.0	0.67
GaP	4.30	2.25
AlSb	3.6	1.6
InP	4.4	1.35
SiC	4.0	2.2
GaSb	4.06	0.7
SiO_2 (amorphous)	~0.9	~9

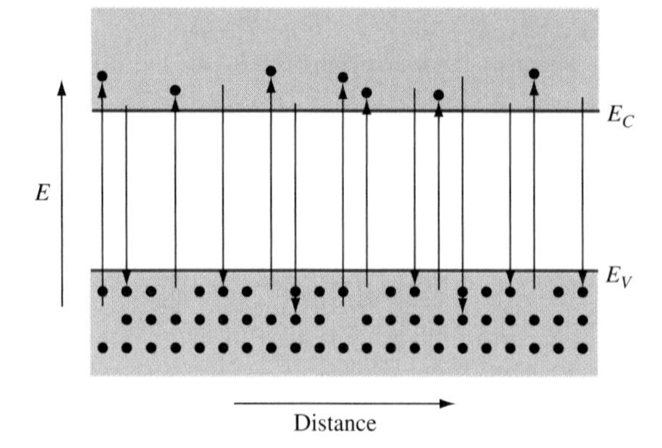

Figure 1.9 At room temperature, electrons are being excited up to the conduction band and relaxing back to the valence band. At any given moment there is some number of electrons in the conduction band.

Let us now discuss the occupancy of these energy bands. All things tend to seek their lowest energies, so in general one would expect to find most of the electrons in the valence band or lower at any given time. In fact, at absolute zero, every electron is in the lowest possible state, implying that all the lowest allowed bands are full, up to and including the entire valence band.

> At absolute zero (0 kelvins), every electron is in the lowest possible energy state. In a perfect semiconductor, every state in the valence band is occupied by an electron. Every state in the conduction band is empty.

Assuming the valence band to be filled, and no states in the forbidden band, the minimum energy that can be absorbed by a valence electron is E_g, enough to excite an electron from E_V to E_C. If the energy available is not enough to span the gap, the electron cannot absorb the energy—its final state would be a forbidden one. Thus, the electron will remain in the valence band. Note that current cannot flow at absolute zero—the electrons cannot move because all the states are filled; there is nowhere to go.

At room temperature, because of thermal agitation, a few[4] electrons are excited into the conduction band, as seen in Figure 1.9. Each one eventually falls back down to a vacant state in the valence band, re-emitting the excess energy as heat or light. The average time an electron spends in the conduction band is called the "electron lifetime" or just "lifetime" and is on the order of 10^{-10} to 10^{-3} seconds, depending on the material.

[4]"Few" is relative—in silicon, the number is on the order of 10 billion electrons per cubic centimeter in the conduction band at room temperature, about 1 out of every 10^{15} electrons in the crystal.

Electrons in the conduction band are free to move around within the crystal. They travel at constant energy (between collisions), but now there are many empty states at the same energy into which an electron can move. This band is called the *conduction band* because the moving electrons carry current.

Notice that in Figure 1.9, we show only the top of the valence band and the bottom of the conduction band on the energy band diagram. Because everything tends to seek its lowest energy, and because the valence band is essentially full, we would very rarely expect to see any empty states in the valence band except near the very top. If an electron deep inside the valence band were excited into the conduction band, a valence electron of higher energy would immediately (within about 10^{-12} seconds) fall into the resultant vacant state. Similarly, if an electron were excited to an energy high up in the conduction band, it would very quickly find a lower energy state. Therefore, all of the interesting activity is occurring near the top of the valence band and near the bottom of the conduction band.

At nonzero temperatures, there are a few empty states in the valence band. We call these empty states *holes,* and interestingly, they can move around too. The state doesn't actually move, but if an electron adjacent to an empty state moves into that state, it leaves an empty state behind. From Figure 1.10 one can see that if an electron moves to a vacant state to the left, that has the same net effect as one hole moving one step to the right.

If, however, one electron represents a unit of current, that one hole could also be thought of as carrying current. If a negatively charged electron moves to the left, the current it "carries" goes to the right. If we think of the same current as resulting from the hole moving to the right, the hole must be positively charged. As we will see, it turns out to be very useful to think of holes as positively charged current carriers in semiconductors. The concept of holes is developed more completely in Chapter 3.

Polycrystalline and Amorphous Materials Polycrystalline materials have small regions (grains) of single-crystal material with different crystalline orientations. These grains have dimensions on the order of a few nanometers to

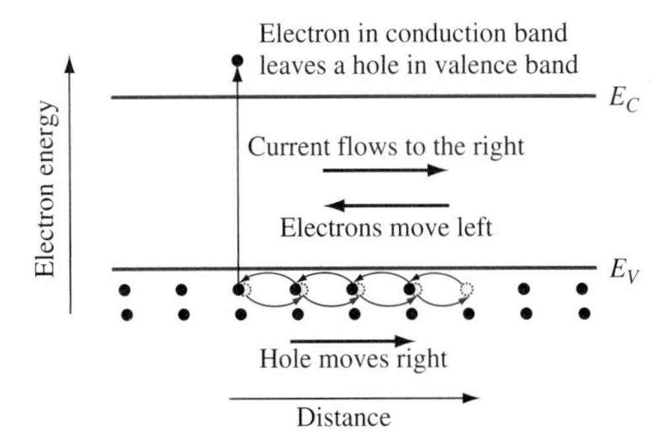

Figure 1.10 Movement of many electrons is treated as the movement of one positively charged "hole."

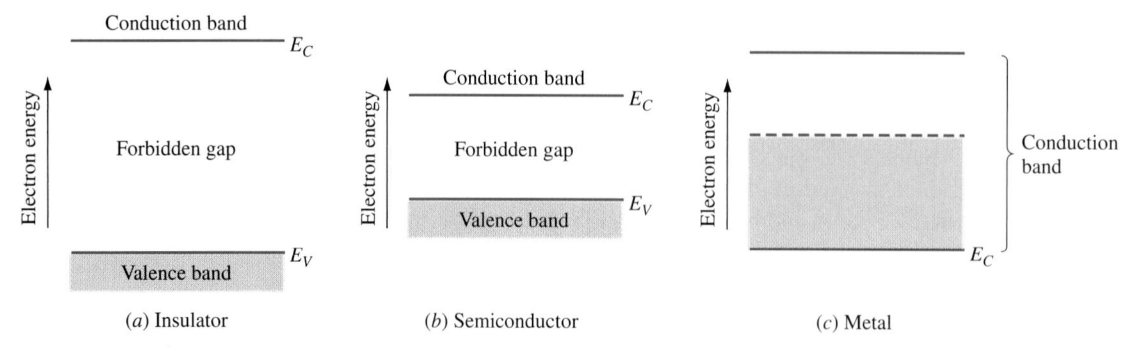

Figure 1.11 Energy band diagrams for (a) an insulator, (b) a semiconductor, and (c) a metal. The energies in the shaded regions are in general occupied.

a few millimeters. Because of the different crystalline orientations of the grains, the crystal periodicity at the grain boundaries is interrupted. This in turn affects the band structure near the grain boundaries.

Amorphous materials have some short-range order, but no long-range periodicity. Consequently, the band structure departs significantly from that of a crystalline material.

Insulators, Semiconductors, and Metals Silicon is called a semiconductor because it conducts better than an insulator but not as well as a metal. The reasons for this can be found, in part, from the energy band diagrams for these various materials, as seen in Figure 1.11. An insulator has a very wide forbidden gap,[5] and thus few valence electrons have sufficient energy to become excited to the conduction band. Thus, it is unlikely even at room temperature for electrons to get excited into the conduction band, and insulators conduct poorly.

Compare this situation with a metal. In a metal, the highest occupied band, referred to as the conduction band, is only partially filled. Therefore, all of the electrons in this band can contribute to current, even at very low temperatures.

A semiconductor has an intermediate-size band gap, Figure 1.11b. At a finite temperature it has more electrons in the conduction band and holes in the valence band than an insulator, so it conducts better. On the other hand, it has nowhere near the conductivity of a metal. We will examine other important ramifications of semiconductors' energy band structure in future chapters.

1.4 WAVE-PARTICLE DUALITY

Classically, energy is considered to be transported either by waves (e.g., sound waves, water waves, electromagnetic waves) or by particles (e.g., electrons, bullets). However, particles such as electrons also exhibit wavelike behavior.

[5]The case is shown for a crystalline insulator. Many insulators (e.g., SiO_2) are amorphous and the value of the forbidden gap is not well defined since a small concentration of states exist throughout the "forbidden band."

For electrons, some of their behaviors are more easily explained by considering them to be waves while other behaviors lend themselves to the particle description. In practice, the particle description is more convenient for solving some kinds of problems while the wave explanation is better for other problems. Both descriptions will be used to explain the operation of semiconductor devices such as transistors and lasers.

Similarly, what are classically considered to be waves can sometimes be considered to be particles. There are a number of experiments that demonstrate the wave-particle duality.[6] Here we simply state the results important in the study of electron devices.

1. Classical waves have energies that are quantized. Each quantum of energy can be considered a particle. For electromagnetic radiation (e.g., light), these particles are called *photons*. For acoustic waves (e.g., sound), the particles are called *phonons*. Each such particle has energy

$$E = h\nu = \frac{h}{2\pi} \cdot 2\pi\nu = \hbar\omega \tag{1.24}$$

where h is Planck's constant and ν (Greek letter nu) is the frequency of the classical wave. In this book we distinguish between photons and phonons by the expressions

$$E = h\nu = \hbar\omega_{\text{pht}} \qquad \text{(photons)}$$

$$E = \hbar\omega_{\text{phn}} \qquad \text{(phonons)}$$

where the subscripts pht and phn refer respectively to photons and phonons.

2. Classical particles can be considered to be waves possessing energy and wavelength. The electric field of an electromagnetic wave traveling in the x direction can be expressed by a simple sinusoidal function of amplitude A:

$$\mathscr{E}(x,t) = A\cos\left[2\pi\left(\frac{x}{\lambda} - \nu t\right)\right] = A\cos(Kx - \omega t)$$

where λ is the wavelength, t is time, and K is called the *wave vector*. Since the photon energy is $E_{\text{pht}} = h\nu = \hbar\omega_{\text{pht}}$, and $\hbar = h/2\pi$,

$$\mathscr{E}(x,t) = A\cos\left(Kx - \frac{E}{\hbar}t\right) \tag{1.25}$$

The wave vector K is also called the propagation constant. (In one-dimensional problems, it is often called the *wave number*.) It is related to the wavelength by

$$\boxed{K = \frac{2\pi}{\lambda}} \tag{1.26}$$

---

[6]The interested reader is referred to texts on atomic physics [1].

Matter can also be described using waves. An electron is an example of a particle that has mass, but can be described by a wave as well. A matter wave is expressed by a wave function, $\Psi(x, t)$

$$\Psi(x, t) = A \sin\left(Kx - \frac{E}{\hbar}t\right) \tag{1.27}$$

where the wavelength of the matter wave is

$$\lambda = \frac{2\pi}{K} \tag{1.28}$$

Matter waves are normally written as

$$\Psi(x, t) = Ae^{j[Kx-(E/\hbar)t]} \tag{1.29}$$

and Equation (1.27) is the imaginary part of Equation (1.29).

These expressions are for waves in a vacuum where the amplitude A is constant with position. They will be modified slightly within a material, e.g., for electrons in a semiconductor.

1.5 THE WAVE FUNCTION

Equation (1.29) is an example of a one-dimensional wave function of a matter wave in vacuum. In general, the wave function is a function of three spatial dimensions and time; i.e.,

$$\Psi = \Psi(x, y, z, t) \tag{1.30}$$

For simplicity, unless the three-dimensional formulation is important for the concepts being introduced, the one-dimensional formulation will be used in this book.

1.5.1 PROBABILITY AND THE WAVE FUNCTION

A basic connection between the properties of the wave function $\Psi(x, t)$ and the behavior of the associated particle is the *probability density* $P(x, t)$. The probability density is given by

$$\boxed{P(x, t) = \Psi^*(x, t)\Psi(x, t)} \tag{1.31}$$

where $\Psi^*(x, t)$ is the complex conjugate of $\Psi(x, t)$. The probability that the particle will be found in region dx near the coordinate x at time t is

$$P(x, t)\,dx = \Psi^*(x, t)\Psi(x, t)\,dx \tag{1.32}$$

Since the particle must be somewhere, the probability of finding it in space is unity, or integrating over all space,

$$\int P(x, t)\,dx = \int \Psi^*(x, t)\Psi(x, t)\,dx = 1 \tag{1.33}$$

in which case the wave function is said to be *normalized*.

1.6 THE ELECTRON WAVE FUNCTION

As will be discussed in Supplement A to Part 1, the wave function $\Psi(x, t)$ for an electron can be written as the product of a space-dependent part and a time-dependent part. In Supplement A to Part 1 it is demonstrated that for the potentials that do not vary with time, the time-dependent wavefunction is $e^{-j(E/\hbar)t}$, so that

$$\boxed{\Psi(x, t) = \psi(x)e^{-j(E/\hbar)t}} \tag{1.34}$$

where $\Psi(x, t)$ is the time-dependent wave function, $\psi(x)$ is the time-independent wave function, and E is the particle energy. The time-independent wave function $\psi(x)$ depends on the potential energy variation with position, $E_P(x)$. The time-independent wave function can be obtained from the time-independent Schroedinger's equation as indicated in Supplement A:

$$\frac{-\hbar^2}{2m}\frac{d^2\psi(x)}{dx^2} + E_P(x)\psi(x) = E\psi(x) \tag{1.35}$$

Time-independent Schroedinger equation

Knowing the potential energy as a function of x, we can find $\psi(x)$ from Equation (1.35). Once the wave function is known, the behavior of the electron can be predicted. For example, in Example 1.1 we saw that a potential energy distribution that formed a potential well trapped an electron in a region of space. The probability of finding the electron in a particular region can be found by inserting the wave function into Equation (1.32).

For semiconductor devices, we are primarily interested in the behavior of electrons in semiconductor crystals. We will work up to this gradually.

1.6.1 THE FREE ELECTRON IN ONE DIMENSION

In the simplest model, we consider the (artificial) case of an electron of mass m_0 in a "crystal" in which the electron potential energy is a constant. Figure 1.12a shows the physical picture. Figure 1.12b shows the simplified energy diagram in the x direction. Note that there are still energy barriers at the surfaces—but we will simplify the problem by assuming the electron is nowhere near a surface, so that the problem has now been reduced to an electron in a universe of constant potential energy. This is called the *free-electron approximation.*

Any electron, including this one, has a total energy $E = E_P + E_K$, which because of conservation of energy is a constant. That is, the electron must move horizontally on the energy diagram in Figure 1.12b. To change energies, the electron would have to give up energy or acquire it from somewhere—for example, by colliding with an atom or a photon. We'll keep the problem simple for the moment by making the further assumption that such collisions don't occur. The results then will be valid only during the time between collisions.

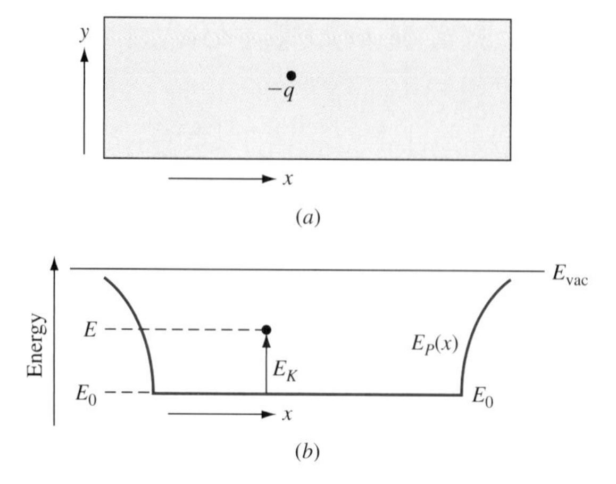

Figure 1.12 The free electron model for an electron in a crystal; (a) the physical picture; (b) the potential is assumed constant everywhere inside the crystal.

Since E_P is equal to some constant E_0, we can write the time-independent Schroedinger's equation in one dimension [Equation (1.35)] as

$$-\frac{\hbar^2}{2m_0}\frac{d^2\psi(x)}{dx^2} + E_0\psi(x) = E\psi(x) \tag{1.36}$$

We have substituted the constant potential energy (independent of position) E_0 for $E_P(x)$. Both E_0 and the total energy E are constants, so rewriting this differential equation in the standard form results in

$$\frac{d^2\psi(x)}{dx^2} + \frac{2m_0}{\hbar^2}(E - E_0)\psi(x) = 0 \tag{1.37}$$

The solution to Equation (1.37) is

$$\psi(x) = Ae^{jKx} + Be^{-jKx} \tag{1.38}$$

or alternatively

$$\psi(x) = C\sin(Kx) + D\cos(Kx) \tag{1.39}$$

Here A and B or C and D are some constants to be determined from the boundary conditions, and

$$K = \sqrt{\frac{2m_0(E - E_0)}{\hbar^2}} = \sqrt{\frac{2m_0E_K}{\hbar^2}} \tag{1.40}$$

The total energy E minus the potential energy E_0 is just the kinetic energy E_K, as indicated in this equation. The time dependence of the wave function will be that shown in Equation (1.34). The total wave function for the free electron is therefore

$$\Psi(x, t) = Ae^{j[Kx-(E/\hbar)t]} + Be^{j[-Kx-(E/\hbar)t]} \tag{1.41}$$

This equation has the form of two plane waves, one traveling in the $+x$ direction, and one traveling in the $-x$ direction. Certainly, the electron could be going either way. Also, note that the quantity $E/\hbar$ appears where the angular frequency ω usually appears in an equation for an electromagnetic plane wave. It would be great if we could say that the electron "wave" has a frequency of $\omega = E/\hbar$ but, unfortunately, the value of E depends on the choice of potential energy reference. Since the reference energy is arbitrary, $E/\hbar$ is not unique and so the "frequency" of the electron is not a physically measurable quantity.

We have established that $E/\hbar$ is not an easily interpreted quantity for the electron wave. What about K? What is its meaning?

We consider the first term of Equation (1.41), $Ae^{j[Kx-(E/\hbar)t]}$. If we consider a point of constant phase of this wave, for positive K the phase term $Kx - (E/\hbar)t$ remains constant with increasing time only if x also increases. In other words, the wave is going in the positive x direction if K is positive. Similarly for negative K, this term represents a wave traveling in the negative x direction. Equation (1.41) can be written as

$$\Psi(x, t) = Ae^{j[Kx-(E/\hbar)t]} \tag{1.42}$$

and the direction of propagation is given by the sign of K. As indicated in connection with Equations (1.25) and (1.27), the quantity K is called the *wave vector*.

We use the wave vector to describe the wavelike properties of the electron. For example, the velocity of a point of constant phase of the wave is called the phase velocity, and is given by

$$v_p = \frac{x}{t} = \frac{E}{\hbar K}$$

The phase velocity is not unique because the total energy E is dependent on the choice of potential energy reference.

The velocity associated with the center of mass of the particle is the "group velocity" v_g of the wave. The group velocity is given by

$$v_g = \frac{dx}{dt} = \frac{1}{\hbar}\frac{dE}{dK} \tag{1.43}$$

To further discuss the concept of wave vector we can rearrange Equation as follows:

$$E = E_0 + \frac{\hbar^2 K^2}{2m_0} \tag{1.44}$$

This is the E-K relation for the free electron, shown in Figure 1.13. Note that the second term in Equation (1.44) is the kinetic energy (since $E = E_P + E_K = E_0 + E_K$) and looks a lot like $p^2/2m_0$, the kinetic energy of a classical particle. Here, though, the classical momentum p is replaced by $\hbar K$.

1.6.2 THE DE BROGLIE RELATIONSHIP

Since we can consider the free electron to be either a wave or a particle, it would be useful to have a way to connect the two different descriptions. Examining

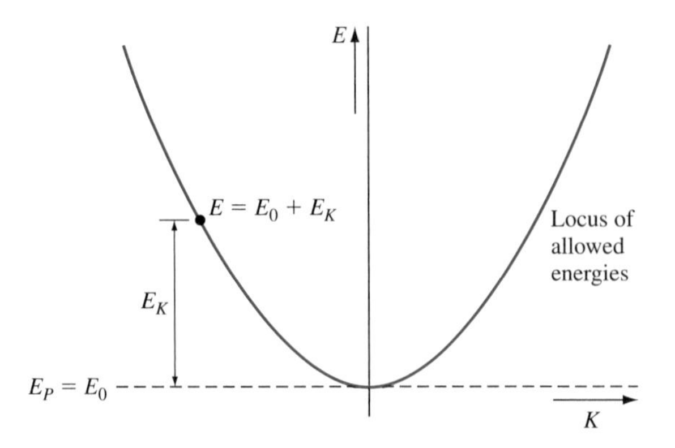

Figure 1.13 The *E-K* diagram for the free electron.

Equation (1.42) we see that at constant t, Ψ repeats itself whenever $Kx = n2\pi$. That is, x increases by one wavelength λ for every integer n, or

$$K = \frac{2\pi}{\lambda} \tag{1.45}$$

If we take the derivative of Equation (1.44) we get

$$\frac{dE}{dK} = \frac{\hbar^2 K}{m_0}$$

Inserting this into Equation (1.43) and multiplying both sides by m_0, we find that $m_0 v_g = \hbar K$ which is the classical momentum p of the free electron.

Since $K = 2\pi/\lambda$, and $\hbar = h/2\pi$, we obtain the de Broglie relationship

$$\lambda = \frac{h}{p} \tag{1.46}$$

It relates the wavelength (quantum mechanics) to the momentum of the particle (classical mechanics). It is important to remember that the de Broglie relationship holds *only* when the potential energy is constant over the path of the electron. It does not hold for electrons in solids.

We can get still more out of Equation (1.44). If we take the second derivative, we find that

$$\frac{\partial^2 E}{\partial K^2} = \frac{\hbar^2}{m_0} \tag{1.47}$$

This implies that the curvature (the second derivative) of the *E-K* locus is inversely proportional to the mass. For the parabolic *E-K* relation of the free electron, the curvature is constant so the mass is constant, a reassuring result.

*1.6.3 THE FREE ELECTRON IN THREE DIMENSIONS

If the potential energy is a constant in three-dimensional space $[E_P(x, y, z) = E_P(r) = E_0]$, then the wave function becomes

$$\Psi(\vec{r}, t) = \Psi(x, y, z, t) = Ae^{j[\vec{K}\cdot\vec{r} - (E/\hbar)t]} \tag{1.48}$$

where

$$\vec{K} = \hat{i}K_x + \hat{j}K_y + \hat{k}K_z \tag{1.49}$$

and

$$\vec{r} = \hat{i}x + \hat{j}y + \hat{k}z \tag{1.50}$$

Here x, y, and z are the coordinates of any position in space at which the wave function is being evaluated.

The magnitude of the wave vector $\vec{K}$ in three dimensions is

$$|\vec{K}| = \sqrt{K^2} = \sqrt{\frac{2m}{\hbar^2}(E - E_0)} = \sqrt{\frac{2m}{\hbar^2}E_K} \tag{1.51}$$

or

$$E - E_0 = E_K = \frac{\hbar^2 K^2}{2m} \tag{1.52}$$

as we found for the free electron in one dimension.

For a three-dimensional system with constant potential energy, then,

1. There is one allowed energy "band"—the range of energies between E_0 and infinity represents the allowed energy states for the free electron.[7] The energy band has a minimum at $E = E_0,\ K = 0$.
2. The velocity of the electron in a given direction is given by

$$v_x = \frac{1}{\hbar}\frac{\partial E}{\partial K_x} \qquad v_y = \frac{1}{\hbar}\frac{\partial E}{\partial K_y} \qquad v_z = \frac{1}{\hbar}\frac{\partial E}{\partial K_z} \tag{1.53}$$

3. From Equation (1.52), the E-K relation is a paraboloid in three dimensions. The curvature of the paraboloid is given by its second derivative $\partial^2 E/\partial K^2 = \partial^2 E_0/\partial K^2 + \partial^2 E_K/\partial K^2$. For the free electron, E_0 is constant everywhere, so the curvature of the E-K curve is the same in any direction in K space (the total energy is also a constant):

$$\frac{\partial^2 E}{\partial K^2} = \frac{\partial^2 E_K}{\partial K^2} = \hbar\frac{1}{m_0} \tag{1.54}$$

   The mass m_0 is a constant for the free electron.
4. De Broglie's relation $\lambda = h/p$ is valid for this case, and $p = mv$.

Although the free electron model is a poor approximation for electrons in semiconductors, the wave-particle duality discussed above leads to somewhat similar results for *quasi-free electrons* in real semiconductors, as discussed in the next section.

1.6.4 THE QUASI-FREE ELECTRON MODEL

Although the free-electron model (E_P = constant) is sometimes used for metals because of its simplicity, it isn't appropriate for semiconductors because it

---

[7]Here we ignore the relativistic effects that limit the electron energy to mc^2. Such high energies are not important in semiconductor electronics.

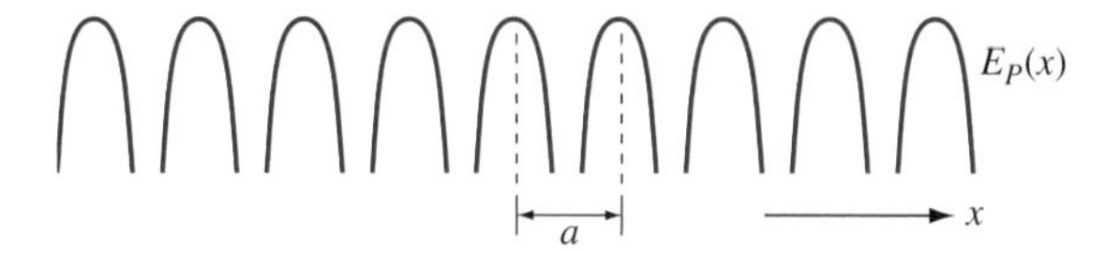

Figure 1.14 The electron potential energy in a crystal is a periodic function.

doesn't predict the existence of discrete energy bands. More realistic models that do predict multiple bands do so because they take into account the periodicity of the potential energy for an electron in a crystal lattice. Since the exact functional relation of the potential energy isn't accurately known, these models assume different periodic functions for the potential energy. Such a model is known as a *quasi-free electron model.* We examine one such case, starting with a one-dimensional universe and then extending the result to three dimensions.

Consider an electron deep within a solid, a one-dimensional crystal of lattice constant a. The corresponding potential energy function for the electron is shown in Figure 1.14, where the electron is considered to be a point charge. By assuming the electron is not close to any surface, we can neglect the effects of the surfaces, and the potential energy looks like an infinite, strictly periodic function.

Let the potential in the crystal be described by some periodic function $E_P(x)$. Since it is periodic in x with periodicity a, then $E_P(x) = E_P(x \pm na)$, where n is some integer. Then Schroedinger's equation [Equation (1.35)] becomes

$$\frac{d^2\psi(x)}{dx^2} + \frac{2m_0}{\hbar^2}[E - E_P(x)]\psi(x) = 0 \tag{1.55}$$

Compare this with Equation (1.37) for the electron in a constant potential.

Since we don't know the exact form of the potential energy $E_P(x)$, we can't solve Schroedinger's equation exactly. However, since we do know that the potential is periodic, we can use a theorem known as the *Bloch theorem* [2], which states that for an electron in a periodic potential, the time-independent wave function is

$$\psi(x) = U_K(x)e^{jKx} \tag{1.56}$$

where $U_K(x)$ is some function that is also periodic in x with the periodicity of the crystal. Beyond that, we do not know the exact form of $U_K(x)$. The complete Bloch wave function is

$$\Psi(x,t) = U_K(x)e^{j[Kx-(E/\hbar)t]} \tag{1.57}$$

which represents a unit amplitude plane wave ($e^{j[Kx-(E/\hbar)t]}$), modulated by some periodic function $U_K(x)$ with period a. Since it is periodic,

$$U_K(x) = U_K(x + na) \tag{1.58}$$

where n is an integer. Even though we do not know everything about U_K, since it is periodic, we can deduce some things about the general characteristics of the E-K relationship.

For example, we know that the crystalline forces acting on the electron are independent of the direction of propagation (sign of K). Therefore

$$\boxed{E(K) = E(-K)} \tag{1.59}$$

That in turn implies that $E(K)$ has an extremum (either a relative maximum or minimum) at $K = 0$.

The time-independent wave function, Equation (1.56), can be multiplied by

$$1 = e^{j(2\pi nx/a)}e^{-j(2\pi nx/a)} \tag{1.60}$$

giving

$$\psi(x) = U_K(x)e^{-j(2\pi nx/a)}e^{jKx}e^{j(2\pi nx/a)} \tag{1.61}$$

We write this as

$$\psi(x) = U_K e^{-j(2\pi nx/a)}e^{j(K+2\pi n/a)x} \tag{1.62}$$

Observe that both U_K and $e^{-j(2\pi nx/a)}$ are periodic in x with the periodicity a of the lattice. We can therefore combine them into one Bloch modulation factor

$$U'_K = U_K e^{-j(2\pi nx/a)} \tag{1.63}$$

If we define

$$\boxed{K' = K + \frac{2\pi n}{a}} \tag{1.64}$$

we can rewrite Equation (1.62) as

$$\psi(x) = U'_K e^{jK'x} \tag{1.65}$$

Since Equation (1.65) is the same as Equation (1.62), and has the same periodicity, both equations must represent states corresponding to the same energy, or

$$\boxed{E(K) = E\left(K + \frac{2\pi n}{a}\right)} \tag{1.66}$$

Therefore, the $E(K)$ relation is periodic in K, with period $2\pi/a$.

A possible E-K relation illustrating these points is indicated in Figure 1.15a. We have shown that $E(K)$ is an even function [Equation (1.59)], and that $E(K) = E(K + 2\pi n/a)$, for $n = 0, \pm 1, \pm 2, \ldots$. From this we can conclude that the E-K curve, whatever its shape actually is, is symmetric about the points $K = 0, \pm 2\pi/a, \pm 4\pi/a, \ldots$, and thus it must have either equivalent maxima or equivalent minima at these points. In addition, because of this symmetry, the E-K relation must have other equivalent extrema midway between these values, or at $K = \pm\pi/a, \pm 3\pi/a, \ldots$.

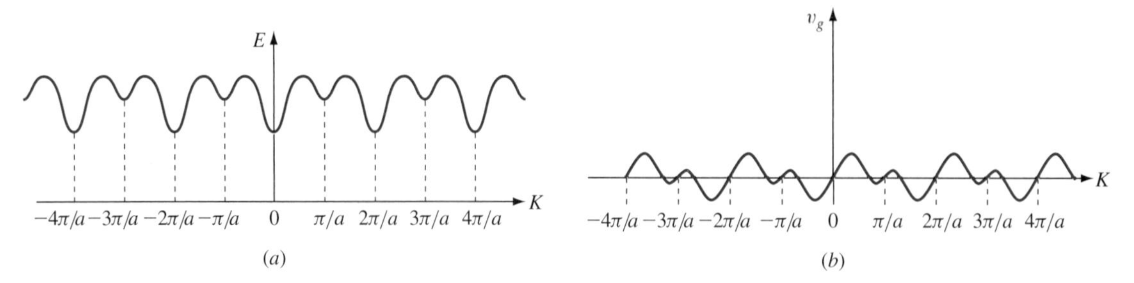

Figure 1.15 One possible E versus K diagram for the periodic potential. (a) E versus K; (b) the group velocity v_g versus K.

From Figure 1.15a, we see that:

1. $E(K)$ is periodic in K space, with period $2\pi/a$.
2. Equivalent extrema in E exist at $K = 0, \pm 2\pi/a, \pm 4\pi/a, \ldots$.
3. Equivalent extrema exist at $K = \pm\pi/a, \pm 3\pi/a, \ldots$.
4. The slope of the E-K curve is zero at $K = 0, \pm\pi/a, \pm 2\pi/a, \pm 3\pi/a, \ldots$.
5. The group velocity $v_g = (1/\hbar)\, dE/dK$ is periodic in K space with the same periodicity as the E-K curve. This is shown in Figure 1.15b.

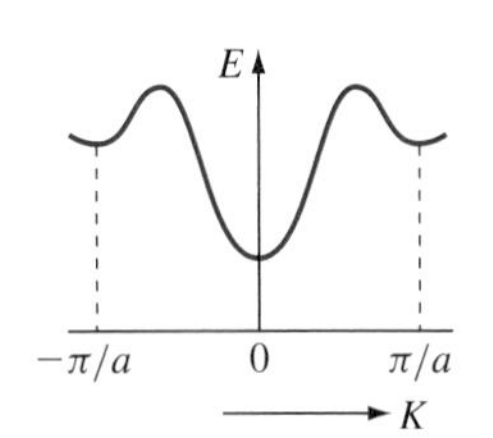

Figure 1.16 The reduced, or first Brillouin, zone.

In fact, it turns out that *all* of the measurable quantities are periodic with the same periodicity, $2\pi/a$ in K space. Since everything repeats, it is customary to consider just the region from $-\pi/a \leq K \leq \pi/a$. This is known as the *reduced zone* or the *first Brillouin zone* [3, 4], as shown in Figure 1.16 (compared with the extended zone of Figure 1.15). In fact, since E versus K is symmetric about $K = 0$, sometimes only half of the reduced zone is drawn.

For this example, a relative minimum was assumed to exist at $K = 0$ and thus also at $K = 2\pi n/a$. We could just as well have chosen a relative maximum, and there may be other extrema as well (we chose a case for which there is also a maximum in the interior of the zone). The exact shape and number of maxima and minima cannot be predicted from the general considerations presented here. The actual E-K diagram could be calculated in principle if $E_P(x)$ were known.

We now return to the E-K diagram of Figure 1.16. This is actually another form of the energy band diagram, since it shows the maximum and minimum energies of a particular band. Figure 1.16 is redrawn in Figure 1.17a to illustrate this point. The allowed range of energies is between E_2 and E_0. In Figure 1.17b the energy band diagram in real space E-x is plotted, and the energies of the maxima and minima are plotted as functions of position in the crystal. The maxima and minima are the same everywhere in the crystal, as long as the electron is not near a surface. Also plotted against x is the potential energy function (the periodic function) for the electron, with the electron considered to be a point.

In a nearly empty band, electrons will tend to occupy the lower energy states and thus have properties associated with the minimum of the E-K plot in Figure 1.17a. Likewise, in a nearly filled band, the vacant states (holes) will be near the band maximum.

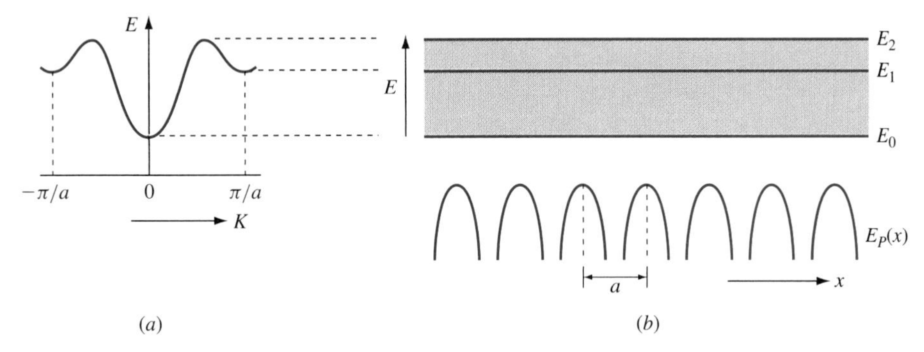

Figure 1.17 (a) the *E-K* diagram; (b) the corresponding energy band (*E-x*) diagram.

The *E-K* diagram and the regular energy band diagram complement each other. In the *E-K* diagram, no information about the position of a given electron is known, but the energy band diagram does not reveal anything about the wave vector (wavelength, direction of motion). In effect, the two descriptions, *E-x* and *E-K*, are Fourier transforms of each other. Both relationships are required for an understanding of electronic behavior in crystals, and we will use both types of diagram.

Since the value of $U_K(x)$ changes periodically with position, Equation (1.56) for the wave function is a modulated plane wave, or Bloch wave function. Again, the wave vector K is related to the wavelength by

$$K = \frac{2\pi}{\lambda} \tag{1.67}$$

but this time

$$K \neq \sqrt{\frac{2m_0}{\hbar^2}(E - E_P(x))} \tag{1.68}$$

Since we didn't assume we knew anything about the potential energy function $E_P(x)$ other than that it is periodic, no more is known about the exact form of $U_K(x)$ except that it is periodic.

We saw that for a free electron, the momentum $p = \hbar K$. This is not the case for the quasi-free electron model because the assumption of constant potential energy everywhere does not apply. De Broglie's relation does not hold. We note, however, that $\hbar K$ is sometimes referred to as "crystal momentum" since it is analogous to classical momentum of the free electron.

In a three-dimensional crystal the potential energy $E_P(r)$ is periodic in all crystallographic directions and

$$\psi(r) = U_K(r)e^{j\vec{K}\cdot\vec{r}} \tag{1.69}$$

where $U_K(r)$ has the same periodicity as $E_P(r)$ and $|K| = 2\pi/\lambda$.

In the free electron and quasi-free electron models, collisions were neglected and so an electron particle could be represented by a single Bloch wave. In a real crystal, however, the electron mean free path is on the order of a few nanometers, and realistically such collisions must be considered. As a result, an electron particle must be considered as a superposition of wave functions. This topic is considered in more detail in Supplement A to Part 1.

1.6.5 REFLECTION AND TUNNELING

There is an interesting consequence to considering the electron to be a wave. Consider an electron wave approaching the potential barrier shown in Figure 1.18a. The electron has some energy lower than the height of the barrier. We expect the electron to be reflected from the barrier, but we cannot say that the wave stops abruptly at the barrier and turns around. Instead, the electron wave actually penetrates the barrier a short distance, as shown in the figure. This means that part of the electron charge is in the (classically) forbidden region. For example, if the barrier in the figure represents the surface of the crystal, some fraction of the electron charge is actually found outside the crystal.

This tunneling process is analogous to the penetration of electromagnetic waves a short distance (the *skin depth*) into a conductor. To reduce this penetration, *shielding* is employed to protect electronic systems. These systems are

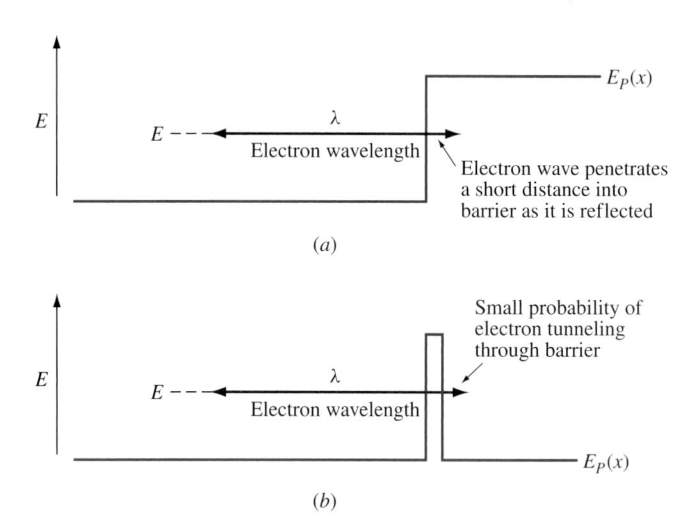

Figure 1.18 An electron wave is extended in space. (a) When the wave reflects from the potential barrier, the electron wave function extends a short distance into the forbidden region. Thus some fraction of the electron charge is found to the right of the barrier. (b) If the barrier is very thin, the electron wave function Ψ may extend all the way through it. Since the probability density $\Psi^*\Psi$ is not zero on the far side of the barrier, there is some (small) chance that the electron will cross through the barrier and emerge on the other side.

enclosed by a metal of high conductivity (e.g., copper or aluminum) and sufficient thickness to permit negligible electromagnetic penetration. There is an analogy for electrons: To prevent electrons from penetrating, a barrier must be of high enough potential and sufficient thickness. The barrier in Figure 1.18a is infinitely thick, so the electron penetrates a short distance but ultimately is reflected.

Next, suppose the barrier is very thin, as shown in Figure 1.18b. Since the electron wave penetrates the barrier a short distance, if the barrier is thinner than the penetration distance, part of the electron wave could be found on the other side. A physical electron cannot be divided in two, so it must end up either on one side or the other. The wave function of the electron extends through the barrier, however. Recall that the wave function multiplied by its complex conjugate is a probability density function [Equation (1.32)]. The quantum mechanical interpretation is that since the wave function extends into the allowed region on the far side of the barrier, there is some probability that the electron will "tunnel" through the barrier and end up on the other side. The probability is quite small, and usually the electron will be reflected from the barrier. Although the probability of a given electron tunneling through the barrier can be miniscule, the number of electrons striking the barrier can be large, with a resulting tunneling current that is not necessarily negligible.

The penetration of the electron into a barrier and the tunneling of electrons through thin potential barriers are both important quantum mechanical effects that very much affect the operation (and reliability) of semiconductor devices, as we see throughout this book. These effects are discussed in more detail in the Supplements to Part 1 and in the chapters on device operation.

1.7 A FIRST LOOK AT OPTICAL EMISSION AND ABSORPTION

We have talked about electrons being in the conduction band or in the valence band of a semiconductor, and noted that electrons can change energy states, moving from one band to the other. When an electron makes a transition, e.g., from the valence band to the conduction band, it has to acquire the extra energy from somewhere. The extra energy can come from vibration of the crystal lattice (phonons), or from an optical source (photons).

As indicated earlier, although we usually think of light as electromagnetic wave energy, light can also be thought of as consisting of particles or photons. The energy of a photon is related to the frequency of the light wave:

$$E = h\nu = \hbar\omega_{\text{pht}} \tag{1.70}$$

where h is Planck's constant and ν is the frequency.

Light energy must be absorbed or emitted in integer multiples of $h\nu$. That is, only an integral number of photons can be absorbed or emitted. For example, consider a beam of red light whose wavelength is 600 nm. The frequency of the light (wave) is given by:

$$\nu = \frac{c}{\lambda} \tag{1.71}$$

where c is the speed of light and λ is the wavelength. Thus for $\lambda = 600$ nm, the corresponding frequency is

$$\nu = \frac{3.0 \times 10^8 \text{ m/s}}{600 \times 10^{-9} \text{ m}} = 5 \times 10^{14} \text{ Hz} \tag{1.72}$$

or 500 terahertz. The energy of a single photon is then

$$E = h\nu = (6.63 \times 10^{-34} \text{ J} \cdot \text{s})(5 \times 10^{14} \text{ s}^{-1}) = 3.3 \times 10^{-19} \text{ J}$$
$$= 3.3 \times 10^{-19} \text{ J} \cdot \frac{1 \text{ eV}}{1.6 \times 10^{-19} \text{ J}} = 2.06 \text{ eV} \tag{1.73}$$

Light whose wavelength is 600 nm can be emitted or absorbed only in multiples of 2.06 eV.

EXAMPLE 1.3

An electron in a hydrogen atom can switch from one allowed energy level to another by either absorbing or emitting energy. Scientists use the spectra of the light emitted by various materials to determine their energy structure. For each of the following spectral lines, determine the initial and final energy states.

■ Solution

a. $\lambda_1 = 656.28$ nm (red). From Equations (1.70) and (1.71), the energy corresponding to this wavelength is

$$E = \frac{hc}{\lambda} = \frac{(6.63 \times 10^{-34} \text{ J} \cdot \text{s})(2.998 \times 10^8 \text{ m/s})}{656.28 \times 10^{-9} \text{ m}} = 3.02 \times 10^{-19} \text{ J} = 1.89 \text{ eV}$$

We have to try some combinations of allowed energy levels for the hydrogen atoms to find which difference corresponds to this value. From Table 1.1, the answer is $E_3 - E_2 = (E_{\text{vac}} - 1.51) - (E_{\text{vac}} - 3.4) = 1.89 \text{ eV}$. Thus the initial state of the electron is state 3, and the final state is state 2.

b. $\lambda_2 = 486.13$ nm (blue). The energy corresponding to this wavelength is

$$E = \frac{hc}{\lambda} = 2.55 \text{ eV}$$

This energy results only from a transition from E_4 to E_2, or

$$E_4 - E_2 = (E_{\text{vac}} - 0.85) - (E_{\text{vac}} - 3.4) = 2.55 \text{ eV}$$

c. $\lambda_3 = 434.05$ nm (violet). The energy corresponding to this wavelength is

$$E = 2.86 \text{ eV}$$

This corresponds to

$$E_5 - E_2 = (E_{\text{vac}} - 0.544) - (E_{\text{vac}} - 3.4) = 2.86\,\text{eV}$$

d. Which energy level do all of these transitions have in common? This group of transitions is called the Balmer series. The Balmer series describes optical transitions ending on E_2. There is another point to be made here. The energy of a photon depends on the wavelength (frequency) of the light ($E = h\nu$). A more energetic (more intense) beam of 600 nm would consist of more photons, but each photon still has to have 2.06 eV of energy. The power of a light beam shining on an object depends on both the energy per photon and the number of photons striking the object per second.

For example, suppose that a beam of photons of $\lambda = 600$ nm (2.06 eV) is incident on a semiconductor. Let the semiconductor have a band gap of 2.5 eV as shown in Figure 1.19a. This particular material cannot absorb a photon of

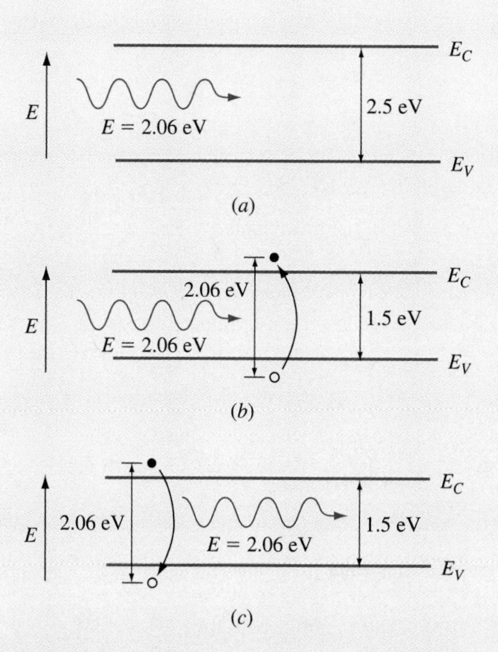

Figure 1.19 (a) A photon of energy 2.06 eV is incident on a material of energy gap 2.5 eV. The photon cannot be absorbed. (b) The band gap is small enough that allowed states separated by 2.06 eV exist, thus the photon can be absorbed. The photon's energy is given to the electron. (c) In emission, the electron goes to a lower energy state, releasing the extra energy in the form of a photon.

2.06 eV. Why not? Consider an electron near the top of the valence band. To absorb the photon's energy, the electron would have to end up in a new energy state 2.06 eV higher than where it started. There is no allowed energy state there, however, because it would be in the forbidden energy band. Thus, the material is transparent to this particular wavelength of light.[8]

In Figure 1.19b, a new material is chosen whose band gap is smaller than the energy of the photon. Now the photon's energy can be absorbed, in which case the energy of the photon is transferred to the electron and the photon is annihilated.

In optical emission, the process is reversed. An electron initially at a high energy state (i.e., in the conduction band) can "fall" down to the valence band—but it must release the extra energy. This energy could be heat or light, or some combination of the two. Here the law of conservation of energy is used to discuss emission and absorption. Another restriction, conservation of wave vector, analogous to the conservation of momentum in classical mechanics, must also be observed, as discussed in Supplement A to Part 1.

EXAMPLE 1.4

Light of 100 μW is incident on a photodetector.

a. If the light is green (500 nm), how many photons are striking the surface per second?
b. If the power remains at 100 μW, but the wavelength of the light is changed to infrared ($\lambda = 1\ \mu$m), now how many photons strike the surface per second?

■ Solution

a. The energy per photon is $E = h\nu$. The frequency ν for the green light is given by

$$\nu = \frac{c}{\lambda} = \frac{2.99 \times 10^8\ \text{m/s}}{500 \times 10^{-9}\ \text{m}} = 5.98 \times 10^{14}\ \text{s}^{-1} = 598\ \text{THz}$$

The energy per photon is thus

$$E = h\nu = (6.62 \times 10^{-34}\ \text{J} \cdot \text{s})(5.98 \times 10^{14}\ \text{s}^{-1})$$
$$= 3.95 \times 10^{-19}\ \text{J} \cdot \left(\frac{1}{1.6 \times 10^{-19}\ \text{J/eV}}\right) = 2.47\ \text{eV}$$

[8] It is, of course, possible for two photons to strike the electron simultaneously and thus give $2 \times 2.06 = 4.12$ eV to the electron. This is more than enough energy to excite an electron from the valence band to the conduction band, but three particles (two photons and one electron) colliding simultaneously is statistically unlikely.

Power is $P = E/t = 100\,\mu\text{W}$, so

$$
\begin{aligned}
100\,\mu\text{W} &= \text{(energy per second)} \\
&= \text{(energy per photon)} \cdot \text{(number of photons striking surface/second)} \\
&= N \cdot 3.95 \times 10^{-19}\,\text{J}
\end{aligned}
$$

$$
\begin{aligned}
N &= \frac{100 \times 10^{-6}\,\text{W}}{3.95 \times 10^{-19}\,\text{J/photon}} = \frac{100 \times 10^{-6}\,\text{J/s}}{3.95 \times 10^{-19}\,\text{J/photon}} \\
&= 2.54 \times 10^{14}\,\text{photons/s}
\end{aligned}
$$

b. For the infrared case, the energy per photon is

$$E = h\nu = \frac{hc}{\lambda} = \frac{(6.62 \times 10^{-34}\,\text{J} \cdot \text{s})(2.99 \times 10^{8}\,\text{m/s})}{1 \times 10^{6}\,\text{m}} = 1.98 \times 10^{-19}\,\text{J}$$

and the number of photons per second is

$$N = \frac{100 \times 10^{-6}\,\text{W}}{1.98 \times 10^{-19}\,\text{J/photon}} = \frac{100 \times 10^{-6}\,\text{J/s}}{1.98 \times 10^{-19}\,\text{J/photon}} = 5.05 \times 10^{14}\,\text{photons/s}$$

Thus, the number of photons required to deliver a given amount of power depends on the wavelength of the light.

Because we often convert wavelength to energy in our calculations, it is helpful to multiply the constants together once and remember the result:

$$
\begin{aligned}
E = h\nu = \frac{hc}{\lambda} &= \frac{(6.62 \times 10^{-34}\,\text{J} \cdot \text{s})(2.99 \times 10^{8}\,\text{m/s})}{\lambda} \cdot \frac{1\,\text{eV}}{1.6 \times 10^{-19}\,\text{J}} \\
&= \frac{1.24 \times 10^{-6}\,\text{eV}}{\lambda}
\end{aligned}
$$

or

$$\boxed{E(\text{eV})\lambda(\mu\text{m}) = 1.24 \qquad \text{Golden Rule}} \tag{1.74}$$

This is known as the *Golden Rule*. The energy per photon in eV times the photon wavelength in μm is 1.24. Thus, a photon whose wavelength is 500 nm (0.5 μm) has an energy of $1.24/0.5 = 2.48$ eV.

EXAMPLE 1.5

An optical communication system (discussed in Chapter 11) is shown schematically in Figure 1.20a. It consists of a semiconductor laser diode optical source, a semiconductor photodetector, and a fiber optic cable to transmit the optical signal between source and detector. As will be discussed in Chapter 11 and indicated in Figure 1.20b, modern fiber has a minimum absorption coefficient at $\lambda = 1.55\ \mu$m. Thus to obtain a maximum fiber length between repeaters, the laser should emit light at this wavelength.

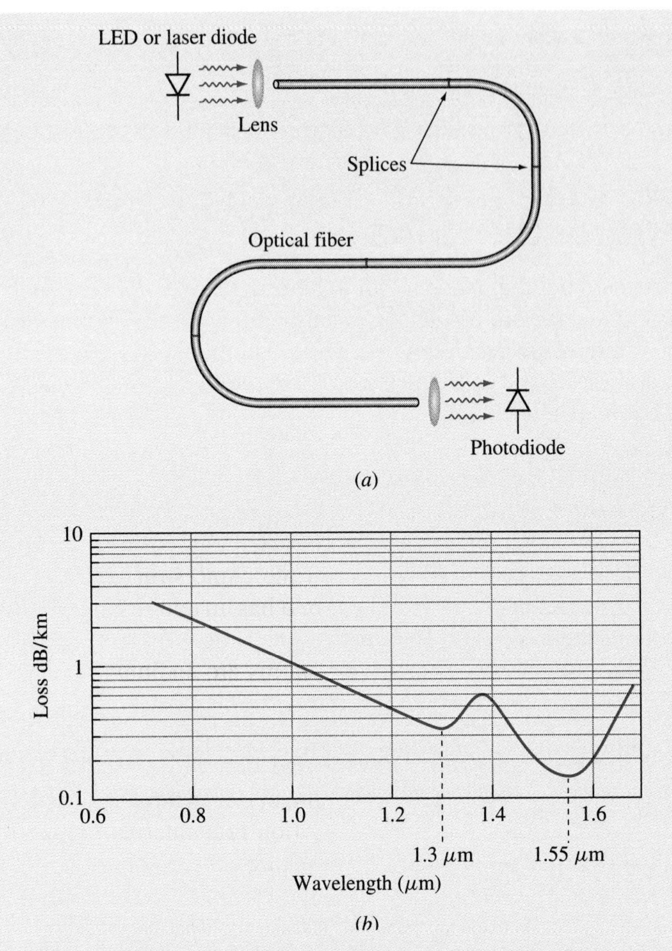

Figure 1.20 (a) A communication fiber optic link contains a light source, a fiber, and a photodetector. (b) Typical absorption spectrum for optical fiber.

a. Assuming the laser emits photons having energy equal to its band gap, what must be its band gap?
b. Since most photodetectors are sensitive to photons having energy greater than their band gaps, what is the minimum band gap of the detector?

■ **Solution**

a. From the Golden Rule [Equation (1.74)],

$$E(\text{eV}) = \frac{1.24}{\lambda(\mu\text{m})} = \frac{1.24}{1.55} = 0.80 \text{ eV}$$

The laser's band gap should be 0.8 eV.

b. Similarly, the minimum band gap of the photodetector should be 0.8 eV.

Note: To obtain a material with a specific band gap, appropriate alloys of three or four materials are used. Choosing the ratios of the various elements in these alloys to produce the desired band gap is referred to as *band-gap engineering.*

1.8 CRYSTAL STRUCTURES, PLANES, AND DIRECTIONS

We have pointed out that silicon, gallium arsenide, and other semiconductors, as well as some metals, are crystals. Crystallography is of great interest to people who fabricate semiconductor devices. For the purposes of appreciating the operating physics of diodes and transistors, some minimal understanding of crystal planes and directions is useful.

Crystals are regular structures in which the atoms are arranged in a pattern that repeats throughout the material. For example, consider the crystal in Figure 1.21a. It is termed *simple cubic.* The atoms are arranged in cubes, with an atom at each vertex. All six faces are square with dimension a, called the *lattice constant.* There are several variations on the cubic structure. One is the face-centered cubic (FCC) lattice, Figure 1.21b. It has, in addition to the corner atoms, an atom in the center of each face of the cube. Common table salt (NaCl) has a face-centered cubic structure—alternate atoms are sodium and the ones in between are chlorine. The body-centered cubic structure, Figure 1.21c, has an atom in the center of each cube.

Silicon's crystal structure is the diamond structure, which is actually a variation on the cubic structure. The cubic element has an atom in the center of every face, like the FCC lattice, but has in addition four internal atoms as shown in Figure 1.22a. In actuality, the diamond structure consists of two FCC lattices that interpenetrate.

GaAs is also a diamond lattice in the sense that the atoms are arranged in the same way as in Figure 1.22a. Here, however, half of the atoms are gallium and

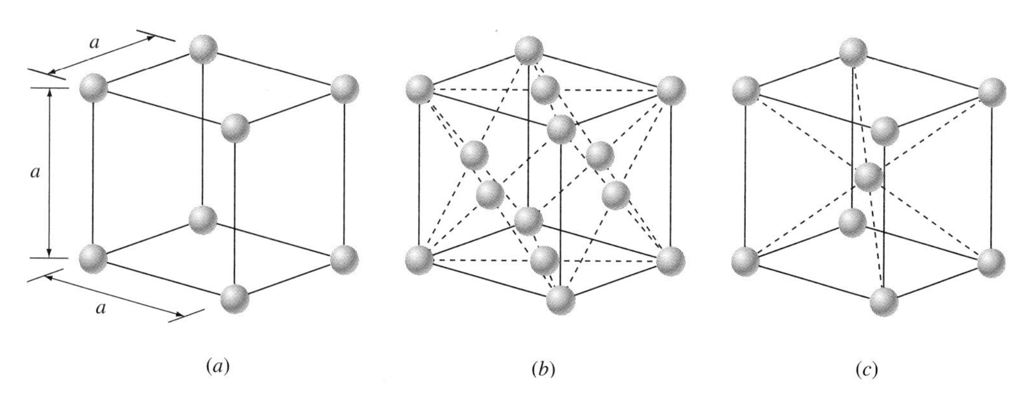

Figure 1.21 Cubic crystals: (a) simple cubic; (b) face-centered cubic, an atom in the center of every face, and (c) body-centered cubic.

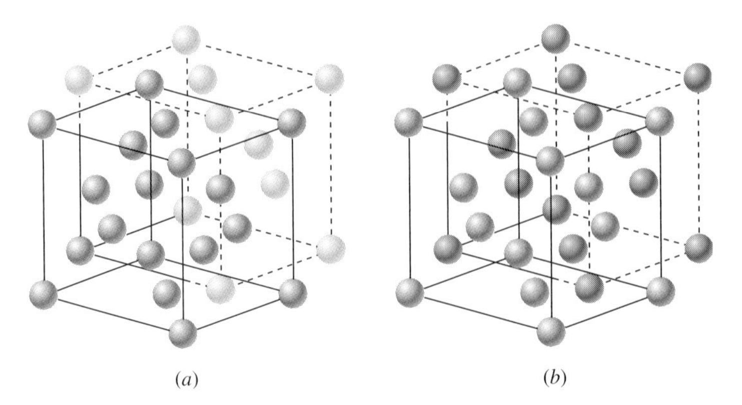

Figure 1.22 (a) The diamond structure consists of two interpenetrating FCC lattices. The second FCC cube is offset by one-quarter of the longest diagonal. The dashed lines indicate the part of the second FCC lattice that is outside the unit diamond cell. (b) A zinc blende material has the same structure, but two types of atoms. The black atoms are one type (for example, gallium) and the colored atoms are the other (arsenic).

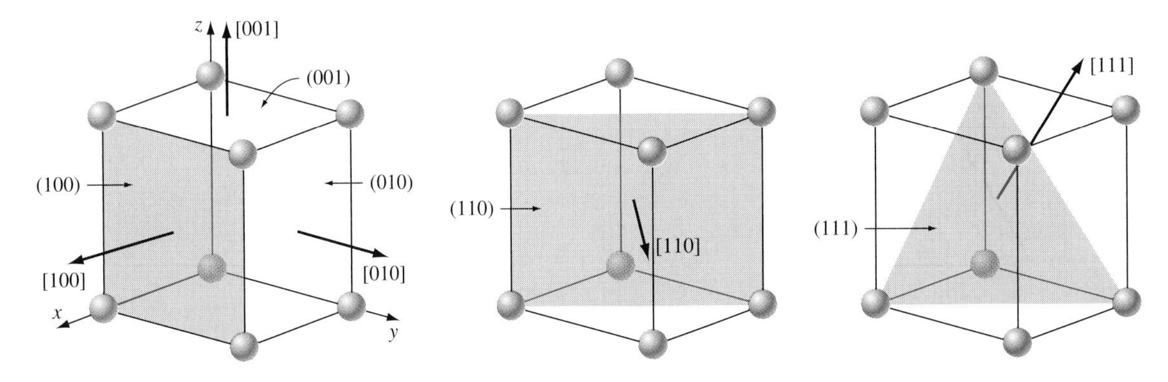

Figure 1.23 The three most important crystallographic planes (in parentheses) and the corresponding crystallographic directions (square brackets).

half are arsenic, as shown in Figure 1.22b. This results in the *zinc blende* structure, named after a mineral that exhibits this arrangement. It is interesting to note that the zinc blende structure is actually two interpenetrating FCC lattices; in GaAs, one FCC is of gallium and the other is of arsenic. As we will see later on in the book, the properties of semiconductor devices depend on the crystal plane of the surface of the semiconductor and in which crystallographic direction the electrons (or holes) travel.

There are three surfaces and three crystallographic directions that are the most important in semiconductors. These are shown in Figure 1.23 for a cubic lattice. The (100) plane is the plane of any one of the faces of the cube. The (110) plane includes two edges of the cubes and cuts the cube diagonally. The (111) plane results when the cube is cut from one corner to the farthest. The directions [100], [110], and [111] are perpendicular to the corresponding planes. Thus, any of the <100> directions[9] is along a cube edge, any <110> direction is diagonal across the cube but parallel to one plane of the crystal, and the <111> direction cuts the cube along a diagonal. These directions and planes are important to those fabricating the devices, and familiarity with the terminology is useful.

1.9 SUMMARY

We have laid some groundwork for understanding the fundamental role of electrons in materials. From the Bohr model, we saw that the allowed energy states of an electron are actually quantized—only certain discrete values are allowed. This is a fundamental concept in the field of quantum mechanics, which will be discussed in more detail in the next chapter and throughout the book.

These states were determined analytically for an isolated hydrogen atom. We then discussed covalent bonding in molecules and in crystalline solids, in which electrons are shared equally between atoms to hold the solid together. In bonding that is partially ionic, the electrons are also shared but in addition there is a Coulombic attraction between nuclei that helps to hold the solid together.

When many atoms are brought close together as in a solid, the discrete allowed states smear into energy bands, in accordance with the Pauli exclusion principle. We saw that for a given material, there are certain ranges of energy that are allowed and other forbidden bands in which no states exist (at least not for the ideal crystal).

We also learned how to tell, from the energy band diagram, which electrons are bound to individual atoms—those whose bands do not extend as far as the next atom. Those electrons that are involved in covalent bonding occupy the valence band that extends across the crystal so that electrons can be shared. The higher energy conduction band also extends across the crystal, and contains electrons that are quasi-free. These electrons can travel around the crystal easily, conducting electricity.

The concept of wave-particle duality was briefly discussed. What are classically thought of as particles (e.g., electrons) also possess wave properties (e.g., wavelength), while what are classically considered waves (e.g., photons) can also be considered to be particles. Matter waves are characterized by a wave function, which can be calculated from Schroedinger's equation if the potential energy of the particle is known. Two examples were considered for an electron within a crystal: the free electron, in which the electron potential energy is considered to be constant, and the quasi-free electron, where the potential is a periodic function with position within the crystal. These two cases will be referred to repeatedly throughout this book.

---

[9] A specific plane specification is enclosed in parentheses, while equivalent planes, e.g., (100), (010), (001), (−100), etc., are enclosed in curly brackets, {100}. Similarly, specific directions are enclosed in brackets, [100], [010}, etc., and equivalent directions are enclosed in angle brackets, <100>.

The interactions of classical particles (electrons) and waves (light) were briefly discussed in terms of the conservation of energy associated with the interaction of the particles involved.

Because the electrical properties of semiconductors depend on their crystal structure, the structures of some semiconductors of interest (silicon and gallium arsenide) were briefly discussed.

Finally, we learned to determine some electrical properties of material directly from the energy band diagram. A material with a wide band gap will generally not conduct well and can be an insulator. On the other hand, in a metal, the band containing the conduction electrons is partly full even at low temperatures, and these materials conduct very well. Semiconductors have intermediate band gaps, and their conductivity is between that of insulators and metals.

It may seem intuitive that electrical engineers should want materials that either conduct very well, or insulate very well. The usefulness of semiconductors is certainly not obvious from what we've learned so far. We will spend the next few chapters developing an understanding of the special properties that these intermediate band-gap energies confer. The rest of the book will be devoted to showing how to use these properties to make useful semiconductor devices.

1.10 READING LIST

The following numbers refer to articles and books relevant to Chapter 1 listed in Appendix G: [1–18].

1.11 REFERENCES

1. Robert Eisberg and Robert Resnick, *Quantum Physics of Atoms, Molecules, Solids, Nuclei and Particles,* 2nd ed., John Wiley and Sons, New York, 1985.
2. Charles Kittel, *Introduction to Solid State Physics,* 4th ed., Chap. 9, John Wiley and Sons, New York, 1971.
3. R. A. Smith, *Wave Mechanics of Crystalline Solids,* Chap. 4, Chapman and Hall, London, 1961.
4. Leon Brillouin, *Wave Propagation in Periodic Structures,* 2nd ed., Dover, New York, 1953.

1.12 REVIEW QUESTIONS

1. Define an electron volt.
2. Define a photon.
3. State the Pauli exclusion principle.
4. Explain why the energy levels of an atom become energy bands in a solid.
5. What is the difference between the band structure of a crystalline insulator and of a semiconductor?

6. What is the difference between the band structure of a semiconductor and of a metal?
7. Explain why a semiconductor acts as an insulator at 0 K and why its conductivity increases with increasing temperature.
8. Define a hole in a semiconductor.
9. Indicate pictorially how a hole contributes to conduction.
10. Draw the potential energy picture of a metal. Explain qualitatively the existence of the potential barrier at the surface.
11. We saw that it is possible, from a quantum mechanical point of view, for an electron to tunnel through a thin barrier, even though such tunneling is forbidden classically. The thicker the barrier or the higher the barrier, the lower the probability of tunneling. A barrier on the order of tenths of a nanometer is thin enough for an electron to tunnel through. Comment on the probability of an entire person, composed of an astronomical number of electrons, protons, etc., tunneling through a brick wall.
12. Explain why an electron's energy increases when it absorbs a photon. What happens to the photon?

1.13 PROBLEMS

1.1 Show that Equation (1.6) follows from Equation (1.3).

1.2 Consider a lithium nucleus, of charge $+3q$. Calculate the first three electron energies for an electron in a Li^{++} ion, using the Bohr model.

1.3 Show that Equations (1.12) and (1.13) follow from (1.8) and (1.11).

1.4 In each of the potential energy distributions in Figure 1P.1, sketch the magnitude and direction of the force on the electron.

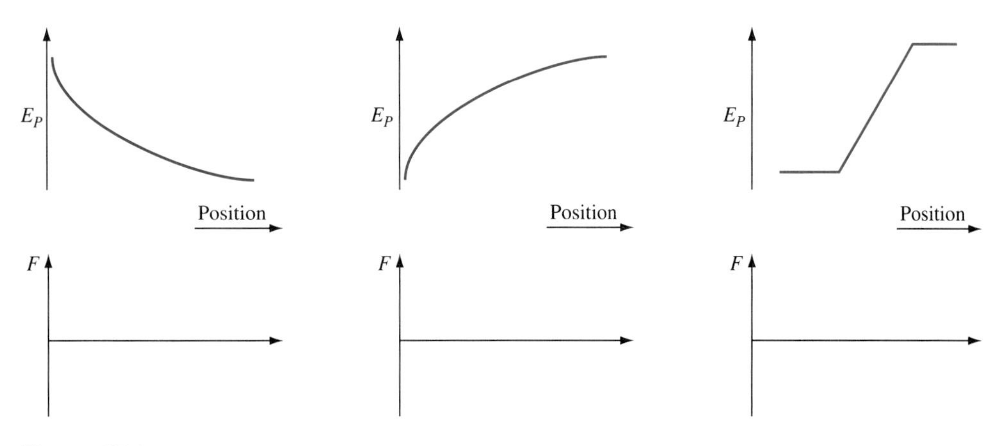

Figure 1P.1

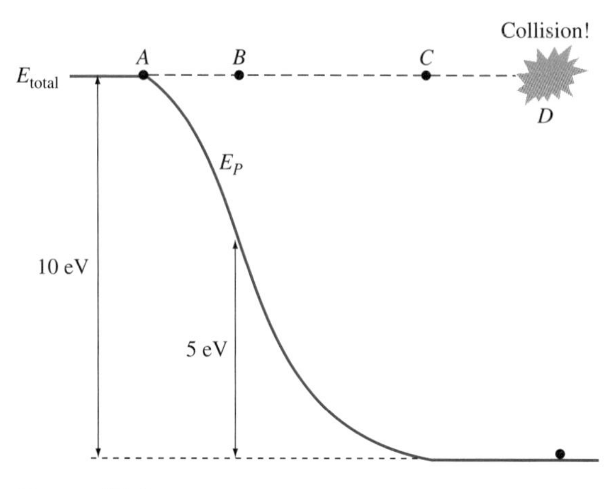

Figure 1P.2

1.5 Consider the electron in the energy diagram of Figure 1P.2. Taking the energy the electron has at Point A as E_{total}, at each of the indicated positions, find the total energy, the kinetic energy, the potential energy, and the electron's velocity. Indicate the direction of force (if any). Recall that total energy is conserved. At point *D* the electron collides inelastically with something (perhaps an atom in the crystal). After the collision, the electron's energy is equal to its potential energy, and its kinetic energy is zero. Its total energy is much less than before the collision; where did the extra energy go?

1.6 Find the kinetic energies in each of the following. Express all your answers in electron volts.

a. An electron in the lowest allowed energy states of the hydrogen atom [according to the Bohr model, Eq. (1.14)].

b. A free electron, initially at rest at the back of a cathode ray tube in your television, accelerated through a potential of 10 kV to strike the phosphor layer.

c. A tiny, drifting dust particle, of mass 1 μg and velocity a leisurely 1 mm/s.

1.7 For the following semiconductor materials, indicate to what degree you expect covalent or ionic bonding, and why:

Ge

GaP

InGaAsP

HgCdTe

1.8 For each of the semiconductors below, draw (to scale) the energy band diagrams:

Silicon: $E_g = 1.12$ eV, $\chi = 4.05$ eV

Germanium: $E_g = 0.67$ eV, $\chi = 4.0$ eV

Gallium arsenide: $E_g = 1.43$ eV, $\chi = 4.07$ eV

Indium phosphide: $E_g = 1.35$ eV, $\chi = 4.35$ eV

1.9 What minimum energy must an electron at the bottom of the conduction band in aluminum antimonide gain to become free of the crystal? Repeat for an electron at the top of the valence band.

1.10 A nondegenerate semiconductor cannot conduct current at absolute zero (degeneracy will be discussed in Chapter 2). How much energy must at least one electron obtain in silicon before conduction is possible?

1.11 At room temperature in a cubic centimeter of intrinsic silicon, there will be about 10 billion electrons in the conduction band.

a. How many holes are in the valence band?

b. If electrons are constantly seeking lower energies and recombining with holes (empty states at lower energies), then how can the number 10 billion remain constant?

1.12 Suppose the electron in Figure 1.12 is traveling to the right at constant energy. What happens to it as it approaches the surface of the material? Explain your answer, using the energy diagram.

1.13 Show that Equation (1.38) is a solution to Equation (1.37). What is the significance of the positive and negative signs of *K*?

1.14 a. Calculate the de Broglie wavelength of

i. A free electron with 1 eV of kinetic energy

ii. An electron with 10 keV of kinetic energy

iii. A tiny, drifting dust particle of mass 1 μg and a leisurely velocity of 1 mm/s

iv. Yourself, walking at 4 mph on your way to class.

b. What is a typical size of an atom? You begin to see why quantum mechanics and the wave description are not useful for large objects.

1.15 What is the wavelength of an electron at the bottom of the *E-K* relationship of Figure 1.13? What is its kinetic energy there?

1.16 Consider the *E-K* diagram shown in Figure 1P.3.

a. Verify that it meets the required criteria:

i. $E(K)$ is periodic in *K* space with period $2\pi/a$.

ii. Equivalent extrema exist at $K = 0, \pm 2\pi/a, \pm 4\pi/a, \ldots$.

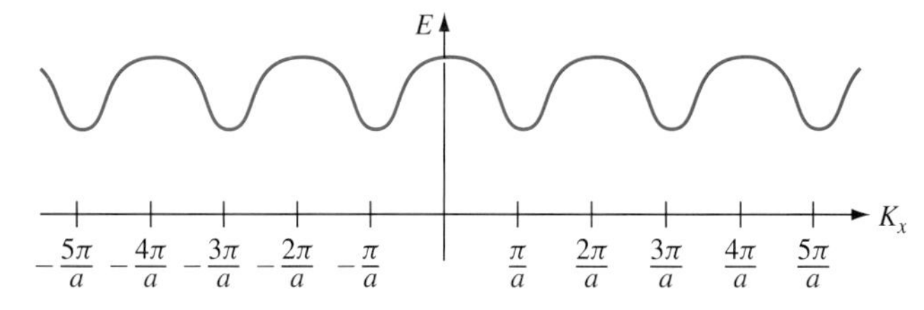

Figure 1P.3

iii. Equivalent extrema exist at $K = \pm\pi/a, \pm 3\pi/a, \pm 5\pi/a, \ldots$.
iv. The slope of the E-K curve is zero at $K = 0, \pm\pi/a, \pm 2\pi/a, \ldots$.

b. Indicate the first Brillouin zone.

c. Sketch the corresponding v_g-K diagram.

d. In what regions of the E-K diagram are electrons most likely to be found for this material?

1.17 Explain the analogy between using a conductor thicker than the skin depth to shield a region of space from electromagnetic waves, and the ability of an electron to penetrate a potential barrier.

1.18 The infinitely thick potential barrier of Figure 1.18a can be considered a crude approximation to the potential barrier at the surface of a semiconductor (see Figure 1.12).

a. How, then, might you construct a thin potential barrier like that in Figure 1.18b? Thin potential barriers are used in a wide variety of semiconductor devices, including tunnel diodes, contacts, and field-effect transistors.

b. How would you construct a potential well (thin region of lower potential energy bounded by region of higher potential energy)? Potential wells are widely used in lasers, photodectors, and heterojunction bipolar transistors.

1.19 a. From the Bohr model, what emission wavelength would you expect for a transition in hydrogen from E_2 to E_1? Transitions ending at E_1 are collectively called the Lyman series and are generally found in the ultraviolet region of the spectrum.

b. What emission wavelength would you expect from a transition from E_4 to E_3? This is the first emission line in the Paschen series, and is in the infrared.

1.20 a. What wavelength of light should you shine on hydrogen to cause electrons to go from E_1 to E_2 by optical absorption?

b. What would happen if you passed a beam of $\lambda = 430$ nm through a hydrogen gas? Explain your answer.

1.21 In discussing Figure 1.19a, we pointed out that, in a material with a band gap of 2.5 eV, an electron near the top of the valence band could not absorb a photon of energy 2.06 eV, since it would have to end up at a forbidden energy state.

a. What about an electron deep in the valence band, more than 2.06 eV below the band edge E_V? Why is it unlikely for this electron to absorb the photon?

b. Why is unlikely for a photon of 2.06 eV to be absorbed by an electron in the conduction band?

1.22 For a simple cubic crystalline structure of lattice constant $a = 0.50$ nm,

a. How many atoms are there per unit volume? (*Hint:* an easy way to proceed is to calculate the volume of the unit cell and the number of atoms per unit cell. Although there are eight atoms involved in any given unit cell, each atom in the simple cubic structure is part of eight different cells, one corner of each. Thus, there are 8 atoms $\times \frac{1}{8}$ atom per corner or 1 atom per unit cell.)

b. How many atoms per unit *area* are there in the (100) plane? The (110) plane? The (111) plane?

c. What if the lattice is FCC instead (still with $a = 0.50$ nm)? Now how many atoms per unit volume are there?

2 CHAPTER

Homogeneous Semiconductors

2.1 INTRODUCTION AND PREVIEW

In this chapter, we extend our understanding of electron motion in solids to investigate some of the electronic properties of homogeneous semiconductors. Homogeneous semiconductors are those that consist of one uniform material; for example, pure (intrinsic) silicon, or silicon with impurities uniformly distributed. In later chapters, devices that combine different materials and nonuniformly distributed impurities are discussed.

Semiconductors conduct current in a unique manner. Instead of the current being carried by electrons as it is in metals, two types of particles or charge carriers can carry current in semiconductors: electrons and holes. The hole is an artificial concept, as indicated in Chapter 1, but it is very convenient for describing the electronic properties of semiconductors.

To discuss the behavior of electrons and holes, it is necessary to use quantum mechanics, since the motion of electrons in a semiconductor crystal cannot be accurately described by using conventional classical mechanics. Quantum mechanics is discussed somewhat further in Supplement A to Part 1.

As discussed in Chapter 1, an electron can be described by its wave function. The wave function is found by solving Schroedinger's equation [Equation (1.35)], for which one has to know the potential energy function E_P of the electron as a function of position. Unfortunately, this is not completely nor accurately known for an electron in a solid. Two approximations were discussed in Chapter 1: the free electron approximation, in which E_P was assumed constant with position, and the quasi-free electron approximation, in which E_P is a periodic function of position.

Fortunately, for the case of semiconductors it is possible to use "pseudo-classical mechanics" to describe the motion of most electrons of interest, which is to say those that carry electrical current. Pseudo-classical mechanics is a way of applying Newton's laws and other familiar equations, using "effective" quantities that give the correct results. For example, we can use an effective mass for electrons, which, when inserted into classical equations such as $F = ma$, will predict the actual motion of the electron. Newton's law then becomes

$$F = m^* a \tag{2.1}$$

where m^* is referred to as the *effective mass* of the electron and takes into account the interaction of the electron with the forces resulting from the periodic nature of the lattice. The force F in Equation (2.1) is then the resulting *external* force acting on the electron. In this way, we can treat the electron as if it were a normal free particle, even though it is in reality nothing close to that. This approach works only up to a point, but is very convenient where applicable.

The electrons of interest in semiconductors usually are near the top or bottom of a band. Those near the bottom of a band behave as the "quasi-free" electrons discussed briefly in Chapter 1. It will turn out that holes, found near the top of the valence band, can also be treated as particles.

The relationship between energy and wave vector K for electrons in crystals is discussed in this chapter. The wave vector arose when we chose to consider electrons as waves, but the energy-wave vector relationship allows us to determine some of the parameters needed to treat the electrons as particles.

The effects of the impurity and temperature dependencies of the quasi-free electron (and hole) concentrations and their distributions with energy are also discussed in this chapter.

2.2 PSEUDO-CLASSICAL MECHANICS FOR ELECTRONS IN CRYSTALS

We have seen that for electrons in crystals, we can't use classical mechanics, with familiar quantities such as mass, velocity, kinetic energy, and potential energy, and familiar equations such as $F = ma$. It would be convenient, however, to be able to use classical mechanics, as opposed to quantum mechanics, because of the simpler equations and because of the physical insight that classical mechanics provides. We show in this section that it is possible to define a pseudo-classical mechanics in which we use the classical equations, but the true electron mass is replaced by an effective mass m^*, which incorporates the electron interaction with the periodic potential of the crystal. This effective mass, when used in the classical equations, correctly predicts the behavior of the electron in a crystal provided the force F on the electron is taken as the external, or applied, force. We start with the easiest case, an electron in a one-dimensional crystal.

2.2.1 ONE-DIMENSIONAL CRYSTALS

The Free Electron In Chapter 1, the energy-wave vector relation for an electron in a one-dimensional crystal was discussed for two cases. In the free-electron approximation, the potential energy was assumed constant with the value E_0. Then the E-K relation has the form

$$E = E_0 + \frac{\hbar^2 K^2}{2m_0} \tag{2.2}$$

where the second term is the kinetic energy

$$E_K = \frac{\hbar^2 K^2}{2m_0} \tag{2.3}$$

For this free-electron case, de Broglie's relation is valid and $m_0 v = P = h/\lambda$, where v is the electron velocity. The kinetic energy has the classical value

$$E_K = \frac{m_0 v^2}{2} = \frac{P^2}{2m_0}$$

For this case Newton's law, $F = m_0\, dv/dt$, is valid. Note that for the free electron case, there are no complicating additional forces from the crystal, and thus we can use the actual classical equations and the actual free electron mass.

The velocity of the electron in the crystal is given by the group velocity [Equation (1.43)]

$$v = v_g = \frac{1}{\hbar}\frac{dE}{dK} \tag{2.4}$$

and is proportional to the slope of the E-K diagram. From Equation (2.2), the electron mass m_0 can be expressed as

$$m_0 = \hbar^2 \left(\frac{d^2 E}{dK^2}\right)^{-1} \tag{2.5}$$

and since the E-K curve is parabolic, d^2E/dK^2 is a constant and thus m_0 is constant in energy.

The Quasi-Free Electron Recall that the quasi-free electron model takes the electron to be in a region in a crystal whose potential energy is modeled as a periodic function extending to infinity in both directions (Figure 1.14). To obtain a physical picture of the behavior of quasi-free electrons, it is convenient to express their properties in forms analogous to those of a free electron, Equations (2.2) to (2.5).

To find the effective mass, we will examine the E-K diagram of a material. Consider first the case of an electron near the bottom of the conduction band as shown in Figure 1.16 and repeated in Figure 2.1a. We have chosen a material for which the bottom of the conduction band is at $K = 0$; GaAs is one such material. We will expand the E-K relation in a power series. Since we are considering an electron near the bottom of the band where $E = E_C$, we expand the energy about $K = 0$:

$$E = E_C + \left(\frac{dE}{dK}\right)\bigg|\, K + \left(\frac{1}{2}\frac{d^2E}{dK^2}\right)\bigg|\, K^2 + \text{HOTs} \tag{2.6}$$

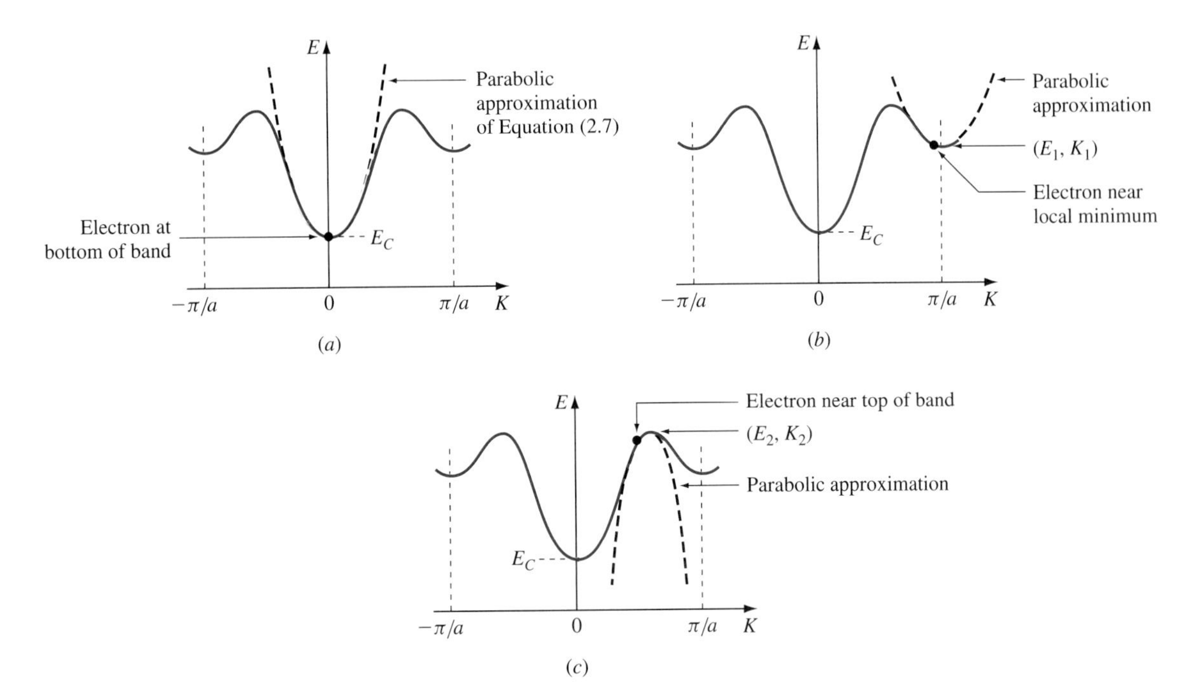

Figure 2.1 The *E*-*K* diagram for (a) an electron at the bottom of the conduction band at $K = 0$, where the velocity and kinetic energy are both zero and thus the total energy is equal to the potential energy; (b) an electron in a local minimum, where it has a different effective mass, and (c) an electron near the top of the band.

where HOTs refers to higher-order terms. The derivatives are evaluated at $K = 0$ where $dE/dK = 0$. Neglecting the HOTs gives

$$E = E_C + \left(\frac{1}{2}\frac{d^2E}{dK^2}\right)K^2 \tag{2.7}$$

This is the expression for a parabola. We can justify neglecting the higher order terms provided we stay in a region where the *E*-*K* relation is parabolic, in this case near the bottom of the band.

For a free electron, the group velocity was given by

$$v = \frac{1}{\hbar}\frac{dE}{dK}$$

[Equation (2.4)]. Here, too, the group velocity has the same form. That is, the velocity of the electron in the crystal is the group velocity, given by

$$\boxed{v = v_g = \frac{1}{\hbar}\frac{dE}{dK}} \tag{2.8}$$

or the group velocity is proportional to the slope of the *E*-*K* diagram. From

Figure 2.1a, an electron at the very bottom of the band ($K = 0$, $dE/dK = 0$) has zero velocity. Since the kinetic energy is that energy associated with motion, at $K = 0$ we can deduce that the kinetic energy $E_K = 0$. The total energy of an electron is the sum of the kinetic energy and potential energy, so at the bottom of the conduction band, the total energy is the same as the potential energy E_P. In other words,

$$E_P = E_C \tag{2.9}$$

Thus, the potential energy for an electron near the bottom of a band is the energy at the bottom of the band.

Compare Equation (2.7) for the electron in a one-dimensional crystal with Equation (2.2) for a free electron (both repeated here):

$$E = E_0 + \frac{\hbar^2}{2m_0}K^2 \qquad \text{free electron} \tag{2.2}$$

$$E = E_C + \left(\frac{1}{2}\frac{d^2E}{dK^2}\right)K^2 \qquad \text{quasi-free electron} \tag{2.7}$$

For the free electron, the quantity E_0 is the potential energy and the second term in Equation (2.2) is its kinetic energy. For the electron near the bottom of the conduction band in a crystal [Equation (2.7)], we have substituted E_C for the potential energy ($E_P = E_C$), and thus the quantity $\frac{1}{2}(d^2E/dK^2)K^2$ is the kinetic energy, or the energy associated with the motion of the quasi-free electron. Therefore

$$E_K = \frac{1}{2}\frac{d^2E}{dK^2}K^2 \tag{2.10}$$

Since the term $\frac{1}{2}(d^2E/dK^2)$ of Equation (2.10) is analogous to $\hbar^2/2m_0$ of Equation (2.2), we can define an effective mass to be

$$\boxed{m^* = \hbar^2\left[\frac{d^2E}{dK^2}\right]^{-1}} \tag{2.11}$$

analogous to the case of a free electron [Equation (2.5)],

$$m_0 = \hbar^2\left[\frac{d^2E}{dK^2}\right]^{-1} \tag{2.5}$$

The effective mass is inversely proportional to the curvature of the *E*-*K* diagram.

Recall that by neglecting the higher-order terms, we effectively restrict ourselves to regions where the *E*-*K* curve approximates a parabola, in this case near the bottom of the conduction band. In such a region, the effective mass is constant. Notice, however, that the *E*-*K* curve also approximates a parabola in the vicinity of $K = \pm\pi/a$, and the concept of constant effective mass can be used here also. Near this minimum (E_1, K_1), we can expand the energy about $K = K_1$ in a manner analogous to the above procedure. An electron in the parabolic region near this minimum has a different potential energy (where $v_g = 0$ and thus $E_K = 0$) at $E_P = E_1$, and

$$E_K = \frac{1}{2} \left. \frac{d^2E}{dK^2} \right|_{K=K_1} (K - K_1)^2$$

and

$$m^* = \hbar^2 \left[\left. \frac{d^2E}{dK^2} \right|_{K=K_1} \right]^{-1}$$

Since the curvature can be different here than at $K = 0$, however, in general so is the value of m^*.

Recall that for a classical particle of mass M, the kinetic energy is $E_K = Mv^2/2 = P^2/2M$, where $Mv = P$, the particle momentum. Similarly, for an electron near the minimum at $K = 0$ in a periodic crystal, from Equations (2.10) and (2.11),

$$E_K = \frac{m^* v^2}{2} = \frac{\hbar^2 K^2}{2m^*} \tag{2.12}$$

where v is the group velocity, and thus the quantity $\hbar K$ is often referred to as the *crystal momentum,* as pointed out in Chapter 1.

$$\text{Crystal momentum} = \hbar K$$

The kinetic energy near a minimum at K_1 becomes

$$E_K = \frac{\hbar^2 (K - K_1)^2}{2m^*} \tag{2.13}$$

where m^* is the electron effective mass at this minimum. Note that in this case the electron momentum is not the same as the crystal momentum, or $m^* v \neq \hbar K$. In an electron transition it is the crystal momentum $\hbar K$, rather than the particle momentum, that is conserved.

Now consider an electron near E_2 in Figure 2.1c. Note that the *E-K* diagram looks parabolic here also, although the curvature is negative. The slope of the *E-K* curve is zero at the top of the band, so from Equation (2.8) the electron velocity at E_2 is zero, and the kinetic energy is also zero. The potential energy of the electron near the local maximum at the top of the band is therefore E_2. In that case, the total energy of the electron in Figure 2.1c is less than its potential energy. This implies that the kinetic energy is negative. In the parabolic region near a maximum, the curvature of the *E-K* curve is negative, or from Equation (2.11), m^* is actually negative. This may seem nonphysical, and we will return to address this later. For now we note that a negative effective mass producing a negative kinetic energy is consistent with the expression for E_K in Equation (2.12) or (2.13).

Between the regions near the extrema, the effective mass as defined by Equation (2.11) is a function of energy since the HOTs of Equation (2.6) cannot be neglected. Since the development here is based on neglecting the HOTs, the results are valid only in the parabolic regions where m^* is constant.

Next, we consider the force on an electron. Recall from classical mechanics

$$F = -\frac{dE_P}{dx} \tag{2.14}$$

In an analogous manner, in pseudo-classical mechanics we have E_P equal to E_C, for an electron near the bottom of the band, and

$$F = -\frac{dE_C}{dx} \tag{2.15}$$

An example of a case in which a force is applied is shown in Figure 2.2a. Here a voltage is applied across a semiconductor sample, generating an electric field $\mathscr{E}$. From classical mechanics, we know that $F = -q\mathscr{E}$. From Equation (2.15), we can see that if a force is applied, there must be a slope or gradient in the conduction band edge. This idea is illustrated in Figure 2.2b. In this case, the electric field is in the negative x direction, and the force on the electrons accelerates them to the right ($F = -dE_C/dx$). Remember that by conservation of energy, the total energy (between collisions) is constant and the electrons travel horizontally on the energy band diagram. From the diagram, we observe that with increasing x, the potential energy (E_C) decreases and the kinetic energy (the difference between the total energy and the potential energy) increases. This is consistent with our expectation that as the electron is accelerated by the electric field, its kinetic energy increases.

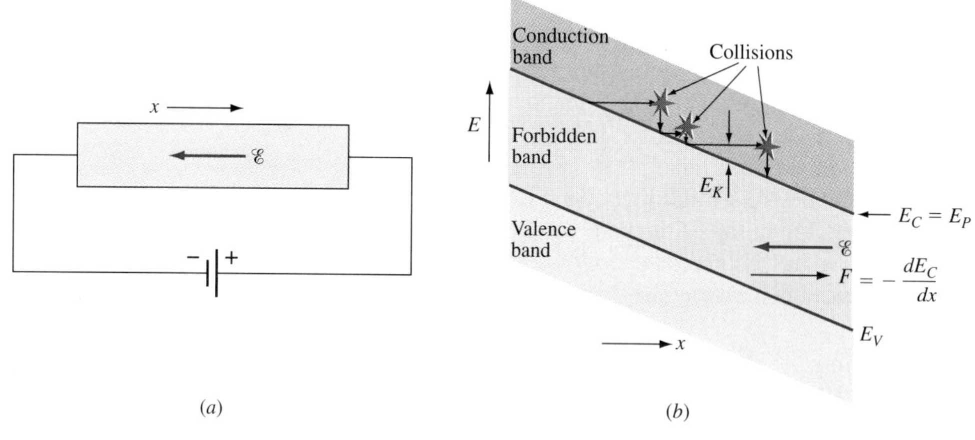

Figure 2.2 An external electric field is applied across a bar of semiconductor. (a) The physical picture; (b) the energy band diagram. Electrons in the conduction band are accelerated to the right; they travel at constant energy between collisions.

As it travels, the electron will collide with atoms, defects, or impurities. Energy can be transferred to the other particle during the collision, in which case the electron loses some of its total energy, as shown in Figure 2.2b. The electron continues to be accelerated by the force of the applied electric field, however. The acceleration is

$$a = \frac{dv}{dt} \tag{2.16}$$

Since $F = m^*a$, near a minimum, where m^* is positive, the electron is accelerated in the direction of the applied force. Near a maximum, however, the effective mass is negative. This means the electron is accelerated opposite to the applied force—a counterintuitive concept. This behavior results because the total force on the electron is the combination of the externally applied force plus the force exerted by all the atoms in the crystal. The effective mass accounts for the internal atomic forces so that pseudo-classical mechanics can be used. In this case, the effects of the internal forces combine to accelerate the electron the "wrong" way.

*2.2.2 THREE-DIMENSIONAL CRYSTALS

In pseudo-classical mechanics, we use the *E-K* diagram to find the effective mass, from which we predict the behavior of an electron, under an applied field (for example). Recall from Chapter 1 that we developed our understanding of the general shape and properties of the *E-K* diagram from a discussion of the Bloch wave function (i.e., it is periodic in *K*, the slope is zero at the center of the Brillouin zone, etc.). The discussion of the *E-K* diagram and the effective parameters used in pseudo-classical mechanics in one dimension can be extended to two and three dimensions. We state here the results for three-dimensional crystals:

1. The Bloch wave in three dimensions is given by

$$\begin{aligned} \Psi(\vec{r}, t) &= U_K(\vec{r})e^{j[\vec{K}\cdot\vec{r} - (E/\hbar)t]} \\ \psi(\vec{r}) &= U_K(\vec{r})e^{j(\vec{K}\cdot\vec{r})} \end{aligned} \tag{2.17}$$

   where $\vec{r}$ is a position vector and $U_K(\vec{r})$ is periodic in $\vec{r}$ with the periodicity of the crystal.
2. The E-$\vec{K}$ relation in any band is periodic in K space with the periods $2\pi/a$, $2\pi/b$, and $2\pi/c$, where a, b, and c are the periodicities in the x, y, and z directions.
3. All of the information in the E-$\vec{K}$ curve, and parameters derived from it, is contained in the reduced zone or the first Brillouin zone.
4. The (group) velocity of an electron in three dimensions is

$$\vec{v} = \frac{1}{\hbar}\left(\hat{i}\frac{\partial E}{\partial K_x} + \hat{j}\frac{\partial E}{\partial K_y} + \hat{k}\frac{\partial E}{\partial K_z}\right) \tag{2.18}$$

where $\hat{i}$, $\hat{j}$, and $\hat{k}$ are the unit vectors in the x, y, and z directions respectively.

5. There are relative extrema in K space at $K = 0$ and at the edge of the reduced zone in every principal crystallographic direction. Stated differently, at the edge of the reduced zone (in three dimensions, the edge is a surface), the slope of the E-K curve is zero in every direction in K space. There is one exception in cases where two bands coincide at the boundary of the reduced zone. In that case, for cubic crystals the slopes of the two E-K curves do not each have to be zero, but their sum must be zero.
6. The E-$\vec{K}$ relation is difficult to plot for a three-dimensional crystal; in practice, E is plotted as a function of K along the principal crystallographic directions. Because of symmetry about $K = 0$, it is necessary to plot E for only half of the Brillouin zone.
7. Near a relative extremum of a band, the effective mass m^* can be defined such that Newton's laws can be used. This effective mass is positive near a minimum and negative near a maximum.
8. Near an extremum, the curvature of the E-K plot may depend on the direction of K, and thus the direction in which the electron is traveling. In such a case, the effective mass is direction dependent:

$$m_x^* = \hbar^2 \left(\frac{\partial^2 E}{\partial K_x^2} \right)^{-1}$$
$$m_y^* = \hbar^2 \left(\frac{\partial^2 E}{\partial K_y^2} \right)^{-1} \tag{2.19}$$
$$m_z^* = \hbar^2 \left(\frac{\partial^2 E}{\partial K_z^2} \right)^{-1}$$

9. For a semiconductor having a cubic unit cell structure and having an extremum at $K = 0$, the curvature is direction independent and

$$m^* = m_x^* = m_y^* = m_z^* \qquad (K = 0) \tag{2.20}$$

or the effective mass is a scalar.

2.3 CONDUCTION BAND STRUCTURE

We are now ready to examine the individual conduction band structures for some common semiconductor materials. The E-K relations for the conduction bands of GaAs, Si, and Ge are plotted in Figure 2.3, along the direction of K with the lowest minimum for that material.

The absolute minimum in the conduction band for GaAs occurs at $K = 0$. Because of the symmetry of its cubic structure, the curvature of the E-K plot near that minimum, and thus m^*, is the same for any direction of motion. For GaAs,

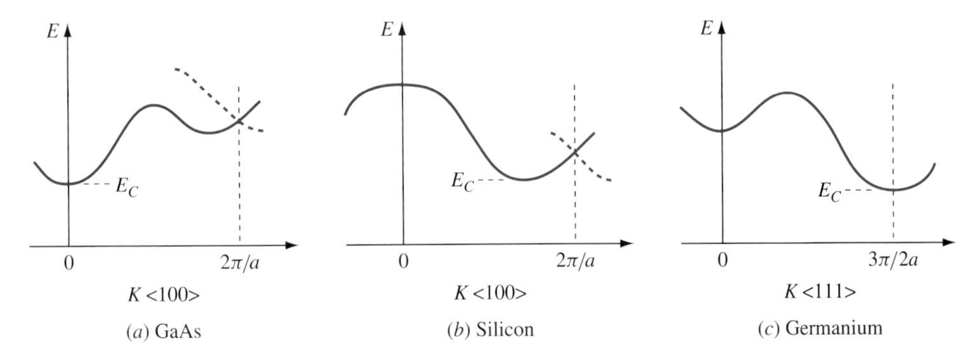

Figure 2.3 The *E*-*K* diagrams for three common semiconductors. The crystallographic direction is shown on the *K* axis. The slope of the *E*-*K* curve must be zero at the Brillouin zone edge, unless multiple bands coincide there.

then, the effective mass for an electron in the conduction band is a scalar [Equation (2.20)]. Its value has been measured to be $0.067m_0$, where m_0 is the rest mass of a free electron. GaAs is an example of a material with two bands having the same energy at the edge of the zone in the <100> directions[1]; thus, the *E*-*K* slope there is not zero, but rather the sum of the slopes is zero.

Silicon (Figure 2.3b), on the other hand, has conduction band minima in the <100> directions at a value of K of about 0.85 of the K value at the zone edge. This is the bottom of the conduction band, where most of the electrons of interest are found. Unfortunately, the effective mass is not a scalar here, because the curvature is different in different directions (in K space) at this minimum. For an electron traveling in a [100] direction in silicon, there is one effective mass, called the longitudinal effective mass $m^*_\parallel$, and the two transverse effective masses, used for travel in the other two directions, are equal and denoted $m^*_\perp$.

In Figure 2.3c, we see that germanium has absolute minima in the conduction band at the zone edges in the <111> directions. The effective mass is not a scalar here, either. Another minimum at $K = 0$ exists at a higher energy. Electrons in this higher-energy minimum have scalar effective mass.

When the effective mass near the bottom of the conduction band is not a scalar, as in Si and Ge, its value depends on the direction in which the electron is traveling. Therefore, some sort of weighted average of $m_\parallel$ and $m_\perp$ is required to obtain a value of m^*. The particular averaging method and the resulting values of m^* depend on the particular type of problem being solved. The two most common averaging techniques result in the *conductivity effective mass* m^*_{ce} for electrons, used for conductivity and related calculations, and the *density-of-states effective mass* m^*_{dse} for electron concentrations. Both of these averaging techniques

[1]The edge of the zone in diamond and zinc blende structures is at $K = 2\pi/a$, where a is the periodicity of equivalent planes. In the <100> directions the spacing between atomic planes is $a/2$. In the <111> directions the spacing is $2a/3$ where a is the lattice constant.

Table 2.1 Effective masses for electrons near the bottom of the conduction band in units of m_0, the rest mass of the free electron at 300 K

Material	$m^*_{(K=0)}$	$m^*_{\parallel}$	$m^*_{\perp}$	m^*_{ce}	m^*_{dse}
Si	—	0.92	0.197	0.26	1.09
GaAs	0.067	—	—	0.067	0.067
Ge	—	1.64	0.082	0.12	0.56
InP	0.077	—	—	0.077	0.077

are discussed in Appendix D. Table 2.1 lists the values for the effective mass components and average values for some common semiconductors. [1]

2.4 VALENCE BAND STRUCTURE

Although the semiconductor band structure varies from material to material for the conduction band, the valence bands are qualitatively similar for most semiconductors important in electronics. They also tend to be simpler than the conduction bands. The valence bands consist of three overlapping bands. These bands all have absolute maxima at $K = 0$ but with different curvatures, resulting in different effective masses, as shown in Figure 2.4.

Recall that the effective mass of an electron, being inversely proportional to the curvature of the *E-K* diagram, is negative near the top of the valence band. Furthermore, the valence band is normally almost entirely full of electrons, with just a few empty states near the top of the band. If we consider these empty states to be the holes, then we can consider the holes to have a *positive* charge along with a *positive* effective mass. Later in the chapter, it is shown that this move is legitimate.

Two of the bands have the same energy maxima at $K = 0$. The two bands h and l refer to *heavy holes* (smaller curvature) and *light holes* (higher curvature)

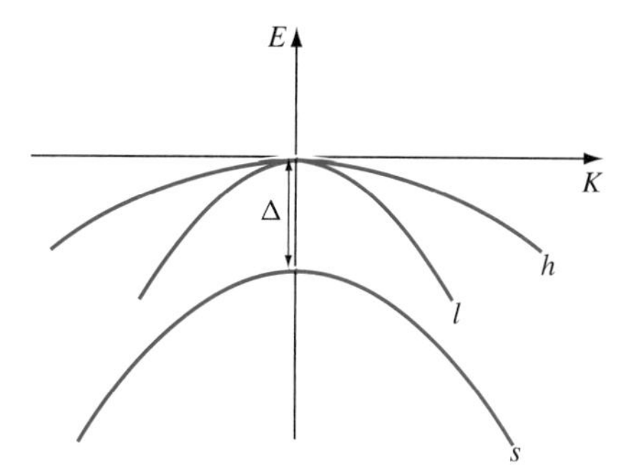

Figure 2.4 The *E-K* diagram for the valence band in most semiconductors, indicating (*h*) the heavy hole band, (*l*) the light hole band, and (*s*) the split-off band.

Table 2.2 Effective masses for holes in the valence bands of several semiconductors (masses are given as multiples of the electron rest mass m_0)

Material	m^*_{lh}	m^*_{hh}	m^*_{sh}	Δ, eV	m^*_{ch}	m^*_{dsh}
Si	0.16	0.48	0.24	0.044	0.36	1.15
GaAs	0.082	0.45	0.15	0.34	0.34	0.48
Ge	0.044	0.28	0.08	0.29	0.21	0.292
InP	0.08	0.4	0.15	0.11	0.3	0.42

respectively. A third band is split off by an energy Δ, due to spin-orbit coupling. This spin-orbit interaction results from magnetic forces that are influenced by the individual spins of the electrons and the magnetic fields resulting from their orbits. The split-off band is designated s.

The values of effective mass for holes associated with each of the bands, along with the split-off energy Δ, are given in Table 2.2. Also given are values for the conductivity effective mass and density-of-states effective mass for holes, m^*_{ch} and m^*_{dsh}, as discussed in Appendix D. Again, these result from two different ways of averaging, except that for electrons in the conduction bands, we were averaging the longitudinal and transverse effective masses. For holes in the valence band we combine the light hole and heavy hole. Normally, the split-off band is enough removed in energy from the other bands that it contains few holes and can be ignored. As for electrons, conductivity effective masses are used for calculations involving electrical conduction while the density-of-states effective mass is used in calculations related to carrier concentrations.

2.5 INTRINSIC SEMICONDUCTORS

A semiconductor is said to be intrinsic if it contains no impurities and no crystalline defects. As we will see, this implies that the concentration (number per unit volume) of electrons in the conduction band is equal to the concentration of holes in the valence band.

In an intrinsic semiconductor at absolute zero, electrons occupy all of the electronic energy states in the valence band, and all the states in the conduction band are empty. This is to be expected since, at absolute zero, every electron will be found at the lowest possible energy state.

At higher temperatures, the electrons acquire some thermal energy, which is transferred to them from the crystal lattice. The atoms in the crystal lattice vibrate, and these lattice vibrations can be transmitted through the crystal as waves. These acoustic waves are called *phonons,* and associated with each phonon is an energy and a wave vector. Like electrons and photons, phonons can be treated as waves or as particles, as discussed in Supplement B to Part 1.

The phonons can excite electrons from the valence band to the conduction band. This leaves an empty state or hole in the valence band. The electron is now quasi-free, and there is a quasi-free hole in the valence band. The electron and hole together are referred to as an *electron-hole pair*. When the electron-hole pair is created by absorption of phonons, we call the process *thermal generation,* since the phonons or lattice vibrations carry the thermal energy of the crystal. If a photon provides the energy, the process is termed *optical generation*.

Figure 2.5 indicates the generation process in physical space using a *bond diagram* (a) and on the *energy band diagram* (b). In the bond diagram (here represented in two dimensions), each silicon atom shares its four outer electrons (the valence electrons) with four neighboring silicon atoms. Thus, the electrons make up the covalent bonds between the atoms, and each atom has a complete outer shell of eight electrons.

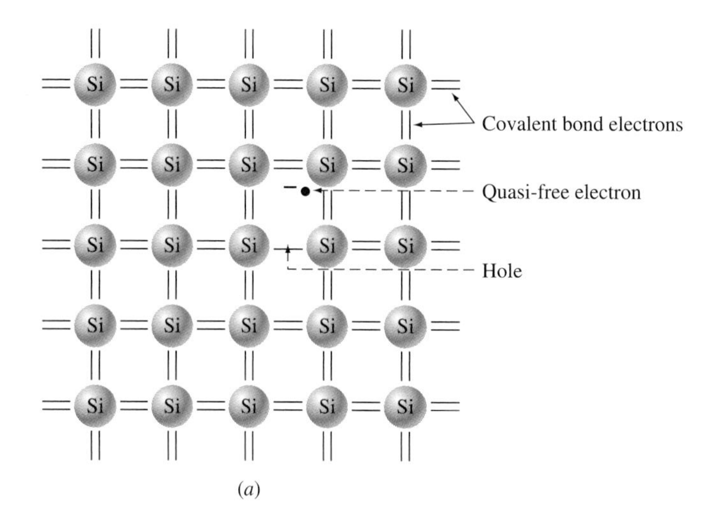

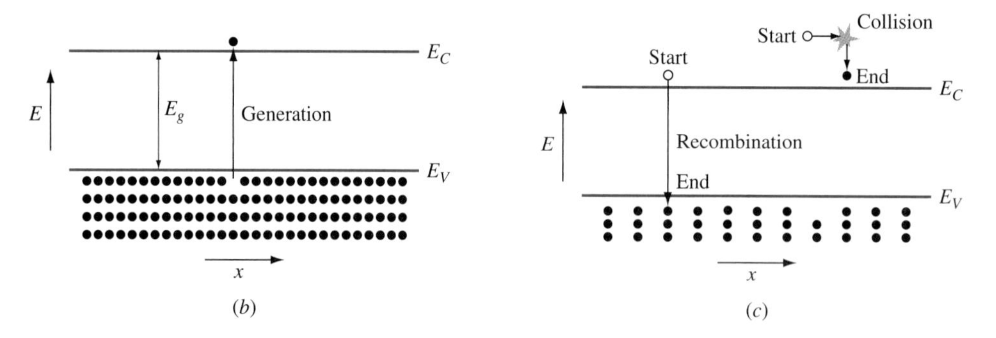

Figure 2.5 In thermal generation, a valence electron acquires some extra energy and moves into the conduction band. (a) Physical picture or bond diagram; (b) energy band diagram. In recombination (c), an electron from the conduction band falls into a hole in the valence band, and both the conduction-band electron and the hole disappear. It is also possible for an electron to lose energy by colliding with something (c, right) in which case it may remain in the conduction band.

If one electron in the valence band should acquire some extra energy (e.g., by absorbing photons or phonons) it may break out of the bond and go to a higher energy state in the conduction band. Here, the electron no longer contributes to the covalent bonding. This electron now can move about the crystal. If the electron, now in the conduction band, should move, then that moving charge represents current.

Similarly, the vacant state in the valence band can be considered a quasi-free hole, which is free to move throughout the crystal. If an electron from a nearby bond shifts into the empty state, then a new empty state appears in that nearby bond. The hole, in other words, has moved.

Note that for an intrinsic semiconductor, which we are considering, the next highest state for a bound (valence) electron is in the conduction band and so the valence electron must gain energy equal to or greater than the energy gap E_g. Since E_g is on the order of 1 eV for semiconductors of interest, and since phonon energies are less than about 0.1 eV, to make the transition from valence band to conduction band by thermal generation, the electron must absorb several phonons. From a particle point of view, these several phonons must simultaneously collide with one electron, a process that becomes less probable as E_g, and thus the number of phonons required, increases. Thus we expect that the larger the semiconductor band gap, the smaller the number of electrons and holes at any given temperature.

For every electron that is thermally excited to the conduction band there is necessarily a hole left behind in the valence band. Therefore, in an intrinsic semiconductor the equilibrium concentration n_0 of electrons in the conduction band is the same as the equilibrium concentration of holes p_0 in the valence band. That is, for intrinsic material

$$\boxed{n_0 = p_0 = n_i \qquad \text{intrinsic}} \tag{2.21}$$

where the subscript 0 indicates the concentration at equilibrium and n_i is defined as the equilibrium concentration of electrons or holes in an intrinsic semiconductor.

Eventually a given electron will recombine if a hole is available. That is, the electron will move into the empty state in the valence band, filling the hole as shown on the left of Figure 2.5c. When recombination occurs, both the quasi-free electron and the hole disappear. The energy lost by the recombining electron is given off as a photon (for example, in lasers), as phonons (heat), or a combination of the two.

When the electron returns to the valence band, there is no longer a quasi-free electron, and there is also no longer a hole. This process is called recombination. When recombination occurs, both the quasi-free electron and the hole disappear.

Electrons will remain in the conduction band an average time τ_n before recombining with holes in the valence band. This average time between generation and recombination is called the electron lifetime or carrier lifetime. (Electrons and holes are considered carriers because they carry current.) Typical values of τ range from 10^{-10} to 10^{-3} seconds at room temperature and are material dependent.

An electron in the conduction band can also lose energy by colliding with another particle (e.g., a phonon), as seen on the right-hand side of Figure 2.5c. In this case, the electron may move to a lower energy in the conduction band. The mean free time between collisions for the electron is appreciably shorter than its lifetime, the mean free time being about 10^{-13} to 10^{-12} seconds. Thus, the electron will make many collisions before recombining.

2.6 EXTRINSIC SEMICONDUCTORS

We have seen that in intrinsic semiconductors, the concentration of conduction band electrons is equal to the concentration of holes in the valence band. By adding *dopant* atoms, it is possible for the numbers of electrons and holes to be unequal, in which case the material is said to be *extrinsic*. A semiconductor is said to be extrinsic if $n_0 \neq p_0$. Extrinsic semiconductors are created by incorporating *impurity* atoms into the intrinsic material (a process called *doping*). The dopant atoms can be either donors or acceptors, as developed in the following sections.

If $n_0 > p_0$, a semiconductor is said to be n type, meaning current is carried predominantly by negatively charged electrons. If $p_0 > n_0$, the material is p type, and current is carried predominantly by positively charged holes.

2.6.1 DONORS

Assume that an atom with five electrons in its outer shell, such as phosphorus, is substituted for a Si atom in an otherwise pure crystal of silicon, as shown in Figure 2.6a for the bond representation. Silicon has a valence of four. The extra

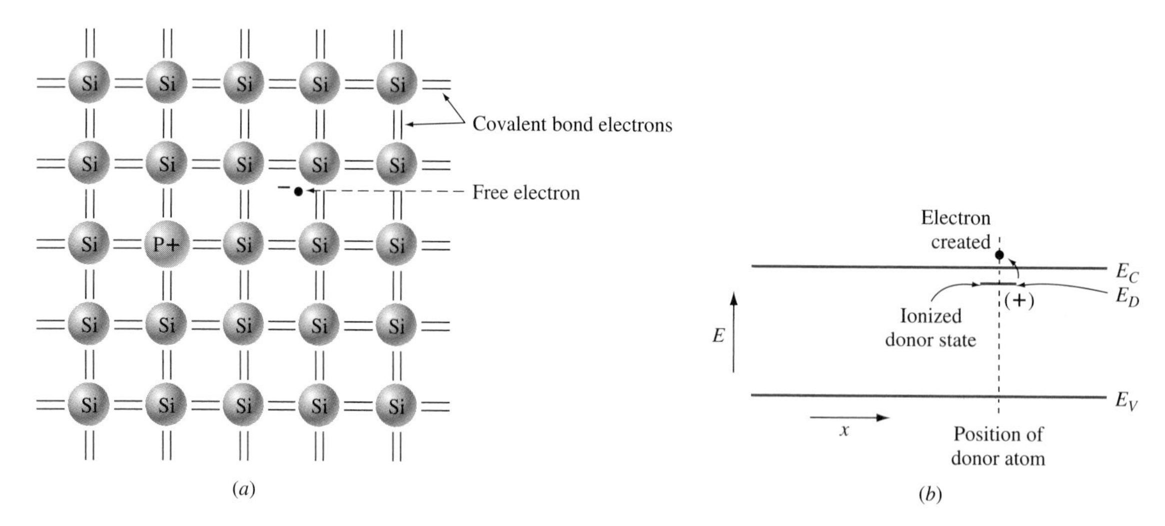

Figure 2.6 Donors in a silicon crystal: (a) bond diagram of the crystal; (b) energy band diagram of silicon doped with one phosphorus (donor) atom.

electron of the phosphorus atom is not needed for the covalent bonding, since there are enough other electrons to fill the bonds. Thus, the extra electron is more loosely bound to the phosphorus atom than the valence electrons. The extra electron can be easily "donated" to the conduction band, as we will see, and thus phosphorus is called a *donor* atom.

The energy band diagram for a Si crystal doped with a phosphorus donor is shown in Figure 2.6b. The phosphorus atom has a slightly different set of energy levels than the surrounding silicon atoms, and some of these are actually in silicon's forbidden gap. The lowest energy (ground) state associated with the donor atom exists only near that donor atom, so the donor state of energy E_D does not exist throughout the crystal. Also, the donor state is close to the conduction band edge. This indicates that it takes very little energy to excite the donor's fifth electron into the conduction band. That is consistent with our previous idea that the electron is loosely bound to the donor atom because it is not participating in bonding. Notice that there is no hole created in this case. The phosphorus atom left behind is ionized positively, but it cannot move since it is "chained down" to the crystal by the four valence bonds. Only the electron contributes to conduction, not the ion.

Now we examine donors more quantitatively, and try to determine their energy levels. Earlier we said that the extra electrons contributed by the donor atoms appeared in the conduction band. When that is the case, the donor atom is positively ionized since it has lost one electron. The positive charge of the ion produces a Coulombic force on the negative electron. This problem resembles the hydrogen atom problem, in which an electron is bound to the single-positive-charge hydrogen nucleus. From the Bohr model (as well as from quantum mechanics) for the hydrogen atom the nth energy level is given by

$$E_n = E_{\text{vac}} - \frac{m_0 q^4}{2(4\pi\varepsilon_0)^2\hbar^2 n^2} = E_{\text{vac}} - \frac{13.6}{n^2}\text{ eV} \qquad (2.22)$$

and the Bohr radius (in quantum mechanical terms the most probable distance between the electron and the nucleus) is

$$r_n = \frac{4\pi\varepsilon_0 n^2\hbar^2}{m_0 q^4} = 0.053n^2\text{ nm} \qquad (2.23)$$

For the hydrogen atom in vacuum, E_{vac} is the minimum total energy required to remove an electron from the influence of the hydrogen core (i.e., to create a free electron). A single phosphorus atom in space similarly has a set of discrete energy levels, as shown in Figure 2.7a, and an electron of energy E_{vac} or higher is free of the effect of the atom. Analogously, for a donor atom in a semiconductor, E_C is the total energy an electron must attain to be removed from the influence of the donor ion, Figure 2.7b. When the electron has energy $E > E_C$ it is free to move around the crystal in the conduction band. An electron at elevated energy states also has a large Bohr radius, so at sufficiently high energy the distance between the electron and bound donor ion is large enough that the

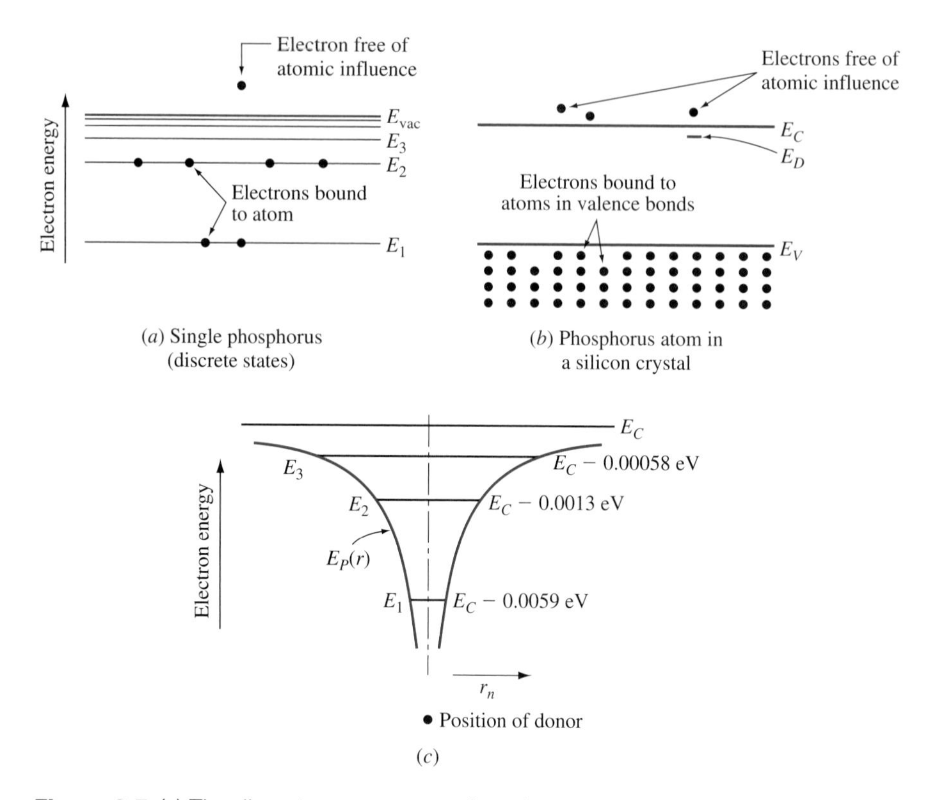

Figure 2.7 (a) The discrete energy states in a single phosphorus ion. (b) The energy band diagram for a semiconductor crystal containing a donor atom. For the discrete atom, an electron must have an energy equal to E_{vac} or higher to escape the influence of the nucleus. In the semiconductor, the electron must have an energy of E_C or greater to escape the influence of the donor ion. (c) Energy diagram for a donor in GaAs.

Coulombic force between them is negligible. In Figure 2.7b only one donor level is shown (the lowest one), but actually there are several allowed states for a donor in a semiconductor. For example, Figure 2.7c shows the band structure for a donor atom in GaAs. There are discrete states below the conduction band edge, analogous to the single hydrogen atom in vacuum. If any of these states is occupied, it will most likely be the lowest one. We call that level the donor ground state energy E_D.

To continue our analysis, we assume the following:

1. The average distance between the electron and the ion is large enough that the material between them more closely resembles the crystal than free space. This permits us to replace the permittivity ε_0 of free space with the macroscopic permittivity ε of the material.
2. The electron in the bound state of the donor atom can be treated as though it has a conductivity effective mass m^*_{ce} equal to that of an electron near the

bottom of the conduction band. (Although restricted to the region near the donor ion, the bound electron is moving in the periodic potential of the crystal, as are conduction electrons.)

Under these assumptions, Equations (2.22) and (2.23) become

$$E_n = E_C - \frac{m_{ce}^* q^4}{2(4\pi\varepsilon)^2 n^2 \hbar^2} = E_C - \frac{13.6}{n^2}\left(\frac{m_{ce}^*/m_0}{\varepsilon^2/\varepsilon_0^2}\right) \text{eV} \tag{2.24}$$

and

$$r_n = \frac{4\pi\varepsilon n^2 \hbar^2}{m_{ce}^* q^2} = 0.053 \frac{\varepsilon/\varepsilon_0}{m_{ce}^*/m_0} n^2 \text{ nm} \tag{2.25}$$

We consider the case of GaAs in which an As atom (valence 5) is replaced by a tellurium atom (valence 6) and so there is an extra electron not required for bonding. For GaAs, $m_{ce}^*/m_0 = 0.067$ and $\varepsilon/\varepsilon_0 = 13.2$. This gives

$$E_n = E_C - \frac{0.0052}{n^2} \text{ eV} \tag{2.26}$$

$$r_n = 10.4 n^2 \text{ nm} \tag{2.27}$$

We see that all values of r_n are large compared with the lattice constant of GaAs (0.565 nm) so assumption 1 is reasonably valid. That is, the lowest energy at which a bound electron can be found corresponds to an orbit with a radius of about 20 lattice constants.

The energy calculated for the ground state (the nonionized donor atom) from Equation (2.26) is 0.0052 eV below the conduction band edge, which agrees reasonably well with experiment (0.0059 eV). Thus, assumption 1 can be taken to be reasonably accurate. Note that for the excited states, r_n is large enough that assumption 1 is quite good and the energies calculated from Equation (2.26) agree very closely with experiment.

For the case of silicon (valence 4), doped with P (valence 5), the agreement is not as good. Using $m_{ce}^*/m_0 = 0.26$ and $\varepsilon/\varepsilon_0 = 11.8$ results in

$$E_n = E_C - \frac{0.026}{n^2} \text{ eV} \tag{2.28}$$

$$r_n = 2.4 n^2 \text{ nm} \tag{2.29}$$

For the ground state, from Equation (2.28), $E_C - E_1 = 0.026$ eV, which is appreciably smaller than the measured value of 0.045 eV for phosphorus in silicon. This discrepancy is explained by noticing that $r_1 = 2.4$ nm [Equation (2.29)], which is not sufficiently large compared with silicon's lattice constant of 0.543 nm to justify assumption 1. Notice, however, that for all of the *excited* states, r_n is large enough to reasonably satisfy assumption 1, and for these states the model agrees quite well with experiment.

For semiconductors (or insulators) with large effective mass or small dielectric constant, however, Equations (2.24) and (2.25) are not valid and donor states are at energies appreciably below the conduction band.

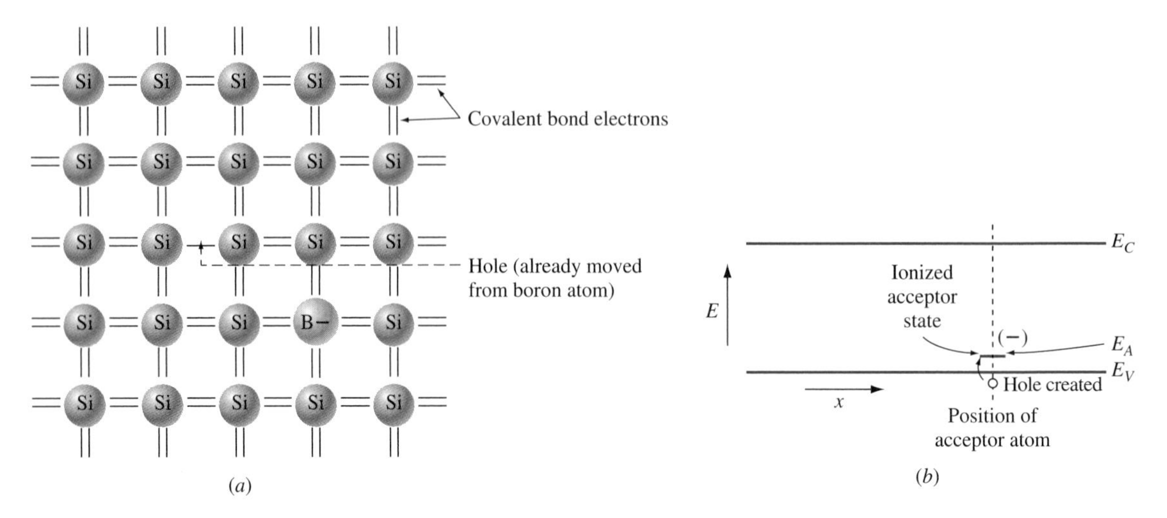

Figure 2.8 Acceptors in a semiconductor: (a) bond diagram; (b) energy band diagram. An electron is excited from the valence band to the acceptor state, leaving behind a quasi-free hole.

2.6.2 ACCEPTORS

Instead of doping a semiconductor with donors, one could add impurities that have fewer electrons in the outer shell than the replaced atoms. Consider the case of Si (four valence electrons) doped with boron, which has three valence electrons. Figure 2.8a shows how doping Si with boron will produce a p-type material. Boron is called an acceptor in silicon. We know that atoms in general like to fill their outer shells, even if it means sharing electrons. The silicon atoms near the boron atom would like to have eight electrons apiece, but they are collectively one electron short. If an electron from a nearby silicon atom should occupy that state, the bond becomes full, but the boron atom becomes negatively charged since it now has more electrons than protons. Since the electron most likely came from some nearby covalent bond, a hole was left behind. The energy band diagram for a semiconductor material doped with one acceptor is shown in Figure 2.8b. Note that in (a), the hole has already moved a short distance from the boron atom.

It does not require much energy for an electron to be excited from the valence band to the acceptor state, at energy E_A. When that does happen, a hole is created in the valence band (a hole that is free to move around) and the acceptor atom becomes negatively ionized. Again, the acceptor atom is locked into the crystal and cannot move or carry current; only the hole can move.

For the acceptors, to find the energy of the acceptor states E_A, we look at the valence band rather than the conduction band. As indicated earlier, in most semiconductors of interest to electronics there are three valence bands, all of which have a maximum at $K = 0$. Two of these have maxima at $E = E_V$, while the third is split off by an energy Δ. Following a similar procedure to the calculations we used for the donor states, and making the same assumptions, for the bound states of holes we obtain

$$E_n = E_V + \frac{13.6}{n^2}\left(\frac{m^*_{ch}/m_0}{\varepsilon^2/\varepsilon_0^2}\right) \text{eV} \tag{2.30}$$

$$r_n = 0.053n^2\left(\frac{\varepsilon}{\varepsilon_0}\right)\left(\frac{m_0}{m^*_{ch}}\right) \text{nm} \tag{2.31}$$

where m^*_{ch} is the conductivity effective mass for holes. By "bound state for a hole," we mean that the acceptor atom is occupied by a hole (is un-ionized).

Recall that earlier we indicated that if a III-V material such as GaAs is the host material, it could be made n type by substituting atoms from the sixth column of the periodic table (e.g., tellurium) for arsenic atoms (five valence electrons). [GaAs could also be made n type by substituting atoms from the fourth column, such as silicon, for the gallium atoms (three valence electrons). In either case there is one more electron added than is needed to complete the valence bond.] Similarly, we can obtain p-type GaAs by replacing column III (gallium), with column II (zinc), or by replacing column V, arsenic, with column IV, silicon. The system is then one electron short of the number needed to complete the covalent bonding, and that atom becomes an acceptor.

In GaAs, column IV elements such as germanium and silicon can be either donors or acceptors, depending on which type of atom is replaced. These are called *amphoteric impurities*. Silicon preferentially occupies Ga sites and thus is normally a donor.

2.7 THE CONCEPT OF HOLES

2.7.1 HOLE CHARGE

We mentioned earlier that states in the valence band that are *not* occupied by electrons are referred to as holes, and that these holes can move around in the valence band. In this section, we will develop this idea further, and show more rigorously that holes can be thought of as carrying current and possessing a positive charge. We approach this discussion by considering the current.

Consider an n-type semiconductor with an electric field applied. There are n electrons per unit volume in the conduction band, and p holes per unit volume in the valence band ($n > p$, for n type). The charge density in the conduction band is therefore $-qn$, the charge per electron times the density of electrons. The electron current density (amperes per unit area) is

$$J = -qn\langle v\rangle \tag{2.32}$$

where $\langle v\rangle$ is the average velocity of the electrons. We could also have written Equation (2.32) in the form

$$J = -\frac{q}{\text{volume}}\sum_i v_i \tag{2.33}$$

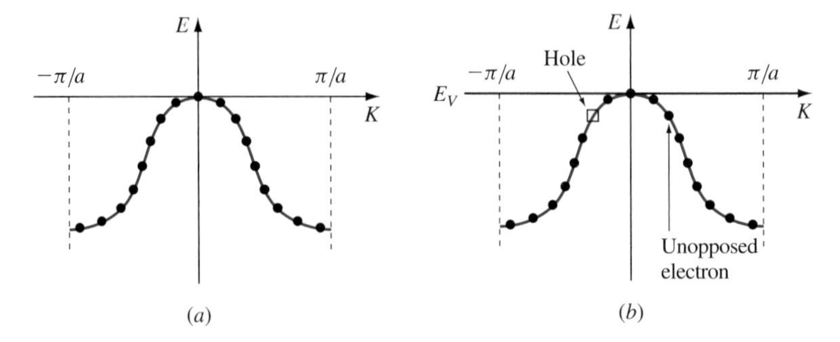

Figure 2.9 States in the valence band on the *E-K* diagram. (a) All states are full; (b) one state is empty (hole), meaning there is an electron with no opposing electron at the same energy but opposite *K* vector and opposite velocity.

where the volume is the volume of the crystal and the summation is taken over the velocities of the individual electrons.

Equation (2.33) is also valid for electrons in the valence band, but the valence band is essentially filled. A completely filled valence band is shown for the one-dimensional case in the *E-K* diagram in Figure 2.9a. Recall that the velocity of an electron is given by

$$v = \frac{1}{\hbar}\frac{dE}{dK} \tag{2.34}$$

From the figure, the *E-K* diagram is symmetrical, so for each electron with velocity v_i there is an opposing electron with velocity $-v_i$. Thus for electrons in the valence band, the total current density $J = 0$ when every state is filled. Note that the electrons are moving about in real space, but at a given energy, at every value of E there are two states in the figure, with opposite velocities.

At temperatures above absolute zero, the valence band is largely occupied by electrons, but there is still a significant population of holes (for example in p-type material the holes represent 10^{-8} to 10^{-4} of the valence band population of electrons). Most of the electrons in the valence band are opposed by electrons with opposite velocity, but not all, as shown in Figure 2.9b. It is therefore convenient to consider the unopposed electrons:

$$J = \frac{-q}{\text{volume}}\sum_i v_{ui} \tag{2.35}$$

where the subscript u refers to the unopposed electrons. Equation (2.35) can be rewritten

$$J = \frac{-q}{\text{volume}}\sum[v_i - v_{hi}] = \frac{-q}{\text{volume}}\left[\sum v_i - \sum v_{hi}\right] \tag{2.36}$$

where the first term in the brackets is the summation over all the electrons in the (full) band while the second (v_{hi}) is the summation over the vacant states (holes), $\sum v_{ui} = \sum(v_i - v_{hi})$.

Since

$$\sum_i v_i = 0 \tag{2.37}$$

it follows that

$$J = +\frac{q}{\text{volume}} \sum_i v_{hi} \tag{2.38}$$

If we consider the holes to be particles, from Equation (2.38) we see that they must be each considered to have a charge of $+q$. Thus for a hole density p, the total hole current density is

$$J = +qp\langle v_h \rangle \tag{2.39}$$

It therefore makes no difference whether we consider the current to be due to all the electrons in the valence band of charge $-q$ or whether we consider the current to be due to the number of holes of charge $+q$. Generally we talk about the holes when discussing conduction current in the valence band.

*2.7.2 EFFECTIVE MASS OF HOLES

Earlier we showed that when an electron is near the top of the valence band, its effective mass is negative, since the curvature of the band is negative in that region. We now show that if we consider the vacant states to be holes with positive charge, they must also be considered to have a positive effective mass, equal in magnitude to the (negative) effective mass of the unopposed electron.

We do this by considering a p-type semiconductor with a current flowing to the right (electrons flowing to the left), as shown in Figure 2.10a. A magnetic field is also applied such that the direction of $\vec{B}$ is into the page. The Lorentz force $\vec{F} = Q[\vec{v} \times \vec{B}]$ on an unopposed electron is given by

$$\vec{F}_n = -q[\vec{v}_{ue} \times \vec{B}] = m^*_{ue}\vec{a}_{ue} \tag{2.40}$$

where m^*_{ue} is the electron effective mass and $\vec{a}_{ue}$ is the resulting acceleration on the unopposed electron. Since $\vec{v} \times \vec{B}$ is directed downward in this example, the force on the electron is directed upward. The unopposed electron is necessarily near the top of the valence band (because the states at lower energies are very likely to be occupied) so its effective mass m^*_{ue} is negative. This means that the electron is, in fact, accelerated *downward,* the result being that the bottom of the sample becomes negatively charged. A voltage thus generated from top to bottom of the sample is shown in the figure. This charge buildup is known as the Hall effect and is used frequently to characterize materials as discussed in Supplement B to Part 1.

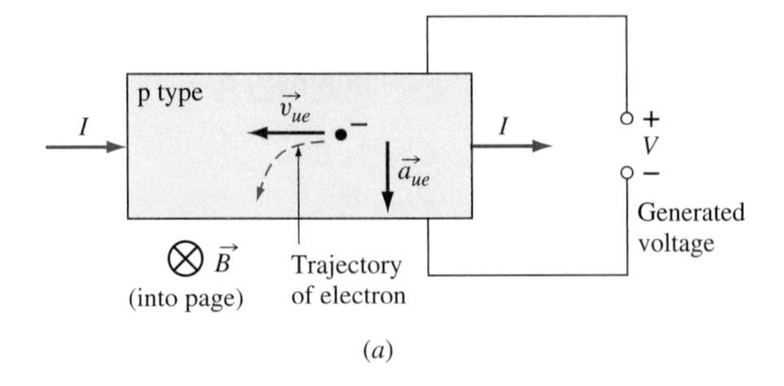

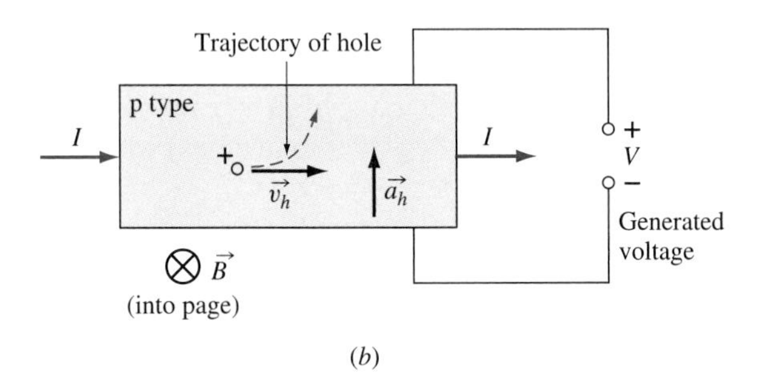

Figure 2.10 Lorentz force on an electron in a p-type sample. (a) The unopposed electrons (negative charge, negative effective mass) are accelerated downward; (b) holes, which have positive charge and positive effective mass, are accelerated upward. In either case the net result is the same: The sample becomes more negatively charged at the bottom with respect to the top.

If, instead of considering the electrons, we consider a hole as shown in Figure 2.10b, the Lorentz force is

$$\vec{F}_p = +q[\vec{v}_h \times \vec{B}] = m_h^* \vec{a}_h \tag{2.41}$$

where m_h^* and a_h are the effective mass and acceleration of the hole. The holes, charged positively, have a velocity v_h to the right, because the current is flowing to the right. The net Lorentz force is then of the same magnitude and direction as for the unopposed electron. For the same *physical* result, that is, a sample whose top is positively charged with respect to its bottom, each hole must have actually gone upward, implying that its effective mass is in fact positive, but equal in magnitude to that of the electron in the opposing state.

We conclude, therefore, that in a nearly filled band (the valence band), the vacant states can be treated as particles, called holes. These holes have positive charge and positive effective mass, and thus can respond to external forces. The effective mass of a given hole is equal in magnitude but opposite in sign to that of an electron near the top of the valence band (i.e., positive).

2.8 DENSITY-OF-STATES FUNCTIONS FOR ELECTRONS IN BANDS

We have established that there are bands of energy states that electrons can occupy in a semiconductor. We have also pointed out that most electrons of interest will be near the bottom of the conduction band, and that most holes will be near the top of the valence band.

To describe current flow in semiconductors, however, we need more information. We need to know the density of quasi-free electrons and of holes and their distributions with energy more precisely. It is not evident yet, but in most electronic devices, the device current is determined by the number of electrons or holes that have enough energy to surmount various energy barriers that are built into the devices. To determine the current-voltage relations in these devices, then, we have to know the distribution of electrons (and holes) with energy.

There are two things that go into that determination. The first is that we must know how the available states are distributed in energy. The second factor we need to know is the probability that a state at a given energy is occupied. For example, we expect intuitively that the higher the energy state, the less likely it is to be occupied.

We begin with the distribution of states in the next section, and in the following section discuss the probability of occupancy.

2.8.1 DENSITY OF STATES AND DENSITY-OF-STATES EFFECTIVE MASS

We begin by examining the density-of-states function, which is derived in Appendix D. For a free electron ($E_P = E_0$ and is constant), the available states are distributed in energy according to

$$S(E) = \frac{1}{2\pi^2}\left(\frac{2m_0}{\hbar^2}\right)^{3/2}\sqrt{E - E_0} \tag{2.42}$$

where $S(E)$, the *density-of-states function,* is the number of states per unit volume per unit energy. We would like to use an analogous formula to describe the density of states in a semiconductor. We confine ourselves to the parabolic regions of the *E-K* diagram, where the effective mass is constant. We can then write for an electron near the bottom of the conduction band

$$\begin{aligned} S(E) &= \frac{1}{2\pi^2}\left(\frac{2m^*_{dse}}{\hbar^2}\right)^{3/2}\sqrt{E - E_C} \\ &= \frac{1}{2\pi^2}\left(\frac{2m^*_{dse}}{\hbar^2}\right)^{3/2}\sqrt{E_K} \qquad \text{conduction band} \end{aligned} \tag{2.43}$$

where m^*_{dse} is referred to as the density-of-states effective mass for electrons. The kinetic energy E_K is the difference between the total energy E and the potential energy E_C for an electron in the parabolic region near the bottom of the conduction band: $E_K = E - E_C$.

Near the bottom of the conduction band of a material such as GaAs, m_e^* is a scalar. As we mentioned earlier, however, and discuss in some detail in Appendix D, for a material like Si, in each equivalent minimum the effective mass is direction dependent, so some type of average of longitudinal and transverse electron masses ($m_\parallel^*$ and $m_\perp^*$ respectively) must be used, and the number of equivalent minima must be taken into account to arrive at a value of m_{dse}^*. The details of this are also handled in Appendix D. Values for some semiconductors of interest were given in Table 2.1.

Similarly, in the parabolic region near the top of the valence band, the density of states function is

$$S(E) = \frac{1}{2\pi^2}\left(\frac{2m_{dsh}^*}{\hbar^2}\right)^{3/2}\sqrt{E_V - E} \qquad \text{valence band} \qquad (2.44)$$

This equation is for the density of states of electrons or *holes,* the valence band edge E_V is the potential energy for holes in the valence band, and m_{dsh}^* is some combination of the light hole and heavy hole effective masses as discussed in Appendix D. Some results were listed in Table 2.2.

The density-of-states functions for electrons in the conduction band and in the valence band are plotted schematically in Figure 2.11 where the $S(E)$ plots are superimposed on the energy band diagram (energy versus position). Note that there are no states for electrons or holes in the forbidden band (for purely intrinsic material), so $S(E)$ is zero there. We emphasize that Equations (2.43) and (2.44) for $S(E)$ are valid only in the parabolic regions of the E-K curves.

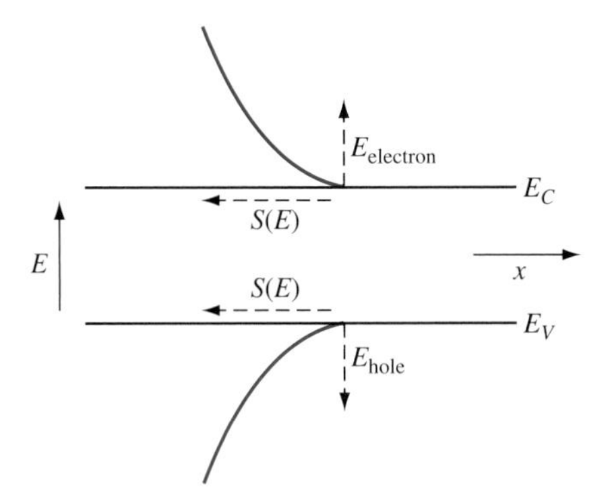

Figure 2.11 The density-of-states functions for electrons in the conduction band and the valence band. The density of states versus energy plot is superimposed on the energy band diagram (energy versus position x).

2.9 FERMI-DIRAC STATISTICS

Now that we know how the available states are distributed in energy, we need to examine the probability that a given energy state is occupied. Given one state at energy E_a, let $f(E_a)$ be the probability that an electron occupies this particular state. If there are S_a states per unit volume at energy E_a, then the electron concentration n_a at E_a is

$$n_a(E_a) = S_a f(E_a) \tag{2.45}$$

We can generalize for states distributed in energy:

$$n(E) = S(E)f(E) \tag{2.46}$$

where $n(E)$ is the number of electrons per unit volume per unit energy. In a small energy range dE,

$$n(E)\,dE = S(E)f(E)\,dE \tag{2.47}$$

In a given band, then, at equilibrium, the total number of electrons per unit volume, n_0, in the entire energy band is

$$n_0 = \int_{\text{band}} S(E)f(E)\,dE \tag{2.48}$$

where the integration is taken over the total band of allowed energies.

To determine n_0 exactly, both $S(E)$ and $f(E)$ must be known. In Section 2.8.1, we gave equations for $S(E)$, but they are valid only near the band extrema, because these equations involve the density of states effective mass. As long as we consider carriers near the band extrema, $S(E)$ is known. We then turn to the probability of occupancy $f(E)$. The term $f(E)$ can be determined for all energies, but has a different form depending on whether the electrons under consideration are free to move through the crystal (as within a band) or bound to localized states (as for impurities).

The derivation for $f(E)$ is somewhat involved; we state and discuss the results in the following sections. [2]

2.9.1 FERMI-DIRAC STATISTICS FOR ELECTRONS AND HOLES IN BANDS

The probability that an electron occupies a given state at energy E in an allowed band is given by

$$f(E) = \frac{1}{1 + e^{(E-E_f)/kT}} \qquad \text{Fermi-Dirac probability function} \tag{2.49}$$

where the energy E_f is a reference energy called the *Fermi energy* or *Fermi level,* and Equation (2.49) is known as the *Fermi-Dirac probability function.*

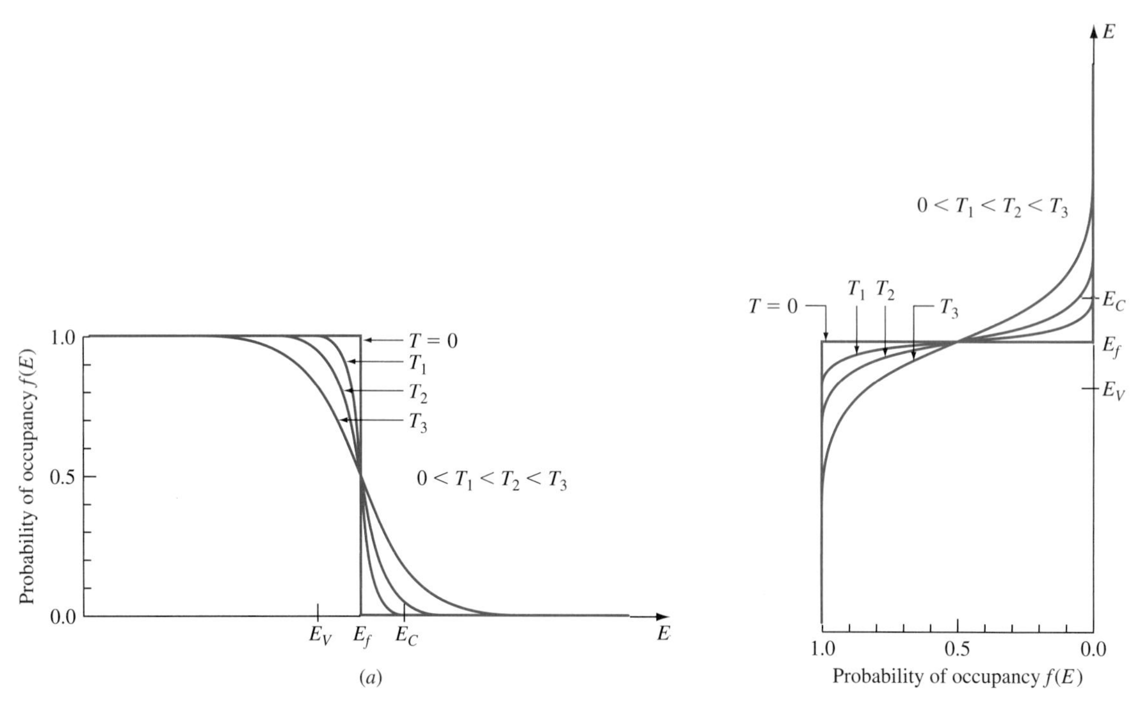

Figure 2.12 The Fermi-Dirac distribution function gives the probability of occupancy of an energy state E if the state exists.

Figure 2.12a shows the Fermi-Dirac distribution as a function of energy for several temperatures. We also indicate possible values for E_C and E_V. In Figure 2.12b we have rotated by 90° the conventional plot of dependent variable on the y axis, independent variable on the x axis. We do this to make this plot agree with other plots, such as energy band diagrams, in which energy is plotted vertically. Probability increases to the left on the plot in Figure 2.12b.

The Fermi level is chosen to be the particular energy level for which $f(E) = \frac{1}{2}$, or the probability of occupancy of a state at that energy (if the state exists) is 50 percent. States at energies below the Fermi level are more likely to be occupied than empty. States at energies above the Fermi level have a probability of occupancy less than $\frac{1}{2}$, meaning they are more likely to be empty than full.

The figure shows the Fermi-Dirac function for several different temperatures. At absolute zero, every electron is at its lowest possible energy. In a semiconductor, that means that every state in the valence band is occupied. The probability of occupancy of any state in the valence band is therefore unity, and the probability of occupancy of any state in the conduction band is zero. The Fermi level happens to be very close to the middle of the forbidden gap for intrinsic materials. Note that the probability of occupancy of a state at the Fermi level is $\frac{1}{2}$, but that there are, in fact, no states there.

Now consider a somewhat higher temperature, T_1. Since the crystal has some thermal energy, some electrons have enough energy to be in the conduction band. The probability of occupancy of a state near the bottom of the conduction band is less than $\frac{1}{2}$, but is not zero either. This means that at any given time, a given state is *probably* empty, but there is a small probability that an electron occupies it. The higher the energy, the less likely the state is to be occupied. Similarly, a state near the top of the valence band is *probably* occupied. Some of these states will be empty, indicating that there are some holes. The states most likely to be unoccupied in the valence band are those near the top, meaning that most of the holes will be found near the top of the valence band.

The probability of a state being occupied by a *hole* is one minus the probability of a state being occupied by an electron, since the state must either be occupied or not. Thus the Fermi-Dirac distribution for holes is

$$f_p(E) = 1 - f(E) \tag{2.50}$$

From Equation (2.49), this gives

$$f_p(E) = 1 - \frac{1}{1 + e^{(E-E_f)/kT}} \tag{2.51}$$

or

$$f_p(E) = \frac{1}{1 + e^{(E_f-E)/kT}} \tag{2.52}$$

If in Equation (2.49) the term $e^{(E-E_f)/kT} \gg 1$, then $f(E) \ll 1$ and

$$f(E) \approx e^{-(E-E_f)/kT} \tag{2.53}$$

Similarly, in Equation (2.52), for $f_p(E) \ll 1$,

$$f_p(E) \approx e^{-(E_f-E)/kT} \tag{2.54}$$

Equations (2.53) and (2.54) are referred to as *the Boltzmann approximations to the Fermi-Dirac probability function,* or simply the *Boltzmann probability function.*

EXAMPLE 2.1

When can the Boltzmann approximation safely be used?

■ **Solution**

The expression $f(E) = e^{-(E-E_f)/kT}$ is approximately equal to the true probability

$$f(E) = \frac{1}{1 + e^{(E-E_f)/kT}}$$

if the quantity $e^{(E-E_f)/kT}$ is large compared with unity. Let us take "large compared with" to mean greater by a factor of 10. Then $e^{(E-E_f)/kT} \approx 10$ and

$E - E_f > kT \ln(10)$. Thus for the Boltzmann approximation to apply, we require that $E - E_f > kT \ln(10) = 2.3kT$. That is, if we are calculating the probability of occupancy of a state that is greater than $2.3kT$ away from the Fermi level, we can use the simpler, approximate form [Equation (2.53) for electrons or Equation (2.54) for holes].

Returning to Figure 2.12, we see that with increasing temperature, the probability of occupancy of a given state in the conduction band increases. Likewise, the probability of occupancy of a given state in the valence band decreases. This means that at higher temperatures, there are more electrons and more holes available to carry current.

EXAMPLE 2.2

Estimate the probability of occupancy of a state at the bottom of the conduction band in intrinsic Si at room temperature.

■ Solution

The band gap of silicon at room temperature is 1.12 eV. The electron is at the bottom of the conduction band, or at $E = E_C$. As we will show later, the Fermi level is approximately at midgap for intrinsic materials, so $E_C - E_f \approx E_g/2 = 0.56$ eV. From Equation (2.49), we have

$$f(E) = \frac{1}{1 + e^{(E-E_f)/kT}} = \frac{1}{1 + e^{(E_C-E_f)/kT}}$$

$$= \frac{1}{1 + e^{(0.56\,\text{eV})/(0.026\,\text{eV})}} = \frac{1}{1 + 2.26 \times 10^9} = 4.4 \times 10^{-10}$$

or about one state in two billion is occupied. Note that in this example the Boltzmann probability distribution function is valid.

The Fermi-Dirac probability function for electrons in bound states is somewhat different, but for most of the cases in this book, we will assume that all donors and acceptors are ionized. This assumption is justified in Supplement B to Part 1.

2.10 ELECTRON AND HOLE DISTRIBUTIONS WITH ENERGY

We now know the distribution of available states $S(E)$ near the bottom of the conduction band and near the top of the valence band, and we know the probability of occupancy $f(E)$ of a state of energy E. In this section, we will put this information together to find the distribution of electrons with energy in the conduction band and the distribution of holes with energy in the valence band. The total concentrations of electrons in the conduction band and holes in the valence band are then obtained.

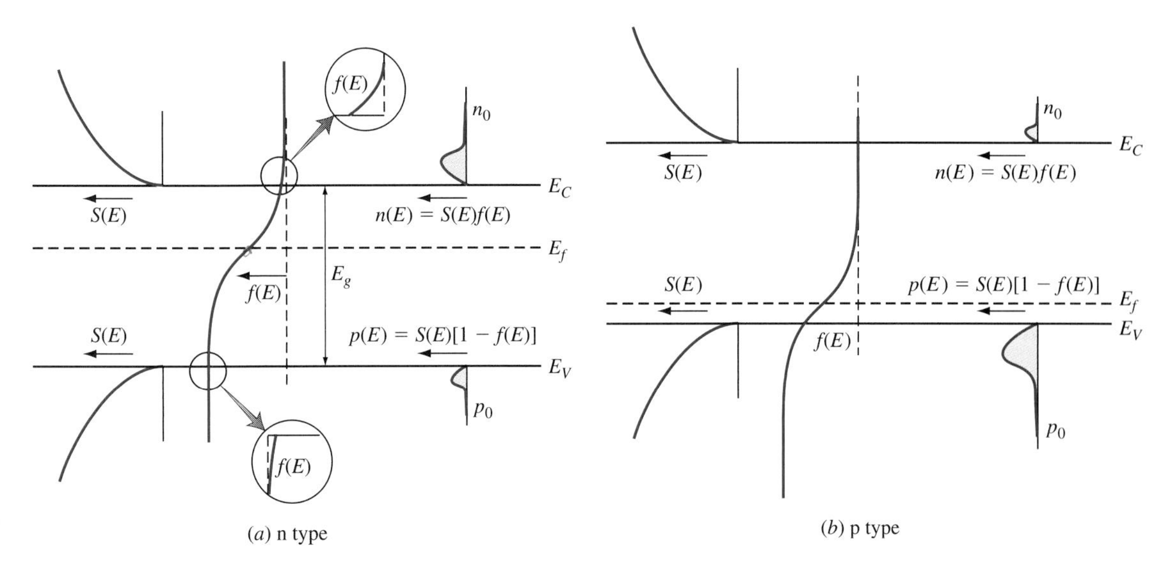

Figure 2.13 The distribution of the electrons near the bottom of the conduction band, *n*(*E*), is the product of the density-of-states distribution *S*(*E*) times the probability of occupancy of states *f*(*E*) at a particular energy. The distribution of holes near the top of the valence band *p*(*E*) is the product of the density-of-states distribution times the probability of vacancy of states at a particular energy. (a) n type, (b) p type.

Figure 2.13 shows a plot of the energy band diagram of a semiconductor. Part (a) depicts the case of an n-type semiconductor and (b), p-type. Plots for *S(E)* and *f(E)* are superimposed onto the energy band diagram. The electron and hole distribution functions, $n(E)$ and $p(E)$, are also shown shaded, where

$$n(E) = S(E)f(E) \qquad \text{conduction band}$$
$$p(E) = S(E)[1 - f(E)] = S(E)f_p(E) \qquad \text{valence band}$$

The area under the respective curves represents the total electron concentration n_0 in the conduction band and hole concentration in the valence band p_0. The shaded area for electrons is larger than for holes in n-type material, because the Fermi level is closer to the conduction band than to the valence band. In p-type material, the hole concentration is larger than the electron concentration because the Fermi level is closer to the valence band.

The total number of electrons in the conduction band at equilibrium is obtained by integrating over the band:

$$n_0 = \int_{E_C}^{\text{top}} S(E)f(E)\,dE = \int_{E_C}^{\text{top}} n(E)\,dE \tag{2.55}$$

where n_0 is the equilibrium concentration of electrons in the conduction band and the integration is taken from the bottom of the conduction band E_C to the top of the band.

Similarly, the equilibrium hole distribution function in the valence band becomes

$$p_0 = \int_{\text{bottom}}^{E_V} S(E) f_p(E)\, dE = \int_{\text{bottom}}^{E_V} p(E)\, dE \tag{2.56}$$

where the integration is taken over the range of valence band energies.

Recall that most of the carriers are concentrated near the extrema of their respective bands. This is fortunate, because these are also the regions where the effective mass is constant, and thus where $S(E)$ is known. Expressions for $S(E)$ were given by Equations (2.43) and (2.44), and are repeated here:

$$\boxed{S(E) = \frac{1}{2\pi^2}\left(\frac{2m^*_{dse}}{\hbar^2}\right)^{3/2}\sqrt{E - E_C}} \qquad \text{electrons} \tag{2.43}$$

$$\boxed{S(E) = \frac{1}{2\pi^2}\left(\frac{2m^*_{dsh}}{\hbar^2}\right)^{3/2}\sqrt{E_V - E}} \qquad \text{holes} \tag{2.44}$$

Note that the effective masses in the equations above are the "density of states" effective masses. Further, the energies at the top of the conduction band and the bottom of the valence band are not accurately known, meaning Equations (2.55) and (2.56) cannot be solved exactly.

We will now solve Equations (2.55) and (2.56) (approximately) for *nondegenerate semiconductors*. By definition, a semiconductor is said to be nondegenerate if the probability of any state in the conduction band being occupied by an electron is small, and the probability that any state in the valence band being occupied by a hole is small.[2] If by "small" we take a probability of 10 percent, then this implies that the Fermi level is at least $2.3kT$ below the conduction band edge and at least $2.3kT$ above the valence band edge. When $E_C - E_f > 2.3kT$ and $E_f - E_V > 2.3kT$, then the Boltzmann approximation to the Fermi-Dirac probability function is valid and can be used in both bands.

To solve the equations we need to make two more simplifications:

1. Assume $S(E)$ has the form given in Equations (2.43) and (2.44) anywhere that $E \geq E_C$ or $E \leq E_V$ respectively. The density of states function $S(E)$ is zero in the forbidden gap.
2. The upper limit of integration in the conduction band is extended to $+\infty$ and the lower limit of integration in the valence band is extended to $-\infty$.

[2]Conversely, a semiconductor is said to be degenerate if the Boltzmann approximation is not valid for one of the bands. An alternative definition of a degenerate semiconductor is that the Fermi level is inside the conduction band (degenerate, n type) or inside the valence band (degenerate, p type).

The last assumption can be justified since the probability factors $f(E)$ and $f_p(E)$ decrease rapidly (exponentially) with energy away from the band edges. In those regions, the product of the density of states and the probability will be so small it contributes little to the integration. The major contribution to the integration will be within the first few kT of the band edges, where Equations (2.43) and (2.44) are valid (and the density-of-states effective masses are known).

We can now write Equations (2.55) and (2.56) as

$$n_0 = \frac{1}{2\pi^2}\left(\frac{2m^*_{dse}}{\hbar^2}\right)^{3/2} \int_{E_C}^{\infty} \sqrt{E - E_C}\, e^{-[(E-E_f)/kT]}\, dE \tag{2.57}$$

and similarly

$$p_0 = \frac{1}{2\pi^2}\left(\frac{2m^*_{dsh}}{\hbar^2}\right)^{3/2} \int_{-\infty}^{E_V} \sqrt{E_V - E}\, e^{-[(E_f-E)/kT]}\, dE \tag{2.58}$$

The results are

$$\boxed{n_0 = N_C e^{-[(E_C-E_f)/kT]}} \tag{2.59}$$

$$\boxed{p_0 = N_V e^{-[(E_f-E_V)/kT]}} \tag{2.60}$$

where

$$\boxed{N_C = 2\left(\frac{m^*_{dse}kT}{2\pi\hbar^2}\right)^{3/2}} \tag{2.61}$$

$$\boxed{N_V = 2\left(\frac{m^*_{dsh}kT}{2\pi\hbar^2}\right)^{3/2}} \tag{2.62}$$

Here N_C is called the *effective density of states in the conduction band*. If there were N_C states all located at $E = E_C$, the probability of one state being occupied would be $e^{-[(E_C-E_f)/kT]}$ and the density of electrons in the conduction band, n_0, would be given by Equation (2.59). Likewise, N_V is referred to as the effective density of states in the valence band.

The effective densities of states N_C and N_V can be expressed for a nondegenerate semiconductor as

$$N_C = 2.54 \times 10^{19}\left(\frac{m^*_{dse}}{m_0}\right)^{3/2}\left(\frac{T}{300}\right)^{3/2} \text{ cm}^{-3} \tag{2.63}$$

$$N_V = 2.54 \times 10^{19}\left(\frac{m^*_{dsh}}{m_0}\right)^{3/2}\left(\frac{T}{300}\right)^{3/2} \text{ cm}^{-3} \tag{2.64}$$

These forms are useful for comparing materials or temperatures. For example, for a material of smaller effective mass, the effective density of states is correspondingly reduced.

We can derive a very useful relation by multiplying Equations (2.59) and (2.60):

$$n_0 p_0 = N_C N_V e^{-[(E_C - E_f)/kT]} e^{-[(E_f - E_V)/kT]} = N_C N_V e^{-(E_g/kT)} = n_i^2 \tag{2.65}$$

where $E_g = E_C - E_V$ or

$$\boxed{n_0 p_0 = n_i^2 \qquad \text{nondegenerate semiconductors}} \tag{2.66}$$

The quantity n_i is the electron concentration in the conduction band for intrinsic material. This shows that the electron-hole product at equilibrium is strongly dependent on temperature and band-gap. Equation (2.65) holds for all nondegenerate semiconductors at equilibrium, but for intrinsic materials it is also true that

$$n_0 = p_0 = n_i \qquad \text{intrinsic} \tag{2.22}$$

Equation (2.66) is often referred to as the *law of mass action* and is valid only for nondegenerate semiconductors. It is analogous to a similar relation in chemistry.

EXAMPLE 2.3

Find the intrinsic concentration n_i for the case of Si ($n_0 = p_0 = n_i$) at room temperature (300 K). The effective masses are[3] $m^*_{dse} = 1.09 m_0$ and $m^*_{dsh} = 1.15 m_0$.

■ Solution

From Equation (2.65),

$$n_i = \sqrt{N_C N_V}\, e^{-E_g/2kT} \tag{2.67}$$

From Equations (2.63) and (2.64), we find that at 300 K, $N_C = 2.86 \times 10^{19}$ cm^{-3} and $N_V = 3.10 \times 10^{19}$ cm^{-3}.

Since the result depends strongly (exponentially) on $E_g/2kT$, we use the accurate values of $E_g = 1.1242$ eV for Si at $T = 300$ K and $kT = 0.02586$ eV to avoid round-off error. Then

$$n_i(300\,\text{K}) = \sqrt{(2.86 \times 10^{19})(3.01 \times 10^{19})}\, e^{-1.1242/(2\times 0.02586)}$$
$$= 1.08 \times 10^{10}\ \text{electrons/cm}^3$$

From Equations (2.59) and (2.60), we can write for an intrinsic semiconductor

$$n_0 = p_0 = n_i = N_C e^{-[(E_C - E_i)/kT]} = N_V e^{-[(E_i - E_V)/kT]} \tag{2.68}$$

[3]The parameters for Si used here are from reference [1]. In some of the earlier literature slightly different values are used for effective masses, resulting in $n_i \approx 1.5 \times 10^{10}$ cm^{-3}.

where E_i is defined as the Fermi energy for an intrinsic semiconductor. Then

$$\frac{N_C}{N_V} = e^{(E_C - E_i)/kT} e^{-(E_i - E_V)/kT} = e^{(E_C + E_V - 2E_i)/kT} \tag{2.69}$$

Solving for E_i gives

$$E_i = \frac{(E_C + E_V)}{2} + \frac{kT}{2} \ln \frac{N_V}{N_C} \tag{2.70}$$

where $(E_C + E_V)/2$ is the energy at midgap.

We can use this to locate the intrinsic Fermi level E_i. From Equations (2.61) and (2.62),

$$\frac{N_V}{N_C} = \left(\frac{m^*_{dsh}}{m^*_{dse}}\right)^{3/2} \tag{2.71}$$

and

$$E_i = \left(\frac{E_C + E_V}{2}\right) + \frac{3}{4} kT \ln \left(\frac{m^*_{dsh}}{m^*_{dse}}\right) \tag{2.72}$$

This means that E_i is offset from midgap by the term $\frac{3}{4} kT \ln(m^*_{dsh}/m^*_{dse})$, which is usually small.

EXAMPLE 2.4

Find the energy by which E_i is offset from midgap for Si at room temperature.

■ **Solution**

From Equation (2.72),

$$E_i - E_{\text{midgap}} = \frac{3}{4} kT \ln \left(\frac{m^*_{dsh}}{m^*_{dse}}\right) = \frac{3}{4}(0.026) \ln \left(\frac{1.15}{1.09}\right) = 1.05 \text{ meV}$$

which is small compared with $E_g = 1.12$ eV.

It is often convenient to use alternative expressions to those of Equations (2.59) and (2.60) for n_0 and p_0. Since for an intrinsic semiconductor

$$\boxed{\begin{aligned} n_0 &= n_i = N_C e^{-[(E_C - E_i)/kT]} \\ p_0 &= n_i = N_V e^{-[(E_i - E_V)/kT]} \end{aligned}} \tag{2.73}$$

we can write

$$\boxed{\begin{aligned} N_C &= n_i e^{(E_C - E_i)/kT} \\ N_V &= n_i e^{(E_i - E_V)/kT} \end{aligned}} \tag{2.74}$$

and (2.59) and (2.60) become

$$\boxed{\begin{aligned} n_0 &= n_i e^{(E_f - E_i)/kT} \\ p_0 &= n_i e^{(E_i - E_f)/kT} \end{aligned}} \tag{2.75}$$

We emphasize that the above equations are valid only for nondegenerate semiconductors (E_f more the $2.3kT$ away from either band edge).

When the material has donors or acceptors, they contribute to the overall concentration of electrons and holes. For example, if a given semiconductor is doped with N_D donors per cubic centimeter, more electrons are added to the conduction band. We can show that this has the effect of raising the Fermi level. Rearranging Equation (2.75) gives

$$E_f - E_i = kT \ln\left(\frac{n_0}{n_i}\right) \tag{2.76}$$

As n_0 increases, so does $E_f - E_i$.

When a sample is doped with N_D donors per cubic centimeter, it is not necessarily the case that $n_0 = N_D + n_i$. This is because when the donors are ionized, and their electrons are given to the conduction band, some of those electrons will fall to the valence band and recombine with holes. This is illustrated in Figure 2.14. This also implies that the number of holes in n-type material is something less than the intrinsic value p_i. Typically, however, the doping concentration is much greater than the intrinsic concentration. This means that the number of holes available for recombination is small and *most* of the donated electrons remain in the conduction band. Thus for an n-type semiconductor, as long as $N_D \gg n_i$, we can make the approximation that

$$n_0 \approx N_D \qquad \text{when } N_D \gg n_i \tag{2.77}$$

Similarly, if a p-type sample is doped with N_A acceptors, some electrons from the conduction band may fall into some of the acceptor states, but if $N_A \gg n_i$, the effect on the concentration of holes is small and

$$p_0 \approx N_A \qquad \text{when } N_A \gg n_i \tag{2.78}$$

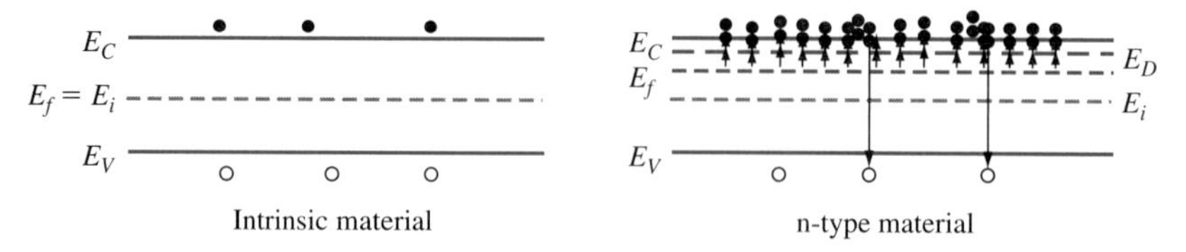

Figure 2.14 Electrons and holes in intrinsic and doped n-type material. In the doped case, some of the electrons from the donors recombine with holes.

Whether the material is doped n type or p type, however, as long as the doping level is not degenerate, it is still true that

$$n_0 p_0 = n_i^2$$

EXAMPLE 2.5

Find the electron and hole concentrations in GaAs doped with $N_A = 10^{16}\ \text{cm}^{-3}$, and locate the Fermi level. The band gap of GaAs is 1.43 eV.

■ Solution

We first find n_i for GaAs, from Equation (2.67), for which we need to know the effective densities of states N_C and N_V. Assuming room temperature, and using the values of the density-of-states effective masses for GaAs from Tables 1.1 and 1.2, we have from Equations (2.63) and (2.64)

$$\begin{aligned} N_C &= 2.54 \times 10^{19} \left(\frac{m_{dse}^*}{m_0}\right)^{3/2} \left(\frac{T}{300}\right)^{3/2} \text{cm}^{-3} \\ &= 2.54 \times 10^{19} \left(\frac{(0.067 m_0)}{m_0}\right)^{3/2} (1)^{3/2} \\ &= 4.4 \times 10^{17}\ \text{cm}^{-3} \end{aligned}$$

Similarly,

$$\begin{aligned} N_V &= 2.54 \times 10^{19} \left(\frac{m_{dsh}^*}{m_0}\right)^{3/2} \left(\frac{T}{300}\right)^{3/2} \text{cm}^{-3} \\ &= 2.54 \times 10^{19} (0.48)^{3/2} \\ &= 8.3 \times 10^{18}\ \text{cm}^{-3} \end{aligned}$$

Therefore, from Equation (2.67),

$$\begin{aligned} n_i &= \sqrt{N_C N_V} e^{-(E_g/2kT)} \\ &= \sqrt{(4.4 \times 10^{17})(8.3 \times 10^{19})}\, e^{-[1.43/2(0.026)]} \\ &= 2.2 \times 10^6\ \text{cm}^{-3} \end{aligned}$$

Since the doping concentration N_A is significantly larger than n_i, we can assume $p_0 \approx N_A$:

$$p_0 = N_A = 1 \times 10^{16}\ \text{cm}^{-3}$$

We can locate the Fermi level by rearranging Equation (2.60) to give

$$E_f - E_V = -kT \ln\left(\frac{p_0}{N_V}\right) = -0.026\ \text{eV} \cdot \ln\left(\frac{10^{16}}{8.3 \times 10^{18}}\right) = 0.17\ \text{eV}$$

Note that we must express the Fermi level relative to some reference position; in this case we have chosen the valence band edge.

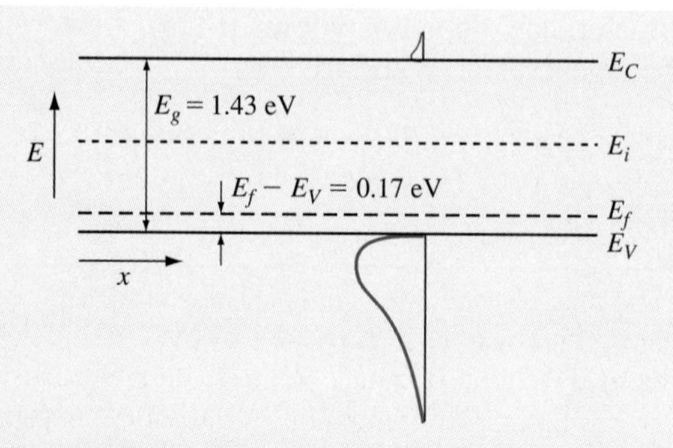

Figure 2.15 Energy band diagram for the p-type GaAs of Example 2.5.

The energy band diagram is drawn in Figure 2.15. This tells us that the Fermi level is below the intrinsic level, consistent with our knowledge that the material is p type, having more holes than electrons.

We indicated that these equations apply only when the material is nondegenerate, so we should check that. The material is nondegenerate if the Fermi level is at least $2.3kT$ about the valence band edge. Since $2.3kT = 0.06$ eV, and $E_f - E_V = 0.17$ eV, the Fermi level is sufficiently removed from the valence band edge that the semiconductor is nondegenerate.

Finally, we can find the concentration of electrons from Equation (2.66):

$$n_0 = \frac{n_i^2}{p_0} = \frac{(2.2 \times 10^6)^2}{10^{16}} = 4.8 \times 10^{-4}\ \text{cm}^{-3}$$

It is common to have materials that contain both acceptors and donors. For example, a diode consists of a p-type region and an n-type region that abut. To fabricate these, one might start with a p-type material, and add extra donors to part of it. This is called *compensation.* If the number of donors exceeds the number of acceptors, the material becomes n type. One has to include the effects of both dopant types in computing the numbers of electrons and holes. For example, when both donors and acceptors are present, some of the electrons from donor states will fall into acceptor states, ionizing both dopants but not contributing any charge carriers. Thus, assuming all donors and acceptors are ionized,

$$\text{If } N_D > N_A, \quad \text{then } n_0 = N_D - N_A \text{ and the material is n type} \tag{2.79}$$

and

$$\text{If } N_A > N_D, \quad \text{then } p_0 = N_A - N_D \text{ and the material is p type} \tag{2.80}$$

Both of these equations apply only when $N_D - N_A$ or $N_A - N_D \gg n_i$.

EXAMPLE 2.6

A sample of silicon is doped everywhere with a background concentration of $N_A = 4 \times 10^{16}\ \text{cm}^{-3}$. Then $10^{17}\ \text{cm}^{-3}$ donors are added. Find the room temperature concentrations of electrons and holes in the original and final materials and draw the energy band diagram for each.

■ Solution

Let us begin with the p-type material. We have $N_A = 4 \times 10^{16}\ \text{cm}^{-3}$, and we found earlier that for silicon $n_i = 1.08 \times 10^{10}\ \text{cm}^{-3}$. We have therefore satisfied the condition that $N_A \gg n_i$, and so the hole concentration in the p-type material is $p_0 = N_A = 4 \times 10^{16}\ \text{cm}^{-3}$. We can solve Equation (2.60) to find the Fermi level:

$$E_f - E_V = -kT \ln\left(\frac{p_0}{N_V}\right) = -0.026\ \text{eV} \cdot \ln\left(\frac{4 \times 10^{16}}{3.1 \times 10^{19}}\right)$$
$$= 0.17\ \text{eV}$$

This is greater than $2.3kT = 0.06$ eV, so the material is nondegenerate and our use of Equation (2.73) is valid.

The concentration of the electrons in the p-type material is

$$n_0 = \frac{n_i^2}{p_0} = \frac{(1.08 \times 10^{10})^2}{4 \times 10^{16}} = 2.9 \times 10^3\ \text{cm}^{-3} \qquad \text{p type}$$

For the n-type material, there are both donors and acceptors. Again assuming room temperature and nondegeneracy, we can write

$$n_0 = N_D - N_A = 10^{17} - 4 \times 10^{16} = 6 \times 10^{16}\ \text{cm}^{-3}$$

From Equation (2.59) the Fermi level is located at

$$E_C - E_f = -kT \ln\left(\frac{N_D}{N_C}\right) = -0.026\ \text{eV} \cdot \ln\left(\frac{6 \times 10^{16}}{2.86 \times 10^{19}}\right)$$
$$= 0.16\ \text{eV}$$

and the material is thus nondegenerate. The minority carrier concentration is then found from $n_0 p_0 = n_i^2$, giving

$$p_0 = \frac{n_i^2}{n_0} = \frac{(1.08 \times 10^{10})^2}{6 \times 10^{16}} = 1.9 \times 10^3\ \text{cm}^{-3}$$

The energy band diagrams for these two materials are shown in Figure 2.16. The p-type material is shown in part (a) and the n-type material is in part (b). Notice the positions of the Fermi level in each case and the relative sizes of the carrier distributions (not shown to scale). Also note that in the p-type material, the acceptor energy level E_A is shown, but there are no donors. In the n-type material, both E_A and E_D exist.

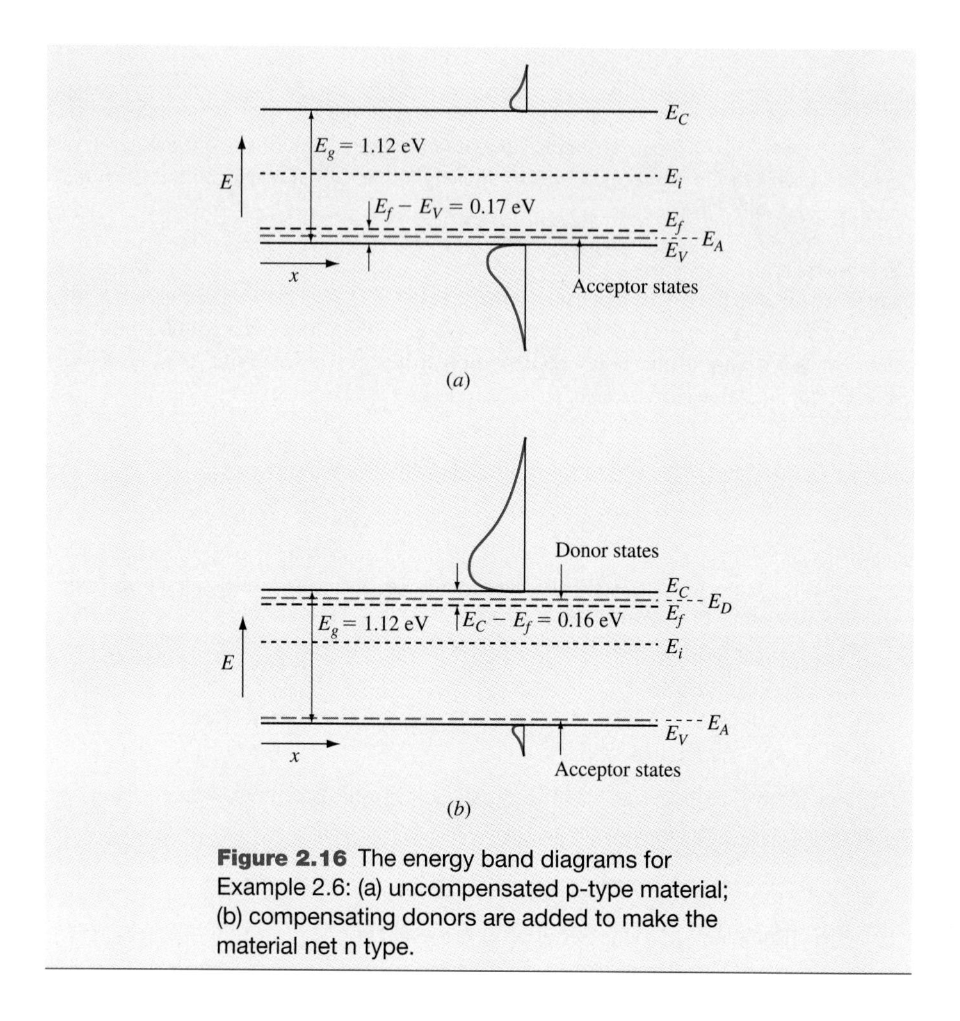

Figure 2.16 The energy band diagrams for Example 2.6: (a) uncompensated p-type material; (b) compensating donors are added to make the material net n type.

Now, let us go back and examine more closely the expression for the electron density distribution function with energy, $n(E)$ for electrons in the conduction band. From Equation (2.46), we know that

$$n(E) = S(E)f(E)$$

where, from Equation (2.43),

$$S(E) = \frac{1}{2\pi^2}\left(\frac{2m^*_{dse}}{\hbar^2}\right)^{3/2}\sqrt{E - E_C}$$

We also recall that for a nondegenerate semiconductor, the probability of occupancy $f(E)$ of a state of energy E can be expressed as

$$f(E) = e^{-[(E-E_f)/kT]} \tag{2.81}$$

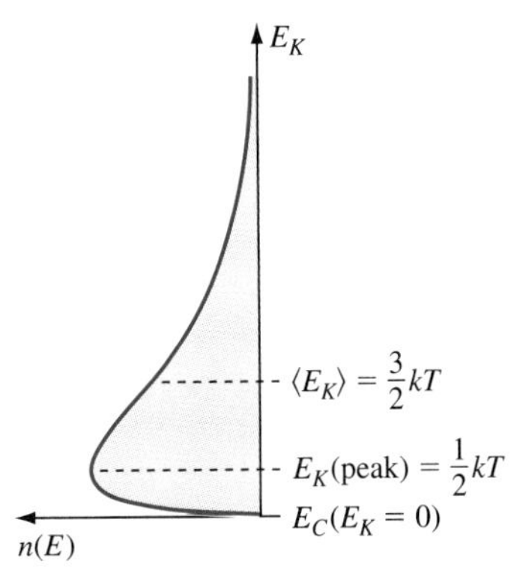

Figure 2.17 The electron distribution function $n(E)$ as a function of energy (energy on the vertical axis).

But, since $E - E_f = (E - E_C) + (E_C - E_f) = E_K + (E_C - E_f)$, we can write

$$n(E) = \frac{1}{2\pi^2}\left(\frac{2m^*_{dse}}{\hbar^2}\right)^{3/2} e^{-[(E_C-E_f)/kT]}\sqrt{E_K}\, e^{-E_K/kT} \tag{2.82}$$

where $E_K = (E - E_C)$ is the kinetic energy of the electron at energy E.

By letting a constant C be defined as

$$C = \frac{1}{2\pi^2}\left(\frac{2m^*_{dse}}{\hbar^2}\right)^{3/2} e^{-[(E_C-E_f)/kT]} \tag{2.83}$$

we can simplify this to

$$n(E_K) = C\sqrt{E_K}\, e^{-E_K/kT} \tag{2.84}$$

This function was shown schematically in Figure 2.13, and it is shown enlarged in Figure 2.17. We see that with increasing E_K, the electron concentration $n(E_K)$ increases, reaches a maximum, and then decreases approximately exponentially.

EXAMPLE 2.7

a. Find the kinetic energy at which the peak electron concentration occurs, $E_K(\text{peak})$.
b. Find the average kinetic energy $\langle E_K \rangle$ for an electron in the conduction band.
c. Find the wavelength of an electron in the conduction band of GaAs having this average kinetic energy.

■ **Solution**

a. Since $n(E_K)$ reaches a peak at $E_K(\text{peak})$, we know that the derivative is zero. From Equation (2.84),

$$\frac{dn(E_K)}{dE_K} = 0 = C\left[\sqrt{E_K}\left(-\frac{1}{kT}\right)e^{-E_K/kT} + e^{-E_K/kT}\left(\frac{1}{2\sqrt{E_K}}\right)\right]$$

which gives the result

$$n(E_K(\text{peak})) = \tfrac{1}{2}kT$$

or, in other words, the kinetic energy for the maximum value of $n(E) = \frac{1}{2}kT$ above the conduction band edge. At room temperature this is $\frac{1}{2}(0.026)\,\text{eV} = 0.013\,\text{eV}$.

b. We can find the average value of the kinetic energy from $n(E_K)$, the energy distribution function for electrons in the conduction band,

$$\langle E_K \rangle = \frac{\int_0^\infty E_K n(E_K)\, dE_K}{\int_0^\infty n(E_K)\, dE_K}$$

Substituting for $n(E_K)$ from Equation (2.84) gives

$$\langle E_K \rangle = \frac{\int_0^\infty E_K C E_K^{1/2} e^{-E_K/kT}\, dE_K}{\int_0^\infty C E_K^{1/2} e^{-E_K/kT}\, dE_K} = \frac{\int_0^\infty E_K^{3/2} e^{-E_K/kT}\, dE_K}{\int_0^\infty E_K^{1/2} e^{-E_K/kT}\, dE_K}$$

Evaluating the integrals, using Appendix E, we get

$$\langle E_K \rangle = \tfrac{3}{2}kT$$

This is a familiar number from chemistry and physics classes. The key points here are:

- For E_K greater than about $3kT$, $S(E)$ varies slowly compared with $f(E)$.
- To good approximation, we can consider $n(E_K)$ to decrease exponentially with E_K.

c. To find the wavelength, from Chapter 1 we know that the wave vector,

$$K = \sqrt{\frac{2m^*}{\hbar^2} E_K} = \frac{2\pi}{\lambda}$$

This implies that the wavelength of an electron with the average kinetic energy of $\frac{3}{2}kT$ is

$$\lambda = \frac{2\pi}{K} = \frac{2\pi}{\sqrt{\frac{2m^*}{\hbar^2} E_K}} = \frac{2\pi\hbar}{\sqrt{2m^*\left(\frac{3}{2}kT\right)}}$$

Since for electrons in GaAs, $m^* = m^*_{dse} = 0.067m_0$,

$$\lambda = \frac{2\pi(1.055 \times 10^{-34}\,\text{J}\cdot\text{s})}{\sqrt{2(0.067)(9.11 \times 10^{-31}\,\text{kg})\left[\frac{3}{2}(1.38 \times 10^{-23}\,\text{J/K})(300\,\text{K})\right]}} = 24\,\text{nm}$$

The electron wavelength is on the order of 40 lattice constants. We confirm again that the electron is spread out over the crystal, as indicated in Chapter 1.

*2.11 TEMPERATURE DEPENDENCE OF CARRIER CONCENTRATIONS IN NONDEGENERATE SEMICONDUCTORS

In discussing the equilibrium carrier concentrations up to this point, we assumed that the semiconductor was at or near room temperature (300 K). Therefore, the band gap of Si was assumed to be constant (1.12 eV) and the intrinsic carrier concentration is $1.08 \times 10^{10}\ \text{cm}^{-3}$ (300 K). It was further assumed that $N_D \gg n_i$ and $N_A \gg n_i$ and that all donors and acceptors were ionized such that at equilibrium the majority carrier concentration is

$$n_0 = N_D - N_A \qquad N_D > N_A \qquad \text{n type}$$

$$p_0 = N_A - N_D \qquad N_A > N_D \qquad \text{p type}$$

The minority carrier concentration was then determined from the relation

$$n_0 p_0 = n_i^2 \tag{2.66}$$

These assumptions, however, must be modified at high temperatures, where thermally generated electron-hole pairs contribute to the carrier concentration, and at low temperatures, where the impurities are not completely ionized.

*2.11.1 CARRIER CONCENTRATIONS AT HIGH TEMPERATURES

In Section 2.10 we assumed that all of the donors had given up their extra electrons to the conduction band, and that all the acceptors had likewise become occupied by electrons from the valence band. In Supplement B to Part 1, it is shown that in Si these assumptions are valid above about 200 K. The concentration of electrons still bound to donors or holes still bound to acceptors is small compared with the impurity concentration and thus, to good approximation, all impurities can be considered to be ionized. Further, since for semiconductors used in devices the impurity concentration is normally much larger than n_i, the contribution of thermally generated majority carriers can be neglected. At high enough temperatures, however, n_i is comparable to the net doping level and must be considered.

The intrinsic carrier concentration n_i can be expressed [Equation (2.67)] as

$$n_i = \sqrt{N_C N_V}\, e^{-E_g/2kT} \tag{2.85}$$

and so should be highly temperature dependent. The temperature dependence of N_C, N_V, and E_g must be considered when calculating $n_i(T)$. The temperature dependence of N_C and N_V was discussed in Section 2.10 (equations repeated here).

$$N_C = 2.54 \times 10^{19} \left(\frac{m^*_{dse}}{m_0}\right)^{3/2} \left(\frac{T}{300}\right)^{3/2} \text{ cm}^{-3} \tag{2.63}$$

$$N_V = 2.54 \times 10^{19} \left(\frac{m^*_{dsh}}{m_0}\right)^{3/2} \left(\frac{T}{300}\right)^{3/2} \text{ cm}^{-3} \tag{2.64}$$

The temperature dependence of E_g for Si can be expressed

$$E_g(T) = 1.170 - \frac{4.73 \times 10^{-4} T^2}{T + 636} \text{ eV} \tag{2.86}$$

This is plotted in Figure 2.18 (solid line).

The dashed line in Figure 2.18 shows that above 250 K, E_g can be approximated by the simpler straight-line expression

$$E_g = 1.206 - 2.73 \times 10^{-4} T \text{ eV} \quad (T > 250\,\text{K}) \tag{2.87}$$

Next, let us discuss the intrinsic carrier concentration. For silicon, n_i can be expressed as [1]

$$n_i = 5.71 \times 10^{19} \left(\frac{T}{300}\right)^{2.365} e^{-(6733/T)} \text{ cm}^{-3} \tag{2.88}$$

At high enough temperatures, thermally generated carriers can contribute to the concentration of majority carriers. To find out at what temperatures this effect

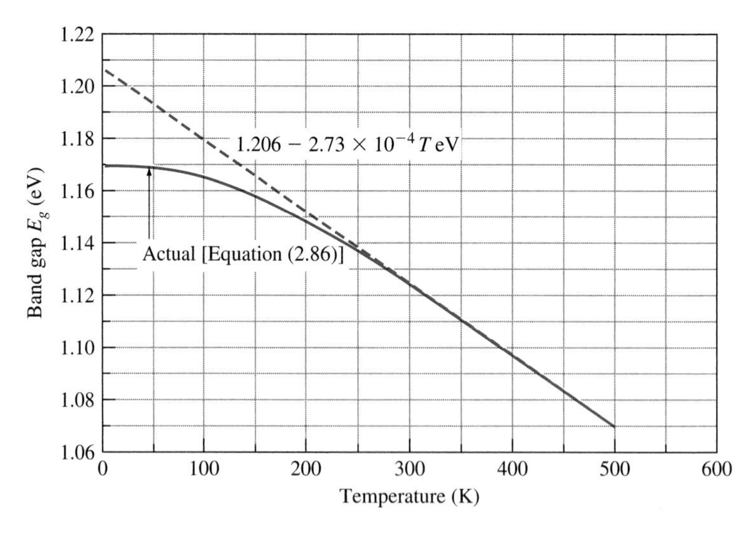

Figure 2.18 Energy band-gap dependence of silicon on temperature.

becomes important, we find the electron and hole concentrations employing the technique of space charge neutrality. That is, the number of positive charges must be equal to the number of negative charges in any macroscopic region. Electrons and ionized acceptors are negatively charged, while holes and ionized donors are positively charged. Thus for a uniformly doped semiconductor, at equilibrium, for space charge neutrality and assuming all impurities are ionized,

$$p_0 + N_D = n_0 + N_A \tag{2.89}$$

or

$$\boxed{p_0 - n_0 + N_D - N_A = 0 \qquad \text{space-charge neutrality}} \tag{2.90}$$

For nondegeneracy such that

$$n_0 p_0 = n_i^2$$

Equation (2.90) can be expressed

$$\frac{n_i^2}{n_0} - n_0 + N_D - N_A = 0 \tag{2.91}$$

or

$$n_0^2 - n_0(N_D - N_A) - n_i^2 = 0 \tag{2.92}$$

Solving for n_0,

$$n_0 = \frac{N_D - N_A}{2} + \left[\left(\frac{N_D - N_A}{2}\right)^2 + n_i^2\right]^{1/2} \tag{2.93}$$

The hole concentration can then be found from Equation (2.66):

$$p_0 = \frac{n_i^2}{n_0}$$

Note that for $(N_D - N_A) \gg n_i$ (the case assumed in Section 10, where high-temperature effects are not significant), Equation (2.93) reduces to our previous result of

$$n_0 = N_D - N_A \qquad (N_D - N_A) \gg n_i \tag{2.94}$$

Figure 2.19 shows the variation of n_0 with temperature, with net donor doping as a parameter, above 200 K, where complete ionization is a good approximation. It can be seen that for n_i greater than about 20 percent of $(N_D - N_A)$, n_0 increases appreciably with increasing temperature.

Similarly, for a nondegenerate p-type semiconductor,

$$p_0 = \frac{N_A - N_D}{2} + \left[\left(\frac{N_A - N_D}{2}\right)^2 + n_i^2\right]^{1/2} \tag{2.95}$$

and $n_0 = n_i^2 / p_0$.

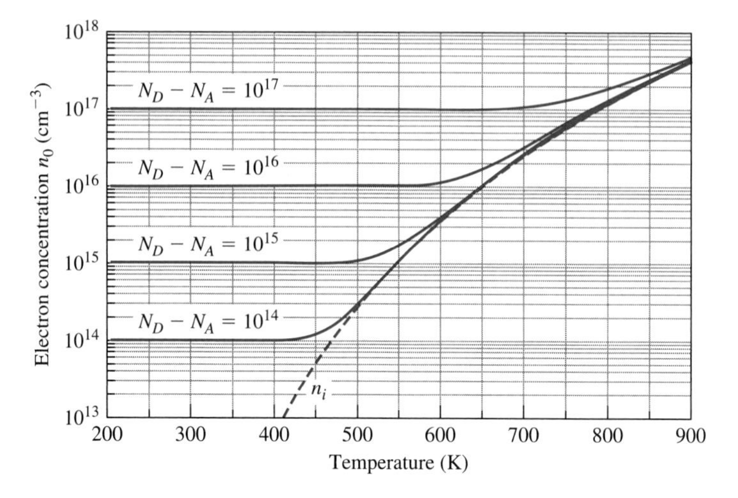

Figure 2.19 Plot of electron concentration n_0 as a function of temperature in n-type silicon for four values of net doping. Also indicated is the temperature dependence of n_i.

EXAMPLE 2.8

For $N_D \gg N_A$, find N_D such that n_0 is 10 percent greater than N_D.

■ Solution

From Equation (2.93), neglecting N_A since it is small,

$$n_0 = 1.1N_D = \frac{N_D}{2} + \left[\left(\frac{N_D}{2}\right)^2 + n_i^2\right]^{1/2}$$

$$N_D(1.1 - 0.5) = \left[\frac{N_D^2}{4} + n_i^2\right]^{1/2}$$

$$0.6N_D = N_D\left[\frac{1}{4} + \frac{n_i^2}{N_D^2}\right]^{1/2}$$

$$(0.6)^2 = 0.36 = \frac{1}{4} + \frac{n_i^2}{N_D^2}$$

$$N_D = \frac{n_i}{\sqrt{0.11}} = 3.0n_i$$

From Figure 2.19, it is seen that n_0 depends on N_D as well as temperature. It is convenient to have a normalized plot in which N_D is incorporated. Assuming $N_D \gg N_A$, from Equation (2.93)

$$n_0 = \frac{N_D}{2} + \left[\left(\frac{N_D}{2}\right)^2 + n_i^2\right]^{1/2} \tag{2.96}$$

which can be expressed as

$$\frac{n_0}{N_D} = \frac{1}{2}\left\{1 + \left[1 + \left(\frac{2n_i}{N_D}\right)^2\right]^{1/2}\right\} \tag{2.97}$$

For the temperature dependence of n_i given by Equation (2.88), n_0/N_D is plotted as a function of temperature in Figure 2.20 on a linear scale for $N_D = 10^{16}\,\text{cm}^{-3}$. As discussed next, at low temperatures $n_0 < N_D$, since some electrons are still bound to donor atoms in this case. This is also shown in Figure 2.20.

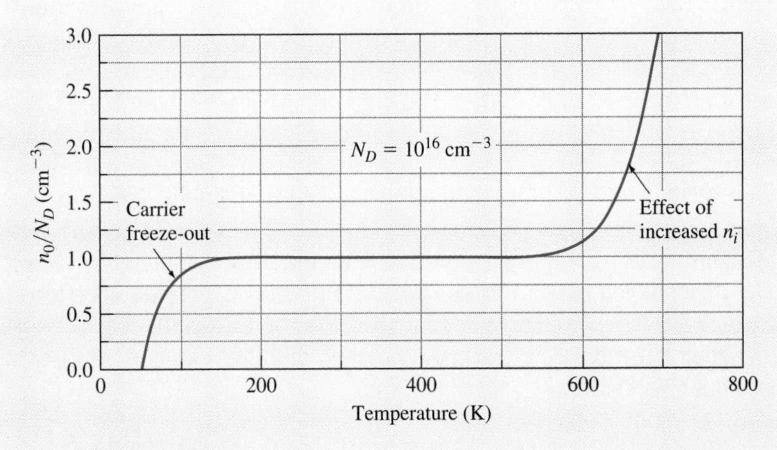

Figure 2.20 Normalized plot of n_0/N_D as a function of temperature. This plot is for $N_D = 10^{16}$ cm^{-3}.

*2.11.2 CARRIER CONCENTRATIONS AT LOW TEMPERATURES (CARRIER FREEZE-OUT)

Electrons bound to donor atoms or holes bound to acceptor atoms are said to be "frozen out." Carrier freeze-out, which can be significant at low temperatures, will affect the carrier concentrations since the dopants are not all ionized under these conditions. The equilibrium concentration of carriers frozen out is discussed in Supplement B to Part 1. Here, we point out that, to find the equilibrium carrier concentrations at low temperatures (below about 200 K), we find the number of dopants that are not ionized. Each donor that is not ionized contains an electron that would otherwise be in the conduction band, so the concentration of "missing" electrons n_D is equal to the concentration of un-ionized donors. Similarly there are p_A holes still attached to the acceptors. The space charge neutrality equation [Equation (2.90)] becomes

$$p_0 + p_A - n_0 - n_D + N_D - N_A = 0 \tag{2.98}$$

The result for a doping concentration of $N_D = 10^{16}\,\text{cm}^{-3}$ of phosphorus in silicon is shown at the low-temperature end of Figure 2.20.

We point out that, with increased doping, a decreasing fraction of the carriers will be frozen out. In Si doped greater than about 4×10^{18} cm^{-3} (degenerately doped), the ground states of the dopant atoms overlap. In this case, there are no bound states and no carrier freeze-out occurs.

2.12 DEGENERATE SEMICONDUCTORS

Up to now we have discussed the properties of semiconductors with small doping levels (i.e., nondegenerate semiconductors). The values for the semiconductor bandgaps we have discussed are for intrinsic materials at equilibrium. It was assumed that this value of E_g is independent of doping levels for nondegenerate semiconductors. For most electron devices, however, there are at least some regions that are degenerately doped. The high doping levels change the electronic properties of the materials enough to significantly affect some device characteristics. A primary effect is known as *impurity-induced band-gap narrowing.* We will discuss this effect first, then discuss a related effect, impurity-induced *apparent band-gap narrowing,* and then examine the effect on equilibrium carrier concentrations. As before, we will discuss the band gap at equilibrium.[4]

The dependence of E_g on doping concentration has a minor effect on the electrical characteristics of semiconductor diodes and field-effect transistors, but it has a major effect on the electrical properties of bipolar junction transistors, as discussed in Chapter 9.

2.12.1 IMPURITY-INDUCED BAND-GAP NARROWING

In Chapter 1, we showed that as the atoms in a crystal get closer together, their electronic orbits overlap in space, and the discrete energy levels of the individual atoms spread out in energy and form bands. A similar situation exists for impurities in semiconductors—the discrete energy levels of the dopants can smear into *minibands.* This happens when the doping concentration is high enough that the dopant atoms are near enough to each other that they can influence one another. This is reflected in the energy band diagram of Figure 2.21. On the left is a nondegenerately doped n-type semiconductor with no acceptors. The donor atoms are sparse enough that they act as isolated atoms in a sea of silicon. Under heavy enough doping, shown on the right, the discrete donor states spread into a band.

The dominant effect behind the minibands is shown schematically in Figure 2.22. Here we use the Bohr model and consider two donor atoms that are close enough together that their first excited states overlap. If the first excited states overlap, then so do all of the higher-energy impurity states. All excited states overlap in energy with each other, and they also overlap with the conduction band of the host silicon. This implies that the bottom of the conduction

---

[4]In some nonequilibrium cases, the electrostatic energy associated with electron-hole interactions can further reduce the bandgap. [3]

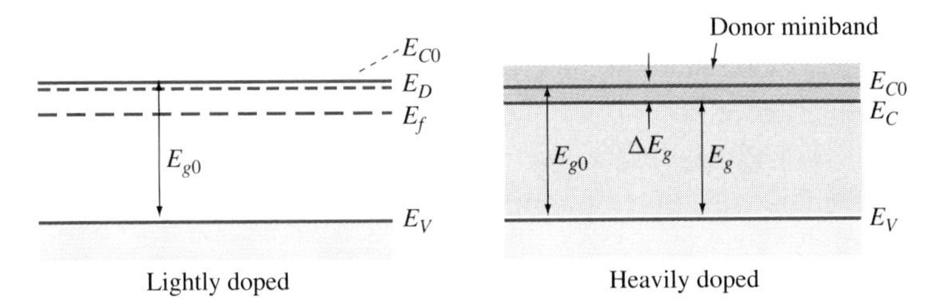

Figure 2.21 Under high doping concentrations, the formerly discrete donor levels smear into a band, effectively narrowing the band gap by an amount ΔE_g.

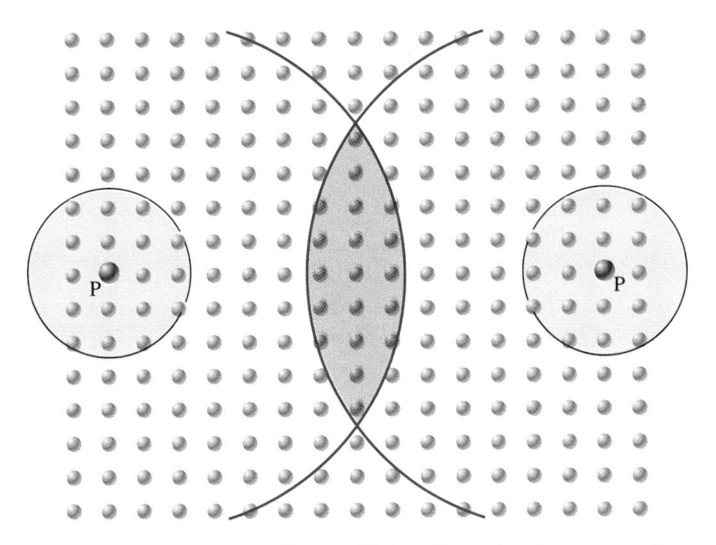

Figure 2.22 The states for the higher donor levels can overlap if the doping concentration is high enough (dopant atoms close enough together).

band, E_C, is actually lowered from its intrinsic value E_{C0} to the bottom of the overlapping impurity bands, as shown in Figure 2.21. The result is that the band gap is reduced. [4]

As the donor concentration increases, the shift in the conduction band, ΔE_C, also increases, where

$$\boxed{\Delta E_C = E_{C0} - E_C} \tag{2.99}$$

This increase in ΔE_C with increasing doping N_D is largely due to an increase in the number of spatially overlapping equi-energy impurity states and also from the broadening in energy of these impurity bands. At a sufficiently high doping level, even the ground states overlap. At that point, ΔE_C continues to increase, but now the downward shift of the conduction band edge results primarily from the broadening of the ground state band. For Si the ground states of the donors overlap at an impurity concentration of about $4 \times 10^{18}\,\text{cm}^{-3}$. This concentration is high compared with some other materials. For semiconductors with small electron conductivity effective mass, the ground states overlap at much smaller doping levels. For example, GaAs with $m^* = m^*_{ce} = 0.067m_0$, the donor ground states overlap at $N_D \approx 2 \times 10^{16}\,\text{cm}^{-3}$.

Normally, for device considerations, it is more useful to talk in terms of the reduction in the band gap, rather than that of the conduction band edge. We thus define the *impurity-induced band-gap reduction* ΔE_g, where for n-type material,

$$\Delta E_g = E_{g0} - E_g(N_D) \qquad \text{n-type} \tag{2.100}$$

where E_{g0} is the band gap of intrinsic material and $E_g(N_D)$ is the bandgap for a donor concentration of N_D. The band-gap narrowing for n-type Si has been fit to the empirical expression [5]

$$\Delta E_g = \left[\left(4.372 \times 10^{-11} N_D^{0.5}\right)^{-4} + \left(1.272 \times 10^{-6} N_D^{0.25}\right)^{-4}\right]^{-1/4} \text{eV} \tag{2.101}$$

where N_D is the donor concentration in cm^{-3}. This expression is plotted in Figure 2.23.

From Figure 2.23 it is seen that the band-gap reduction is negligible for $N_D < 10^{16}\,\text{cm}^{-3}$ and increases to 128 meV at $N_D = 10^{20}\,\text{cm}^{-3}$. For $N_D \ll N_C$, the concentration of states in this impurity band is small and can be neglected. Then for N_D less than about $10^{18}\,\text{cm}^{-3}$, the electron concentration in the impurity band can be neglected, and to good approximation

$$\boxed{n_0 = N_C e^{-(E_{C0} - E_f)/kT}}$$

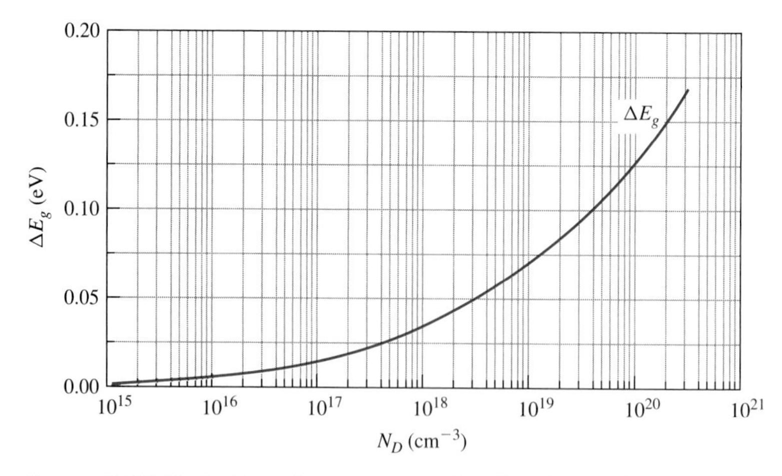

Figure 2.23 Reduction of room-temperature band gap ΔE_g as a function of donor density in phosphorus-doped silicon.

2.12.2 APPARENT BAND-GAP NARROWING

We have seen how the conduction band edge, and thus the band gap, decrease with increased donor doping. As the doping increases, however, the Fermi level moves closer to the band edge. At some doping level the Fermi level will actually cross into the conduction band.

In this case, it is usually the energy of the Fermi level that is of more interest than that of the conduction band edge. For example, the apparent band gap E_g^* may be considered to be the energy that must be overcome for an electron in the valence band to be promoted into the conduction band. In a nondegenerate semiconductor, the energy required is the energy difference between the top of the valence band and the bottom of the conduction band, $E_g(0)$, in Figure 2.24. In a degenerately doped material, however, if the Fermi level is inside the conduction band, the states below E_f are occupied. Thus, the energy required to promote an electron is now $E_f - E_V = E_g^*$ instead of $E_C - E_V = E_{g0}$. The impurity-induced apparent band-gap narrowing ΔE_g^* is then

$$\boxed{\Delta E_g^* = E_{g0} - E_g^*} \tag{2.102}$$

The value of E_g^* can be measured by optical transmission spectroscopy. In this experiment the material is illuminated with light of different photon energies. For absorption to occur, the photons must have sufficient energy to excite electrons from the valence band to the lowest empty states in the conduction band. For photon energies less than this, the material is optically transparent. The condition for transparency is

$$h\nu < E_g^* = E_{g0} - \Delta E_g^*$$

The value of E_g^* is difficult to measure, and there is considerable scatter in the reported data. However, the doping concentration dependence of ΔE_g^* has

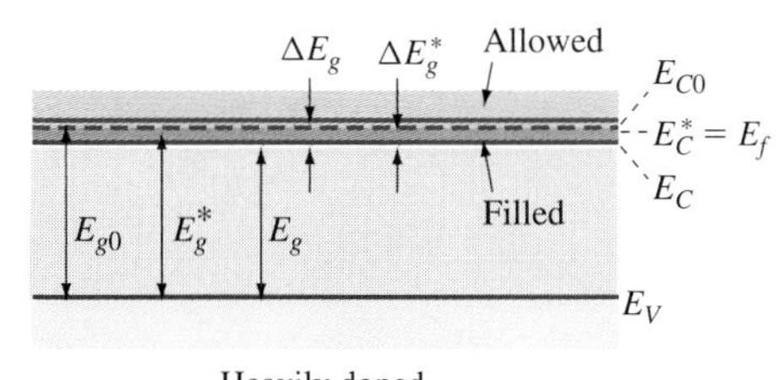

Figure 2.24 Plot of energy band diagram of a degenerate semiconductor. The intrinsic energy gap E_{g0}, the actual energy band gap E_g, and the apparent band gap E_g^* are indicated.

been fit to the following expressions:

$$\Delta E_g^*(N_D) = 18.6 \ln\left(\frac{N_D}{7 \times 10^{17}}\right) \text{ meV} \qquad \text{n type} \tag{2.103}$$

for $N_D \geq 7 \times 10^{17}$ cm^{-3} and zero for $N_D < 7 \times 10^{17}$ cm^{-3}. [6, 7] Similarly,

$$\Delta E_g^*(N_A) = 9\left[F + \sqrt{F^2 + 0.5}\right] \text{ meV} \qquad \text{p type} \tag{2.104}$$

for $N_D > 10^{17}$ cm and zero for $N_D < 10^{17}$ cm^{-3}. [8] Here the factor F has the form

$$F = \ln\left(\frac{N_A}{10^{17}}\right) \tag{2.105}$$

The above expressions are plotted in Figure 2.25 up to $N_D,\ N_A = 10^{20}$ cm^{-3}. Above this doping concentration there is some evidence that for n-type Si, ΔE_g^* saturates at about 100 meV. [5]

Note that the above data were obtained for noncompensated Si. For heavily compensated material, because of the spread in both donor and acceptor impurity bands, it would be expected that ΔE_g^* would be larger than indicated here.

A comparison of Figures 2.23 and 2.25 indicates that in Si the Fermi level crosses into the (reduced) conduction band at $N_D \approx 4 \times 10^{18}$ cm^{-3}.

As will be discussed in Chapter 9, the emitters of bipolar junction transistors are very heavily doped, and the apparent band-gap narrowing reduces the current gain of the transistor.

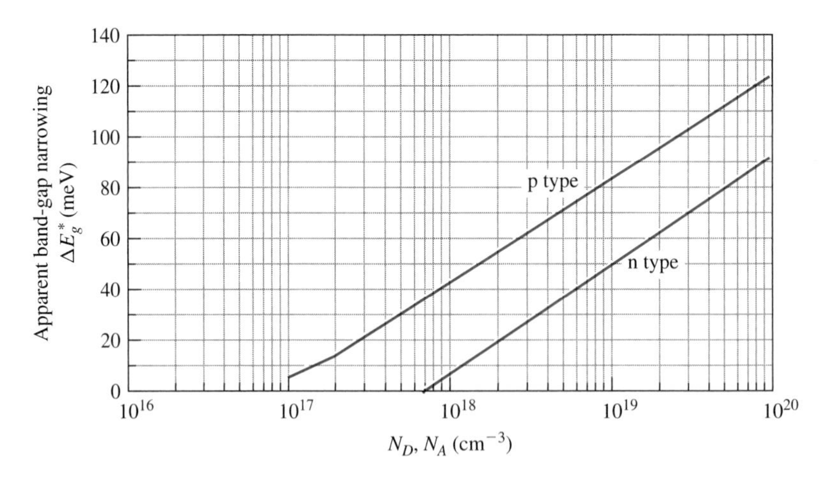

Figure 2.25 Apparent band-gap narrowing as a function of impurity concentration for uncompensated n-type and p-type Si.

2.12.3 CARRIER CONCENTRATIONS IN DEGENERATE SEMICONDUCTORS

Throughout this chapter, we have discussed and used formulas for finding the equilibrium carrier concentrations n_0 and p_0 for nondegenerately doped semiconductors. We now investigate how to handle those that are degenerately doped.

In Section 2.10 we obtained an expression for the electron-hole product at equilibrium in a nondegenerate semiconductor

$$n_0 p_0 = n_i^2 = N_C N_V e^{-E_{g0}/kT} \qquad \text{nondegenerate} \tag{2.106}$$

where E_{g0} is the band-gap energy of the intrinsic semiconductor. Since most semiconductor devices have at least one region that is degenerate, however, it is useful to have an expression for the $n_0 p_0$ product for such cases.

Consider an n-type degenerate semiconductor whose energy band diagram is indicated in Figure 2.24. Here E_{C0} is the intrinsic conduction band edge, E_C is the actual conduction band edge, E_C^* is the apparent conduction band edge (the Fermi energy), and ΔE_C^* is the impurity-induced effective conduction band change. Neglecting any affect of N_A on E_V,

$$\Delta E_g^* \approx \Delta E_C^* = E_{C0} - E_C^* \tag{2.107}$$

Since in this material all donor states overlap with each other and overlap with the conduction band, there are no bound states, and

$$n_0 = N_D \tag{2.108}$$

(or $n_0 = N_D - N_A$ in compensated material).

Since the Fermi level is far removed from the valence band, the Boltzmann approximation can be used to calculate p_0.

$$p_0 = N_V e^{-(E_f - E_V)/kT} = N_V e^{-E_g^*/kT} = N_V e^{-(E_{g0} - \Delta E_g^*)/kT} \tag{2.109}$$

Then the $n_0 p_0$ product is

$$n_0 p_0 = N_D N_V e^{-E_{g0}/kT}\, e^{\Delta E_g^*/kT} \tag{2.110}$$

Multiplying and dividing by N_C and rearranging gives

$$\begin{aligned} n_0 p_0 &= \left[\frac{N_D}{N_C} e^{\Delta E_g^*/kT}\right]\left[N_C N_V e^{-E_{g0}/kT}\right] \\ &= \frac{N_D}{N_C} e^{\Delta E_g^*/kT} n_i^2 \qquad \text{degenerate n type} \end{aligned} \tag{2.111}$$

That is, the $n_0 p_0$ product n_i^2 for nondegenerate semiconductors is modified by the factor $(N_D/N_C)e^{\Delta E_g^*/kT}$ in degenerately doped n-type materials. For p-type materials, the result is

$$n_0 p_0 = \frac{N_A}{N_V} e^{\Delta E_g^*/kT} n_i^2 \qquad \text{degenerate p type} \tag{2.112}$$

EXAMPLE 2.9

Consider a silicon sample doped with 5×10^{19} cm^{-3} phosphorus donors and no acceptors. Find the $n_0 p_0$ product.

■ Solution

From Figure 2.25, $\Delta E_g^* = 0.08$ eV for this n-type material. Then from Equation (2.111),

$$n_0 p_0 = \frac{5 \times 10^{19}}{2.86 \times 10^{19}} e^{0.08/0.026} n_i^2 = 38 n_i^2$$

The $n_0 p_0$ product is increased by a factor of about 40 above its nondegenerate (intrinsic) value.

Note that Equations (2.111) and (2.112) are valid for degenerate semiconductors in which the Fermi level is within the conduction and valence bands respectively.

The $n_0 p_0$ product is sometimes expressed as

$$n_0 p_0 = n_i^2 e^{\Delta E_g^{**}/kT} \qquad \text{alternative form} \tag{2.113}$$

Comparing Equations (2.111) and (2.113), we see that the definition of ΔE_g^{**} is

$$\Delta E_g^{**} - \Delta E_g^* = kT \ln \frac{N_D}{N_C}$$

For the case of Example 2.9, the difference is $\Delta E_g^{**} - \Delta E_g^* = 1.2$ meV, but increases with increased doping.

2.13 SUMMARY

In this chapter, we saw that we could define a carrier effective mass in such a way that if it were used in classical mechanical equations (pseudo-classical mechanics), the results would predict what actually happens to the electron or hole in response to an external force. Electrons were found to have negative effective mass in some cases (e.g., near the top of the valence band), but in those cases it makes more sense to talk about holes having positive effective mass and positive charge $+q$. Electrons near the bottom of the conduction band can be treated as particles of charge $-q$ and with a positive effective mass.

Most of the electrons available for conduction are found near the bottom of the conduction band, and similarly most of the holes are found near the top of the valence band. This is fortuitous, since those happen to be the regions where the pseudo-classical mechanics applies.

For a one-dimensional crystal, near the conduction band minimum E_C at $K = 0$, the electron energy can be expressed

$$E = E_C + \left(\frac{1}{2}\frac{d^2E}{dK^2}\right) K^2$$

where E_C is its potential energy and its kinetic energy is

$$E_K = \frac{1}{2}\frac{d^2E}{dK^2} K^2$$

The electron (group) velocity is

$$v = v_g = \frac{1}{\hbar}\frac{dE}{dK}$$

and the effective mass m^* is related to the curvature of the E-K curve:

$$m^* = \hbar^2 \left(\frac{d^2E}{dK^2}\right)^{-1}$$

Analogous expressions hold for holes in the valence band.

For a three-dimensional semiconductor having a single minimum at $K = 0$ (e.g., GaAs, InP), the effective mass is a scalar and single-valued. For cases in which the effective mass is not a scalar or single-valued, we differentiate between two different averages for effective mass. One is the density-of-states effective mass m^*_{ds}, used in calculations involving the density-of-states function. The other is the conductivity effective mass m^*_c. For other processes (e.g., tunneling, discussed in Supplement A to Part 1) different effective masses are used. It is important to recognize which effective mass to use in any given calculation.

The values of electron and hole density-of-states effective masses and conductivity effective masses are given in the following table.

Material	m^*_{dse}/m_0	m^*_{dsh}/m_0	m^*_{ce}/m_0	m^*_{ch}/m_0
Si	1.09	1.15	0.26	0.36
GaAs	0.067	0.48	0.067	0.34
Ge	0.56	0.29	0.12	0.21
InP	0.077	0.42	0.077	0.30

We also saw how the electrons and holes are distributed in energy. This result comes from two parameters:

1. The density of states $S(E)$ in each band, which varies parabolically with energy (near the band edges) for electrons in the conduction band and for holes in the valence band, where

$$S(E) = \frac{1}{2\pi^2}\left(\frac{2m^*_{ds}}{\hbar^2}\right)\sqrt{|E_K|}$$

2. The probability $f(E) = 1/(1 + e^{(E-E_f)/kT})$ of occupancy of a given state of energy E, where E_f is the Fermi energy.

The electron concentration distribution function with energy is then

$$n(E) = S(E)f(E)$$

2.13.1 NONDEGENERATE SEMICONDUCTORS

A material is defined as nondegenerate if the Fermi level is inside the forbidden gap and at least $2.3kT$ away from either band edge.

Substitutional impurities can affect the carrier concentrations in semiconductors. Donors have more electrons available than required for bonding, and these electrons are easily excited into the conduction band. Acceptors have too few electrons, and electrons from the valence band are easily excited into these acceptor states, leaving holes in the valence band.

To find the concentrations of carriers in a semiconductor, the steps are:

1. Assume all donors and acceptors are ionized (valid at room temperature as long as the net doping is sufficiently large that $|N_D - N_A| \gg n_i$, the intrinsic concentration. Under these conditions, in n-type material $n_0 = N_D - N_A$, and in p-type material $p_0 = N_A - N_D$. This gives us the majority carrier concentration.
2. To find the minority carrier concentration, if the material is nondegenerate, we can use

$$n_0 p_0 = n_i^2 = N_C N_V e^{-E_g/kT}$$

where n_i is the electron concentration of intrinsic material and

$$N_C = 2.54 \times 10^{19} \left(\frac{m_{dse}^*}{m_0}\right)^{3/2} \left(\frac{T}{300}\right)^{3/2} \text{ cm}^{-3}$$

$$N_V = 2.54 \times 10^{19} \left(\frac{m_{dsh}^*}{m_0}\right)^{3/2} \left(\frac{T}{300}\right)^{3/2} \text{ cm}^{-3}$$

The equilibrium carrier concentrations are

$$n_0 = N_C e^{-[(E_C - E_f)/kT]}$$

$$p_0 = N_V e^{-[(E_f - E_V)/kT]}$$

or alternatively

$$n_0 = n_i e^{-[(E_f - E_i)/kT]}$$

$$p_0 = n_i e^{-[(E_i - E_f)/kT]}$$

At high temperatures, thermally excited electron-hole pairs contribute to the carrier concentrations. At very low temperatures, the carrier concentrations are reduced as a result of electrons becoming bound to donor atoms or holes to acceptor atoms.

2.13.2 DEGENERATE SEMICONDUCTORS

If the material is sufficiently degenerate, i.e., the Fermi level is within the conduction band (n type) or within the valence band (p type), we use

$$n_0 p_0 = \frac{N_D}{N_C} e^{\Delta E_g^*/kT} n_i^2 \qquad \text{degenerate n type}$$

$$n_0 p_0 = \frac{N_A}{N_V} e^{\Delta E_g^*/kT} n_i^2 \qquad \text{degenerate p type}$$

where ΔE_g^* is the apparent or effective impurity-induced band-gap shrinkage, (the energy between the intrinsic band edge and the Fermi level).

2.14 READING LIST

The following refer to articles and books relevant to Chapter 2 listed in Appendix G. [1–21]

2.15 REFERENCES

1. M. A. Green, "Intrinsic concentration, effective density of states and effective mass in silicon, *J. Appl. Phys.,* 67, pp. 2944–2954, 1990.
2. J. S. Blakemore, *Semiconductor Statistics,* App. A, Dover Publications, Toronto, 1987.
3. H. P. D. Lanyon and R. A. Tuft, "Bandgap Narrowing in Heavily Doped Silicon," *IEDM Technical Digest,* p. 316, 1978.
4. Yaun Taur and Tak H. Ning, *Fundamentals of Modern VLSI Devices,* Chap. 2, Cambridge University Press, New York, 1998.
5. S. Sokolic' and S. Amon, "Modeling heavily doping effects for low temperature device simulation," *Journal de Physique IV,* Colloque C6, pp. C6-133–C6-138, 1994.
6. J. del Alamo, S. Swirhun, and R. M. Swanson, "Measuring and modeling minority carrier transport in heavily doped silicon," *Solid State Electronics,* 28, pp. 47–54, 1985.
7. J. del Alamo, S. Swirhun, and R. M. Swanson, "Simultaneous measurement of hole lifetime, hole mobility and bandgap narrowing in heavily doped n-type silicon," *IEDM Technical Digest,* pp. 290–293, 1985.
8. S. E. Swirhun, Y.–H. Kwark, and R. M. Swanson, "Measurement of electron lifetime, electron mobility and band-gap narrowing in heavily doped p-type silicon, *IEDM Technical Digest,* pp. 24–27, 1986.

2.16 REVIEW QUESTIONS

1. What is the distinction between an intrinsic and extrinsic semiconductor?
2. a. Define intrinsic concentration of holes.
   b. What is the relationship between this density and the intrinsic concentration of electrons?
   c. What do these equal at 0 K?
3. Define:
   a. Donor.
   b. Acceptor.

4. A semiconductor is doped with both donors and acceptors of concentrations N_D and N_A respectively. Write the equation or equations from which to determine the equilibrium electron and hole concentrations n_0 and p_0.
5. Explain physically the meaning of the following statement: An electron and a hole recombine and disappear.
6. Explain when and why the Boltzmann approximation can be used.
7. Explain what is meant by a distribution function. Use as an example the distribution in age of people in the United States.
8. Plot the Fermi-Dirac distribution function $f(E)$ versus energy E for $T = 0$ K and $T = 2500$ K. What are the meanings of these plots?
9. The electron energy distribution function is given by the product of two factors. What is the interpretation to be given to each of these factors?
10. Sketch the energy band diagrams for (a) an intrinsic, (b) an n-type, and (c) a p-type semiconductor. Indicate the positions of the Fermi level, the conduction band edge, the valence band edge, the donor level, and the acceptor level and label both axes.
11. Explain what it means for a semiconductor to be degenerate.
12. Under what conditions do discrete donor states expand into minibands?
13. Draw an energy band diagram for a degenerately doped p-type semiconductor. Indicate the band-gap narrowing and the apparent band gap narrowing.

2.17 PROBLEMS

Unless otherwise specified, assume $T = 300$ K.

2.1 Of the two materials whose E-K diagrams are shown in Figure P2.1, which will have the lowest effective mass for electrons? Which will have the lowest effective mass for holes? Explain how you arrived at your conclusion.

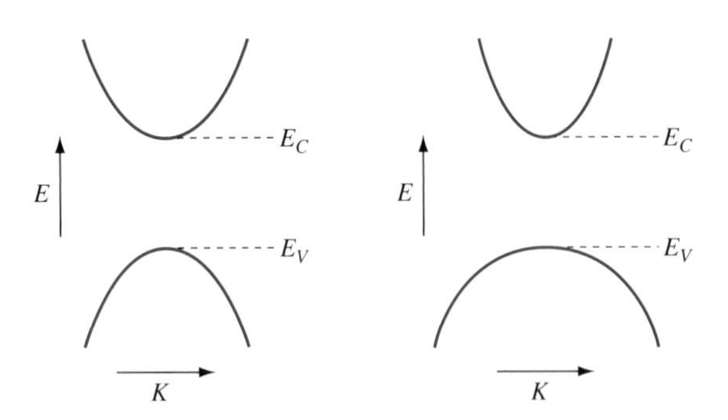

Figure P2.1

2.2 Assume a material has an *E-K* diagram given by

$$E(K)_{\text{conduction}} = E_C + E_1 \sin^2(Ka)$$
$$E(K)_{\text{valence}} = E_V - E_2 \sin^2(Ka)$$

Let $a = 0.5$ nm, $E_1 = 5$ eV, and $E_2 = 4$ eV.

a. Sketch the *E-K* diagram for the first Brillouin zone. Label the axes completely.

b. What is the effective mass for an electron near the bottom of the conduction band?

c. What is the effective mass for holes near the top of the valence band?

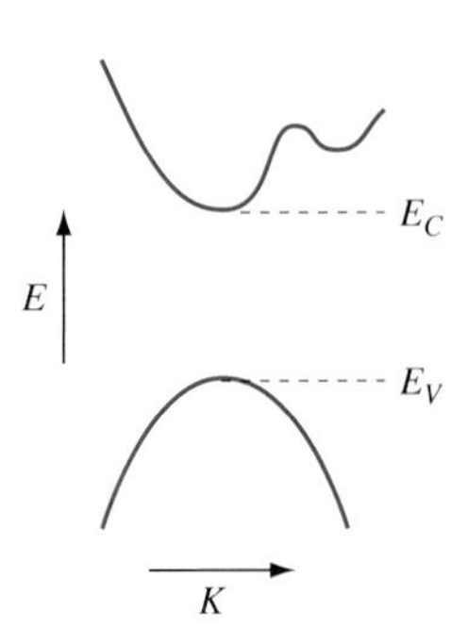

Figure P2.2

2.3 Explain in your own words why the effective mass of an electron in a semiconductor is different from its mass in vacuum.

2.4 For the material of Figure P2.2, there are two different effective masses for electrons. Which effective mass will be the one displayed by most of the electrons in the conduction band? Why?

2.5 Consider the energy band diagram in Figure P2.3.

a. Find the electric field, and express the result in V/m. In what direction does it point?

b. Find the force on the electron. In what direction is the electron accelerated?

c. Find the force on the hole. In what direction is the hole accelerated?

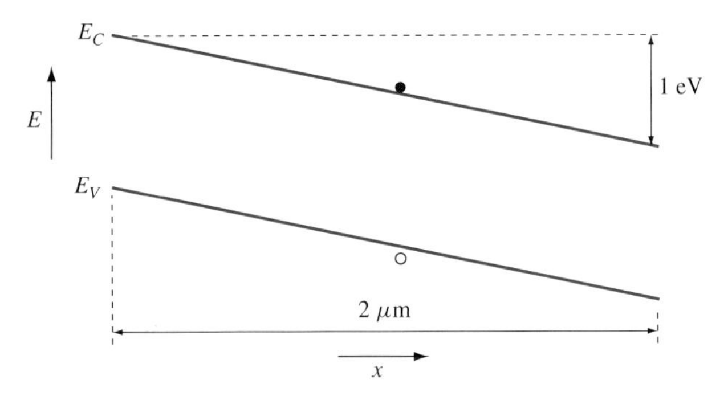

Figure P2.3

2.6 For an optical emission to occur, four things are needed:

a. An electron must be at an elevated energy state.

b. There must be an empty state at a lower energy for the electron to go to.

c. Energy must be conserved (energy of electron before emission = energy of electron after emission + energy of photon).

d. *K* must be conserved.

It turns out that photons have negligible *K* compared with electrons of interest in semiconductor electronics. That implies an electron making an

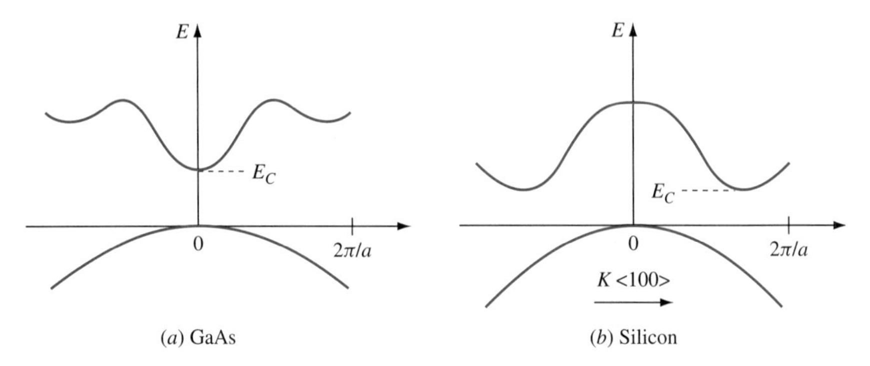

Figure P2.4

optical transition must end up at the same value of K as when it started (the energy can change, since the energy difference goes to the photon). On the basis of your expectation of where electrons and holes are likely to be found, and the four conditions above, which material will be a more efficient light emitter, Si or GaAs? Their E-K diagrams are shown in Figure P2.4. Explain your reasoning. Silicon is an indirect gap material, meaning the conduction band minimum is not at the same value of K as at the maximum in the valence band. GaAs is a direct gap material.

2.7 Consider the energy band diagram in Figure P2.5.

a. What is the potential energy of the electron?
b. What is the kinetic energy of the electron?
c. What is the potential energy of the hole?
d. What is the kinetic energy of the hole?
e. The vertical axis is energy—meaning the electron energy. What direction represents increasing hole energy, up or down?

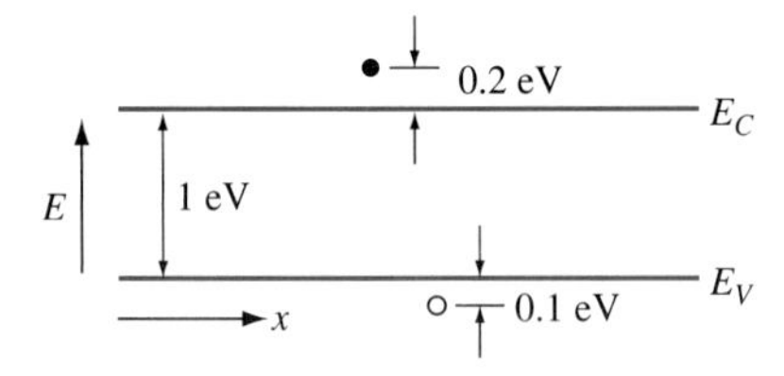

Figure P2.5

2.8 A sample of GaAs is subjected to an electric field of 100 V/cm.

a. What is the velocity of an electron in the conduction band if it is initially at rest and is accelerated by this field for a period of 0.1 ps? (Use the conductivity effective mass since this problem is related to conduction.)
b. What is the terminal velocity for a hole under the same conditions?

2.9 Near the bottom of the conduction band in a semiconductor, the electron energy can be expressed as $E = E_C + AK^2$, where A is independent of K and is positive.

a. Find an expression for the electron effective mass in terms of A.

b. Find an expression for the electron velocity as a function of E.

2.10 Consider a semiconductor material whose three-dimensional *E-K* relationship has the same curvature in the x and z directions but a different curvature in y. Let us consider the x and y directions, and let the relationship be $E(x, y) = AK_x^2 + BK_y^2$.

a. Using a computer program such as MATLAB, produce a mesh plot of this function for $B = 2A$. Rotate the perspective to get an understanding of the shape of this function.

b. In which direction is the effective mass smaller?

c. What determines which effective mass a given electron actually has in this crystal?

2.11 For each of the following, identify whether the impurity will produce n-type or p-type material, or be amphoteric:

a. Phosphorus replacing a gallium atom in gallium arsenide.

b. Zinc in germanium.

c. Carbon in indium phosphide.

d. Silicon in zinc selenide.

2.12 Show that the difference between the intrinsic level and midgap is small, using GaAs as an example. Compare the result with that of silicon.

2.13 What is the probability that a state at the conduction band edge of intrinsic GaAs is occupied at room temperature?

2.14 The donor ground states for tellurium in GaAs are 5.9 meV below the conduction band. (There are two of them because of spin.) At room temperature, what is the probability that a given ground state is occupied if the Fermi level is 0.1 eV below E_C?

2.15 What is the probability of occupancy of a state 0.1 eV above the Fermi level? Below the Fermi level?

2.16 Verify the definition of nondegeneracy; show that if the Fermi level is $2.3kT$ below the conduction band edge, the probability of occupancy of the lowest energy state in the conduction band is 10 percent.

2.17 Calculate the effective densities of states for electrons and holes at room temperature for

a. Silicon.

b. GaAs.

c. Ge.

2.18 Compare the effective density of states for electrons, N_C, and for holes, N_V, for GaAs.

2.19 Show that Equation (2.52) follows from Equation (2.51), that is:

$$f_p(E) = 1 - \frac{1}{1 + e^{(E-E_f)/kT}}$$

becomes

$$f_p(E) = \frac{1}{1 + e^{(E_f-E)/kT}}$$

2.20 Recall that the equations $n_0 = N_C e^{-[(E_C-E_f)/kT]}$ and $p_0 = N_V e^{-[(E_f-E_V)/kT]}$ are valid only for nondegenerately doped semiconductors. One reason is that the effective densities of states are defined only for nondegenerately doped materials. What is the other reason?

2.21 a. Silicon is doped with $N_D = 5 \times 10^{15}$ cm^{-3}. Find n_0 and p_0, and locate the Fermi level. Draw the energy band diagram.

b. A new batch of silicon is doped with boron to $N_A = 4 \times 10^{15}$ cm^{-3}. Find n_0 and p_0 and locate the Fermi level. Draw the energy band diagram.

2.22 Silicon is doped with $N_A = 4 \times 10^{17}$ cm^{-3}. Find n_0 and p_0. Draw the energy band diagram indicating the position of the Fermi level.

2.23 GaAs is doped with $N_D = 10^{15}$ cm^{-3} and $N_A = 4 \times 10^{14}$ cm^{-3}. Draw the energy band diagram and locate the Fermi level.

2.24 Ge is doped with 10^{15} Sb/cm^{-3}. Assuming Sb to be a shallow donor, find n_0 and p_0 at room temperature. Draw the energy band diagram.

2.25 a. Calculate the intrinsic carrier concentration for Ge at room temperature.

b. Make a semilogarithmic plot of n_i versus temperature over the range 200 to 500 K. The intrinsic carrier concentration should be on the vertical (logarithmic) axis. Add curves for silicon and GaAs.

2.26 A diode is made by using an n-type Si substrate ($N_D = 10^{15}$ cm^{-3}) and then adding acceptors to a concentration of $N_A = 10^{16}$ cm^{-3} to one region. What are p_0 and n_0 in the p-type region?

2.27 A sample of InP is doped with $N_D = 10^{16}$ cm^{-3} and $N_A = 10^{15}$ cm^{-3}.

a. Find n_0 and p_0.

b. Locate E_f.

c. Sketch the energy band diagram. Label both axes and identify E_C, E_V, E_f, E_i, and E_{vac} on your drawing. Indicate the electron affinity and the band gap.

2.28 A sample of silicon is doped with $N_D = 4 \times 10^{16}$ cm^{-3} and $N_A = 8 \times 10^{15}$ cm^{-3}. Find the equilibrium concentrations of electrons and holes, and locate the Fermi level.

2.29 How heavily would you need to dope silicon with donors to violate the assumption of nondegenerate doping? How many acceptors would be needed to just cause a degenerately doped type material?

2.30 What donor concentration is required to elevate the Fermi level in Si to $2.3kT$ above E_i? What acceptor concentration will lower E_f $2.3kT$ below E_i?

2.31 Given that for a particular sample of silicon $N_D = 10^{15}$ cm^{-3} and $n_0 = 10^6$ cm^{-3}, find N_A and p_0.

2.32 Manganese makes a donor trap state 0.53 eV below the conduction band edge in silicon. If the silicon is doped with $N_D = 4 \times 10^{16}$ cm^{-3}, what is the probability of occupancy of the trap state? Assume the concentration of Mn is small enough not to affect the overall doping and that the trap state is single-valued.

2.33 A sample of Si doped with phosphorus has its Fermi level 0.2 eV below E_C and the donor ground state is 0.045 eV below E_C. Find n_0.

2.34 The probability that an energy state in the conduction band edge, E_C, of Si is 10^{-4}:

a. Is this Si n type, p type, or intrinsic?

b. Find $N_D - N_A$.

2.35 The effective density-of-states function N_C for a given semiconductor is 7×10^{17} cm^{-3}. What is the electron density-of-states effective mass?

2.36 Given two semiconductors A and B, let them have the same density-of-states effective masses. Let $E_g = 1$ eV for A and 2 eV for B. Find the ratios of the intrinsic carrier concentrations.

2.37 Semiconductor devices for many applications must be able to withstand and operate over a wide range of temperatures, to operate from Antarctica or deep space to a tropical climate in a hot truck. For example, military specifications for semiconductor devices cover the range -55°C to $+150$°C. Repeat Problem 2.26 at these two temperature extremes.

2.38 In a nondegenerate semiconductor, the electron distribution peaks at $E_C + kT/2$ as indicated in Figure 2.17, and the average electron kinetic energy is at $E_C + 3kT/2$. For intrinsic material, find the probability that a state at these two energies is occupied. Repeat for $E_C + 10kT$.

2.39 Calculate the electron concentrations in Si and GaAs if the Fermi level is 0.2 eV below the conduction band edge.

2.40 Complete the mathematical steps in Example 2.7 and verify the results.

2.41 At what doping concentration does apparent band-gap narrowing become significant in n-type Si? (We will define significant as a narrowing of 0.03 eV.)

2.42 Suppose you were to dope some silicon with $N_D = N_A = 10^{16}$ cm^{-3}.

a. Where do you expect the Fermi level to be?

b. What do you expect n_0 and p_0 to be?

c. Verify (or adjust) your expectations by performing the calculations.

What do you think is going on physically? What is happening to the electrons in the donor and acceptor states?

2.43 Plot and compare n_i for Si as a function of temperature from 250 to 500 K as calculated from Equation (2.88) with that calculated from Equations (2.85), (2.86), (2.63), and (2.64). Neglect the small temperature dependences of the density-of-states effective masses.

2.44 A silicon sample doped with $N_A = 10^{15}$ and $N_D = 10^{14}$ cm^{-3} is at a temperature of 600 K. Find n_0 and p_0.

2.45 The sample of Problem 2.44 is cooled to 250 K. What are the equilibrium carrier densities n_0 and p_0 now?

2.46 In this chapter we differentiated between conductivity effective mass and density-of-states effective mass. This problem illustrates the calculation of the conductivity of a semiconductor by using the conductivity effective mass. (A more thorough discussion appears in Chapter 3.) The current density is the current passing a plane of unit area per second. Consider an n-type semiconductor. The current density is

$$J = -qn_0 \langle v \rangle = -\sigma \mathscr{E}$$

where n_0 is the electron concentration, $\langle v \rangle$ is the average electron velocity, and σ is the conductivity. The negative sign indicates that the current is in the opposite direction to the electron velocity, since the electron is charged negatively. Since this is an n-type semiconductor, $n_0 = N_D$ and the conductivity can be expressed as

$$\sigma = -\frac{qN_D \langle v \rangle}{\mathscr{E}}$$

Find the conductivity of a semiconductor with $N_D = 10^{17}$ cm^{-3} if the average time between collisions of electrons is 10^{-13} seconds.

CHAPTER 3

Current Flow in Homogeneous Semiconductors

3.1 INTRODUCTION

In Chapter 2 we learned how to determine how many carriers are available for conduction in a given semiconductor—holes in the valence band and electrons in the conduction band. In this chapter, we consider the mechanisms by which current flows.

There are two basic mechanisms by which charge carriers move in semiconductors: *drift* and *diffusion*. We begin with the more intuitive drift current, which results when the electrons and holes are in an electric field. The diffusion current, which is discussed later, arises when there is a variation in the concentration of carriers with position.

3.2 DRIFT CURRENT

A semiconductor at equilibrium has a net total current of zero. Although the electrons and holes are moving because of their thermal energy, the direction of movement is completely random, so all the currents contributed by the individual charge carriers add up to zero. Consider a single electron in the conduction band. Its path when there is no applied field is shown in Figure 3.1a. After each collision, the new direction is random, so that, averaged over time, the electron makes no overall progress in any particular direction and the net current is zero.

Recall from Figure 2.2, however, that when an external electric field is applied to a semiconductor, the negatively charged electrons are accelerated and the positively charged holes are accelerated in the opposite direction. Figure 3.1b shows the progress of a particular electron under the influence of a high electric field. Averaged over a time long enough to include many collisions, there is a

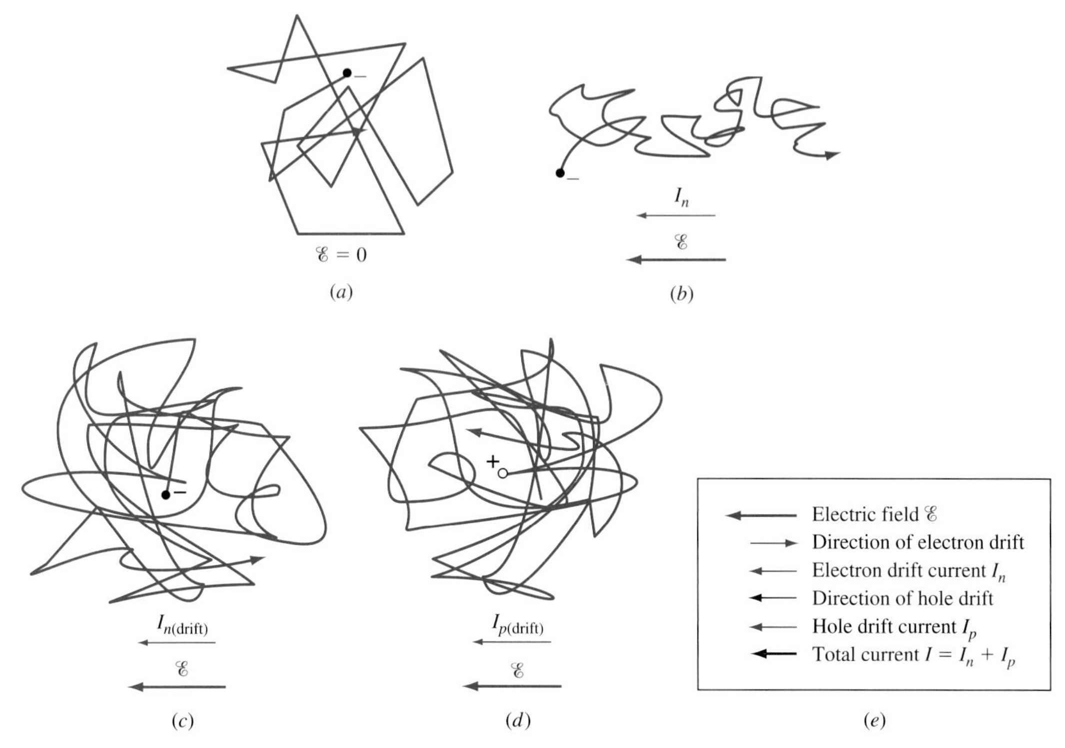

Figure 3.1 The motion of an electron in a crystal. The electron changes direction randomly whenever it makes a collision. (a) Under no applied field there is no net progress in any particular direction. (b) When a field is applied, the electron tends to drift in some particular direction. A trajectory such as this would be found only under very high fields. (c) Under low fields, the drift velocity is much smaller than the thermal speed. It takes many collisions before any appreciable progress is made. (d) A hole undergoes similar motion, but being positively charged, is accelerated in the opposite direction. (e) Electrons and holes drift in opposite directions but the resulting currents are in the same direction.

general tendency for the electron to drift toward the right. Any net motion of a charge results in electric current, so this motion is termed "drift current."

For more typical fields in a bulk semiconductor, the velocity induced by the applied field is actually quite small compared with the thermal speed as indicated in Figure 3.1c. For a field on the order of 10 V/cm, the average drift velocity is on the order of 10^4 cm/s. Compare this with the average electron speed between collisions (thermal speed) on the order of 10^7 cm/s at room temperature.

Electric current is defined as the amount of charge crossing some plane per unit time. The charge for electrons is negative; in Figures 3.1b and c, therefore, the electron current $I_{n(\text{drift})}$ is going to the left. We are using *drift* in the subscript to differentiate this current mechanism, driven by the presence of an electric field, from the diffusion current we will discuss later. Holes can also travel in these types of paths, as shown in Figure 3.1d. The overall drift of the holes is in the opposite direction to that of the electron. This is because the holes are positively charged, so the same field accelerates them in the opposite direction to the electrons. The current the holes produce is, however, in the same direction as their motion, and the hole current $I_{p(\text{drift})}$ is also to the left, as shown in Figure 3.1e. The total drift current is the sum of these two:

$$I_{(\text{drift})} = I_{n(\text{drift})} + I_{p(\text{drift})} \tag{3.1}$$

Current I is a convenient quantity to define for a wire, but for a semiconductor, it is often more useful to talk about current density J, which is the amount of charge crossing a plane of unit area per unit time. That is,

$$J = \frac{I}{\text{area}} \tag{3.2}$$

We now derive an expression for current density. We recall from Ohm's law that the resistance R of a uniform sample of length L and cross-sectional area A (Figure 3.2) is

$$R = \frac{V}{I} = \frac{\rho L}{A} \tag{3.3}$$

where ρ is referred to as *resistivity* and has common dimensions of ohm-centimeters ($\Omega \cdot \text{cm}$). Since the semiconductor is uniform, $V = \mathscr{E}L$, where $\mathscr{E}$ is the electric field and L is the sample length. Using $I = JA$, we can express Equation (3.3) as

$$\frac{\mathscr{E}L}{JA} = \frac{\rho L}{A} \tag{3.4}$$

or

$$\boxed{J_{(\text{drift})} = \frac{\mathscr{E}}{\rho} = \sigma\mathscr{E}} \tag{3.5}$$

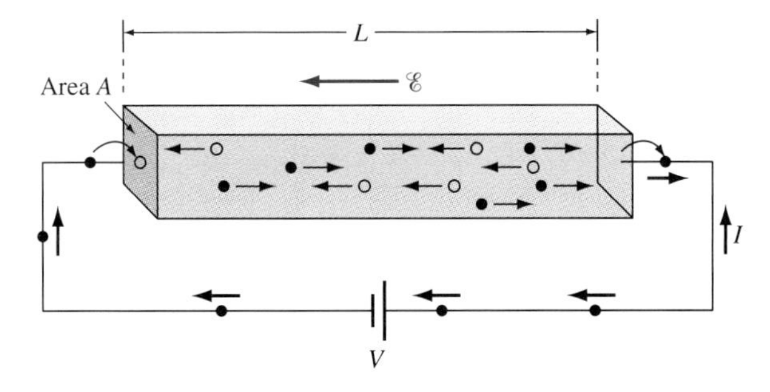

Figure 3.2 Current is carried by electrons in a wire, but in a semiconductor current can be carried by both electrons and holes.

where $\sigma = 1/\rho$ and is referred to as conductivity, with dimensions of reciprocal ohm-centimeters $[(\Omega \cdot \text{cm})^{-1}]$ or siemens per centimeter (S/cm), and we have added the subscript *drift* again since there will also be a diffusion current density component discussed later. Equation (3.5) is often referred to as *Ohm's law in point form*.

Since, in a semiconductor, electrons and holes can carry current, the total current density is

$$J_{(\text{drift})} = J_{n(\text{drift})} + J_{p(\text{drift})} \tag{3.6}$$

where $J_{n(\text{drift})} = \sigma_n \mathcal{E}$ and $J_{p(\text{drift})} = \sigma_p \mathcal{E}$. Here σ_n is the conductivity due to electrons and σ_p the conductivity due to holes. In an n-type semiconductor, electrons carry most of the current, since there are more electrons than holes, while in p-type material most of the current is carried by holes.

Before we continue, we have an observation to make about Figure 3.2. Both holes and electrons carry the current in the semiconductor, but in the wire, the current (the same amount) is carried by electrons alone. What happens to the holes? At the left end of the semiconductor bar, we see electrons arriving at the semiconductor-wire boundary from the wire, and at the same time, holes are arriving there from the bar. The electron, in effect, enters the semiconductor at the ohmic contact between wire and semiconductor and annihilates the hole. An opposing electron cancels every hole that gets lost. The hole is annihilated, but remember, it was only an empty state to begin with. When the electron moves into the state, the hole is lost but the electron keeps going. Stated differently, at the ohmic contact the equilibrium concentrations of electrons and holes are maintained in the semiconductor.

We next obtain expressions for σ_n and σ_p. We first consider hole conduction.

Consider a p-type semiconductor of hole concentration p as shown schematically in Figure 3.3. We have used the more general notation p instead of the equilibrium concentration p_0, because once a field is applied the sample is no

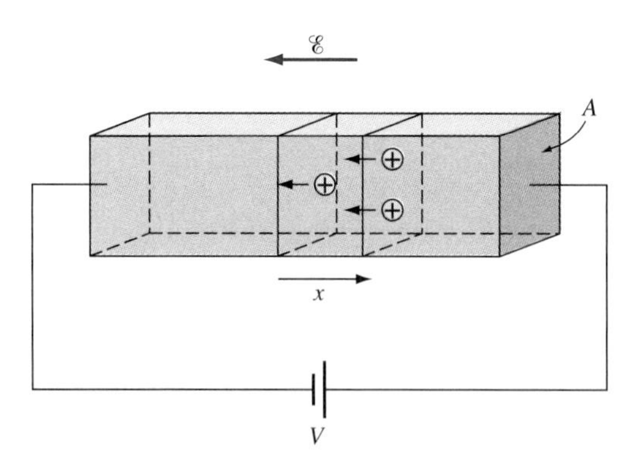

Figure 3.3 We find the conductivity of a sample by considering the flow of (in this case) holes across an area *A*.

longer at equilibrium and in some cases, $p \neq p_0$. Here an electric field is applied in the negative x direction. The force on the holes, $F = +q\mathscr{E}$, will move them in the direction of the field. Assume that all the holes have an average field-induced (drift) velocity v_{dp}. The total charge crossing the area A in time dt is equal to $Q = qpAv_{dp}\,dt$. Since current is defined as the charge passing a plane in unit time,

$$I_p = \frac{dQ}{dt} = qpAv_{dp} \tag{3.7}$$

The hole current density is $J_p = I_p/A$. But from Equations (3.7), (3.2), and (3.5),

$$J_{p(\text{drift})} = qpv_{dp} = \sigma_p\mathscr{E} \tag{3.8}$$

and

$$\sigma_p = qp\frac{v_{dp}}{\mathscr{E}} = qp\mu_p \tag{3.9}$$

where μ_p is called the *hole mobility* and is the hole drift velocity per unit field:

$$\boxed{\mu_p \equiv \frac{v_{dp}}{\mathscr{E}}} \tag{3.10}$$

The mobility is a measure of how easily a carrier moves in a particular material. It is normally expressed in centimeters squared per volt-second ($cm^2/V \cdot s$). Similarly the electron mobility is

$$\boxed{\mu_n = -\frac{v_{dn}}{\mathscr{E}}} \tag{3.11}$$

and the electron conductivity σ_n is

$$\boxed{\sigma_n = qn\mu_n} \tag{3.12}$$

Finally, the total drift current density is

$$J_{(\text{drift})} = J_{n(\text{drift})} + J_{p(\text{drift})} = (q\mu_n n + q\mu_p p)\mathscr{E} = \sigma\mathscr{E} \tag{3.13}$$

and the total conductivity is

$$\boxed{\sigma = qn\mu_n + qp\mu_p} \tag{3.14}$$

3.3 CARRIER MOBILITY

The electron and hole mobilities are dependent on the impurity concentrations of donors and acceptors, on temperature, and on whether the carriers are minority carriers or majority carriers. (Majority carriers are electrons in n-type material and holes in p-type material.) The physics of these effects are discussed in the next section, but here we give some results for silicon at room temperature.

First, the mobility varies with the doping concentration. While there is considerable scatter in the experimental data,[1] for noncompensated material, the mobility in silicon is often characterized by the empirical formula:

$$\mu = \mu_0 + \frac{\mu_1}{1 + \left(\dfrac{N}{N_{\text{ref}}}\right)^{\alpha}} \tag{3.15}$$

where N is the doping concentration (either N_D or N_A) and N_{ref} and α are fitting parameters.

The fitting parameters depend on whether the carriers of interest are majority or minority carriers.[2] At room temperature Equation (3.15) becomes:

Majority carriers [1]:

$$\mu_n(N_D) = 65 + \frac{1265}{1 + \left(\dfrac{N_D}{8.5 \times 10^{16}}\right)^{0.72}} \tag{3.16}$$

$$\mu_p(N_A) = 48 + \frac{447}{1 + \left(\dfrac{N_A}{1.3 \times 10^{16}}\right)^{0.76}} \tag{3.17}$$

Minority carriers [2–4]:

$$\mu_n(N_A) = 232 + \frac{1180}{1 + \left(\dfrac{N_A}{8 \times 10^{16}}\right)^{0.9}} \tag{3.18}$$

$$\mu_p(N_D) = 130 + \frac{370}{1 + \left(\dfrac{N_D}{8 \times 10^{17}}\right)^{1.25}} \tag{3.19}$$

These equations apply only to silicon, and only under low field. (We will address high-field effects in Section 3.3.5.) They are plotted in Figure 3.4. From these plots we can see:

1. At low impurity concentrations, majority carrier and minority carrier electron mobilities approach the same values: $\mu_n \approx 1330 \text{ cm}^2/\text{V} \cdot \text{s}$.
2. A similar result holds for holes: $\mu_p \approx 495 \text{ cm}^2/\text{V} \cdot \text{s}$.

---

[1] The majority carrier mobilities have been extensively measured in uncompensated semiconductors. Minority carrier mobility is much more difficult to measure and fewer results have been reported.

[2] It may surprise the student to know that in some devices, such as bipolar junction transistors, the minority carriers influence the device operation more than the majority carriers. In some field-effect transistors, the majority carriers are more important.

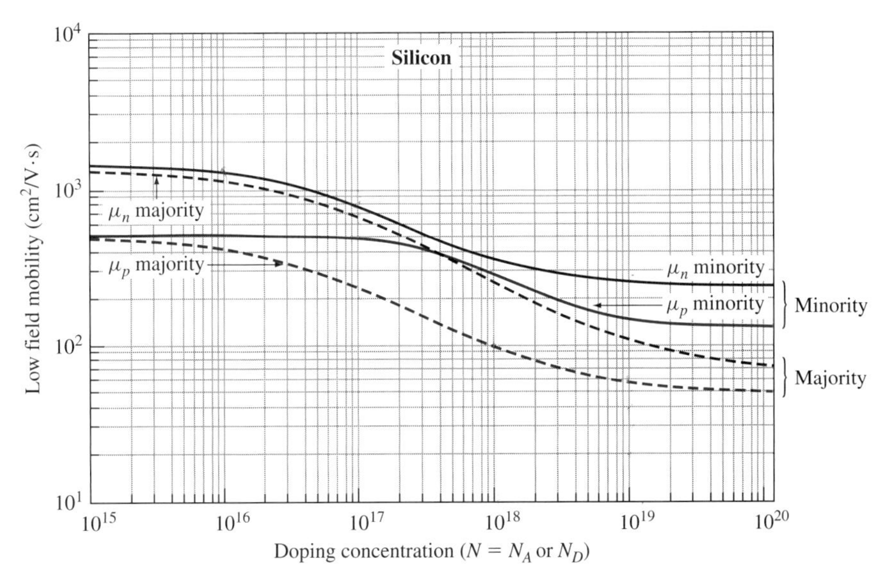

Figure 3.4 Room temperature majority and minority carrier mobility as functions of doping in p-type and n-type silicon. Solid lines: minority carriers; dashed lines: majority carriers.

3. Electron and hole mobilities (both majority carrier and minority carrier) reduce monotonically with increasing impurity concentration.
4. For a given doping level, minority carrier mobilities for electrons and holes are greater than corresponding majority carrier mobilities.
5. These fractional differences increase with increased doping.

EXAMPLE 3.1

Compute the room temperature resistivity of intrinsic silicon, and compare that with uncompensated n- and p-type silicon doped with 10^{17} cm^{-3} donors (or acceptors). Assume $n = n_0$ and $p = p_0$.

■ **Solution**

The resistivity of a sample is given by the reciprocal of the conductivity:

$$\boxed{\rho = \frac{1}{\sigma}}$$

and from Equation (3.14),

$$\sigma = q(\mu_n n + \mu_p p)$$

a. *Intrinsic silicon:* We already know the room temperature equilibrium carrier concentrations for intrinsic silicon are $n_0 = p_0 = n_i = 1.08 \times 10^{10}$ cm^{-3}.

From Figure 3.4 we find the mobilities for electrons and holes for undoped (intrinsic) silicon:

$$\mu_n = 1330 \frac{\text{cm}^2}{\text{V} \cdot \text{s}}$$

$$\mu_p = 495 \frac{\text{cm}^2}{\text{V} \cdot \text{s}}$$

We can find the conductivity:

$$\begin{aligned}\sigma &= q(\mu_n n + \mu_p p) = q(\mu_n n_0 + \mu_p p_0)\\ &= 1.6 \times 10^{-19}\ \text{C}\left(1330 \frac{\text{cm}^2}{\text{V} \cdot \text{s}} \cdot 1.08 \times 10^{10}\ \text{cm}^{-3} + 495 \frac{\text{cm}^2}{\text{V} \cdot \text{s}} \cdot 1.08 \times 10^{10}\ \text{cm}^{-3}\right)\\ &= 3.15 \times 10^{-6}\ (\Omega \cdot \text{cm})^{-1}\end{aligned}$$

From the conductivity, we can find the resistivity:

$$\rho = \frac{1}{\sigma} = \frac{1}{3.15 \times 10^{-6}} = 3.17 \times 10^5\ \Omega \cdot \text{cm}$$

b. *n-type silicon:* We expect, since this material is more heavily doped, and since it has many electrons to carry current, that the result will be higher conductivity or lower resistivity than that of intrinsic material. Since $N_D \gg n_i$ and since virtually all impurities are ionized,

$$n_0 = N_D = 10^{17}\text{cm}^{-3}$$

$$p_0 = \frac{n_i^2}{n_0} = \frac{(1.08 \times 10^{10}\ \text{cm}^{-3})^2}{10^{17}\ \text{cm}^{-3}} = 1.17 \times 10^3\ \text{cm}^{-3}$$

From Figure 3.4, for a donor concentration of 10^{17}, we find the electron mobility and hole mobilities to be:

$$\mu_n \approx 650 \frac{\text{cm}^2}{\text{V} \cdot \text{s}} \qquad \text{majority carrier mobility}$$

$$\mu_p \approx 480 \frac{\text{cm}^2}{\text{V} \cdot \text{s}} \qquad \text{minority carrier mobility}$$

Using these values, we obtain

$$\begin{aligned}\sigma &= q(\mu_n n_0 + \mu_p p_0)\\ &= 1.6 \times 10^{-19}\ \text{C}\left(650 \frac{\text{cm}^2}{\text{V} \cdot \text{s}} \cdot 10^{17}\ \text{cm}^{-3} + 480 \frac{\text{cm}^2}{\text{V} \cdot \text{s}} \cdot 1.17 \times 10^3\ \text{cm}^{-3}\right)\\ &= 1.6 \times 10^{-19}(6.5 \times 10^{19} + 5.62 \times 10^5) = 10.4\ (\Omega \cdot \text{cm})^{-1}\end{aligned}$$

Note that the second term, the hole term, is negligible in this case. The resistivity is then

$$\rho = \frac{1}{\sigma} = \frac{1}{10.4} = 0.096\ \Omega \cdot \text{cm}$$

The doped Si is considerably more conductive than the intrinsic material.

c. *Same doping level as the previous example but p type instead of n type:* It is worthwhile to start a problem by thinking ahead to what kind of result we might expect. We expect the p-type material to be more conductive than the intrinsic, again because it has more carriers. On the other hand, holes are generally less mobile than electrons, so the p-type material may not be as conductive as the equivalent n-type. We find the carrier concentrations first:

$$p_0 = N_A = 10^{17}\ \text{cm}^{-3}$$

$$n_0 = \frac{n_i^2}{p_0} = 1.17 \times 10^3\ \text{cm}^{-3}$$

Since

$$\mu_n = 750\ \frac{\text{cm}^2}{\text{V}\cdot\text{s}} \qquad \text{minority carriers}$$

$$\mu_p = 230\ \frac{\text{cm}^2}{\text{V}\cdot\text{s}} \qquad \text{majority carriers}$$

Then

$$\begin{aligned}\sigma &= q(\mu_n n_0 + \mu_p p_0)\\ &= 1.6 \times 10^{-19}\ \text{C}\left(750\ \frac{\text{cm}^2}{\text{V}\cdot\text{s}} \cdot 1.17 \times 10^3\ \text{cm}^{-3} + 230\ \frac{\text{cm}^2}{\text{V}\cdot\text{s}} \cdot 10^{17}\ \text{cm}^{-3}\right)\\ &= 3.7\ (\Omega\cdot\text{cm})^{-1}\end{aligned}$$

In this case, the electrons are so few they contribute negligibly to conduction. The resistivity for the p-type sample is

$$\rho = \frac{1}{\sigma} = \frac{1}{3.7\ (\Omega\cdot\text{cm})^{-1}} = 0.27\ \Omega\cdot\text{cm}$$

The p-type material is slightly less conductive than the n-type sample at the same doping levels because the current is carried primarily by holes and the holes are less mobile than the electrons.

3.3.1 CARRIER SCATTERING

Recall that by definition [Equations (3.10) and (3.11)], mobility is dependent on the drift velocity:

$$\boxed{\mu = \left|\frac{v_d}{\mathscr{E}}\right|}$$

The carrier drift velocity is influenced by scattering events, i.e., change in direction and/or energy of a carrier by collisions with a particle. Minority carrier mobility is dependent on these scattering mechanisms. As discussed later, the drift velocity is also dependent on the time the carrier spends in impurity states. This impurity band conduction, along with scattering, contributes to majority carrier scattering. Since both majority and minority carrier mobilities are influenced by scattering, we will examine the physics of scattering first.

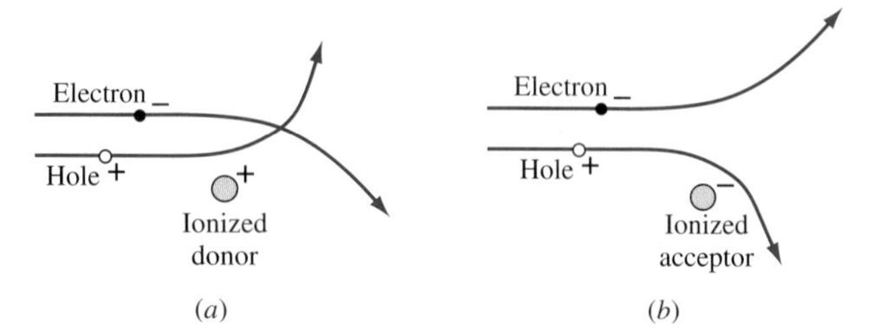

Figure 3.5 (a) An electron approaching an ionized donor is deflected toward it, but a hole is deflected away from the donor. (b) Electrons deflect away from the negatively charge ionized acceptors but holes deflect toward them.

Carriers can be scattered by interactions (collisions) with "particles" such as ionized impurity atoms and by phonons. The drift velocity and thus the mobility are dependent on the mean free time between collisions.

Ionized Impurity Scattering The scattering effect of the impurities can be understood by recognizing that the impurities (donors and acceptors) are typically ionized, so they are charged. The Coulombic forces will deflect an electron or hole approaching the ionized impurity. This is known as *ionized impurity scattering* and is shown schematically in Figure 3.5. The situation is shown for a positive ion (a) and a negative ion (b). Holes are deflected away from the positive ion and toward the negative ion; electrons are deflected oppositely. The amount of deflection depends on the speed of the carrier and its proximity to the ion. It follows, then, that the more heavily a material is doped, the higher the probability that a carrier will collide with an ion in a given time and the smaller will be the mean free time between collisions and the smaller the mobility.

Furthermore, since the ionized donors and acceptors both scatter carriers approximately equally, the mobility due to ionized impurity scattering in compensated material depends on the *total* doping concentration $N_D + N_A$. Although Figure 3.4 is for uncompensated material, in practice the minority carrier mobilities are often estimated from it since usually $N_D \gg N_A$ or $N_A \gg N_D$.

Lattice (Phonon) Scattering Another influence on the carrier mobility results from lattice scattering, often called phonon scattering. We know that at any temperature the vibrating atoms create pressure (acoustic) waves in the crystal. As discussed in the Supplement B to Part 1, these pressure waves are called *phonons*. Like electrons, phonons can be considered to be particles (wave-particle duality), each with energy

$$E_{\text{phonon}} = \hbar\omega \tag{3.20}$$

and wave vector

$$K = \frac{2\pi}{\lambda} \tag{3.21}$$

where ω is the angular frequency of the lattice vibration and λ is the corresponding wavelength. Thus, considering phonons as particles, a phonon can collide with an electron (or hole) and scatter it. Phonon energies extend over a small range—usually less than 0.1 eV. The energy distribution of phonons and the concentration of phonons at a given energy depend on the amplitude of the pressure wave, and thus on temperature. With increasing temperature, the stronger lattice vibrations cause an increase in the concentration of phonons and thus increased scattering or reduced mobility.

3.3.2 SCATTERING MOBILITY

Let us explore how the scattering affects the mobility. We start by considering the force on an electron due to the electric field:

$$F = -q\mathscr{E} = m_{ce}^* a = m_{ce}^* \frac{dv}{dt} \tag{3.22}$$

where a is the electron acceleration due to the field $\mathscr{E}$ and v is the electron velocity. Note that, since we are considering conduction, we use the conductivity effective mass for electrons. Here we consider the mobility associated with a single scattering mechanism, e.g., ionized impurity scattering.

From Equation (3.22) we can write

$$dv = -\frac{q\mathscr{E}}{m_{ce}^*} dt \tag{3.23}$$

Integrating both sides of Equation (3.23) (between collisions) gives

$$v(t) - v(t_0) = -\frac{q\mathscr{E}}{m_{ce}^*}(t - t_0) \tag{3.24}$$

where t_0 is the time of the previous collision. This gives us the velocity reached by a particular electron just before its next collision at time t. If we consider a large number of collisions and take the average, we get

$$v_{dn} = \langle v(t) - v(t_0) \rangle = -\frac{q\mathscr{E}}{m_{ce}^*}\langle t - t_0 \rangle \tag{3.25}$$

We define the drift velocity of electrons as $v_{dn} = \langle v(t) - v(t_0) \rangle$. We also define the mean free time between collisions as

$$\bar{t}_n = \langle t - t_0 \rangle \tag{3.26}$$

Then the drift velocity for electrons can be written

$$v_{dn} = -\frac{q\mathscr{E}}{m_{ce}^*}\bar{t}_n \tag{3.27}$$

From Equations (3.11) and (3.27)

$$\boxed{\mu_n = \frac{q\bar{t}_n}{m_{ce}^*}} \tag{3.28}$$

For holes, a similar calculation with charge $+q$ and conductivity effective mass m^*_{ch} gives

$$v_{dp} = \frac{q\bar{t}_p}{m^*_{ch}}\mathscr{E} = \mu_p \mathscr{E} \tag{3.29}$$

$$\boxed{\mu_p = \frac{q\bar{t}_p}{m^*_{ch}}} \tag{3.30}$$

where μ_p is the hole mobility.

From the above it can be seen that the carrier mobility for any given scattering mechanism depends on the carrier conductivity effective mass and on the scattering time associated with that scattering mechanism. The scattering time is dependent on the frequency of carrier collisions with other "particles."

Because the scattering times for carrier-ion and for carrier-phonon collisions are independent, by Matthiessen's rule,

$$\frac{1}{t_n} = \frac{1}{t_{nii}} + \frac{1}{t_{nl}} \tag{3.31}$$

and

$$\frac{1}{\mu_n} = \frac{1}{\mu_{nii}} + \frac{1}{\mu_{nl}} \tag{3.32}$$

where the subscripts ii and l refer to ionized impurity and lattice (phonon) scattering respectively.

EXAMPLE 3.2

Find the mean free times between scattering, $\bar{t}_n$ and $\bar{t}_p$, for intrinsic Si at room temperature.

■ **Solution**

For intrinsic Si, $N_D = N_A = 0$. Then from Equation (3.28), $\bar{t}_n = (m^*_{ce}\mu_n)/q$. From Figure 3.4, $\mu_n = 1330\ \text{cm}^2/\text{V}\cdot\text{s} = 0.133\ \text{m}^2/\text{V}\cdot\text{s}$. The conductivity effective mass for electrons is, from Table 2.1, $m^*_{ce} = 0.26m_0 = 0.26 \times 9.11 \times 10^{-19}$ kg.

The mean scattering time is

$$\bar{t}_n = \frac{0.26 \times 9.11 \times 10^{-31} \times 0.133}{1.60 \times 10^{-19}} = 2 \times 10^{-13}\ \text{s}$$

Similarly for holes, $\mu_p = 495\ \text{cm}^2/\text{V}\cdot\text{s}$, $m^*_{ch} = 0.36m_0$, and $\bar{t}_p = 1 \times 10^{-13}$ s

3.3.3 IMPURITY BAND MOBILITY

The above two scattering mechanisms, ionized impurity scattering and phonon scattering, apply to both minority carriers and to majority carriers. There is an additional scattering mechanism associated with majority carriers traveling

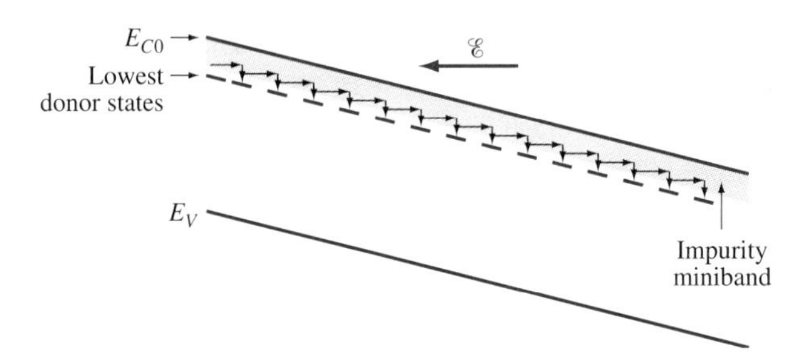

Figure 3.6 Majority carriers (in this case, electrons) drift in an n-type semiconductor. Electrons spend some time in the impurity band, where their mobility is reduced relative to that in the conduction band.

within the dopant impurity states. As discussed in Chapter 2, with increased doping the states associated with the dopant atoms overlap in space and energy to form an impurity band extending from the bottom of the normal conduction band E_{C0} into the normally forbidden band. This impurity bandwidth increases with increased doping. This is illustrated for electrons in an uncompensated n-type semiconductor by Figure 3.6.

Earlier we discussed the deflection of electrons in the conduction band by ionized impurities. Electrons in the impurity band, however, are in states associated with the donor atoms. These electrons can be temporarily captured by the lowest donor state, as shown in the figure. These electrons are repeatedly captured and re-released, slowing down their progress. Additionally, some of the electrons in the miniband are in orbit around the dopant atom. While they can move from one dopant atom to another since these orbits overlap, they cannot do this as readily as a Bloch wave (that is, as they would in the conduction band). With increased donor concentration, this impurity band becomes wider (in energy and in concentration of states). A greater fraction of the electrons travel in the impurity band where their drift velocity and thus mobility are reduced [Equation (3.11)].

We emphasize that in *uncompensated* material this mechanism (the slowing of majority carriers by the donor or acceptor states) is important only for majority carriers. In the n-type semiconductor of Figure 3.6, holes travel only in the normal valence band. In compensated material, however, in which both donors and acceptors are present, electrons will travel in the donor band while holes travel in analogous acceptor bands if the minority doping is great enough. The minority carrier hole (electron) mobility is then a function of the acceptor (donor) concentration.

Note that the (majority) electron mobility μ_n *associated with this mechanism of impurity band scattering* is dependent on N_D but independent of N_A, while μ_p is a function of N_A and independent of N_D. This is in contrast to ionized impurity scattering, which we mentioned before, is a function of $N_D + N_A$ for both majority and minority carriers.

We emphasize that Figure 3.4 reflects measurements in uncompensated material. In compensated semiconductors, minority carrier current will be partially carried in the impurity bands, which will reduce the minority carrier mobilities from their values in Figure 3.4. Often however, $N_D \gg N_A$ or $N_A \gg N_D$, in which case the minority carrier band transport does not have a large effect on mobility.

From Figure 3.4, it can be seen that for small impurity concentrations such that the impurity band transport effect is negligible, the majority and minority carrier mobilities approach each other for electrons and for holes. With increasing doping level, this effect becomes more prominent and the majority carrier mobility becomes increasingly smaller than the minority carrier mobility.

In this section, we developed the concept of mobility using silicon. For comparison, the low-field doping dependence of majority carrier mobilities is plotted in Figure 3.7 for GaAs and Ge. Minority carrier mobility data are not available for these materials. It is seen that the low-field electron mobility of GaAs is appreciably larger than for Si or Ge, primarily because of its small effective mass.

3.3.4 TEMPERATURE DEPENDENCE OF MOBILITY

The mobility is also sensitive to temperature. With increasing temperature, the carrier mobility due to lattice scattering decreases because the increased phonon concentration causes increased scattering. For silicon, the mobility due to lattice scattering varies as $T^{-2.6}$ for electrons and as $T^{-2.3}$ for holes.

The effect of ionized impurity scattering, however, decreases with increasing temperature because the average thermal speeds of the carriers are increased. Thus, the carriers spend less time near an ionized impurity as they pass and the scattering effect of the ions is thus reduced, or the mobility due to ionized impurity scattering increases with increasing temperature.

These two effects operate simultaneously on the carriers. The temperature dependence of mobility for a typical semiconductor is shown schematically in Figure 3.8. At lower temperatures, ionized impurity scattering dominates, while at higher temperatures, phonon scattering dominates.

3.3.5 HIGH-FIELD EFFECTS

We assumed in the previous section that the velocity imparted by the applied electric field (drift velocity) was small compared with the thermal speed, or that the field had negligible effect on the scattering time $\bar{t}$. This assumption of constant $\bar{t}$ results in the drift velocity being proportional to the field, or

$$v_d = \frac{q\bar{t}}{m_c^*}\mathscr{E} \qquad \text{low field} \tag{3.33}$$

At high enough fields, however, the mean free time between collisions $\bar{t}$ decreases with increasing $\mathscr{E}$, resulting in reduced mobility. The dependence

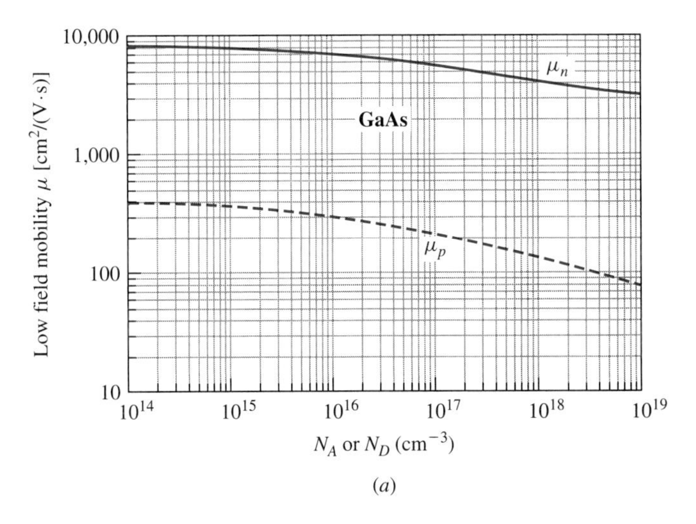

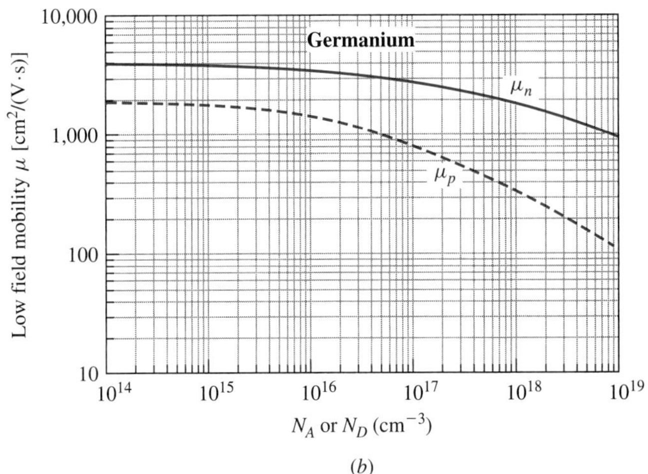

Figure 3.7 Room temperature electron and hole majority carrier mobilities (low field) as functions of uncompensated dopant concentration N in (a) GaAs and (b) Ge.

of v_d on $\mathscr{E}$ has been measured experimentally and is shown in Figure 3.9 for electrons and holes in high-purity bulk Si and Ge, and for electrons in GaAs at room temperature.

At low fields, the drift velocity v_d is proportional to $\mathscr{E}$, so μ is constant. This value of μ is called the low-field mobility μ_{lf}, and that is what was plotted in Figures 3.4 and 3.7. As the field is increased, v_d increases sublinearly and appears to approach a limiting velocity v_{sat}. The value of v_{sat} is on the order of 1×10^7 cm/s for both electrons and holes in Si. It is on the order of 6×10^6 cm/s for electrons in GaAs and for both electrons and holes in Ge.

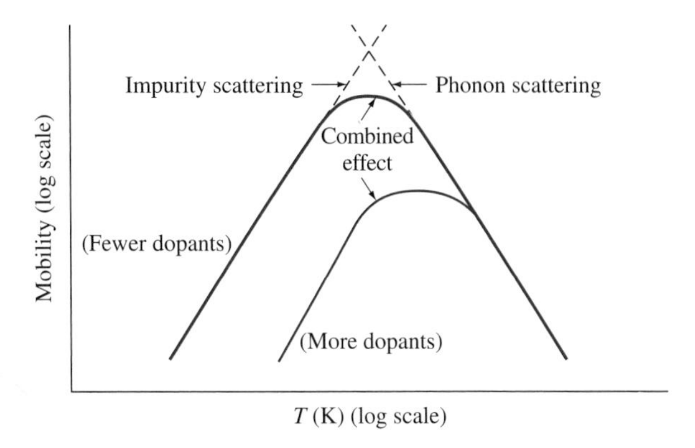

Figure 3.8 Mobility as a function of temperature. At low temperatures, impurity scattering dominates, but at high temperatures, lattice vibrations dominate.

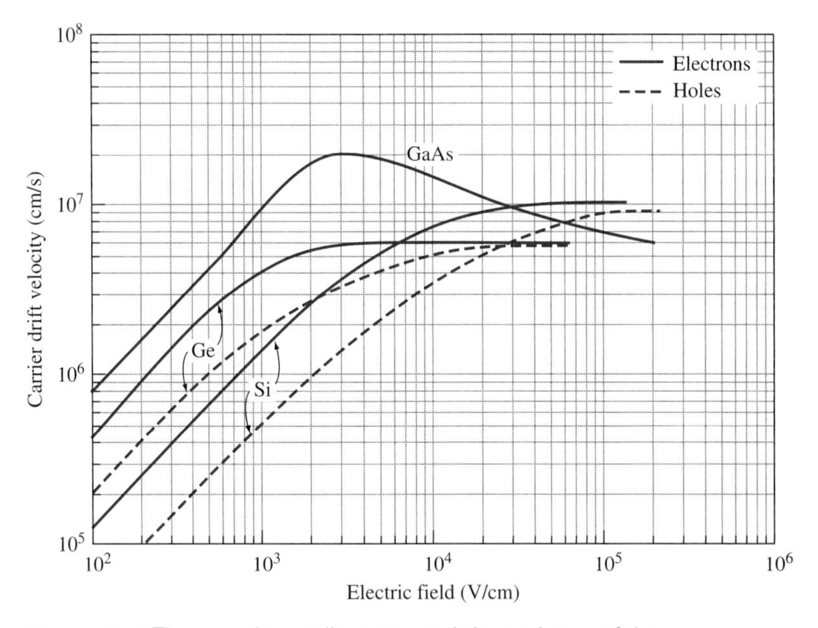

Figure 3.9 The experimentally measured dependence of the drift velocity on the applied field.

This velocity saturation results from a process called *optical phonon scattering*. An optical phonon, as discussed in the Supplement B to Part 1, refers to a lattice vibration in which adjacent planes of atoms vibrate out of phase. At high fields, carriers are accelerated enough to gain sufficient kinetic energy between collisions that, when they collide with the lattice, they can impart enough energy to create an optical phonon. The probability that a carrier having this high an energy will create a phonon is quite high, and thus the maximum carrier velocity is limited by the relation

$$\frac{m^* v_{\text{max}}^2}{2} \approx E_{\text{phonon (opt.)}} \tag{3.34}$$

where $E_{\text{phonon (opt.)}}$ is the optical phonon energy and m^* is the carrier effective mass in the direction of $\mathscr{E}$. The value of $E_{\text{phonon (opt.)}}$ is 0.063 eV for Si and 0.034 eV for GaAs and Ge.

EXAMPLE 3.3

Estimate the value of the saturation velocity for electrons in Si. Make the following assumptions:

1. After each collision, an electron loses all of its kinetic energy.
2. Electrons gain the kinetic energy $E_{\text{phonon (opt.)}}$ without intermediate collisions; i.e., the creation of optical phonons is the only scattering mechanism.
3. Between $E_K = 0$ and $E_K = m^* v_{\text{max}}^2/2$, the electrons are in the parabolic region of the E-K curve and so the concept of constant effective mass is valid.

■ Solution

From assumptions (1) and (2) and Equation (3.34), the average velocity is $v_{\text{max}}/2$, or

$$v_{\text{sat}} = \frac{v_{\text{max}}}{2} = \sqrt{\frac{E_{\text{phonon (opt.)}}}{2m_{ce}^*}} \tag{3.35}$$

where we use m_{ce}^* for m^*, since velocity is related to conduction.

For Si, $E_{\text{phonon (opt.)}} = 0.063$ eV and $m_{ce}^* = 0.26m_0$, and

$$v_{\text{sat}} = \sqrt{\frac{0.063 \times 1.6 \times 10^{-19}}{2 \times 0.26 \times 9.11 \times 10^{-31}}} = 1.46 \times 10^7 \text{ cm/s}$$

This value is somewhat above the experimental value on the order of 1×10^7 cm/s. The discrepancy results primarily from assumption 2. Since electrons do make intermediate collisions, the average saturation velocity is less than that calculated above.

The data in Figure 3.9 were measured on high-purity materials at room temperature. For more highly doped semiconductors, the scattering time is reduced, with a corresponding reduction in the low-field velocity and the saturation velocity. In doped materials, the mean free time between collisions $\bar{t}$ is smaller, and assumption 2 becomes less accurate; i.e., intermediate collisions are increasingly important. Because of the dependence of $\bar{t}$ on impurity concentration and temperature, v_{sat} also depends on these parameters.

Notice from Figure 3.9 that the drift velocity v_d increases monotonically with $\mathscr{E}$ for electrons and holes in Si and Ge. For electrons in GaAs, however, v_d increases, reaches a maximum, and then decreases, approaching v_{sat}. This can be explained with the aid of the E-K for electrons in GaAs (Figure 2.3a). At low field, most of the electrons are in the minimum at $K = 0$, where $m^* = 0.067m_0$. When they make collisions, they are scattered back into this same minimum. At higher fields, they can gain enough kinetic energy between collisions to be scattered into the higher-energy minima, where the effective mass has a higher value, $m^* \approx 0.55m_0$. This reduces the drift velocity [Equation (3.27)] and the corresponding saturation velocity [Equation (3.35)].

We will see in Chapter 7 that this velocity saturation effect has a significant influence on the performance of field-effect transistors, and cannot be ignored.

3.4 DIFFUSION CURRENT

We discussed earlier how applying an electric field resulted in drift current. Now we look into a second type of current in a semiconductor, the diffusion current. Diffusion is the process of mobile particles moving from regions of high concentrations to regions of low concentrations. This diffusion results from the random (thermal) motion of particles (often referred to as *random walk*).

As an example, consider a tray of ping-pong balls in which someone has very carefully placed all the balls in a neat group in the center of the tray. If we set the tray to vibrating (analogous to giving the lattice thermal energy) then the balls do not stay neatly grouped for very long. They diffuse, and will eventually be uniformly distributed throughout the tray.

This same diffusion process occurs for mobile electrons and holes, and since they are charged particles, their motion results in current, referred to as *diffusion current*.

Consider again a bar of semiconductor material of cross-sectional area A at room temperature, as shown in Figure 3.10a. There is no electric field. We assume that at some time t the electrons are distributed nonuniformly with position, as shown in Figure 3.10b. Since the temperature of the semiconductor is nonzero, each electron has some thermal energy, and therefore some speed. The direction of the electron's motion, however, is random since there is no electric field. Each electron can go either to the right or to the left, with 50 percent probability either way.

Consider the plane $x = x_0$. The electrons pass this plane in either direction. On the left of the plane, there is a high concentration n_L of electrons, with half traveling to the right and half traveling to the left. On the right, there is a smaller

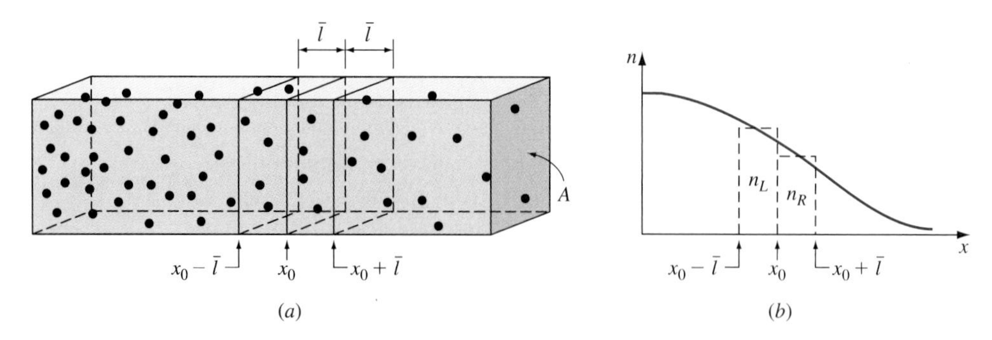

Figure 3.10 Diffusion. (a) A sample of semiconductor with a nonuniform distribution of electrons; (b) the electron density distribution.

concentration n_R of electrons, and half of those are going to the left. Thus, more electrons cross the $x = x_0$ plane from left to right than from right to left. There is a net flux of electrons to the right, and thus net electron diffusion current to the left. (Holes diffuse in exactly the same manner,[3] always from regions of high concentration to regions of low concentration. The hole diffusion current is in the same direction as the hole flux because of the positive charge of the holes.)

We next calculate the diffusion current density. We start by finding the electron flux density, which is the number of electrons crossing unit area at x_0 per unit time. Let $\bar{t}$ be the mean free time between collisions and $\bar{l}$ be the electron mean free path, which is the average distance a carrier progresses in its mean free time.[4] We consider two regions, each of volume $\bar{l}A$, on either side of x_0. On the average, half the electrons in this volume will arrive at the plane x_0 without colliding (the other half are going the other way), and they'll arrive there in time $\bar{t}$. In time $\bar{t}$, one-half of the electrons on the left will cross x_0 to the right and one-half of the electrons on the right will cross x_0 to the left. The net number of electrons crossing x_0 in time $\bar{t}$ is one-half the difference in the number on either side within a distance $\bar{l}$ from x_0. The electron flux density is therefore

$$F_n = \frac{(n_L - n_R)\bar{l}}{2\bar{t}} \tag{3.36}$$

where n_L and n_R are the electron densities in the left and right regions respectively. If the change in n is small in distance $\bar{l}$, we can write

$$(n_L - n_R) \approx -\frac{dn}{dx}\bar{l} \tag{3.37}$$

and the electron flux density is

$$F_n = \frac{-\bar{l}^2}{2\bar{t}}\frac{dn(x)}{dx} = -D_n\frac{dn(x)}{dx} \tag{3.38}$$

where

$$D_n \equiv \frac{\bar{l}^2}{2\bar{t}} \tag{3.39}$$

The quantity D_n is called the *diffusion coefficient* for electrons. The minus sign in Equation (3.38) indicates that the electron flux density is in the direction of decreasing n, that is, always toward regions of lower concentration.

The electron diffusion *current* density is equal to the flux density multiplied by the charge of the electron:

$$J_{n(\text{diff})} = -qF_n = qD_n\frac{dn(x)}{dx} \tag{3.40}$$

---

[3]And so do any other particles with random motion, for example, the ping-pong balls mentioned earlier or impurity atoms in a crystal.

[4]We saw that $\bar{t}$ is on the order of 10^{-13} s and thermal speed $\bar{v}$ is about 10^7 cm/s at room temperature. Then to a reasonable approximation, $\bar{l} = \bar{v}\bar{t}$ and $\bar{l}$ is on the order of 10 nm.

Similarly for holes,

$$J_{p(\text{diff})} = +qF_p = -qD_p \frac{dp(x)}{dx} \tag{3.41}$$

Note that, for diffusion, there is no force at work—the only thing causing the currents is the thermal energy of the carriers and the variation in concentration.

As will be discussed shortly, in many devices $dn(x)/dx$ and $dp(x)/dx$ vary with position and thus $J_{n(\text{diff})}$ and $J_{p(\text{diff})}$ [Equations (3.40) and (3.41)] also vary with position. To keep the total current constant, the drift currents change accordingly. This topic is discussed further in Section 3.7, where we consider the *continuity equations*.

We note that both diffusion coefficient and mobility are measures of how easily the particles move through a material, and one might suspect that they are related. In the next chapter, we will derive this relation, known as the *Einstein relation*. We state the results here

$$\boxed{\frac{D_n}{\mu_n} = \frac{kT}{q}} \tag{3.42}$$

$$\boxed{\frac{D_p}{\mu_p} = \frac{kT}{q}} \tag{3.43}$$

From the above, the diffusion coefficients are proportional to the mobilities. Figure 3.11 shows plots of room temperature minority carrier and majority carrier diffusion coefficients for electrons and holes in Si as functions of doping. (Compare with Figure 3.4 for mobilities.)

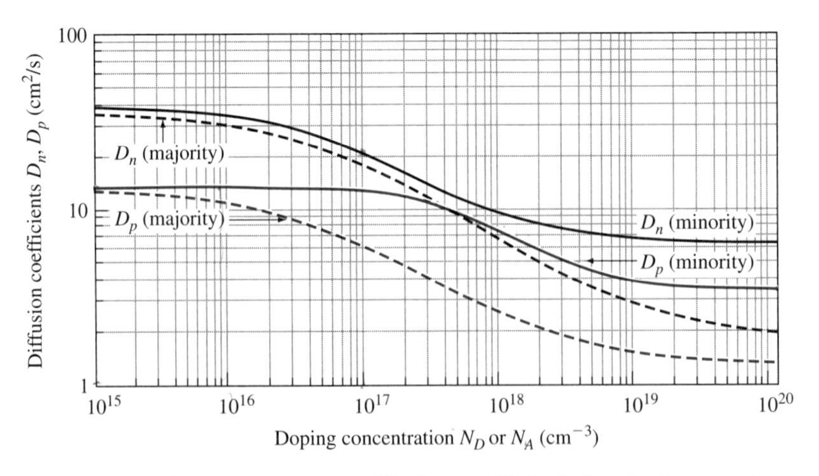

Figure 3.11 Room temperature diffusion coefficients for electrons and holes as a function of doping concentration for silicon.

It is possible to apply a force to the electrons and holes as well, by applying an electric field. For the case in which electrons are subject to both an electric field and a variation in concentration, the total electron current is

$$J_n = J_{n(\text{drift})} + J_{n(\text{diff})} = q\mu_n n(x)\mathscr{E} + qD_n \frac{dn(x)}{dx} \tag{3.44}$$

and the hole current also has drift and diffusion components:

$$J_p = J_{p(\text{drift})} + J_{p(\text{diff})} = q\mu_p p(x)\mathscr{E} - qD_p \frac{dp(x)}{dx} \tag{3.45}$$

Using the Einstein relation, we can express Equations (3.44) and (3.45) as

$$J_n = q\mu_n \left[n\mathscr{E} + \frac{kT}{q}\frac{dn}{dx} \right] \tag{3.46}$$

$$J_p = q\mu_p \left[p\mathscr{E} - \frac{kT}{q}\frac{dp}{dx} \right] \tag{3.47}$$

The two components of each of these are the drift and the diffusion currents. The total current is

$$\begin{aligned} J &= J_{n(\text{drift})} + J_{p(\text{drift})} + J_{n(\text{diff})} + J_{p(\text{diff})} \\ &= q\mu_n n\mathscr{E} + q\mu_p p\mathscr{E} + qD_n \frac{dn}{dx} - qD_p \frac{dp}{dx} \end{aligned} \tag{3.48}$$

where in p-type material, μ_n and D_n are the minority carrier mobilities and diffusion coefficients, and μ_p and D_p are their majority carrier counterparts. The opposite is the case for n-type material.

In many devices (e.g., resistors), there are regions where both n and p are uniform in position (i.e., $dn/dx = dp/dx = 0$). In that case Equation (3.48) becomes

$$J = (q\mu_n n + q\mu_p p)\mathscr{E} = (\sigma_n + \sigma_p)\mathscr{E} = \sigma\mathscr{E} \tag{3.49}$$

which is Ohm's law in point form as expressed in Equation (3.5).

3.5 CARRIER GENERATION AND RECOMBINATION

At a finite temperature, there are always some electrons in the conduction band and some holes in the valence band. These carriers arise from ionized impurities and from the excitation of electrons from valence to conduction band, which create electron-hole pairs. Near and above room temperature, to good approximation, all impurities are ionized and so here we consider only those processes involving electron-hole pairs. By definition, a process of creating electron-hole pairs or exciting an electron from valence band to conduction band is referred to as *generation*. A process by which an electron from the conduction band is moved to the valence band, thus annihilating an electron-hole pair, is called *recombination.*

At equilibrium, generation and recombination in a semiconductor occur at equal rates and thus the equilibrium concentration of electrons and holes (n_0 and p_0) are constant. However, semiconductor devices operate under nonequilibrium conditions, in which case this is not necessarily true, and $n \neq n_0$ and $p \neq p_0$. In many

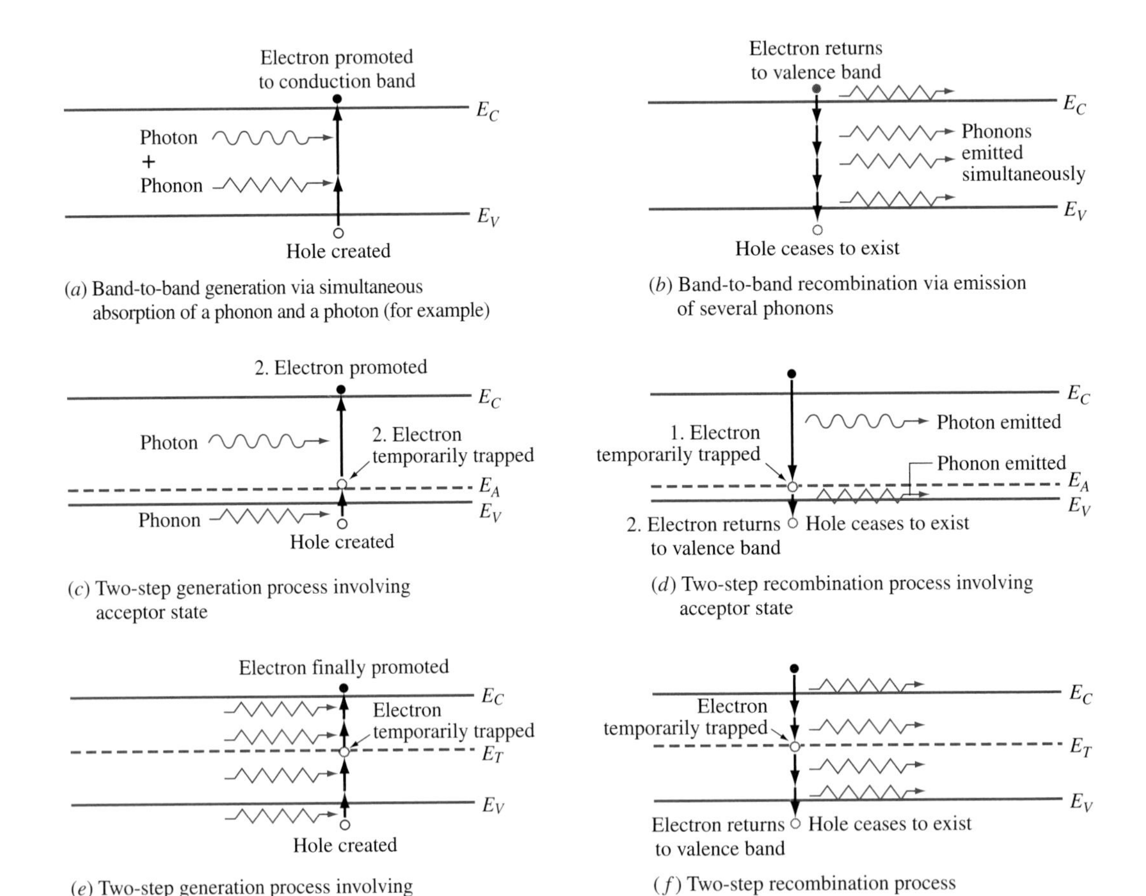

Figure 3.12 Various generation and recombination processes. (a) An electron-hole pair is generated when an electron absorbs (in this case) a phonon plus a photon. This generation could also occur by the absorption of a single photon or multiple phonons. The photons and phonons are absorbed simultaneously. (b) Band-to-band recombination via the simultaneous emission of multiple phonons. (c) A two-step generation process, in which, for example, the electron absorbs a phonon to promote it to the acceptor state, then in the next step it absorbs a photon to go to the conduction band. (d) A typical recombination event in p-type material involves emission of a photon to take the electron temporarily to the acceptor level, then the subsequent emission of the phonon returns it to the valence band, annihilating a hole. (e) and (f) Recombination and generation via trap states.

devices the processes of generation and recombination are important in determining their electrical characteristics. There are a number of physical processes involved in generation and recombination. Here we discuss the more common processes.

3.5.1 BAND-TO-BAND GENERATION AND RECOMBINATION

An electron can be excited from valence band to conduction band by the absorption of a photon having energy greater than the band gap (e.g., GaAs), by the simultaneous absorption of a photon and a phonon (e.g., Si), or by the simultaneous absorption of several phonons. The process of band-to-band transitions is illustrated in Figure 3.12a. In this process an electron-hole pair is generated. Figure 3.12b illustrates the inverse process. When an electron recombines with a hole, then a photon, a photon plus a phonon, or multiple phonons (as illustrated) are generated. In either case, whether generation or recombination, energy and wave vector (crystal momentum) must be conserved.

3.5.2 TWO-STEP PROCESSES

Generation and recombination can also occur by a two-step process involving electronic states within the forbidden band. Figure 3.12c indicates the case for generation in p-type GaAs. Phonons excite electrons from the valence band to acceptor levels. Photons (or multiple phonons) can then excite the electrons from acceptor levels to the conduction band. Figure 3.12d indicates the recombination process. The transitions from conduction band to acceptor states emit photons while the transition from acceptor to valence band emits phonons.

Generation and recombination in Si is often by a two-step process [5], involving an energy level, called a *trap level,* of energy E_T, near the center of the forbidden band as indicated in Figures 3.12e and f. In silicon, each step is induced by the absorption (e) or emission (f) of multiple phonons. While in principle one might expect that each step could result from absorption or emission of photons, this is not experimentally observed in Si.

3.6 OPTICAL PROCESSES IN SEMICONDUCTORS

Until now, we have been primarily considering semiconductors for which the carrier concentrations have their equilibrium values. In devices, however, often carrier concentrations have nonequilibrium values. For example, when a semiconductor is exposed to and absorbs light, extra electrons and holes are produced in excess of their equilibrium concentrations. In this section, we discuss the optical processes of absorption and emission in semiconductors and the effects of these excess carrier concentrations on current.

*3.6.1 ABSORPTION

In Chapter 1, we discussed absorption of photons by an atom. A single atom has discrete energy states, and for a photon to be absorbed, its energy must be equal to the difference in energy between two allowed states. In a semiconductor, the

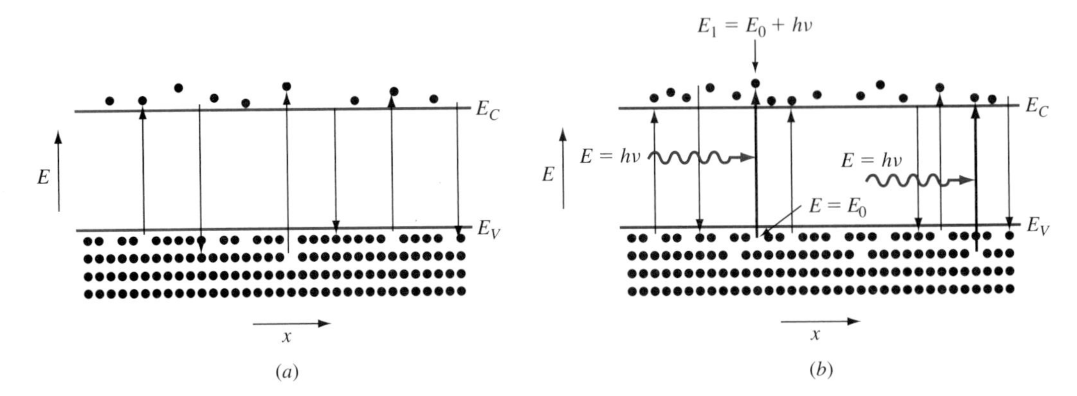

Figure 3.13 (a) At equilibrium, electrons and holes are generated and destroyed at equal rates, thus maintaining some constant equilibrium n_0 and p_0. (b) When light shines on the sample, the photons can be absorbed, producing extra electron-hole pairs.

process is similar, except that, instead of isolated energy states, there exist allowed energy bands. Thus, there will be a range of photon energies that can be absorbed. We will see that there are further restrictions, however, based on the E-K diagrams.

Consider a semiconductor material whose energy band diagram is shown in Figure 3.13a. The bottom of the conduction band and the top of the valence band are shown. At equilibrium, there are n_0 electrons in the conduction band and p_0 holes in the valence band. Electron-hole pairs are constantly being generated thermally, and electrons are constantly recombining with holes as they seek lower energies. At thermal equilibrium, these two processes happen at exactly the same rate.

Now assume a photon of energy

$$E = h\nu \tag{3.50}$$

arrives at the semiconductor as in Figure 3.13b, where ν is the frequency of the light and h is Planck's constant. We already know that if the photon is to be absorbed, by conservation of energy the energy of the *system* must be the same before and after the absorption event. Before the collision of the electron and the photon, their combined energy is

$$E_{\text{electron + photon}} = E_0 + h\nu \tag{3.51}$$

where E_0 is the initial electron energy. When the photon collides with the electron, it is possible that the electron will absorb the energy of the photon. When the photon energy is transferred to the electron, the photon is annihilated. Therefore, after the event the electron, by conservation of energy, now has energy

$$E_1 = E_0 + h\nu \tag{3.52}$$

where $E_1 > E_C$.

Notice that if a photon of energy $E = h\nu$ less than the band gap is incident, that photon cannot be absorbed. For it to be absorbed, the electron would have to end up at an energy state somewhere inside the forbidden gap. Except through the help of traps or impurities that might provide the occasional state within the forbidden gap, this is not possible. Therefore, light of photon energy less than the band gap is not absorbed, and the semiconductor appears transparent to that radiation.

We have not told the whole story, however. Aside from conservation of energy, there is also the law of conservation of wave vector K (analogous to conservation of momentum in classical mechanics).[5] Photons of interest in semiconductor electronics[6] have wave vectors that are small compared with those at the edge of the Brillouin zone, and for many cases can be considered to be essentially zero. Therefore, when the electron makes the energy transition, it must do so at virtually constant K, which is to say it must make a vertical transition on the E-K diagrams. This is shown in Figure 3.14.

The particular semiconductor whose E-K diagram is shown in Figure 3.14a is called a direct gap material. *Direct* means that the minimum in the conduction band is at the same value of K as the maximum in the valence band, here at $K = 0$. Thus, the transition results from the direct interaction of a photon with an electron. Direct gap materials are generally efficient emitters and absorbers of optical energy because it is easy for electrons to move between the conduction band and valence band without having to acquire or give off K. GaAs and InP are good examples of direct gap materials.

An indirect gap material, on the other hand, is one like that shown in Figure 3.14b. In an indirect material, the minimum in the conduction band is not at the same value of K as the top of the valence band. An electron cannot go from one band to the other simply by absorbing a photon of energy close to the band gap, because the photon cannot supply adequate wave vector. The electron needs to acquire both energy and wave vector to make the transition in indirect materials. Examples of indirect materials are Si, Ge, and GaP.

This raises an interesting point. Silicon is the material most commonly used in photodetectors and solar cells; yet it is an indirect gap material and thus not an efficient absorber of light. Furthermore, its primary band gap is 1.12 eV, which corresponds to a photon wavelength of just over 1 μm—in the near infrared. According to Figure 3.14b it should not be able to absorb light at all until considerably higher energies. Photons of energy less than that of the direct band gap can be absorbed, however, via a three-particle collision involving an electron, a photon, and a phonon. Phonons have adequate K vector but have little energy, while the opposite is true for photons. If a photon and a phonon collide with an electron at the same time, the electron can get both enough energy and enough K to get across the forbidden gap. Such a three-body collision is statistically unlikely, and as a result, silicon is not a very efficient absorber at wavelengths near the band gap. Still, it is widely used because of its low cost.

---

[5]As indicated earlier, the quantity $\hbar K$ is often referred to as *crystal momentum* (or sometimes just *momentum*).

[6]By this we mean radiation in the ultraviolet to infrared range—not X-rays.

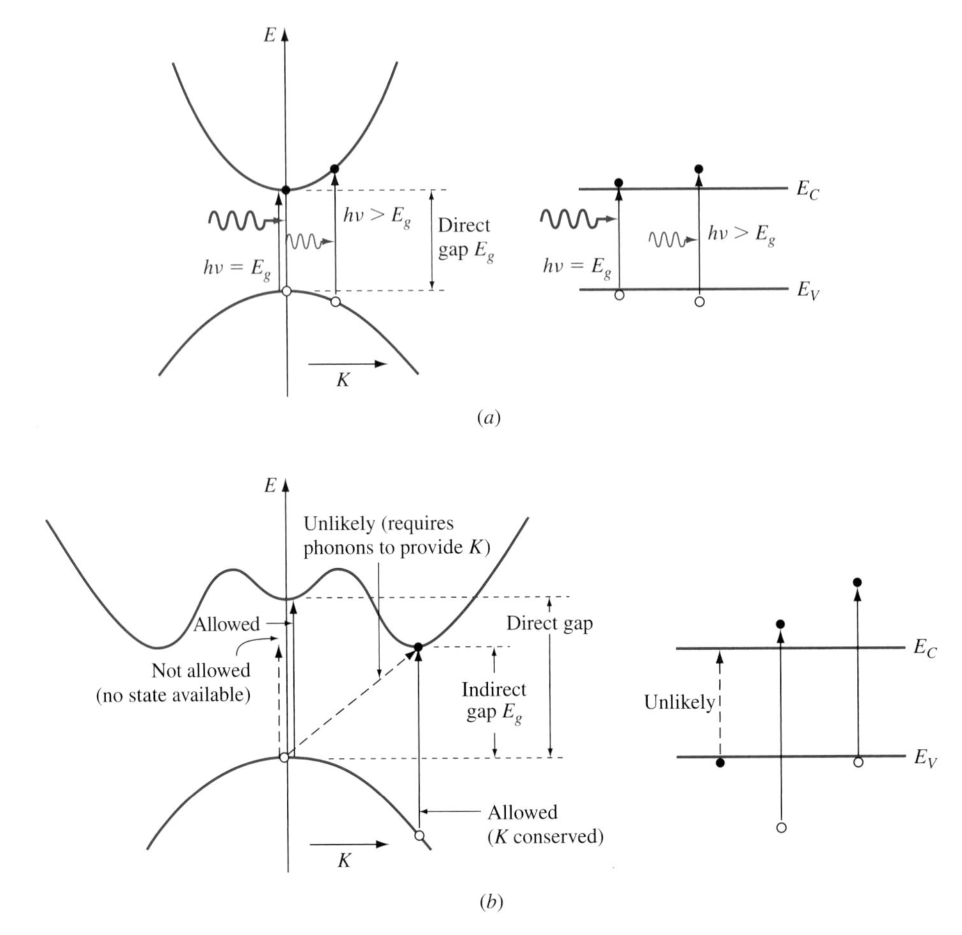

Figure 3.14 For absorption to occur, K must be conserved as well as E. (a) A direct gap semiconductor; on the left is the E-K diagram, and on the right the conventional energy band diagram. (b) An indirect gap material (so called because conduction band minimum and the valence band maximum do not occur at the same value of K and thus the photon-electron interaction is indirect).

EXAMPLE 3.4

Verify that photons of interest have negligible wave vector compared with that of electrons at the edge of the Brillouin zone.

■ Solution

We consider a photon of wavelength 620 nm (energy 2 eV) (orange). It has wave vector

$$K_{\text{pht}} = \frac{2\pi}{\lambda} = \frac{6.28}{620 \times 10^{-9}\ \text{m}} = 10.1 \times 10^6\ (\text{m})^{-1}$$

We recall from Chapter 2 that an electron at the edge of the first Brillouin zone has a wave vector of π/a. For a lattice constant of $a = 0.5$ nm,

$$K_e = \frac{\pi}{a} = \frac{3.14}{0.5 \times 10^{-9}} = 6.28 \times 10^9 \text{ m}^{-1}$$

Then the ratio of the K vectors is

$$\frac{K_{\text{pht}}}{K_e} = \frac{10.1 \times 10^6}{6.28 \times 10^9} = 1.6 \times 10^{-3}$$

Thus, the K vector of the photon is about $1/1000$ that of the electron. Note, however that high-energy (X-ray) photons can have wave vectors comparable to those of electrons at the edge of the Brillouin zone.

When light shines on a semiconductor, equilibrium is disturbed. When a given photon is absorbed, an extra electron (beyond the equilibrium number) is produced in the conduction band, and an extra hole is simultaneously produced in the valence band. An electron-hole pair is thus produced for every photon that is absorbed. These electrons and holes are termed *excess carriers*—excess above the equilibrium concentrations. The excess electron concentration is Δn and the excess hole concentration is Δp. The total carrier concentrations are the equilibrium values plus the excess concentrations:

$$\boxed{\begin{aligned} n &= n_0 + \Delta n \\ p &= p_0 + \Delta p \end{aligned}} \tag{3.53}$$

These excess carriers can diffuse if they are not uniformly distributed, and they can drift if there is an electric field. Even if the drift and diffusion currents are zero (or add to zero), as long as there are excess carriers, the material is not at equilibrium. We will see shortly that if the light is turned off, the excess carriers disappear with time and the material returns to equilibrium. Before we discuss that, however, we will examine optical emission in semiconductors.

*3.6.2 EMISSION

The inverse optical process of absorption is optical emission. In this process, an electron initially in a state of energy E_1 makes the transition to a state of lower energy E_0 as shown in Figure 3.15. By conservation of energy, the excess energy must be given off, either as a photon (light) or phonons (heat), or both.

In the absence of phonon creation, the energy of the photon emitted must be equal to the energy lost by the electron. Therefore, we expect to see emission only at photon energies equal to the difference between allowed electron energies. We consider the case of a p-type direct gap semiconductor as shown in Figure 3.16a. At energies greater than that of the gap, emission is possible, but K must still be conserved. Therefore, optical transitions must still occur vertically

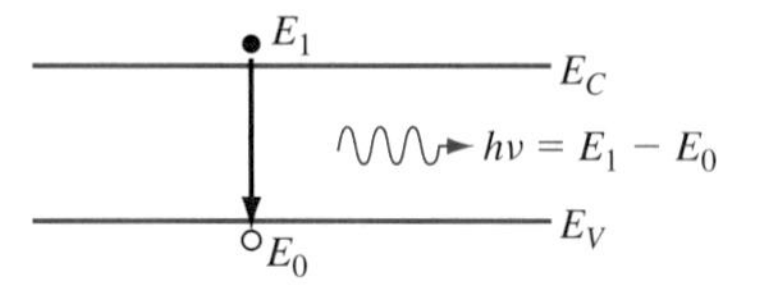

Figure 3.15 Optical emission. The electron loses energy, giving off the excess as a photon of $E = h\nu$.

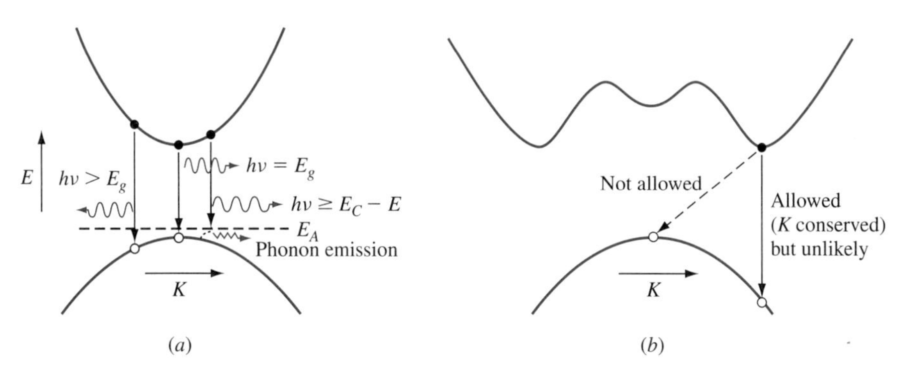

Figure 3.16 Emission on the *E-K* diagram. Both *K* and *E* must be conserved. (a) A direct gap material; (b) an indirect semiconductor.

on the *E-K* diagram. For energies close to, but greater than, the band gap, optical emission can occur only in direct gap semiconductors. Optical transitions are also possible, however, from conduction band to acceptor level, producing lower-energy photons. In this case, the accompanying transition from E_A to valence band produces phonons. For narrow-gap materials (e.g., InSb, $E_g = 0.17$ eV), band-to-band transitions are the more probable. For wider-gap materials (e.g., GaAs, $E_g = 1.43$ eV) conduction band-to-acceptor state transitions are more probable.

An indirect energy band diagram for a material such as GaP is shown in Figure 3.16b. Notice that one transition is marked as allowed but unlikely. This is because, although *K* would be conserved, the probability of a hole existing at this low energy is remote. The other transition shown is not permitted since *K* is not conserved. As a result, indirect gap materials are less efficient light emitters than direct gap materials.[7]

[7]There are exceptions. For example, by doping an indirect gap semiconductor appropriately, producing *isoelectronic traps,* as discussed in Chapter 11, substantial light emission from some indirect gap materials (e.g., GaP) can be obtained.

3.7 CONTINUITY EQUATIONS

The continuity equations are mathematical statements of the conservation of particles. They are fundamental to an understanding of the variations of carrier concentrations (electrons and holes) with time and position and their effects on the electrical and electro-optical characteristics, both static and transient. For simplicity, we consider the time dependence of the carrier concentrations in the x direction for a one-dimensional case. It is thus assumed that at a given time the carrier concentrations are uniform in the y and z directions.

In a metal wire, we find the current by examining the number of electrons entering or leaving a volume per unit time. In a semiconductor, both electrons and holes can enter a volume at one end and leave at the other, and we must account for both. Carriers may also pile up or disappear (recombine) in the volume. Therefore their numbers leaving at any given time may not be the same as the numbers entering. The continuity equation takes into account all the sources and sinks of electrons and holes.

Consider the semiconductor differential volume of unit cross-sectional area and length dx, as shown in Figure 3.17. An electron flux density F_n, composed of drift and diffusion, is flowing into the volume, and the flux density flowing out will in general be different. The difference is made up by the generation and recombination and the pile-up or decrease of carriers in the volume. In this differential volume, then, the rate of increase in conduction band electron concentration is

$$\frac{\partial n}{\partial t}\, dx = -\frac{\partial F_n}{\partial x}\, dx + (G_n - R_n)\, dx \tag{3.54}$$

where G_n is the electron generation rate and R_n is the electron recombination rate (number of carriers per unit volume per unit time). The term $(\partial F_n/\partial x)\, dx$ represents the difference in electron flux at either end of the region dx.

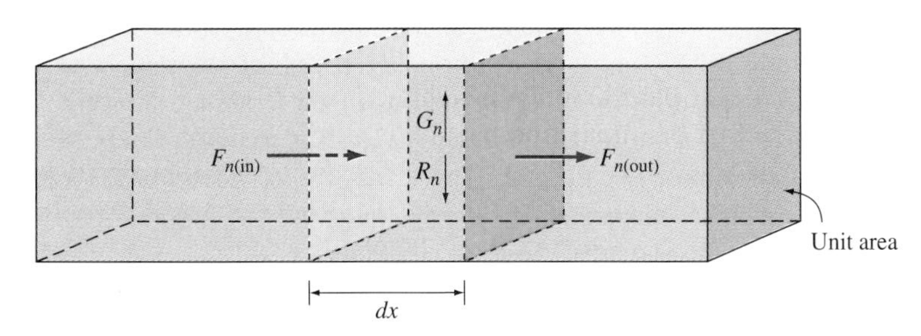

Figure 3.17 The geometry for determining the continuity equation. The rate at which carriers accumulate in the incremental volume depends on the incoming and outgoing currents as well as the recombination and generation within the region dx.

We saw there are several mechanisms for generation and recombination. Let G_{th} be the thermal (phonon-induced) generation rate, G_{op} be the optical (photon induced) rate and G_{other} be the generation rate due to other processes (e.g., trapping, detrapping). Then the total electron generation rate is

$$G_n = G_{n(\text{th})} + G_{n(\text{op})} + G_{n(\text{other})} \tag{3.55}$$

measured in electrons generated per unit volume per second. Since the electron flux density is related to the current density by $F_n = -J_n/q$, Equation (3.54) becomes, after the dx terms are canceled,

$$\boxed{\frac{\partial n}{\partial t} = \frac{1}{q}\frac{\partial J_n}{\partial x} + (G_n - R_n)} \quad \text{continuity equation for electrons} \tag{3.56}$$

This equation states that the rate of increase in electron density at a point is equal to the increase in n due to unequal electron currents, plus the net electron generation rate $G_n - R_n$. We reiterate that J_n is the total electron current density and in general consists of drift and diffusion current. Equation (3.56) is called the *continuity equation for electrons*. Similarly, the continuity equation for holes is

$$\boxed{\frac{\partial p}{\partial t} = -\frac{1}{q}\frac{\partial J_p}{\partial x} + (G_p - R_p)} \quad \text{continuity equation for holes} \tag{3.57}$$

Next, returning to Equation (3.56), we recall that $n = n_0 + \Delta n$ [Equation (3.53)], and that the equilibrium electron concentration n_0 is time-independent, or $\partial n_0/\partial t = 0$. Thus

$$\frac{\partial n}{\partial t} = \frac{\partial n_0}{\partial t} + \frac{\partial \Delta n}{\partial t} = \frac{\partial \Delta n}{\partial t} \tag{3.58}$$

Further, neglecting other processes,

$$G_{\text{n}} = G_{\text{th}} + G_{\text{op}} \tag{3.59}$$

Recombination also occurs. An electron has some lifetime associated with it, which is the average time an electron spends in the conduction band, before recombining. We will discuss the lifetimes in detail in Section 3.8, but for now, for a well-defined[8] minority carrier (electron) lifetime τ_n, the recombination rate is proportional to the number of electrons available for recombination (meaning electrons in the conduction band) by

$$R = \frac{n}{\tau_n} = \frac{n_0}{\tau_n} + \frac{\Delta n}{\tau_n} \tag{3.60}$$

[8]By *well-defined* we mean that the lifetime is the average time that a minority carrier spends in its respective band, $\bar{t}_n = \tau_n$ or $\bar{t}_p = \tau_p$. This is usually the case for semiconductors used in electronic devices.

Substituting these results back into Equation (3.56) gives

$$\frac{\partial n}{\partial t} = \frac{\partial \Delta n}{\partial t} = \frac{1}{q}\frac{\partial J_n}{\partial x} + \left(G_{\text{th}} + G_{\text{op}} - \frac{n_0}{\tau_n} - \frac{\Delta n}{\tau_n}\right) \tag{3.61}$$

We will consider the special case of equilibrium. Equilibrium means that there is no net current ($J = 0$), there are no external fields applied, there are no temperature gradients, and no light is shining on the sample. Equilibrium therefore means there are no excess carriers. For equilibrium, then, $\Delta n = 0$, $J_n = 0$, and $G_{\text{op}} = 0$. Thus the left-hand side of Equation (3.61) is equal to zero, resulting in

$$G_{\text{th}} = \frac{n_0}{\tau_n} \tag{3.62}$$

Equation (3.61) then becomes, with the aid of Equation (3.62),

$$\boxed{\frac{\partial n}{\partial t} = \frac{\partial \Delta n}{\partial t} = \frac{1}{q}\left(\frac{\partial J_n}{\partial x}\right) + \left(G_{\text{op}} - \frac{\Delta n}{\tau_n}\right)} \tag{3.63}$$

Similarly for holes,

$$\boxed{\frac{\partial p}{\partial t} = \frac{\partial \Delta p}{\partial t} = -\frac{1}{q}\left(\frac{\partial J_p}{\partial x}\right) + \left(G_{\text{op}} - \frac{\Delta p}{\tau_p}\right)} \tag{3.64}$$

where, in general, G_{op}, Δn, and Δp are functions of position.

The current density consists of drift and diffusion

$$\begin{aligned} J_n &= J_{n(\text{drift})} + J_{n(\text{diff})} = qn\mu_n\mathscr{E} + qD_n\frac{dn}{dx} \\ J_p &= J_{p(\text{drift})} + J_{p(\text{diff})} = qp\mu_p\mathscr{E} - qD_p\frac{dp}{dx} \end{aligned} \tag{3.65}$$

and the continuity equations become

$$\boxed{\begin{aligned} \frac{\partial n}{\partial t} &= \frac{\partial \Delta n}{\partial t} = n\mu_n\frac{\partial \mathscr{E}}{\partial x} + \mu_n\mathscr{E}\frac{\partial n}{\partial x} + D_n\frac{\partial^2 n}{\partial x^2} + G_{\text{op}} - \frac{\Delta n}{\tau_n} \\ \frac{\partial p}{\partial t} &= \frac{\partial \Delta p}{\partial t} = p\mu_p\frac{\partial \mathscr{E}}{\partial x} + \mu_p\mathscr{E}\frac{\partial p}{\partial x} - D_p\frac{\partial^2 p}{\partial x^2} + G_{\text{op}} - \frac{\Delta p}{\tau_p} \end{aligned}} \tag{3.66}$$

The use of these continuity equations will be illustrated in the next two sections for simple one-dimensional problems. We will also use the continuity equations to calculate transient effects for some diodes in Chapter 5, to understand the operation of bipolar junction transistors in Chapters 9 and 10, and to find the current in photodetectors and solar cells in Chapter 11. In the more complicated cases, such as transistor analysis, the continuity equations in two or three (spatial) dimensions are often required. Often these cannot be solved in closed form, but are solved numerically in device simulators.

In three dimensions the continuity equations are

$$\frac{\partial n}{\partial t} = \frac{1}{q}\nabla \cdot J_n + (G_n - R_n)$$

$$\frac{\partial p}{\partial t} = -\frac{1}{q}\nabla \cdot J_p + (G_p - R_p)$$

3.8 MINORITY CARRIER LIFETIME

We have seen that electrons and holes can be generated thermally or optically and that there is also a restoring force: recombination. When electrons are found at elevated energies (e.g., in the conduction band), they tend to seek lower energies, and after some time they will recombine with holes in the valence band. In this section, we are interested in the time that takes.

We define the *minority carrier lifetime* τ as a measure of the average time, (τ_n), an electron spends in the conduction band in a p-type semiconductor or a hole (τ_p) spends in the valence band in an n-type semiconductor. One method used to determine the lifetime is to measure the time dependence of photoconductivity when a semiconductor is illuminated.

We consider a uniform semiconductor sample connected in the circuit of Figure 3.18. The semiconductor is thin, and the wavelength of the light is chosen such that the absorption coefficient is small—this way the illumination can be considered uniform throughout the sample. If light pulses are applied, they create electron-hole pairs through absorption. The absorption creates excess carriers, and since there are more carriers to conduct, the conductivity increases. The resulting variation in current with time is measured with an oscilloscope as a time variation of voltage across the load resistance R_L. If the value of R_L is chosen to be small compared with the resistance R_S of the semiconductor, then the measured current is V_A/R_S. Recalling that the resistance of a semiconductor

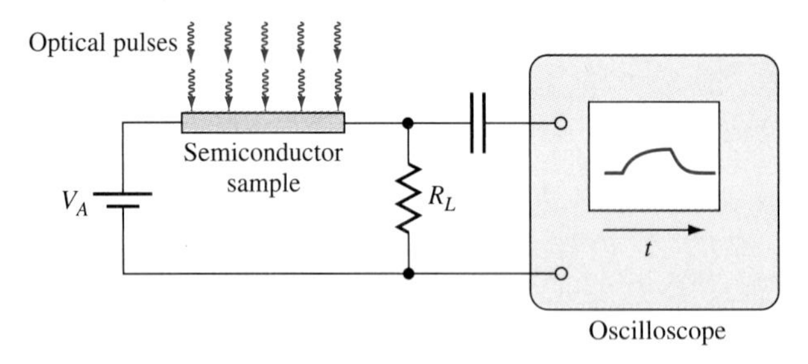

Figure 3.18 Schematic of a circuit used to measure minority carrier lifetime in semiconductors.

sample is $R_S = \rho L/A$, and $\sigma = 1/\rho$, we have

$$i(t) = \frac{V_A}{\dfrac{\rho(t)L}{A}} = \frac{V_A A\sigma(t)}{L} \tag{3.67}$$

The time-dependent voltage is then proportional to $i(t)$, which in turn is proportional to the conductivity $\sigma(t)$, which changes in time as the number of carriers changes.

Recall that the conductivity of a semiconductor sample is given by

$$\sigma = \sigma_n + \sigma_p = q(\mu_n n + \mu_p p) \tag{3.14}$$

where

$$\begin{aligned} n &= n_0 + \Delta n \\ p &= p_0 + \Delta p \end{aligned} \tag{3.53}$$

Again, n_0 and p_0 are the equilibrium concentrations of electrons and holes while Δn and Δp are the excess (in this case, photoinduced) concentrations. We can combine Equations (3.14) and (3.53) to obtain

$$\sigma = q(\mu_n n_0 + \mu_n \Delta n + \mu_p p_0 + \mu_p \Delta p) \tag{3.68}$$

or

$$\sigma = \sigma_0 + \sigma_{pc} \tag{3.69}$$

Here the conductivity in the dark is

$$\sigma_0 = q(\mu_n n_0 + \mu_p p_0) \tag{3.70}$$

and the photoconductivity is

$$\sigma_{pc} = q(\mu_n \Delta n + \mu_p \Delta p) \tag{3.71}$$

As an example, we consider a p-type direct gap semiconductor in which recombination is band to band; i.e., electrons in the conduction band recombine directly with holes in the valence band. This is shown schematically at the right-hand end of Figure 3.19. The thermal generation (due to phonon absorption) and optical generation (photon absorption) are also shown.

Since each photon creates an electron-hole pair, and since each recombining electron annihilates an electron-hole pair, the concentration of excess electrons and holes is the same, or $\Delta n = \Delta p$. The rates at which carriers are being generated may not be the same as the rate at which they recombine, in which case carriers can accumulate or deplete.

The time dependence of the photocurrent is proportional to the time dependence of Δn and Δp. Since $\Delta n = \Delta p$, we solve the continuity equation [Equation (3.63) or (3.66)] for Δn. Because the semiconductor is uniform and uniformly illuminated, $\mathscr{E}$ is constant and $d\mathscr{E}/dx$, dn/dx, and thus $\partial J_n/\partial x = 0$. Then

$$\frac{d\Delta n}{dt} = G_{\text{op}} - \frac{\Delta n}{\tau_n} \tag{3.72}$$

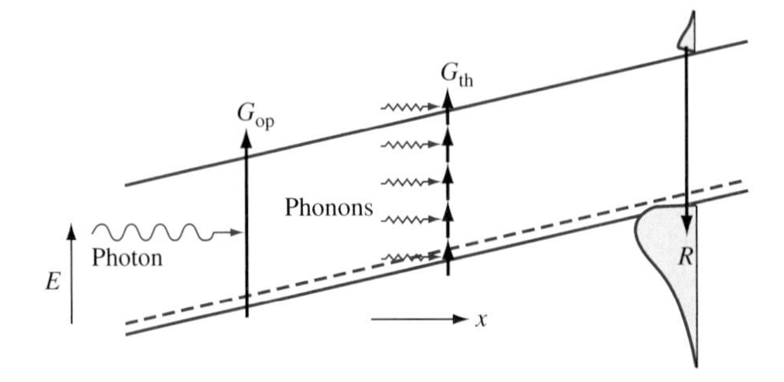

Figure 3.19 Energy band diagram of the semiconductor of Figure 3.18, under electrical bias and optical illumination. The combination rate R, thermal generation rate G_{th}, and the optical generation rate G_{op} are illustrated.

where the partial derivative is replaced by the derivative. Equation (3.72) must be solved for Δn with the initial conditions appropriate to the particular problem.

The photoconductivity is determined from Equation (3.71) or, since $\Delta p = \Delta n$,

$$\sigma_{pc}(t) = q\Delta n(t)(\mu_n + \mu_p)$$

Then from Equation (3.67), the time dependence of the current depends on $\Delta n(t)$. We will start with the case in which the semiconductor is initially at equilibrium, and the light is then turned on at $t = 0$.

3.8.1 RISE TIME

Before the light is turned on, the semiconductor is at equilibrium. Thus at $t = 0$, $\Delta n = 0$. In this case, the solution to Equation (3.72) for $t > 0$ is

$$\Delta n = G_{\text{op}}\tau_n(1 - e^{-t/\tau_n}) \tag{3.73}$$

This is an increasing function with time, and increases exponentially with some characteristic time τ_n.

Figure 3.20a shows a plot of $\Delta n(t)$ for this case. We see that Δn increases with a time constant τ_n and reaches a maximum of

$$\Delta n_{\max} = G_{\text{op}}\tau_n \tag{3.74}$$

The maximum is proportional to G_{op}, which is related to the light intensity and absorption rate, and to τ_n. When the light is turned on, it takes some time for the steady-state excess carrier concentration to be established (the rise time).

3.8.2 FALL TIME

Now let us assume the light is extinguished at some time $t = t_0$. After t_0, the optical generation rate goes to zero. Inserting $G_{\text{op}} = 0$ into Equation (3.72) results in the solution

$$\Delta n(t) = \Delta n(t_0)e^{-(t-t_0)/\tau_n} \tag{3.75}$$

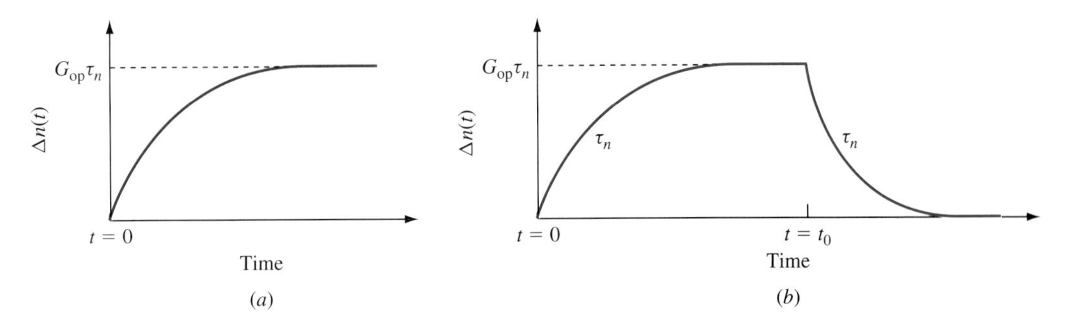

Figure 3.20 Variation of excess carriers in a semiconductor under pulsed illumination. (a) When the light is turned on, the excess carrier concentration increases exponentially. For the complete pulse (b), the rise and fall time constants are equal to the minority carrier lifetimes.

or Δn decays with the same time constant τ_n. For an n-type semiconductor, with similar approximations, Δn and τ_n are replaced by Δp and τ_p respectively.

From Equation (3.60), the recombination rate is given by $R = n/\tau_n$. For a p-type direct gap semiconductor with $n \ll N_A$, which is usually the case,

$$R = n\beta p = n\beta N_A$$

where β is the probability that a given electron will recombine with a hole in unit time. Then

$$\tau_n = \frac{1}{\beta N_A}$$

or τ_n varies inversely with N_A. The heavier the doping, the shorter the lifetime. In a p-type indirect gap semiconductor (e.g., Si), however, recombination is via trap states within the forbidden band. Thus, R is limited by the density of the trap states N_T.

We have seen that while the light is on, the excess carrier concentration (and along with it the photoconductivity and photocurrent) reaches some maximum value that depends on G_{op}, which is proportional to the light intensity. The maximum current is also proportional to the minority carrier lifetime, from Equations (3.67), (3.71), and (3.74). Thus in a photodetector, the sensitivity increases as the carrier lifetime gets longer. The rise and fall times are also proportional to τ_n, however, so for high speed of response, one sacrifices sensitivity.

EXAMPLE 3.5

Show that for a p-type semiconductor the average time an electron spends in the conduction band is equal to the time constant τ_n.

■ Solution

Consider the photoconductivity experiment of Figure 3.18, and suppose that at $t = t_0 = 0$ the illumination is turned off. Then Equation (3.75) becomes

$$\Delta n(t) = \Delta n(0)e^{-t/\tau_n}$$

The average time $\bar{t}_n$ that an electron spends in the conduction band is given by

$$\bar{t}_n = \frac{\int_0^\infty t \frac{dn}{dt} dt}{\int_0^\infty \frac{dn}{dt} dt}$$

We can substitute for dn/dt by using $dn/dt = d\Delta n/dt$. We then have that

$$\frac{dn}{dt} = \frac{d\Delta n}{dt} = \Delta n(0) \left(\frac{-1}{\tau_n}\right) e^{-t/\tau_n},$$

and thus in the denominator

$$\int_0^\infty \frac{dn}{dt} dt = \Delta n(0) \left[\frac{-1}{\tau_n}\right] \int e^{-t/\tau_n} dt = \Delta n(0)$$

For the integral in the numerator,

$$\int_0^\infty t \frac{dn}{dt} dt = \int_0^\infty t \frac{d\Delta n}{dt} dt = \Delta n(0)\tau_n$$

Thus

$$\bar{t}_n = \frac{\int_0^\infty t \frac{dn}{dt} dt}{\int_0^\infty \frac{dn}{dt} dt} = \frac{\Delta n(0)\tau_n}{\Delta n(0)} = \tau_n$$

As indicated above, the minority carrier lifetime is the average time an electron spends in the conduction band in p-type material or, in n-type material, the average time a hole spends in the valence band. Minority carrier lifetime has been measured as a function of doping in uncompensated Si; there is considerable scatter in the data. At high doping levels where the lifetimes are short, minority carrier lifetimes become difficult to measure. At low doping, the lifetime is strongly dependent on the trap concentration. Minority carrier lifetimes have been fitted to the empirical expressions [3, 4]

$$\tau_n = \left[3.45 \times 10^{-12} N_A + 9.5 \times 10^{-32} N_A^2\right]^{-1} \quad (3.76)$$

$$\tau_p = \left[7.8 \times 10^{-13} N_D + 1.8 \times 10^{-31} N_D^2\right]^{-1} \quad (3.77)$$

where N_A and N_D are expressed in cm^{-3} and τ_n and τ_p are in seconds. These lifetimes are plotted against doping level from Equations (3.76) and (3.77) in Figure 3.21. We can see that for high-purity Si, the minority carrier lifetimes are in the millisecond range.[9] They decrease with increased doping and for a concentration of 10^{20} cm^{-3} are on the order of a nanosecond.

[9]There is considerable scatter in the measured lifetimes of lightly doped Si because of the wide variations in trap concentrations. The plots in Figure 3.21 represent the highest measured values (lowest trap densities).

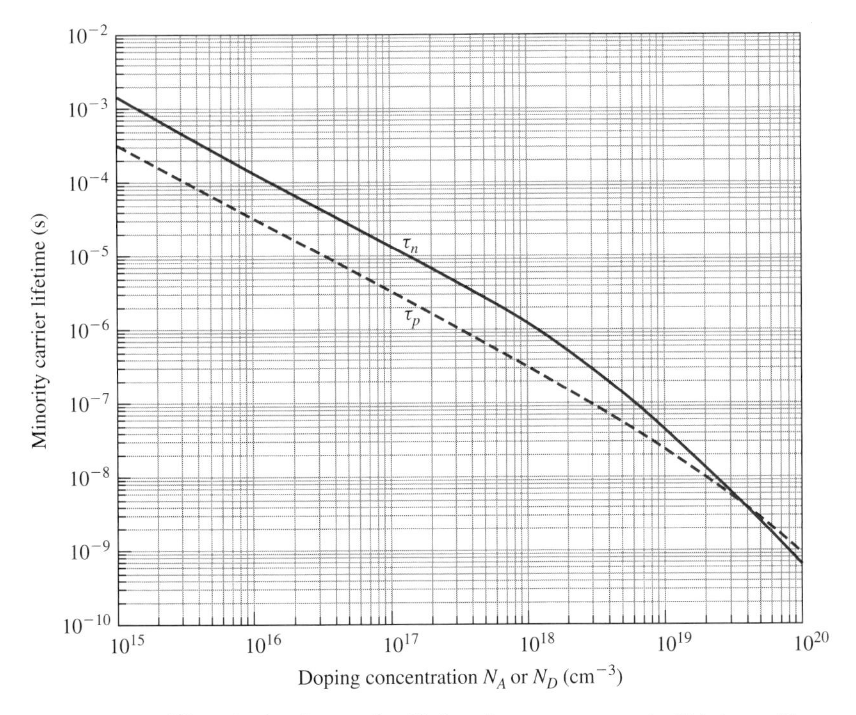

Figure 3.21 Plot of minority carrier lifetime in uncompensated high quality Si as a function of doping concentration N_A or N_D.

3.9 MINORITY CARRIER DIFFUSION LENGTHS

We have seen that recombination causes the carriers to have a finite lifetime. During that lifetime, they can also diffuse, since they will always have some kinetic energy. In this section, we will see how far a carrier diffuses, on the average, before it recombines.

Here we consider a p-type semiconductor that is illuminated from one side as shown in Figure 3.22a. The illumination is steady state, with photons of absorption coefficient high enough that for practical purposes they all can be considered to be absorbed at the surface. We also assume that there is no electric field in the semiconductor. The excess electrons generated at the surface then will diffuse into the regions of lower electron concentration—deeper into the semiconductor. As they penetrate, they will recombine with holes. Figure 3.22b indicates the variation of the excess electrons as a function of position with time as a parameter. With increasing time after the light is turned on, the electron concentration at the surface increases. The electrons diffuse into the bulk and recombine. The steady-state condition is reached when the generation rate at the surface is equal to the recombination rate in the bulk.

We can determine Δn as a function of x and t by solving the continuity equation for electrons [Equation (3.63)]. This can be simplified for the steady-state condition since then $\partial n/\partial t = 0$. Further, since all excess electrons are generated

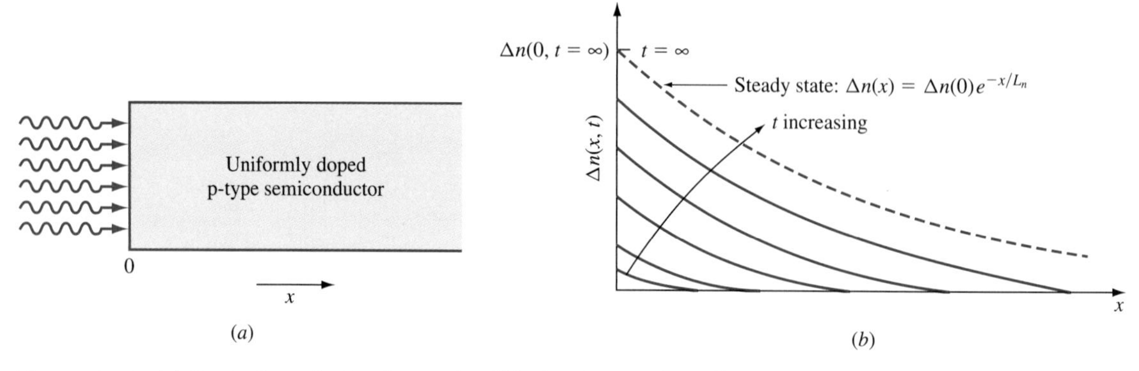

Figure 3.22 (a) Illustration of minority carrier diffusion in a surface-illuminated p-type semiconductor. The absorption is assumed to occur at the surface. (b) Plots of the excess minority carrier concentration as a function of distance into the bar with increasing time. As the excess carriers are generated at the surface, they diffuse to regions of lower concentration, where they recombine.

at the surface, for $x > 0$, $G_{op} = 0$, and since $\mathscr{E} = 0$, $J_{n(\text{drift})} = 0$. Equation (3.63) then becomes

$$\frac{dJ_n}{dx} = q\frac{\Delta n}{\tau_n} \tag{3.78}$$

where the derivative replaces the partial derivative because there is no variation with time. Since

$$J_{n(\text{diff})} = qD_n\frac{dn}{dx} = q\left(\frac{dn_0}{dx} + \frac{d\Delta n}{dx}\right)$$

and n_0 is constant, Equation (3.78) becomes

$$qD_n\frac{d^2(\Delta n)}{dx^2} = q\left(\frac{\Delta n}{\tau_n}\right) \tag{3.79}$$

or

$$\frac{d^2(\Delta n)}{dx^2} = +\frac{\Delta n}{D_n\tau_n} \tag{3.80}$$

This is a second-order differential equation, which can be solved with the boundary conditions $\Delta n = \Delta n(0)$ at $x = 0$ and $\Delta n = 0$ at $x = \infty$. The solution is

$$\Delta n(x) = \Delta n(0)e^{-x/\sqrt{D_n\tau_n}} = \Delta n(0)e^{-x/L_n} \tag{3.81}$$

Here we have introduced a new quantity called the diffusion length:

$$\boxed{L_n = \sqrt{D_n\tau_n}} \tag{3.82}$$

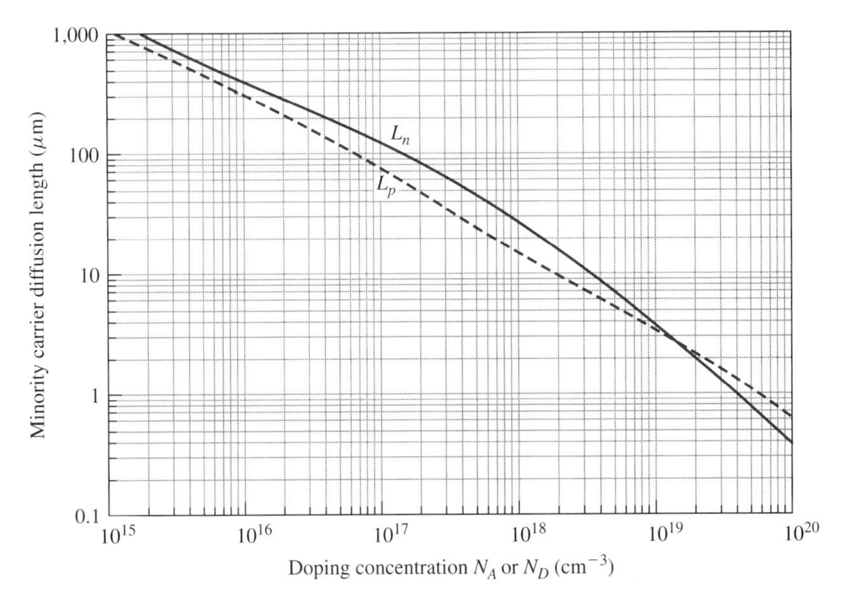

Figure 3.23 Minority carrier diffusion lengths L_n and L_p as functions of impurity concentration N_A or N_D in uncompensated high quality Si.

It is the average distance an electron diffuses before it recombines. Similarly the diffusion length for holes is

$$\boxed{L_p = \sqrt{D_p \tau_p}} \qquad (3.83)$$

In more heavily doped Si, both D_n and τ_n are reduced below their intrinsic values, and thus so is L_n. The minority carrier diffusion lengths for Si can be calculated from Equations (3.82) and (3.83), using the minority carrier diffusion coefficients as expressed by the Einstein relation $D/\mu = kT/q$ and Equations (3.18) and (3.19), as plotted in Figure 3.11. The minority carrier lifetimes come from Equations (3.76) and (3.77). These minority carrier diffusion lengths L_n and L_p are plotted as functions of doping in Figure 3.23 for Si. It can be seen that the diffusion lengths reduce from about 1 mm at 10^{15} impurities per cm^3 to less than 1 μm at 10^{20} cm^{-3}.

The diffusion lengths in direct gap semiconductors are appreciably smaller than this, since their carrier lifetimes are much less.

3.10 QUASI FERMI LEVELS

For a semiconductor at equilibrium, the Fermi level is constant and its value can be used as a measure of the equilibrium concentrations of electrons and holes. In this section, we will introduce the *quasi Fermi level,* which can be used as a reference to find the concentrations of electrons and holes when a material is *not* at equilibrium.

Recall that for a nondegenerate semiconductor

$$\left.\begin{aligned} n_0 &= N_C e^{-(E_C - E_f)/kT} = n_i e^{(E_f - E_i)/kT} \\ p_0 &= N_V e^{-(E_f - E_V)/kT} = n_i e^{(E_i - E_f)/kT} \end{aligned}\right\} \quad \text{equilibrium} \tag{3.84}$$

It is useful to have similar expressions for electrons and holes for the nonequilibrium case. We can write analogues to Equations (3.84):

$$\left.\begin{aligned} n &= N_C e^{-(E_C - E_{fn})/kT} = n_i e^{(E_{fn} - E_i)/kT} \\ p &= N_V e^{-(E_{fp} - E_V)/kT} = n_i e^{(E_i - E_{fp})/kT} \end{aligned}\right\} \quad \text{nonequilibrium} \tag{3.85}$$

where n and p are respectively the total electron and hole concentrations, including the equilibrium and excess carriers. Equation (3.85) in effect defines the quasi Fermi levels for electrons, E_{fn}, and for holes, E_{fp}.

Solving for the quasi Fermi levels, we find that Equation (3.85) becomes

$$\begin{aligned} E_{fn} &= E_C - kT \ln \frac{N_C}{n} = E_i + kT \ln \frac{n}{n_i} \\ E_{fp} &= E_V + kT \ln \frac{N_V}{p} = E_i - kT \ln \frac{p}{n_i} \end{aligned} \tag{3.86}$$

As an example, consider the illuminated p-type semiconductor of Figure 3.22 in steady state. At the dark end of the bar, $n = n_0$ and $p = p_0$, the material is in equilibrium, and the Fermi level is defined. At equilibrium $E_{fn} = E_{fp} = E_f$, as shown in Figure 3.24. Toward the illuminated end of the sample, both n and p increase from their equilibrium values, and the quasi Fermi levels separate as indicated. Since the material is p type, $p_0 \gg n_0$, and since

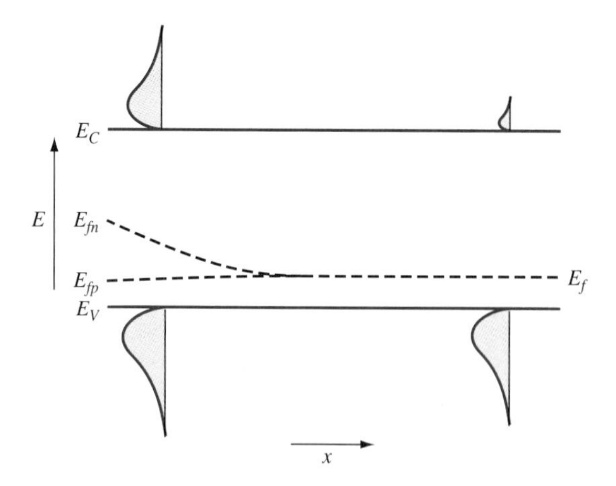

Figure 3.24 Illustration of quasi Fermi levels for electrons and holes for the steady-state nonequilibrium case of Figure 3.22, with $\mathscr{E} = 0$.

$\Delta n = \Delta p$ at any position, there are more holes than electrons, so the addition of a fixed number of excess carriers has a smaller fractional effect on the total concentration of holes than on the electrons. Thus the change in E_{fp} relative to E_f is less than for E_{fn}.

Just as we need to know the equilibrium electron and hole concentrations to determine the Fermi energy, to obtain the quasi Fermi levels we must know the electron and hole concentrations in the nonequilibrium case. The use of quasi Fermi levels on an energy band diagram is a convenient way to represent the electron and hole concentrations as a function of position.

We emphasize that, just as Equation (3.84) is valid for nondegenerate semiconductors, Equations (3.85) and (3.86) are valid only for small Δn such that the quasi Fermi levels are greater than about $2.3kT$ from their respective band edges.

In a nondegenerate semiconductor, the current densities J_n and J_p can be expressed in terms of the gradient of the quasi Fermi levels E_{fn} and E_{fp} respectively. The electron current can be expressed as in Equation (3.46):

$$J_n = q\mu_n \left[n(x)\mathscr{E}(x) + \frac{kT}{q}\frac{dn}{dx} \right] \tag{3.87}$$

But from Equation (3.85),

$$\begin{aligned} \frac{dn}{dx} &= \frac{d}{dx}[N_C e^{-(E_C - E_{fn})/kT}] \\ &= \frac{1}{kT} N_C e^{-(E_C - E_{fn})/kT} \left[\frac{dE_{fn}(x)}{dx} - \frac{dE_C(x)}{dx} \right] \\ &= \frac{n(x)}{kT} \left[\frac{dE_{fn}(x)}{dx} - \frac{dE_C(x)}{dx} \right] \end{aligned} \tag{3.88}$$

Since $dE_C(x)/dx = q\mathscr{E}(x)$, however, Equation (3.87) becomes

$$J_n(x) = \mu_n(x) n(x) \frac{dE_{fn}}{dx} \tag{3.89}$$

which can be expressed as

$$J_n(x) = q\mu_n(x) n(x) \frac{d}{dx}\left(\frac{E_{fn}}{q}\right) = \sigma_n(x) \frac{d}{dx}\left(\frac{E_{fn}}{q}\right) \tag{3.90}$$

This resembles Ohm's law in point form, with $\mathscr{E}$ replaced by $d/dx(dE_{fn}/q)$.

Similarly,

$$J_p(x) = \sigma_p(x) \frac{d}{dx}\left(\frac{E_{fp}}{q}\right) \tag{3.91}$$

The above relations for J_n and J_p are valid for any combinations of drift and diffusion currents. For $dn/dx = dp/dx = 0$, there is no diffusion current and

$$\frac{d}{dx}\left(\frac{E_{fn}}{q}\right) = \frac{d}{dx}\left(\frac{E_{fp}}{q}\right) = \mathscr{E} \qquad \text{no diffusion}$$

It should be emphasized again that the above use of quasi Fermi levels is valid only for nondegenerate semiconductors.

3.11 SUMMARY

We discussed two mechanisms of current flow in semiconductors. Current flows by carrier drift in an electric field, and by diffusion in the presence of a carrier concentration gradient.

In general, current flow is by a combination of drift and diffusion. The electron and hole current densities are

$$J_n = q\mu_n n\mathscr{E} + qD_n \frac{dn}{dx} = q\mu_n \left[n\mathscr{E} + \frac{kT}{q}\frac{dn}{dx} \right]$$

$$J_p = q\mu_p p\mathscr{E} - qD_p \frac{dp}{dx} = q\mu_p \left[p\mathscr{E} - \frac{kT}{q}\frac{dp}{dx} \right]$$

In the above, the mobilities and diffusion coefficients decrease with increasing doping concentrations. For either carrier type their values are larger if they are minority carriers than if they are majority carriers. The fractional difference increases with increased doping. For uncompensated Si, the values of μ can be found from Figure 3.4 and the values of D can be found from Figure 3.11. Mobilities and diffusion coefficients are related by the Einstein relation $D/\mu = kT/q$.

The carrier mobility is proportional to the mean free time between collisions, $\bar{t}$

$$\mu = \frac{q\bar{t}}{m_c^*}$$

at low fields, the mobility is independent of field. At high fields, it decreases with increasing field.

Illuminating a semiconductor with photons of energy greater than the band gap produces hole-electron pairs. Since photons have little wave vector, this produces a near-vertical transition in the electron *E-K* diagram for a direct gap semiconductor. For an indirect gap semiconductor a phonon is required to furnish the change in wave vector required by the electron transition, making optical transitions less probable in indirect materials.

Minority carrier lifetimes and diffusion lengths are material dependent, but in any given material, both decrease with increasing impurity concentrations.

The continuity equations for electrons and for holes were introduced:

$$\boxed{\frac{\partial n}{\partial t} = \frac{1}{q}\frac{\partial J_n}{\partial x} + (G_n - R_n)}$$

$$\boxed{\frac{\partial p}{\partial t} = -\frac{1}{q}\frac{\partial J_p}{\partial x} + (G_p - R_p)}$$

In three dimensions they become

$$\frac{\partial n}{\partial t} = \frac{1}{q}\nabla \cdot J_n + (G_n - R_n)$$

$$\frac{\partial p}{\partial t} = -\frac{1}{q}\nabla \cdot J_p + (G_p - R_p)$$

These equations are mathematical statements of the conservation of particles. While they are quite general, they can be solved in closed form for only relatively simple cases.

The minority carrier lifetime was defined as the average time τ_n an electron spends in the conduction band in a p-type semiconductor before recombining with a hole, or the average time τ_p a hole spends in the valence band in an n-type semiconductor. Analogously, the minority carrier diffusion lengths L_n and L_p are defined as the average distance a minority carrier will diffuse before recombining.

$$L_n = \sqrt{D_n \tau_n}$$

$$L_p = \sqrt{D_p \tau_p}$$

For a well-defined carrier lifetime, the continuity equations become, for electrons,

$$\boxed{\frac{\partial n}{\partial t} = \frac{\partial \Delta n}{\partial t} = \frac{1}{q}\left(\frac{\partial J_{n(\text{drift})}}{\partial x} + \frac{\partial J_{n(\text{diff})}}{\partial x}\right) + \left(G_{\text{op}} - \frac{\Delta n}{\tau_n}\right)}$$

and similarly for holes,

$$\boxed{\frac{\partial p}{\partial x} = \frac{\partial \Delta p}{\partial t} = \frac{-1}{q}\left(\frac{\partial J_{p(\text{drift})}}{\partial x} + \frac{\partial J_{p(\text{diff})}}{\partial x}\right)\left(G_{\text{op}} - \frac{\Delta p}{\tau_p}\right)}$$

Just as the Fermi level is a useful concept for describing the electron and hole concentrations in semiconductors at equilibrium, the quasi Fermi levels E_{fn} and E_{fp} are useful for nonequilibrium cases. For nondegenerate semiconductors

$$n = N_C e^{-(E_C - E_{fn})/kT} = n_i e^{(E_{fn} - E_i)/kT}$$

$$p = N_V e^{-(E_{fp} - E_V)/kT} = n_i e^{(E_i - E_{fp})/kT}$$

From knowledge of the quasi Fermi levels, the electron and hole currents can be expressed

$$J_n(x) = \sigma_n(x)\frac{d}{dx}\left(\frac{E_{fn}}{q}\right)$$

$$J_p(x) = \sigma_p(x)\frac{d}{dx}\left(\frac{E_{fp}}{q}\right)$$

3.12 READING LIST

Items 1, 2, 4, 6, 8, 10 to 12, 14 to 19, and 22 in Appendix G are recommended.

3.13 REFERENCES

1. D. M. Caughey and R. Thomas, "Carrier mobilities in silicon empirically related to doping and field," *Proc. IEEE,* 55, pp. 2192–2193, 1967.
2. J. del Alamo, S. Swirhun, and R. M. Swanson, "Measuring and modeling minority carrier transport in heavily doped silicon," *Solid State Electronics,* 28, pp. 47–54, 1985.
3. J. del Alamo, S. Swirhun, and R. M. Swanson, "Simultaneous measurement of hole lifetime, hole mobility and bandgap narrowing in heavily doped n-type silicon, *IEDM Technical Digest,* pp. 290–293, 1985.
4. S. E. Swirhun, Y.–H. Kwark, and R. M. Swanson, "Measurement of electron lifetime, electron mobility and band gap narrowing in heavily doped p-type silicon, *IEDM Technical Digest,* pp. 24–27, 1986.
5. Chi-Tang Sah, Robert W. Noyce, and Williiam Shockley, "Carrier Generation and Recombination in *P-N* Junctions and *P-N* Junction Characteristics," *Proc. IRE,* 45, pp. 1228–1242, 1957.

3.14 REVIEW QUESTIONS

1. What are the two basic mechanisms by which current flows in semiconductors? Explain the physics of each.
2. What is the difference between current and current density?
3. Define conductivity. Give its dimensions.
4. Explain how both electrons and holes contribute to conductivity.
5. There are no holes in a metal. Explain what happens to the holes carrying current in a semiconductor when that current reaches and continues on into the wire connecting the semiconductor with the rest of the circuit.
6. Define mobility. Give its dimensions.
7. Explain why the majority and minority carriers have different mobilities.
8. Why is a doped semiconductor more conductive than an intrinsic one?
9. What is ionized impurity scattering? Is it stronger under low doping or high doping? Under low temperature or high temperature? Why?
10. What is phonon scattering? Is it stronger under low temperature or high temperature? Why?
11. Explain how the presence of an impurity miniband affects the mobility of majority carriers. Under what doping conditions (heavy, light, any) will this effect be appreciable?
12. What is velocity saturation? What causes it?

13. Electrons always diffuse to regions of lower concentration. Is the same true for holes?

14. What is the difference between optical generation and thermal generation?

15. What is a trap state?

16. Why are direct gap materials better light emitters and absorbers than indirect gap materials?

17. Do photons have large K vector or small K vector? What does that imply about optical transitions from one band to the other on the E-K diagram?

18. What are excess carriers?

19. Define carrier lifetime.

20. Explain in words the meaning of the continuity equations. What are the various terms and what do they mean?

21. Explain in words how the continuity equation can be used to find the carrier lifetimes. You may wish to use Figure 3.17 in your explanation.

22. Define equilibrium.

23. What is the difference between dark conductivity and photoconductivity?

24. What is a quasi Fermi level? How is it different from the Fermi level?

3.15 PROBLEMS

3.1 Calculate the thermal speed of an electron in Si with kinetic energy 0.013 eV. Draw an energy band diagram for silicon and indicate where this electron will be. Compare your calculated thermal speed with the typical drift velocities cited in the text of 10^4 cm/s. How does it compare with typical saturation velocities?

3.2 Calculate the resistivity for a uniformly doped silicon sample with 10^{17} donors per cubic centimeter.

3.3 A lightly doped Si sample ($N_D = 10^{14}$ cm^{-3}) is heated from 300 to 400 K. Is its resistivity expected to increase or decrease? Explain your answer. Repeat for Si with $N_D = 10^{18}$ cm^{-3}.

3.4 Germanium is an interesting semiconductor because it has a small band gap ($E_g = 0.67$ eV). (In fact, for a while it was not considered to be a semiconductor but was classified as a metal. Now it is a semiconductor again.) As a result, it has a higher intrinsic concentration n_i than either silicon or GaAs. Do you expect the conductivity of intrinsic germanium to be less than or greater than that of intrinsic silicon? How about compared with GaAs? Why?

3.5 Find the conductivity of silicon doped with $N_D = 10^{16}$ cm^{-3} and $N_A = 10^{19}$ cm^{-3}. Be sure to take into account the minority and majority carriers.

3.6 A voltage of 2.5 V is applied to a sample of silicon whose cross-sectional area is 0.1 μm × 1 μm. The length of the path is 0.1 μm. If the material is doped n type with $N_D = 10^{18}$ cm^{-3}, what is the current in the sample? What is the current density? These dimensions could represent the channel of a field effect transistor.

3.7 Calculate the electron drift velocity for $N_D - N_A = 10^{16}$ cm^{-3} in a bar of Si of cross-sectional area 1.0 mm^2 for a current of 50 mA.

3.8 Explain how an electron can "collide" with a crystal defect. *Hint:* the normal periodic potential of the lattice is disturbed by a crystalline defect. A poor-quality crystal will have lower carrier mobilities than a good one.

3.9 Explain why, for a given noncompensated semiconductor with a given doping level N, the electron mobility is larger in p-type Si than in n-type Si. Refer to Figure 3.4.

3.10 With reduced N in Problem 3.9, the electron mobilities approach each other. Explain.

3.11 Compare the mean free time between collisions for electrons and for holes in intrinsic GaAs. How do these values compare with those for silicon?

3.12 For Problem 3.11, find an average gain in kinetic energy between collisions for electrons and holes for an applied field of 100 V/cm.

3.13 Estimate the saturation velocity of electrons in intrinsic GaAs. How does your estimate compare with the experimental data?

3.14 The electron velocity in Si has its saturation value ($v_{sat} \approx 1 \times 10^7$ cm/s) over the range of 5×10^4 to 2×10^5 V/cm. Plot the mobility-field (μ-$\mathscr{E}$) and $\bar{t}$-$\mathscr{E}$ relations over this range of fields.

3.15 Consider a p-type Si sample of $N_A = 10^{18}$ cm^{-3} and $N_D = 0$. Over a length of 1 μm the electron concentration drops linearly from 10^{16} cm^{-3} to 10^{13} cm^{-3}. Calculate the electron diffusion current density.

3.16 A material is doped such that the electron concentration varies linearly across the sample. The sample is 0.5 mm thick. The donor concentration varies from $N_D = 0$ at $x = 0$ to $N_D = 10^{16}$ cm^{-3} at $x = 0.5$ μm.

a. Write equations for $n(x)$ and $p(x)$.

b. Find the electron diffusion current density.

c. Find the hole diffusion current density at $x = 0$ and $x = 0.5$ μm. Can the minority carriers contribute significant diffusion current?

d. Find an expression for $E_C(x) - E_f$ as a function of x. Sketch the energy band diagram. (The Fermi level is constant at equilibrium, so draw the Fermi level as a flat line and adjust $E_C - E_f$ appropriately.)

e. At equilibrium, the total current must be zero. Show that there must therefore be an internal electric field present in this sample. (The field is generated by the variation in doping, as we will see in the next chapter.)

3.17 Comment on the probability of absorption (zero, low, medium, high) by a photon of $\lambda = 600$ nm (red) by the following materials:

Si

Ge

GaAs

InAs

SiC

GaN

CdS

3.18 Explain physically why the carrier lifetimes are much less in direct gap materials than in indirect.

3.19 Light of $h\nu = 1.5$ eV (this in the near infrared, at $\lambda = 826$ nm) at a power level of 10 mW/cm^2 shines on an intrinsic sample of GaAs of area 1 cm^2. Let the electron lifetime be 10 ps.

a. What is the number of photons arriving at the semiconductor surface per sec? (Recall energy = power $\times$ time.)

b. Verify that photons of this energy can be absorbed.

c. Assuming every photon is absorbed and creates an electron-hole pair, and assuming the GaAs sample is 1 mm thick, what is the optical generation rate?

d. What are the equilibrium electron and hole densities (in the dark)?

e. What are the excess carrier concentrations when the light is on?

f. What are the recombination rates for electrons and for holes when the light is off? When the light is on?

g. What are the steady-state carrier densities n and p?

h. How much does the conductivity of this sample change compared with its dark value?

i. Suppose the power level is kept the same, but the wavelength of the light is shifted further into the infrared, at $E = h\nu = 1$ eV ($\lambda = 1240$ nm). What is the generation rate now?

3.20 Solve Equation (3.72) to prove that Equation (3.73) is a solution under the appropriate initial conditions. Repeat for Equation (3.75).

3.21 A direct gap semiconductor sample is illuminated at one end with light of $\lambda = 500$ nm (green), with an intensity of 1 mW/cm^2. The area of the illuminated surface is 1 cm^2. Assume the carrier lifetimes are 10 ns.

a. Find the number of photons striking the sample per second.

b. If every photon is absorbed uniformly (with x) within 1 μm of the surface, what are the excess carrier concentrations Δn and Δp in this region?

3.22 As we saw, it is possible to produce a nonuniform spatial distribution of charge carriers by shining a light on one part of a semiconductor, producing more electrons and holes in that part than elsewhere. What might be another way to produce more electrons and holes in one location, without applying an electric field? Would there be a net diffusion current?

3.23 Consider a bar of semiconductor illuminated as shown in Figure P3.1.

a. Sketch the concentrations of electrons and holes as functions of position.

b. In which direction(s) will the electrons diffuse? Holes?

c. In what directions do the electron and hole diffusion currents go?

d. Explain why a plot of voltage versus position along the length of the sample is as shown in the figure.

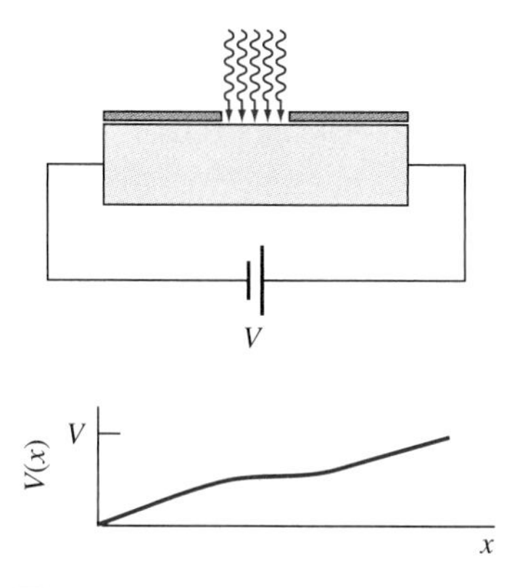

Figure P3.1

3.24 a. Find the conductivity (in the dark) of a sample of GaAs doped with $N_A = 10^{17}$ cm^{-3}.

b. If the sample is illuminated such that the excess electron concentration is 10^{16} cm^{-3}, what is the excess hole concentration?

c. What is the conductivity of this sample when the light is on?

3.25 Consider an n-type Si sample with $N_D - N_A = 10^{17}$ cm^{-3}. Under illumination, $\Delta n = \Delta p = 5 \times 10^{16}$ cm^{-3}. Find the quasi Fermi levels for electrons and holes. Compare these locations with the Fermi levels when the light is off.

3.26 A sample of InP is doped such that $E_f - E_V = 0.2$ eV. It is also illuminated such that $\Delta n = \Delta p = 10^3$ cm^{-3}. Find the quasi Fermi levels and sketch the energy band diagram. Repeat for $\Delta n = \Delta p = 10^{10}$.

CHAPTER 4

Nonhomogeneous Semiconductors

So far, we have examined *homogeneous* semiconductors. By homogeneous we mean that the entire semiconductor is made of the same material and doped uniformly. Examples of the other case, nonhomogeneous semiconductors, are semiconductors whose doping varies with position or where the semiconductor composition varies. A classic example of the former is the pn junction diode (the subject of Chapter 5), which is a junction between n-type and p-type material. We will lead up to the diode gradually, starting with a semiconductor in which the doping level is nonuniform with position. We then discuss the case in which the composition of the semiconductor itself is graded.

4.1 CONSTANCY OF THE FERMI LEVEL AT EQUILIBRIUM

First, we will establish a very important result:

In a system at equilibrium, the Fermi level is at constant energy.

To show this, we consider two materials in intimate contact as indicated in Figure 4.1a. In this example, the two semiconductors have different band gaps (meaning they are not the same material), and they are doped differently as well. Assume these materials have different Fermi levels as indicated in (b). When the materials are first joined, they are in a state we call *electrical neutrality,* as we shall see later. This state is not stable, because there are electrons at higher energies in A than in B, and we expect that these will flow from A to available lower energies in B until the system of joined materials reaches some kind of equilibrium. At equilibrium, the total current must be zero. Therefore, if carriers are moving from A to B, the same number of carriers must be moving from B to A.

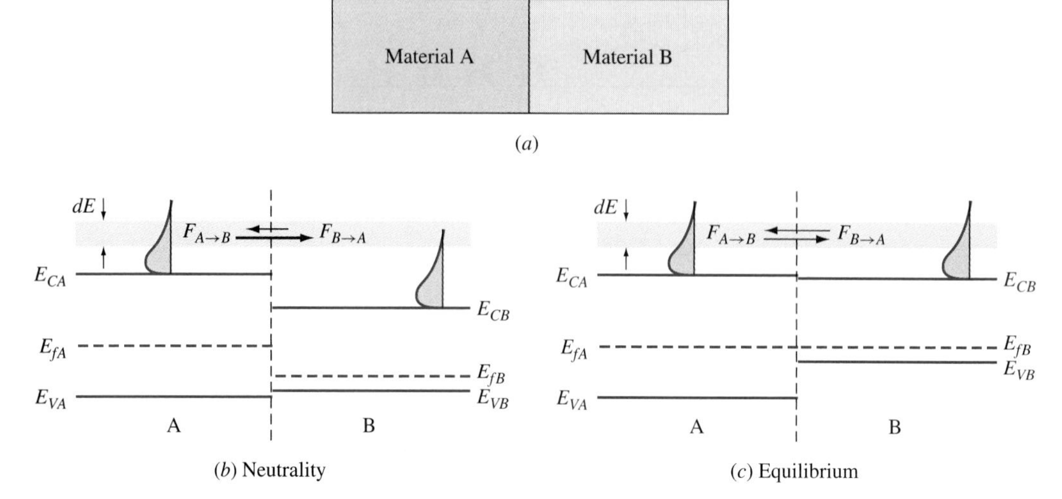

Figure 4.1 (a) When two different materials are in contact at equilibrium the net current must be zero. (b) When the materials are first joined, they are both electrically neutral. In the text it is shown that at equilibrium, the Fermi levels must be equal, as shown in (c).

To find the number of carriers available in A, let the density-of-states function in material A be $S_A(E)$, and the probability of occupancy be $f_A(E)$. The electron concentration at some energy E is $S_A(E)f_A(E)$. Similarly, $S_B(E)$ is the density-of-states function at energy E in material B and $f_B(E)$ is the Fermi probability at E in B. Let $F_{A \to B}$ be the rate of electron transfer (electron flux) from A to B. The movement of electrons requires electrons at a given state and nearby empty states to be at the same energy. For carriers to move from A to B, there must be empty states in B. The concentration of the *empty* states in B is $S_B(E)(1 - f_B(E))$. We can see from part (b) of the figure that, as drawn, in a given energy range dE (shaded in the figure) the concentration of electrons in material A is much higher than in material B. This implies that there are many empty states in B for electrons to go to.

If we consider the transfer flux of carriers in some small energy range dE, then

$$F_{A \to B} = C S_A(E) f_A(E) S_B(E)\,(1 - f_B(E))\,dE \tag{4.1}$$

where C is a constant. Similarly, if $F_{B \to A}$ is the transfer flux from B to A,

$$F_{B \to A} = C S_B(E) f_B(E) S_A(E)\,(1 - f_A(E))\,dE \tag{4.2}$$

At equilibrium, however, the two fluxes must be equal (no current), so $F_{A \to B} = F_{B \to A}$ and

$$S_A(E) f_A(E) S_B(E)\,(1 - f_B(E)) = S_B(E) f_B(E) S_A(E)\,(1 - f_A(E)) \tag{4.3}$$

which reduces to

$$f_A(E) = f_B(E) \tag{4.4}$$

The probability of occupancy is the Fermi function, so we can substitute Equation (2.49):

$$f_A(E) = \frac{1}{1 + e^{(E - E_{fA})/kT}}$$
$$f_B(E) = \frac{1}{1 + e^{(E - E_{fB})/kT}} \tag{4.5}$$

into Equation (4.4), producing

$$E_{fA} = E_{fB} \tag{4.6}$$

Thus, the Fermi levels are equal at equilibrium. This is shown in Figure 4.1c. The position of the Fermi level with respect to the conduction and valance band edges within a given material is a function of the doping, so if the Fermi levels are adjusted so that they line up in the figure, the entire energy band diagram must adjust along with it. This is still idealized to some extent, and in Part 2 we will discuss the details of constructing the energy band diagrams for heterojunctions (junctions between two different materials, as in this example).

While the discussion above shows that at equilibrium the Fermi levels of two dissimilar materials are equal, it can be generalized to state that the Fermi level is constant throughout a system at equilibrium. The materials can be semiconductors, nondegenerate or degenerate, or insulators, metals, or alloys. This important result will be used extensively in describing the operation of electronic devices.

4.2 GRADED DOPING

Up to now we have always taken the net doping concentration in a semiconductor—either N_D or N_A, to be a constant with position. This is often not the case in semiconductor devices. We now consider the case of a semiconductor with nonuniform doping. First, the influence of the doping profile on the equilibrium energy band diagram is investigated and how this, in turn, affects carrier concentrations.

Many semiconductors used in devices use compensated materials, or those in which both acceptors and donors are present. The quantity of interest in computing electron and hole concentrations, locating Fermi levels, and drawing energy band diagrams is actually the *net* doping concentration:

$$N'_D = N_D - N_A \qquad \text{n type} \tag{4.7}$$

or

$$N'_A = N_A - N_D \qquad \text{p type} \tag{4.8}$$

As an example of graded doping, we choose the case of compensated p-type Si shown in Figure 4.2 in which the net acceptor concentration, $N'_A = N_A - N_D$, decreases with position from left to right. In this discussion, we assume that the

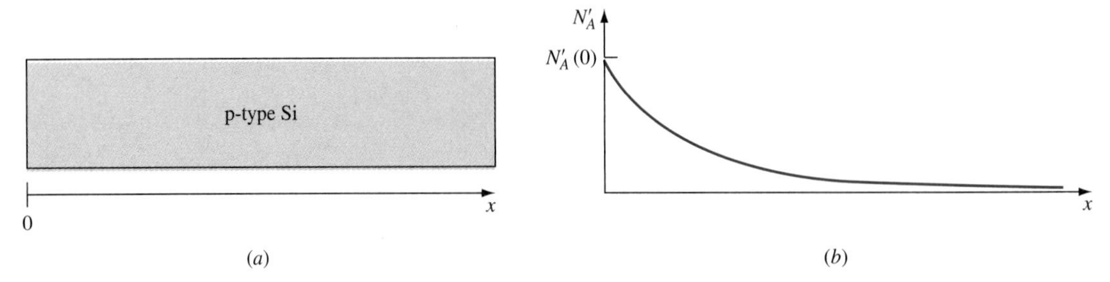

Figure 4.2 Nonuniformly doped semiconductor. (a) The bar of semiconductor material; (b) the net acceptor concentration as a function of distance.

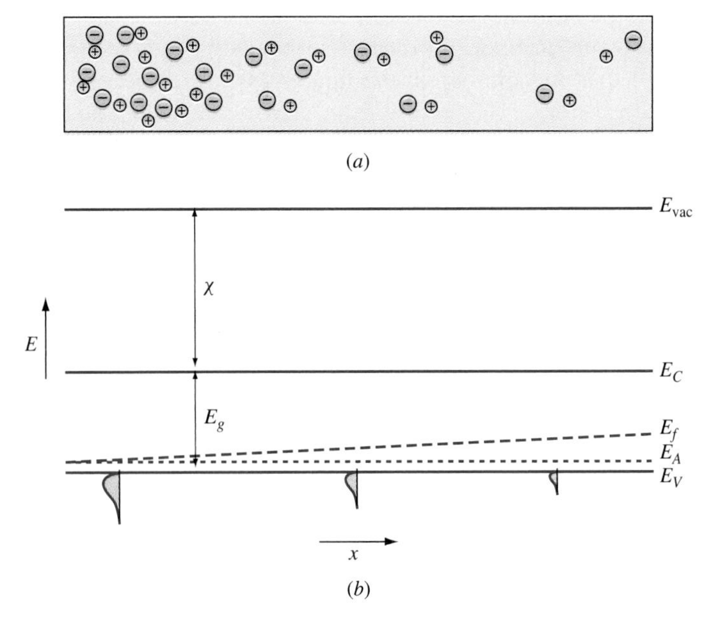

Figure 4.3 (a) The energy band diagram before equilibrium for the nonuniformly doped semiconductor (b) Neutrality exists in every macroscopic region.

material is everywhere nondegenerate and band-gap narrowing is therefore negligible, or that the band gap is independent of doping. This situation is typical in the base region of an npn bipolar transistor, for example.

We know from Chapter 2 that, where the net acceptor doping is heavier, there will be more holes. In the sample in Figure 4.2, we expect the Fermi level to be close to the valence band edge near the left end where the doping is heavy, and we expect the Fermi level to be closer to midgap at the right end where the material is more lightly doped. We therefore might expect an energy band diagram similar to that in Figure 4.3, in which electrical neutrality is assumed to exist in every macroscopic region. By electrical neutrality, we mean that in any macroscopic region, the concentration of negative charge in the region equals the positive charge concentration, as shown in part (a) of the figure. In part (b), the vacuum level E_{vac} and the acceptor state energy level E_A are shown, and the electron affinity χ is also indicated.

Assuming that all the impurities are ionized, the hole concentration is equal to the net acceptor concentration, or

$$p(x) = N'_A(x) = N_V e^{-(E_f - E_V)/kT} \tag{4.9}$$

Since the left side of Equation (4.9) is a function of x, the right side must be also. The quantities N_V, k, and T are all constants, so it follows that $E_f - E_V$ varies with position. But recall that the Fermi level is the energy level at which the Fermi probability function is $\frac{1}{2}$. The Fermi function is independent of the distribution of states, the doping concentrations, etc., and depends solely on temperature. We know, however, that at equilibrium the Fermi level is at constant energy independent of position. Therefore, we can anticipate that the edge of the valence band will shift in energy as the doping varies. We also now realize that Figure 4.3 is wrong. We now need to determine how to correct it.

Returning to Figure 4.3, the case of the assumed neutrality, we notice that the electric field

$$\mathscr{E} = \frac{1}{q}\frac{dE_{vac}}{dx} \tag{4.10}$$

is zero but that there is a gradient in the concentration of holes. We know from Chapter 3 that holes, being mobile, will diffuse to regions of lower concentration, in this case to the right. There is also a gradient in the concentration of ionized acceptors from Figure 4.2. The acceptor ions are negatively charged, but they can't diffuse because they are locked into the crystal lattice. Therefore, there is a net diffusion of positive charges (holes) from left to right, leaving uncompensated negative ions on the left. This is shown in Figure 4.4a. The number of positive charges (holes) per unit volume at the right-hand end of the sample then increases, leaving the negatively charged ions on the left. This separation of charges, however, creates an electric field in the semiconductor, and we no longer have a condition of electrical neutrality. The polarity of the field, though, is such that it tends to move the holes back toward the left. Thus, the diffusion of carriers to the right produces an electric field that tends to move them back to the left.

We can determine what the final situation at equilibrium will be. We know that at equilibrium, the net hole current density is zero, and that therefore the drift current due to the built-in electric field must balance the diffusion current caused by the concentration gradient:

$$J_p = J_{p(\text{drift})} + J_{p(\text{diff})} = 0 = q\mu_p p_0(x)\mathscr{E} - qD_p\frac{dp_0(x)}{dx} \tag{4.11}$$

where the hole concentration $p = p_0$ at equilibrium is a function of position. Because a relatively small shift in charge is required to establish this field, the

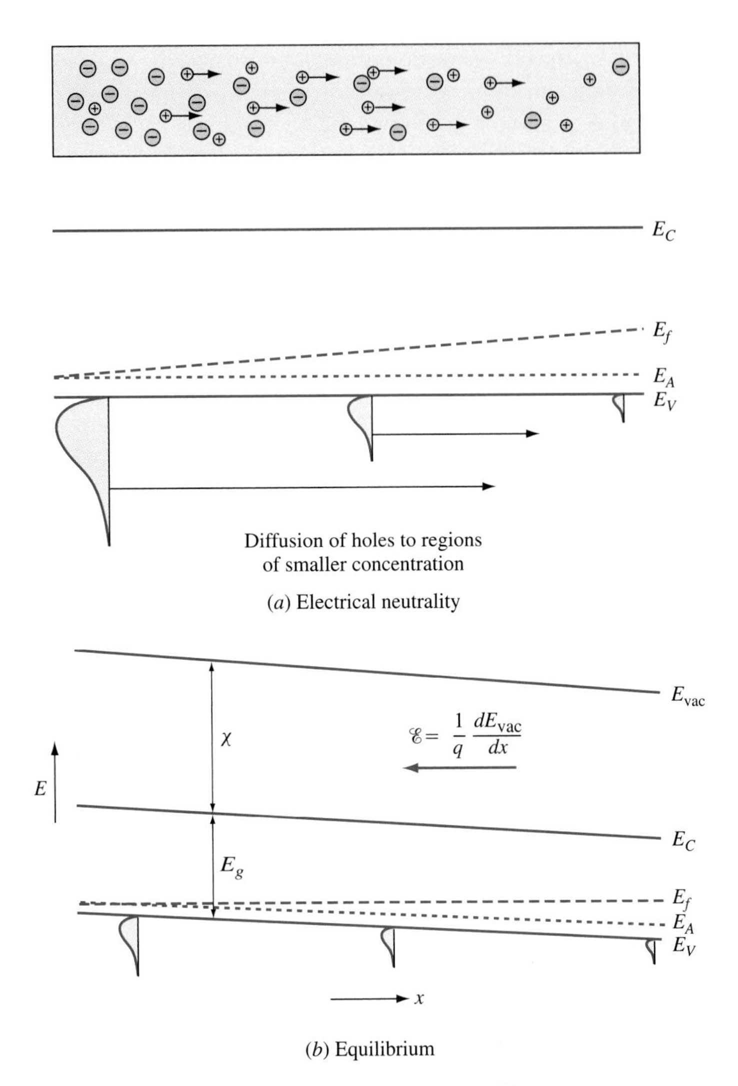

Figure 4.4 (a) The holes diffuse to regions of lower concentration, leaving behind ionized acceptors. This sets up an electric field. (b) The correct energy band diagram for a nonuniformly doped semiconductor at equilibrium.

approximation $p(x) = p_0(x) = N'_A(x)$ is still good, and Equation (4.11) becomes, with the aid of Equation (4.9),

$$q\mu_p p_0(x)\mathscr{E} = -q\frac{D_p N_V}{kT}e^{-(E_f - E_V)/kT}\left[\frac{d(E_f - E_V)}{dx}\right]$$

$$= -q\frac{D_p}{kT}p_0(x)\frac{d(E_f - E_V)}{dx} \qquad (4.12)$$

Since the Fermi level is a constant, $dE_f/dx = 0$, and

$$\mu_p \mathscr{E} = \frac{D_p}{kT}\frac{dE_V}{dx} \tag{4.13}$$

Next, since the semiconductor species (e.g., Si) is constant, so are the electron affinity, band-gap and ionization potential. Thus, E_V and E_C are parallel to E_{vac}. This means, from Equation (4.10),

$$\mathscr{E} = \frac{1}{q}\frac{dE_{\text{vac}}}{dx} = \frac{1}{q}\frac{dE_C}{dx} = \frac{1}{q}\frac{dE_V}{dx} \tag{4.14}$$

The electric field is therefore proportional to the slopes of the valence band edge and conduction band edge. The corrected energy band diagram at equilibrium is shown in Figure 4.4b.

Note that at equilibrium, there is *not* a state of electrical neutrality—to establish equilibrium the positive and negative charges were separated to some extent. However, since a small charge transfer is required to establish equilibrium, this region is normally referred to as a *quasi-neutral region,* or often as a *neutral region.*

To summarize, a convenient method to draw an energy band diagram at equilibrium is:

1. Assume electrical neutrality in every macroscopic region.
2. Using the vacuum level as a reference, i.e., with E_{vac} constant with position, draw the energy band diagram taking the electron affinity and band gap into account.
3. From a knowledge of the net doping, find the Fermi level with respect to the appropriate band edge (E_V for p-type, E_C for n-type material).
4. Adjust (tilt) the neutrality band diagram such that the Fermi level is at constant energy, keeping electron affinity and band gap constant.

4.2.1 THE EINSTEIN RELATION

From Equations (4.13) and (4.14),

$$\mu_p \mathscr{E} = \frac{D_p}{kT} q\mathscr{E} \qquad \text{or} \qquad \frac{D_p}{\mu_p} = \frac{kT}{q}$$

or for either type of carrier,

$$\boxed{\frac{D}{\mu} = \frac{kT}{q}} \tag{4.15}$$

This is the familiar Einstein relation of Chapter 3.

4.2.2 A GRADED-BASE TRANSISTOR

As an example of graded doping, we consider the p-type base of a Si npn bipolar junction transistor (BJT), shown schematically in Figure 4.5. This device will be discussed in considerable detail in Chapter 9. Here it suffices to indicate that the active transistor is the vertical region under the emitter contact (E), along the dotted line. The transistor consists of the heavily doped n-type emitter (labeled n^+), the p-type base (p), and the n-type collector (n well). The buried n^+ layer, being heavily doped, is highly conductive and electrically connects the n well to the collector contact.

The doping profile as obtained experimentally for a BJT from a specific process is shown in Figure 4.6a. Here the concentrations of various dopants are plotted as functions of depth (position) in the crystal. Figure 4.6b shows the region of interest on an expanded horizontal scale. We see that the base region is 0.14 μm in depth with net p-type doping varying from about 6×10^{17} cm^{-3} at the emitter edge ($x = 0.13$ μm, $N_D(As) = N_A(B)$) to 4×10^{16} cm^{-3} at the collector edge ($x = 0.27$ μm, $N_A(B) = N_D(P)$). On the semilogarithmic plot of Figure 4.6, the base doping can be approximated as a straight line, meaning the net acceptor concentration varies approximately exponentially with position:

$$N'_A(x) = N'_A(x_0)e^{-(x-x_0)/\lambda} \tag{4.16}$$

where $N'_A(x_0)$ is the straight line extrapolation of N'_A at x_0. Here λ is a constant that describes the doping variation with position and x_0 is a reference position, which we will locate at $x_0 = 0$. Equation (4.16) can be rewritten:

$$\frac{x-0}{\lambda} = \ln\left(\frac{N'_A(0)}{N'_A(x)}\right) \tag{4.17}$$

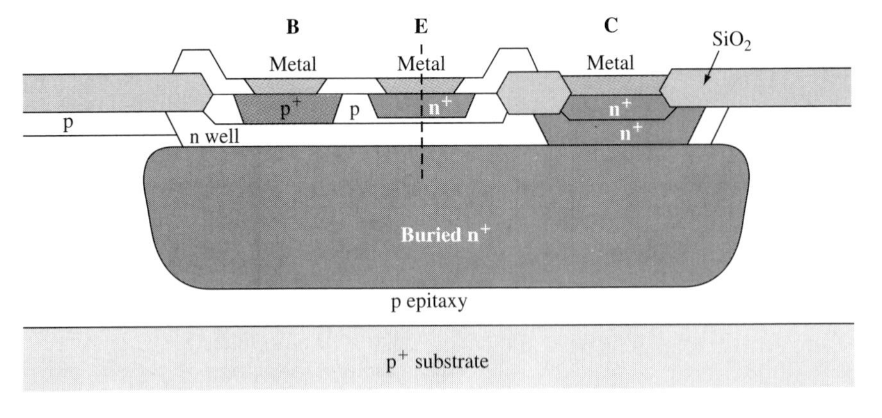

Figure 4.5 Schematic diagram of a silicon npn bipolar junction transistor.

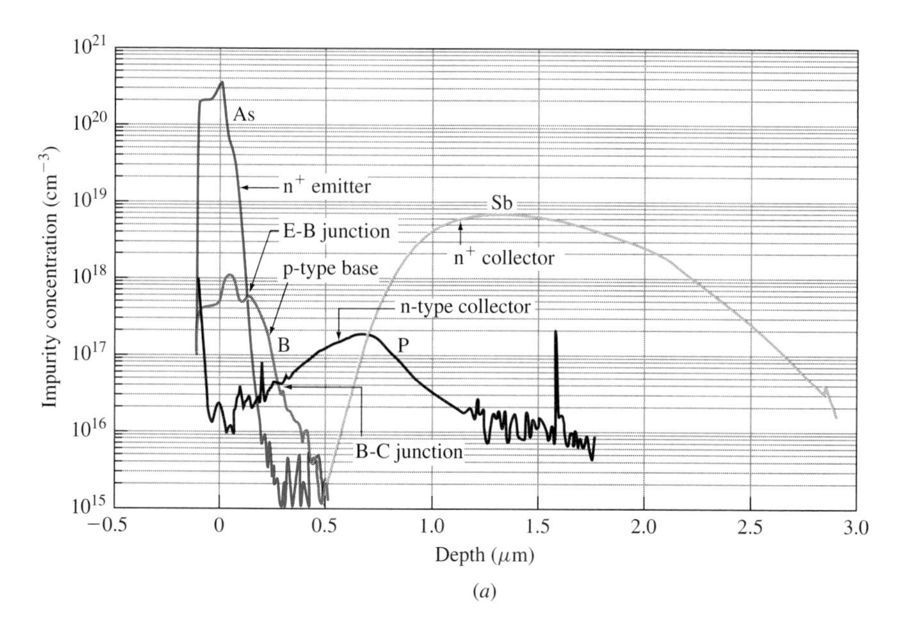

(*a*)

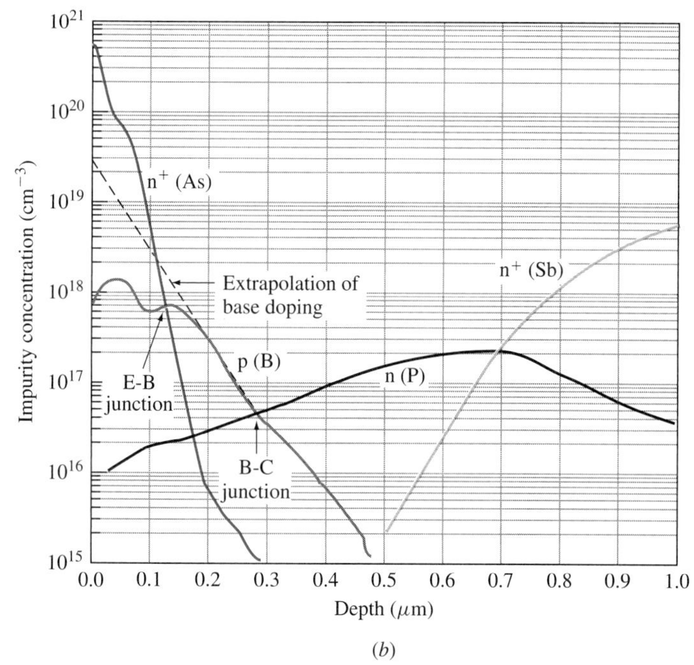

(*b*)

Figure 4.6 Experimentally measured [by secondary ion mass spectroscopy (SIMS)] plot of the doping profile normal to the emitter of the device of Figure 4.5. (Courtesy IBM.)

Evaluating Equation (4.17) at the position $x = 0.27\ \mu$m gives a value for the constant λ:

$$\frac{0.27 \times 10^{-4}\ \text{cm}}{\lambda} = \ln\left(\frac{2.5 \times 10^{19}\ \text{cm}^{-3}}{4 \times 10^{16}\ \text{cm}^{-3}}\right)$$

or $\lambda = 4.2 \times 10^{-6}$ cm $= 0.042\ \mu$m for this device.

Assuming complete ionization, we can write

$$p(x) = N'_A(x_0)e^{-(x-x_0)/\lambda} = N_V e^{-(E_f - E_V(x))/kT} \tag{4.18}$$

or

$$e^{-(x-x_0)/\lambda + (E_f - E_V(x))/kT} = \frac{N_V}{N'_A(x_0)} \tag{4.19}$$

Solving for $E_f - E_V$ gives

$$E_f - E_V(x) = kT\left[\frac{x}{\lambda} + \ln\frac{N_V}{N'_A(x_0)}\right] \tag{4.20}$$

This tells us the functional form of the valence band edge as a function of position. But, by Equation (4.14), the electric field is given by

$$\mathscr{E} = \frac{1}{q}\frac{dE_V}{dx} \tag{4.21}$$

Taking the derivative of Equation (4.20), and remembering that E_f is constant, we obtain an expression for the electric field as a function of position for this case:

$$\mathscr{E} = -\frac{kT}{q}\frac{1}{\lambda} \tag{4.22}$$

For this (exponential) doping profile, then, the electric field is constant with distance. At room temperature, where $kT/q = 0.026$ eV, the electric field $\mathscr{E}$ is constant with x and equal to

$$\mathscr{E} = -\frac{kT}{q}\frac{1}{\lambda} = -0.026\ \text{V}\left(\frac{1}{4.2 \times 10^{-6}\ \text{cm}}\right) = -6.2 \times 10^{3}\ \text{V/cm}$$

or 6.2 kV/cm (0.62 V/μm). This built-in field is used to decrease the electron transit time across the base and thus increase the switching speed of transistors, as discussed in Part 4 of this book.

We emphasize that the electric field is built-in, not applied. It arises naturally when the doping concentration is not uniform, because the charged carriers diffuse. The semiconductor can still be at equilibrium as long as no electric field is applied *externally.* The total current is still zero at equilibrium.

Figure 4.7 shows all of the electron and hole fluxes in this sample, and the resulting currents. Electrons diffuse to regions of lower concentration (in this case to the left), resulting in electron diffusion current to the right. Electrons drift

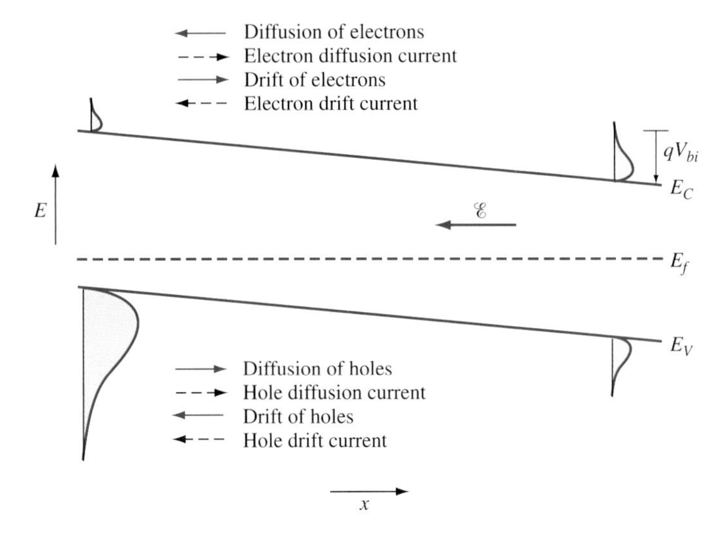

Figure 4.7 The currents that arise in the nonuniformly doped semiconductor must still sum to zero at equilibrium.

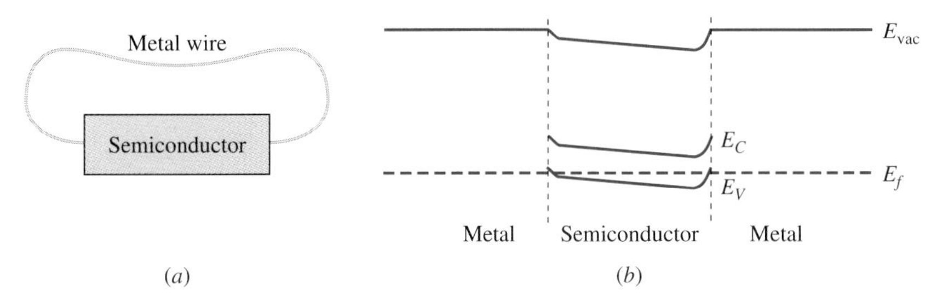

Figure 4.8 (a) A wire is used to connect one end of the graded-doping semiconductor to the other. (b) The energy band diagram. There are built-in fields at the ends of the semiconductor where it is attached to the outside environment (a metal wire in this case), but the voltages add up to zero around the loop.

in the direction opposite to the electric field (to the right) resulting in a net electron drift current to the left. Convince yourself that the directions of the arrows in the figure are correct for holes.

Note that because of the built-in field, there is also a built-in voltage across this sample. The left side is at a higher electron potential energy than the right, meaning that there is a net voltage difference V_{bi} across the sample.[1] That does not imply that a nonuniformly doped semiconductor can be a battery, however. Figure 4.8 shows why. Suppose we connect a metal wire from one end of the

[1]The q in the figure is needed because the figure shows energy in eV and the voltage V_{bi} is in volts. The factor of q keeps the units correct.

semiconductor bar to the other. The system is still at equilibrium because we have not applied any external voltages. The energy band diagram for this situation shows that the vacuum level is the same everywhere in the wire. In the semiconductor, the vacuum level, and along with it the band edges, bends to meet the vacuum level of the metal. The band bending is greater at one end of the semiconductor bar than the other. Recalling that where the bands bend there is an electric field, you immediately deduce that there are some new built-in electric fields at the junctions between the semiconductor and the metal. The built-in voltages add up to zero around the circuit in the figure (Kirchhoff's voltage rule). The electric fields are all inside the semiconductor, and thus they cannot be accessed from outside. This will happen regardless of the material chosen to connect the two ends.

*4.3 NONUNIFORM COMPOSITION

Next, consider a semiconductor whose *composition* is varied gradually from one end to the other. For example, perhaps the crystal is grown such that it starts out as pure Si ($E_g = 1.12$ eV), but Ge ($E_g = 0.67$ eV) is introduced in gradually increasing quantities until the final alloy is $\mathrm{Si}_y\mathrm{Ge}_{1-y}$, where y denotes the fractional concentration of Si. Since Si and Ge each have a valence of four, and have the same (diamond) crystal structure, the resultant Si:Ge alloy is also a semiconductor with the diamond structure, and it has some intermediate band gap. Let us also assume that the alloy is doped with acceptors such that the Fermi level is equidistant from the valence band. Note that since N_V varies only slightly with position (because of the small variation in electron effective mass as the composition changes [Equation (2.62)]), to keep $E_f - E_V$ constant, the net acceptor concentration is not quite uniform [Equation (4.9)].

To determine the equilibrium energy band diagram of the graded semiconductor alloy, we follow the procedure used in the previous section for graded doping. We first draw the diagram for the case in which electrical neutrality exists in every macroscopic region, as indicated in Figure 4.9, in which the grading in the band gap is exaggerated for clarity. With the vacuum level as a reference, we see that the conduction band edge E_C is not a function of position. This is because for the Si:Ge alloy, the electron affinity χ is essentially independent of alloy composition. The affinities are 4.05 eV (Si) and 4.0 eV (Ge). The band gap, however, *is* a noticeable function of alloy composition, since Si has a significantly different band gap (1.12 eV) than Ge (0.67 eV). Therefore as E_g decreases, E_V moves upward. It is known that E_g decreases linearly with the Ge concentration, by an amount 7.5 meV per each atomic percent of Ge. [1] In this example, the doping level is essentially uniform, so the free hole concentration p_0 is also uniform (approximately N_A'). The electron concentration, on the other hand, is not uniform, since the band gap is not constant (recall that n_i increases with decreasing E_g, and that $n_0 p_0 = n_i^2$).

Since the hole concentration p_0 is constant, the hole diffusion is negligible. However, since the valence band edge is tilted, there is a force on the holes. The

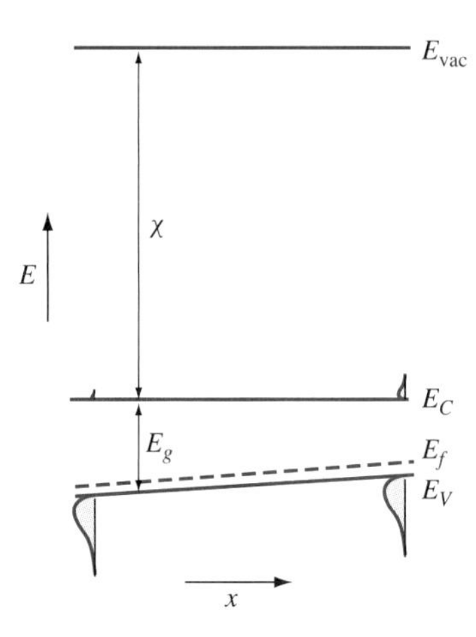

Figure 4.9 The energy band diagram for a Si:Ge alloy of nonuniform Si composition (exaggerated for clarity), under electrical neutrality.

force exerted on the holes is minus the gradient of the hole's potential energy:

$$F_h = -\frac{dE_{P(\text{holes})}}{dx} \tag{4.23}$$

and for the holes, the potential energy is the valence band edge. Thus

$$\frac{dE_{P(\text{holes})}}{dx} = -\frac{dE_V}{dx} \tag{4.24}$$

The minus sign arises from the fact that we use the energy band diagram for negatively charged electrons and thus the hole energy increases downward, since holes are positively charged. Equation (4.23) then becomes

$$F_h = \frac{dE_V}{dx} \tag{4.25}$$

This force tends to redistribute the holes, until the sample reaches equilibrium, or the Fermi level is constant with position. The case at equilibrium is illustrated in Figure 4.10.

The alert student will observe from Figure 4.10 that a hole in the valence band will not be subjected to any force at equilibrium ($F_h = dE_V/dx \approx 0$), but that an electron in the conduction band *will* experience a force accelerating it to the right:

$$F_e = -\frac{dE_{P(\text{electron})}}{dx} = -\frac{dE_C}{dx} \tag{4.26}$$

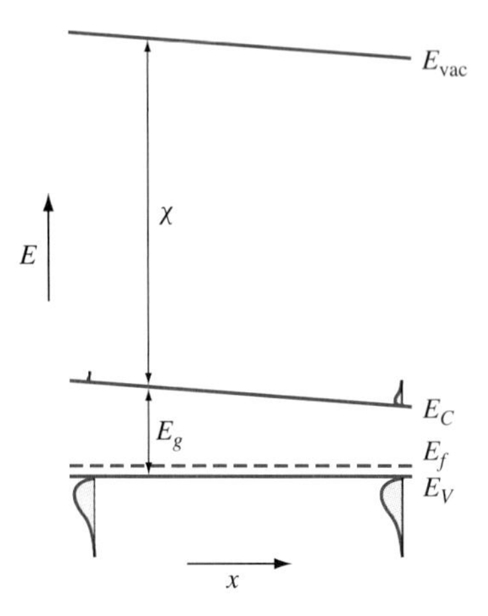

Figure 4.10 The energy band diagram for the semiconductor of nonuniform composition at equilibrium (again, exaggerated).

At equilibrium, there *is* a force on the electrons that causes drift and a gradient in the electron concentration that causes diffusion. These drift and diffusion currents cancel at equilibrium.

Since the forces on electrons and holes are different, we define *effective electric fields* $\mathscr{E}_e^*$ and $\mathscr{E}_h^*$ [2] by the equations:

$$\boxed{F_e = -q\mathscr{E}_e^* = -\frac{dE_C}{dx}} \tag{4.27}$$

for electrons and

$$\boxed{F_h = q\mathscr{E}_h^* = \frac{dE_V}{dx}} \tag{4.28}$$

for holes.

The true electric field is equal to $(1/q)(dE_{\text{vac}}/dx)$ and by definition is equal to the force on a unit positive test charge. This can be understood by imagining a small hole drilled inside the semiconductor from left to right, with a small test charge Q_T on the end of an insulating rod inserted into the hole. Since the test charge is far enough from the semiconductor surface, its potential energy is E_{vac} (its kinetic energy is zero). The force exerted on this test charge is

$$F = Q_T\mathscr{E} = -\frac{dE_P}{dx} = -\frac{dE_{\text{vac}}}{dx} \tag{4.29}$$

EXAMPLE 4.1

Consider a transistor with a Si:Ge base in which the band gap varies linearly by 0.1 eV across the 0.05-μm base width. Assume that the net acceptor concentration is constant. Find the effective electric field and force for holes and electrons, and the true electric field.

■ **Solution**

Since ΔE_g varies 7.5 meV per atomic percent Ge, this transistor has a linear grading of the Ge content in its base from zero to 13 percent. We saw earlier that for Si:Ge alloys, if the acceptor concentration is constant, the valence band level is essentially constant. Therefore, from Equation (4.28) the force (and the effective electric field) for holes is virtually zero:

$$\mathscr{E}_n^* = \frac{F_h}{q} = \frac{1}{q}\frac{dE_V}{dx} = 0$$

Since the valence band edge is flat, then as the band gap decreases, the conduction band edge energy decreases. For a base width of 0.05 μm, the resulting effective field for electrons is, from Equation (4.27),

$$\mathscr{E}_e^* = \frac{1}{q}\frac{dE_C}{dx} = \frac{1}{q}\frac{dE_g}{dx} = \frac{1}{1.6 \times 10^{-19}\ \text{C}}\frac{(-0.1\ \text{eV})(1.6 \times 10^{-19}\ \text{V/eV})}{0.05\ \mu\text{m}}$$

$$= -\frac{0.1\ \text{V}}{5 \times 10^{-8}\ \text{m}} = -2 \times 10^6\ \text{V/m} = -20\ \text{kV/cm}$$

As we will see in Part 4, such a graded-composition Si:Ge alloy is sometimes employed in the base region of high-performance bipolar junction transistors. Electrons injected from the emitter into the base are accelerated to the collector by this high field, decreasing the transit time across the base and effectively decreasing the switching time or increasing the operating frequency of the transistor.

The true electric field is given by Equation (4.29). In this case, though, the electron affinity is almost constant across the material, so E_{vac} is nearly parallel to E_C. Since their slopes are the same, in this example the true electric field is the same as the effective electric field for electrons.

$$\mathscr{E} = \frac{1}{q}\frac{dE_{\text{vac}}}{dx} \approx \frac{1}{q}\frac{dE_C}{dx} = -20\ \text{kV/cm}$$

The negative sign on the expressions for both the effective and true fields means the field is in the negative x direction.

*4.4 GRADED DOPING AND GRADED COMPOSITION COMBINED

In an actual Si:Ge transistor the field in the base is further increased by grading the acceptor concentration [3] as described in Section 4.2, which in turn speeds up the device even more, increasing the operating frequency.

EXAMPLE 4.2

A Si:Ge transistor has a base that is compositionally graded as in Example 4.1, but in addition, the net acceptor density decreases exponentially from 10^{18} cm^{-3} to 3×10^{16} cm^{-3} over this region. Find the effective electric field for electrons.

■ **Solution**

Figure 4.11 shows the energy band diagram in the base region (a) for the case of neutrality and (b) for equilibrium. Since $\Delta E_g = 0.1$ eV and the base width W_B is 0.05 μm as in Example 4.1, the effective field due to compositional grading alone is -20 kV/cm. In the previous case, E_V was considered constant with position. Here, however, there is an additional field resulting from the varying doping. Since

$$p(x) = N_A(x) = N_V e^{-(E_f - E_V(x))/kT}$$

$$E_V(x) - E_f = kT \ln \frac{N_A(x)}{N_V}$$

then the additional tilt ΔE_V to E_V due to the varying doping concentration is

$$\Delta E_V = (E_V(W_B) - E_V(0)) = (E_V(W_B) - E_f) - (E_V(0) - E_f)$$

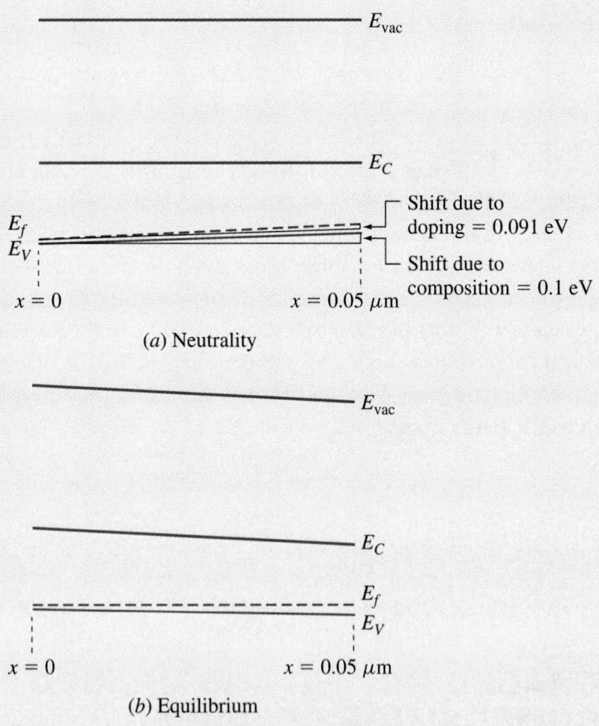

(*a*) Neutrality

(*b*) Equilibrium

Figure 4.11 The energy band diagram for the sample of Example 4.3, with both graded composition and varying doping. (a) Electrical neutrality; (b) equilibrium.

or

$$\Delta E_V = kT\left(\ln\frac{N_A(W_B)}{N_V} - \ln\frac{N_A(0)}{N_V}\right) = kT\ln\frac{N_A(W_B)}{N_A(0)}$$

$$\Delta E_V = 0.026\ln\frac{3\times10^{16}}{10^{18}} = -0.091\text{ eV}$$

The negative sign indicates the valence band edge moves down away from the Fermi level (in Figure 4.11a the Fermi level is shown moving up). Combining the field induced by the doping gradient (which moves the conduction band edge down) with the effect of the compositional grading (which moves the valence band and the conduction band down) yields

$$\Delta E_C = 0.1\text{ eV} + (-\Delta E_V) = 0.191\text{ eV}$$

Since the linear compositional grading and the exponential doping gradient each contribute to constant fields, the net quasi-electric field for electrons is

$$\mathscr{E}_e^* = \frac{0.191\text{ V}}{0.05\ \mu\text{m}} = 3.8\text{ V}/\mu\text{m} = 38\text{ kV/cm}$$

4.5 SUMMARY

In this chapter, we have extended our understanding of the physics of materials to predict what happens in nonuniform semiconductors. The results are interesting indeed. Internal electric fields can be generated in a semiconductor simply by varying the doping concentrations across the sample or by varying the material composition, or both. These fields are utilized in the base regions of bipolar junction transistors to improve their performance, as will be discussed in Part 4.

The built-in electric fields cannot be accessed externally; the sample can't be used as a battery. Furthermore, we saw that at equilibrium the drift currents produced by the built-in electric fields are balanced by opposing diffusion currents; the net electron and hole currents are still zero.

We obtained the useful concept that for a system in intimate contact, at equilibrium the Fermi level is constant throughout the system. While the concept of a Fermi level is valid only at equilibrium, analogous quasi Fermi levels can be used for nonequilibrium cases.

4.6 READING LIST

Items 1, 2, 4, 8–12, and 17–19 in Appendix G are recommended.

4.7 REFERENCES

1. John D. Cressler, "Re-Engineering Silicon: Si-Ge heterojunction bipolar technology," *IEEE Spectrum,* 32, pp. 49–55, 1995.
2. H. Kroemer, "Quasielectric and quasimagnetic fields in nonuniform semiconductors," *RCA Review,* 18, pp. 332–347, 1957.

3. D. L. Harame, J. H. Comfort, J. D. Cressler, E. F. Crabbe, J. Y. -C. Sun, B. S. Meyerson, and T. Tice, "Si-SiGe epitaxial-base transistors—Part I: Materials, physics, and circuits," *IEEE Trans. Electron Devices,* 42, pp. 455–568, 1995.

4.8 REVIEW QUESTIONS

1. Explain why the Fermi level is constant at equilibrium.
2. How does varying the doping across a sample create a built-in electric field?
3. What is meant by *quasi-neutral region?*
4. What are the steps for drawing the energy band diagram for a new system?
5. How does grading the base of a transistor improve its operation?
6. When a material has a built-in electric field, what is the current through it at equilibrium?
7. How does grading the composition create a built-in electric field?
8. Why can't a material with a built-in electric field be used as a battery?
9. What is meant by the term *effective electric field* (as for electrons)? How is it different from the true electric field?

4.9 PROBLEMS

4.1 Consider a p-type Si sample of length 0.1 μm in which the net doping varies exponentially from 5×10^{17} to 5×10^{15}. Find the electric field.

4.2 Consider the graded-composition Si:Ge alloy discussed in Section 4.3. The hole concentration is assumed constant (N'_A is essentially constant and all dopants are ionized). There is no electric field or diffusion for holes at equilibrium but there is a field for electrons in the conduction band. Since at equilibrium there is no net current, the drift electron current must be offset by an opposing diffusion current. Identify and explain the source of the varying electron concentration that produces the diffusion current.

4.3 Find the time required for an electron to traverse the p region of Example 4.2 due to drift alone.

4.4 A graded alloy is manufactured in the AlGaAs system. At $x = 0$, the material is pure GaAs ($\chi = 4.07$ eV, $E_g = 1.43$ eV), and over a distance of 2 μm the composition changes to $Ga_{0.6}Al_{0.4}As$ ($\chi = 3.84$ eV and $E_g = 1.92$ eV). The material is intrinsic. Find the effective electric fields for electrons and holes, and find the true electric field.

4.5 Consider the equilibrium energy band diagram of Figure P4.1.

a. Find the effective electric field for electrons.

b. Find the effective electric field for holes.

c. Sketch the carrier concentrations.

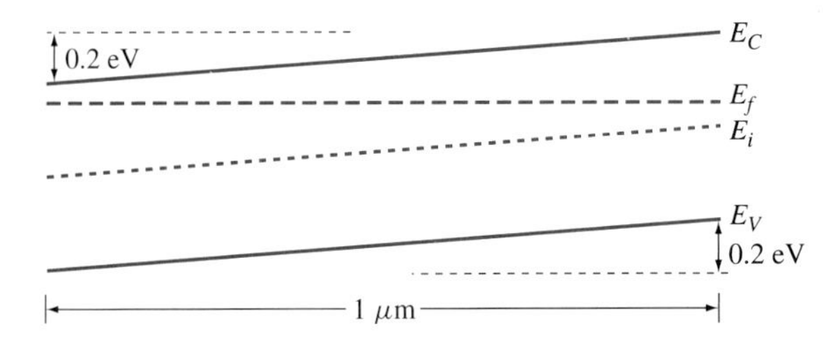

Figure P4.1

d. Indicate the directions of each of the following:
   i. Hole diffusion
   ii. Hole diffusion current
   iii. Hole drift
   iv. Hole drift current
   v. Electron diffusion
   vi. Electron diffusion current
   vii. Electron drift
   viii. Electron drift current
   ix. Electric field for electrons

4.6 For the sample whose energy band diagram is shown in Figure P4.2:
a. Find the effective electric field for electrons.
b. Find the effective electric field for holes.
c. Sketch the carrier concentrations.
d. Indicate the directions of each of the following:
   i. Hole diffusion
   ii. Hole diffusion current
   iii. Hole drift
   iv. Hole drift current
   v. Electron diffusion
   vi. Electron diffusion current
   vii. Electron drift
   viii. Electron drift current

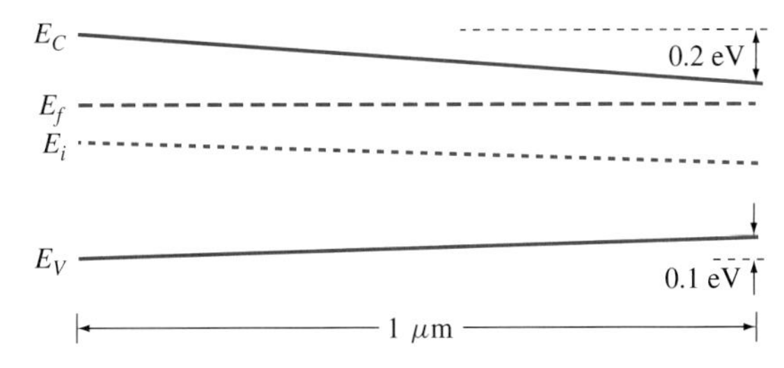

Figure P4.2

ix. Electric field for electrons

x. Electric field for holes

4.7 Draw the energy band diagram for GaAs that is doped such that $E_c - E_f = 0.2 + (0.8\ \text{eV}/\mu\text{m})x$ for $0 < x < 0.5\ \mu\text{m}$.

a. Find the effective electric field for electrons.

b. Indicate the directions of $J_{n(\text{diff})}$, $J_{n(\text{drift})}$, $J_{p(\text{diff})}$, and $J_{p(\text{drift})}$.

c. Indicate the direction of the electric field.

4.8 A sample of $Al_yGa_{1-y}As$ has a composition that is graded linearly from pure GaAs to $Al_{0.4}Ga_{0.6}As$ over a length of 10 μm. The sample is not doped. Find the value of the built-in electric field for electrons and holes. The band gap of $Al_yGa_{1-y}As$ is given by $E_g = 1.43 + 1.425y$ (in eV) for $y < 0.43$. Note the y here is composition fraction, not distance. Assume that the electron affinity is constant and E_i is at midgap.

4.9 An optoelectronic device will be made of layers of $Ga_xIn_{1-x}As_yP_{1-y}$ grown on an InP substrate. To make sure the layers will have the same lattice constant as the substrate, the gallium fraction x is kept equal to $0.47y$. The arsenic fraction is varied among the layers from $y = 0$ to $y = 0.5$. The band gap varies as $E_g = 1.35 - 0.72y + 0.12y^2$ eV. Assuming for convenience that the effective masses are constant at m^*_{dse} and m^*_{dsh}, find the ratio of $n_i(y = 0)$ to $n_i(y = 0.5)$ from the pure InP layers to the $y = 0.5$ layer.

4.10 Consider the Si:Ge alloy discussed in Example 4.2. If the acceptor concentration varies exponentially from 10^{18} cm^{-3} to 2×10^{16} cm^{-3}, show that the total effective electric field for electrons is $\mathscr{E}^*_e = \mathscr{E}^*_{\text{doping}} + \mathscr{E}^*_{\text{composition}}$.

4.11 We will discuss pn junctions in Chapter 5. However, you already have enough knowledge to deduce some things from the energy band diagram, shown in Figure P4.3.

a. In what region is there an electric field?

b. What is the value of the built-in voltage V_{bi}?

c. Sketch the electron and hole concentrations. Are the directions of the drift and diffusion components of the electron and hole currents correct for equilibrium?

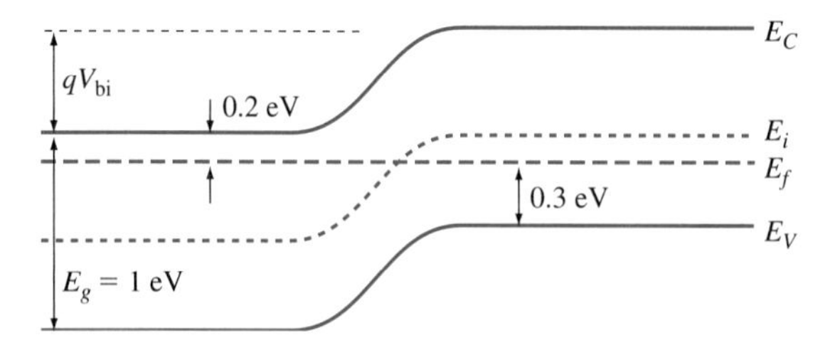

Figure P4.3

Supplement to Part 1: Materials

Throughout this book, supplements are used to expand on the material in each part of the book. In this one, several items are discussed.

In Chapter 1, quantum mechanics was used to solve two problems of interest for electrons in semiconductor crystals: the simplified free-electron model and the more realistic quasi-free electron model. In the first section of this supplement (A), quantum mechanics is discussed in slightly more detail and some specific (though simplified) problems are investigated for electron behavior in semiconductor devices. Some importance is placed on quantum mechanical tunneling of electrons since it has considerable influence on the electrical characteristics of several electronic devices and is a factor in some device failure mechanisms.

In the second part, supplement (B), carrier concentrations and mobilities are revisited. In Chapter 2, we indicated how to *calculate* semiconductor properties from their carrier concentrations and mobilities. So, we will first show one method used to measure the majority carrier concentrations and mobilities in the laboratory.

Second, Fermi-Dirac statistics for bound states are discussed. Recall that in our previous discussion of the probability of occupancy and the resulting carrier distributions, the assumption was made that, whether a state was occupied or not occupied, its occupancy had no effect on the probability of occupancy of any other states. For the dopants, however, this does not apply. Consider donors as an example. Each donor atom has several possible energy states in the forbidden band, but if an electron occupies any one of them, the donor is no longer ionized and, in effect, the other states cease to exist. Thus, in this supplement it is shown how to handle bound states.

Third, we illustrate that except at low temperatures ($T < 200$ K) in common semiconductors, essentially all donors and acceptors are ionized.

Finally, the role of phonons on electronic behavior is discussed.

1A SUPPLEMENT

Introduction to Quantum Mechanics

S1A.1 INTRODUCTION

A reasonable question in the mind of the student might be, "Why should I know quantum mechanics?"

You saw in Chapter 1 how an understanding of the quantization of the electron states in an atom led eventually to the understanding of energy bands in a crystal. The energy band concept is used repeatedly throughout this book. You also saw how different materials could be expected to have different band structures. The energy band structure can also be changed by the presence of electric fields, crystal defects, impurities, etc. You will want to be able to predict the behavior of electrons and holes when they encounter these disturbances, because it is precisely these disturbances that make transistors, lasers, and all semiconductor devices work.

It turns out that the classical (Newtonian) mechanics you learned in freshman physics does not always reliably predict the behavior of the charge carriers (electrons and holes). This is because those carriers have such small sizes, and these nonclassical effects become even more important when the disturbances are also on a scale on the order of nanometers, as occurs very frequently. A qualitative understanding of electron behavior based on quantum mechanics is essential for an understanding of device operation. In Supplement 1A, a few principles are presented that will allow you to predict what electrons (and holes) are going to do in the presence of various energy band structures.

S1A.2 THE WAVE FUNCTION

All matter can be thought of as consisting of either particles or waves. Since the particle description of electrons is intuitive, let us discuss in more detail the other case, in which we consider electrons to be behaving as waves. In quantum mechanics, we say the electron can be described by a wave function $\Psi(x, y, z, t)$. For simplicity, unless the three-dimensional spatial formulation is required for the concepts being introduced, the one-dimensional formulation will be used here. The concept of wave function is somewhat disturbing the first time one comes across it, partly because it cannot be directly measured. We will see, however, that knowledge of the wave function along with the *operators* discussed below permits us to predict the behavior of electrons.

Inherent in quantum mechanics is the philosophy that only properties of particles (waves) that can be measured, called *observables,* are meaningful. The rule for calculating the average (of several measurements) or expected value for some observable quantity O is:

$$\langle O \rangle = \frac{\int \Psi^*(x,t) O_{\text{op}} \Psi(x,t)\, dx}{\int \Psi^*(x,t)\Psi(x,t)\, dx} \tag{S1A.1}$$

where Ψ^* is the complex conjugate of the wave function, O_{op} is an operator associated with the observable O and operates on the wave function, and the integration is taken over all space. Table S1A.1 shows the operators for various

Table S1A.1 Quantum mechanical operators

Observable	Operator
x	x
y	y
z	z
r	r
$f(r)$	$f(r)$
v_x	$\frac{\hbar}{jm}\frac{\partial}{\partial x}$
v_y	$\frac{\hbar}{jm}\frac{\partial}{\partial y}$
v_z	$\frac{\hbar}{jm}\frac{\partial}{\partial z}$
$p_x = mv_x$	$\frac{\hbar}{j}\frac{\partial}{\partial x}$
$\vec{p} = \hat{\imath}p_x + \hat{\jmath}p_y + \hat{k}p_z$	$\vec{p}_{\text{op}} = \hat{\imath}p_{x_{\text{op}}} + \hat{\jmath}p_{y_{\text{op}}} + \hat{k}p_{z_{\text{op}}}$
v_x^2	$-\frac{\hbar^2}{m^2}\frac{\partial^2}{\partial x^2}$
p_x^2	$-\hbar^2\frac{\partial^2}{\partial x^2}$
$E_K = \frac{mv^2}{2} = \frac{p^2}{2m} = \frac{\vec{p}\cdot\vec{p}}{2m}$	$-\frac{\hbar^2}{2m}\left(\frac{\partial^2}{\partial x^2} + \frac{\partial^2}{\partial y^2} + \frac{\partial^2}{\partial z^2}\right) = -\frac{\hbar^2}{2m}\nabla^2$
E	$-\frac{\hbar}{j}\frac{\partial}{\partial t}$

observable quantities. Some of the operators, for example, those for position observables, are merely multiplication factors. Others perform an operation such as taking a derivative. We will do an example in the next section, after we have further developed the idea of wave functions.

S1A.3 PROBABILITY AND THE WAVE FUNCTION

As indicated in Section 1.5.1, the probability that a particle will be found in region dx near the coordinate x at time t is

$$\boxed{P(x,t)\,dx = \Psi^*(x,t)\Psi(x,t)\,dx} \tag{S1A.2}$$

where the function $P(x,t)$ is called the *probability density*. Since the particle must be somewhere, the probability of finding it in space at any given time is unity. That is, integrating over all space,

$$\boxed{\int P(x,t)\,dx = \int \Psi^*(x,t)\Psi(x,t)\,dx = 1} \tag{S1A.3}$$

in which case the wave function is said to be *normalized*. While Equation (S1A.1) is valid for any wave function, for normalized wave functions, the denominator is equal to unity and Equation (S1A.1) has the simpler form

$$\boxed{\langle O\rangle = \int \Psi^* O_{\text{op}} \Psi\,dx} \tag{S1A.4}$$

*S1A.3.1 PARTICLE IN A ONE-DIMENSIONAL POTENTIAL WELL

As an example of applying the operators, we consider an electron of energy E in a potential energy configuration in which the potential energy is some constant $E_P = E_0$ in a region of length L, as shown in Figure S1A.1a, and infinitely high everywhere else. Such a structure is called a *potential well,* and the electron inside the well cannot escape. The electron in the well is traveling at some constant and finite energy, so it cannot cross the infinite potential barriers. This is a crude approximation for the potential energy of an electron in an atom (See Figure 1.2), except that here the sides are vertical and go to infinity. From physical intuition, then, we expect that the electron is reflected at each wall and oscillates back and forth. Later, after we have discussed how to find the wave function for a particle, we will show how to find the non-normalized wave functions of the electrons in

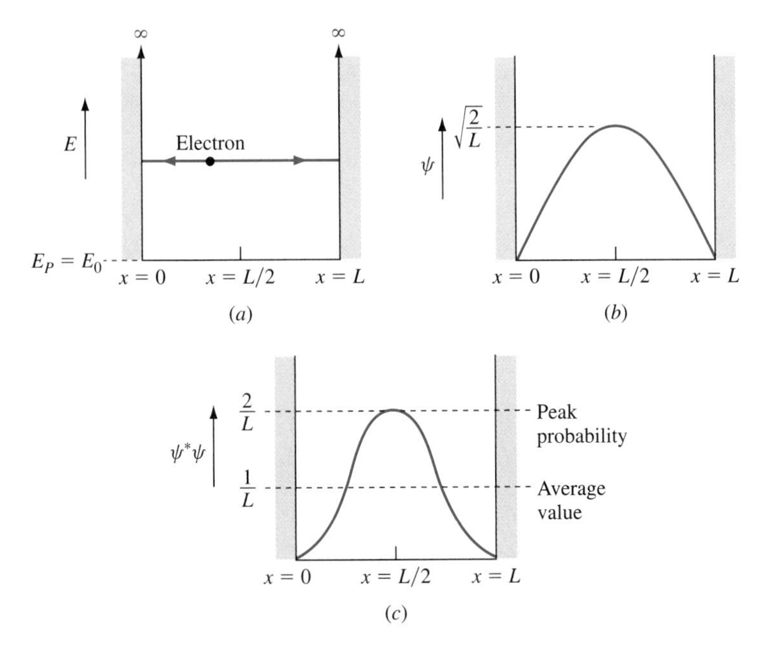

Figure S1A.1 Particle in a one-dimensional well. (a) The potential energy diagram. The electron oscillates back and forth between the walls at constant energy. (b) The wave function for the lowest allowed energy state. (c) The probability distribution function.

this potential well. For now, let us accept that the result for the lowest energy state in this problem is[1]

$$\Psi(x,t) = C \sin\left(\frac{\pi x}{L}\right) e^{-j(E/\hbar)t} \qquad 0 < x < L$$

$$\Psi(x,t) = 0 \qquad \text{elsewhere}$$

where C is some constant. Since we know that the electron has to be inside the well (it does not have enough energy to escape under any circumstance), the probability is unity that at any time it is somewhere in the well, or

$$\int_0^L \Psi^* \Psi \, dx = \int_0^L \psi^* \psi \, dx = 1$$

Here we introduce ψ, the spatial part of wave function Ψ. That is, ψ depends only on x, and the time dependence is treated separately. That is,

$$\Psi(x,t) = \psi(x) e^{-j(E/\hbar)t}$$

The time dependence will be derived explicitly later.

[1]This results from the complete reflection of the traveling waves at the boundaries of the well where $E_p = \infty$, producing the standing wave as indicated.

We can determine the value of C from the equation by recalling the probability function integrated over all space must be unity:

$$\int_0^L \Psi^*\Psi\, dx = \int_0^L \psi^*\psi\, dx = \int_0^L C^* \sin\left(\frac{\pi x}{L}\right) C \sin\left(\frac{\pi x}{L}\right) dx = 1$$

We simplify our work with a substitution of variables by letting $y = (\pi x/L)$. Now the above equation becomes

$$\frac{C^*CL}{\pi} \int_0^\pi \sin^2 y\, dy = 1$$

Since this integral is equal to $\pi/2$, we obtain $C^*C = 2/L$, or

$$C = \sqrt{\frac{2}{L}}$$

Figure S1A.1b shows a plot of the wave function $\psi(x)$, and the probability density function $\psi^*\psi$ is shown in Figure S1A.1c for this case. Note that the maximum value of $\psi^*\psi$ is $2/L$. The average value of $\psi^*\psi$ is $1/L$ as indicated. This average value, times the well length L, gives the expected unity probability that the particle is in the well.

Now, we find the average position of the electron. Using the normalized wave function and Equation (S1A.4) [we have already normalized the wave function, so it is proper to use Equation (S1A.4) instead of Equation (S1A.1)], we find

$$\langle x \rangle = \int_0^L \Psi^*(x,t) \cdot x \cdot \Psi(x,t)\, dx = \int_0^L \left(\frac{2}{L}\right)^{1/2} \sin\frac{\pi x}{L} \cdot x \cdot \left(\frac{2}{L}\right)^{1/2} \sin\frac{\pi x}{L}\, dx$$

$$= \frac{2}{L} \int_0^L x \sin^2\left(\frac{\pi x}{L}\right) dx$$

This integral is slightly different from the one we used to find C, but we proceed the same way. Again letting $y = \pi x/L$, and using an integral table to find that $\int_0^\pi y \sin^2 y\, dy = \pi^2/4$, we obtain $\langle x \rangle = L/2$ as expected. Thus, the particle's average position of the electron is in the center of the well.

S1A.4 SCHROEDINGER'S EQUATION

We saw that if the wave function for a quantum mechanical particle is known, one can calculate the expected (or average) value of the physical observables, such as position, speed, and momentum. The challenge in quantum mechanics is finding out what the wave functions are for a particular particle in some given situation. Once the wave functions are known, we can determine everything (observables) about the behavior of the particle, with the use of Equation (S1A.1) and the appropriate operator for the observable from Table S1A.1.

The wave functions are found from Schroedinger's equation, which can be expressed for a one-dimensional problem as

$$-\frac{\hbar^2}{2m}\frac{\partial^2\Psi(x,t)}{\partial x^2} + E_{P_{op}}\Psi(x,t) = -\frac{\hbar}{j}\frac{\partial\Psi(x,t)}{\partial t} \qquad \text{Schroedinger's equation}$$

(S1A.5)

Notice that $-(\hbar^2/2m)(\partial^2/\partial x^2)$ is the kinetic energy operator (Table S1A.1), that $-(\hbar/j)(\partial/\partial t)$ is the total energy operator, and that $E_{P_{op}}$ is the potential energy operator. Then Schroedinger's equation can be expressed

$$E_{K_{op}}\Psi + E_{P_{op}}\Psi = E_{op}\Psi \tag{S1A.6}$$

This is analogous to the relation in classical mechanics that the kinetic energy plus the potential energy is equal to the total energy:

$$E_K + E_P = E$$

Now, the physical problem to be solved is described through the potential energy, for example the potential energy of a free electron, an electron in an atom, or an electron in a crystal. The potential energy operator depends on the particular problem being solved. The potential energy might vary in space, for example, the way it does in the vicinity of an atom, as seen before in Figure 1.2. For many problems, E_P can be considered to be independent of time to first approximation. In this case

$$-\frac{\hbar^2}{2m}\frac{\partial^2\Psi(x,t)}{\partial x^2} + E_P(x)\Psi(x,t) = -\frac{\hbar}{j}\frac{\partial\Psi(x,t)}{\partial t} \tag{S1A.7}$$

where from Table S1A.1, $E_{P_{op}}(x) = E_P(x)$.

S1A.5 APPLYING SCHROEDINGER'S EQUATION TO ELECTRONS

We now have the elements of a procedure for finding the values of observable quantities for electrons. The procedure consists of three steps:

1. Determine the potential energy function. This can be done from knowledge of the forces acting on an electron and considering it to be a point charge.

Since the force is minus the gradient of the potential energy,

$$F = -\nabla E_P \rightarrow -\frac{dE_P}{dx} \tag{S1A.8}$$

2. Use the potential energy function in Schroedinger's equation and solve to find the wave function Ψ.
3. Insert the wave function Ψ into Equation (S1A.1) along with the appropriate operator to find the value of the corresponding observable.

While in principle this appears straightforward, only a few physical problems have simple analytical solutions in closed form. There is some value, however, in considering simplified cases—they may not be quantitatively realistic, but they can provide qualitative insight.

For the case in which the potential energy is not a function of time, Schroedinger's equation can be solved in part by using the technique of separation of variables. The result is that the wave function can be expressed as the product of two functions, one time-independent $\psi(x)$ and one space-independent $T(t)$:

$$\boxed{\Psi(x, t) = \psi(x)T(t)} \tag{S1A.9}$$

where the lowercase ψ represents the time-independent wave function as before, and uppercase Ψ the complete wave function. Equation (S1A.9) is inserted into Schroedinger's equation, which is then solved. The result is two separate ordinary differential equations, each with the same constant of separation, E.

The time-independent equation is

$$-\frac{\hbar^2}{2m}\frac{d^2\psi(x)}{dx^2} + E_P(x)\psi(x) = E\psi(x) \tag{S1A.10}$$

and the time-dependent equation is

$$-\frac{\hbar}{j}\frac{\partial T(t)}{\partial t} = ET(t) \tag{S1A.11}$$

We can solve Equation (S1A.11) immediately. The solution to the time-dependent equation is

$$\boxed{T(t) = e^{-j(E/\hbar)t}} \tag{S1A.12}$$

where E is the constant of separation. This constant E is the total energy of the electron.

The time-independent wave function $\psi(x)$ can be found from the time-independent Schroedinger equation, Equation (S1A.10). The solution to this one, however, depends on the particular form of $E_P(x)$, which in turn depends on the particular problem being solved.

Once we have the appropriate expression for $E_P(x)$, Schroedinger's equation can be solved by using the boundary conditions appropriate to the problem.

There are two restrictions on the solutions of wave functions $\Psi(x, t)$ and $\psi(x)$:

1. The wave function must be finite, single-valued, and continuous in space.
2. The gradient of the wave function must be continuous.

We will present the results of some interesting (and important) problems in the next section.

S1A.6 SOME RESULTS FROM QUANTUM MECHANICS

In Section 1.6, Schroedinger's equation was solved for two cases of interest: the free electron, in which the electron potential energy was considered constant with position (E_0), and the more realistic model for electrons in crystals, the quasi-free electron model in which the potential energy was considered to be a periodic function of position with the periodicity of the lattice. Some of the major results for these two cases are repeated here.

S1A.6.1 THE FREE ELECTRON

The wave function for the free electron can be expressed as [recall Equation (1.42)]

$$\Psi(x, t) = Ae^{j[Kx-(E/\hbar)t]} \quad \text{one dimension} \tag{S1A.13}$$

$$\Psi(\vec{r}, t) = Ae^{j[\vec{K}\cdot\vec{r}-(E/\hbar)t]} \quad \text{three dimensions} \tag{S1A.14}$$

where A is a constant and represents the amplitude of the wave function. The total electron energy is represented by E and the wave vector by K, where the magnitude of K is [Equation (1.44)]

$$K = \frac{2\pi}{\lambda} = \sqrt{\frac{2m_0(E - E_0)}{\hbar^2}} = \sqrt{\frac{2m_0 E_K}{\hbar^2}} \tag{S1A.15}$$

and the direction of K represents the direction of motion. Here, the kinetic energy is $E_K = (E - E_0)$.

The group velocity, $v_g = v$ of the electron wave is the velocity of its center of mass:

$$v_g = \frac{1}{\hbar}\frac{dE}{dK} \quad \text{one dimension}$$

$$v_g = \frac{1}{\hbar}\nabla_K E \quad \text{three dimensions} \tag{S1A.16}$$

where $\nabla_K E$ represents the gradient of E in K space.

The total energy can be expressed as

$$E = E_0 + \frac{\hbar^2 K^2}{2m_0} \qquad \text{one dimension}$$

$$E = E_0 + \frac{\hbar^2}{2m_0}\left(K_x^2 + K_y^2 + K_z^2\right) = E_0 + \frac{\hbar^2 K^2}{2m_0} \qquad \text{three dimensions}$$

(S1A.17)

Equations (S1A.17) are parabolas in one dimension (recall Figure 1.13) and parababoloids (three dimensions) respectively. All are with minima at $K = 0$ and $E = 0$. The mass of the free electron is related to the curvature of the E-K plot:

$$m_0 = \hbar^2 \left(\frac{d^2 E}{dK^2}\right)^{-1} \tag{S1A.18}$$

EXAMPLE S1A.1

Show that the momentum of the free electron is given by $\hbar K$. We will do this by calculating the average momentum of the free electron using quantum mechanics.

■ Solution

We use Equation (S1A.1) and the momentum operator from Table S1A.1. For an electron traveling in the positive x direction, the non-normalized wave function in one dimension is $\Psi(x, t) = Ae^{j[Kx-(E/\hbar)t]}$. The average momentum is then

$$\langle p_x \rangle = \frac{\displaystyle\int [A^* e^{-j[Kx-(E/\hbar)t]}]\frac{\hbar}{j}\frac{d}{dx}[Ae^{j[Kx-(E/\hbar)t]}]\,dx}{\displaystyle\int [A^* e^{-j[Kx-(E/\hbar)t]}][Ae^{j[Kx-(E/\hbar)t]}]\,dx}$$

$$= \frac{\hbar}{j}(jK)\frac{\displaystyle\int [A^* e^{-j[Kx-(E/\hbar)t]}][Ae^{j[Kx-(E/\hbar)t]}]\,dx}{\displaystyle\int [A^* e^{-j[Kx-(E/\hbar)t]}][Ae^{j[Kx-(E/\hbar)t]}]\,dx} \qquad \text{free electron}$$

$$= \hbar K$$

Thus, for the free electron, the quantity $\hbar K$ is the momentum of the particle.

Similarly, for a free electron in three dimensions,

$$\vec{p} = \hbar \vec{K} \qquad \text{free electron in three dimensions}$$

since the curvature is independent of direction of K.

S1A.6.2 THE QUASI-FREE ELECTRON

Recall from Chapter 1 that the solution to Schroedinger's equation for an electron in a periodic potential such as a crystal is a Bloch wave function

[Equation (1.56)]:

$$\Psi(x,t) = U_K(x)e^{j[Kx-(E/\hbar)t]} \tag{S1A.19}$$

$$\Psi(\vec{r},t) = U_K(\vec{r})e^{j[\vec{K}\cdot\vec{r}-(E/\hbar)t]} \tag{S1A.20}$$

where U_K is a function with the periodicity of the crystal. Here the E-K relation cannot be determined unless $E_P(x)$ or $E_P(\vec{r})$ is known. However, as for a free electron (recall Section 1.6.4),

$$|K| = \frac{2\pi}{\lambda} \tag{S1A.21}$$

and the group velocities are again

$$\begin{aligned} v_g &= \frac{1}{\hbar}\frac{dE}{dK} && \text{one dimension} \\ v_g &= \frac{1}{\hbar}\nabla_K E && \text{three dimensions} \end{aligned} \tag{S1A.16}$$

The quasi-free electron model predicts multiple allowed energy bands, each with its own E-K relation.

Within any band, the E-K relation is periodic in K space with periodicity

$$\frac{2\pi}{a} \qquad \text{one dimension}$$

$$\left(\frac{2\pi}{a}, \frac{2\pi}{b}, \frac{2\pi}{c}\right) \qquad \text{three dimensions}$$

S1A.6.3 THE POTENTIAL ENERGY WELL

We have so far considered the free electron and the quasi-free electron, which were covered in Chapter 1. In both models, we considered the case of an electron deep within the crystal such that the surface effects were neglected. Here we will consider the effects of the surfaces, which constitute high potential barriers to the electron. The electrons are essentially trapped in the resulting potential well. This situation is of considerable importance in many semiconductor devices.

In Figure S1A.2a we show the case of a one-dimensional crystal indicating the periodic potential within the crystal and the potential energy at the surface. Physically, we expect that an electron approaching the surface will be decelerated and repelled, because the force is equal to minus the gradient of the potential energy. In Figure S1A.2a, the slope of the left potential energy barrier, for example, is negative, so the force is in the positive x direction, or to the right. An electron traveling toward the left surface will be decelerated and turned around.

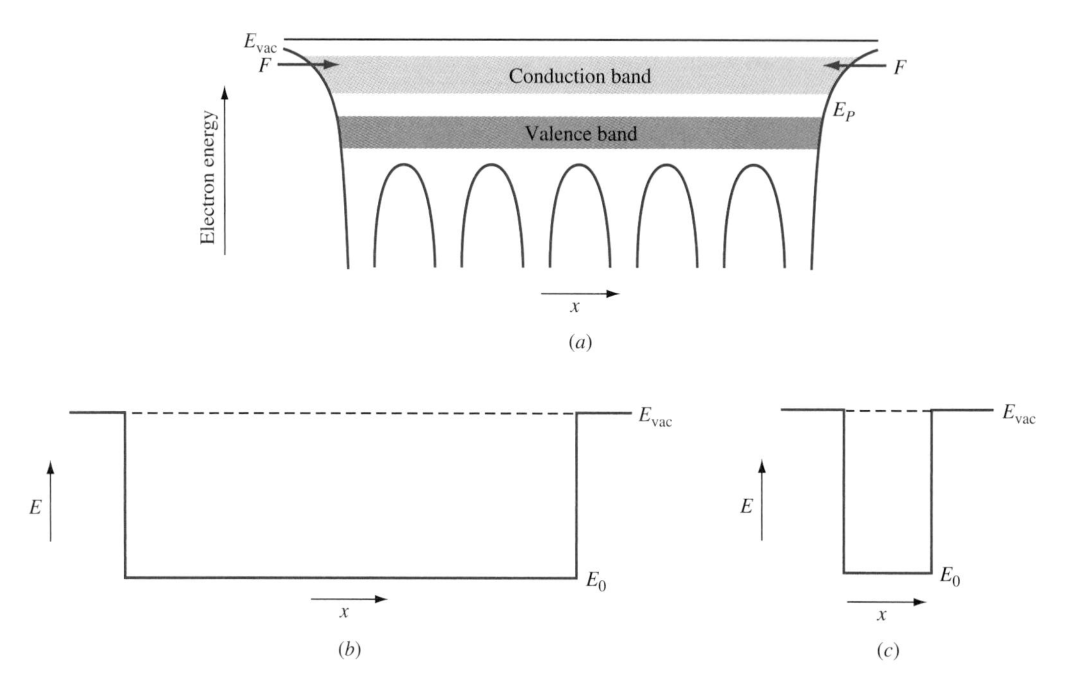

Figure S1A.2 The potential energy for an electron in a crystal. (a) The actual potential, including the effects of the surfaces; (b) the idealized potential, which is easier to solve. (c) The potential energy diagram for an electron in an idealized but thin crystal.

At the right edge, the slope is positive and therefore the force is in the negative x direction. The electron is trapped in the well. Between collisions, electrons further than about a mean free path from the surface will behave as quasi-free particles.

The potential energy distribution in Figure S1A.2a would be hard to describe analytically, but we could simplify the problem as shown in Figure S1A.2b. Here we replace the surface potentials with abrupt potential energy barriers, and replace the periodic crystal potential with a constant potential energy.

When we considered the quasi-free electron, we assumed the material was so thick that the electron never came near enough to the surfaces to be influenced by those barriers. We are now going to complicate the problem by letting the material be so thin that the electron is definitely influenced by the interfaces at the surface; i.e., the electron mean free path is much larger than the width of the well. Consider a very thin sheet of this crystalline material, Figure S1A.2c. There are two symmetrical energy barriers here, but now they are very close together (in space), on the order of a few lattice constants. These structures are approximations to quantum wells in light-emitting diodes and lasers, and to some field-effect transistors to be discussed later.

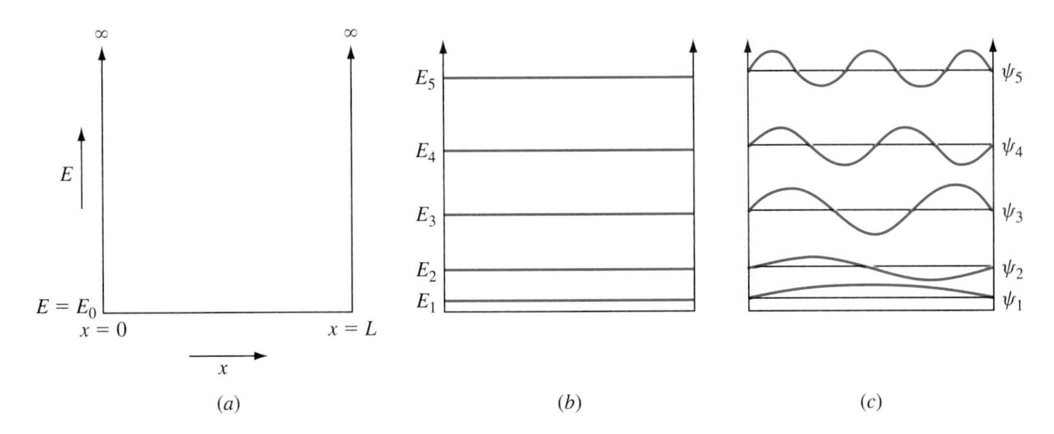

Figure S1A.3 The infinite potential well. (a) The potential energy is E_0 everywhere except at $x \leq 0$ and $x \geq L$, where the potential is infinite; (b) the first five energy levels for the electron in the well; (c) the corresponding wave functions.

S1A.6.4 THE INFINITE POTENTIAL WELL IN ONE DIMENSION

In Example S1A.1, we presented the infinite potential well and gave the result for the wave function of the electron in the lowest allowed state. We revisit that problem now, but this time we will solve it. Recall that we assumed the barriers are so high in energy that they appear infinite. This problem is not very realistic but it is easy to solve.

The potential energy diagram for the one-dimensional infinite potential well is shown in Figure S1A.3a. We can write the potential energy as

$$E_P(x) = E_0 \qquad 0 \leq x \leq L \tag{S1A.22}$$

$$E_P(x) = \infty \qquad x \leq 0 \text{ and } x \geq L \tag{S1A.23}$$

Now that we have an expression for the potential energy, we can write the time-independent Schroedinger's equation [Equation (S1A.10)]:

$$-\frac{\hbar^2}{2m^*}\frac{d^2\psi(x)}{dx^2} + E_0\psi(x) = E\psi(x) \qquad 0 \leq x \leq L \tag{S1A.24}$$

$$-\frac{\hbar^2}{2m^*}\frac{d^2\psi(x)}{dx^2} + (\infty)\psi(x) = E\psi(x) \qquad x \leq 0 \text{ and } x \geq L \tag{S1A.25}$$

The time dependence is $e^{-(jE/\hbar)t}$ as usual [Equation (S1A.12)], since the potential energy is not changing with time.

The electron cannot possibly be found outside the well, since it would have to have infinite potential energy. Since the probability distribution function for the electron is $\psi^*\psi$, this quantity must be zero outside the well. Therefore, the only solution in these two regions is $\psi = 0$.

Schroedinger's equation for the region inside the well, Equation (S1A.24), is the same as the equation for the free electron [Equation (1.36)] because the potential is constant in this region. The only difference is that here we have used the effective mass m^* instead of m_0 since this electron is not strictly free. The solution was found earlier [Equation (1.38)] as:

$$\psi(x) = Ae^{jKx} + Be^{-jKx} \tag{S1A.26}$$

where

$$K = \pm\sqrt{\frac{2m^*}{\hbar^2}(E - E_0)} \tag{S1A.27}$$

Note that in this case this solution applies only inside the well.

We can expect that the electron will oscillate back and forth inside the well, so we must keep both the positive-traveling and the negative-traveling solutions in Equation (S1A.27).

Next, we invoke the boundary conditions. Recall that $\psi(x)$ must be continuous. Since we have established that the wave function $\psi(x)$ is zero outside the well, $\psi(x)$ must also go to zero at $x = 0$ and $x = L$ for the solution inside the well. With these boundary conditions on Equation (S1A.26) we find

$$\psi(x = 0) = 0 = A + B \tag{S1A.28}$$

$$\psi(x = L) = 0 = Ae^{jKL} + Be^{-jKL} \tag{S1A.29}$$

From the first of these,

$$A = -B \tag{S1A.30}$$

meaning

$$\psi(x) = A(e^{jKx} - e^{-jKx}) = -2jA\sin(Kx) = C\sin(Kx) \tag{S1A.31}$$

where C is a constant $(C = -2jA)$, and we have used the Euler relation $\sin\theta = (e^{j\theta} - e^{-j\theta})/2j$. The second condition, (S1A.29), can be satisfied only for

$$KL = n\pi \qquad n = \pm 1, \pm 2, \pm 3, \ldots \tag{S1A.32}$$

This means that the wave vector K is quantized (takes on discrete values) and can have only the particular values

$$\boxed{K = \frac{n\pi}{L}} \tag{S1A.33}$$

Now, from Equation (S1A.27), we know that K is related to the kinetic energy E_K of the electron, since $E_K = E - E_0$, so therefore the kinetic energy is also quantized:

$$E_K = (E - E_0) = \frac{\hbar^2 K^2}{2m^*} = \frac{\hbar^2\pi^2}{2m^* L^2}n^2 \tag{S1A.34}$$

Only discrete energies are allowed, just as we found for the atom. The first few energies are shown in Figure S1A.3b. The actual energies are different from the case for the hydrogen atom, since the basic problems are different, but the quantization of the energy levels is a common feature.

Since n can be any integer, there are an infinite number of solutions to the time-independent Schroedinger's equation in the infinite potential well:

$$\psi_n(x) = C \sin(K_n x) = C \sin\left(\frac{n\pi x}{L}\right) \tag{S1A.35}$$

each of which has the form of a standing wave. The first five solutions are shown in Figure S1A.3c.

The complete solution contains the time dependence as well. From Equations (S1A.9) and (S1A.12),

$$\Psi_n(x, t) = C \sin(K_n x) e^{-j(E_n/\hbar)t} \tag{S1A.36}$$

Note that the wave vector K_n can be positive or negative, depending on the sign of n. From Equation (S1A.34) then, we see that a given magnitude of n ($n = |\pm 1|, |\pm 2|, \ldots$) results in the same value of E_K.

Also, notice that the wider the well gets (L), the closer together the energy levels get, as indicated in Figure S1A.4. For an infinite L [part (c) of the figure], we'd have a continuum of allowed energies. As the well gets narrow enough, the states separate into discrete energy levels [(a) and (b)]. For such a case, the potential well is referred to as a *quantum well.*

The subscript n is a quantum number, because it identifies which quantum state is being discussed. Each wave function ψ_n is called an *eigenfunction,* describing a particular state (*eigenstate*); each quantized parameter such as K_n and E_n is called an *eigenvalue.*

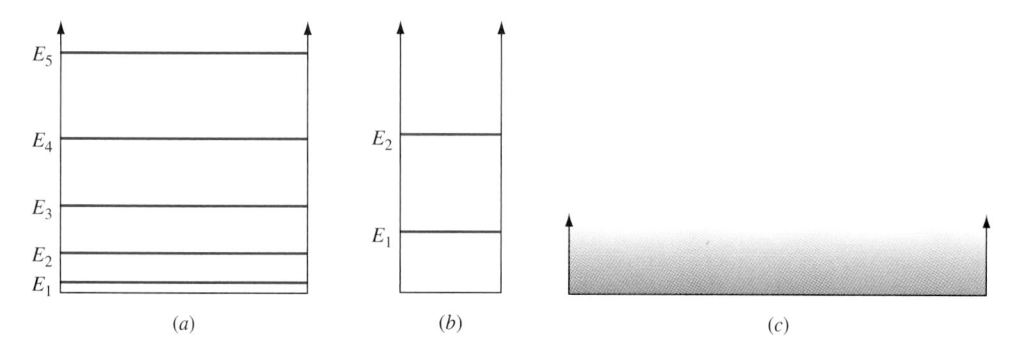

Figure S1A.4 The effect of the well width on the energy levels in an infinite potential well. (a) Some discrete states; (b) as the well gets narrowed the energy levels are spaced further apart; (c) for a very wide well the energy levels are so close together they appear to be quasi-continuous (an energy band).

S1A.6.5 REFLECTION AND TRANSMISSION AT A FINITE POTENTIAL BARRIER

We saw that the electron wave in the infinite well reflected off the two infinite barriers and oscillated back and forth in the well. We will now consider a case where the electron approaches a potential barrier that is finite. There are two possible configurations for this problem. The electron approaching the barrier can have enough energy to go over the barrier, or the electron's energy can be smaller than the barrier.

We begin by assuming the electron has energy *higher* than the barrier. Consider an electron in a material, traveling toward one of the surface barriers, as in Figure S1A.5. We assume an abrupt barrier of finite height. Classical mechanics would predict that the electron would continue past the barrier, although with lower kinetic energy.

As would be suspected, the results will be different when we consider very small particles like electrons, where classical mechanics doesn't apply. In quantum mechanics, we find that sometimes the electron will go past the barrier, and sometimes it will turn back. For any given approach to the barrier, we cannot say for certain whether the electron will be reflected or continue on. Using Schroedinger's equation, however, we can predict the probability of reflection or transmission for any given attempt at the barrier by the electron.

As always, we need to know the wave function to predict the behavior of the electron. The incident electron is in a region of constant potential energy. Until it gets close enough to be influenced by the barrier, it looks like a free electron. We can write the wave function ψ_i for the incident electron, having positive K since it is traveling to the right, as:

$$\psi_i = Ae^{jK_i x} \tag{S1A.37}$$

where

$$K_i = \sqrt{\frac{2m}{\hbar^2} E_{Ki}} \tag{S1A.38}$$

If the electron wave is reflected, we can write

$$\psi_r = Be^{-jK_r x} \tag{S1A.39}$$

where $K_r = K_i$ since $E_{Kr} = E_{Ki}$.

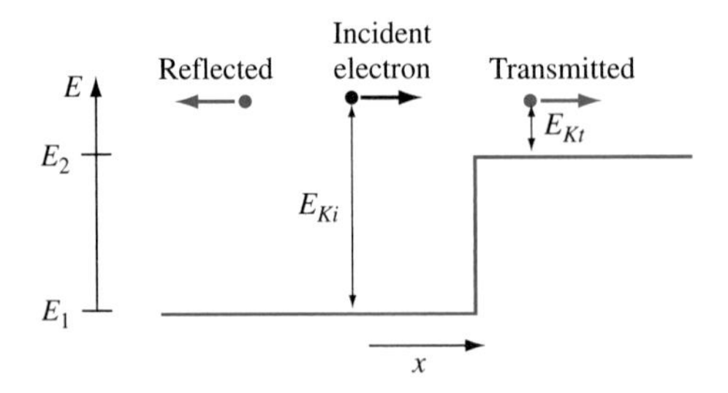

Figure S1A.5 Reflection and transmission of an electron by a finite potential barrier.

For the transmitted wave, the wave function reflects the changed potential and kinetic energies:

$$\psi_t = Ce^{jK_t x} \tag{S1A.40}$$

where the subscripts i, r, and t refer to incident, reflected, and transmitted waves respectively.

We have expressions now for the wave functions for the electron whether it is incident, reflected, or transmitted. The relationships between coefficients A, B, and C will tell us the probability of reflection and transmission. These can be found by using the boundary conditions, and the analysis is the subject of a homework problem. We give the results here. The probability of the electron reflecting from the barrier is

$$R = \left[\frac{K_i - K_t}{K_i + K_t}\right]^2 \tag{S1A.41}$$

and the probability of its being transmitted across the barrier is

$$T = 1 - R = \frac{4K_i K_t}{(K_i + K_t)^2} \tag{S1A.42}$$

Note that for an electron at energy E crossing the barrier, from Equations (S1A.41) and (S1A.42), the reflection and transmission probabilities are independent of the direction of the electron. That is, it doesn't matter whether the electron approaches the barrier from the low-energy or high-energy side.

EXAMPLE S1A.2

Obtain an expression for the reflection coefficient in terms of total energy and the potential energy barrier ΔE_P.

■ **Solution**

Since $E_K = E - E_P = \hbar^2 K^2/2m^*$,

$$K_i = \sqrt{\frac{2m^*}{\hbar^2}(E - E_{Pi})}$$

$$K_t = \sqrt{\frac{2m^*}{\hbar^2}(E - E_{Pt})}$$

where E_{Pi} and E_{Pt} are respectively the potential energies of the incident and transmitted electrons. Letting $\Delta E_P = (E_{Pt} - E_{Pi})$ gives, from Equation (S1A.41),

$$R = \left[\frac{1 - \sqrt{1 - \dfrac{\Delta E_P}{E}}}{1 + \sqrt{1 - \dfrac{\Delta E_P}{E}}}\right]^2$$

S1A.6.6 TUNNELING

Recall from Section 1.6.5 that, in quantum mechanics, an electron may tunnel through to the other side of a potential barrier if the barrier is thin enough. Here we consider three cases of tunneling. First we consider the case in which the tunneling barrier has a finite width, the case we saw in Chapter 1. We then consider the case for infinite width and lastly the case in which the electron tunnels into the forbidden gap of a semiconductor, a common occurrence in semiconductor devices.

Case 1: Finite Barrier Width We repeat the problem of the previous section, but this time we take the energy of the electron to be *less* than the height of the barrier, and the barrier to have width L, as shown in Figure S1A.6. From a classical point of view, this should turn out to be a pointless exercise because, of course, the particle cannot cross the barrier; it does not have enough energy so it must be reflected at the barrier. From a quantum mechanical point of view, we will obtain very different results. The result is a concept that has no analog in classical mechanics: the ability of a particle to pass through a region where its total energy is less than its potential energy.

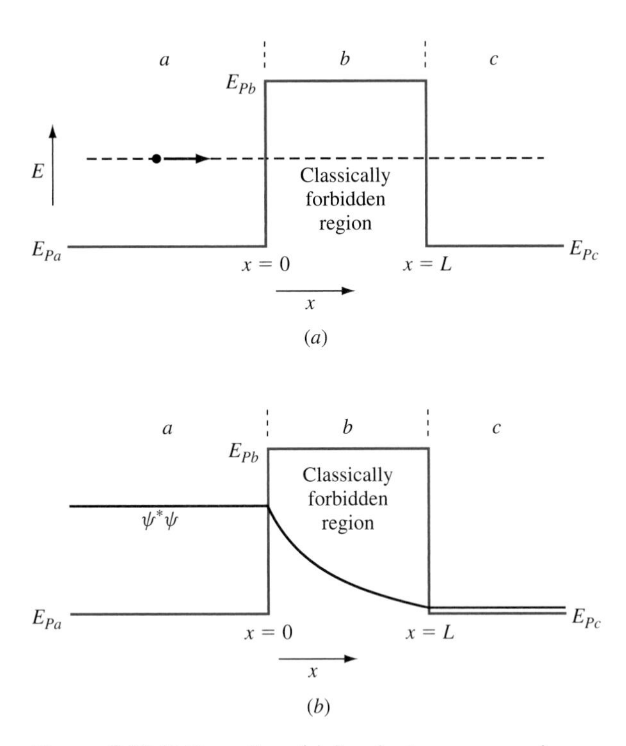

Figure S1A.6 Tunneling. (a) An electron approaches a finite potential barrier; (b) the probability density function shows there is a small chance the electron will appear on the far side of the barrier.

In Figure S1A.6, assume the electron starts out in region a, with a total energy E that is less than the barrier height. We will show that the electron can travel through the barrier, even though this region of space is classically forbidden. Keep in mind that the energy of the electron is conserved, so it travels at constant energy.

We begin by writing the wave function of the incident electron in region a as

$$\psi_a = Ae^{jKx} = Ae^{j\sqrt{(2m^*/\hbar^2)(E-E_{Pa})}x} \tag{S1A.43}$$

This is the usual solution for an electron in a region of constant potential energy, and could have been written as sines and cosines. The point is it has an oscillatory wavelike nature. (The figure shows not the wave function, but $\psi^*\psi$, which is constant in region a, as we will see later.) Let us neglect reflection at the a/b boundary ($x = 0$). Then the amplitude of the wave function at $x = 0$ is A.

In the forbidden region, since $K = \sqrt{(2m^*/\hbar^2)(E - E_{Pb})}$, and $(E - E_{Pb})$ is negative, K is imaginary in region b. We therefore have

$$\psi_b = Ae^{-\sqrt{(2m^*/\hbar^2)(E_{Pb}-E)}x} \tag{S1A.44}$$

In this case the exponent in the wave function is real, so this function does not oscillate but decays exponentially with x.

Both ψ_a and ψ_b have the same coefficient because ψ must be continuous across the boundary $x = 0$ (since reflection at the boundary is neglected). At $x = L$, just inside the far side of the barrier, we have

$$\psi_b(L) = Ae^{-\sqrt{(2m^*/\hbar^2)(E_{Pb}-E)}L} \tag{S1A.45}$$

We also neglect the reflection at this boundary. Then in region c we have

$$\psi_c(x) = Ce^{-j\sqrt{(2m^*/\hbar^2)(E-E_{Pc})}x} \tag{S1A.46}$$

where

$$C = Ae^{-\sqrt{(2m^*/\hbar^2)(E_{Pb}-E)}L} \tag{S1A.47}$$

Let us examine these results. In regions a and c, the wave functions are oscillatory. The probability density function $\psi^*\psi$ is $\psi^*\psi = A^2$ in region a and $\psi^*\psi = C^2$ in region c. These are both constant, as shown in Figure S1A.6b.

In region b, however, the wave function is a decaying exponential function. The probability density function $\psi^*\psi$ will also be a decaying exponential, given by

$$\boxed{\psi^*\psi = A^2e^{-2\sqrt{(2m^*/\hbar^2)(E_{Pb}-E)}x}} \tag{S1A.48}$$

We interpret the figure, then, to suggest that there is some reduced, but non zero, probability of finding the electron on the opposite side of the barrier. The electron will be reflected part of the time, but not always.

It seems from the figure that the thicker the barrier, the smaller the probability of finding the electron on the other side. We can calculate this probability by using the boundary conditions. The wave function ψ must be continuous across both boundaries. Since we are neglecting reflection at the boundaries, this becomes

$$\frac{\psi_b^* \psi_b(L)}{\psi_b^* \psi_b(0)} = e^{-2\sqrt{(2m^*/\hbar^2)(E_{Pb}-E)}L} \tag{S1A.49}$$

EXAMPLE S1A.3

Let us consider how to make a tunneling structure in semiconductors. Consider a semiconductor material of band gap E_{g1}. We grow a thin (2 nm) layer of another material whose band gap is larger, E_{g2}. Then we add a thick layer of the first material. The resulting energy band diagram is shown in Figure S1A.7.

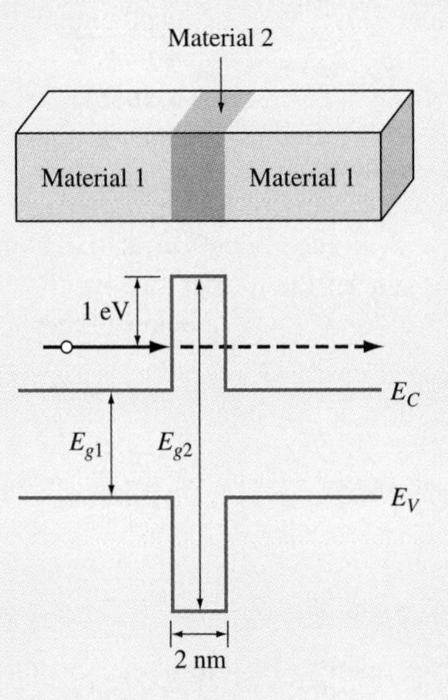

Figure S1A.7 A potential barrier is formed by sandwiching a large band gap material between two lower band gap materials. Electrons in the conduction band can tunnel through the barrier.

Now suppose an electron in the conduction band approaches the thin layer at energy 1 eV below the top of the barrier. Neglecting reflection and the influence of the valence band edge in the wide-gap material, what is the probability that the electron will penetrate the barrier?

■ **Solution**

From Equation (S1A.49), the probability of finding the electron at the far side of the barrier, $\psi^*\psi(L)$, is

$$e^{-2\sqrt{(2m^*/\hbar^2)(E_{Pb}-E)}L}$$

$$= \exp\left(-2\sqrt{\frac{2(9.1\times10^{-31}\text{ kg})}{(1.05\times10^{-34}\text{ J}\cdot\text{s})^2}(1\text{ eV}\cdot1.6\times10^{-19}\text{ J/eV})}\cdot2\times10^{-9}\text{ m}\right)=1.1\times10^{-9}$$

where we have taken $m^* = m_0$. This says that about one electron in a billion tunnels in this case.

While the probability of tunneling in this example may seem small, the number of incident electrons in a real material can be large enough such that this thin insulating region can, in fact, be a good conductor. In Chapter 6, we will see that tunneling through barriers similar to this is used extensively in integrated circuits to provide low-resistance contacts.

Let us do another example. While this text is about semiconductor materials and devices, it is instructive to consider a case of tunneling in superconductors.

Superconductor (Giaever) Tunneling ***EXAMPLE S1A.4**

As another example of tunneling, we consider two superconductors (Sn) separated by a thin semiconductor (SnO_2). This structure, shown schematically in Figure S1A.8a, is fabricated by oxidizing the surface of a sample of tin, producing about 2 nm of SnO_2. Then another layer of Sn is deposited, and electrical contacts are made to each end. While Sn is normally considered a metal, below 3.7 K it becomes a superconductor (zero electrical resistance). Furthermore, in a superconductor the normally continuous conduction band splits into two bands separated by a small forbidden energy E_g (1.15 meV for Sn). The equilibrium energy band diagram of this structure is shown schematically (not to scale) in Figure S1A.8b. The applied voltage is zero. The lower energy band is completely filled with electrons while the upper band is completely empty because of the low temperature. The thin layer of SnO_2 is a semiconductor with a forbidden band on the order of four electron volts.

At equilibrium, we do not expect current to flow. The bottom band is filled but there are no empty states at the same energy as a filled state, so the electrons cannot move since they must move at constant energy.

We now apply a small voltage (e.g., 0.5 mV) across the device. The energy band diagram will appear as shown in Figure S1A.8c. The band tilts in the SnO_2 because the applied voltage is dropped across the semiconductor (the superconductors, having infinite conductivity, cannot support a voltage), and a change in voltage implies a change in potential energy. Remember that the energy band diagrams are for the negatively charged electrons and so on these diagrams positive voltage corresponds to a reduced electron potential energy. Tunneling occurs at constant energy, and when tunneling does occur, an electron tunnels from a filled state to an empty state. In this case,

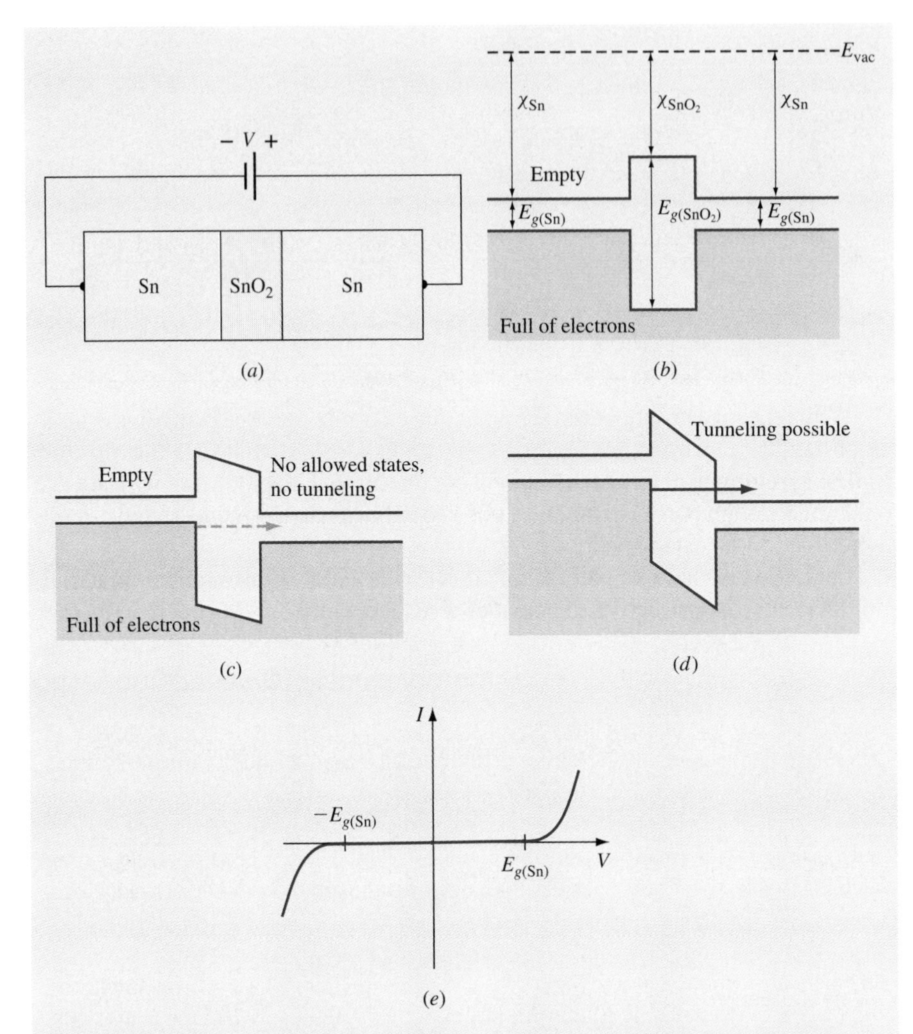

Figure S1A.8 Giaever tunneling. (a) The physical structure consists of two layers of superconductor (Sn is a superconductor below 3.7 K) with a thin layer of semiconductor (SnO_2) between them. (b) The energy band diagram at equilibrium. (c) The energy band diagram under small bias. Electrons cannot tunnel because empty states are not available at the same energies as those occupied by electrons. (d) Under larger bias, electrons from the valence band on the left tunnel through the forbidden region to empty states in the conduction band on the right. (e) A plot of current versus voltage for this device.

as at equilibrium, there are no empty states on the far side of the barrier at the same energies as filled states on the near side, so current still cannot flow.

Figure S1A.8d shows the energy band diagrams for an applied voltage just greater than the band gap of Sn. Now electrons can tunnel from the filled states on the left to the empty states on the right and current flows. The current-voltage characteristic of this device is shown in (e). Note that since the structure is symmetrical, so is the *I-V* characteristic.

While this structure is not very useful as an electronic device, it does provide a means of measuring the band gap of superconducting materials. Incidentally, this experiment with its interpretation, along with another type of tunneling (Josephson tunneling) in superconductors and tunneling in semiconductors (Esaki tunneling) resulted in a Nobel Prize in 1973, shared by Ivar Giaever, Brian D. Josephson, and Leo Esaki.

Case 2: Infinite Barrier Width Next, suppose the region b in Figure S1A.6 is infinitely thick. For $L = \infty$, from Equation (S1A.49), $\psi^*\psi$ decays exponentially to zero. This implies, however, that the probability of finding the electron on the other side of the barrier (in the classically forbidden region) is not zero, at least not close to the barrier. But the electron cannot penetrate infinitely far, either, since its probability density function decays to zero. We thus conclude that the electron penetrates some distance past the finite height barrier, but is reflected back to the low potential energy side.

Since the electron has some charge associated with it, we can say that some of that charge is actually within the classically forbidden region.

Case 3: Tunneling into the Forbidden Band of a Semiconductor Next, consider the case of Si, which has a band gap of $E_g = 1.12$ eV. An electron is incident to the Si surface at energy E where $E_V < E < E_C$, as indicated in Figure S1A.9a. The electron can tunnel some distance into the silicon, even though classically its energy is forbidden. This situation actually resembles reflection from a barrier, because the electron will penetrate some distance into the silicon and then be repelled back. While the electron is in the silicon, however, its charge can affect the potentials near the surface and alter the energy band structure, as will be discussed in Chapter 6. Here we investigate the distance that an electron can tunnel into the silicon, and calculate the associated charge concentration.

For this situation (tunneling into a forbidden region of a material) Equation (S1A.48) has the form

$$\psi^*\psi = A^2 e^{-2\sqrt{(2m^*/\hbar^2)(E_P^* - E)}x} \tag{S1A.50}$$

where m^* represents the electron effective mass. We know that an electron in the conduction band has a particular effective mass, and a hole in the valence band has a (different) effective mass, but what about an electron in the forbidden gap? Its effective mass is some combination of electron and hole effective masses (tunneling effective mass). Next, the term E_P^* is the effective potential energy of the tunneling electron. We have the same difficulty here—the potential energy for an electron in the conduction band is E_C and the potential energy for a hole in the valence band is E_V. The electron in the forbidden region sees two potential energies, E_C and E_V. The term $(E_P^* - E)$ in Equation (S1A.50) then has the form

$$(E_P^* - E) = \frac{(E_C - E)(E - E_V)}{(E_C - E) + (E - E_V)} = \frac{(E_C - E)(E - E_V)}{E_g} \tag{S1A.51}$$

similar to the effective resistance of two resistors in parallel. Since $(E_P^* - E)$ is a function of E, so is $\psi^*\psi$.

EXAMPLE S1A.5

We define the characteristic tunneling distance x_T as the value of x such that $\psi^*\psi(x_T)/\psi^*\psi(0) = e^{-1}$. Estimate the characteristic tunneling distance into the forbidden band of Si as a function of energy. Assume $m^* = m_0/2$, and is independent of energy.

■ Solution

From the definition of characteristic tunneling distance in the problem statement,

$$\frac{\psi^*\psi(x_T)}{\psi^*\psi(0)} = e^{-2\sqrt{(2m^*/\hbar^2)(E_P^*-E)}x_T} = e^{-1}$$

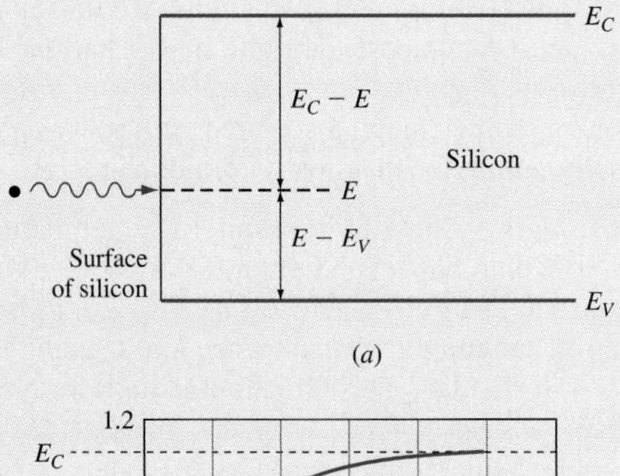

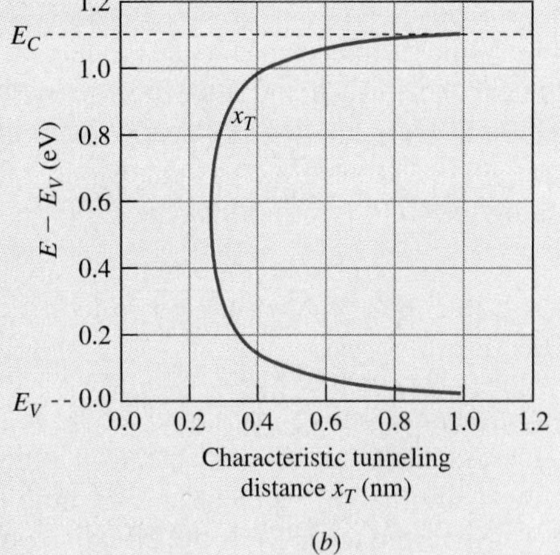

Figure S1A.9 (a) Illustration of an electron tunneling from a Si surface into its forbidden band. (b) Plot of the characteristic tunneling distance as a function of electron energy. At any energy, 63 percent of the tunneling charge is in the region $0 < x < x_T$.

or

$$x_T = \frac{1}{2\sqrt{\dfrac{2m^*}{\hbar^2}(E_P^* - E)}} \tag{S1A.52}$$

Figure S1A.9a shows the two terms that appear in Equation (S1A.51). Letting $m^* = m_0/2$, and setting $E_V = 0$ such that $E_C = 1.12$ eV, we obtain a plot of x_T as a function of E, as shown in Figure S1A.9b.

Note that the classically forbidden region contains charge. In this region, from Equation (S1A.50) and Figure S1A.6,

$$\psi^*\psi(x) = A^2 e^{-2\sqrt{(2m^*/\hbar^2)(E_P^* - E)}x}$$

Since the probability of finding an electron in a region dx is $\psi^*\psi\,dx$ and since the electron has charge q, the forbidden region will contain electronic charge density $Q_{VT}(x)$ (tunneling charge density) proportional to $\psi^*\psi(x)$; i.e., the charge density will decrease exponentially with distance into the classically forbidden region.

Since for a given electron $Q_{VT}\,dx = -q\psi^*\psi\,dx$, the fractional tunneling charge within one tunneling characteristic length x_T is

$$\frac{Q_{VT}(x < x_T)}{Q_{VT}(x < \infty)} = \frac{\displaystyle\int_0^{x_T} \psi^*\psi\,dx}{\displaystyle\int_0^{\infty} \psi^*\psi\,dx} = (1 - e^{-1}) = 0.63 = 63\%$$

We can see that most of the tunneling charge is concentrated within x_T of the surface, a distance on the order of a few tenths of a nanometer, as Figure S1A.9b shows.

The above result will be of considerable importance in Chapter 6 for metal-semiconductor diodes, in which the tunneling electrons originate from the conduction band in the metal.

S1A.6.7 THE FINITE POTENTIAL WELL

In this section we consider what would happen if a layer of narrow band-gap semiconductor were sandwiched between two materials of wide band gap. Figure S1A.10 shows an idealized one-dimensional energy band diagram for the resulting structure, called a *quantum well.* Quantum wells made this way are widely used in lasers, as discussed in Chapter 11. We will consider only the conduction band here. It appears to the electron as a finite potential well.

We consider the case in which the electron is at energy lower than the top of the well, as shown in Figure S1A.10b. We intuitively expect the particle to be trapped in the well, and oscillate back and forth. We will not go through the

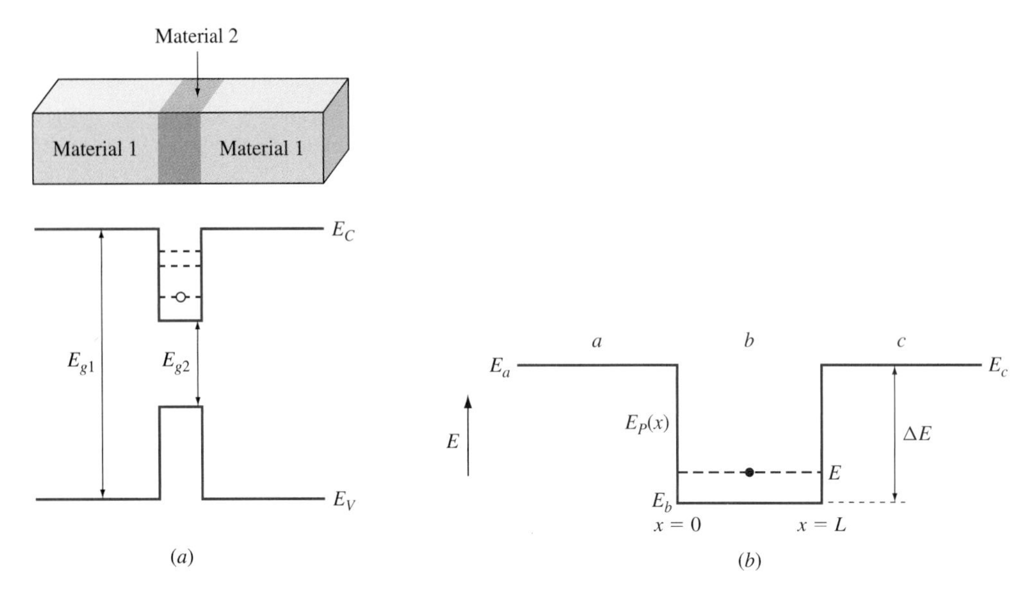

Figure S1A.10 A quantum well is formed when a layer of narrow band gap material is sandwiched between two layers of wide band gap material.

derivation, but rather outline the steps and discuss the results. Again, there are three regions of space, each with a constant potential energy. We have solved this problem many times now. The wave functions are

$$\Psi_{a,b,c}(x) = A_{a,b,c}e^{jK_{a,b,c}x} + B_{a,b,c}e^{-jK_{a,b,c}x} \tag{S1A.53}$$

where

$$|K_{a,b,c}| = \sqrt{\frac{2m^*}{\hbar^2}(E - E_{P_{a,b,c}})} = \sqrt{\frac{2m^*}{\hbar^2}E_{K_{a,b,c}}} \tag{S1A.54}$$

and the *K*'s are different for the different regions because the E_P's are different.

Inside the well, (region b) *K* is real, resulting in solutions of the form $\psi(x) = C\sin(Kx)$, as in the case of the infinite potential well. The solutions outside the finite well where *K* is imaginary are decaying exponentials as in the case for tunneling. The first three solutions are shown in Figure S1A.11, where we have plotted both the wave functions and the probability distribution functions.

Looking at the probability distribution functions, we can see that now there is a small, but finite, probability that the electron can be found outside the well. Another way to express that is to say that the electron spends some small amount of time outside the well as it oscillates back and forth between the barriers. The electron actually penetrates the barrier on either side for a short distance. The higher the step in the potential well, the shorter this distance is, or the more tightly confined the electron is to the well.

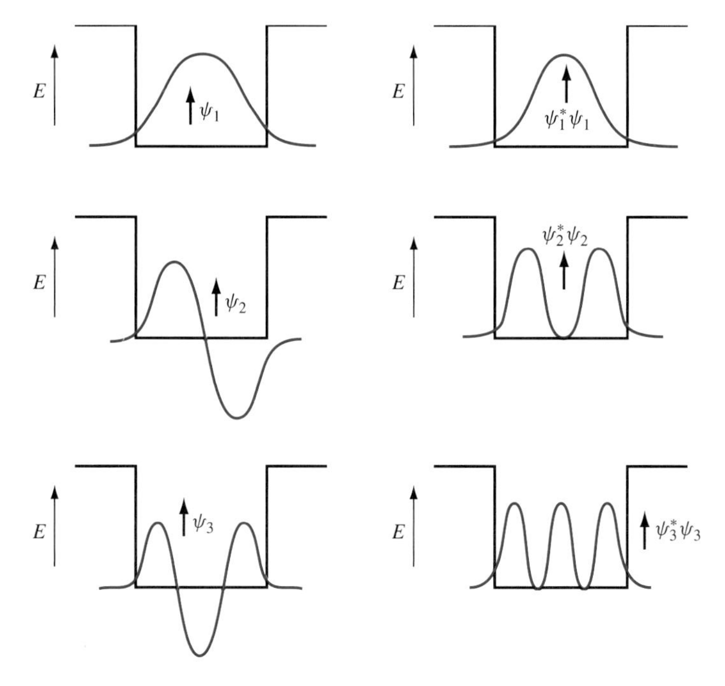

Figure S1A.11 The first three wave functions in the finite potential well and the associated probability density functions $\psi^*\psi$.

As was the case with the infinite well, the allowed energy states get farther apart as the well becomes narrower. For example, the well shown in Figure S1A.11 happened to have three discrete states. Above the top of the well is a continuum of allowed states—the rest of the conduction band. Thus, designers of quantum well devices can control the number and spacing of allowed energy levels by adjusting the thickness of the well and by choice of materials and thus barrier heights. This is called *band-gap engineering*. One example where band-gap engineering is used is in lasers—by controlling the well width and depth, the energies of the discrete states in the potential well for electrons (in the conduction band) and the potential well for holes (in the valence band) can be adjusted to produce emission at a particular wavelength of light.

S1A.6.8 THE HYDROGEN ATOM REVISITED

You will recall that in Chapter 1 we used the Bohr model to find the allowed energy levels in the hydrogen atom. Here we briefly outline the steps used to apply quantum mechanics to this problem.

The hydrogen atom is necessarily a three-dimensional object, and best described in spherical coordinates. We will need to follow the usual steps:

1. Describe the potential energy as a function of position (in spherical coordinates this time), assuming the electron to be a point charge.

2. Solve Schroedinger's equation (also in spherical coordinates) for the allowed wave functions. As with every problem, once we know the wave function(s), we can use Equation (S1A.1) to find the average or expected value of any observable.

We begin with the potential energy of an electron in the presence of a single positive charge, the nucleus of the atom. Solving Schroedinger's equation with $E_P(r)$ given by Equation (1.6) results in quantized energies of values

$$E_n = E_{\text{vac}} - \frac{m_0 q^4}{2(4\pi\varepsilon_0)^2 n^2 \hbar^2} \tag{S1A.55}$$

This is the same result obtained from the Bohr model.

S1A.6.9 THE UNCERTAINTY PRINCIPLE

An important concept in quantum mechanics is the *uncertainty principle,* often referred to as the *Heisenberg uncertainty principle*. "Uncertainty" is a way of saying that a quantity is not exactly known. For example, an electron oscillating back and forth in a potential well is known to be in the well somewhere, but since its wave function is spread out in space we can't say it is located precisely at a given point. The uncertainty principle states that for certain pairs of observables, the more accurately one of them is known, the less accurately the other is. Examples are

$$\boxed{\Delta p \Delta x \sim 2h} \tag{S1A.56}$$

$$\boxed{\Delta E \Delta t \sim 2h} \tag{S1A.57}$$

where Δx is the uncertainty in position and so forth, and the symbol $\sim$ is taken to mean "on the order of." The observables x and p are said to be conjugate variables.

From Equation (S1A.56), then, the more accurately a particle's momentum is known, the less accurately you can determine its position. Equation (S1A.57) says that the more accurately the electron energy is known, the less is known about the amount of time it spends at that energy. Let us explore this idea.

We will start with an example familiar to most electrical engineers: a classical voltage pulse of amplitude V_T and time duration T centered at $t = 0$ as shown in Figure S1A.12a. This pulse can be expressed in the time domain as

$$\begin{aligned} V(t) &= V_T \qquad -\frac{T}{2} < t < \frac{T}{2} \\ V(t) &= 0 \qquad \text{elsewhere} \end{aligned} \tag{S1A.58}$$

If you have studied Fourier transforms, you will recall that this voltage pulse can also be expressed in the frequency domain—as a superposition of terms

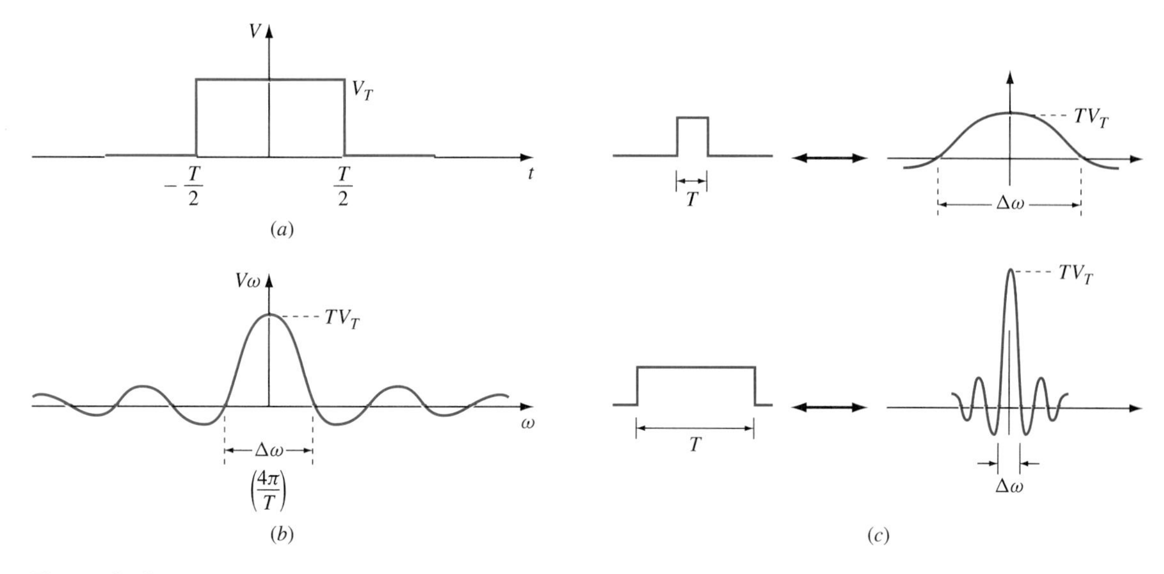

Figure S1A.12 Illustration of the uncertainty principle. A square voltage pulse (a) of duration T and amplitude V_T can be expressed in the frequency domain (b) by its Fourier transform, a modulated sinusoidal function.

$V(\omega)\sin(\omega t)$. The angular frequency ω is a variable and $V(\omega)$, the amplitude of the function at ω, is found by taking the Fourier transform of $V(t)$:

$$V(\omega) = \int_{-\infty}^{\infty} V(t)e^{-j\omega t}\,dt \tag{S1A.59}$$

Substituting Equation (S1A.58) into (S1A.59),

$$V(\omega) = \int_{-T/2}^{T/2} V_T e^{-j\omega t}\,dt \tag{S1A.60}$$

This can be solved to obtain

$$V(\omega) = TV_T \frac{\sin\left(\frac{\omega T}{2}\right)}{\frac{\omega T}{2}} \tag{S1A.61}$$

In other words, we have expressed the voltage pulse in the frequency domain. The frequency representation is plotted in Figure S1A.12b.

Similarly, given an expression in the frequency domain, we can express that function in the time domain by using the inverse Fourier transform:

$$V(t) = \frac{1}{2\pi}\int_{-\infty}^{\infty} V(\omega)\,e^{j\omega t}\,d\omega \tag{S1A.62}$$

If $V(\omega)$ is expressed by Equation (S1A.61), Equation (S1A.62) results in the original voltage pulse in the time domain as given in Equation (S1A.58) and Figure S1A.12a. This, of course, requires an infinite range of frequency ω, since the integration limits are $\pm\infty$. An approximation can be made, however, by using only a finite range $\Delta\omega$ in ω. We arbitrarily choose to define the uncertainty in ω as the width in frequency space between the first zeros of the function, $V(\omega)$. This is $\Delta\omega \sim 4\pi/T$ as shown in (b). If we integrated over this range of ω with the aid of Equation (S1A.62), we would get a reasonable approximation to the voltage pulse of Figure S1A.12a. Then, since the uncertainty in the time domain $\Delta t = T$,

$$\Delta\omega\Delta t = \frac{4\pi}{T}T \sim 4\pi \qquad \text{(S1A.63)}$$

We could have chosen a different definition of the size of $\Delta\omega$—for example, taking the next set of zeros—but the exact number isn't important. The principle here is that the narrower the pulse in one domain, the broader it is in the other domain, as indicated in Figure S1A.12c. It is still, however, the same physical pulse. Mathematically ω and t are thus conjugate variables. Relations similar to Equation (S1A.63) exist for all such conjugate variable pairs.

Now let us do a quantum mechanical example. We consider the (artificial) case of a free or quasi-free electron that is localized such that its probability function $\psi^*\psi$ is constant in a finite region. For mathematical simplicity, we choose the region $-L/2 < x < L/2$ as indicated in Figure S1A.13. In the case of the voltage pulse, we represented the time domain pulse as a superposition of an infinite number of different sine waves [Equations (S1A.61) and (S1A.62)] centered about $\omega = 0$, its value for infinite T. Analogous to that case, we can in this instance superpose wave functions of different wave vectors to achieve this particular probability density. Here, K will be centered about K_0 since for infinite L, $K = K_0$.

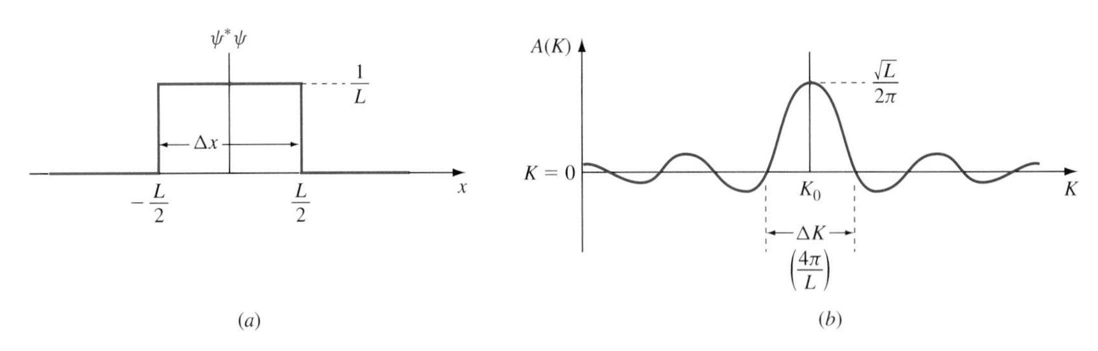

Figure S1A.13 Illustration of the uncertainty principle for an electron in a conduction band. In (a) the electron of wave vector K_0 is confined to the region $-L/2 \leq x \leq L/2$. In (b), the corresponding distribution in K for the value of $\psi^*\psi$ of (a) is shown.

The amplitude of the plane wave component at a given K is

$$A(K) = \frac{\sqrt{L}}{2\pi}\left[\frac{\sin\dfrac{(K - K_0)\,L}{2}}{\dfrac{(K - K_0)\,L}{2}}\right] \tag{S1A.64}$$

and is plotted in Figure S1A.13b.

Since the electron is restricted to a region of length L, we represent the uncertainty in its position as $\Delta x = L$. As in the previous problem, we can take $\Delta K = 4\pi/L$. Then

$$\Delta K\,\Delta x = \frac{4\pi}{L}L \sim 4\pi \tag{S1A.65}$$

Although the uncertainty principle was illustrated above for K and x with E_P constant, it holds for any two conjugate quantities. These include

$$\begin{aligned}\Delta K_x \Delta x &\sim 4\pi\\ \Delta K_y \Delta y &\sim 4\pi\\ \Delta K_z \Delta z &\sim 4\pi\\ \frac{\Delta E}{\hbar}\Delta t &\sim 4\pi\end{aligned} \tag{S1A.66}$$

Note that in the case of a free particle, the momentum $p = \hbar K$ and

$$\Delta p \Delta x \sim 2h \tag{S1A.67}$$

Let us do an example to relate the uncertainty principle to real life.

EXAMPLE S1A.6

Find the spread of the photon energy spectrum for electron transitions from an excited state E_2 to a ground state E_1.

■ Solution

In a laser, light is emitted when an electron makes the transition from a state of energy E_2 to a state of lower energy E_1, emitting a photon of angular frequency, $\omega = 2\pi\nu$. An electron in the upper state E_2 has a lifetime—the amount of time that it spends in that state on the average. If Δt is the time the electron spends in the upper state, then from Equation (S1A.57),

$$\frac{\Delta E}{\hbar}\Delta t \sim 4\pi \tag{S1A.68}$$

The spread in the emitted photon energy spectrum can be measured with a spectrometer. The width of the photon energy spread (Figure S1A.14) is a measure of Δt.

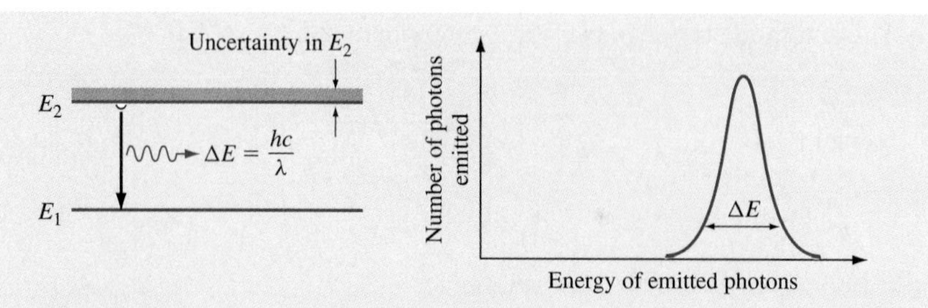

Figure S1A.14 The uncertainty in the energy levels of a laser creates some uncertainty in the energy of the emitted photons.

For a lifetime of $\Delta t = 1$ ns,

$$\Delta E = \frac{4\pi\hbar}{\Delta t} = \frac{4\pi \cdot 6.58 \times 10^{-16}\,\text{eV} \cdot \text{s}}{10^{-9}\,\text{s}} = 8.27 \times 10^{-6}\,\text{eV}$$

This corresponds to a spectral width of 2 GHz.

S1A.7 SUMMARY

We have expanded our understanding of quantum mechanics as described by Schroedinger's equation. The solutions to Schroedinger's equation are called wave functions, and we can use them to tell us everything about the observable quantities, or quantities that can, in principle, be measured. The value of an observable is found by using the operator corresponding to the observable, and the wave function representing the particle whose observable needs to be known.

We saw that once we know the wave function for a particle in a given physical situation, then we can calculate all of the observables for that particle. We also found that for most realistic problems, the potential energy cannot be accurately described. We therefore resorted to simplified models, which we used to get some physical intuition about certain kinds of structures. For example, we saw that for an electron in a potential well, the energy levels are quantized, consistent with the electron in the hydrogen atom. The hydrogen nucleus is a potential well for electrons.

We found that in quantum mechanics, particles can tunnel through potential barriers that classical physics predicted would reflect them every time.

We also began to explore the uncertainty principle. This nonintuitive concept predicts (for example) that the more accurately one knows a particle's position, the less accurately one will know its momentum. The uncertainty principle prevents us from making precise statements. In quantum mechanics, we tend to express everything as a probability—the electron has a probability of tunneling through this barrier, or this electron is, on the average, here. We can never say where the electron is, but we can discuss where it's likely to be.

In this section, we began applying the principles of quantum mechanics to some simple but common device structures: potential barriers and potential wells.

S1A.8 REVIEW QUESTIONS

1. Explain why the wave function is important if it can't be measured.
2. Consider the electron reflecting off a finite potential barrier of finite width L, as shown in Figure S1A.6, but suppose this time that electron has a total energy greater than the barrier height. In Section S1A.6.5, we saw that when the electron approaches the a/b boundary, it has some probability of being reflected. Suppose it is not reflected, however, but travels across the barrier. What happens when it reaches the b/c boundary of the barrier? What happens after that?

S1A.9 PROBLEMS

S1A.1 Assume the wave function Ψ is separable, as shown in Equation (S1A.9). Insert that into Schroedinger's equation and show that Equations (S1A.10) and (S1A.11) result. If the procedure is not obvious, review separation of variables from your differential equations course.

S1A.2 Solve Equation (S1A.11) to show that Equation (S1A.12) is a solution.

S1A.3
a. Find the average value of the position x for the $n = 2$ state of the infinite potential well, where $\Psi_2 = A\sin(2\pi x/L)e^{j(E_2/\hbar)t}$ for $0 \le x \le L$.
b. What is the most probable location for the electron? Explain how you drew your conclusion.

S1A.4 For the electron in the lowest energy state of the infinite potential well, find
a. The average momentum p_x
b. The average velocity v_x
c. The average energy E

S1A.5 Consider a plane wave $\Psi(x, t) = e^{j[Kx-(E/\hbar)t]}$. What is the average velocity v_x?

S1A.6 Consider a free electron with a kinetic energy of 1 eV. It has a wave function $\Psi(x, t) = Ae^{j[(5.14\times10^9\ \text{m}^{-1})x-(1.5\times10^{15}\ \text{s}^{-1})t]}$.
a. Find the group velocity classically (i.e., using momentum, energy, etc.). (*Hint:* This is a plane wave.)
b. Find the group velocity using quantum mechanics [i.e., using the observable operator for group velocity, $v_{\text{op}} = (\hbar/jm)(\partial/\partial x)$].

S1A.7 Consider the (artificial) wave function shown in Figure PS1A.1. The wave function ψ is zero for $x < 0, x > L$.
a. Sketch the probability distribution function.
b. Indicate the region(s) in which the electron is most likely to be found.

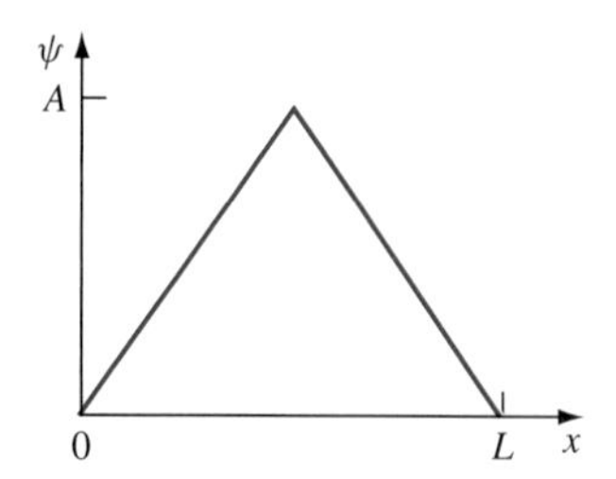

Figure PS1A.1

c. To find the probability of an electron being in a certain region, one integrates $\Psi^*\Psi$ over the region of interest, and divides by the integral over all space [see Equation (S1A.1)]. Calculate the probability of finding the electron in the region $L/4 < x < L/2$.

S1A.8 Find the eigenfunctions for the infinite potential well, but let the well begin at $x = -L/2$ and end at $x = L/2$. Since the electron can't read, it doesn't know you have changed coordinates and should end up with exactly the same states as before. Sketch the first three eigenfunctions and compare with Figure S1A.3. *Hint:* Make sure that you have a complete set of solutions.

S1A.9 Consider an electron approaching a potential barrier with an energy lower than the barrier, as shown in Figure S1A.6. Explain why neglecting reflections at the two sides of the barrier in Figure S1A.6 is a reasonable approximation. Take the reflection coefficient at either side of the barrier to be on the order of 20 to 30 percent, and take the barrier to be thick. *Hint:* Compare the transmission probabilities, neglecting reflection, with the reflection probabilities.

S1A.10 Sketch the electrical characteristics of a tunneling junction in which the superconductor Pb ($E_g = 2.73$ meV) is used instead of Sn (i.e., the structure is $Pb/Sn0_2/Pb$). Label the significant points on your horizontal axis with numeric values.

S1A.11 Consider the potential energy distribution function shown in Figure PS1A.2. It consists of two finite wells separated by a finite barrier. Sketch what you expect the wave function to look like. Can the electron move from one well to the other? If so, by what process? Can the electron escape from the wells?

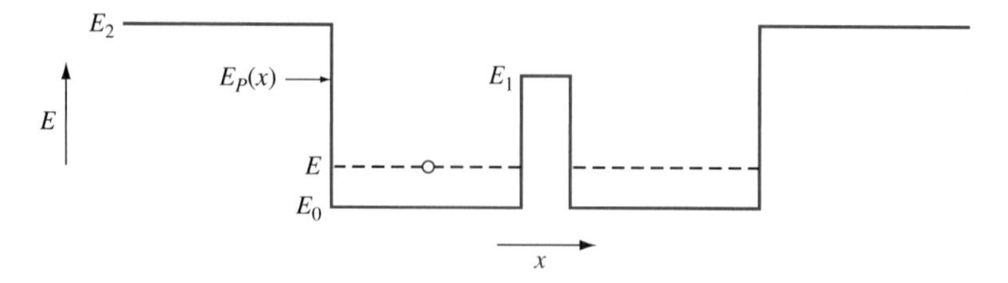

Figure PS1A.2

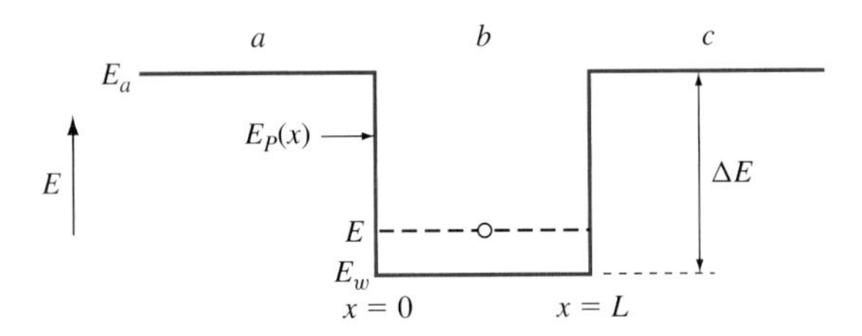

Figure PS1A.3

S1A.12 (Lengthy.) In this problem you will solve the problem of the finite potential well, using results presented in this supplement. The geometry of the problem is shown in Figure PS1A.3. The energy of the bottom of the well is E_w, and the energy of the barriers is E_a. The difference is $E_a - E_w = \Delta E$. The well extends from $x = 0$ to $x = L$. We will call the three regions a, b, and c as shown in the figure.

a. Write the time-independent Schroedinger's equation for this problem (it will require three equations, one for each region).

b. Solve Schroedinger's equation for each of the three regions. The results should be of the form:

$$\psi_a(x) = Ae^{ax}$$

$$\psi_b(x) = B\cos(Kx) + C\sin(Kx)$$

$$\psi_c(x) - De^{-a(x-L)x}$$

Express the constants a and K in terms of the energies in the regions.

c. Use the boundary conditions to express the constants B, C, and D in terms of A. You will need to apply (1) continuity of ψ at both boundaries and (2) continuity of $d\psi/dx$ at $x = 0$. The results are: $A = B$, $C = A(a/K)$, and

$$D = A\left[\cos(KL) + \frac{a}{K}\sin(KL)\right]$$

d. We need to find the value(s) of K. Use the final boundary condition, continuity of $d\psi/dx$ at $x = L$, to show that

$$\tan(KL) = \frac{2a}{K\left(1 - \dfrac{a^2}{K^2}\right)}$$

e. The result of (d) is known as the characteristic equation, and cannot be solved in closed form. You can solve it graphically, however, by plotting the left hand side versus K and the right hand side versus K and seeing where they cross.

i. To do this, first find the maximum value that K can have for an electron trapped in the well. (*Hint:* Its kinetic energy must be less than $E_a - E_w$.)

ii. Given that $a = \sqrt{(2m/\hbar^2)(E_a - E)}$ and $K = \sqrt{(2m/\hbar^2)(E - E_w)}$, show that $a = \sqrt{K_{\max}^2 - K^2}$.

iii. Write a program that plots the two sides of the characteristic equation from $K = 0$ to $K = K_{\max}$. As an example, use $L = 2$ nm and $\Delta E = 0.5$ eV.

iv. How many states are in this well, and what are their energies?

f. Plot the wave function for the lowest energy state, from $x = -0.2L$ to $1.2L$. The wave function should be symmetrical.

g. Use your program to find the number of allowed states for

$$L = 3\,\text{nm},\ \Delta E = 0.5\,\text{eV}$$
$$L = 5\,\text{nm},\ \Delta E = 0.5\,\text{eV}$$
$$L = 2\,\text{nm},\ \Delta E = 1\,\text{eV}$$
$$L = 3\,\text{nm},\ \Delta E = 0.2\,\text{eV}$$

SUPPLEMENT 1B

Additional Topics on Materials

In this supplement we discuss some additional topics of interest in semiconductor materials. In Chapter 3 we saw how the carrier concentration and mobility affect conductivity. Here we will start by showing how these are measured, using the Hall effect. Then, we will go on to the Fermi-Dirac statistics for bound states; in Chapter 2 we covered only the probability of occupancy of states in a band, but for donors and acceptors there are some differences.

Then, we will explore phonons in more detail. We saw earlier how they influence mobility by scattering electrons, and we also learned that phonons are involved in many transitions—electrons moving from one energy state to another may emit or absorb phonons. In Section S1B.3 we will find that phonons have energy bands and E-K characteristics similar to those of electrons.

S1B.1 MEASUREMENT OF CARRIER CONCENTRATION AND MOBILITY

In Chapter 2, we saw how to predict the carrier concentration from the doping levels. In practice, since no doping process can be perfectly controlled, we must measure the sample to see exactly what doping resulted. One measurement used is known as the *Hall effect.*

Also in Chapter 2, we gave the mobility of a sample of known doping as an experimental parameter. In practice, it depends not only on doping but also on the quality of the crystal. The Hall effect can be combined with resistivity measurements to find the carrier mobility of a sample. The resistivity measurement is covered first, and then the Hall effect.

S1B.1.1 RESISTIVITY MEASUREMENT

To see how resistivity is measured in a semiconductor, consider the sample shown in Figure S1B.1. A current flows through the uniformly doped sample in

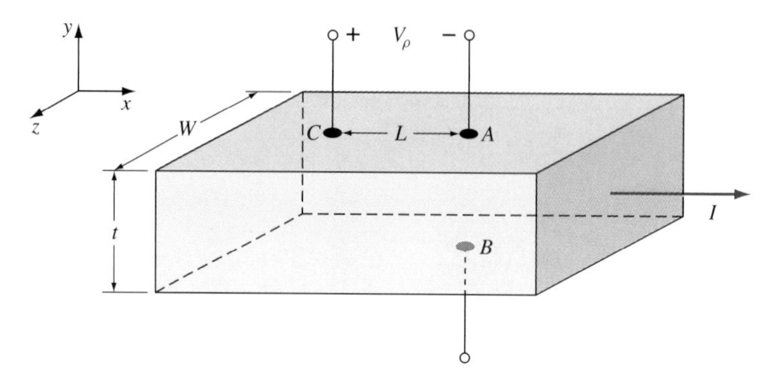

Figure S1B.1 Structure used to measure resistivity and Hall effect.

the x direction. The resistivity of a material is obtained from a measurement of the voltage V_ρ between contacts A and C on the surface of the sample. The resistance of the sample between A and C is

$$R_{C\text{-}A} = \frac{V_\rho}{I} = \frac{\rho L}{A} = \frac{\rho L}{tW} \tag{S1B.1}$$

where $V_\rho = V_C - V_A$ and $t \times W$ is the cross-sectional area. Then the resistivity is

$$\rho = \frac{V_\rho}{I}\frac{tW}{L} \tag{S1B.2}$$

which is normally expressed in ohm-centimeters.

The conductivity is the reciprocal of the resistivity:

$$\sigma = \frac{1}{\rho} = \frac{I}{V_\rho}\frac{L}{tW} \tag{S1B.3}$$

S1B.1.2 HALL EFFECT

There is a third measurement point, B, shown in Figure S1B.1. We do not expect any voltage between points A and B, since these contacts are opposite each other, or on an equipotential plane.

Next, consider a magnetic field applied in the negative z direction, perpendicular to the direction of current flow, as shown in Figure S1B.2a. The sample is n type in this case, so the majority carriers are electrons.

Since the current is flowing to the right, the electrons are moving to the left. The magnetic field produces a Lorentz force, F_{Hn}, on a charge Q and of velocity v_n, $\vec{F}_{Hn} = Q\vec{v}_n \times \vec{B}$ where $\vec{v}_n \times \vec{B}$ is the cross product of velocity and magnetic field. For an electron:

$$\vec{F}_{Hn} = -q(\vec{v}_n \times \vec{B}) \tag{S1B.4}$$

where the charge Q on the electron is $-q$. This force accelerates the electron toward the top of the sample. The top of the sample then becomes negatively charged. A negative Hall voltage, V_H, is produced.

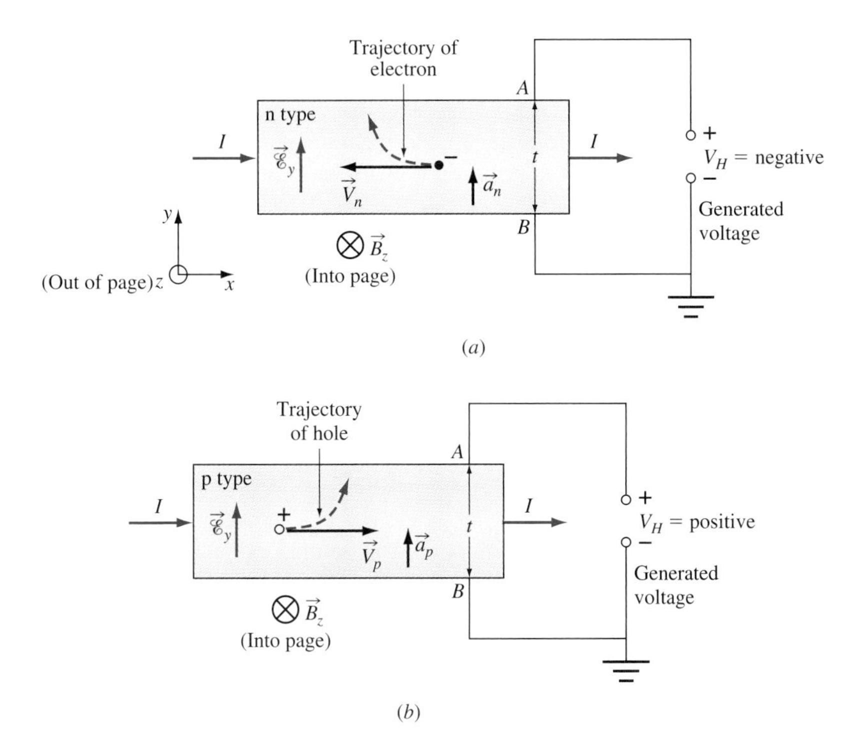

Figure S1B.2 The Hall effect can be used to determine the doping type and carrier concentration of a sample. (a) In n-type material, the Lorentz force causes the electrons to accumulate at the top of the sample, producing a negative Hall voltage. (b) In p-type material, the holes move upward, producing a positive Hall voltage.

Figure S1B.2b shows the case for a p-type material. In this case, holes flow to the right, and the Lorentz force on the holes is

$$\vec{F}_{Hp} = +q(\vec{v}_p \times \vec{B}) \tag{S1B.5}$$

This force is again in the upward direction, but since the holes are positive, the Hall voltage V_H becomes positive. Thus the sign of the measured Hall voltage, $V_H = V_A - V_B$ indicates whether the material is n type or p type.

The Hall voltage sets up a field in the y direction:

$$\mathscr{E}_y = \frac{dV_y}{dy} = \frac{V_H}{t} \tag{S1B.6}$$

The carriers are confined to the sample however (note V_H is an open-circuit voltage), and so they accumulate at the surface. Assuming that all carriers have their average velocity, the net force in the y direction on each electron (or hole) is zero, which in turn means that the magnetic force is equal and opposite to the electric force set up by the Hall voltage. Then for electrons we have

$$F_y = -q\mathscr{E}_y + qv_x B = 0 \tag{S1B.7}$$

where the electric field is pointing in the positive y direction (upward in the figure) and the Lorentz force is pointing upward (v_x is negative), or

$$\mathscr{E}_{yn} = v_x B \tag{S1B.8}$$

For electrons, the field points in the positive y direction, and for holes, it points in the negative y direction.

The current I is given by

$$I = -qnv_x tW \tag{S1B.9}$$

Combining Equations (S1B.8) and (S1B.9) gives

$$\mathscr{E}_y = -\frac{IB}{nqtW} = +\frac{R_H IB}{tW} \tag{S1B.10}$$

where, for electrons, the Hall coefficient R_H is defined as

$$\boxed{R_{Hn} = \frac{-1}{nq}} \qquad \text{electrons} \tag{S1B.11}$$

Since the Hall voltage is

$$V_H = \mathscr{E}_y t \tag{S1B.12}$$

we can combine Equations (S1B.10), (S1B.11), and (S1B.12) to write

$$R_{Hn} = \frac{-1}{nq} = \frac{V_{Hn} W}{IB} \tag{S1B.13}$$

Note that for n-type material, both R_H and V_H are negative.

Therefore, from knowledge of V_H, I, B, and W, the Hall coefficient and thus the electron density n can be determined.

Similarly for a p-type semiconductor,

$$\boxed{R_{Hp} = \frac{+1}{pq} = \frac{V_{Hp} W}{IB}} \tag{S1B.14}$$

where for holes both R_H and V_H are positive.

That gives us the electron or hole concentrations n or p. Next, we want to know the carrier mobilities. The carrier mobility can be obtained from the product of σR_H. For electrons, $\sigma_n = nq\mu_n$ where μ_n is the majority carrier mobility and

$$\boxed{\sigma_n R_{Hn} = nq\mu_n \frac{-1}{nq} = -\mu_n} \tag{S1B.15}$$

For holes

$$\boxed{\sigma_p R_{Hp} = pq\mu_p \frac{1}{pq} = \mu_p} \tag{S1B.16}$$

The mobility found in this manner is known as the Hall mobility μ_H, but while it is simple to measure, it is not really the best value to use. In the above discussion, we assumed that all carriers had the same (average) velocity and so the magnetic and electric forces were equal and opposite for each carrier. While the electric force ($F_y = Q\mathscr{E}_y$) is the same for each carrier, the magnetic force ($Qv_x \times B$) depends on the carrier velocity. Instead of equating electric and magnetic forces, the condition that should be used is that the current in the y direction be zero. For this approach, the calculation of the Hall effect is modified. In the modified approach, the Hall coefficients become, for electrons and holes respectively, [1]

$$\begin{aligned} R_{Hn} &= \frac{-1}{nq} \cdot \frac{\overline{t_n^2}}{(\bar{t}_n)^2} \\ R_{Hp} &= \frac{1}{pq} \cdot \frac{\overline{t_p^2}}{(\bar{t}_p)^2} \end{aligned} \tag{S1B.17}$$

where $\bar{t}_n$ and $\bar{t}_p$ are the mean free times between collisions for electrons and holes respectively. This modifies Equations (S1B.15) and (S1B.16).

For those cases in which $\bar{t}_n$ and $\bar{t}_p$ are independent of velocity, the drift mobilities μ are related to the Hall mobilities by the factor $8/3\pi$:

$$\mu = \mu_H \cdot \frac{8}{3\pi} \tag{S1B.18}$$

or on the order of 15 percent less than that obtained from the Hall effect.

For more heavily doped semiconductors, where ionized impurity scattering is significant, the proportionality factor is different, but less than unity.

S1B.2 FERMI-DIRAC STATISTICS FOR ELECTRONS IN BOUND STATES

In this section, we will address the probability of occupancy of bound states associated with the dopant atoms. In Chapter 2, we considered the probability of occupancy of a state in the conduction band or valence band. Although we did not state it explicitly there, because the states of interest were in these large bands of states, comprising many energies, the occupancy of one state did not affect the existence of the remaining states. That is not true, however, for localized states such as those associated with donors or acceptors.

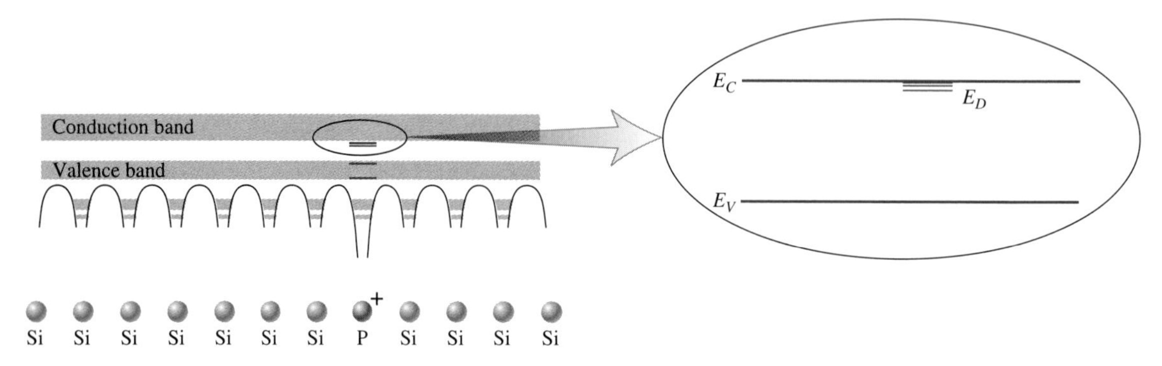

Figure S1B.3 A donor atom has multiple states analogous to the hydrogen atom. When one state is occupied by an electron, the atom is no longer ionized and the other states effectively disappear.

Consider, for example, a donor atom. At room temperature, the atom is most likely ionized (the extra electron has gone into the conduction band and left the vicinity of the atom). At low temperatures, the concentration of donors that are not ionized (are still occupied by electrons) can be significant. Carrier freeze-out will change the concentrations of electrons and holes as discussed in Chapter 2.

If we apply the Bohr model to this positively charged ion, the result is that there are many unoccupied states associated with it, as shown in Figure S1B.3. This is analogous to the many possible energy states of the hydrogen atom. If, however, an electron should take up residence in one of those states, the other states "disappear" as far as other electrons are concerned, because the atom is no longer ionized.[1]

The statistics governing the probability of occupancy are therefore altered from the Fermi function we used for carriers in bands. For the bound state case, the probability that an electron is bound to a donor (a donor state is occupied) is [2]

$$\boxed{f(E_D) = \frac{1}{1 + \left[\sum_i \frac{1}{g_i} e^{(E_{Di} - E_f)/kT}\right]}} \qquad \text{(S1B.19)}$$

where E_{Di} is the energy of the ith bound state and g_i is the number of states at this energy, or the number of ways in which the same state can be occupied; for example a given energy might have two different states including spin. The summation is taken over all the bound states.

This expression is somewhat unwieldy, since all bound states and their energies must be known. Further, states may overlap (and thus not be bound)

[1]While it is possible for a second electron to be bound to the donor, creating a negative ion, the probability is small because its binding energy is small.

depending on the interimpurity spacing, which in turn depends on the doping concentration. It is convenient to define an effective nondegenerate donor energy E_D^* such that an equation similar to Equation (2.49) can be used for bound states:

$$\boxed{f(E_D) = \frac{1}{1 + e^{(E_D^* - E_f)/kT}}} \tag{S1B.20}$$

where $f(E_D)$ is the probability that one of the donor states is occupied. By equating Equations (S1B.19) and (S1B.20), and taking the summation over all bound (non-overlapping) states for a given donor concentration, E_D^* can be calculated. [3] Figure S1B.4a shows the value of E_D^* with respect to the bottom of

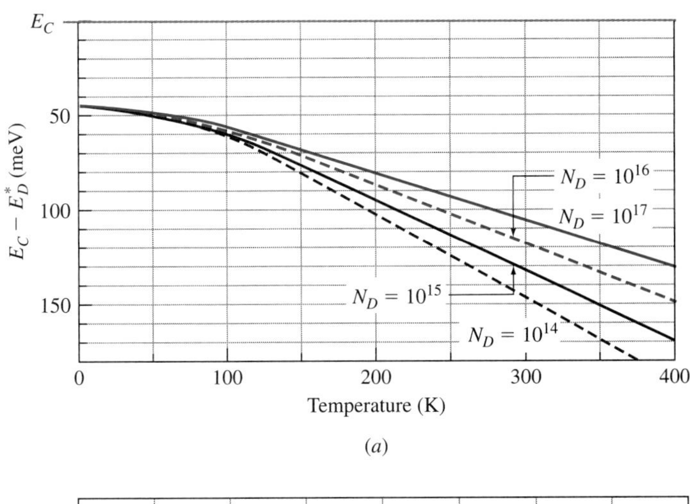

(a)

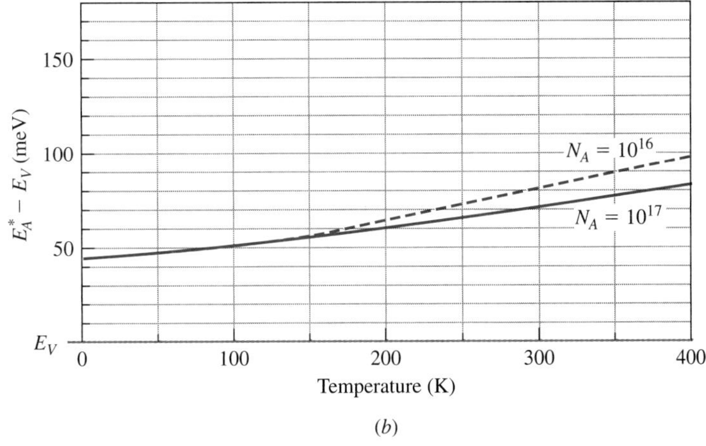

(b)

Figure S1B.4 (a) The effective donor energy for phosphorus in silicon as a function of temperature. (b) The effective acceptor energy for boron in silicon.

the conduction band as a function of temperature, for phosphorus in silicon, over a range of doping levels.

At absolute zero, all electrons would be at the ground state energy of the donor atom, which for phosphorus in Si is at $E_C - E_D = 45\,\text{meV}$. With increasing temperature, the Fermi-Dirac function spreads out in energy so that excited states and conduction band states are occupied.

The case for "bound holes" is similar:

$$\boxed{f_p(E_A) = \frac{1}{1 + e^{(E_f - E_A^*)/kT}}} \tag{S1B.21}$$

where $f_p(E_A)$ is the probability that an acceptor state is *not* occupied by an electron, or is occupied by a hole. Figure S1B.4b shows a plot of E_A^* as a function of temperature for two doping concentrations of boron in Si. The reduced dependence of E_A^* on temperature and doping compared with that of E_D^* results from the fact that each acceptor energy has four states (due to two valence bands, with the same energies at $K = 0$, and spin) while donors have 12 states at each energy (six equivalent minima and spin).[2] While the effective donor and acceptor energies of Figure S1B.4 were calculated for P and B in Si, they are expected to differ little for impurities with approximately equal ground state energies.

S1B.3 CARRIER FREEZE-OUT IN SEMICONDUCTORS

In n-type material, the concentration of electrons in donor states n_D can be expressed as

$$n_D = N_D f(E_D) = \frac{N_D}{1 + e^{(E_D^* - E_f)/kT}} \tag{S1B.22}$$

which requires a knowledge of the Fermi energy. This can be determined from the free carrier concentration

$$n_0 = N_C e^{-(E_C - E_f)/kT} \tag{S1B.23}$$

and the charge neutrality condition

$$n_0 + n_D = N_D - N_A \tag{S1B.24}$$

where at low temperatures the hole concentration is negligible.

The above three equations are solved simultaneously for n_0 and n_D in non-compensated n-type (P-doped) Si with $N_D = 10^{16}\,\text{cm}^{-3}$ and plotted as functions of temperature in Figure S1B.5a. It can be seen that above 200 K, more than 90 percent of the donors are ionized. Above room temperature (300 K), more than 95 percent are ionized.

---

[2]Each set of donor states in Si is split into three sets of states with slightly different energies.

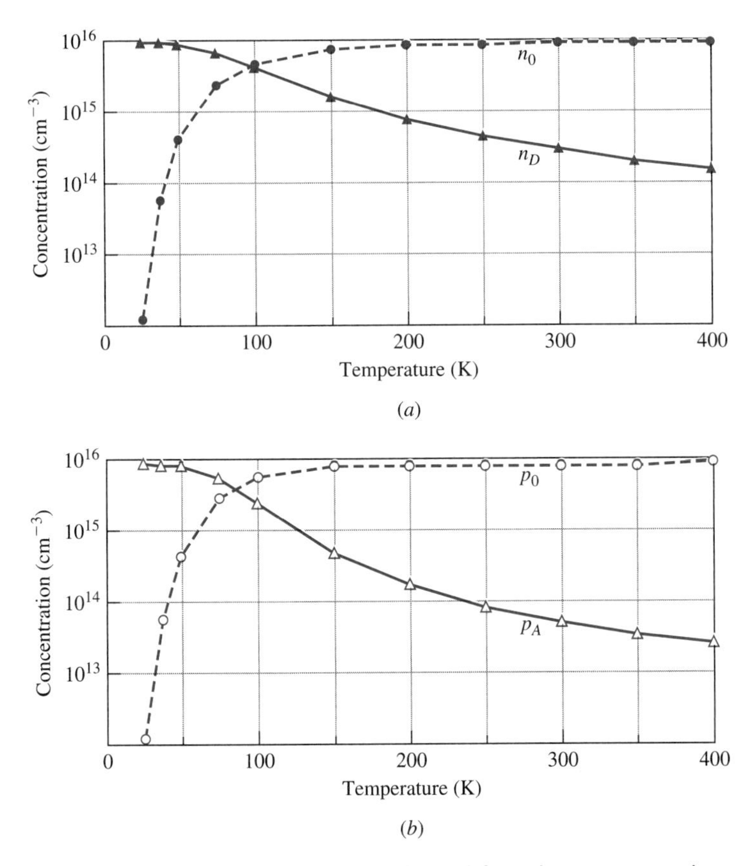

Figure S1B.5 (a) Concentrations of quasi-free electrons n_0 and occupied donor states n_D as a function of temperature for P in Si at a doping concentration of $N_D = 10^{16}$ cm^{-3}. (b) Concentrations of quasi-free holes p_0 and occupied acceptor state p_A for B in Si ($N_A = 10^{16}$ cm^{-3}).

Similar plots for free and bound holes for boron-doped Si are shown in (b). For $T > 300$ K more than 99 percent of the acceptors are ionized.

While the fractional ionization depends slightly on doping level, we can conclude that for semiconductors at room temperature and above, the assumption of complete ionization of impurities is valid.

S1B.4 PHONONS

In Chapter 3, the term *phonon* was used to describe lattice vibrations in crystals. Phonons influence the mean free time between electron or hole collisions and thus the mobilities of the carriers. In Chapter 3, we mentioned that at high electric fields, the drift velocity is limited by carrier interactions with *optical*

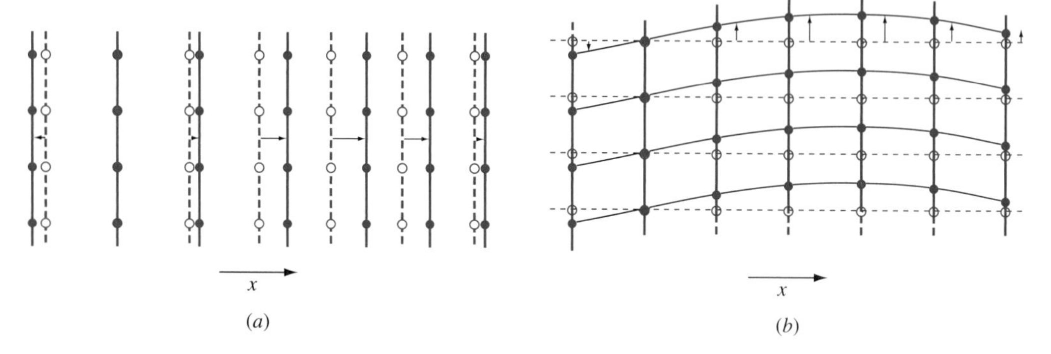

Figure S1B.6 Displacement of atomic planes under the influence of a pressure wave. For a longitudinal wave (a), the displacement is in the direction of motion. For a transverse wave (b), the displacement is transverse to the direction of motion. For a three-dimensional crystal, for each longitudinal wave there are two transverse waves. The dashed lines represent the equilibrium positions, and the solid lines indicate the deflected positions at a given time.

phonons. At low fields, the *acoustical phonons* influence the mobility. In this section, we discuss these phonons in slightly more detail. We will see that phonons can be treated as particles or waves, and since the phonons travel in a periodic structure, they develop energy bands.

Figure S1B.6 shows the longitudinal and transverse pressure waves in a crystal in which the wave is traveling in the x direction. The dashed lines represent planes of atoms at their equilibrium positions, while the solid lines represent the planes displaced by the pressure wave. A longitudinal wave is one in which the displacement is in the direction of motion as indicated in (a). In a transverse wave (b) the displacement is perpendicular to the direction of motion.

We will develop the idea of energy bands for phonons by analogy with electrons. In Chapter 1, electrons traveling in a periodic structure were treated as waves (Bloch waves), with wavelength $\lambda = 2\pi/|K|$, and also as particles of energy E and crystal momentum $\hbar K$. These electrons exist in energy bands, each band with its own E-K relation. All the important information for electrons (effective mass, allowed energies, potential energy, etc.) is contained in the energy band diagrams (E-K relations) of the first Brillouin zone. The first Brillouin zone is in the region $-\pi/a \leq K \leq \pi/a$, where a is the periodicity of the lattice.

Phonons and photons are also waves that can move through the lattice. We are interested in phonons in this section. Phonon waves have some angular frequency ω and some wave vector $|K| = 2\pi/\lambda$.

The energy associated with lattice vibrations of angular frequency ω is given by

$$\boxed{E = \left(n + \tfrac{1}{2}\right)\hbar\omega} \qquad \text{(S1B.25)}$$

where n is an integer. Defining $\hbar\omega$ as the energy of one phonon, E_{phonon},

$$\boxed{E_{\text{phonon}} = \hbar\omega} \tag{S1B.26}$$

The integer n in Equation (S1B.25) indicates the number of phonons at frequency ω. A phonon thus has the characteristics of a wave and a particle, like electrons and photons. The frequencies of the lattice vibrations are on the order of 10^{13} Hz, while the phonon energies are a fraction of an electron volt. Since we consider the interaction of phonons with electrons and photons whose energies are expressed in electron volts, it is convenient to consider phonon energies rather than their frequencies.

A detailed treatment of these lattice vibrations [4] is beyond the scope of this text, but to summarize the major results:

1. There exist, for phonons, Brillouin zones in K space, periodic with periodicity $2\pi/a$ where a is the periodicity of equivalent planes.
2. All information exists in the reduced zone (first Brillouin zone) extending from $-\pi/a \leq K \leq +\pi/a$.
3. This zone is symmetric around $K = 0$.
4. There are as many allowed energy bands as there are atoms per primitive unit cell. Most semiconductors of interest (e.g., Si, Ge, GaAs) have two atoms per primitive unit cell.[3] Thus in these materials there are two energy bands for phonons.
5. In each band, there are three branches, one due to the longitudinal wave and two transverse branches corresponding to the two transverse directions.
6. In a cubic crystal, from symmetry the two transverse branches are the same.

Expressions for the longitudinal branches can be easily calculated. [4] We will not do the calculation here, but the procedure is to treat the primitive unit cell as having two atoms of masses M_1 and M_2 (in silicon these would be the same since all atoms are silicon atoms) connected by a "spring" of spring constant C representing the attractive and repulsive force between successive planes. The resulting E-K diagram is plotted in Figure S1B.7 for the case of unequal masses (e.g., GaAs).

The "acoustical phonons" correspond to the lower phonon energy band, and the "optical phonons" occupy the upper energy band. The reasons for these names will emerge shortly.

We make the following observations from the figure:

1. The group velocity of a phonon at a given frequency is proportional to the slope of the E-K plot, since $v = (1/\hbar)\, dE/dK$.

---

[3]The "primitive unit cell" of a crystal is the smallest arrangement of atoms that when repeated and translated will produce the entire lattice. There are four primitive unit cells in one conventional unit cell in a diamond or zinc blende lattice, where the conventional unit cell is a cube.

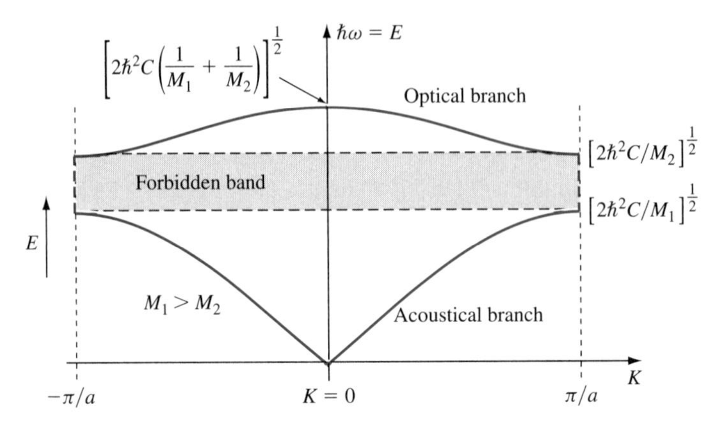

Figure S1B.7 Phonon *E-K* diagram for longitudinal waves in a crystal having two different atoms of masses M_1 and M_2. The maximum frequency is on the order of 10^{13} Hz. The force constant between adjacent planes is *C*.

2. In the lower branch, for small *K* (large λ) the velocity is constant and equal to the velocity of sound, hence the term *acoustical branch.*
3. In the upper branch at $K = 0$, the slope is zero and thus the velocity $v = 0$. Physically, this corresponds to the two adjacent planes in the crystal vibrating out of phase. Since the center of mass is constant in time, the wave does not propagate in the crystal.
4. The top of the upper branch at $K = 0$ corresponds to a frequency on the order of 10^{13} Hz. This is comparable to the frequencies of photons in the infrared. We will see later that this allows phonons to interact with photons. For now it suffices to say that this is why this branch is referred to as the *optical branch.*
5. The forbidden energy band increases with the difference in mass of the two atoms. For elemental semiconductors (e.g., Si, Ge, diamond) $M_1 = M_2$ and the forbidden band disappears.
6. The energy at $K = 0$ in the optical band is dependent on the sum of the reciprocal masses. The smaller the masses, the larger the energy.

Figure S1B.8 shows the phonon *E-K* curves for Si, Ge, and GaAs. The transverse as well as the longitudinal branches are shown in the (100) direction for $K \geq 0$. For Si and Ge, where $M_1 = M_2$, the longitudinal branches have the same value at the zone edge. The masses of Ga and As are only slightly different and so in GaAs a small forbidden energy band exists.

We indicated that the upper band is called the optical band since it can interact directly with optical photons. Consider the situation in which a photon produces a phonon. To transform the optical energy into acoustic energy, both energy and wave vector must be conserved. On the E_{phonon}-*K* diagram of Figure S1B.9, the *E-K* diagram of the photon is superimposed on the phonon

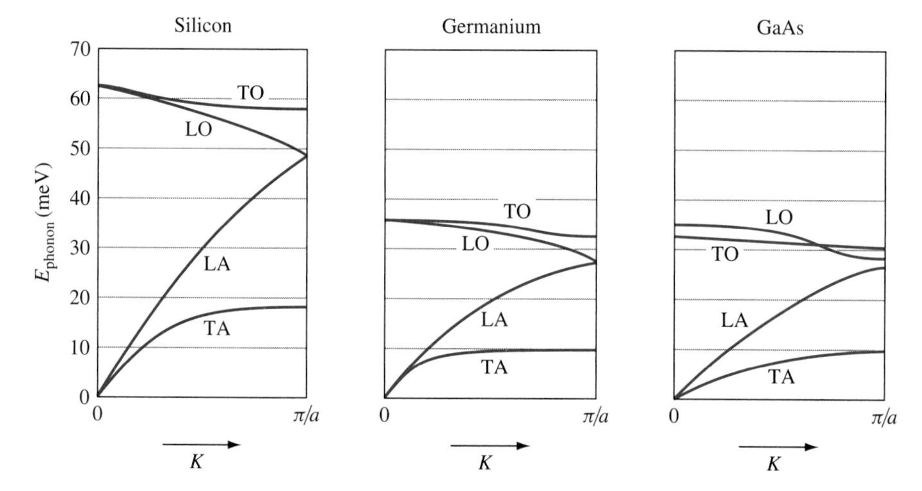

Figure S1B.8 Energy-wave vector diagrams for Si, Ge, and GaAs in the <100> directions. Shown are transverse acoustical (TA), longitudinal acoustical (LA), longitudinal optical (LO) and transverse optical (TO) branches.

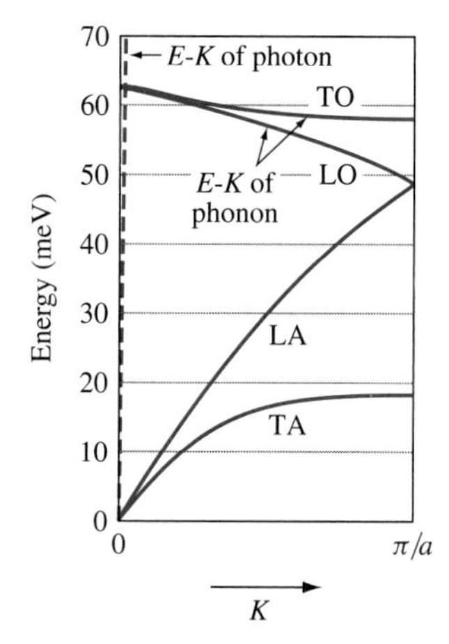

Figure S1B.9 Interaction of a photon and a phonon requires conservation of both E and K. Thus a photon can create a phonon only in the optical band of the phonon's E-K curve.

diagram. The two intersect at a point in the optical branch of the phonon curve. Thus, only phonons in the optical branch can interact directly with photons. Since for a wave, $v = (1/\hbar)\, dE/dK$, and the velocity of light is on the order of 4 orders of magnitude greater than the velocity of sound, the photon E-K characteristic is almost vertical on the phonon E-K characteristic.

Let us compare the energies and K vectors for the photons and phonons. Since $v = (1/\hbar)\, dE/dK$, for a photon of small K (large λ), the velocity of light and thus dE/dK are constant. Thus the photon wave vector is

$$K = \frac{E_{\text{photon}}}{\hbar c}$$

where c is the velocity of light in the material. In silicon, at $K = 0$, $E_{\text{phonon}} = 0.063$ eV. Assuming $c = 3 \times 10^8$ m/s, a photon of light of the same energy, 0.063 eV, has a wave vector $K_{\text{photon}} \approx 2 \times 10^5\ \text{m}^{-1}$. Compare this with the K of a phonon at the edge of the zone: $K_{\text{phonon}} = \pi/a \approx 5.6 \times 10^9\ \text{m}^{-1}$. Thus only phonons near $K = 0$ are able to interact with light.

What about phonon interaction with electrons? Can a phonon excite an electron from the valence band to the conduction band, for example? The range of wave vector for phonons is the same as that for electrons, so from a conservation of K point of view such an event is possible. The phonon energies, however, are much smaller than that needed to excite an electron across the forbidden gap.[4] The phonons can, however, interact with electrons via scattering, as illustrated next.

*S1B.4.1 CARRIER SCATTERING BY PHONONS

Recall from Section 3.3 that electrons can be scattered by lattice vibrations (phonon scattering). There the scattering was described in terms of particle-particle collisions. This scattering can also be described by electron wave–phonon wave interaction.

We know that, at any temperature, the vibrating atoms create pressure waves in the crystal. The pressure waves cause periodic compression and dilation of the atomic spacing, as shown in Figure S1B.10a. Since the band gap is pressure sensitive (it is sensitive to atomic spacing), the regions of compression cause E_g to increase, while regions of dilation tend to decrease E_g. This results in the energy band structure shown in Figure S1B.10b. Electrons in the conduction band are partially reflected (scattered) off these changes in potential energy E_C. This is easier to see for the extreme cases in which E_C is assumed to change abruptly as shown in the idealized model of (c). As the electron travels, it encounters a series of potential barriers, each of which has some probability of reflecting it as indicated in Section S1A.2.6. Similarly, holes are reflected (scattered) by the changes in E_V.

The temperature also has an effect. With increasing temperature, the stronger lattice vibrations cause an increased variation in band gap and thus increased scattering or reduced mobility.

In addition, high-energy electrons could be scattered by the creation of optical phonons. This last effect limits the electron's kinetic energy and thus its velocity (velocity saturation).

[4]However, a number of phonons can simultaneously collide with a valence band electron and excite it to the conduction band.

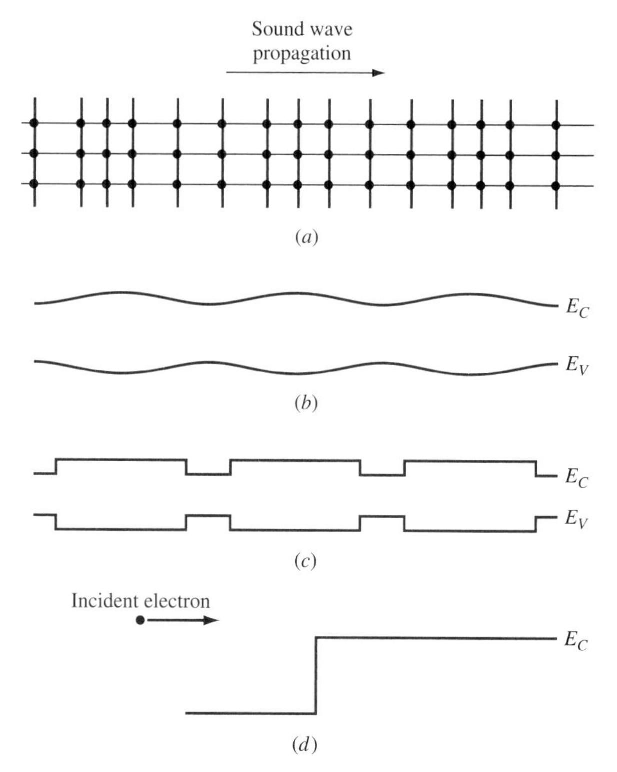

Figure S1B.10 Phonon scattering. Phonons are the particle description of acoustic or compressional waves in the crystal. (a) The phonon causes compression and dilation of the crystal. (b) The varying lattice spacing causes periodic band-gap narrowing and widening. (c) These band-gap variations can be modeled as abrupt steps, in which case they resemble the reflection-from-a-potential-barrier problem as shown in (d).

These two interactions (electron with optical phonons and electrons with acoustical phonons) are illustrated in Figure S1B.11, in which the kinetic energy–wave vector (E_K-K) diagrams of electrons and phonons are superimposed. An electron at position A has a high probability of creating an acoustical phonon of energy E_A and wave vector K_A. To conserve energy the electron must lose this energy, and to remain on its allowed E-K diagram it must simultaneously lose wave vector.

In the above process, an acoustical phonon was created, to carry off the energy and wave vector given off by the electron. In the reverse process, an acoustical phonon can be annihilated, giving its energy and wave vector to an electron, thus increasing its energy and wave vector. In either case the electron is scattered—its energy (and direction) is changed. At equilibrium, the creation and

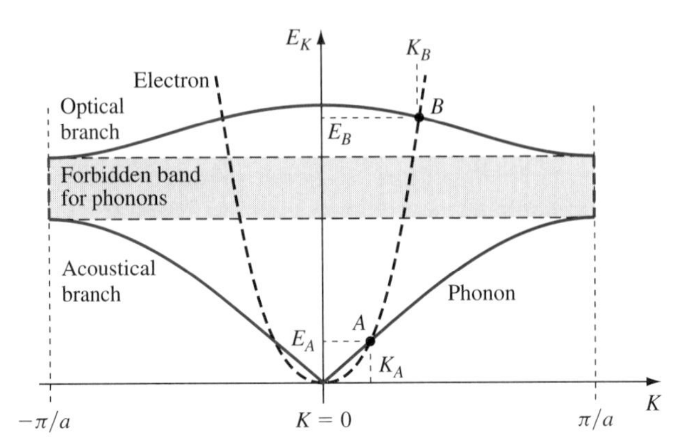

Figure S1B.11 Kinetic energy-wave vector relations for phonons and conduction band electrons in a direct-gap semiconductor. An electron can create an optical phonon of energy E_B and wave vector K_B, or an acoustical phonon of energy E_A and wave vector K_A.

annihilation rates of phonons are equal, and result in mean free times between scattering events on the order of 10^{-13} seconds.

Assume an electron can gain enough kinetic energy to reach the point B in Figure S1B.11. This could happen if the electron is accelerated strongly by a high electric field. An electron at B can lose its energy and wave vector, this time creating an optical phonon of energy E_B and wave vector K_B. Since this interaction is highly probable, it effectively puts a limit on the kinetic energy an electron can achieve, and this is what causes the electron velocity to saturate.

S1B.4.2 INDIRECT ELECTRON TRANSITIONS

Next, we examine the role of phonons in indirect materials. In Chapter 3, we indicated that optical absorption in an indirect gap semiconductor requires a three-particle process involving an electron, a photon, and a phonon. This is illustrated in Figure S1B.12 for the case of Si, where again the E-K diagrams of electrons, photons, and phonons are superimposed. We saw earlier that a single phonon does not have enough energy to excite an electron from the valence band to the conduction band. Similarly, for energies of interest in semiconductor electronics, photons have significant energy $h\nu$ but negligible wave vector. An electron can, however, be excited from the valence band to the conduction band by simultaneously absorbing a photon and a phonon. The phonon furnishes the necessary wave vector K_C and energy $\hbar\omega_{\text{phonon}}$, while the energy furnished by the photon is $h\nu$. The photon energy required to excite an electron across the energy gap is at least:

$$\boxed{h\nu = E_g - \hbar\omega_{\text{phonon}}}$$

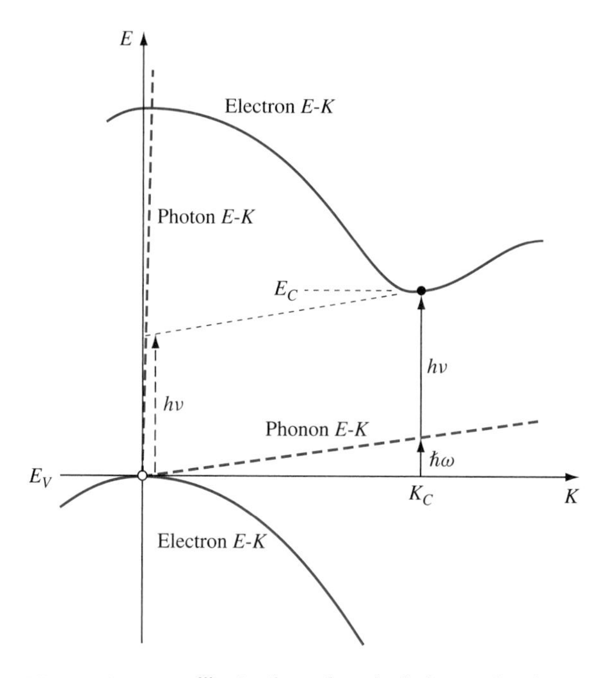

Figure S1B.12 Illustration of optical absorption in an indirect semiconductor involving an electron, a phonon, and a photon. The wave vector needed to make the transitions comes almost entirely from the phonon; the phonon contributes a small amount of energy, $\hbar\omega$, as well, but most of the energy is supplied by the photon.

to conserve energy. To conserve wave vector, we require

$$\boxed{K_{\text{phonon}} \approx K_C}$$

where K_C is the electron wave vector at the bottom of the conduction band.

It is also possible for the band-to-band transition to result from the absorption of a photon accompanied by the *emission* of a phonon (as opposed to absorption of a phonon). The minimum photon energy required in this case is

$$\boxed{h\nu = E_g + \hbar\omega_{\text{phonon}}}$$

and

$$\boxed{K_{\text{phonon}} \approx -K_C}$$

S1B.5 SUMMARY

In this supplement, we have investigated some additional topics in semiconductor theory. First, we discussed the physics of the Hall effect and showed how it can be combined with resistivity measurements to measure the carrier concentration of a sample and find the majority carrier mobilities. We saw that the values of mobility, called the *Hall mobilities,* are not quite equal to the drift mobilities because of the assumption of uniform velocity of all carriers.

We also examined the probability of occupancy of the bound states. We verified our earlier assumptions that at room temperature all the dopants can be assumed to be ionized, to within a few percent accuracy. At lower temperatures, however, carrier freeze-out occurs, reducing the number of carriers and thus the sample conductivity.

We spent some time discussing the role of phonons in semiconductors. Phonons can scatter electrons, thus influencing the mean free time between collisions (and therefore mobility), and they can also assist with electron transitions.

In Part 2 of this book, we will apply our understanding of basic electrical processes in semiconductors to the operation of two-terminal devices, the diodes.

S1B.6 READING LIST

Items 2, 9, 17, and 23 to 25 in Appendix G are recommended.

S1B.7 REFERENCES

1. J. P. McKelvey, *Solid State and Semiconductor Physics,* Harper and Row, New York, 1966, Chapters 7 and 9.
2. J. S. Blakemore, *Semiconductor Statistics,* Dover Publications, New York, 1987, Chapter 3.
3. R. G. Pires, R. M. Dickstein, S. Titcomb, and R. L. Anderson, "Carrier Freezeout in Silicon," *Cryogenics,* 1990, pp. 1004–1068.
4. See for example, C. Kittel, *Introduction to Solid State Physics,* 4th ed., John Wiley and Sons, New York, 1986, Chapter 4.

S1B.8 REVIEW QUESTIONS

1. Explain how a current flowing in the x direction in the presence of a magnetic field in the z direction produces a voltage in the y direction.
2. Why is the probability of occupancy for a bound state different from that for a state in the conduction or valence band?
3. Why does the effective donor state (or acceptor state) get progressively closer to the conduction band edge (valence band edge) as the doping gets larger?

4. Explain the difference between a transverse acoustic wave and a longitudinal acoustic wave. Can an electromagnetic wave (for example, a photon) be longitudinal?
5. Explain how an electron scatters from an acoustic wave.
6. With the help of Figure S1B.12, explain how an electron in an indirect material can make a band-to-band transition. What is the probability of this transition compared with that for a direct gap material?

S1B.9 PROBLEMS

S1B.1 a. A sample of silicon is subjected to a resistivity test as in Figure S1B.1. The sample dimensions are $t = 0.001$ cm, $W = 0.01$ cm, and $L = 0.1$ cm. The current between points A and C is 1 mA and the measured voltage is 6.25 V. What is the resistivity of the sample?

b. The same sample is now placed in a Hall effect setup. The applied magnetic field is $B_z = 0.5\,\text{T} = 5 \times 10^{-5}$ Wb/cm^2. The current is 10 mA and the measured Hall voltage is $-3.12\,\mu$V.

i. What is the net carrier concentration?

ii. Is the sample n type or p type?

iii. Find the Hall mobility of the majority carriers.

S1B.2 a. What is donor ground-state energy $E_C - E_D$ for phosphorus in silicon at zero kelvins?

b. What is the room temperature effective donor energy for lightly doped Si, e.g., $N_D = 10^{14}\,\text{cm}^{-3}$?

c. Repeat for Si doped n type with $N_D = 10^{17}\,\text{cm}^{-3}$?

d. What is the effective acceptor state energy $(E_A^* - E_V)$ at room temperature for boron in silicon with $N_A = 10^{17}\,\text{cm}^{-3}$?

e. A sample is doped p type with $N_A = 10^{17}$ and $N_D = 10^{16}$. What is the effective donor state energy?

S1B.3 Consider the phonon *E-K* diagram of Figure S1B.7, for a crystal with two types of atoms, of mass M_1 and M_2 in a zinc blende lattice.

a. For a material consisting of a single atom type (e.g., silicon), the lattice becomes a diamond lattice and $M_1 = M_2$. Explain why the optical phonon energy for diamond (carbon) is greater than that for silicon, which in turn is greater than that for germanium.

b. Explain why the saturation velocity of germanium is less than that for silicon.

c. Explain why the saturation velocity of electrons in GaAs is less than its peak value.

S1B.4 Acoustic waves travel faster in materials that are stiffer (harder). The group velocity of phonons is proportional to the slope of the *E-K* diagram for acoustic phonons (as in Figure S1B.7). At the edge of the Brillouin zone, this slope is zero, corresponding to standing waves. Traveling waves would be found nearer to the origin on the *E-K* diagram. Use the phonon *E-K* curve to predict which should be harder, diamond (carbon) or lead, assuming they have the same force constant *C*, and comparable lattice constants.

PART 2

Diodes

We now have the necessary background in the electronics of semiconductor materials to progress to devices. Most electronic devices depend on the electrical characteristics of junctions between different materials. Usually (except for the case of superconductor junctions), at least one of these is a semiconductor. Such junctions are two-terminal devices and are referred to as *diodes.*

In Part 2 of this book, we investigate the electrical characteristics of a variety of junctions. These include:

1. *Homojunctions.* Junctions between two differently doped regions of the same semiconductor material.
2. *Heterojunctions.* Junctions between two different materials. Although technically this includes combinations such as a semiconductor-insulator junction, in common usage *heterojunctions* refers to junctions between two different semiconductors.
3. *Metal-semiconductor junctions.* Junctions between a metal and a semiconductor.

An understanding of semiconductor junctions and their electrical characteristics is essential to the understanding of most semiconductor devices—transistors, lasers, etc.—since these devices are composed of such junctions.

Chapter 5 deals with diodes. A semiconductor diode is a junction. It is also a two-terminal device that acts as a switch: for one polarity of applied voltage it acts as a near open circuit and for the other polarity it acts as a near short circuit. This is illustrated by the current-voltage characteristics of a Si homojunction diode in Figure II.1. For an applied voltage less than about 0.7 V, the current is small enough that the diode can be considered an open circuit. For applied voltages greater than about 0.7 V, the current increases rapidly with voltage and the diode can be approximated by a voltage source of 0.7 V. An approximate equivalent circuit is indicated in Figure II.2. A generalized symbol for a diode is indicated in Figure II.3. Current flows in the direction indicated for any voltage

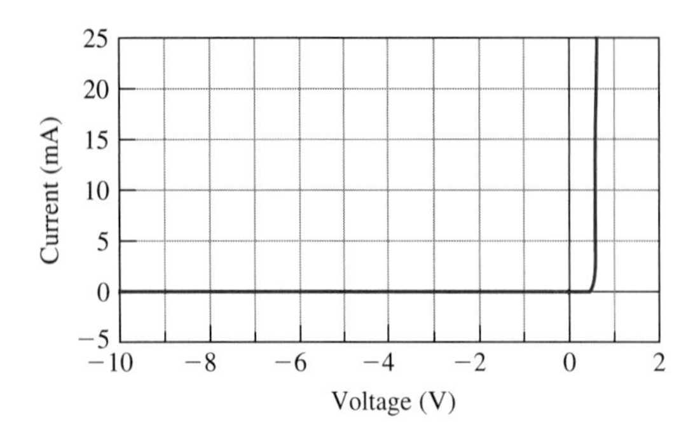

Figure II.1 The current-voltage characteristic of a silicon homojunction diode. For $V < 0.7$ V, the current can be considered negligible. For current flow, the diode voltage is nearly constant and equal to 0.7 V.

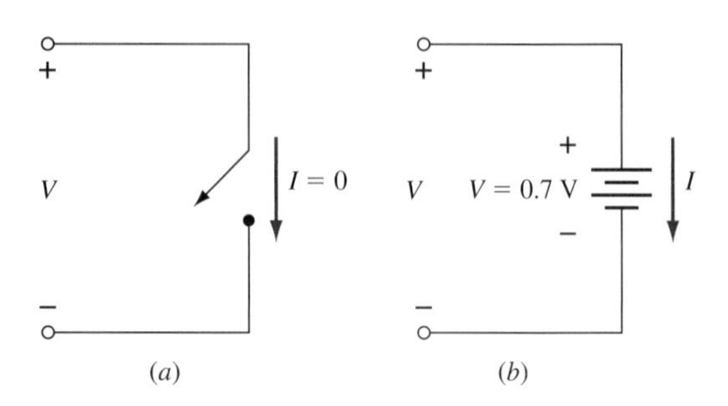

Figure II.2 (a) Approximate equivalent circuit for a silicon homojunction diode. For $V < 0.7$ V, the diode acts as an open circuit. (b) For an applied voltage greater than this, the diode approximates a constant voltage $V = 0.7$ V.

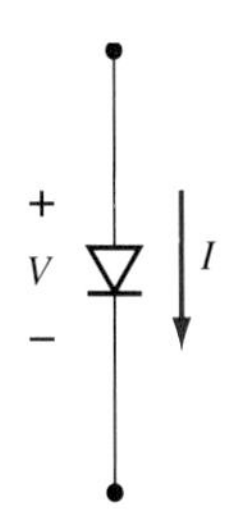

Figure II.3 Symbol commonly used to represent a diode. For forward bias ($V > 0$) current flows in the direction indicated. For reverse bias ($V < 0$) the current is essentially zero.

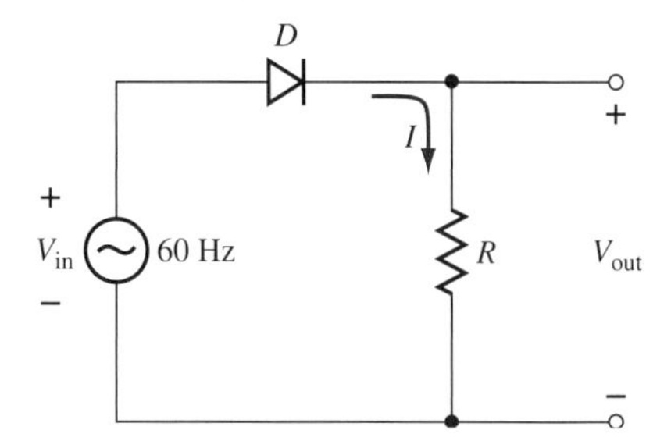

Figure II.4 Circuit diagram for a rectifier. The output voltage is $V_{out} = IR$. The diode conducts, and thus current flows only for forward bias of $V_{in} > 0.7$ V.

greater than zero, but until the voltage reaches about 0.7 V, the current is negligibly small.

A common application of a diode is in a rectifier circuit as indicated in Figure II.4. Here an input 60 Hz sine wave is applied to a series combination of a diode D and a resistor R. The output voltage ($V_{out} = IR$) is taken across the resistor.

EXAMPLE

Consider the rectifier circuit of Figure II.4 in which $V_{in} = 5\sin(2\pi ft)$, where $f = 60\,\text{Hz}$, $R = 2\ \text{k}\Omega$ and the diode is represented by the equivalent circuit of Figure II.2. Plot the input and output waveforms and the current waveform.

■ Solution

The input waveform is given as $V_{in} = 5\sin(2\pi ft) = 5\sin(377t)$, as indicated in Figure II.5. For $V_{in} < 0.7$ V the current is zero and thus $V_{out} = 0$. For $V_{in} > 0.7$ V the

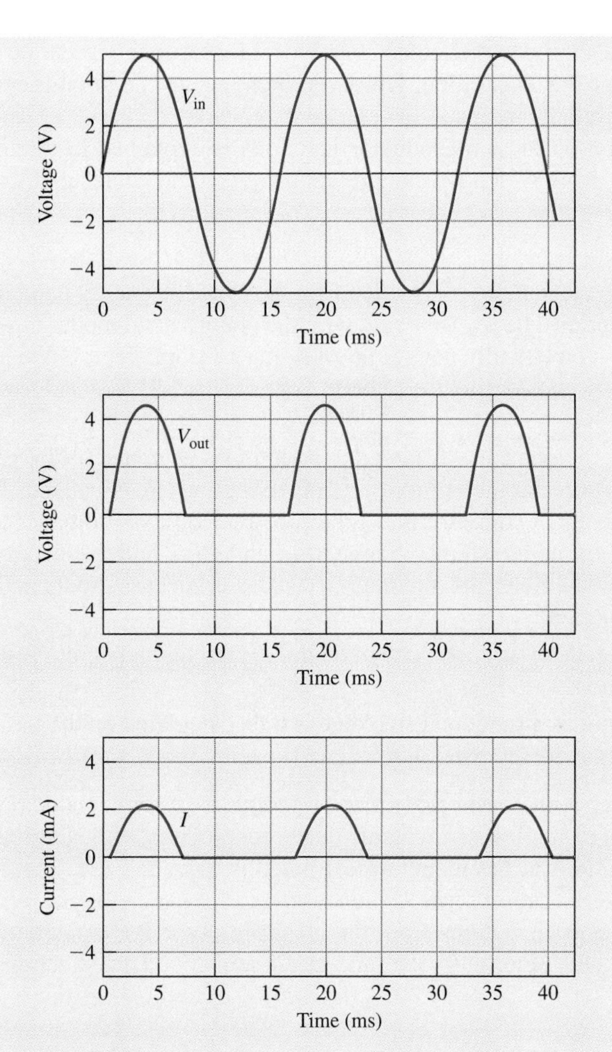

Figure II.5 Input voltage, output voltage and current for the rectifier circuit of Figure II.4, with $V_{in} = 5\sin(\omega t)$, $R = 2\ \text{k}\Omega$, and frequency of 60 Hz ($\omega = 2\pi f = 377$ rad/s).

voltage across the diode is 0.7 V and thus $V_{out} = V_{in} - 0.7$ V. The current then is

$$I = \frac{V_{out}}{R} = \frac{5\sin 377t - 0.7\ \text{V}}{2\ \text{k}\Omega}$$

This particular circuit is referred to as a half-wave rectifier. It converts an ac voltage, which has zero average value, to a pulsating voltage whose average is now nonzero. With additional circuitry this pulsating voltage can be smoothed out to produce a constant dc voltage.

The diode equivalent circuit of Figure II.2 is reasonably accurate for the rectifier circuit of the illustration. However, most electronic circuits require more accurate equivalent circuits as developed in the next chapter.

The operation of semiconductor junctions is explained in terms of energy band models, from which we can predict the electrical characteristics of the devices. To draw the energy band diagrams for junctions the following procedure is used:

1. We use the technique of Chapter 4 for drawing the energy band diagram of each material. That is, we begin with the energy band model in which each material is electrically neutral in every macroscopic region. We normally start out by considering the regions to be electrically isolated from each other.
2. Next, the regions are considered to be in intimate contact. Charge will transfer, such that the Fermi levels in the regions are equalized. The energy band diagram at equilibrium is then constructed, taking into account the locations of the transferred charge, the band gaps, and the electron affinities and work functions and any interface states or electric dipole in the vicinity of the junction.
3. The influence of applied voltage on the energy band diagram is then investigated.
4. The current as a function of voltage is then calculated on the basis of the shape of the energy band diagram and specific models for carrier transport.

In Chapter 5, the basic principles of operation of semiconductor diodes are discussed. As examples, semiconductor homojunctions with step function doping concentrations at the metallurgical junction are analyzed. These structures can be considered as *prototype homojunctions.*

In Chapter 6, deviations from the ideal prototype homojunction model are treated along with heterojunctions and metal-semiconductor junctions. ■

CHAPTER 5

Prototype pn Homojunctions

5.1 INTRODUCTION

A pn homojunction (often called simply a pn junction) consists of a single crystal of a given semiconductor in which the doping level changes from p type to n type at some boundary. The term *homojunction* implies that the junction is between two regions of the same material (e.g., silicon), as opposed to the term *heterojunction* in which the junction is between two different semiconductors (for example, Ge and Si).

There are several methods of fabricating pn junctions. A common technique is to implant phosphorus (donor) atoms into a p-type Si substrate. The resulting structure is shown schematically in Figure 5.1a. In Figure 5.1b a cross-sectional view is shown along the cut A-A′. Since the depth of the n-type region is small compared with its lateral dimension, for most purposes the side effects can be ignored and the properties of the junction can be determined by examining the cross section B-B′. This is indicated schematically in Figure 5.1c.

A typical doping profile for ion implantation is shown schematically in Figure 5.1d. The doping in the p-type substrate (N_A) is assumed constant throughout the crystal. When the donors are implanted, they penetrate the substrate with high energy, causing crystal damage. A donor implantation is followed by a short annealing step (heating the crystal up to eliminate defects). Wherever there are more donors than acceptors ($N_D > N_A$), the semiconductor is n type in that region. It is p type wherever $N_A > N_D$. The metallurgical junction is at x_0, the point at which $N_D = N_A$, or the net doping is zero. It is the net doping profile that determines the energy band diagram, so in Figure 5.1e, $N'_D = N_D - N_A$ and is plotted as a function of position.

To solve for the electrostatic properties of the junction, the N'_D profile must be known, but it is typically not a simple mathematical function. Thus, an

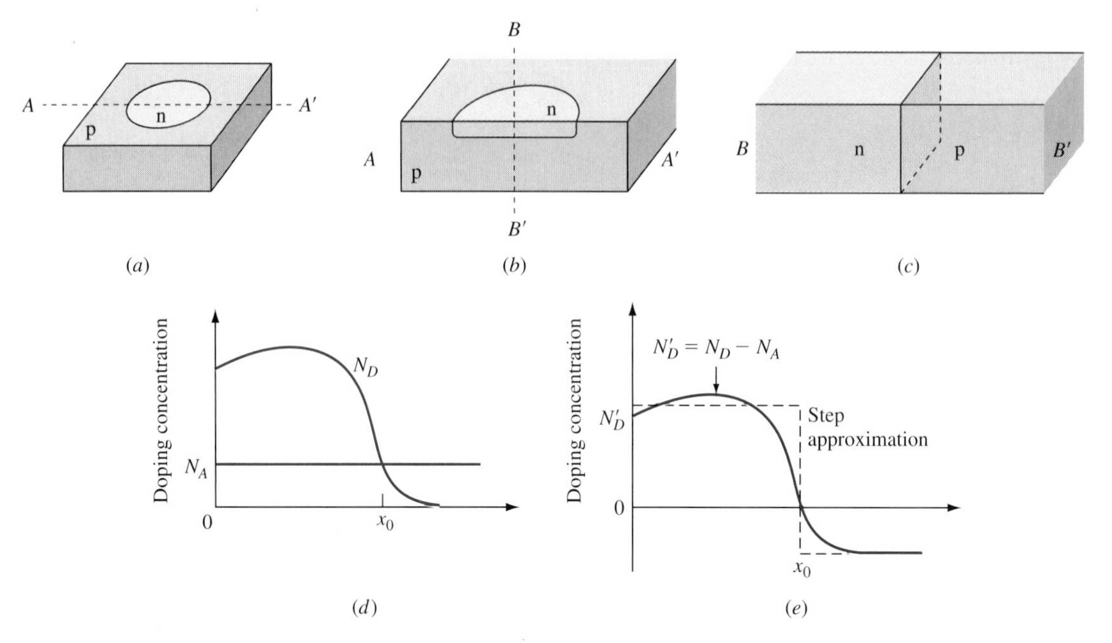

Figure 5.1 (a) The physical picture of a planar pn junction; (b) cross section through $A - A'$; (c) schematic representation of the pn junction; (d) typical doping profile showing a p-type substrate with implanted donors (the junction occurs where $N_D = N_A$); (e) the net doping concentration $N_D - N_A$ for this junction, and the step approximation (dashed line).

approximation to the $N_D - N_A$ profile is often used.[1] The most common of these is the step approximation, in which the net doping is assumed to be a step function, as shown by the dashed line in Figure 5.1e.

To develop some physical understanding of diodes, we will analyze this simplified model (step junction) of a semiconductor homojunction. We refer to this as the *prototype homojunction*. The purpose is to illustrate the basic principles of operation of semiconductor diodes. In practice, examples of the prototype junction are seldom encountered, but the simplifying approximations permit analytical calculations for many of the device properties. These results, then, can be used as first approximations for more realistic junctions, which are considered in Chapter 6.

The approximations used in the step-junction model are:

1. The doping profile is a step function. On the n-type side, $N'_D = N_D - N_A$ and is constant. On the p side, $N'_A = N_A - N_D$ and is constant.
2. All impurities are ionized. Thus the equilibrium electron concentration on the n side is $n_{n0} = N'_D$. The equilibrium hole concentration on the p side is $p_{p0} = N'_A$.

[1]However, in device simulators using numerical solution methods, the more complicated profiles can be used.

3. Impurity-induced band-gap narrowing effects are neglected. Therefore, for the purposes of this simple model, if one side of the junction is degenerate (N_D, $N_A > 4 \times 10^{18}$ cm^{-3} in Si) the Fermi level is assumed to be at E_{C0}, the bottom of the intrinsic conduction band (n type) or at E_{V0} for p type.[2]

5.2 PROTOTYPE pn JUNCTIONS (QUALITATIVE)

The first step to understanding any junction is to draw its energy band diagram.

5.2.1 ENERGY BAND DIAGRAMS OF PROTOTYPE JUNCTIONS

Electrical Neutrality To draw the energy band diagram for the prototype pn junction, we begin by imagining that the n and p regions are physically separated, and are electrically neutral in every macroscopic region. Again, by *electrically neutral* we mean that there is no region having more positive charges than negative, a situation that will change when the materials are in contact. The energy band diagram of each of the two isolated semiconductors is given in Figure 5.2. The electron affinity χ, the ionization potential γ, and the energy gap E_g are indicated for each material. The subscript n indicates n-type semiconductor while the subscript p indicates p-type semiconductor. Also shown is an

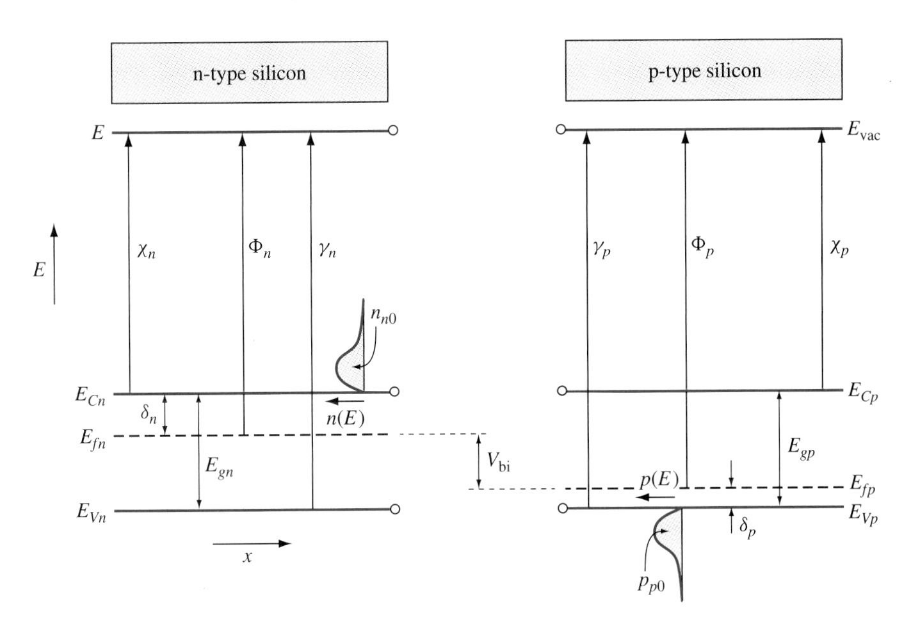

Figure 5.2 Two similar semiconductors, doped differently, before being joined in a junction.

[2]Actually, the Fermi level is generally within the (reduced) conduction band in degenerately doped materials, as discussed in Section 2.11.

additional parameter, the work function Φ. The work function is equal to the energy difference between the vacuum level and the Fermi level, $\Phi = E_{\mathrm{vac}} - E_f$.

Because of the assumption of space charge neutrality everywhere, the energy required for an electron to escape the material is the same in any region. Therefore, the vacuum level is the same for either material at any position. It is convenient to choose as reference the vacuum level for each material on the edge facing the other material. The circles in the vacuum level represent this.

Since the material is silicon on both sides, $\chi_n = \chi_p$, $E_{gn} = E_{gp}$, and $\gamma_n = \gamma_p$. This implies that the bottom of the conduction band is (for neutrality) at the same energy for both materials, and $E_{Cn} = E_{Cp}$. Similarly $E_{gn} = E_{gp}$ and $E_{Vn} = E_{Vp}$. Since electron affinities and ionization potentials are constant, E_C and E_V at the material edges are secondary references as indicated by the additional circles in Figure 5.2. However, because the doping is different in the two materials, the positions of the Fermi levels are not the same, and thus $\Phi_n \neq \Phi_p$.

Equilibrium Upon contact between the two materials,[3] electrons flow (diffuse) from the n-type semiconductor to the p-type semiconductor because there are more quasi-free electrons on the n side than on the p side. As the electrons move toward the p-type region, they leave behind ionized donors (charged positively) that are locked into the crystal lattice. At the same time, holes flow from the p semiconductor to the n semiconductor, leaving behind negatively charged acceptors. This separation of charges sets up an electric field, as shown in Figure 5.3. This is the situation at equilibrium. The presence of the electric field is evident because there is a gradient in the vacuum level and in the band edges. We also observe that the Fermi level is now continuous across the entire sample (recall Section 4.1). We will discuss this figure in more detail shortly, but first let us examine the currents. Because the concentrations of electrons and holes are different on either side of the junction, we expect diffusion current to flow across the junction. At the same time, the presence of an electric field sets up drift current as well. In the transition region between n and p, the electron and hole currents are, from Equations (3.46) and (3.47),

$$\boxed{\begin{aligned} J_n &= q\mu_n n\mathscr{E} + qD_n\frac{dn}{dx} = q\mu_n\left[n\mathscr{E} + \frac{kT}{q}\frac{dn}{dx}\right] \\ J_p &= q\mu_p p\mathscr{E} - qD_p\frac{dp}{dx} = q\mu_p\left[p\mathscr{E} - \frac{kT}{q}\frac{dp}{dx}\right] \end{aligned}}$$

At equilibrium there is no net current, so $J_n = J_p = 0$ and the Fermi levels are equalized. A built-in field $\mathscr{E}$ is generated in what is referred to as the *transition region*. In the transition region, the field produces a drift current for electrons that at every position exactly compensates the diffusion current caused by the electron concentration gradient. A similar balance of hole drift and diffusion currents exists.

[3]Here we must assume that in contact the atomic periodicity is continued across the pn interface, or that the entire structure is a single crystal.

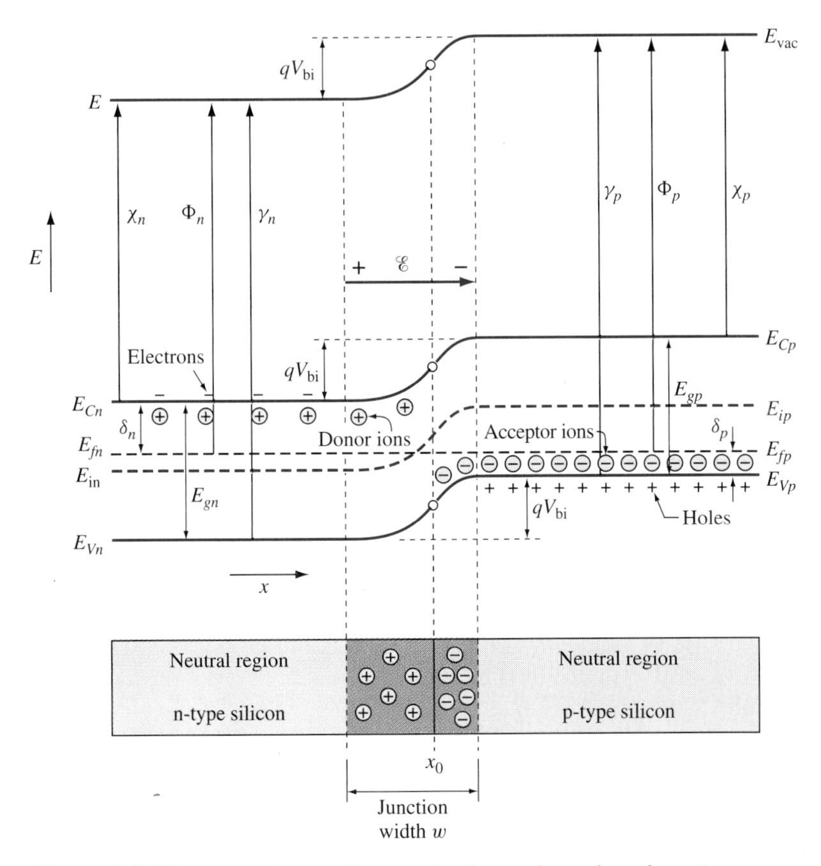

Figure 5.3 The energy band diagram for the pn homojunction at equilibrium.

The built-in field at the junction, however, alters the bands. Since we are using the vacuum levels *at the interface* as a reference, this point will remain unchanged. Also, χ and γ are fundamental properties of silicon, so they will remain unchanged as well. The result is that the vacuum level, conduction band and valence band all pivot about the positions indicated by the circles at the interface, as Figure 5.3 shows. The details of the shape of this energy band diagram will be discussed later. Notice, however, that:

1. Near the interface, the electrons from the n-type semiconductor fill the holes on the p side resulting in a region of negative charge (non-neutralized acceptors) on the p side. Similarly, electrons annihilate holes that diffused into the n-type region, giving a region of positive charge (non-neutralized donors) on the n side. This creates an electric field at the interface between non-neutralized (positive) donor ions in the n region and non-neutralized (negative) acceptor ions in the p region. By *non-neutralized,* we mean that there are no corresponding free electrons or free holes in the same region to neutralize the charge of the ions.

2. The energies E_{vac}, E_C, and E_V are everywhere parallel. This is a direct result of the quantities χ, γ, and E_g being constant.
3. The electron affinity is unchanged across the device. However, an electron escaping from the bottom of the conduction band in the n region to the vacuum level on the right side must overcome the electrostatic potential of the built-in field in addition to the electron affinity of the material.
4. On each side of the junction, there is a region of uncompensated charge. This *space charge region*[4] extends on both sides of the interface, and contains non-neutralized impurity ions. Its width depends on the concentration of impurities on each side and the charge transfer required to align the Fermi levels.
5. The concentration of free carriers in the space charge region is negligible. The carriers are swept out by the electric field. The space charge region is said to be depleted of carriers, and thus is often also referred to as the *depletion region.*
6. The total space charge on either side of the junction is the same (but has opposite sign).
7. A built-in potential energy barrier qV_{bi} exists across the junction, where V_{bi} is referred to as the *built-in voltage.* The magnitude of V_{bi} is proportional to the energy difference in the Fermi levels in Figure 5.2 for the case of neutrality:

$$\boxed{qV_{bi} = \Phi_p - \Phi_n} \tag{5.1}$$

   where as indicated earlier, Φ_p and Φ_n are respectively the work functions of the two semiconductors. By convention, V_{bi} is taken as a positive quantity.
8. The potential energy barrier is the same for the conduction band as for the valence band. This implies that the potential energy barrier is the same for electrons as for holes.
9. Because the magnitude of the space charge on either side of the junction is equal, for equal doping levels, the width of the space charge region is the same on each side of the junction. For unequal doping levels most of the space charge region is on the side with the lighter doping.
10. The built-in voltage increases with increased doping level on either side, resulting from the dependence of work function on doping level [Equation (5.1), Figure 5.2].
11. The electric field, which is proportional to the slope of the vacuum level:

$$\boxed{\mathscr{E} = \frac{1}{q}\frac{dE_{vac}}{dx}} \tag{5.2}$$

[4]Another name for the transition region.

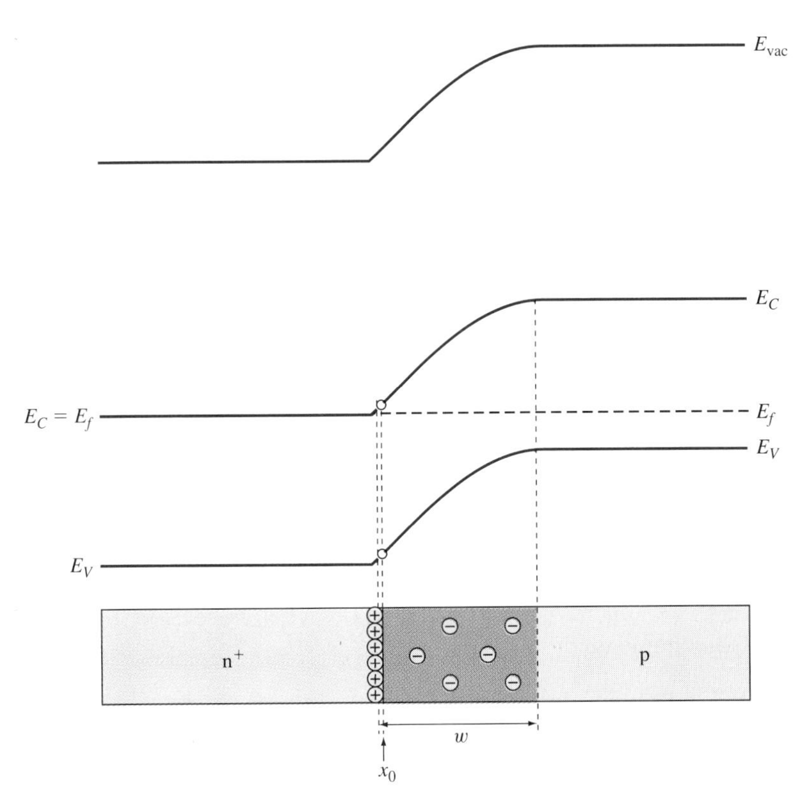

Figure 5.4 The one-sided n$^+$p junction has one heavily doped side. The designation n$^+$ indicates degenerately doped n type. On the degenerately doped side, we approximate that the Fermi level is at the conduction band edge.

has its maximum value at the metallurgical junction, x_0. This will be shown in Section 5.3. The field given by Equation (5.2) is the true electric field, as discussed in Chapter 4.

12. The built-in voltage is mostly across the more lightly doped region.

Recall from item 9 that the transition region is primarily on the more lightly doped side. For a junction with one side degenerate and the other side nondegenerate, essentially all of the depletion region will be on the lightly doped side, as indicated in Figure 5.4. This is referred to as a *one-sided step junction*. In this class are n^+p and p^+n junctions. The notation n^+ indicates degenerately or heavily doped n type, and p^+ indicates heavily doped p-type material.

Energy Band Diagram under Bias We have seen how to determine the energy band diagram for a pn homojunction at equilibrium. Now we investigate what happens to the energy band diagram when a voltage is applied. Consider the case of Figure 5.5, in which the p region is made negative with respect to the n region by the applied voltage V_a. By convention, V_a is measured from the p side to the n side, i.e., with the n side as reference, and is considered to be negative (reverse bias) if

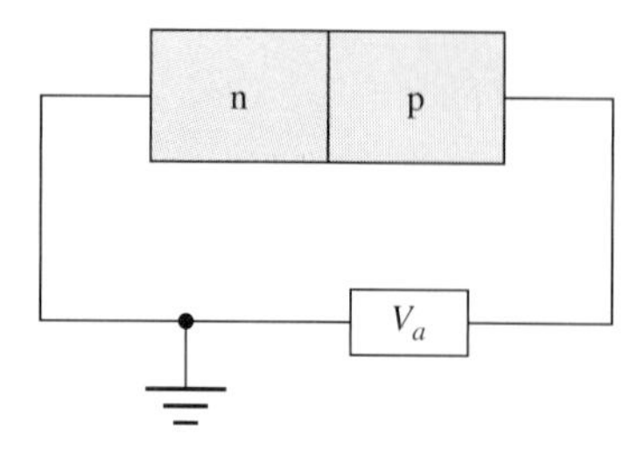

Figure 5.5 A pn homojunction with a bias V_a applied. By convention the applied voltage is measured from p side to n side, or with the n side as a reference. Thus, the diode is forward biased if V_a is positive, and reverse biased if V_a is negative.

it has the same polarity as the built-in voltage and to be positive (forward bias) if it is of opposite polarity. We will consider the case of reverse bias first, so V_a is negative (for example, $V_a = -1$ V).

But how is V_a distributed across the device? Let us consider the device to consist of three regions:

1. The region from the left terminal to the edge of the junction depletion region on the n side. This is called a *quasi-neutral* region (often called a *neutral region*), since in this area the net number of donors is equal (or nearly equal) to the number of electrons in the conduction band.
2. The depletion region itself, which contains ions but virtually no (or a negligible number of) free carriers.
3. The region from the edge of the depletion region on the p side to the right terminal (also quasi-neutral).

We proceed by considering the pn junction as a series connection of the resistances of these three regions. Recall that resistivity is inversely proportional to the concentration of free carriers at a point, i.e., n_{n0} on the n side and p_{p0} on the p side. In the transition region, however, the carrier concentration is much less than either of these values, since virtually all the free carriers are swept out by the built-in electric field. This implies that the resistance of the transition region is much greater than that of the other two, and virtually all of the applied voltage is dropped across this depletion region.

To adjust the equilibrium energy band diagram to reflect the applied bias, we proceed as we did before in constructing the energy band diagrams at equilibrium. Using the vacuum level at the metallurgical junction as a reference, the energy band diagrams will pivot around the circles, as indicated in Figure 5.6. Since the p side is made electrically more negative, that represents higher potential energy for electrons and that side of the energy band diagram moves upward. The n side is more positive, representing a lower potential energy for the electrons, and thus the n side moves downward on the energy band diagram.

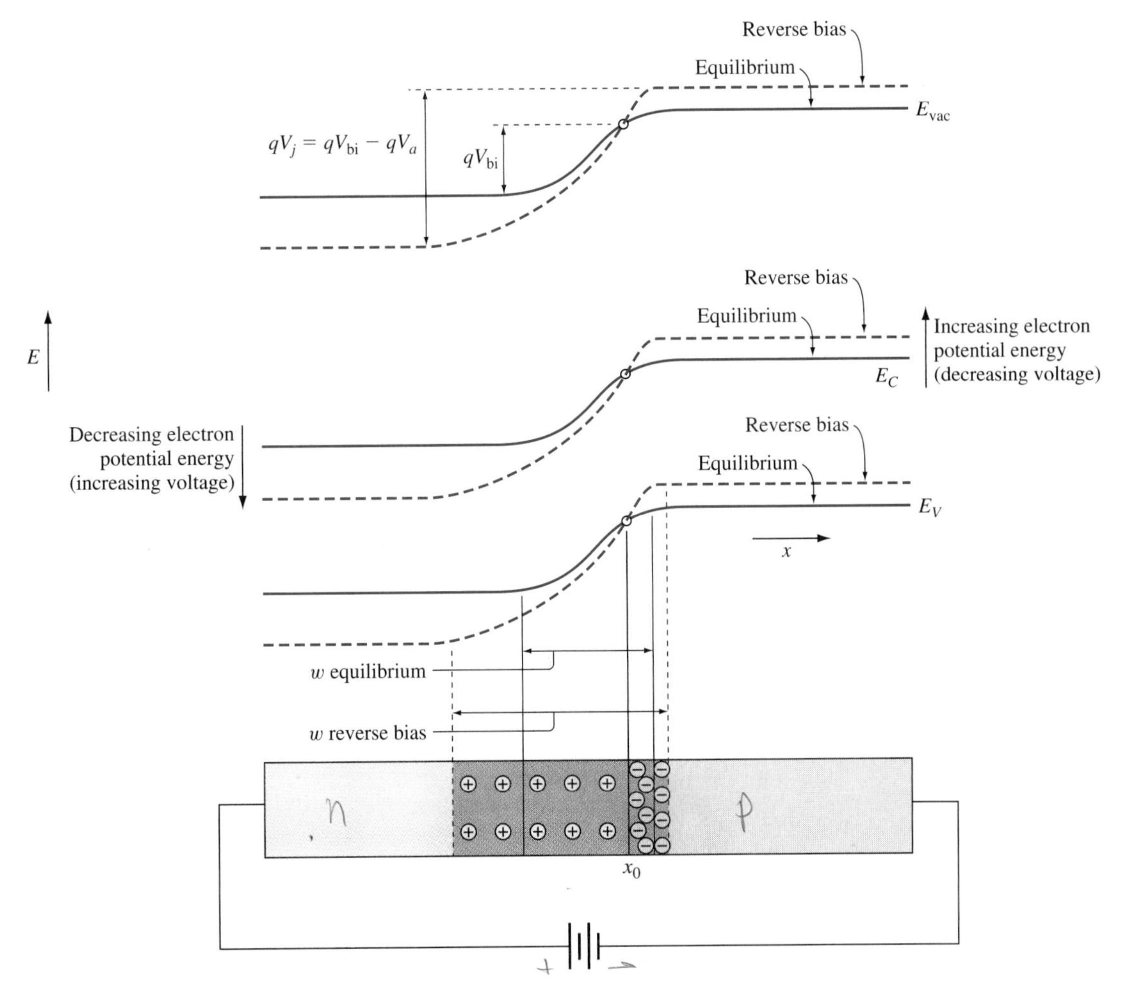

Figure 5.6 The pn homojunction under reverse bias. Solid line: equilibrium energy band diagram; dashed line: energy band diagram under reverse bias. The field increases; this requires more ionized acceptors and donors, so the depletion region gets wider under reverse bias.

Figure 5.6 shows the result. The applied negative bias effectively increases the potential barriers for both electrons and holes.

We can see from Figure 5.6 that the junction voltage V_j is increased:

$$\boxed{V_j = V_{bi} - V_a} \tag{5.3}$$

(with V_a negative because of reverse bias), resulting in a greater field at the metallurgical junction. The internal electric field is generated, however, by the positively charged acceptor and donor ions near the junction, so this increased field requires an increased quantity of charge on either side of the junction. Since the charge is a result of ionized impurities that are fixed in the crystal lattice

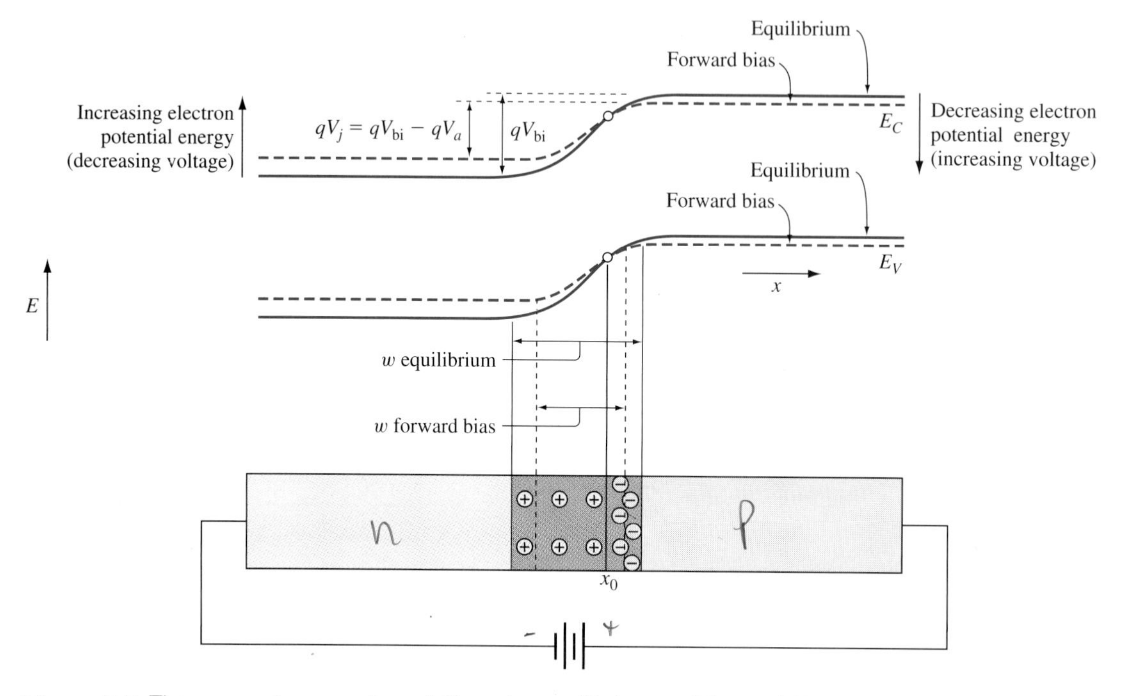

Figure 5.7 The space charge region width under equilibrium and forward bias, and the corresponding energy band diagrams.

(remember that the mobile carriers are swept away from this region by the field), the transition region must expand on each side. Therefore the width of the transition region increases, as shown in the bottom of Figure 5.6.

Similarly, for a positive or forward bias V_a, the voltage across the junction V_j and the transition width both decrease as shown in Figure 5.7. Here the vacuum level is not indicated since it is parallel to E_C and E_V, and thus holds no new information.

5.2.2 DESCRIPTION OF CURRENT FLOW IN A pn HOMOJUNCTION

Now let us qualitatively discuss current flow in a diode. We will start by considering the currents at equilibrium, and then progress to the situations under bias.

Equilibrium The energy band diagram for a diode at equilibrium is shown in Figure 5.8. The electrons are diffusing from their region of high concentration (the n side) to their region of low concentration (the p-type material). There is also a built-in electric field at the junction that produces an electron drift that exactly cancels (at equilibrium) the electron diffusion at every point. Similarly, the holes are diffusing from p to n, but the electric field forces them back toward the p region. Neither a net electron nor hole current flows.

Electron generation and recombination occur throughout the device. In the neutral n region, the generation and recombination rates are equal at equilibrium.

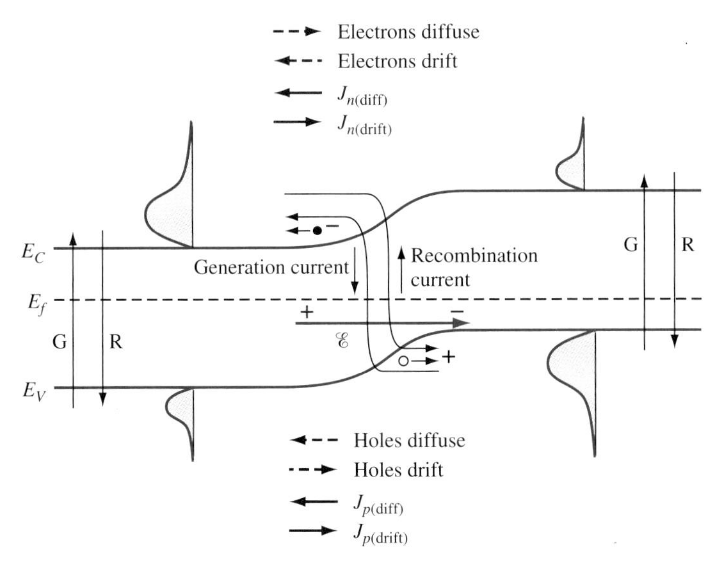

Figure 5.8 The pn homojunction at equilibrium. The built-in field in the transition region causes drift currents that exactly oppose the diffusion current resulting from the different carrier concentrations along the junction. Within the transition region, the generation (G) and recombination (R) currents are equal and opposite at any point.

Thus, since there is no field in this region and no concentration gradients, the generation-recombination processes do not contribute to current. From a similar argument the generation-recombination process in the neutral p region does not produce current.

Within the transition region, however, a field does exist. What is the effect on generated or recombining electron-hole pairs? The electrons generated in the transition region are accelerated toward the n side, producing a current from n to p, as indicated in the figure. Note that the hole produced by the same generation event is accelerated to the p side. This also produces current in the direction n to p. Thus, an electron-hole pair generation event in the transition region produces current across the junction.

Similarly, recombining electrons produce a current in the opposite direction. An electron entering the depletion region from the n side recombines with a hole supplied by the p side. This produces a net current from p to n. At equilibrium, the generation and recombination rates at any position are equal and so the net generation-recombination current is zero.

Reverse Bias Next, we consider the currents that flow under reverse bias. The equilibrium case is shown again in Figure 5.9a, and, for comparison, the reverse-bias case is shown in Figure 5.9b. Under reverse bias, the electrons in the n material (majority carriers) attempt to diffuse toward the p-type material (region of lower concentration). However, there is a potential energy barrier there. A

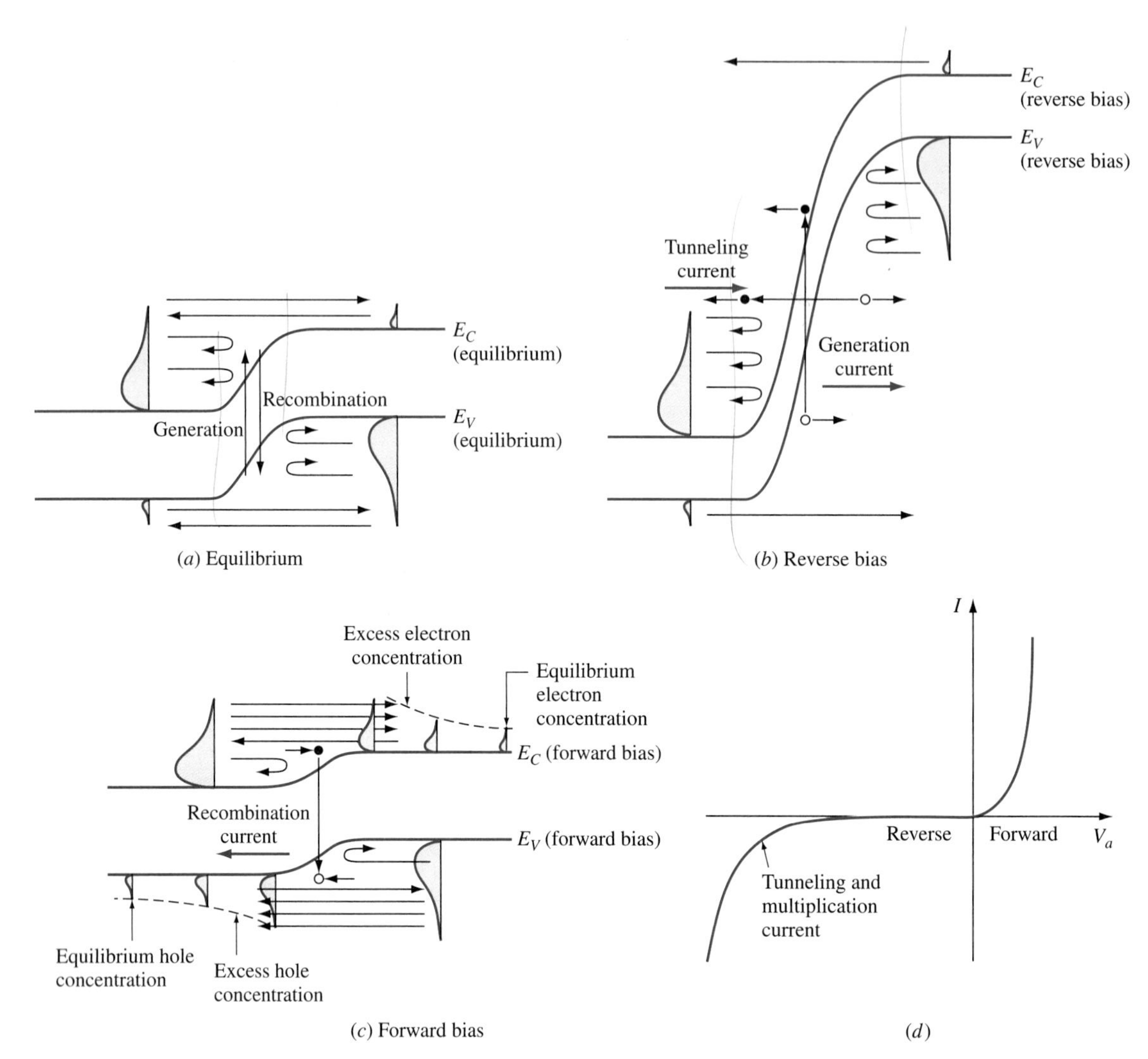

Figure 5.9 Qualitative view of current flow in a pn junction diode. (a) Under equilibrium, both diffusion currents are cancelled by opposing drift currents; (b) under reverse bias, only a small number of carriers are available to diffuse across the junction (once within the junction they drift to the other side). With increasing reverse bias the reverse current increases due to tunneling and carrier multiplication. (c) Under forward bias, the drift current is slightly reduced but the diffusion current is greatly increased; (d) the expected current-voltage characteristic.

negligible number of electrons have enough kinetic energy to surmount the energy barrier. As for the rest of the electrons, as they travel to the right, their kinetic energy decreases and they eventually stop, whereupon the junction's field accelerates them back to the left. Thus from this process there is negligible net electron current across the junction.

The few conduction band electrons on the right of the junction, however, also have some kinetic energy, and at any given moment, 50 percent of them are traveling to the left. If some of those electrons wander into the transition region, the field accelerates them and they cross into the n-type region. Thus, there is a net electron current, but it is very small since there are few electrons available to participate.

Similarly, for holes, those on the p side tend to diffuse to the n side where their concentration is smaller, but they also are prevented from crossing the junction by the barrier. The very small number of holes on the n side that diffuse to the junction, however, are accelerated to the right. The current resulting from electrons and holes diffusing to the junction and being swept across by the field is referred to as *minority carrier diffusion current.*

There are three additional current mechanisms under reverse bias. A generation current is produced by electrons in the valence band in the transition region being thermally excited into the conduction band then swept to the n side. The hole produced by each generated electron is swept to the p side. Note that recombination in the depletion region under reverse bias is negligible, since both the electron and hole concentrations there are miniscule. A second current is a tunneling current. Although electrons on the n side see a high potential barrier, the barrier can be quite thin, as shown in the figure, and some electrons from the valence band on the right can tunnel through the forbidden region into the available states in the conduction band on the left. Tunneling was discussed in the Supplement A to Part 1 and is addressed again in Section 5.3 under "Reverse-Bias Tunneling." This tunneling current, then, is caused by electrons in the valence band of the p side that tunnel through the forbidden region into the conduction band on the n side, and are swept to the left. The generation and tunneling currents under reverse bias are normally referred to as *leakage currents* because of their small size.

A third current mechanism, *carrier multiplication,* results from electron or hole collisions within the depletion region, which produce electron-hole pairs. These additional carriers also contribute to current. This mechanism is considered in more detail in Section 5.3 under "Reverse-Bias Carrier Multiplication and Avalanche."

Forward Bias For forward bias, we refer to Figure 5.9c. Because the potential energy across the junction is reduced in this case, an appreciable number of electrons now have sufficient energy to cross the junction. Thus, there is a net diffusion of electrons across the barrier from the n side to the p side, producing net electron current from p to n.

Similarly, the applied voltage also reduces the barrier to holes; thus there is a net hole flow and a net hole current from p to n. The hole and electron diffusion currents can add up to a significant current under forward bias.

Once the electrons are injected into the p region, their charge is immediately (within the dielectric relaxation time—on the order of 10^{-12} to 10^{-13} seconds) neutralized by an equal number of oppositely charged holes, which are supplied through the ohmic contact to the p material. The dielectric relaxation time is discussed in more detail in the Supplement to Part 2, but the point here is that except in the transition region, the p region is therefore virtually electrically neutral ($\mathscr{E} \approx 0$). Similarly, outside of the transition region the n side is quasi neutral. As indicated earlier, this region is referred to as being *quasi-neutral.*

Under forward bias, there are excess minority carriers near the junction but outside of the transition region that have been injected across the junction. That is, there are excess electrons near the junction on the p side and excess holes near the junction on the n side. The concentrations of excess electrons Δn and holes Δp decay by recombination as they diffuse away from the junction, as shown in Figure 5.9c. But in the quasi-neutral regions, the electric field $\mathscr{E} \approx 0$, so electron flow into the p region is almost entirely by diffusion. A similar argument applies to holes injected into the n-type region. Thus, the minority carrier currents are often referred to as *diffusion currents.*

What about farther from the junction, where there are no excess carriers? What maintains the current flow? In the quasi-neutral regions, away from the junction, the majority carriers carry the current by drift. Although we said the field is zero in these regions, in reality the field is finite but small. Still, the number of majority carriers available to carry current is so large that the diode current can be maintained by drift even with a small electric field.

We noted earlier that with increasing distance from the junction, the concentration of excess minority carriers decreases because of recombination. This results in decreasing electron diffusion current and a corresponding increase in hole drift current to keep the total current constant.

An additional current under forward bias conditions results from electrons recombining with holes *within* the transition region. While at equilibrium the recombination and generation rates are equal, for forward bias the recombination rate predominates because of the increased concentration of both electrons and holes in the depletion region (as they diffuse across).

For the prototype junction considered here, tunneling cannot occur under forward bias. This is because tunneling occurs at constant energy. There are no allowed states at the same energies across the gap from the valence band, as there were in the reverse bias case. Therefore, electrons cannot tunnel from the valence band on the p side to the conduction band on the n side.

To summarize the current mechanisms under forward bias, when a voltage is applied that reduces the potential barrier, electrons and holes are injected across the depletion region. When they reach the other side, they become minority carriers. As they diffuse away from the junction, they recombine, so the excess minority carrier concentration decreases (exponentially) away from the junction, and thus so does the minority carrier diffusion current. The majority carriers make up the difference between the minority carrier diffusion current and the total current. The majority carriers travel by drift. The field is small in the quasi-neutral regions but the majority carrier concentrations are large.

Normally in a pn junction the reverse bias leakage currents are small compared with the currents under forward bias. Thus, to good approximation, the pn junction is a device that permits current to flow in only one direction. The pn junction is often called a *junction diode.*

Current as a function of voltage is discussed quantitatively in the next section. Qualitatively, however, let us consider what to expect. We know that at equilibrium, the number of electrons and holes that have enough energy to diffuse across the barrier is equal to the number that drift back as a result of the built-in field. We also know that electron and hole distribution functions vary exponentially with energy. Therefore, if the barrier is lowered by some amount V_a, the number of carriers that can diffuse across the barrier increases exponentially. (Note that they diffuse against the electric field.) We expect that under forward bias, then, the current across the junction will increase exponentially with V_a, as shown in Figure 5.9d.

Under reverse bias, the barrier is increased, but the barrier height has no influence on the diffusion current from n to p since the number of minority carriers available doesn't change with barrier height. However, with increasing reverse bias the junction width increases, resulting in increased generation current. In addition, the tunneling distance (at constant energy) decreases, causing increased tunnel current. Further, with increasing reverse bias, the field strength within the transition region increases. Thus, the carriers gain more kinetic energy between collisions. During a collision, if the kinetic energy is large enough, an electron-hole pair can be created, resulting in an increase in multiplication current. Thus, we might qualitatively expect a complete current-voltage (I-V_a) characteristic to look something like Figure 5.9d.

Finally, we comment that the diffusion current under forward bias is often referred to as *injection current*. Electrons from the n-type region are injected into the p-type region (over the barrier) and holes from the p region are injected into the n region.

5.3 PROTOTYPE pn HOMOJUNCTIONS (QUANTITATIVE)

We now have a qualitative understanding of the operation of a prototype pn homojunction in terms of its energy band diagram and transport processes. In this section, we will obtain a quantitative description of the energy band diagram of a pn homojunction. We will also obtain quantitative descriptions of the various current mechanisms. Once these characteristics are known and understood, we can find the junction I-V_a characteristics.

5.3.1 ENERGY BAND DIAGRAM AT EQUILIBRIUM (STEP JUNCTION)

Since the junction characteristics depend on the built-in voltage, we will start by finding an expression for V_{bi}. From Equation (5.1),

$$\boxed{qV_{\text{bi}} = \Phi_p - \Phi_n} \tag{5.4}$$

This can be expressed as

$$\boxed{qV_{\text{bi}} = E_g - (\delta_n + \delta_p)} \tag{5.5}$$

where the quantity δ_n is the energy difference between conduction band edge and the Fermi level in the neutral region on the n side of the junction, as indicated in Figure 5.2. It is given by

$$\delta_n = E_C - E_f = kT \ln \frac{N_C}{n_{n0}} = kT \ln \frac{N_C}{N'_D} \qquad \text{n side, nondegenerate} \tag{5.6}$$

If the material is degenerately doped, we take the Fermi level to be *at* the band edge (remember we are neglecting band-gap narrowing for the prototype pn junction). In this case

$$\delta_n = 0 \qquad \text{n side, degenerate} \tag{5.7}$$

Similarly, δ_p is the energy between the Fermi level and the top of the valence band in the neutral p region:

$$\delta_p = E_f - E_V = kT \ln \frac{N_V}{p_{p0}} = kT \ln \frac{N_V}{N'_A} \qquad \text{p side, nondegenerate} \tag{5.8}$$

and

$$\delta_p = 0 \qquad \text{p side, degenerate} \tag{5.9}$$

As indicated earlier, we usually refer to a junction in which both sides are nondegenerate as a *pn junction*. Often, however, one side is degenerately doped, so the nomenclature then is that an n^+p junction has the n side degenerate and the p side nondegenerate. A p^+n junction is the opposite.

Let us now calculate the built-in voltage of the step pn junction of Figure 5.3. From Equations (5.5), (5.6), and (5.8), we get

$$qV_{bi} = \left[E_g - kT\left(\ln \frac{N_C}{N'_D} + \ln \frac{N_V}{N'_A}\right)\right] \qquad \text{pn junction} \tag{5.10}$$

If the junction were n^+p or p^+n, then δ_n or δ_p would be zero, respectively, and Equation (5.10) would be reduced to:

$$qV_{bi} = \left[E_g - kT \ln \frac{N_V}{N'_A}\right] \qquad n^+p \text{ junction}$$
$$qV_{bi} = \left[E_g - kT \ln \frac{N_C}{N'_D}\right] \qquad p^+n \text{ junction} \tag{5.11}$$

We can simplify Equations (5.10) and (5.11) by recognizing that for a nondegenerate semiconductor,

$$\boxed{\begin{aligned} n_i^2 &= N_C N_V e^{-E_g/kT} \\ E_g &= kT \ln \frac{N_C N_V}{n_i^2} \end{aligned}} \tag{5.12}$$

Then from Equations (5.5) through (5.12) we can write

$$\boxed{\begin{aligned} V_{\text{bi}} &= \frac{kT}{q} \ln \frac{N_D' N_A'}{n_i^2} && \text{pn junction} \\ V_{\text{bi}} &= \frac{kT}{q} \ln \frac{N_V N_D'}{n_i^2} && \text{p}^+\text{n junction} \\ V_{\text{bi}} &= \frac{kT}{q} \ln \frac{N_C N_A'}{n_i^2} && \text{n}^+\text{p junction} \end{aligned}} \tag{5.13}$$

For a pn junction, we can also express the built-in voltage as

$$\boxed{V_{\text{bi}} = \frac{E_g}{q} - \frac{kT}{q} \ln \frac{N_C N_V}{N_D' N_A'} \qquad \text{pn junction}} \tag{5.14}$$

This form shows us that the built-in potential approaches the value of the band gap only if both N_D' and N_A' approach N_C and N_V, respectively. Because for Si, $N_C \approx N_V$, the built-in voltages of n^+p and p^+n junctions are virtually the same for a given doping on the lightly doped side.

If impurity-induced band-gap narrowing (see approximation 3 in Section 5.1) were included, the built-in voltage would be somewhat reduced from that in Equations (5.13). We will discuss this in Chapter 9 when we discuss bipolar junction transistors, where the band-gap narrowing has an appreciable effect on the transistor characteristics.

Let us apply our knowledge of how to calculate the built-in voltage of a junction.

EXAMPLE 5.1

Calculate the value of the built-in voltage for a Si pn junction in which the n region is uniformly doped with 10^{16} net donors per cm^3 and the p region has a uniform net acceptor concentration of 10^{15} per cm^3.

■ **Solution**

From Equation (5.13),

$$V_{bi} = \frac{kT}{q} \ln\left[\frac{N_A' N_D'}{n_i^2}\right] = \frac{(0.026\text{ eV})(1.6 \times 10^{-19}\text{ J/eV})}{1.6 \times 10^{-19}\text{ C}} \ln\left[\frac{10^{16}10^{15}}{(1.08 \times 10^{10})^2}\right]$$

$$= 0.026\text{ V} \ln[8.57 \times 10^{10}] = 0.655\text{ V}$$

For a one-sided step junction, the appropriate version of Equation (5.13) would be used.

Because these one-sided step junctions are so common, it is worthwhile to plot Equation (5.13) and thus avoid many repetitive calculations. A plot of V_{bi} for a silicon one-sided step junction in which one side is degenerate is shown in Figure 5.10 as a function of net doping on the lightly doped side. The curves for a p^+n and an n^+p junction are indistinguishable. If impurity-induced band-gap narrowing is considered, the results of Figure 5.10 will be slightly reduced—by the value of the apparent band-gap narrowing, ΔE_g^* in the degenerate material.

5.3.2 ENERGY BAND DIAGRAM WITH APPLIED VOLTAGE

Next we will quantitatively examine the energy band diagram of a pn junction under applied bias. To determine the shape of the energy band diagram, and see how it varies with applied voltage, we start with the variation of charge density Q_V as a function of position. From that we can find the electric field $\mathscr{E}(x)$, and from the electric field we can determine the variation of the voltage $V(x)$ with position in the junction.

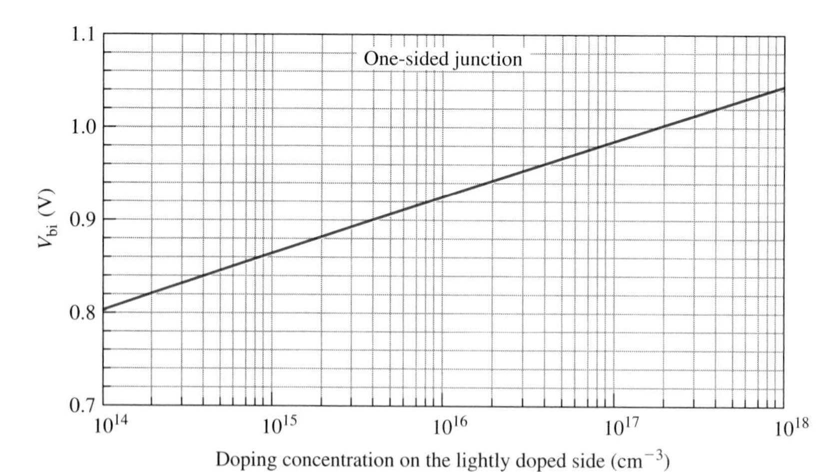

Figure 5.10 The built-in voltage in a one-sided step junction in silicon, as a function of the net doping concentration on the lightly doped side. The curves for the p^+n and n^+p junctions are indistinguishable on the plot.

We will use a coordinate system as shown in Figure 5.11a where x_0 is the position of the metallurgical junction. The material is n type for $x < x_0$ and p type for $x > x_0$. The boundary of the depletion region on the n side is x_n and on the p side is x_p. The width of the depletion region on the n-type side is w_n, and the width on the p side is w_p. The overall width is w. The n side is more heavily doped than the p side in this example.

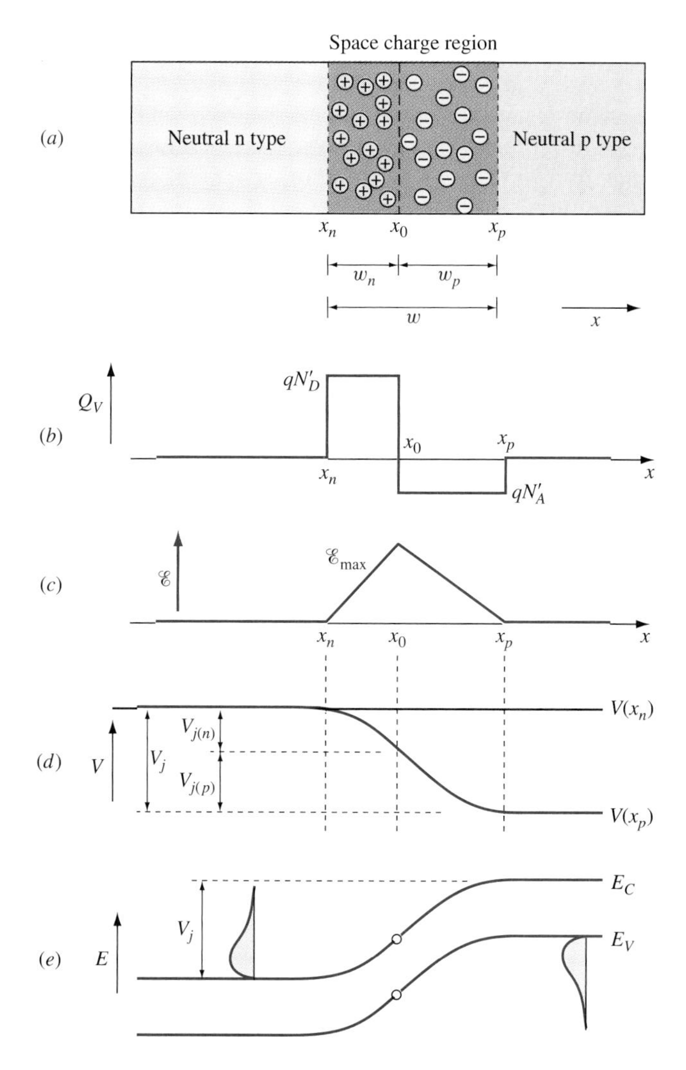

Figure 5.11 A prototype homojunction: (a) the physical diagram; (b) the distribution of charge; (c) the electric field, obtained by integrating the charge; (d) the voltage, obtained by integrating the field; (e) the energy band diagram, with the same shape as the voltage but inverted.

We know the distribution of charge in the device. Under the step approximation, the net doping is constant on the n side and on the p side. In the quasi-neutral regions, the ionized dopants are compensated since there are electrons near the donor ions and holes in the region of acceptor ions. Thus in the quasi-neutral regions there is no uncompensated charge that could give rise to an electric field. In the junction, however, the mobile carriers have been swept out, leaving uncompensated positive donor ions on the n side and uncompensated negative acceptor ions on the p side. These charges, then, will set up the built-in electric field.

Earlier we observed that if a bias is applied, the depletion width changes. Under reverse bias, the junction voltage increases and the depletion width gets wider as more uncompensated charges become "uncovered." Under forward bias the junction voltage decreases and the transition region narrows. We now show this mathematically. The following derivation is general and applies under any bias condition.

To find the electric field $\mathscr{E}(x)$, we solve Poisson's equation:

$$\frac{d\mathscr{E}}{dx} = \frac{Q_V(x)}{\varepsilon} \tag{5.15}$$

along with the appropriate boundary conditions. The charge density $Q_V(x)$ is the charge per unit volume at a given position x (the V subscript is for volume, to remind us that this is a charge density, not a total charge), and ε is the permittivity of the semiconductor.[5] Note that $\mathscr{E}(x)$ is continuous in x.

To solve Poisson's equation we need to know $Q_V(x)$. The possible charges are the electrons, the holes, and the ionized donors and acceptors. We expect, however, very few electrons or holes in the depletion region itself. We therefore use the *depletion approximation* that, in the transition region, $n = p = 0$.[6] Since we are assuming that all dopants are ionized,[7] on the n-type side, where $x_n < x < x_0$,

$$Q_V = qN'_D \tag{5.16}$$

and for $x_0 \leq x \leq x_p$

$$Q_V = -qN'_A \tag{5.17}$$

since the ionized acceptors are negatively charged. We further assume that $Q_V(x)$ is zero and thus $\mathscr{E}(x) = 0$ outside the depletion region:

$$\mathscr{E}(x) = 0 \qquad x \leq x_n \quad \text{and} \quad x \geq x_p \tag{5.18}$$

The charge density under the depletion approximation is plotted in Figure 5.11b.

---

[5]The permittivity is given by $\varepsilon = \varepsilon_r \varepsilon_0$, and ε_r, the relative permittivity (dielectric constant) for silicon is 11.8.

[6]Actually, there are electrons and holes diffusing and drifting across this region, but their numbers are small.

[7]At low (cryogenic) temperatures, it is possible that not all donors and acceptors are ionized. In this case we must subtract the number of un-ionized donors n_D from the total number of donors N_D, and similarly subtract the un-ionized acceptors n_A from the total acceptor concentration N_A. Including these, plus accounting for possible electrons and holes, yields $Q_V(x) = q[N_D - n_D - n - N_A + p + p_A]$. The un-ionized donor and acceptor concentrations are discussed in the Supplement B to Part 1.

Using these expressions in Equation (5.15), on the n side for $x_n \le x \le x_0$ we have

$$\int_0^{\mathscr{E}(x)} d\mathscr{E} = \int_{x_n}^{x} q\frac{N'_D}{\varepsilon}\,dx \tag{5.19}$$

and on the p side for $x_0 < x < x_p$,

$$\int_{\mathscr{E}(x)}^{0} d\mathscr{E} = -\int_{x}^{x_p} q\frac{N'_A}{\varepsilon}\,dx \tag{5.20}$$

On the n side the result is

$$\mathscr{E}(x) = q\frac{N'_D}{\varepsilon}(x - x_n) \qquad x_n \le x \le x_0 \tag{5.21}$$

and on the p side

$$\mathscr{E}(x) = q\frac{N'_A}{\varepsilon}(x_p - x) \qquad x_0 \le x \le x_p \tag{5.22}$$

Since $\mathscr{E}(x)$ is continuous, matching these solutions at $x = x_0$ gives

$$qN'_D(x_0 - x_n) = qN'_A(x_p - x_0) \tag{5.23}$$

The electric field for the step junction increases linearly with x on the n side, and decreases linearly on the p side, as plotted in Figure 5.11c. The maximum of $\mathscr{E}(x)$ occurs at x_0, the metallurgical junction.

Earlier we had predicted that the junction would extend further into the lightly doped side. We will see now whether we were right. Letting w_n and w_p be the transition region widths on the n and p sides respectively, from Equation (5.23) we can write

$$\boxed{\frac{w_n}{w_p} = \frac{(x_0 - x_n)}{(x_p - x_0)} = \frac{N'_A}{N'_D}} \tag{5.24}$$

Therefore if N'_D is the larger doping concentration (on the n side), then w_p is the larger depletion width, verifying our prediction.

Now we find the functional form of $V(x)$, the voltage distribution. We use the expression

$$\mathscr{E} = -\frac{dV}{dx} \tag{5.25}$$

On the n side, we integrate V from $x = x_n$ to $x \le x_0$ and find:

$$\int_{V(x_n)}^{V(x)} dV = -\int_{x_n}^{x} \mathscr{E}(x)\,dx = -\int_{x_n}^{x} \frac{qN'_D}{\varepsilon}(x - x_n)\,dx \tag{5.26}$$

or

$$V(x) - V(x_n) = -\frac{qN'_D}{2\varepsilon}(x - x_n)^2 \qquad x_n \le x \le x_0 \tag{5.27}$$

Likewise for the p side:

$$V(x_p) - V(x) = -\frac{qN'_A}{2\varepsilon}(x_p - x)^2 \qquad x_0 \le x \le x_p \tag{5.28}$$

We can find the voltage drops across the n side of the transition region and across the p side by equating Equations (5.27) and (5.28) at $x = x_0$:

$$V(x_n) - V(x_0) = V_j^n = \frac{qN'_D}{2\varepsilon}(x_0 - x_n)^2 = \frac{qN'_D}{2\varepsilon}{w_n}^2 \tag{5.29}$$

and

$$V(x_0) - V(x_p) = V_j^p = \frac{qN'_A}{2\varepsilon}(x_p - x_0)^2 = \frac{qN'_A}{2\varepsilon}{w_p}^2 \tag{5.30}$$

where $V_j{}^n$ and $V_j{}^p$ are the voltages across the n and p sides of the junction respectively.

Then we can find the total voltage across the junction, V_j:

$$V_j = V_j{}^n + V_j{}^p = \frac{q}{2\varepsilon}[N'_D(x_0 - x_n)^2 + N'_A(x_p - x_0)^2] \tag{5.31}$$

These quantities are indicated on the plot of $V(x)$ shown in Figure 5.11d. Note that the voltage decreases from x_n to x_p, since the electric field is positive.

Since $V(x)$ is continuous, we match Equations (5.27) and (5.28) at x_0:

$$V(x_0) = V(x_n) - \frac{qN'_D}{2\varepsilon}(x_0 - x_n)^2 = V(x_p) + \frac{qN'_A}{2\varepsilon}(x_p - x_0)^2 \tag{5.32}$$

Evaluating Equations (5.29) and (5.30) at x_0 and taking the ratio,

$$\frac{V(x_n) - V(x_0)}{V(x_0) - V(x_p)} = \frac{V_j{}^n}{V_j{}^p} = \frac{N'_D {w_n}^2}{N'_A {w_p}^2} \tag{5.33}$$

With the aid of Equation (5.24), Equation (5.33) becomes

$$\frac{V_j{}^n}{V_j{}^p} = \frac{N'_A}{N'_D} \tag{5.34}$$

indicating that most of the junction voltage is dropped across the more lightly doped region.

Next, let us find expressions for the junction widths. From Equations (5.29), (5.32), and (5.33) we can obtain an expression for w_n:

$$w_n = (x_0 - x_n) = \left[\frac{2\varepsilon V_j{}^n}{qN'_D}\right]^{1/2} = \left[\frac{2\varepsilon V_j}{qN'_D\left(1 + \frac{N'_D}{N'_A}\right)}\right]^{1/2} \tag{5.35}$$

Similarly

$$w_p = (x_p - x_0) = \left[\frac{2\varepsilon V_j{}^p}{qN'_A}\right]^{1/2} = \left[\frac{2\varepsilon V_j}{qN'_A\left(1 + \frac{N'_A}{N'_D}\right)}\right]^{1/2} \tag{5.36}$$

which results in a total junction width

$$w = w_n + w_p = \left[\frac{2\varepsilon V_j(N'_A + N'_D)}{qN'_A N'_D}\right]^{1/2} \qquad \text{pn junction} \tag{5.37}$$

Solving for the junction voltage gives

$$V_j = \frac{qN'_D N'_A w^2}{2\varepsilon(N'_D + N'_A)} \qquad \text{pn junction} \tag{5.38}$$

For a one-sided junction, the junction is almost entirely on the lightly doped side. For an n$^+$p junction, $w \approx w_p$, since in this case $N'_D \gg N'_A$. From Equation (5.37) we have

$$w = \left[\frac{2\varepsilon V_j}{qN'_A}\right]^{1/2} \qquad \text{n}^+\text{p} \tag{5.39}$$

and for a p$^+$n junction

$$w = \left[\frac{2\varepsilon V_j}{qN'_D}\right]^{1/2} \qquad \text{p}^+\text{n} \tag{5.40}$$

We found earlier that the maximum electric field occurs at the junction $x = x_0$; now we can find its value. From Equations (5.21), (5.37), and (5.38),

$$\mathscr{E}_{\max} = \frac{qN'_D}{\varepsilon} w_n = \frac{qN'_A}{\varepsilon} w_p = \left[\frac{2qV_j N'_D N'_A}{\varepsilon(N'_D + N'_A)}\right]^{1/2} = \frac{2V_j}{w} \tag{5.41}$$

Finally, we can obtain the shape of the energy band diagram by recognizing that the potential energy is related to the electric potential:

$$\frac{dE_P}{dx} = -q\frac{dV}{dx} = \frac{dE_C}{dx} = \frac{dE_V}{dx} \tag{5.42}$$

Thus, the conduction band has the same shape as $V(x)$, but inverted, resulting in the energy band diagram of Figure 5.11e.

The width of the transition region is important for determining the various current mechanisms in a pn junction. Figure 5.12 shows the width as a function of junction voltage, $V_j = V_{bi} - V_a$ as calculated from Equation (5.37) for a one-sided step junction. The plot shows this for several values of doping on the lightly doped side. As expected, w increases with reduced doping and increased junction voltage.

In Figure 5.13, the junction width for a one-sided step junction is shown as a function of doping for applied voltages of 0.5 V (forward bias), 0 V, and −5 V (reverse bias).

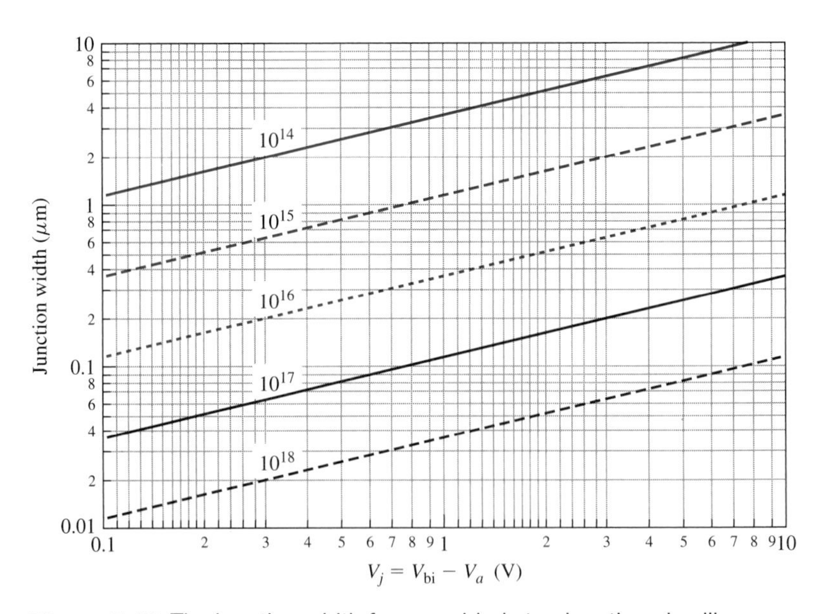

Figure 5.12 The junction width for one-sided step junctions in silicon as a function of junction voltage with the doping on the lightly doped side as a parameter.

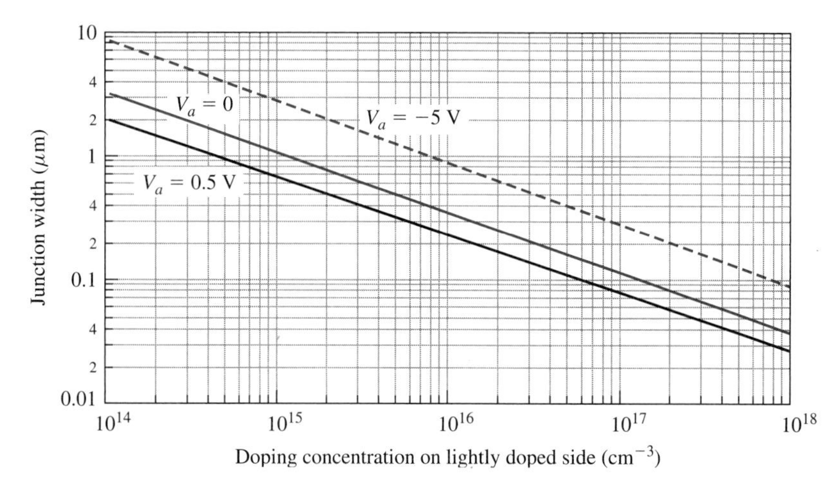

Figure 5.13 Junction width for a one-sided junction is plotted as a function of doping on the lightly doped side for three different operating voltages.

5.3.3 CURRENT-VOLTAGE CHARACTERISTICS OF pn HOMOJUNCTIONS

We now obtain expressions for the steady-state current versus applied voltage (I-V_a) characteristics of prototype pn homojunctions. The case of one-sided step junctions, i.e., n^+p and p^+n junctions, will be discussed specifically later. In addition to the assumptions already made, we assume that:

1. On either side of the junction the minority carrier concentration is everywhere much less than the majority carrier concentration:

$$\begin{aligned} n_p &\ll p_p \approx N'_A \\ p_n &\ll n_n \approx N'_D \end{aligned} \tag{5.43}$$

   This is referred to as the *low-level injection condition.*

2. In the bulk region of a semiconductor the majority carrier concentration is essentially the equilibrium value. This is a result of space-charge neutrality ($\Delta n = \Delta p$) and the low injection condition. This implies that

$$\begin{aligned} n_n &\approx n_{n0} = N'_D \\ p_p &\approx p_{p0} = N'_A \end{aligned} \tag{5.44}$$

3. For minority carriers the drift current can be neglected compared with diffusion current in the quasi-neutral regions. Thus

$$\begin{aligned} J_{np} &= qD_n \frac{dn_p}{dx} \\ J_{pn} &= -qD_p \frac{dp_n}{dx} \end{aligned} \tag{5.45}$$

   where D_n and D_p are the minority carrier diffusion coefficients.

4. The semiconductors are nondegenerate. This implies that Boltzmann's statistics can be used, and at equilibrium

$$n_0 = N_C e^{-[(E_C - E_f)/kT]}$$
$$p_0 = N_V e^{-[(E_f - E_V)/kT]} \tag{5.46}$$

5. We define current to be positive for positive V_a and negative for negative V_a. Thus for positive current, electrons flow from n to p and holes flow from p to n.

To find the I-V characteristic, we first consider a *long-base diode* in which both sides of the quasi-neutral regions are much longer than their minority carrier diffusion lengths, L_n or L_p. After that, we consider the case of a *short-base diode* in which one or both sides are much shorter than a diffusion length.

The most important mechanism for current flow across a pn junction is referred to as minority carrier injection-extraction (I-E) current. It is the fundamental quantity of interest in devices such as bipolar junction transistors as well as in optoelectronic devices such as lasers, light-emitting diodes, and photodetectors. For the prototype junctions considered in this chapter, the I-E current is entirely diffusion current.[8] We will discuss this next. Then we will go on to discuss generation-recombination (G-R) current, tunnel current, and carrier multiplication and avalanche currents.

Diffusion Current Diffusion current consists of minority carriers that diffuse toward or away from the junction. To obtain the I-V characteristic of a diode it is convenient to begin with the continuity equations [Equations (3.63 and 3.64)], which are repeated here:

$$\boxed{\frac{\partial n}{\partial t} = \frac{\partial \Delta n}{\partial t} = \frac{1}{q}\left(\frac{\partial J_n}{\partial x}\right) + \left(G_{\text{op}} - \frac{\Delta n}{\tau_n}\right)} \tag{5.47}$$

$$\boxed{\frac{\partial p}{\partial x} = \frac{\partial \Delta p}{\partial x} = -\frac{1}{q}\left(\frac{\partial J_p}{\partial x}\right) + \left(G_{\text{op}} - \frac{\Delta p}{\tau_p}\right)} \tag{5.48}$$

where Δn and Δp are the excess electron and hole concentrations.

First we consider electrons within the p region of a pn junction, where they are minority carriers. Then from Assumption 3 [Equation (5.45)],

$$J_n = J_{n(\text{diff})} = qD_n \frac{dn}{dx} \tag{5.49}$$

[8]In nonprototype junctions, i.e., those with nonuniform doping on either side, drift current is also important, but the injection-extraction current is often referred to as *diffusion current.*

Combining Equations (5.47) and (5.49) gives

$$\frac{\partial n}{\partial t} = \frac{\partial \Delta n}{\partial t} = D_n \frac{\partial^2 n}{\partial x^2} - \frac{\Delta n}{\tau_n} \tag{5.50}$$

For steady state, $\partial n/\partial t = 0$ and Equation (5.50) can be expressed

$$\frac{\partial^2 n}{\partial x^2} = \frac{d^2 n}{dx^2} = \frac{\Delta n}{D_n \tau_n} = \frac{\Delta n}{{L_n}^2} \tag{5.51}$$

where the minority carrier diffusion length for electrons is, from Equation (3.82),

$$L_n = \sqrt{D_n \tau_n}$$

Since $n_p = n_{p0} + \Delta n_p$ where Δn_p is the excess electron concentration in the p region, and n_{p0} is constant, Equation (5.51) can be expressed

$$\frac{d^2 \Delta n_p}{dx^2} = \frac{\Delta n_p}{{L_n}^2} \tag{5.52}$$

whose solution is

$$\Delta n_p = Ae^{x/L_n} + Be^{-x/L_n} \tag{5.53}$$

The constants A and B are determined from the boundary conditions. For a long-base diode, the p region extends several diffusion lengths and thus the excess carriers will have all recombined by the end of the region. Evaluating Equation (5.53) with $\Delta n = 0$ at $x = \infty$ gives $A = 0$. At $x = x_p$, the boundary condition is that $\Delta n(x_p) = Be^{-(x-x_p)/L_n}$, giving $B = \Delta n(x_p)e^{x_p/L_n}$ and

$$\boxed{\Delta n(x) = \Delta n(x_p)e^{-(x-x_p)/L_n}} \tag{5.54}$$

That is, the excess electron concentration decays exponentially with position from its value at the edge of the junction.

Since

$$J_n = qD_n \frac{dn}{dx} = qD_n \frac{d\Delta n}{dx} \tag{5.55}$$

the electron current is

$$\boxed{J_n = \frac{qD_n}{L_n} \Delta n(x_p)e^{-(x-x_p)/L_n}} \tag{5.56}$$

Note that the electron current density decreases exponentially with x on the p side. But since the total current must remain constant, the hole current increases by the same amount. This hole current is drift current resulting from the small electric field in this quasi-neutral region.

At the edge of the transition region where $x = x_p$, the electron current density is

$$\boxed{J_n(x_p) = q\frac{D_n}{L_n}\Delta n(x_p)} \tag{5.57}$$

It is useful to obtain an expression for $\Delta n(x_p)$ as a function of applied voltage. We first consider the case of equilibrium where $\Delta n(x_p) = 0$ or $n(x_p) = n_{p0}$.

Equilibrium Consider the equilibrium energy band diagram of Figure 5.14. On the n side, the electron concentration is $n_{n0} = N'_D$ while the hole concentration on that side is $p_{n0} = n_i{}^2/N'_D$. Both are uniform in the quasi-neutral n region. On the p side the hole concentration is $p_{p0} = N'_A$ and the electron concentration is $n_{p0} = n_i{}^2/N'_A$. The potential energy barrier for both electrons and holes is qV_{bi}.

We now relate N'_D, n_{p0}, N'_A, p_{n0}, and V_{bi}. In the neutral p region, we have

$$n_{p0} = N_C e^{-[(E_{Cp}-E_f)/kT]} \tag{5.58}$$

Multiplying by $e^{-E_{Cn}/kT}e^{+E_{Cn}/kT}$ allows this to be written

$$n_{p0} = N_C e^{-[(E_{Cn}-E_f)/kT]}e^{-[(E_{Cp}-E_{Cn})/kT]} \tag{5.59}$$

But since the energy step in the conduction band edge is proportional to the built-in voltage:

$$(E_{Cp} - E_{Cn}) = qV_{\text{bi}} \tag{5.60}$$

we can rewrite Equation (5.59) as

$$n_{p0} = [N_C e^{-[(E_{Cn}-E_f)/kT]}]\,e^{-qV_{\text{bi}}/kT} \tag{5.61}$$

The part in the square brackets is the electron concentration on the n side:

$$n_{n0} = N'_D = N_C e^{-q(E_{Cn}-E_f)/kT} \tag{5.62}$$

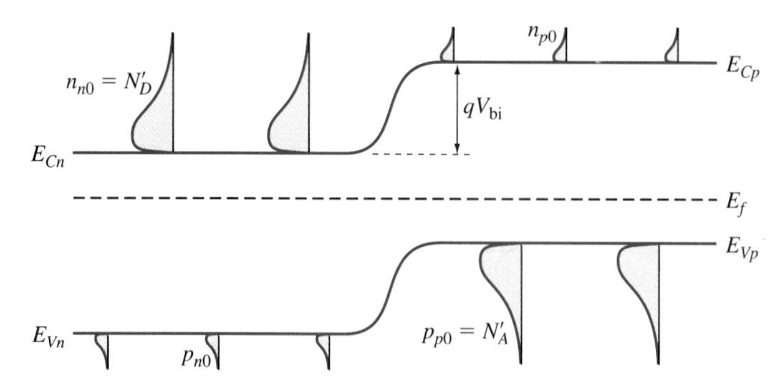

Figure 5.14 Equilibrium energy band diagram for a step junction.

and, therefore, combining Equations (5.61) and (5.62) gives the minority carrier concentration on the p side:

$$n_p(x_p) = n_{p0} = N'_D e^{-qV_{\text{bi}}/kT} = n_{n0} e^{-qV_{\text{bi}}/kT} \tag{5.63}$$

Similarly

$$p_n(x_n) = p_{n0} = N'_A e^{-qV_{\text{bi}}/kT} = p_{p0} e^{-qV_{\text{bi}}/kT} \tag{5.64}$$

These may seem counterintuitive because the carrier concentration on one side appears to depend on the doping on the other side, but remember that the qV_{bi} term takes into account the doping on *both* sides.

Diffusion Current: Forward Bias Next, we consider the case of an externally applied voltage. For example, under a forward bias of V_a (Figure 5.15), the energy barrier is changed to $(E_{Cp} - E_{Cn}) = q(V_{\text{bi}} - V_a)$. The p side has been raised to a higher electric potential, or lowered to a lower electron potential energy, and the n side has increased in potential energy. Equations (5.63) and (5.64) can be adapted to express the concentrations in terms of the doping on the other side and the barrier height. At the edges of the depletion region, x_n and x_p, the minority carrier concentrations are

$$\begin{aligned} n_p(x_p) &= N'_D e^{-q(V_{\text{bi}}-V_a)/kT} \\ p_n(x_n) &= N'_A e^{-q(V_{\text{bi}}-V_a)/kT} \end{aligned} \tag{5.65}$$

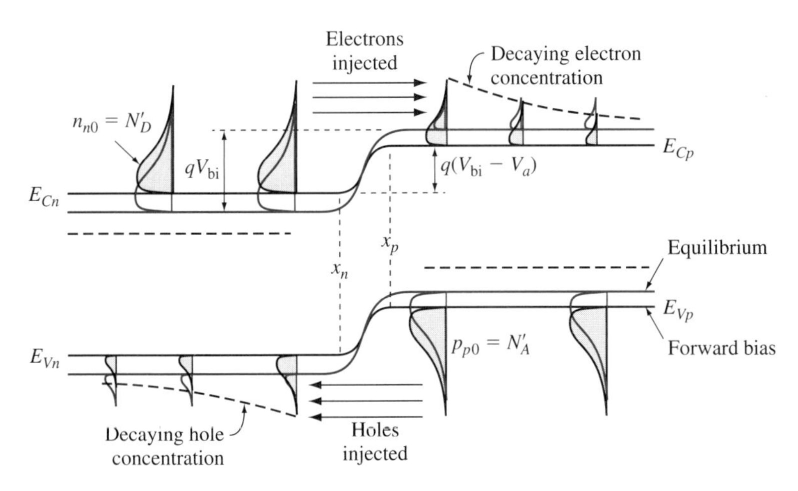

Figure 5.15 Under forward bias, the n side moves up in potential energy and the p side moves down. The colored lines correspond to equilibrium; the black lines represent forward bias. Carriers are injected across the junction by diffusion. They become minority carriers once they get across, and recombine as they continue to diffuse. The result is a varying concentration of minority carriers with distance from the junction.

or from Equations (5.63) and (5.64)

$$\boxed{\begin{aligned} n_p(x_p) &= n_{p0}e^{qV_a/kT} \\ p_n(x_n) &= p_{n0}e^{qV_a/kT} \end{aligned}} \tag{5.66}$$

When forward bias is applied, the excess electrons that are injected into the p region diffuse to the right and recombine with holes on the p side. Holes are also injected into the n side, where they also recombine. Let Δn_p represent the excess electron concentration on the p side and Δp_n represent the excess hole concentration on the n side:

$$\begin{aligned} \Delta n_p(x_p) &= n_p(x_p) - n_{p0} \\ \Delta p_n(x_n) &= p_n(x_n) - p_{n0} \end{aligned} \tag{5.67}$$

Then from Equations (5.66),

$$\begin{aligned} \Delta n_p(x_p) &= n_{p0}(e^{qV_a/kT} - 1) \\ \Delta p_n(x_n) &= p_{n0}(e^{qV_a/kT} - 1) \end{aligned} \tag{5.68}$$

These are excess carriers injected across the junction. We recall from Chapter 3, however, that the excess electrons will diffuse, and as they diffuse in the p material they will recombine because of the abundance of holes. The injected excess electron concentration decreases exponentially with x as

$$\boxed{\Delta n_p(x) = \Delta n_p(x_p)e^{-(x-x_p)/L_n}} \tag{5.69}$$

and on the n side the hole concentration will obey

$$\boxed{\Delta p_n(x) = \Delta p_n(x_n)e^{-(x_n-x)/L_p}} \tag{5.70}$$

These injected carrier concentrations are shown in Figure 5.16. Since the concentrations vary with distance, they set up diffusion currents. For the electron

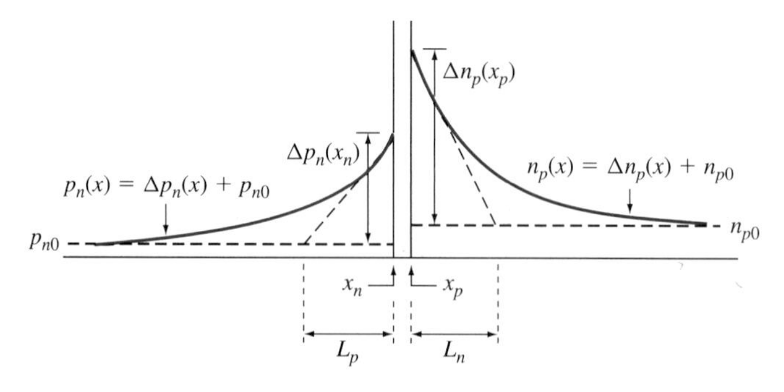

Figure 5.16 The minority carrier concentrations on either side of the junction.

diffusion current on the p side,

$$J_n = qD_n\frac{dn}{dx} = qD_n\frac{d\Delta n}{dx} \tag{5.71}$$

where D_n and L_n are the electron minority carrier diffusion coefficient and the minority carrier diffusion length respectively.

Substituting Equations (5.69) and (5.68) into Equation (5.71) gives us

$$J_n = q\frac{D_n}{L_n}n_{p0}(e^{qV_a/kT} - 1)e^{-(x-x_p)/L_n} \tag{5.72}$$

At the edge of the transition region, where $x = x_p$, the electron diffusion current density is

$$\boxed{J_n(x_p) = q\frac{D_n}{L_n}n_{p0}(e^{qV_a/kT} - 1)} \tag{5.73}$$

Similarily, the hole diffusion current density in the n region at $x = x_n$ is

$$\boxed{J_p(x_n) = q\frac{D_p}{L_p}p_{n0}(e^{qV_a/kT} - 1)} \tag{5.74}$$

where D_p and L_p are hole minority carrier diffusion coefficient and diffusion length respectively.

Equations (5.73) and (5.74) give the minority carrier diffusion currents at the edge on either side of the depletion region. Since recombination and generation in the transition region are neglected, all of the excess electrons on the p side crossing $x = x_p$ had to come from the n side, and therefore also had to cross $x = x_n$. This is shown in Figure 5.17a. Similarly the hole diffusion current $J_p(x = x_n)$ must equal the hole current crossing x_p (Figure 5.17b). We also agreed by Assumption 3 that the drift of minority carriers was neglible, so on either side of the junction the total current of the minority carriers is due to diffusion. Therefore both electron and hole currents are continuous across the transition region and at any point in the transition region, as shown in Figure 5.17c. The total current anywhere between x_n and x_p is given by the sums of the diffusion currents from either side of the junction:

$$J = J_n(x_p) + J_p(x_n) \tag{5.75}$$

The figure also shows the decrease in diffusion current and corresponding increase in majority carrier drift currents in the bulk regions needed to keep the total current constant.

Substituting in Equations (5.73) and (5.74) gives the total current density:

$$\boxed{J = q\left(\frac{D_n n_{p0}}{L_n} + \frac{D_p p_{n0}}{L_p}\right)(e^{qV_a/kT} - 1)} \tag{5.76}$$

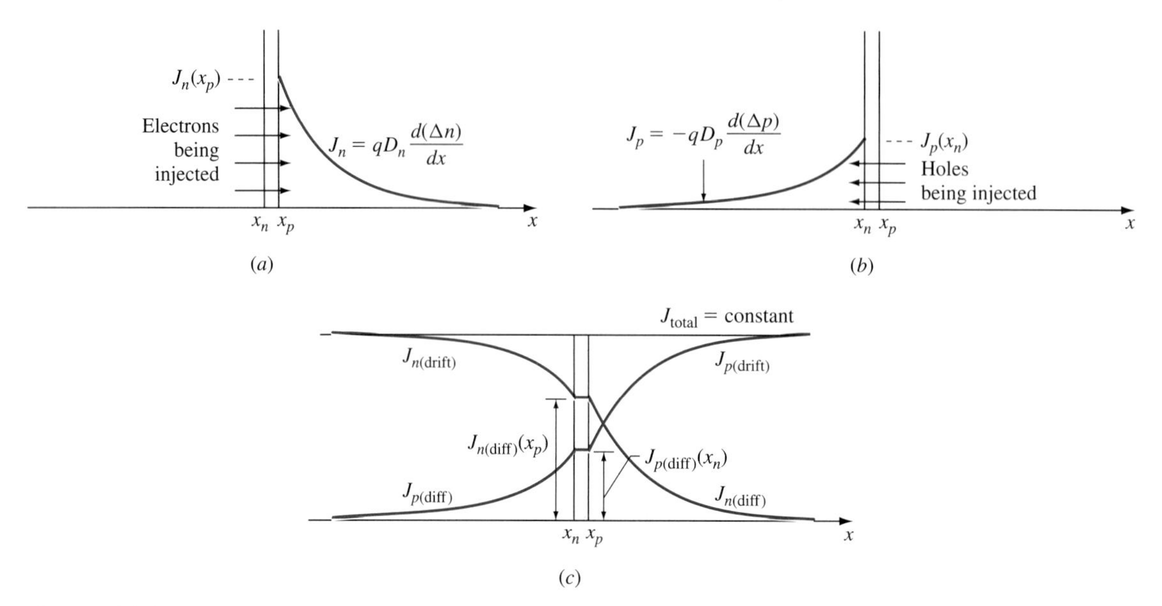

Figure 5.17 (a) The electrons injected into the p region under forward bias must cross the plane $x = x_n$. (b) Similarly, the holes injected into the n region must also cross the plane $x = x_p$. (c) The total current in the junction (neglecting recombination current) is the sum of the two. Outside the junction, the minority diffusion currents decrease and the majority carrier drift current increases to keep the total current constant.

This is normally written in the form

$$\boxed{J = J_0(e^{qV_a/kT} - 1)} \tag{5.77}$$

Note that at x_n and x_p the current is entirely due to diffusion, and

$$\boxed{J_0 = q\left(\frac{D_n n_{p0}}{L_n} + \frac{D_p p_{n0}}{L_p}\right)} \tag{5.78}$$

From the relations $L_n = \sqrt{D_n \tau_n}$, $L_p = \sqrt{D_p \tau_p}$, $n_{p0} = n_i{}^2/N'_A$, and $p_{n0} = n_i{}^2/N'_D$, Equation (5.78) can be expressed

$$J_0 = qn_i{}^2\left(\sqrt{\frac{D_n}{\tau_n}} \cdot \frac{1}{N'_A} + \sqrt{\frac{D_p}{\tau_p}} \cdot \frac{1}{N'_D}\right) \tag{5.79}$$

We can see that J_0 is proportional to $n_i{}^2$.

Equation (5.77) is the familiar diode equation from circuits courses. Now we see that the exponential nature of the diode current is due to lowering the potential barriers.

The ratio of electron to hole current at x_p (or x_n) is

$$\frac{J_n(x_p)}{J_p(x_n)} = \frac{D_n n_{p0}}{L_n} \frac{L_p}{D_p p_{n0}} = \frac{D_n}{D_p} \frac{L_p}{L_n} \frac{N'_D}{N'_A} \tag{5.80}$$

We can see that this ratio depends on N'_D/N'_A. This result will be used in the analysis of bipolar junction transistors in Chapter 9.

From Equation (5.77), for forward voltage ($V_a \gg kT/q$), the diode current can be expressed

$$\boxed{I = I_0 e^{qV_a/kT}} \tag{5.81}$$

or

$$V_a = \frac{kT}{q} \ln \frac{I}{I_0} = \frac{kT}{q}(\ln I - \ln I_0)$$

For a decade change in current, $\ln I = \ln 10 = 2.3$. Thus V_a varies by the factor $\Delta V_a = 2.3kT/q = 60$ mV at room temperature.

Diffusion Current: Reverse Bias The energy band diagram under reverse bias is shown in Figure 5.18. Here the diffusion current results from minority carriers

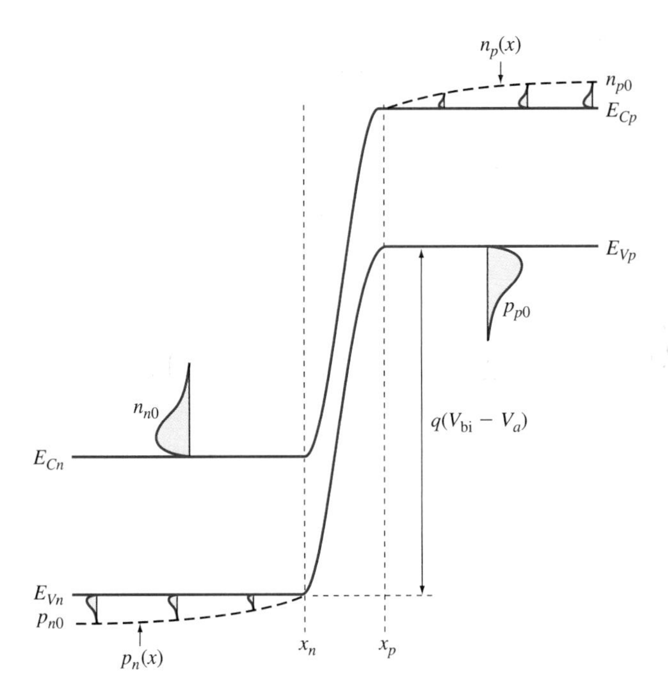

Figure 5.18 Reverse-biased prototype junction. The minority carriers generated within the quasi-neutral regions diffuse to the junction and contribute to the reverse current.

diffusing to the transition region and being swept across by the junction field. At the edge of the depletion region, the minority carrier concentration is essentially zero, and the excess carrier concentration is $\Delta n_p(x_p) = n_p(x_p) - n_{p0} = 0 - n_{p0} = -n_{p0}$. On the n side the minority hole concentration is also zero, and $\Delta p_n(x_n) = p_n(x_n) - p_{n0} = -p_{n0}$. We say that minority carriers are *extracted* (as opposed to injected). Since there are few minority carriers near the junction under reverse bias, the minority carriers diffuse toward the junction. In effect, all of the minority carriers generated within a diffusion length of the transition region contribute to current. There are not many of them so the resulting current is small, as indicated earlier.

Under reverse bias, Equations (5.76) and (5.77) are still valid. However, for $V_a < -3kT$, the exponent becomes large and negative, and the exponential term approaches zero. Therefore, to good approximation the reverse diffusion current density is $-J_0$.

Diffusion Current under Reverse Bias: Short-Base Diode Equation (5.78) for J_0 was derived on the assumption that the thickness on each side of the junction was much larger than a minority current diffusion length. In other words, the neutral region was long enough that the excess minority carrier concentrations could decay by recombination. Often, however, one side has length much less than a diffusion length. This structure is called the *short-base diode,* because this kind of junction occurs in the base region of bipolar transistors, which we will cover in Chapters 9 and 10. We will prepare for that discussion now by considering the pn junction of Figure 5.19. Here the thickness W_B of the "base", in this case the p region thickness excluding the transition region, is much less than L_n, the diffusion length for electrons. In this diode, the n side is long and behaves normally as discussed in the previous section. The p side, however, is different.

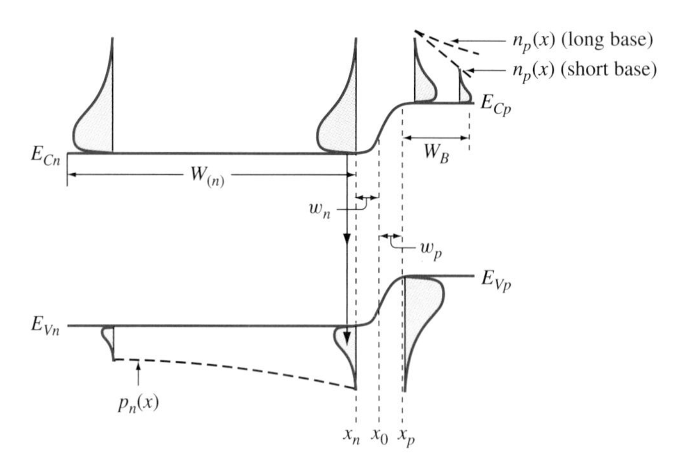

Figure 5.19 Illustration of the diffusion current in a pn junction in which the quasi-neutral region W_B is much shorter than an electron diffusion length.

Under forward bias, the electron carrier injection process is the same. There is negligible recombination in the p region, however; because it is so thin the electrons are not in it long enough to recombine naturally. There is an assumed contact (not shown) on the far end of the region, and that contact supplies the equilibrium electron and hole concentrations, n_{p0} and p_{p0} respectively. That is, instead of decaying exponentially, the carrier concentration is forced to the equilibrium concentration in a distance W_B, the distance between the transition region edge and the end of the quasi-neutral region. Since negligible recombination occurs, the change is very nearly linear. The expression for $d\Delta n/dx$ in Equation (5.71) becomes

$$\frac{d\Delta n}{dx} = \frac{\Delta n_p(x_p)}{W_B} \tag{5.82}$$

and the diffusion current on the p side of the transition region becomes, by Equation (5.73)

$$J_n(x_p) = q\frac{D_n}{W_B} n_{p0}(e^{qV_a/kT} - 1) \tag{5.83}$$

On the n side the diffusion is the same as for the long-base diode, so Equation (5.78) becomes

$$J_0 = q\left(\frac{D_n n_{p0}}{W_B} + \frac{D_p p_{n0}}{L_p}\right) \quad \text{short p-side diode} \tag{5.84}$$

If both sides are short, e.g., on the p side $W_{B(p)} \ll L_n$ and on the n side $W_{(n)} \ll L_p$, as is often the case for the emitter-base junction of a transistor,

$$J_0 = q\left(\frac{D_n n_{p0}}{W_{B(p)}} + \frac{D_p p_{n0}}{W_{(n)}}\right) \quad \text{both sides short} \tag{5.85}$$

and the ratio of electron to hole current at $x = x_p$ or x_n is

$$\frac{J_n}{J_p} = \frac{D_n}{D_p} \cdot \frac{W_{(n)}}{W_{B(p)}} \cdot \frac{N'_D}{N'_A} \quad \text{both sides short} \tag{5.86}$$

We emphasize that the above results are for the case of nondegenerate semiconductors on both sides of the homojunction.

Generation and Recombination Current in pn Homojunctions In the analysis of diffusion current in the previous section, we assumed that all carrier generation and recombination occurred in the quasi-neutral regions, and we ignored generation and recombination in the transition region. However, carrier generation and recombination do occur in the transition region and also contribute to current [generation-recombination (G-R) current].

We saw in Chapter 3 that in many semiconductors (including Si), carrier generation and recombination is primarily via interband (trap) states with energy near the intrinsic level, $E_T \approx E_i$, where E_T is the trap energy. Let us consider such a case. For simplicity we assume $\tau_n = \tau_p = \tau_0$ and $E_T = E_i$. Under these conditions, the net recombination rate is given by

$$R - G = \frac{np - n_i^2}{\tau_0(n + n_i + p + n_i)} \tag{5.87}$$

At equilibrium, $np = n_i^2$ everywhere and the net recombination reduces to zero as expected.

The net recombination rate depends on the numbers of available electrons and holes as indicated in Equation (5.87). At position x within the transition region, the carrier concentrations are given by

$$\begin{aligned} n(x) &= n_{n0} e^{-[E_C(x) - E_C(x_n)]/kT} \\ p(x) &= p_{p0} e^{[E_V(x) - E_V(x_p)]/kT} \end{aligned} \tag{5.88}$$

or the electron concentration decreases rapidly with increasing x and E_C. Likewise, $p(x)$ increases rapidly as x and thus E_V increases. We will use these results in the following sections to examine recombination and generation under reverse bias and under forward bias.

Generation and Recombination: Reverse Bias To consider recombination and generation under reverse bias, we refer to Figure 5.20 in which the energy band diagram is shown. Recall that under reverse bias the transition region widens. When the width is large, there can be significant opportunities for both generation and recombination. Under reverse bias, however, n and p are both small in most of the transition region. To simplify the mathematics, the depletion approximation is used, which assumes n and p can be neglected in the transition region for $x_n \le x \le x_p$. In this case the recombination can be ignored, since there are few electrons to recombine and few holes with which to recombine. Then from Equation (5.87), we have

$$R = 0 \qquad G = \frac{n_i}{2\tau_0} \tag{5.89}$$

This implies that the generation rate is constant within the transition region. The generation current density J_G is

$$J_G = -qGw = -\frac{qn_i w}{2\tau_0} \tag{5.90}$$

However, since the depletion width w for a step junction is given by

$$w = \left[\frac{2\varepsilon(N'_D + N'_A)(V_{bi} - V_a)}{qN'_D N'_A}\right]^{1/2} \tag{5.91}$$

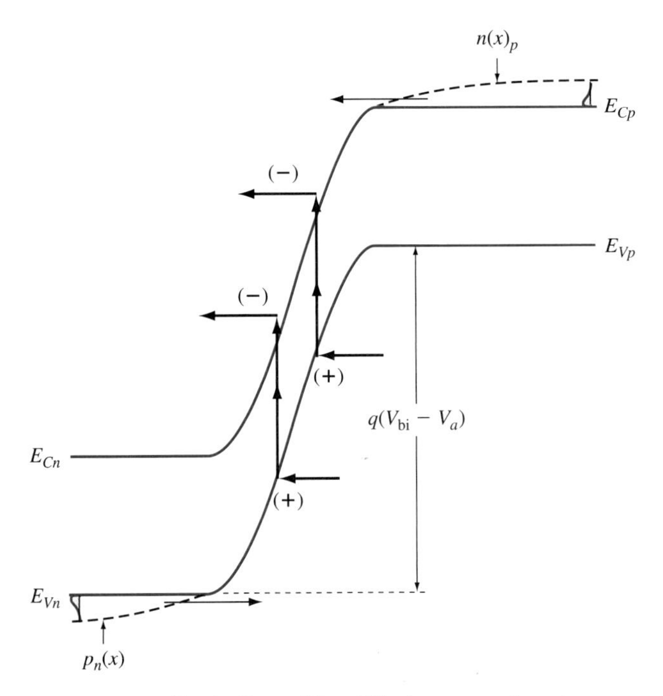

Figure 5.20 Illustration of the diffusion current and generation current in a reverse-based pn junction. The generation current is normally much larger than the diffusion current.

substituting Equation (5.91) into (5.90) results in

$$J_G = -\frac{n_i}{\tau_0}\left[\frac{q\varepsilon(N'_D + N'_A)(V_{bi} - V_a)}{2N'_D N'_A}\right]^{1/2} \qquad \text{reverse bias} \qquad (5.92)$$

EXAMPLE 5.2

Estimate the value of generation current relative to the diffusion current for a typical Si pn junction under reverse-bias conditions.

■ Solution

Let us take for a prototype junction in silicon:

$$N'_A = 10^{17}\ \text{cm}^{-3}$$
$$N'_D = 10^{17}\ \text{cm}^{-3}$$
$$\tau_n \approx \tau_p = \tau_0 \approx 6 \times 10^{-6}\ \text{s}$$
$$V_j = (V_{bi} - V_a) = 5\ \text{V}$$

where τ_0 is taken as the average minority carrier lifetime from Figure 3.21.

From the graph of Figure 3.11, at doping concentrations of $N'_A = N'_D = 10^{17}$ cm^{-3}, the diffusion constants for minority carriers are

$$D_n = 20\,\text{cm}^2/\text{s}$$
$$D_p = 11\,\text{cm}^2/\text{s}$$

The minority carrier diffusion lengths are found from Figure 3.23 to be $L_n = 102\ \mu$m and $L_p = 73\ \mu$m. The minority carrier concentrations are $n_{p0} = p_{n0} = n_i{}^2/10^{17} = 1.16 \times 10^3$ cm^{-3}, and the built-in voltage is, from Equation (5.13)

$$V_{\text{bi}} = \frac{kT}{q}\ln\frac{N'_D N'_A}{n_i{}^2} = 0.026\,\text{V}\ln\left[\frac{10^{17}\times 10^{17}}{(1.08\times 10^{10})^2}\right] = 0.83\ \text{V}$$

We are given the junction voltage, so we can find the applied voltage:

$$V_j = 5\,\text{V} = V_{\text{bi}} - V_a = 0.83 - V_a \qquad \text{or} \qquad V_a = -4.17\ \text{V}$$

For the diffusion current, we find the leakage current from Equation (5.78),

$$J_0 = q\left(\frac{D_n n_{p0}}{L_n} + \frac{D_p p_{n0}}{L_p}\right)$$
$$= 1.6\times 10^{-19}\,\text{C}\left[\frac{(20\,\text{cm}^2/\text{s})(1.16\times 10^3\,\text{cm}^{-3})}{102\times 10^{-4}\,\text{cm}} + \frac{(11\,\text{cm}^2/\text{s})(1.16\times 10^3\,\text{cm}^{-3})}{73\times 10^{-4}\,\text{cm}}\right]$$
$$= 6.7\times 10^{-13}\ \text{A/cm}^2$$

Inserting this into Equation (5.77), we have for the diffusion current

$$J_{\text{diff}} = J_0(e^{qV_a/kT} - 1)$$
$$= 6.7\times 10^{-13}\ \text{A/cm}^2(e^{[(1.6\times 10^{-19}\,\text{C})(-4.17\,\text{V})]/[(0.026)(1.6\times 10^{-19})\,\text{J/eV}]} - 1) \approx -J_0$$
$$= -6.7\times 10^{-13}\ \text{A/cm}^2 = -6.7\times 10^{-21}\ \text{A}/\mu\text{m}^2$$

For the generation current, we calculate the junction width from Equation (5.91) to be $w = 0.36\ \mu$m. Then, using this value in Equation (5.90), we find

$$J_G = -\frac{qn_i w}{2\tau_0} = -\frac{(1.6\times 10^{-19}\,\text{C})(1.08\times 10^{10}\,\text{cm}^{-3})(0.36\times 10^{-4}\,\text{cm})}{2(6\times 10^{-6}\,\text{s})}$$
$$= -5.2\times 10^{-9}\ \text{A/cm}^2 = 5.2\times 10^{-17}\ \text{A}/\mu\text{m}^2$$

or

$$\frac{J_G}{J_{\text{diff}}} \approx \frac{5.2\times 10^{-17}}{6.7\times 10^{-21}} = 7.8\times 10^3$$

The generation current under reverse bias is appreciably larger than the diffusion current. Therefore, under reverse bias the diffusion current can normally be neglected.

Generation and Recombination: Forward Bias Next, let us consider the generation and recombination current under forward bias. The energy band diagram for forward bias is shown in Figure 5.21. In the transition region, the electrons are in thermal equilibrium with the n side of the junction and the holes are in

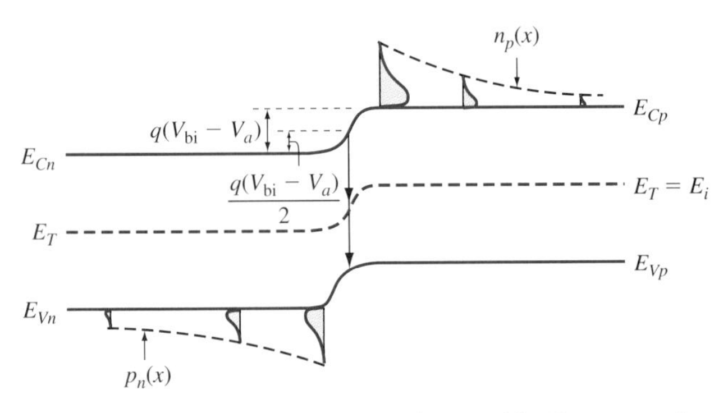

Figure 5.21 Diffusion current and recombination current in a forward-biased pn junction. The barrier for diffusion current is $q(V_{bi} - V_a)$. The barrier for recombination current is about half that value.

thermal equilibrium with the p side. Therefore, from Equation (5.46), one can show that in the transition region,

$$\boxed{\begin{aligned} n(x) &= n_{n0} e^{-[E_C(x) - E_{Cn}]/kT} \\ p(x) &= p_{p0} e^{-[E_{Vp} - E_V(x)]/kT} \end{aligned}} \tag{5.93}$$

where x is position within the transition region.

Therefore, within the transition region the np product is, from Equations (5.93) and (5.13),

$$\boxed{np = n_i^2 e^{qV_a/kT}} \tag{5.94}$$

For reasonable forward bias levels, such that $V_a \geq 3kT/q$, the recombination current is large compared with the generation current, because there are now many electrons and many holes in the same region. Since from Equation (5.94) the np product is much greater than n_i^2 under forward bias, Equation (5.87) then becomes

$$\boxed{R = \frac{np}{\tau_0(n + p)} = \frac{n_i^2 e^{qV_a/kT}}{\tau_0(n + p)}} \tag{5.95}$$

Where in the junction is the recombination rate the greatest? The maximum recombination rate is obtained by setting $dR/dn = 0$ [or more conveniently, $d(1/R)/dn = 0$, since n is in the denominator of Equation (5.95)]. Using $p = (n_i^2/n)e^{qV_a/kT}$ from Equation (5.94) gives the maximum R which is found at the value of x where

$$n = p = n_i e^{qV_a/2kT} \tag{5.96}$$

or

$$R_{\max} = \frac{n_i e^{qV_a/2kT}}{2\tau_0} \tag{5.97}$$

and decreases rapidly on either side.

Since most of the recombination current occurs near where $R \approx R_{\max}$, the barrier for electrons and holes for recombination current is $(V_{bi} - V_a)/2$. This is half the barrier for diffusion current of $(V_{bi} - V_a)$.

The G-R current density is often approximated as

$$\boxed{J_{GR} = J_{GR0}(e^{qV_a/2kT} - 1)} \tag{5.98}$$

where the term J_{GR0} is slightly voltage dependent.[9] Comparing Equation (5.92) for generation-recombination under reverse bias with (5.79) for diffusion under reverse bias, we can conclude that J_{GR} is proportional to n_i while J_{diff} is proportional to n_i^2.

The total current density is then

$$\boxed{J = J_{GR} + J_{diff} = J_{GR0}(e^{qV_a/2kT} - 1) + J_0(e^{qV_a/kT} - 1)} \tag{5.99}$$

where usually $J_{GR0} \gg J_0$ as expected from the analysis for reverse bias.

For forward bias with $V_a > 3kT/q$,

$$J = J_{GR0}e^{qV_a/2kT} + J_0 e^{qV_a/kT} \tag{5.100}$$

At small V_a, recombination current predominates (because $J_{GR} \gg J_0$), but at larger V_a, diffusion current predominates. This is indicated in Figure 5.22[10] for a junction area of 100 μm^2. Values for I_{GR0} and I_0 can be extrapolated from the straight-line regions as indicated. At high currents, the line deviates from linearity because part of the applied voltage (IR) is dropped across the series resistance of the material. Over a range of current, J is often approximated as

$$J = J_s(e^{qV_a/nkT} - 1) \tag{5.101}$$

where J_s is a function of J_{GR0} and J_0, and the *diode quality factor n* (sometimes called the ideality factor) is intermediate between 1 and 2.[11]

Reverse-Bias Tunneling Two other current mechanisms must be considered under reverse bias. These are *tunneling* and carrier *multiplication*. We consider

---

[9]Its value is obtained by integrating over the width of the junction, and the junction width changes with applied voltage.

[10]Because the current is plotted on a common log scale, the slope in Figure 5.22 is $d\log I/dV_a = (d\ln I/dV_a)\log e$.

[11]The diode quality factor of 2 for recombination current is a result of the assumption that the trap level is at the intrinsic level ($E_T = E_i$), and that $\tau_n = \tau_p = \tau_0$. If these conditions are not met, the diode quality factor for recombination current will be less than 2.

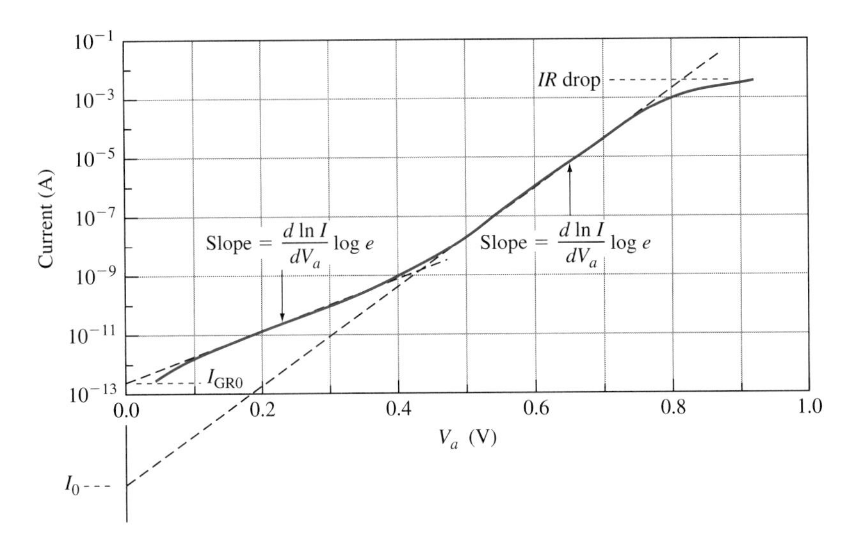

Figure 5.22 The I-V_a characteristic for a forward-biased junction showing the recombination current and diffusion current. At high currents the plot deviates from a straight line by the IR drop in the bulk.

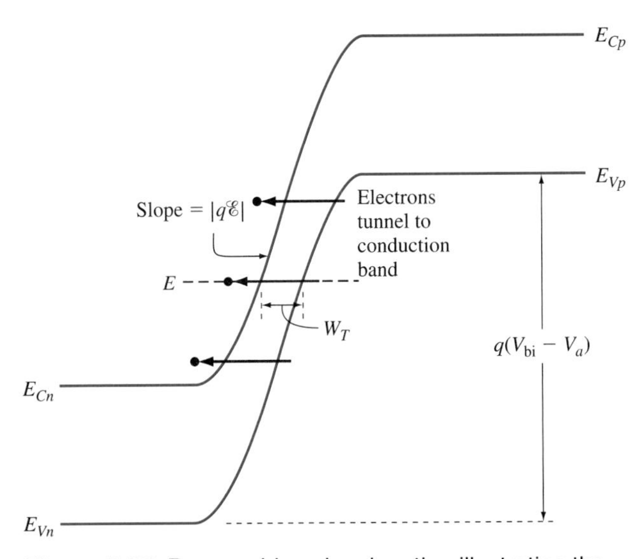

Figure 5.23 Reverse-biased pn junction illustrating the one-step tunneling process.

the reverse-biased junction of Figure 5.23. Tunneling can occur when an electron at a given energy (for example energy E in the figure) sees an empty state *at the same energy* on the other side of the potential barrier. In this case, the barrier is the forbidden gap, and the electrons in the valence band are opposite empty states in the conduction band under reverse bias. This is termed a *one-step tunneling process* since the electron goes directly from the valence band to the conduction band without passing through any intermediate states.

As indicated in the Supplement A to Part 1, an electron has a finite probability of penetrating the classically forbidden band gap. This process is called *tunneling* or sometimes *Zener tunneling* after the person who predicted the effect.

The tunneling probability T for an electron at energy E normally incident to the forbidden region is

$$T = e^{-2\int \alpha\, dx} \tag{5.102}$$

where

$$\alpha = \left[\frac{2m^*}{\hbar^2}(E_P^*(x) - E)\right]^{1/2} \tag{5.103}$$

and the integration is across the forbidden region along the tunneling path. Since the electron effective mass is different in the two bands, m^* in this case represents some average (tunneling) effective mass.

The quantity E_P^* in Equation (5.103) is the "effective potential energy" for the electron. Recall that the potential energy for an electron in the valence band is E_V and the potential energy for an electron in the conduction band is E_C. Both of these are varying with position in this case, as can be seen with the aid of Figure 5.24a. An electron in the forbidden region (i.e., during tunneling) of energy E at position x is affected by two potential energies, $E_C(x)$ and $E_V(x)$. The effective potential energy $E_P^*(x)$ seen by the tunneling electron is analogous to the effective resistance of two parallel resistances:

$$(E_P^*(x) - E) = \frac{(E_C(x) - E)(E - E_V(x))}{(E_C(x) - E) + (E - E_V(x))} = \frac{(E_C(x) - E)(E - E_V(x))}{E_g}$$

The normalized effective potential energy for tunneling as a function of position is indicated in Figure 5.24b. At $x = W_T$ and $x = 0$, $(E_C(0) - E)$ and $[E - E_V(W_T)]$ are equal to zero and thus $(E_P^* - E)$ is equal to zero. Therefore an electron at $x = 0$ (the valence band edge) sees no immediate barrier and enters the forbidden region. As it penetrates deeper into the forbidden region, because $(E_P^* - E)$ increases, the probability of the electron being turned back (reflected) increases. If it tunnels as far as $x = W_T/2$, $(E_P^* - E)$ reaches its maximum value of $E_g/4$. If the electron makes it all the way to the other side, $E_P^* - E$ becomes zero again and the electron enters the conduction band.

After some tedious algebra the tunneling probability becomes

$$T = e^{-\pi W_T \sqrt{m^* E_g}/2^{3/2}\hbar} \tag{5.104}$$

where W_T is the tunneling distance. Tunneling distance depends on the band gap and the slope of the band edges, as shown in Figure 5.24c, where $W_T = E_g/q\mathscr{E}$. The slope increases with increasing applied voltage, reducing W_T and thus increasing the tunneling probability. Since W_T depends on applied voltage and doping level, from Equation (5.104) we can see that the tunneling probability

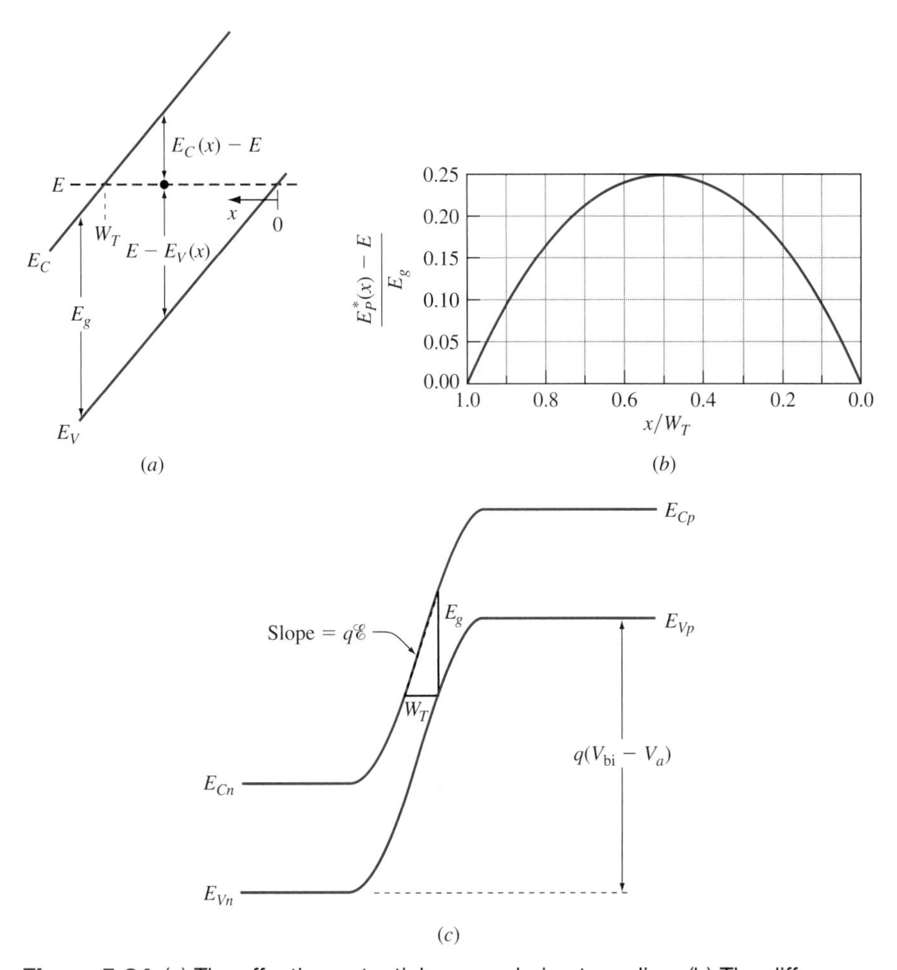

Figure 5.24 (a) The effective potential energy during tunneling. (b) The difference between the total energy and the effective energy, normalized to the band gap, as a function of normalized position of the tunneling electron. At either end of the trip, the electron's energy is entirely potential energy. (c) The geometry used to find the tunneling distance W_T.

and thus the tunnel current depend strongly on effective mass, band gap, doping level and applied voltage.

The I-V_a characteristics resulting from tunneling in a reverse-biased junction are shown in Figure 5.25, along with that for multiplication and avalanche, discussed next.

Reverse-Bias Carrier Multiplication and Avalanche The last current mechanism to be discussed is *multiplication and avalanche current,* also a reverse-bias phenomenon. Consider an electron that is thermally excited to the conduction band on the p side of a reverse-biased junction as shown in Figure 5.26. Let the thermal generation event be called process 1. We recall that electrons (and

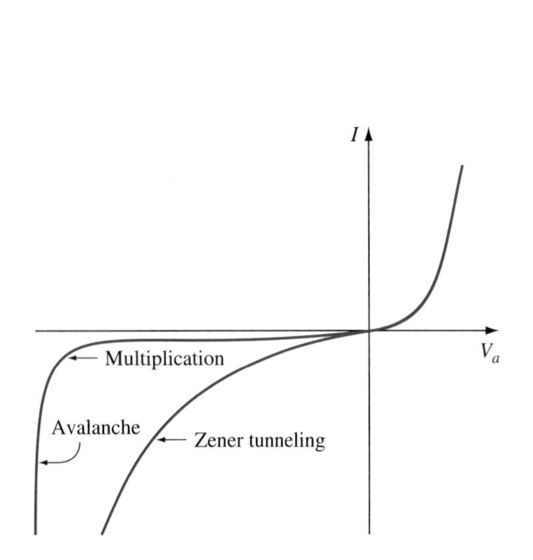

Figure 5.25 The I-V_a characteristics illustrating Zener tunneling, carrier multiplication, and avalanche. The reverse and forward characteristics are not plotted on the same scale.

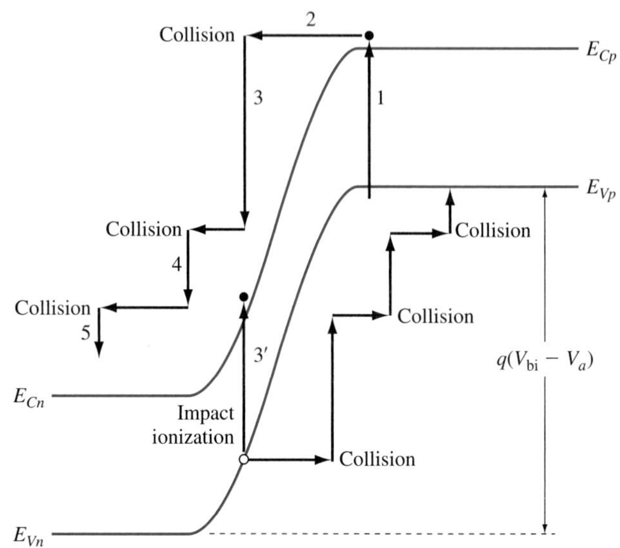

Figure 5.26 The carrier multiplication process under reverse bias. An electron is generated thermally (1), and is accelerated to the left by the field (2). Its kinetic energy increases by more than the band gap, so when the electron collides with another electron in the valence band (3), the second electron can be excited up to the conduction band (3′), a process called *impact ionization.* The ionization results in an electron-hole pair, so one original electron produced two current carriers, effectively multiplying the current.

holes) travel at constant energy between collisions. The thermally generated electron diffuses to the edge of the transition region, where it is accelerated toward the n side until it makes a collision (process 2). It then loses energy (3). It can give this energy to an electron in the valence band, exciting it to the conduction band (3′). This process is called *impact ionization*. There are now three carriers contributing to current—two electrons and one hole. These are swept to their respective sides of the junction. Although one electron started to cross the junction, two carriers finished,[12] meaning the original current was multiplied by a factor of 2. This multiplication process relies on an electron gaining kinetic energy between collisions (in one mean free path), by an amount greater than the band gap, to generate another electron. Collisions 4 and 5 in Figure 5.26 do not meet this requirement and thus do not cause more multiplication. The multiplication factor depends strongly on the electric field in the junction.

[12]It might appear that the current is multiplied by a factor of 3. Note, however, that inpact-generated electron and hole each traverse only a portion of the junction, contributing to one carrier crossing the junction.

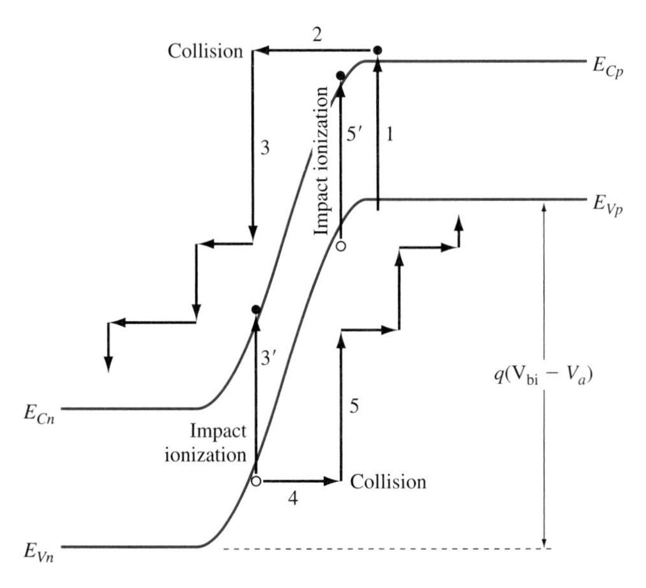

Figure 5.27 Avalanche occurs when the field is large enough to cause the newly excited carriers to in turn create more carriers.

In carrier multiplication, it is also possible for holes to gain enough kinetic energy between collisions to create hole-electron pairs. This is shown in Figure 5.27 as process 5′. If no more carriers were generated, the multiplication factor would be 3. However, above a critical field this generated electron can generate another hole-electron pair. The hole thus generated can do likewise, etc., and the multiplication factor can become infinite. This process is called *avalanche*.

Let us examine the multiplication. Let P be the probability that either a hole or an electron creates an electron-hole pair while it traverses the junction.[13] Let n_{in} be the number of electrons entering the transition region from the p side. Then there will be Pn_{in} ionizing collisions giving $n_{in}(1 + P)$ electrons reaching the n side. But Pn_{in} holes are also generated, which generate $P(Pn_{in}) = P^2 n_{in}$ pairs, etc., or the number of total carriers crossing the junction is

$$n_{in}(1 + P + P^2 + P^3 + \cdots)$$

This can be expressed as

$$\frac{n_{in}}{(1 - P)}$$

The multiplication factor then, is

$$M = \frac{1}{(1 - P)} \tag{5.105}$$

[13]For simplicity, we assume that the probabilities for holes and electrons are equal.

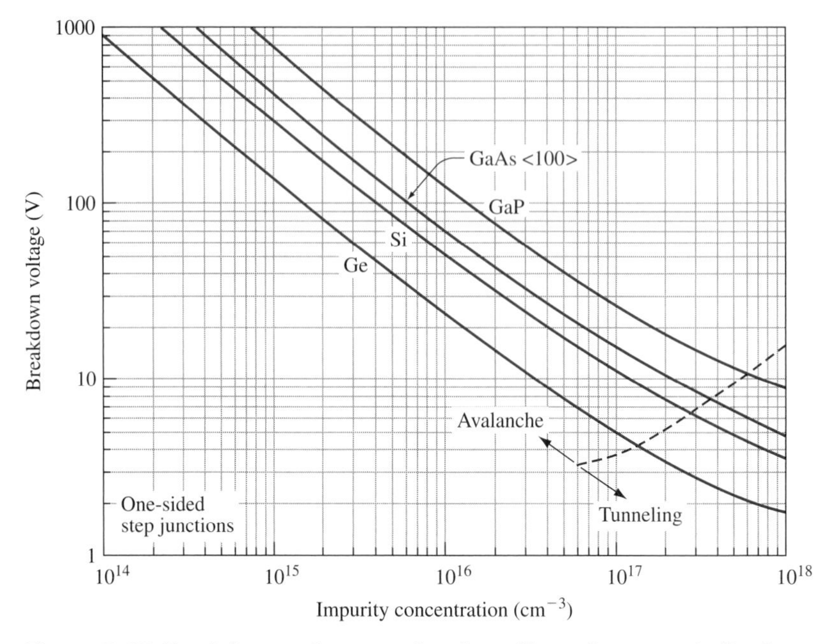

Figure 5.28 Breakdown voltage as a function of impurity concentration for one-sided n^+p or p^+n junctions in Ge, Si, GaAs, and GaP. The dashed line indicates the doping level that separates avalanche from tunneling as the dominant breakdown mechanism. (After S. M. Sze, *Physics of Semiconductor Devices,* 2nd ed., © Wiley, 1981. This material is used by permission of John Wiley & Sons, Inc.)

For $P = 1$, $M = \infty$ and avalanche occurs. The I-V_a characteristic for multiplication and avalanche is *sharp* or *hard,* as was shown in Figure 5.25. If a diode reverse current exceeds a certain value (set by convention, e.g., 10 μA), it is said to have *broken down*. This refers to the I-V_a curve turning sharply downward, but does not imply the device is damaged, since this breakdown is nondestructive. For Si pn junctions with breakdown voltages greater than about 8 V, the breakdown mechanism is primarily avalanche; if it is less than about 6 V, it is Zener tunneling.[14]

5.3.4 REVERSE-BIAS BREAKDOWN

As indicated, with increasing reverse bias the current in a diode increases as a result of tunneling as well as multiplication and avalanche.

The breakdown voltage of one-sided step junctions in each of four common semiconductors is shown in Figure 5.28. The breakdown voltage is plotted as a function of the doping on the lightly doped side. It is seen that for a given semiconductor, the breakdown voltage decreases with increasing doping. For a

[14]*Historical note:* Diodes that break down by the avalanche process are often referred to as *Zener diodes* because it was originally thought that the breakdown was a result of tunneling.

given doping level, the breakdown voltage increases with increasing band-gap energy of the semiconductor.

These observations can be explained with the aid of Figures 5.27 and 5.23 for avalanche breakdown and tunneling breakdown respectively. For avalanche to occur (Figure 5.27) the field in the junction must be large enough that between collisions, a carrier gains sufficient kinetic energy to create hole-electron pairs. With increased doping on the lightly doped side at a given reverse voltage, the junction width decreases and thus the field increases. With increasing band gap, the energy required to create electron-hole pairs increases, and thus the field and reverse voltage required for avalanche breakdown increases.

For tunneling breakdown (Figure 5.23), the electron tunneling probability depends on tunneling distance (determined by the doping and bias) and band gap as indicated in Equation (5.104). Since W_T is inversely proportional to $\mathscr{E}$, maximum tunneling occurs at the energy associated with $\mathscr{E}_{\max}$.

EXAMPLE 5.3

Estimate the tunneling distance for appreciable tunnel current. Consider a p^+n Si junction with $N_D' = 8.0 \times 10^{17}\ \text{cm}^{-3} = 8.0 \times 10^{23}\ \text{m}^{-3}$.

■ Solution

Appreciable tunnel current begins to occur around breakdown. We therefore begin by finding the junction voltage at breakdown, which is the difference between the applied breakdown voltage and the built-in voltage.

From Figure 5.10, the built-in voltage is 1.04 V.

From Figure 5.28, the breakdown voltage is 4 V ($V_a = -4$ V). The junction voltage is then $V_j = (V_{\text{bi}} - V_a) = 5.04$ V.

From Equation (5.41), the maximum field is $\mathscr{E}_{\max} = 2V_j/w$, where from Equation (5.40),

$$w = \left[\frac{2\varepsilon V_j}{qN_D'}\right]^{1/2} = \left[\frac{2 \times 11.8 \times (8.85 \times 10^{-12}\ \text{F/m}) \times (5.04\ \text{V})}{(1.6 \times 10^{-19}\ \text{C}) \times (8 \times 10^{23}\ \text{m}^{-3})}\right]^{1/2}$$

$$= 9.1 \times 10^{-8}\ \text{m} = 91\ \text{nm}$$

Then

$$\mathscr{E}_{\max} = \frac{2 \times 5.04\ \text{V}}{9.1 \times 10^{-8}\ \text{m}} = 1.1 \times 10^{8}\ \text{V/m}$$

Finally, from Figure 5.24c,

$$W_T = \frac{E_g(\text{eV})}{q\mathscr{E}_{\max}} = \frac{1.12\ \text{eV}}{(1.6 \times 10^{-19}\ \text{C})(1.1 \times 10^{8}\ \text{V/m})\left(\dfrac{1\ \text{eV}}{1.6 \times 10^{-19}\ \text{V}}\right)}$$

$$= 1.0 \times 10^{-8}\ \text{m} = 10\ \text{nm}$$

We conclude that for Si, an appreciable tunneling current flows for a tunneling distance on the order of 10 nm or less.

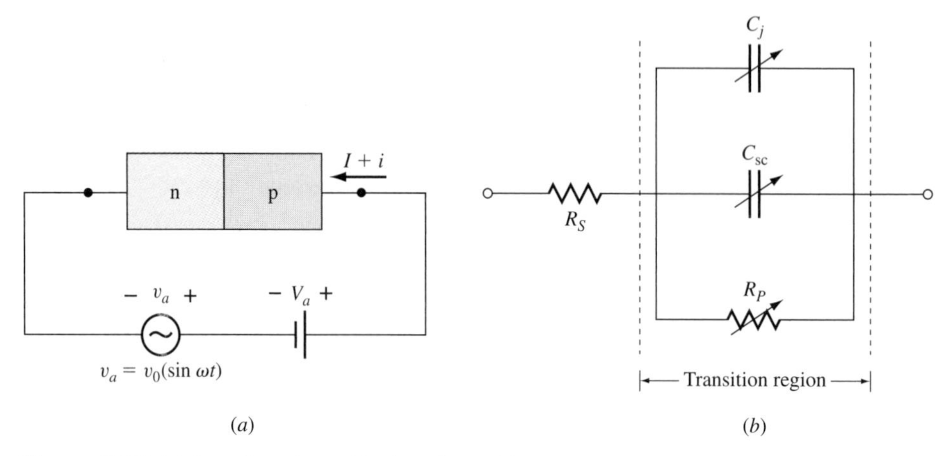

Figure 5.29 (a) A pn junction with dc voltage V_a and ac voltage v_a applied. The resulting dc and ac currents are I and i. (b) Small-signal equivalent circuit for a pn junction indicating the series resistance R_S (of the contacts and quasi-neutral regions, here lumped together into one resistance), the junction resistance R_P, the junction capacitance C_j, and the stored-charge capacitance C_{sc}. The arrows indicate that the parameters change with applied voltage.

5.4 SMALL-SIGNAL IMPEDANCE OF PROTOTYPE HOMOJUNCTIONS

The dc I-V_a characteristics of prototype homojunctions were discussed in Section 5.3. In many applications the small-signal ac response of a diode is important. Consider the circuit of Figure 5.29a, in which a dc voltage V_a and an ac voltage v_a are applied to a diode, producing a dc current I and an ac current i, respectively. The small-signal equivalent circuit of the diode is shown in Figure 5.29b. The series resistance R_S represents the contact resistance plus the resistance of the quasi-neutral regions of the diode. The small-signal (differential) resistance of the transition region is designated as R_p, while C_j and C_{sc} represent the junction and stored charge capacitances, respectively, associated with the junction. We will discuss each of these in turn.

5.4.1 JUNCTION RESISTANCE

The small signal conductance G_P of the transition region is

$$G_P = \frac{dI}{dV_a} \tag{5.106}$$

This is equal to the slope of the dc I-V_a characteristic at any given point.[15] Figure 5.30 shows how the small-signal conductance can be used to determine the

[15] Here, the voltage across R_S is neglected.

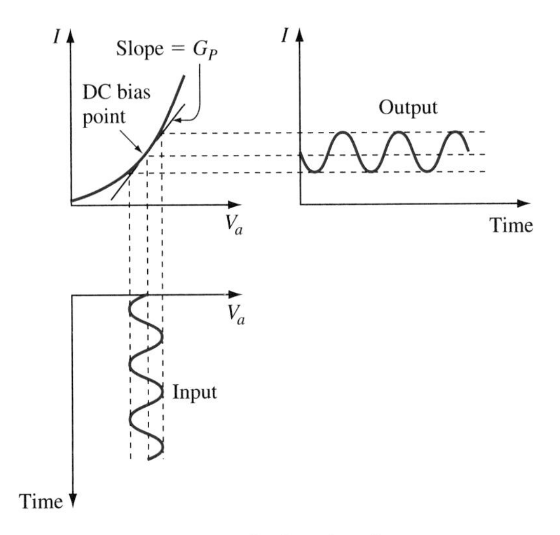

Figure 5.30 The small-signal resistance $R_P = 1/G_P$ is used to find the output current variation for an input voltage variation.

output current for a small, slowly varying input voltage where capacitance current is negligible. The (small-signal) junction resistance is the reciprocal of the slope:

$$R_P = \frac{1}{G_P} = \left[\frac{dI}{dV_a}\right]^{-1} \tag{5.107}$$

EXAMPLE 5.4

Find the junction resistance of a diode at forward currents of 1 mA and 1 μA. Assume the ideality factor is unity and $R_S = 0$.

■ Solution

From Equation (5.107),

$$R_P = \left[\frac{dI}{dV_a}\right]^{-1}$$

The diode current can be expressed

$$I = I_0(e^{qV_a/kT} - 1) = I_0 e^{qV_a/kT} - I_0 \approx I_0 e^{qV_a/kT}$$

where the term $-I_0$ is neglected since it is typically on the order of 10^{-14} A, and so $I \gg I_0$. Then

$$\frac{dI}{dV_a} = \frac{q}{kT} I_0 e^{qV_a/kT} = \frac{qI}{kT}$$

and

$$R_P = \frac{kT}{qI}$$

Since $kT/q = 0.026$ eV,

$$R_P = 26\ \Omega \qquad (I = 1\text{ mA})$$
$$R_P = 26\ \text{k}\Omega \qquad (I = 1\ \mu\text{A})$$

5.4.2 JUNCTION CAPACITANCE

There are two sources of capacitance in pn junctions, the *junction capacitance* and the *stored-charge capacitance*. Both of these limit the speed at which the diode can respond to changes of the input voltage. We will discuss the junction capacitance first.

Recall from Section 5.3.2 that the charge on either side of the metallurgical junction consists of ionized impurities. The amount of charge on each side of the junction is dependent on junction width, which in turn is a function of applied voltage. Thus, a change in applied voltage produces a change in the number of charges on each side of the junction as indicated in Figure 5.31. If the applied voltage is changed by an amount dV_a, the space charge on one side of the junction changes by an amount dQ and the space charge on the other side of the junction changes by $-dQ$. As the applied voltage changes, the electrons and holes must move out of the junction (if the junction width is increased) or into the junction (if the junction width is decreased) to change the number of non-neutralized ions on each side. The electron current resulting from this movement of electrons and holes flows through the external circuit rather than through the junction. Because

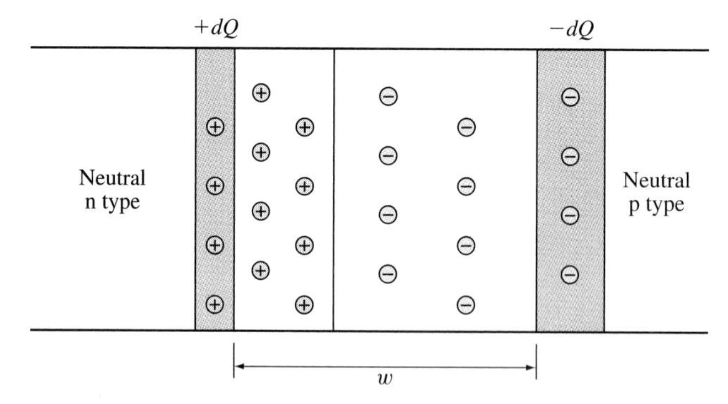

Figure 5.31 Illustration of the junction capacitance C_j. A change in applied voltage V_a produces a change in the number of uncompensated ions on either side of the junction. This produces a change in charge dQ on either side of the junction, making the junction look like a parallel-plate capacitor of width w.

current must be continuous, an equal displacement current flows across the junction, or the junction acts as a capacitor.

The small-signal or differential junction capacitance C_j is

$$\boxed{C_j \equiv \left|\frac{dQ}{dV_a}\right|} \tag{5.108}$$

At equilibrium, there is already some charge on either side of the junction—the ionized dopants in the space charge region. When a voltage is applied, this charge is either increased (reverse bias) or decreased (forward bias). Figure 5.31 emphasizes this point by showing that the space charge changes at the edges of the space charge region. Thus the junction resembles a parallel plate capacitor and we can write for the junction

$$\boxed{C_j = \frac{\varepsilon A}{w}} \tag{5.109}$$

where A is junction area and w is the space charge region width. It is understood that C_j is the differential capacitance defined by Equation (5.108). This capacitance is often called *depletion capacitance*. Since w is dependent on the square root of the voltage, C is nonlinear. With substitution of w from Equation (5.37), the junction capacitance becomes, for the step junction,

$$\boxed{C_j = A\left[\frac{q\varepsilon N_D' N_A'}{2(N_D' + N_A')(V_{\text{bi}} - V_a)}\right]^{1/2} = A\left[\frac{q\varepsilon N_D' N_A'}{2(N_D' + N_A')V_j}\right]^{1/2} \quad \text{pn step junction}} \tag{5.110}$$

For a one-sided step junction,

$$\boxed{C_j = A\left[\frac{q\varepsilon N'}{2(V_{\text{bi}} - V_a)}\right]^{1/2} = A\left[\frac{q\varepsilon N'}{2V_j}\right]^{1/2} \quad \text{one-sided step junction}} \tag{5.111}$$

where N' is the net doping concentration on the lightly doped side. Junction capacitance per unit junction area is plotted in Figure 5.32 as a function of applied voltage V_a for $N' = 10^{16}$ cm^{-3}. One can see that C_j increases with V_a. From Equation (5.110) or (5.111), it would appear that for $V_a = V_{\text{bi}}$, the junction capacitance would be infinite. For large V_a, however, the current and thus the IR_S drop become large. The junction voltage

$$V_j = V_{\text{bi}} - (V_a - IR_S)$$

therefore is always greater than zero.

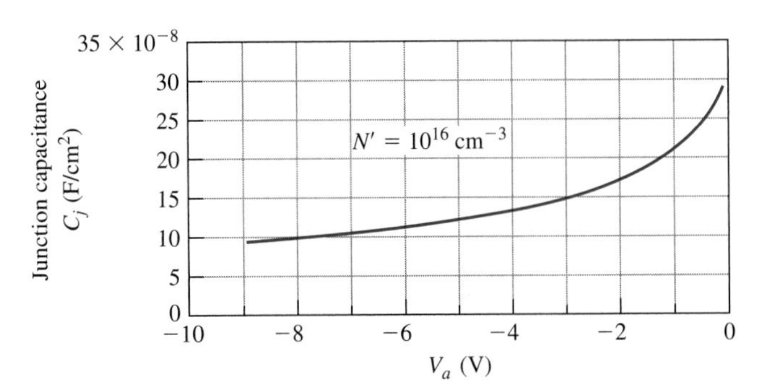

Figure 5.32 Plot of junction capacitance per unit area as a function of reverse bias for a one-sided junction with $N' = 10^{16}$ cm^{-3}.

5.4.3 STORED-CHARGE CAPACITANCE

The other type of capacitance in junctions is called the *stored charge capacitance*. In the case of the previously discussed junction capacitance, the change in charge with a change in voltage was due to a change in the number of non-neutralized ions on each side of the junction. In stored-charge capacitance, the pertinent changing charge on either side of the junction is the change in minority carrier density, as the minority carriers are injected or extracted at the junction edges with changing bias.

We consider the case of an n$^+$p junction under forward bias, Figure 5.33a. We consider the injection of electrons only, since it is large compared with the hole injection in an n$^+$p diode. In steady state, assuming the p-type region is much longer than the electron diffusion length (long-base diode), we have from Equation (5.69), which is repeated here,

$$\Delta n_p(x) = \Delta n_p(x_p)e^{-(x-x_p)/L_n} \tag{5.69}$$

In steady state, this distribution of carriers is maintained constant. We say there is a charge "stored" in this distribution, even though the individual stored electrons or holes are changing from moment to moment. The total stored minority carrier (electron) charge in the p region is the integral of the distribution (times the area of the junction):

$$Q_s = -qA\int_{x_p}^{\infty} \Delta n_p(x)\,dx = qA\Delta n_p(x_p)L_n \tag{5.112}$$

This can be related to the current. We recall that the current across the plane x_p is carried entirely by the diffusion of minority carriers, and that the total current anywhere is equal to the current crossing this plane. The diffusion current is, from Equation (5.69) (again neglecting the small contribution due to holes),

$$I|_{x_p} = qAD_n \left.\frac{d\Delta n_p(x)}{dx}\right|_{x_p} = -qAD_n\frac{\Delta n_p(x_p)}{L_n} \tag{5.113}$$

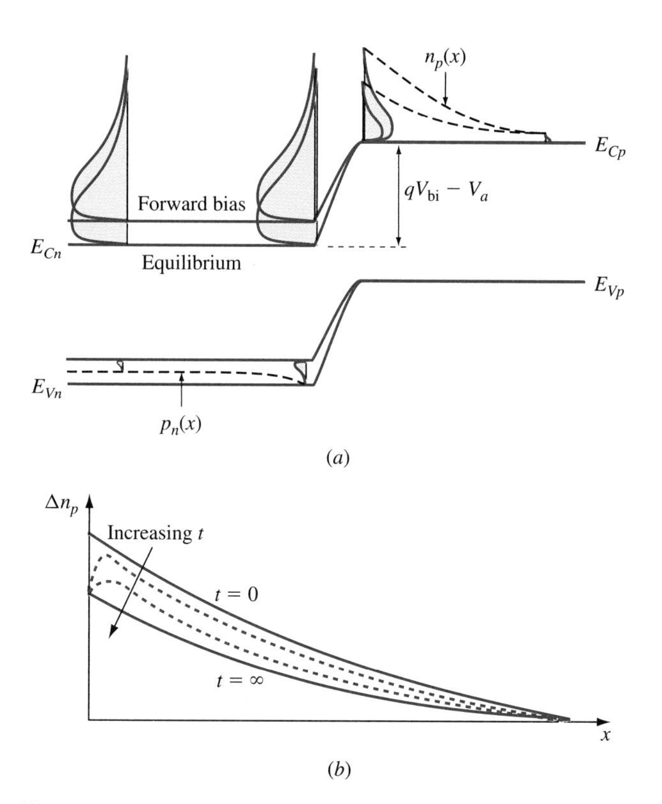

Figure 5.33 Illustration of the stored-charge capacitance in an n$^+$p junction. (a) As the forward bias changes, the number of injected electrons (minority carriers) changes. In a one-sided junction the hole injection is negligible. (b) An abrupt decrease in the forward voltage ΔV_a causes some of the injected ("stored") electrons to return to the n side of the junction and contribute to capacitance.

Combining Equations (5.113) and (5.112) gives

$$\boxed{Q_s = \frac{I L_n{}^2}{D_n} = I\tau_n} \tag{5.114}$$

where we have used $L_n = \sqrt{D_n \tau_n}$. This result shows that the steady-state stored minority carrier charge is proportional to current.

If the applied voltage is changed by dV_a, the steady-state stored charge is changed by dQ_s. The rate of change of Q_s with V_a is

$$\frac{dQ_s}{dV_a} = \frac{dI}{dV_a}\tau_n = \frac{\tau_n}{R_P} \tag{5.115}$$

where R_P is the diode differential resistance given by Equation (5.107).

The expression dQ_s/dV_a has the dimensions of capacitance, but it is not the capacitance as seen from the junction terminals. This can be understood with the aid of Figure 5.33b for the case of V_a being abruptly reduced by dV_a. Since from Equation (5.68), $\Delta n(x_p)$ is proportional to $e^{qV_a/kT}$, the excess carrier concentration at the plane $x = x_p$ also abruptly decreases. The rest of the distribution doesn't change instantaneously, however. Thus the peak concentration of excess electrons is no longer at the plane $x = x_p$ but somewhere to the right of it. Then since $J_n = qD_n(dn/dx)$, the stored electrons will diffuse in both directions to regions of lower concentration. Only those that diffuse to the left into the n^+ region will flow in the external circuit and thus contribute to the capacitance. This charge is referred to as the *reclaimable stored charge* Q_{sr}, and the associated stored-charge capacitance C_{sc} is

$$\boxed{C_{sc} = \frac{dQ_{sr}}{dV_a} = \delta\frac{dQ_s}{dV_a} = \delta I \tau_n} \tag{5.116}$$

where δ is the fraction of the stored charge that is reclaimable.

While this quantity can be calculated, the procedure is tedious. For the case treated here, half of the stored charge is reclaimable, or $\delta = 1/2$. For diodes in which the doping is nonuniform, δ depends on the doping profile. Stored-charge capacitance will be treated again in more detail in the section on bipolar junction transistors.

The stored-charge capacitance in prototype diodes is proportional to diffusion current, which in turn varies exponentially with V_a, while the junction capacitance varies as $\sqrt{V_a}$. As a result, for reverse bias and small forward bias, junction capacitance predominates. For large forward bias, stored-charge capacitance predominates.

EXAMPLE 5.5

Compare the junction capacitance and stored-charge capacitance under reverse bias ($V_a = -5$ V) and forward bias ($V_a = +0.75$ V).

■ Solution

We consider a prototype silicon n^+p junction with $N'_A = N_A = 10^{17}$ cm^{-3}. Let the junction area be 100 μm^2 and the fraction of reclaimable charge $\delta = 0.5$.

■ Junction Capacitance

For an n^+p junction, from Equation (5.111),

$$C_j = A\left[\frac{q\varepsilon N_A}{2(V_{bi} - V_a)}\right]$$

The built-in voltage for this junction is, from Figure 5.10,

$$V_{bi} = 0.98 \text{ V}$$

For $V_a = -5$ V,

$$C_j(-5) = \left(100\,\mu\text{m}^2 \frac{10^{-8}\,\text{cm}^2}{1\,\mu\text{m}^2}\right) \times \left[\frac{(1.6\times10^{-19}\,\text{C})(11.8)(8.85\times10^{-14}\,\text{F/cm})(10^{17}\,\text{cm}^{-3})}{2(0.98+5)}\right]^{1/2}$$

$$C_j(-5) = 0.053\text{ pF}$$

Similarly, for $V_a = 0.75$ V,

$$C_j(0.75) = 0.27\text{ pF}$$

■ Stored-Charge Capacitance

To find the stored-charge capacitance, we will need to find I. This will be the diffusion current from Equation (5.77),[16] and to find that we need $I_0 = AJ_0$. Since $J_n \gg J_p$, holes injected into the n$^+$ material can be ignored and

$$I = I_0(e^{qV_a/kT} - 1) = qA\left(\frac{D_n n_{p0}}{L_n}\right)(e^{qV_a/kT} - 1)$$

where D_n and L_n are the minority carrier diffusion constant and diffusion length respectively.

For the p-type material, we find

$$n_{p0} = \frac{n_i{}^2}{N'_A} = \frac{(1.08\times10^{10}\,\text{cm}^{-3})^2}{10^{17}\,\text{cm}^{-3}} = 1.17\times10^3\,\text{cm}^{-3}$$

From Figure 3.11, we look up D_n (the minority carriers on the p side are the electrons) using the doping level on the *p side.* At $N'_A = 10^{17}$ cm^{-3}, we find $D_n =$ 20 cm^2/s. From Figure 3.23, $L_n = 110\,\mu$m.

Then I_0 becomes

$$\begin{aligned} I_0 &= qA\left(\frac{D_n n_{p0}}{L_n}\right) \\ &= (1.6\times10^{-19}\,\text{C})(100\times10^{-8}\,\text{cm}^2)\left[\frac{(20\,\text{cm}^2/\text{s})(1.17\times10^3\,\text{cm}^{-3})}{11\times10^{-3}\,\text{cm}}\right] \\ &= 3.4\times10^{-19}\,\text{A} \end{aligned}$$

The diffusion current at $V_a = -5$ V is $I = -I_0 = -3.4\times10^{-19}$ A. The current under a forward bias of +0.75 V is $I = I_0(e^{qV_a/kT} - 1) = 3.4\times10^{-19}\,\text{A}(e^{0.75/0.026} - 1) =$ 1.1 μA.

The stored-charge capacitances are:
At $V_a = -5$ V:

$$C_{sc}(-5) = \delta I \tau_n = (0.5)(3.4\times10^{-19}\,\text{A})(3\times10^{-6}\,\text{s}) = 5.1\times10^{-25}\,\text{F} \approx 0$$

where we obtained the minority carrier lifetime from Figure 3.21.

[16] The generation-recombination current does not contribute to stored minority carrier charge.

At $V_a = +0.75$ V:

$$C_{sc}(0.75) = \delta I \tau_n = (0.5)(1.1 \times 10^{-6}\text{ A})(3 \times 10^{-6}\text{ s}) = 1.65\text{ pF}$$

Let us now compare the two capacitances:

$V_a = -5$ V	$C_j = 0.053$ pF	$C_{sc} \approx 0$
$V_a = 0.75$ V	$C_j = 0.27$ pF	$C_{sc} = 1.65$ pF

As expected, the junction capacitance dominates under reverse bias and the stored-charge capacitance dominates under forward bias.

5.5 TRANSIENT EFFECTS

A pn junction is often used as a switch. A voltage or current pulse is applied to change the operating state between forward bias ("on") and reverse bias ("off"). Since a pn junction has capacitance associated with it, it would be expected that some time would be required to make the transition from off to on (turn-on time) and from on to off (turn-off time). These transients are discussed qualitatively with the aid of Figure 5.34. The applied voltage is switched between V_F (forward bias) and V_R (reverse bias). For mathematical simplicity, we assume that both V_F and V_R are much larger in magnitude than V_{bi}. We also assume that $R_1 \gg R_S$ so that the diode series resistance R_S can be neglected (i.e., $R_S = 0$).

Usually the turn-off time is much larger than the turn-on time, so this case is considered first.

5.5.1 TURN-OFF TRANSIENT

For simplicity, we discuss the case of an n^+p junction, so that we can ignore the stored hole charge in the n region. The stored excess electron charge is much larger and dominates in an n^+p junction.

At $t < 0$, the diode of Figure 5.34 is on, with an applied voltage of V_F. The junction is forward biased and excess electrons are being injected into the

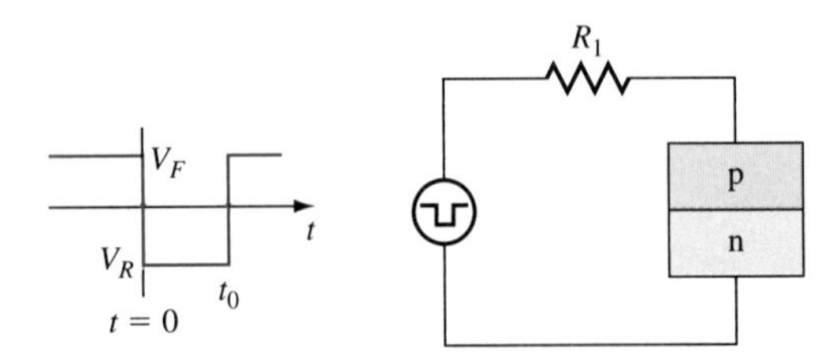

Figure 5.34 The circuit used to illustrate switching turn-off and turn-on transients in a pn junction.

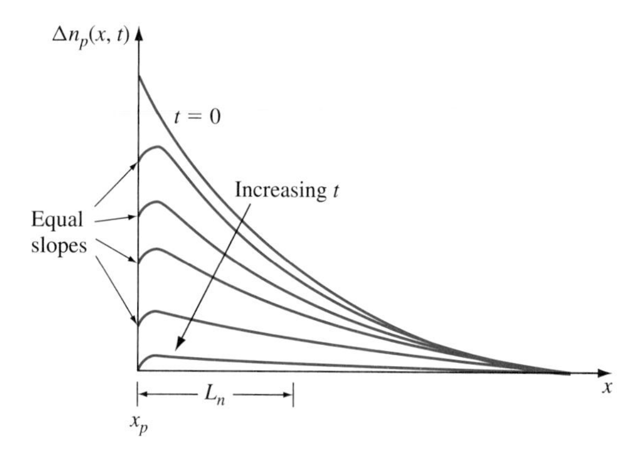

Figure 5.35 Decay of injected electron concentration with time after switching from V_F to V_R.

quasi-neutral p region. At $t = 0$ the excess electron concentration for $x \geq x_p$ varies with position as given by Equation (5.69):

$$\Delta n_p(x, 0) = \Delta n_p(x_p, 0)e^{-(x-x_p)/L_n} \tag{5.117}$$

The junction voltage $V_j(\text{on})$ will be less than V_{bi} (forward bias reduces the barrier). Since $V_F \gg V_{\text{bi}} > V_{\text{on}}$, the forward current up until $t = 0$ is

$$I_F(0) \approx \frac{V_F}{R_1} \tag{5.118}$$

At $t = 0$ the applied voltage is switched from V_F to V_R. The minority carrier (electron) distribution in the quasi-neutral p region is shown in Figure 5.35 as a function of x and t for $t > 0$. Excess carriers are no longer being injected, and as the carriers diffuse away and recombine, the excess carrier concentration decreases. We also observe that some of the carriers will diffuse back into the junction as discussed earlier.

Notice that the slope of Δn_p at x_p is constant for some time as the carrier concentration decays. We have argued before that the current crossing the plane x_p is the same current that flows through the device, but we can evaluate the current easily at this plane because it is entirely due to diffusion here. Since the slope is constant, the current is constant at

$$I_R \approx \frac{V_R}{R_1} \tag{5.119}$$

until $t = t_s$, where t_s is the *storage time*.[17] The current is plotted as a function of time in Figure 5.36. For $t > t_s$ the slope at $x = x_p$ decreases with t and the magnitude of the current decays toward the steady-state reverse current (≈ 0).

---

[17] Note that since V_R is negative, so is I_R.

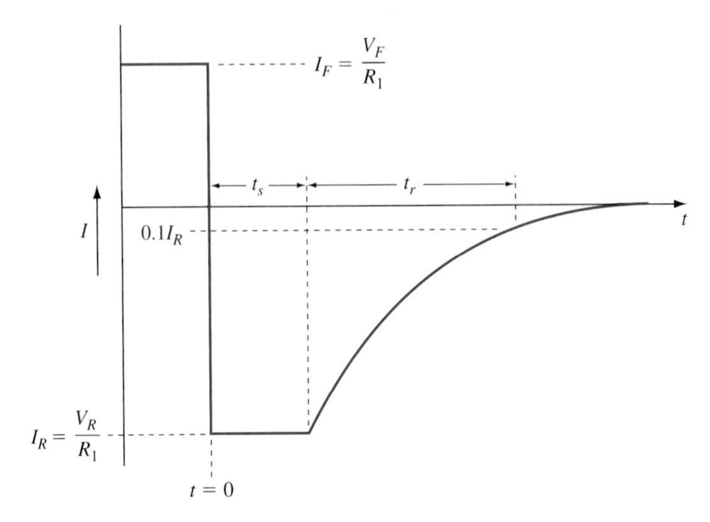

Figure 5.36 Current transient during turn-off of diode, showing the storage time t_s and the rise time t_r.

The storage time can, in principle, be calculated by solving the continuity equation, but this is somewhat involved. An approximate solution[18] gives

$$t_s \approx \tau_n \ln\left(1 + \left|\frac{I_F}{I_R}\right|\right) \tag{5.120}$$

For $V_F = V_R$, $|I_F| \approx |I_R|$, and $t_s = \tau_n \ln 2 = 0.68\tau_n$.

The turn-off time will be shortened if the charge in the p region is reduced, which can be achieved by reducing the electron lifetime or decreasing the $|I_F/I_R|$ ratio. Reducing the carrier lifetime also reduces L_n. This can be accomplished by introducing recombination centers (traps) such as Au or Cu. The traps, however, increase the off current because they provide for more carrier generation in the transition region in the off state. That results in increased power consumption.

The turn-off time can also be reduced by reducing the thickness of the more lightly doped side or by appropriately grading the doping in the more lightly doped side, a topic covered in Chapter 9 for bipolar transistors.

[18]The resulting, more accurate relation,

$$t_s = \tau_n \left[\operatorname{erf}^{-1}\left(\frac{1}{1 + \left|\frac{I_R}{I_F}\right|}\right)\right]^2$$

is used in SPICE simulations.

5.5.2 TURN-ON TRANSIENT

There is also a transient time associated with switching from off back to on (V_R to V_F). Some time is required to reach the steady-state forward voltage. This delay time is a result of the time required to discharge the junction capacitance (decrease the junction width) and then to inject carriers to set up the steady-state electron distribution. Switching to V_F at $t = t_0$ results in a current

$$I_F \approx \frac{V_F}{R_1} \tag{5.121}$$

Figure 5.37 shows the variation of stored electron concentration as a function of position and time for the turn-on transient. Since I_F is constant, so are the slopes of the $n_p(x_p, t)$ curves. Steady state is reached when the recombination rate is equal to the rate of electron injection.

The amount of charge required to discharge the junction capacitance is normally much smaller than the steady-state minority carrier stored charge. This, along with the fact that the stored charge is established by diffusion (a slow process), means that the turn-on time is to good approximation equal to that required to set up the minority carrier steady-state distribution.

Figure 5.38 indicates the diode current and voltage waveforms resulting from a rectangular input voltage waveform. Upon switching of the input voltage from V_F to V_R, the diode current switches from $I_F = V_F/R_1$ to $I_R = V_R/R_1$. During the storage time, while the diode current remains constant, the diode voltage remains at V_{on}. In a real diode, the series resistance R_S is finite. Since the diode terminal voltage is that across the junction plus that across the diode series resistance R_S, at $t = 0$, the terminal voltage reduces from $[V_j(\text{on}) + I_F R_S]$ to $[V_j(\text{on}) + I_R R_S]$, where I_F is positive and I_R is negative. This is indicated in Figure 5.38c. After a time $t = t_s$ the diode current reduces in magnitude and the diode voltage decays toward its steady-state value.

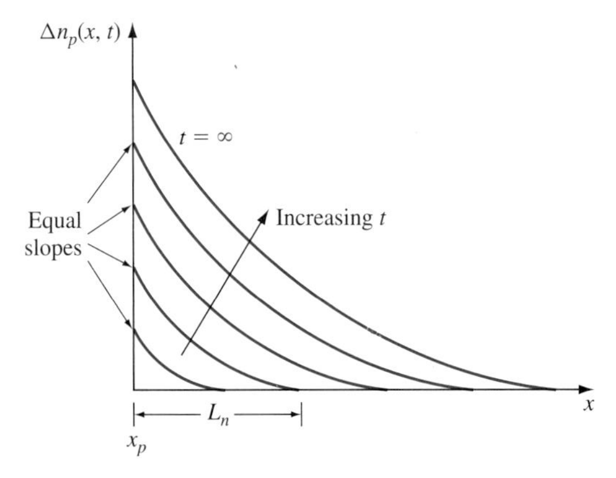

Figure 5.37 The buildup of stored electrons after switching from off to on.

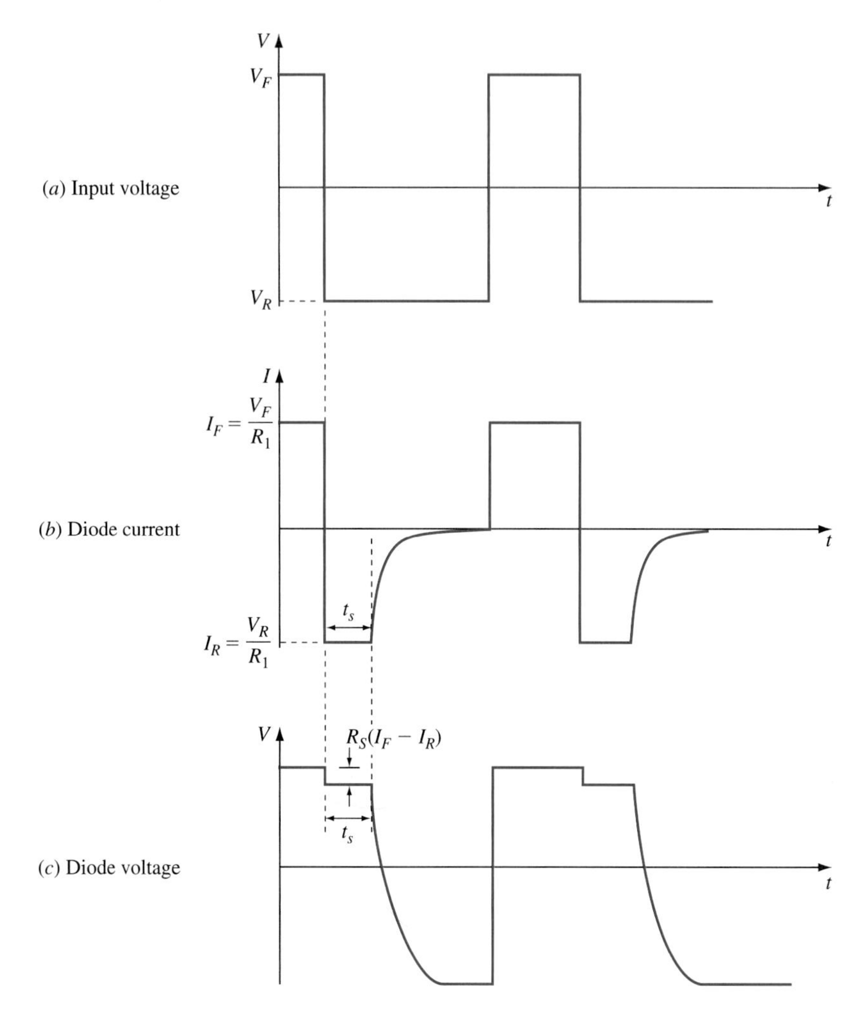

Figure 5.38 The waveform resulting from a diode switched on and off. (a) The input waveform; (b) the diode current; (c) the diode voltage. The turn-off time exceeds the turn-on time significantly.

Note that the turn-on time is much less than the turn-off time, so the maximum switching frequency is limited by the turn-off time.

In the above qualitative discussion of switching transients in a pn junction, approximations were used to illustrate the switching phenomenon resulting from the effect of stored charge. Both forward-bias and reverse-bias voltages were assumed large compared with the voltage V_D across the diode, such that V_D could be neglected in determining I_F and I_R. In many circuits it is convenient to have $V_R = 0$ such that the input voltage switches between V_F and zero as shown in the circuit of Figure 5.39a. As indicated, the diode is represented by its

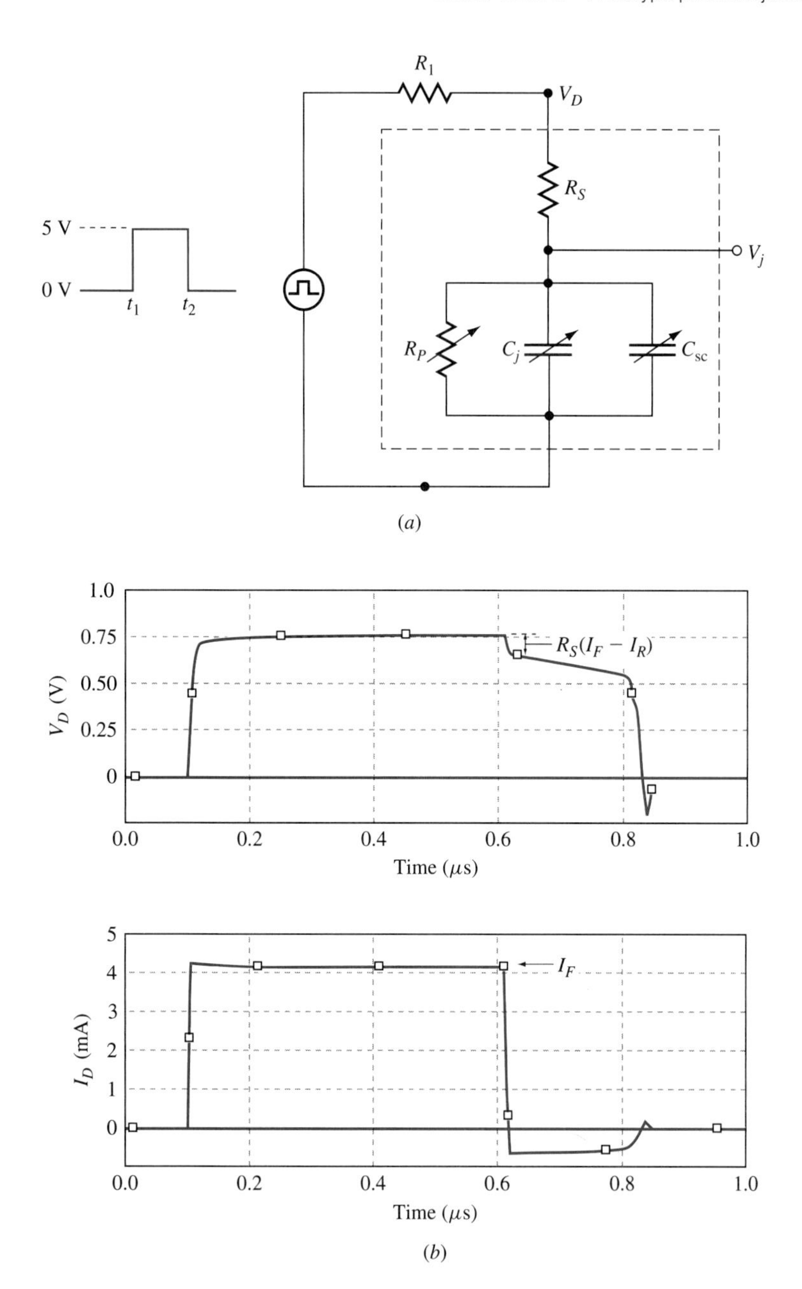

Figure 5.39 (a) The switching circuit model for a pn homojunction diode. (b) The voltage and current response, as determined by SPICE, to an input pulse beginning at $t_1 = 100$ ns and ending at $t_2 = 600$ ns.

equivalent circuit where R_p, C_j and C_{sc} are dependent on voltage and current and the effect of R_S is considered.

It is of interest to determine the diode voltage (V_D) and current (I_D) transients during switching for such a case. This is a complicated problem because of the nonlinearity of R_P, C_j, and C_{sc}. These switching waveforms can be determined, however, with the aid of SPICE (Simulation Program with Integrated Circuit Emphasis). This problem is treated in the Supplement to Part 2. Plots of V_D and I_D as functions of time as simulated by SPICE are given in Figure 5.39b.

The previous discussion was for a long-base diode. For a short-base n^+p diode ($W_B \ll L_n$) there is negligible electron recombination in the quasi-neutral p region and $n_p(x)$ can be written

$$\Delta n_p(x) = \Delta n_p(0)\left(1 - \frac{x}{W_B}\right) \tag{5.122}$$

and

$$I_n = qAD_n \frac{d\Delta n}{dx} = -\frac{qAD_n\,\Delta n(0)}{W_B} \tag{5.123}$$

The total excess electron charge stored in the short base is

$$Q_s = -qA\,\Delta n(0)\cdot\frac{W_B}{2} = I_n t_T \tag{5.124}$$

where t_T is referred to as the *base-transit time* and is given by

$$t_T = \frac{(W_B)^2}{2D_n} \qquad \text{short-base diode} \tag{5.125}$$

This is the average time it takes for the minority carriers (electrons) to traverse the base.

EXAMPLE 5.6

Compare the amount of minority carrier stored charge in a forward-biased short-base diode with that in a long-base diode.

■ Solution

From Equations (5.124), (5.125), and (5.114), and using $L_n{}^2 = D_n\tau_n$,

$$\frac{Q_s(\text{short base})}{Q_s(\text{long base})} = \frac{1}{2}\left(\frac{W_B}{L_n}\right)^2$$

for W_B on the order of 0.1 μm, as in a typical bipolar transistor, and $L_n \approx 31\ \mu$m, we have

$$\frac{Q_s(\text{short base})}{Q_s(\text{long base})} \approx \frac{1}{2}\left(\frac{0.1}{31}\right)^2 \approx 5\times10^{-6}$$

It is clear that the stored charge and thus the turn-off recovery time is greatly reduced by the use of a short-base diode.

5.6 EFFECTS OF TEMPERATURE

At a given voltage the current in a diode is quite temperature sensitive. Consider an n^+p diode. From Equation (5.79),

$$J_0 = \frac{q}{N'_A}\sqrt{\frac{D_n}{\tau_n}}n_i^2$$

$$= \frac{q}{N'_A}\sqrt{\frac{D_n}{\tau_n}}N_C N_V e^{-E_g/kT} \tag{5.126}$$

Neglecting the temperature dependences of D_n, τ_n, N_C, N_V, and E_g compared with the exponential dependence on $1/T$, the fractional variation of J_0 with temperature can be expressed

$$\frac{\dfrac{dJ_0}{dT}}{J_0} = \frac{\dfrac{de^{-E_g/kT}}{dT}}{e^{-E_g/kT}} = \left(\frac{E_g}{kT^2}\right) = \frac{E_g}{kT}\cdot\frac{1}{T} \tag{5.127}$$

For Si at room temperature this becomes

$$\frac{1.12}{0.026}\cdot\frac{1}{300} = 0.14$$

or J_0 varies approximately 14 percent per degree Celsius.

Similarly since J_{GR0} varies as n_i rather than n_i^2, its fractional variation with temperature is about half this value.

An analogous analysis shows that at constant forward current in the typical range of operation (0.1 to 1 mA) the diode forward voltage decreases about 2 mV for an increase of 1°C.

5.7 SUMMARY

In most pn junctions, the doping concentrations are complex functions of position and their electrical characteristics must be calculated numerically. While there are software programs available to do this, their use provides little insight into the physical processes involved. In this chapter, we used a greatly simplified model of a pn junction, the prototype pn homojunction. In this model we assumed that the net doping level on each side is constant with position, or the doping is a step function at the metallurgical junction.

We treated three classes of junctions:

1. pn junctions in which the semiconductor on each side is nondegenerate
2. p^+n junctions in which the p side is degenerate and the n side is nondegenerate
3. n^+p junctions with the n side degenerate and the p side nondegenerate

For the n and p regions, the Fermi energy can be expressed

$$E_f = E_C - kT \ln \frac{N_C}{N_D'} = E_i + kT \ln \frac{N_D'}{n_i} \qquad \text{n region}$$

$$E_f = E_V + kT \ln \frac{N_V}{N_A'} = E_i - kT \ln \frac{N_A'}{n_i} \qquad \text{p region}$$

For a degenerate semiconductor, the approximation is often made that

$$E_f \approx E_C \qquad \text{n}^+ \text{ region}$$

$$E_f \approx E_V \qquad \text{p}^+ \text{ region}$$

5.7.1 Built-in Voltage

An important parameter required to determine the electrical characteristics of any junction is its built-in voltage. In general

$$V_{\text{bi}} = \frac{1}{q}|\Phi_p - \Phi_n|$$

where the work functions Φ_p and Φ_n are evaluated at the edges of the transition region at equilibrium. In general, V_{bi} must be solved numerically. However, for a prototype homojunction,

$$V_{\text{bi}} = \frac{1}{q}\left[E_g - kT \ln \frac{N_C N_V}{N_D' N_A'}\right] = \frac{kT}{q} \ln \frac{N_D' N_A'}{n_i^2} \qquad \text{prototype pn junction}$$

$$V_{\text{bi}} = \frac{1}{q}\left[E_g - kT \ln \frac{N_V}{N_A'}\right] = \frac{kT}{q} \ln \frac{N_C N_A'}{n_i^2} \qquad \text{prototype n}^+\text{p junction}$$

$$V_{\text{bi}} = \frac{1}{q}\left[E_g - kT \ln \frac{N_C}{N_D'}\right] = \frac{kT}{q} \ln \frac{N_V N_D'}{n_i^2} \qquad \text{prototype p}^+\text{n junction}$$

5.7.2 Junction Width

The junction width of a prototype homojunction can be expressed

$$w = w_n + w_p = \left[\frac{2\varepsilon V_j (N_A' + N_D')}{q N_A' N_D'}\right]^{1/2} \qquad \text{pn}$$

$$w = \left[\frac{2\varepsilon V_j}{qN'_A}\right]^{1/2} \qquad \text{n}^+\text{p}$$

$$w = \left[\frac{2\varepsilon V_j}{qN'_D}\right]^{1/2} \qquad \text{p}^+\text{n}$$

where the junction voltage $V_j = V_{\text{bi}} - V_a$ and V_a is the applied voltage.

The ratio of the junction width on the n side to that on the p side is

$$\frac{w_n}{w_p} = \frac{(x_0 - x_n)}{(x_p - x_0)} = \frac{N'_A}{N'_D}$$

and most of the junction width is on the more lightly doped side.

Likewise, the ratio of the junction voltage dropped across the n side to that across the p side is

$$\frac{V_j{}^n}{V_j{}^p} = \frac{N'_A}{N'_D}$$

so most of the junction voltage appears across the more lightly doped side.

5.7.3 Junction Current

The three major current mechanisms for both forward and reverse bias are diffusion current density J_{diff}, drift current J_{drift}, and generation-recombination current density J_{GR}. The total current density can be expressed as $J = J_{\text{GR}} + J_{\text{diff}}$ or

$$J = J_{\text{GR}} + J_{\text{diff}} \approx J_{\text{GR0}}(e^{qV_a/2kT} - 1) + J_0(e^{qV_a/kT} - 1)$$

where J_{diff} is evaluated at the transition region edges, where drift current is negligible.

$$J_{\text{diff}} = J_0(e^{qV_a/kT} - 1)$$

where J_0 increases about 14 percent per degree Celsius. At constant current the diode voltage decreases about 2 mV per degree increase. For forward bias at room temperature, the current increases by a factor of 10 for a voltage change of 60 mV.

The coefficient for the generation-recombination current is

$$J_{\text{GR0}} \approx \frac{qn_i w}{2\tau_0}$$

The transition width w is only slightly voltage dependent (proportional to $\sqrt{V_j}$).

For a device in which the thickness of either side is much longer than a minority carrier diffusion length:

$$J_0 = q\left(\frac{D_n n_{p0}}{L_n} + \frac{D_p p_{n0}}{L_p}\right) \qquad \text{long-base diode}$$

If the p-side thickness W_B is much less than a diffusion length,

$$J_0 = q\left(\frac{D_n n_{p0}}{W_B} + \frac{D_p p_{n0}}{L_p}\right) \qquad \text{short p-region diode}$$

and if both sides are very short,

$$J_0 = q\left(\frac{D_n n_{p0}}{W_{B(p)}} + \frac{D_p p_{n0}}{W_{(n)}}\right) \qquad \text{both sides short}$$

Generally $J_{\text{GR0}} \gg J_0$ and thus J_{GR0} predominates for reverse bias and small forward bias. However, since the diffusion current increases much more rapidly with forward bias, it predominates at higher V_a and thus higher currents.

5.7.4 Junction Breakdown

There are two additional current mechanisms, which can lead to large reverse currents (breakdown) for reverse-biased junctions.

In Zener tunneling, electrons tunnel from the valence band on the p side to the conduction band on the n side. The tunneling current density can be expressed as

$$J \approx e^{-\pi W_T \sqrt{m^* E_g}/2^{3/2}\hbar} \qquad \text{tunneling}$$

Multiplication current results from carrier impact ionization in the high-field depletion region. The carrier multiplication factor M is

$$M = \frac{1}{(1-P)}$$

where P is the probability that one carrier crossing the junction will create an additional hole-electron pair. For unity P, M becomes infinite and avalanche occurs.

While Zener tunneling results in a soft breakdown and predominates in heavily doped junctions (small tunneling distance), avalanche gives a sharp breakdown and predominates for more lightly doped junctions.

5.7.5 Capacitance

There are two effects that contribute to capacitance in a pn junction. The junction capacitance C_j results from the variation of ionized charge in the transition region with applied voltage.

$$C_j = A\left[\frac{q\varepsilon N'_D N'_A}{2(N'_D + N'_A)(V_{bi} - V_a)}\right]^{1/2}$$

Stored-charge capacitance results from the injected minority carrier charge stored in the quasi-neutral regions for forward bias. For an n^+p junction in which the p region is much greater than an electron diffusion length,

$$C_{sc} = \delta I \tau_n \qquad \text{long-base diode}$$

where δ is the reclaimable charge fraction and is equal to $1/2$ for step (prototype) junctions.

For the p region much shorter than L_n, the capacitance is proportional to the p-region thickness, or

$$C_{sc} = \delta I t_T = \delta I\left(\frac{W_B{}^2}{2D_n}\right) \qquad \text{short-base diode with p side short}$$

In this case, $\delta = 2/3$.

For reverse bias and for small forward currents, C_j predominates. For larger forward currents, C_{sc} predominates.

5.7.6 Transient Effects

When operating as a switch, the switching speed is controlled by the time required to charge and discharge the diode capacitances. The major contribution to the time is the time associated with charging and discharging the stored-charge capacitance. Normally, the time required to remove the stored charge in switching from forward bias to reverse bias (turn-off transient) is appreciably greater than in switching from reverse to forward bias.

We emphasize that, while the prototype model is oversimplified for most real devices, it does provide insight to the physical processes involved. In the next chapter, we will refine the model and discuss some more realistic devices.

5.8 READING LIST

Items 1, 2, 4, 8–12, 17–19, and 26 in Appendix G are recommended.

5.9 REVIEW QUESTIONS

1. Referring to Figure 5.8, how can electrons enter the transition region from the left if there is an opposing electric field in the transition region itself?
2. Explain qualitatively why the current in a diode is small under reverse bias and large under forward bias.
3. Why does generation current dominate over recombination current under reverse bias in a pn junction, but recombination current dominates over generation current under forward bias?

5.10 PROBLEMS

5.1 A silicon pn junction is formed between n-type silicon doped with $N_D = 10^{17}\,\text{cm}^{-3}$ and p-type silicon doped with $N_A = 10^{16}\,\text{cm}^{-3}$.

a. Sketch the energy band diagram. Label both axes and all important energy levels.

b. Find n_{n0}, p_{n0}, n_{p0}, and p_{p0}. Sketch the carrier concentrations.

c. What is the built-in voltage?

5.2 Recall that the circuit symbol for a diode is as shown in Figure II.3. Which is the anode (the end labeled +) of a pn junction diode, the p side or the n side? Explain your reasoning.

5.3 A p^+n junction is formed in silicon. On the p side, the Fermi level is at the (intrinsic) valence band edge $E_f = E_{V0}$. The n side is doped with $N_D = 5 \times 10^{16}\,\text{cm}^{-3}$.

a. Sketch the energy band diagram.

b. Sketch the carrier concentrations.

c. What is the built-in voltage?

5.4 Fill in the missing steps to derive Equation (5.35) for the junction width on the n side of the junction.

5.5 A step pn junction diode is made in silicon with the n side having $N'_D = 2 \times 10^{16}\,\text{cm}^{-3}$ and the p side having a net doping of $N'_A = 5 \times 10^{15}\,\text{cm}^{-3}$.

a. Draw, to scale, the energy band diagram of the junction at equilibrium.

b. Find the built-in voltage, and compare with the value measured off your drawing in part (a).

c. Find the junction width.

d. Find the widths of the n side of the depletion region and the p side of the depletion region, and the voltage dropped across each side of the transition region.

e. Plot the electric field. What is its maximum value?

f. Plot the voltage distribution.

g. Plot the potential energy for electrons (E_C).

h. Draw the energy band diagram for $V_a = 0.5$ V.
i. Draw the energy band diagram for $V_a = -5$ V.

5.6 Consider the equilibrium energy band diagram for the pn junction diode shown in Figure P5.1a.

a. Indicate the region(s) where there exists an electric field.
b. What is the value of the built-in voltage?
c. Sketch the electron and hole concentrations. Indicate the directions of the drift and diffusion components of the electron and hole fluxes and currents.
d. The same device is shown in Figure P5.1b, but now a voltage is applied across it. If the total junction voltage $V_j = V_{bi} - V_a$, what is the value of the applied voltage?

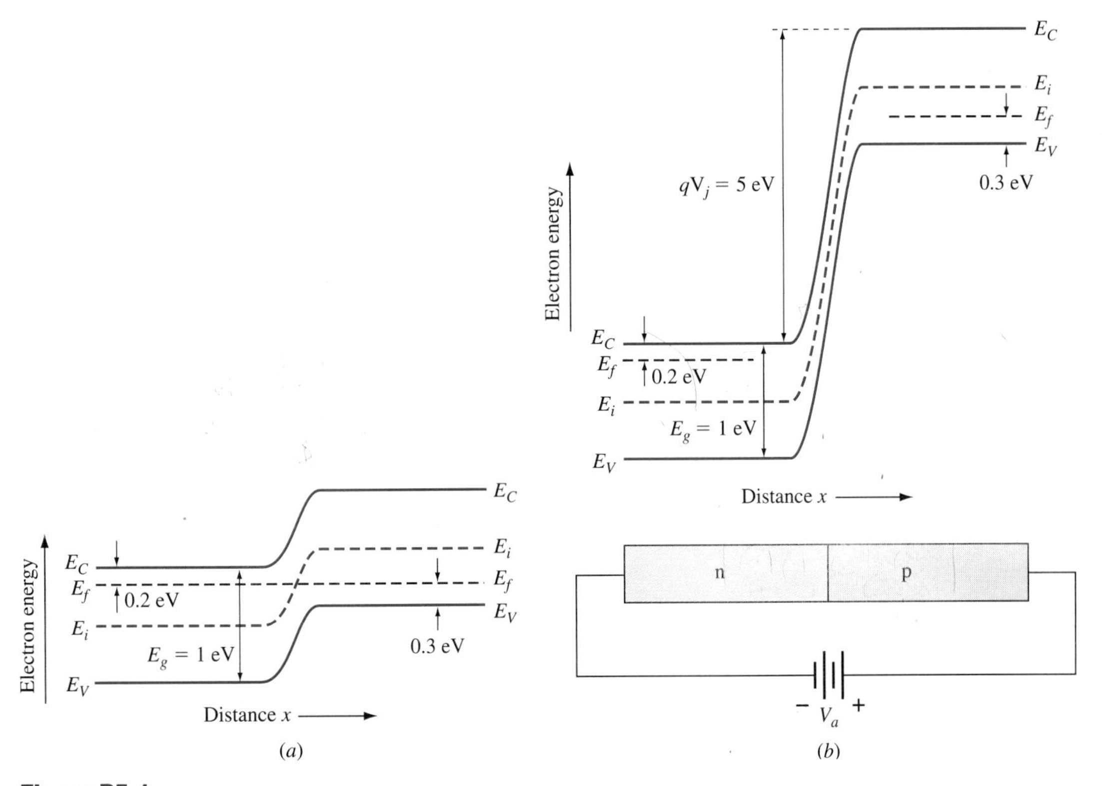

Figure P5.1

5.7 A pn junction is formed in silicon between n-type ($N'_D = 10^{18}$ cm^{-3}) and p-type ($N'_A = 10^{17}$ cm^{-3}) materials. Find, for equilibrium,

a. w_n and w_p and w.
b. The built-in voltage.
c. The maximum electric field.

d. How much of V_{bi} is dropped on the n side? On the p side?

e. Sketch an energy band diagram carefully reflecting your calculations above.

5.8 An n^+p junction has $N'_A = 10^{17}$ cm^{-3}. The Fermi level on the n^+ side is taken to be at the (intrinsic) conduction band edge. Find the junction width at equilibrium.

5.9 A silicon diode has $N'_D = 10^{17}$ on the n side and $N'_A = 10^{16}$ on the p side. It is forward biased at $V_a = 0.5$ V.

a. What is the diffusion current density due to minority carriers at the plane $x = x_p$?

b. What is the minority carrier diffusion density at the plane $x = x_n$?

c. What is the total current density in the junction neglecting recombination and generation?

d. What is the maximum recombination current density in the forward-biased junction? Compare this result with the injection (diffusion current).

e. Which is larger, the injection flux density into the lightly doped side or the injection flux density into the heavily doped side?

f. Repeat part (a) for a reverse bias of $V_a = -5$ V.

g. Repeat part (b) for $V_a = -5$ V.

h. Estimate the generation current density under reverse bias.

i. Compare the generation current density to the diffusion current density in the reverse-biased junction.

5.10 a. Calculate the minority excess carrier concentrations at each edge of the transition region for a silicon diode with $N'_D = 5 \times 10^{17}$ cm^{-3} and $N'_A = 10^{17}$ cm^{-3}. The diode is forward biased with $V_a = 0.5$ V.

b. Sketch the diffusion current as a function of distance on the p side.

c. Keeping in mind that the total current is constant and the difference between the total current and the minority carrier diffusion current is due to drift of majority carriers, sketch the majority carrier drift current as a function of distance on the p side.

5.11 a. Show that Equation (5.93) follows from Equation (5.46).

b. Derive Equation (5.94).

5.12 Consider an n^+p junction under reverse bias of 5 V. Let $N'_A = 5 \times 10^{17}$ cm^{-3} and the junction area be 75 μm^2.

a. Find the reverse current due to diffusion.

b. Find the reverse current due to generation.

5.13 Note that diffusion coefficient, diffusion length, and lifetime depend on doping [Equations (3.18), (3.42), (3.82), and (3.76)]. Consider a one-sided n^+p junction in silicon.

a. Plot the reverse diffusion current as a function of doping for $V_a = -5$ V.

b. Plot the reverse generation current as a function of doping for $V_a = -5$ V.

5.14 A Si junction has $N'_D = 10^{15}$ cm^{-3} and $N'_A = 10^{17}$ cm^{-3}. The junction area is 100 μm^2.

a. Find V_{bi}.

b. Find I_0.

c. Find the current at $V_a = -5$ V, 0 V, and +0.5 V. Remember to consider both diffusion and recombination-generation current. Also bear in mind that most of the junction appears on the lightly doped side (thus lifetime should be chosen for that material).

5.15 Plot the small signal resistance R_P as a function of applied voltage for V_a positive. Let the reverse leakage current be $I_0 = 10^{-15}$ A. Use a logarithmic scale for the resistance axis. Comment on your graph in view of your expectations of the resistance of a diode.

5.16 Consider a symmetrical junction in GaAs in which $N'_D = N'_A = 10^{16}$ cm^{-3}.

a. Find V_{bi}.

b. Calculate the series resistance R_S of the bulk regions if the cross-sectional area of the junction is 50 μm^2 and the lengths of the bulk regions are each 2 μm. Note that the series resistance R_S is due to the finite resistivity of the semiconductor materials; it is not the same as the differential resistance R_P.

5.17 Consider a forward-biased pn junction:

$$I = I_0(e^{q(V_a - IR_S)/kT} - 1)$$

or

$$V_a = \frac{kT}{q} \ln\left(\frac{I}{I_0} + 1\right) + IR_S$$

Let $I_0 = 10^{-16}$ A and $R_S = 20\ \Omega$. Plot I versus V_a for $0.3\text{ V} \leq V_a \leq 1\text{ V}$. (*Hint:* Choose values for I and solve for corresponding values of V_a.)

5.18 A Si junction has $N'_D = 10^{17}$ cm^{-3} and $N'_A = 10^{18}$ cm^{-3}. The junction area is 100 μm^2. What is the junction capacitance at $V_a = -5$ V?

5.19 A junction has $N'_D = 10^{18}$ cm^{-3} and $N'_A = 10^{16}$ cm^{-3}. Both sides are long, and the fraction of reclaimable charge is 0.5. Compare the magnitudes of the junction capacitance and the stored charge capacitance at $V_a = -5$ V, 0 V, and +0.5 V. The junction area is 100 μm^2. Note that generation-recombination current does not contribute to stored charge, only diffusion current contributes.

5.20 A bipolar transistor consists of two pn junctions back to back. In the forward active mode, the emitter-base junction is forward biased and the base-collector junction is reverse biased. The base is generally very thin—much shorter than a diffusion length. Figure P5.2 shows the energy band diagram.

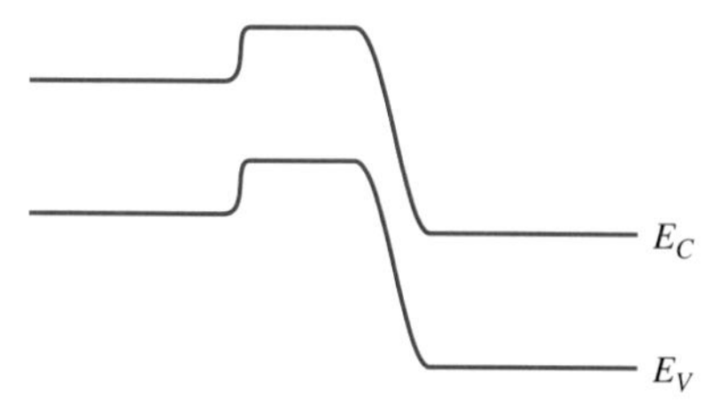

Figure P5.2

a. Identify the emitter, base, and collector on the diagram.
b. Is this an npn or a pnp transistor?
c. At the E-B junction, are electrons injected or extracted?
d. At the C-B junction, are electrons injected or extracted?
e. Sketch the electron concentration distribution you expect across the base.
f. This thin base region is the source of the term *short base diode.* The goal in designing a transistor is to ensure that every electron injected into the base from the emitter eventually ends up in the collector (so that $I_E \approx I_C$). Explain how this short-base structure helps. (*Hint:* What else could happen to the electrons?)

5.21 In Figure 5.39, explain how the diode current can be negative when the applied voltage is zero.

5.22 From Equation (5.120), the storage time is given by

$$t_s = \tau_n \ln\left(1 + \left|\frac{I_F}{I_R}\right|\right)$$

From Figure 5.39 (obtained from a SPICE plot), $I_F = 4.2$ mA and $I_R = -0.6$ mA. Estimate the electron lifetime in the p-type region.

CHAPTER 6

Additional Considerations for Diodes

6.1 INTRODUCTION

In Chapter 5, we treated the case of prototype homojunctions, in which the doping is a step function at the metallurgical junction. While such junctions are seldom encountered in practice, the model is amenable to mathematical analysis and the results are at least indicative of those for a real device.

In this chapter, other, more realistic, junction devices are considered. We will first discuss homojunctions in which the doping profile is not a step function. Then we briefly discuss heterojunctions and metal-semiconductor junctions.

6.2 NONSTEP HOMOJUNCTIONS

As an example of a nonstep homojunction, we consider the case of a silicon bipolar junction transistor (BJT) that consists of two pn junctions back to back. The center (base) region was discussed briefly in Section 4.2.2. The structure of this npn BJT is repeated in Figure 6.1 and the smoothed doping profile along the cut A-A′ near the junction is repeated in Figure 6.2. The doping in each region is nonuniform. The metallurgical junctions (emitter-base and base-collector junctions) where $N_D = N_A$ are indicated. Because the two junctions are very close together (within about 0.15 μm of each other), the current in one junction affects the current in the other. The electrical characteristics of this BJT will be analyzed in Part 4. In this chapter, we investigate the energy band diagrams near the two junctions and discuss some of the consequences of nonuniform doping.

Notice that the doping level in each region varies appreciably with position, and so neither junction can be considered a step junction. Most of the emitter region is degenerate, however, and in this region, to a reasonable approximation, $E_f = E_C$. In the base region the doping decreases with distance from the emitter-base (E-B) junction. This is referred to as a *hyperabrupt doping profile*. At the

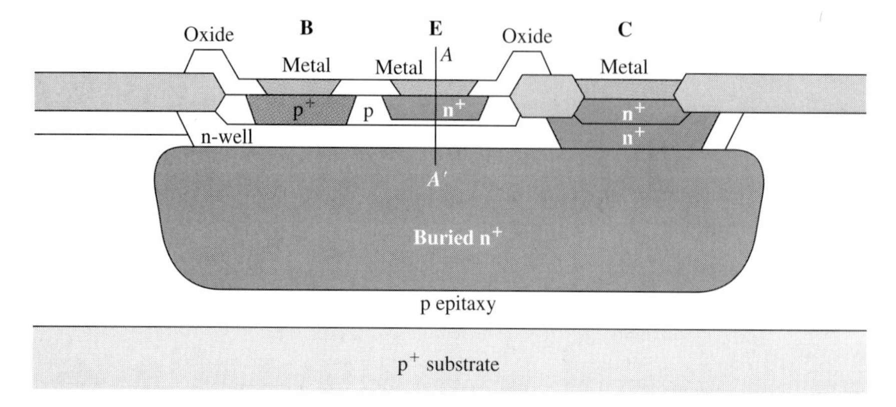

Figure 6.1 Schematic diagram (not to scale) of an npn bipolar transistor.

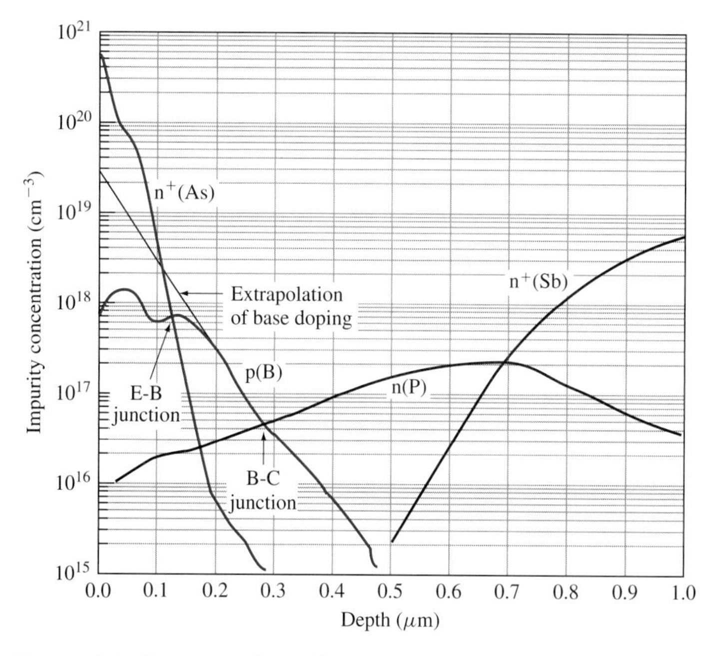

Figure 6.2 Concentration of impurities along the cut *A-A′* of the device in Figure 6.1.

base-collector (B-C) junction, the net doping on each side of the junction increases with distance from the junction. The doping gradient causes an electric field in the base and collector regions.

In Section 4.2.2, we found that the electric field in the base region is on the order of 6.2 kV/cm and in a direction that accelerates electrons in the base toward the collector. This is a substantial field, which indicates that to calculate minority

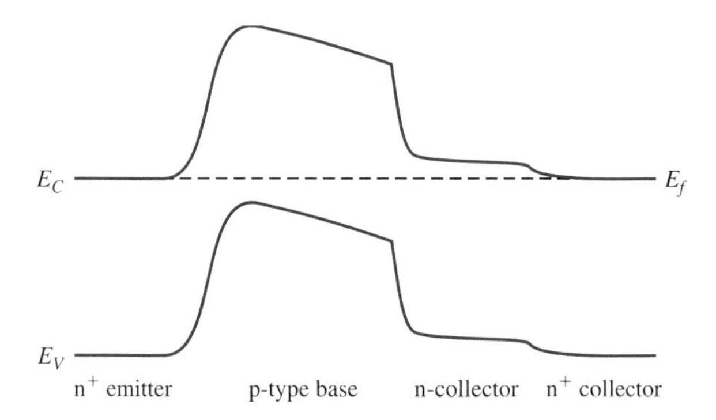

Figure 6.3 Equilibrium energy band diagram corresponding to the impurity profile of Figure 6.2.

carrier (electron) current in the base, drift must be considered in addition to the diffusion in these nonprototype junctions.

The equilibrium energy band diagram of the structure is shown schematically in Figure 6.3. The emitter and n^+ collector are degenerately doped, so there we make the approximation that $E_f = E_C$. The effect of impurity-induced band-gap narrowing is neglected here. We also observe that in the base and collector regions the doping gradients induce electric fields.[1]

While the E-B junction is hyperabrupt, the magnitude of the doping increases with distance on either side of the B-C junction. To obtain the doping profile ($N_D - N_A$ versus distance) near the B-C junction, we observe that, on the semilog plot of Figure 6.2, both acceptor and donor concentrations can be approximated by straight lines. This implies exponential impurity profiles, as we saw in Example 4.1. To describe these profiles on the base side, we use the expression for N_A

$$N_A = N_A(0)e^{-x/\lambda_B} \tag{6.1}$$

where $N_A(0)$ is the (extrapolated) value of N_A at $x = 0$ and λ_B is a constant representing the distance in the base for a change of N_A by a factor of e (2.718). In Chapter 4, λ_B for this case was found to be 0.04 μm. Since, from Equation (4.22), $\mathscr{E} = -(kT/q)(1/\lambda_B)$, $\mathscr{E} = -6.5$ kV/cm in the base.

Performing a similar analysis for N_D on the collector side gives

$$N_D = N_D(0)e^{x/\lambda_C} \tag{6.2}$$

where $\lambda_C = 0.195\,\mu$m and $\mathscr{E} = -133$ V/cm in the collector.

Figure 6.4 shows a plot of $N_D(x)$, $N_A(x)$, and $N_A(x) - N_D(x)$ near the base-collector junction. We approximate this junction as being linearly graded (along the slope indicated in the figure). The linearly graded junction is discussed next.

[1]The doping gradient in the emitter also causes a built-in field to exist there. We will deal with this and with the consequences of band-gap narrowing in the heavily doped emitter in Part 4.

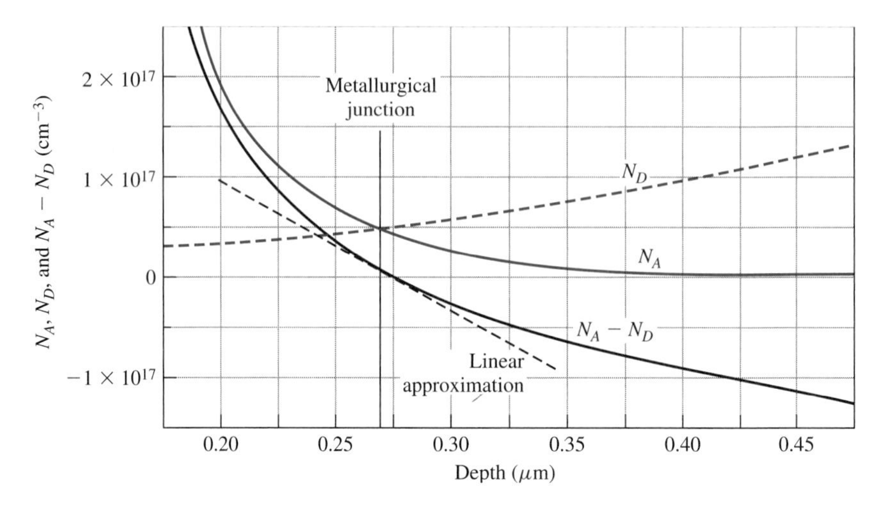

Figure 6.4 Plot of N_A, N_D, $N_A - N_D$ in the vicinity of the base-collector junction. The metallurgical junction is at $x = 0.27\ \mu$m, where $N_A - N_D = 0$.

*6.2.1 LINEARLY GRADED JUNCTIONS

To a first approximation, $N_A - N_D$ of Figure 6.4 can be taken as the tangent to the $N_A - N_D$ function at the metallurgical junction x_0:

$$\boxed{N_A - N_D = -a(x - x_0)} \tag{6.3}$$

where x_0 is the position of the base-collector junction, and a is the magnitude of the slope. This is referred to as the linearly graded approximation. From Figure 6.4 we find $a = 1.2 \times 10^{18}\ \text{cm}^{-3}/\mu\text{m}$.

Figure 6.5a indicates the net doping level as a function of position for the linearly graded approximation with the metallurgical junction at $x = x_0$. For simplicity, we let $x_0 = 0$. Using the depletion approximation (that there are virtually no free carriers in the transition region, so the remaining charges are the ionized impurities on either side), we write the charge volume density Q_V as:

$$Q_V = \begin{cases} qax & -\dfrac{w}{2} \le x \le \dfrac{w}{2} \\ 0 & \text{elsewhere} \end{cases} \tag{6.4}$$

where w is the transition region width. Because of symmetry, the transition region width extends from $-w/2$ to $w/2$ as indicated in Figure 6.5b.

We can find the field in the junction by following the procedure used in Chapter 5 for the step junction. That is, we integrate the charge density to obtain $\mathscr{E}(x)$:

$$\mathscr{E}(x) = \int_{-w/2}^{x} \frac{Q_V}{\varepsilon}\, dx = \frac{qa}{2\varepsilon}\left[x^2 - \left(\frac{w}{2}\right)^2\right] \qquad -\frac{w}{2} \le x \le \frac{w}{2} \tag{6.5}$$

which is quadratic, as seen in Figure 6.5c.

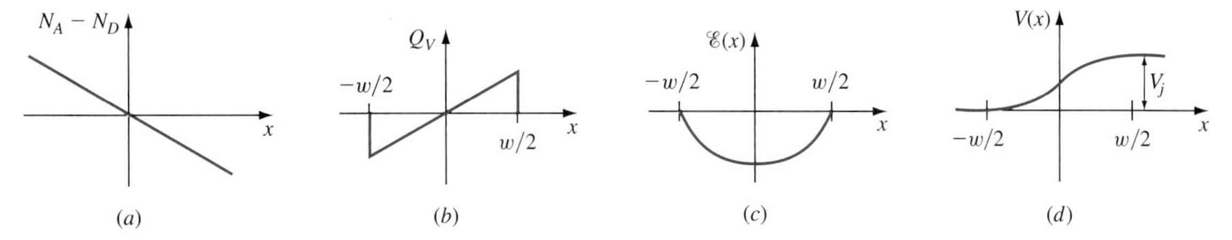

Figure 6.5 In a linearly graded junction the net doping concentration (a) is linear in x. The depletion approximation results in the charge distribution (b). In (c) the electric field is shown, and (d) shows the voltage profile.

Since $\mathscr{E} = -dV/dx$,

$$V(x) = V\left(-\frac{w}{2}\right) + \frac{qa}{6\varepsilon}\left[\frac{w^3}{4} + \frac{3xw^2}{4} - x^3\right] \qquad -\frac{w}{2} \leq x \leq \frac{w}{2} \tag{6.6}$$

as indicated in part (d) of the figure.

Solving for the junction width w yields

$$\boxed{w = \left[\frac{12\varepsilon V_j}{qa}\right]^{1/3}} \tag{6.7}$$

where again $V_j = (V_{\text{bi}} - V_a)$.

Let us now investigate the validity of using the linearly graded approximation for the base-collector junction.

From Equation (6.7) with $a = 1.2 \times 10^{18}\ \text{cm}^{-3}/\mu\text{m}$ and for $V_j = 1$ V, we find $w = 0.088\ \mu\text{m}$, or the transition region extends $0.044\ \mu\text{m}$ on either side of the metallurgical junction. At this distance, the linear approximation in Figure 6.4 is close to the actual doping, so the approximation is reasonable. For reverse bias such that $V_j = V_{\text{bi}} - V_a = 5$ V, however, the depletion region as calculated from Equation (6.7) extends $0.075\ \mu\text{m}$ on each side of the metallurgical junction. At these distances from the junction, the linearly graded approximation is less valid, and the calculated junction width is less credible.

As would be expected, the larger the grading coefficient a, the narrower is the transition region.

The calculation of the built-in voltage V_{bi} is not as straightforward as for the step junction. In Figure 6.6a, the energy band diagram is shown for the case of electrical neutrality. The case for equilibrium is shown in (b). The edges of the transition region are indicated. The built-in voltage is

$$\boxed{V_{\text{bi}} = \frac{1}{q}\left[\Phi_p\left(\frac{-w}{2}\right) - \Phi_n\left(\frac{w}{2}\right)\right]} \tag{6.8}$$

where the work functions are evaluated at the edges of the transition region.

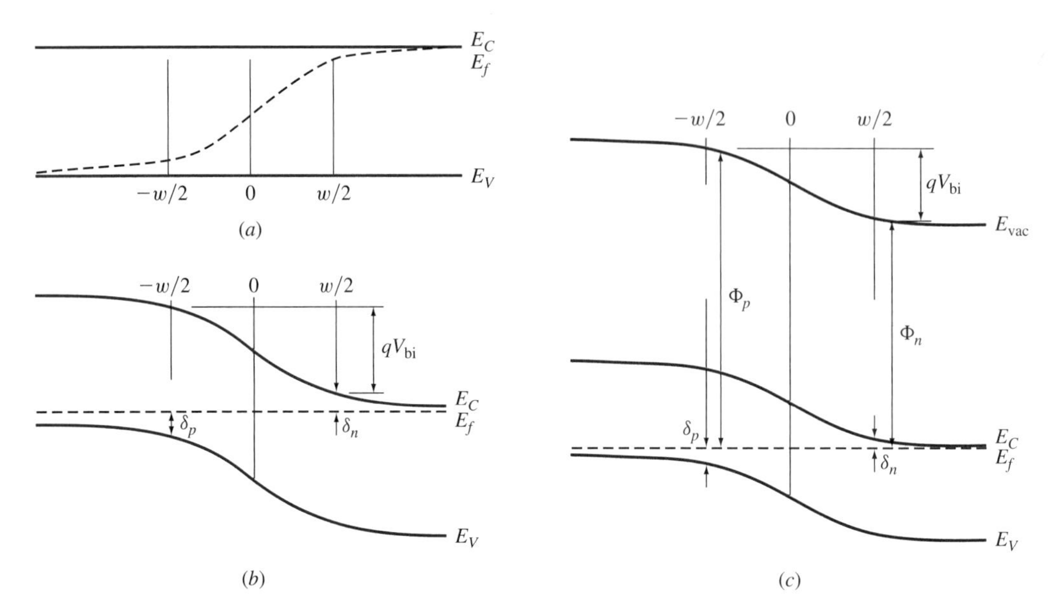

Figure 6.6 Energy band diagrams for the linearly graded junction under (a) neutrality and (b) equilibrium. The built-in voltage can be found from $qV_{bi} = E_g - \delta_n - \delta_p$ or from $qV_{bi} = \Phi_p - \Phi_n$.

Alternatively, V_{bi} can be expressed in terms of δ_n and δ_p:

$$qV_{bi} = E_g - \left[\delta_n\left(\frac{w}{2}\right) + \delta_p\left(-\frac{w}{2}\right)\right] \tag{6.9}$$

Following the procedure for a step junction, we write $\delta_p = kT\ln(N_V/p_{p0})$ and $\delta_n = kT\ln(N_C/n_{n0})$, where n_{n0} and p_{p0} are evaluated at the edges of the depletion region ($-w/2$ and $+w/2$) and

$$V_{bi} = \frac{kT}{q}\ln\left[\frac{\left(\frac{aw}{2}\right)^2}{n_i^2}\right] = \frac{2kT}{q}\ln\left(\frac{aw}{2n_i}\right) \tag{6.10}$$

Substituting w from Equation (6.7), and with $V_j = V_{bi}$, we obtain

$$V_{bi} = \frac{2kT}{q}\ln\left[\frac{a}{2n_i}\left(\frac{12\varepsilon V_{bi}}{qa}\right)^{1/3}\right] \tag{6.11}$$

which can be solved iteratively.

Note that, for an arbitrary doping profile,

$$V_{\text{bi}} = \frac{1}{q}[\Phi_p(x_p) - \Phi_n(x_n)] \tag{6.12}$$

where Φ_p and Φ_n are the work functions evaluated at the edges of the transition region at equilibrium. This is shown in Figure 6.6c.

6.2.2 HYPERABRUPT JUNCTIONS

We treated the base-collector junction as a linearly graded junction, but as indicated earlier, the other junction, the emitter-base n^+p junction, is hyperabrupt. This is because the base doping decreases with increasing distance from the metallurgical junction. If the base doping profile in the vicinity of the junction is known, the junction properties can be calculated numerically. There are software programs for this. Here we simply mention that:

1. The built-in voltage is

$$V_{\text{bi}} = \frac{1}{q}(\Phi_p - \Phi_n)$$

   where the work functions are evaluated at the edges of the transition region at equilibrium.
2. A field exists in the base that accelerates injected minority carriers away from the E-B junction. The effect of this field is to increase the current for a given forward bias, reduce the stored-charge capacitance, and increase the switching speed of the junction. This is particularly important in bipolar transistors.
3. Hyperabrupt junctions exhibit a large fractional variation of junction capacitance with applied voltage. This property is often used in a class of devices called *varactors* (variable-reactance devices).

6.3 SEMICONDUCTOR HETEROJUNCTIONS

A heterojunction, as indicated earlier, is a junction between two dissimilar materials. Such junctions can be between a semiconductor and a metal, for example, or between two different semiconductor materials. In this section we examine semiconductor–semiconductor junctions.

6.3.1 THE ENERGY BAND DIAGRAMS OF SEMICONDUCTOR–SEMICONDUCTOR HETEROJUNCTIONS

We consider junctions formed between two different semiconductor materials of different band gaps, electron affinities, ionization potentials, and work functions. The electrical properties of the junction depend strongly on these parameters. [1–3]

There are several classes of semiconductor heterojunctions, depending on the relative values of χ and E_g. Three cases are shown in Figure 6.7. These are *straddling* or *Type I, staggered* or *Type II,* and *broken-gap* or *Type III*. Because Type I heterojunctions are the most technologically important, we will discuss two examples, in which the forbidden band of the wide-band-gap semiconductor

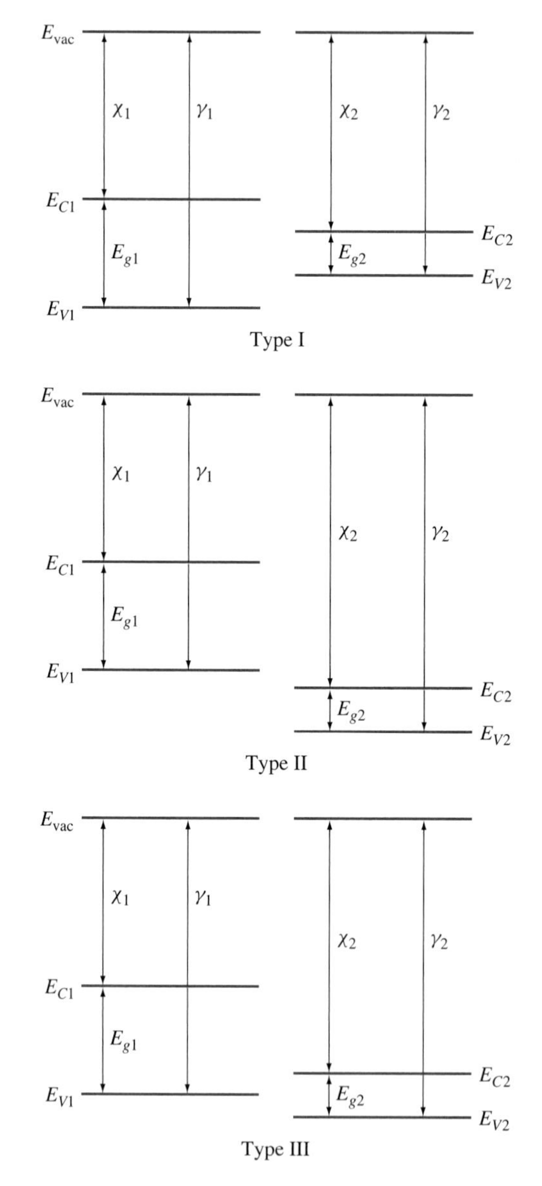

Figure 6.7 Three types of heterojunctions: Type I (straddling), in which the wide-gap energy overlaps that of the narrow gap, is the most important; Type II (staggered); and Type III (broken gap).

(Figure 6.7) overlaps (in energy) that of the narrow-band-gap semiconductor and the doping levels are constant on each side of the junction.

We will first consider the *electron affinity model* (EAM) in which the bulk semiconductor parameters are assumed invariant up to the metallurgical junction and the lattice constants of the two semiconductors are the same. We will then discuss corrections to this model to include charge dipoles near the interface resulting from the difference in valence band energies of the two semiconductors. Also considered are the effects of states within the forbidden band in the vicinity of the metallurgical junction (*interface states*).

Electron Affinity Model (EAM) The first example is a heterojunction between two different semiconductors, as shown in Figure 6.8a. One is an n-type wide-gap semiconductor (subscript 1) and the other is a p-type narrow-gap semiconductor (subscript 2). To obtain the energy band diagram as predicted by the electron affinity model (EAM), we proceed as for homojunctions, assuming electrical neutrality in every macroscopic region and using the vacuum level at the interface as a reference.

Using the convention in which a capital letter is used to indicate the conductivity type of the wide-band-gap material, this would be labeled an Np junction. Notice that because the electron affinities χ are different, there is a difference in the energies of the conduction band edges, ΔE_C. Since the band gaps (and therefore the ionization potentials γ) are also different, the two valence band edges do not line up either, with a difference of ΔE_V. The discontinuities are:

$$\begin{aligned} \Delta E_C &= |\chi_2 - \chi_1| \\ \Delta E_V &= |\gamma_2 - \gamma_1| \end{aligned} \tag{6.13}$$

At the instant the two materials are brought into contact, electrons in the conduction band of the N-type material flow into empty states that exist at lower energies in the conduction band in the p-type material and then recombine with holes in the valence band. This transfer of carriers is by diffusion, as for a

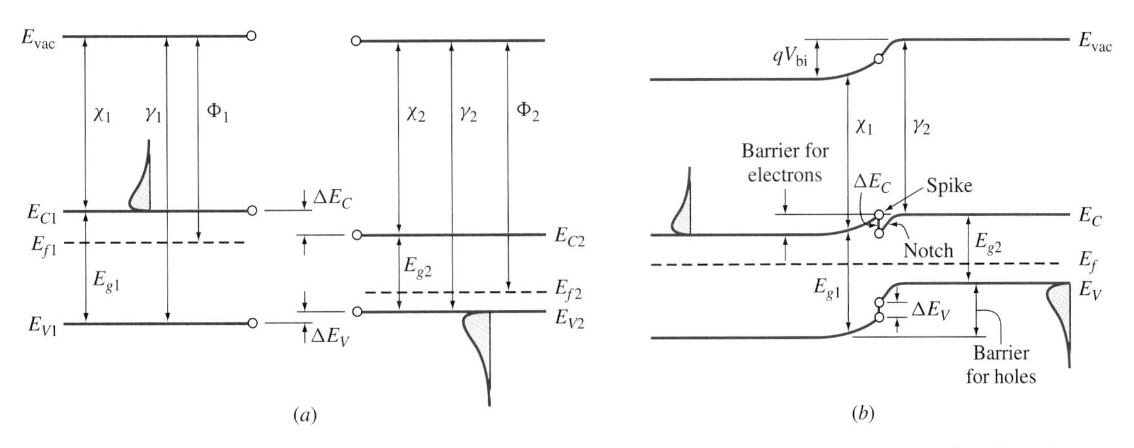

Figure 6.8 Constructing the energy band diagram for an Np Type I heterojunction as predicted by the electron affinity model: (a) electrical neutrality; (b) equilibrium.

homojunction. Similarly, holes diffuse from p to N. As charges migrate, leaving behind the ionized acceptors and donors near the junction, an electric field builds up, again the same as in a homojunction.

The resultant energy band diagram at equilibrium is shown in Figure 6.8b. The energy bands pivot around the reference points E_{vac}, E_{C2}, E_{C1}, E_{V2}, and E_{V1}, all at the interface. When drawing the energy band diagrams, bear in mind that E_{vac} is continuous, and that the electron affinity χ, the band gap E_g, and the ionization potential γ are constants for a given material. Thus, when E_{vac} bends, the conduction band edges must remain parallel to it within each material to maintain constant electron affinity. Similarly, as the conduction band bends, the valence band must remain parallel to it within that material to maintain constant E_g. A potential energy spike and notch (potential energy well) exist in the conduction band. This "glitch" can appear in the valence band instead, depending on the particular semiconductors and their doping levels. The notch or potential well can trap carriers, an effect that can be exploited to improve device performance. For example, in a laser the well is used to trap carriers in the conduction band and increase the probability of stimulated emission.

As was the case for a homojunction, an applied voltage (p side positive, n side negative) tends to decrease the existing barriers and is referred to as a positive bias (Figure 6.9a), while the opposite bias polarity (reverse bias) increases the barriers (Figure 6.9b).

Notice, however, that there are two different barriers—the barrier for electrons in this case is smaller than the barrier for holes. Thus, when the junction is forward biased, electrons will be injected more easily than holes. If we define the *electron injection efficiency* as the electron injection current divided by the total current, this efficiency will be high when the barrier to electrons is less than that for holes. We will see in Part 4 that improving the injection efficiency increases the performance of bipolar transistors.

It is interesting to notice that, for semiconductor heterojunctions, the wide-gap material is transparent to photons of energy less than its band gap and thus

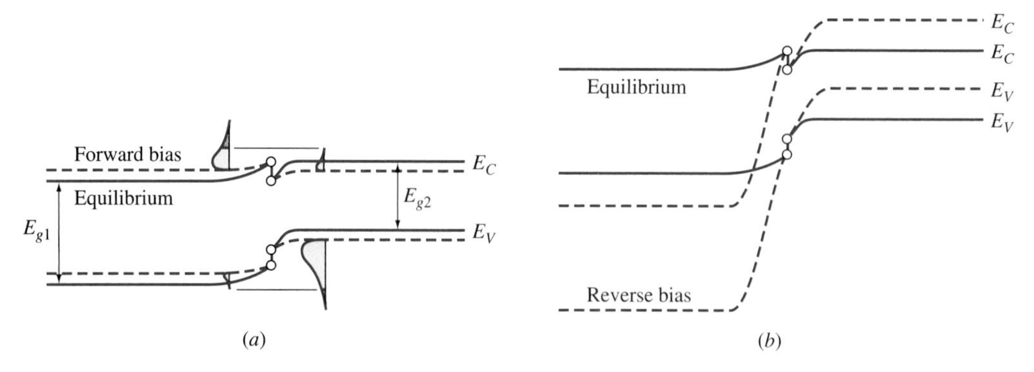

Figure 6.9 Energy band diagram for an Np heterojunction at equilibrium as predicted by the EAM (solid lines) and under bias (dashed lines): (a) forward bias; (b) reverse bias.

serves as an "optical window" to the narrow-gap material. This window effect can be used in solar cells and in light-emitting diodes, where the photons of interest are transmitted through the wide-gap window with little absorption. We will return to this point when we discuss optoelectronic diodes in Chapter 11.

The second type of heterojunction we will analyze using the EAM is an Nn heterojunction. We start as usual with the neutrality diagram as shown in Figure 6.10a. To achieve equilibrium in this case, at the instant the materials are joined, electrons flow across the junction from one semiconductor to the other until equilibrium is reached. The equilibrium case is shown in Figure 6.10b. Here, electrons from the conduction band of the semiconductor with the smaller work function (N) move into the empty states in the conduction band of the material with the larger work function (n). There are a negligible number of holes on either side.

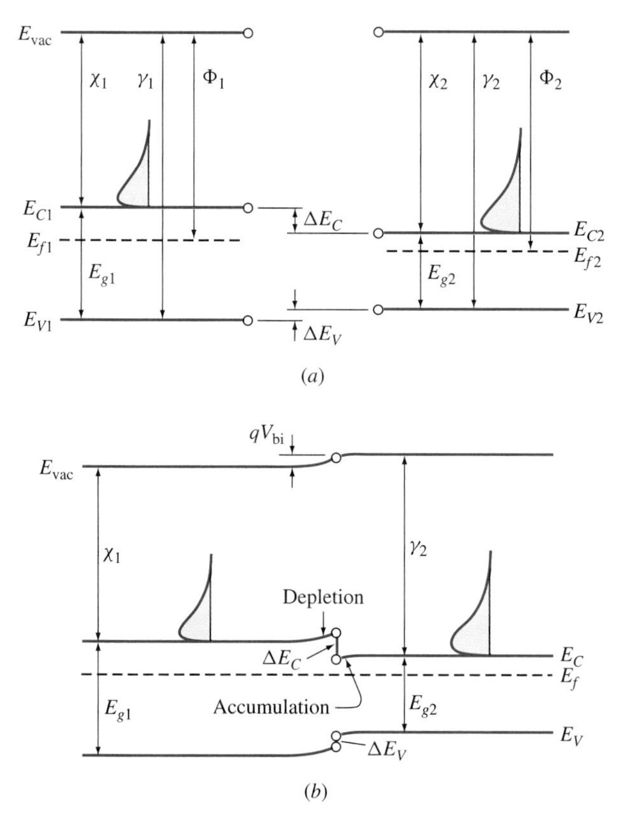

Figure 6.10 The energy band diagram for an Nn Type I heterojunction as predicted by the electron affinity model. (a) Neutrality; to reach equilibrium, electrons flow from the material on the left into the material on the right. (b) Equilibrium.

Since electrons from the left transferred to the right, there is a nonuniform distribution of charge across the junction. The result is a built-in electric field. In this case, however, electrons are depleted from the left-hand side, but they accumulate in the right-hand side. That is, there is a depletion region on the left (carriers are swept out of the region by the field) with charge density qN'_D and an accumulation region on the right (carriers swept into the region by the local field). Since the density of states in the conduction band is large, so is the (negative) charge density. Because the total charge in the transition region is zero, the transition region width and junction voltage exist predominantly on the left side.

The EAM outlined above for drawing the energy band diagrams of heterojunctions provides a first estimate of the shape and nature of the bands. It also assumes, however, that there is an exact match in the lattice constants of the two materials, such that there are no dangling bonds. It also ignores an electric dipole, which can be set up at the interface as a result of the difference in valence band edge of the two materials. These effects, and their influence on the band lineup at the interface as predicted by the EAM, will be discussed qualitatively in the following sections.

***Tunneling-Induced Dipoles** Although the electron affinity model is a simple intuitive model for predicting the band lineup at the interface, the measured values of the induced barriers are not generally in agreement with this model. [1–5] To illustrate this, we take as an example a GaAs:Ge heterojunction. The lattice constants of GaAs and Ge are 0.5653 and 0.5646 nm respectively, a reasonably close match (0.124 percent mismatch). The electron affinity of GaAs is 4.07 eV and that of Ge is 4.0 eV. The electron affinity model then predicts $\Delta E_C = 0.07$ eV and $\Delta E_V = 0.83$ eV, as indicated in Figure 6.11a for the neutrality case.[2] The measured values of ΔE_C and ΔE_V, however, are 0.27 eV and 0.49 eV respectively [1] as indicated in Figure 6.11b.

This discrepancy is explained (at least in part) by the existence of an electronic dipole at the interface resulting from the discontinuity in the valence band edge.[3] To illustrate this, we consider the equilibrium energy band diagram of an arbitrary lattice matched Np heterojunction of Figure 6.12, as predicted by the EAM.

We can see that there are electrons in the valence band of semiconductor B at the same energy as the forbidden band of A. As indicated in the Supplement B to Part 1, these electrons can thus tunnel a short distance (on the order of a nanometer, as illustrated in Figure S1A.9) into the forbidden region before being reflected back by the barrier. This penetration of the electrons into

---

[2]This appears to be a Type II or staggered gap heterojunction, but what is experimentally observed is a Type I junction, as we are about to explain.

[3]Much of the early literature reconciled the differences between the EAM and experiment by the assumption of localized states in the vicinity of the interface. More recently, it has been reported that the discrepancy can be primarily attributed to a tunneling-induced dipole layer [2–4], although there exists controversy concerning the relative importance of interface states and the dipole layer.

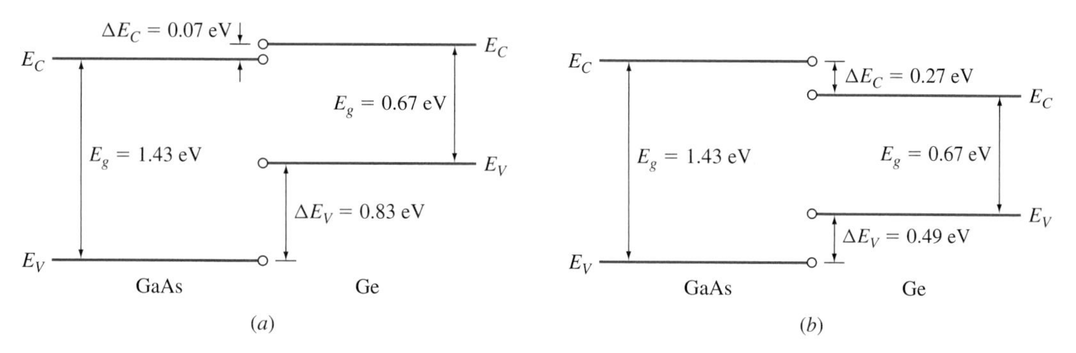

Figure 6.11 Energy band offset for a GaAs:Ge heterojunction (a) as predicted by the electron affinity model and (b) experimentally measured.

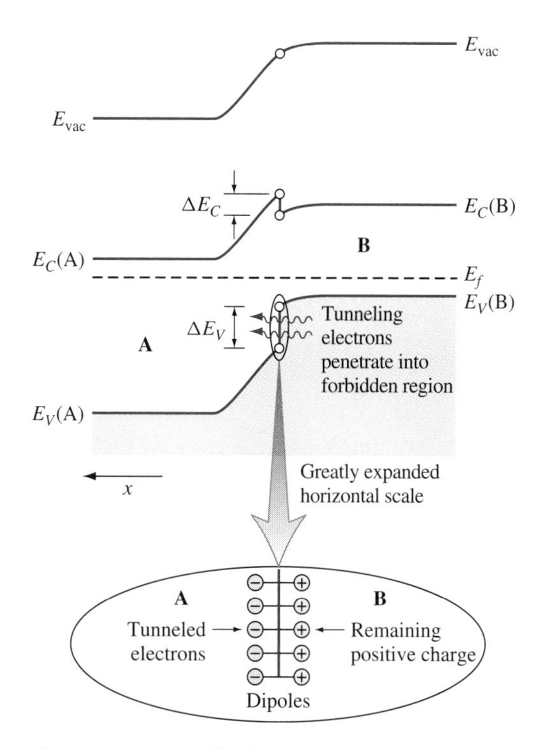

Figure 6.12 Equilibrium energy band diagram of an arbitrary Type I Np heterojunction as predicted by the electron affinity model. Electrons from the valence band of semiconductor B can tunnel a short distance into the forbidden gap of A, thus creating an interfacial dipole.

material A has the effect of placing some negative charge on the A side of the junction, creating a negative space charge in this region. An equal positive space charge must then exist in B, also within about a nanometer of the interface. The positive charge is due to the nuclei of the atoms of semiconductor B remaining fixed. This separation of charge results in a dipole layer being established at the interface.

These interface dipoles influence the energy band diagram. Figure 6.13 shows the junction region of Figure 6.12 on a vastly expanded horizontal scale, to zoom in on the dipole layer. On this scale the bands predicted by the electron affinity model are essentially flat. The circles indicate the band edge location predicted by the EAM. Since negative is "up" on the electron energy band diagram, the tunneled electrons tend to raise the band edges near the surface of material A, pushing the bands upward. Similarly, the positive charge on the surface of B tends to push the bands downward. The squares in the figure indicate the positions of the band edges accounting for the tunneling-induced dipoles.

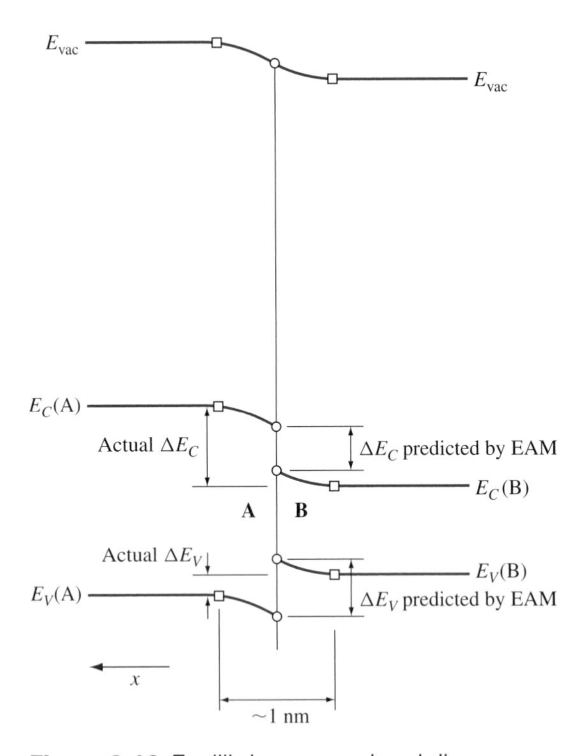

Figure 6.13 Equilibrium energy band diagram within a few nanometers of the interface of the heterojunction of Figure 6.12. The circles indicate the band discontinuities predicted by the electron affinity model; the squares indicate the influence of the tunneling-induced dipoles.

EXAMPLE 6.1

Estimate the tunneling distance of electrons of semiconductor B into the forbidden band of semiconductor A. Assume $m^* = m_0/2$.

■ Solution

Consider an electron that tunnels at an energy $E = E_V(A) + 0.1$ eV. Then from Equation (S1A.50) from Supplement A for Part 1, the electron probability penetration into semiconductor A is

$$\frac{\psi^*\psi(x)}{\psi^*\psi(0)} = e^{-2\left[\sqrt{(2|m^*|/\hbar^2)(E-E_V(A))}\right]x}$$

where E_V is the potential energy of an electron near the valence band edge in semiconductor A, the influence of E_C is neglected, and positive x is taken in the tunneling direction. Defining the penetration distance x_T as the tunneling distance such that $\psi^*\psi(x_T)/\psi^*\psi(0) = e^{-1}$, we have

$$2\sqrt{\frac{2|m^*|[E - E_V(A)]}{\hbar^2}}x_T = 1$$

or

$$\begin{aligned} x_T &= \frac{\hbar}{2\sqrt{2|m^*|[E - E_V(A)]}} \\ &= \frac{1.05 \times 10^{-34}\ \text{J} \cdot \text{s}}{2\sqrt{2(0.5)(9.11 \times 10^{-31}\ \text{kg})(1.6 \times 10^{-19}\ \text{J/eV})(0.1\ \text{eV})}} \\ &= 0.44 \times 10^{-9}\ \text{m} = 0.44\ \text{nm} \end{aligned}$$

Compare this value with that of Figure S1A.9 in Supplement A to Part 1.

Thus the electrons penetrate approximately one-half nanometer into the wide-gap material. The total extent of the dipole interface (about 1 nm) is small compared with the thickness of the transition region, which is generally on the order of a micrometer.

EXAMPLE 6.2

Estimate the electric field caused by the tunneling-induced band bending at the interface of a GaAs:Ge heterojunction.

■ Solution

To find the true electric field, we examine the slope of the vacuum level E_{vac}. We can determine the change in potential by recognizing that the measured value of the discontinuity in the conduction band edge ΔE_C for this junction is 0.27 eV. From Figure 6.11a, the offset in E_C predicted by the electron affinity model is 0.07 in the opposite direction. Thus, the tunneling-induced dipole effect bends the conduction bands a total of $0.07 + 0.27 = 0.34$ eV. The curvature is repeated in the vacuum level.

This band bending occurs over a distance equal to the length of the dipoles, which we will take to be about 1 nm. Since $\mathscr{E} = (1/q)(dE_{vac}/dx)$, the effective electric field is[4]

$$\mathscr{E} \cong \frac{0.34\,\text{eV}\left(\dfrac{1\,\text{eV}}{1.6\times10^{-19}\,\text{J}}\right)}{(1.6\times10^{-19}\,\text{C})(10^{-9}\,\text{m})} = 3.4\times10^{8}\ \text{V/m} = 3.4\times10^{6}\ \text{V/cm}$$

We emphasize again that the tunneling-induced dipole effect occurs over a very short distance. Figure 6.14 compares the equilibrium energy band diagram predicted the by electron affinity model (circles, dashed lines) with that considering the interface dipole effect (squares, solid lines). Since the dipole region is on the order of a nanometer and the transition region is on the order of a micrometer, on this diagram the dipole region is neglected, although it is indicated

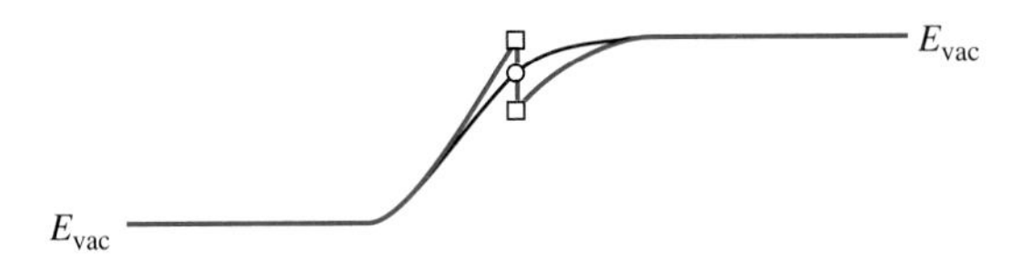

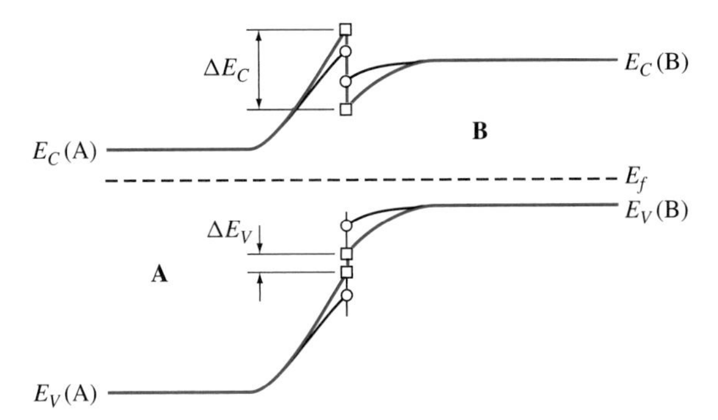

Figure 6.14 Equilibrium diagram of the Np Type I heterojunction considered in Figures 6.12 and 6.13. The lateral scale is reduced by a factor of 10^2 to 10^3 so that the indicated discontinuities appear in E_{vac}, E_C, and E_V. The circles and black lines are for the electron affinity model. The squares and colored lines present the result including the tunneling-induced dipoles.

[4]Note that at such a high field, the electron potential energy departs substantially from being periodic, and in this region the band gap is not well-defined.

as a discontinuity in the vacuum level. The conduction and valence bands also have discontinuities in addition to those predicted by the EAM.

The magnitude of the tunneling-induced dipole effect depends on the size of the discontinuity in the valence band. The larger the discontinuity in E_V at the interface, the larger the number of electrons that can penetrate the barrier into the forbidden region. Thus, the larger the dipole, the greater the error in the EAM. Note that for a homojunction, $\Delta E_V = 0$ and no tunneling-induced dipole exists.

6.3.2 EFFECTS OF INTERFACE STATES

In the above discussion, the presence of any interface states, i.e., localized states near the interface and within the forbidden band, was neglected. The presence of such states can affect the resultant energy band diagram. To examine these effects, we will first discuss surface states, and then extend the discussion to interface states.

Recall that in Chapter 2 it was shown that the existence of well-defined allowed and forbidden energy bands resulted directly from the periodic nature of the semiconductor crystal. When the periodicity is interrupted, for example at or near a surface or defect, the band structure is modified. For defects the result is localized allowed states in the forbidden band that can trap electrons or holes. At a surface, the potential energy is no longer periodic and the resultant band structure discussed in Chapter 2 is no longer applicable. There are two effects: one due to the surface itself and the other due to the nonperiodicity near the surface. These are normally treated by considering the band structure of the bulk to be valid up to the surface, and then surface effects are described by localized *surface states*.

Consider, for example, the Si crystal shown schematically in Figure 6.15a in two dimensions. In the bulk, each Si atom is bound to each of its four neighbors by two electrons (covalent bonding). Surface atoms, however, have only three neighbors. Each surface atom, then, under the condition of neutrality, has one nonbonding electron and one vacant (surface) state. These are referred to as *dangling bonds*. The surface density of surface states (number per unit area) is then comparable to the surface density of the surface atoms. For N atoms/cm^3 in the bulk, there are on the order of $N^{2/3}$ atoms/cm^2 on a surface. For Si, $N \approx 5 \times 10^{22}$ cm^{-3}, giving on the order of 2.7×10^{15} surface atoms/cm^2. The atoms are more closely spaced in the $\{111\}$ faces than in the $\{110\}$ and $\{100\}$ faces, with consequently more surface states. In a three-dimensional crystal, the actual concentration of surface states is appreciably reduced from this value because the surface atoms are displaced from their bulk positions such that they can bond with their neighbors, reducing the number of dangling bands.

The above example is for a "clean" surface.[5] On a real surface, foreign atoms are adsorbed, making chemical bonds with surface Si atoms. Consider, for

---

[5]It has been experimentally found, however, that within about 1 to 3 nm of the surface, the atomic periodicity is disturbed which causes some of the *surface states* to lie within this region.

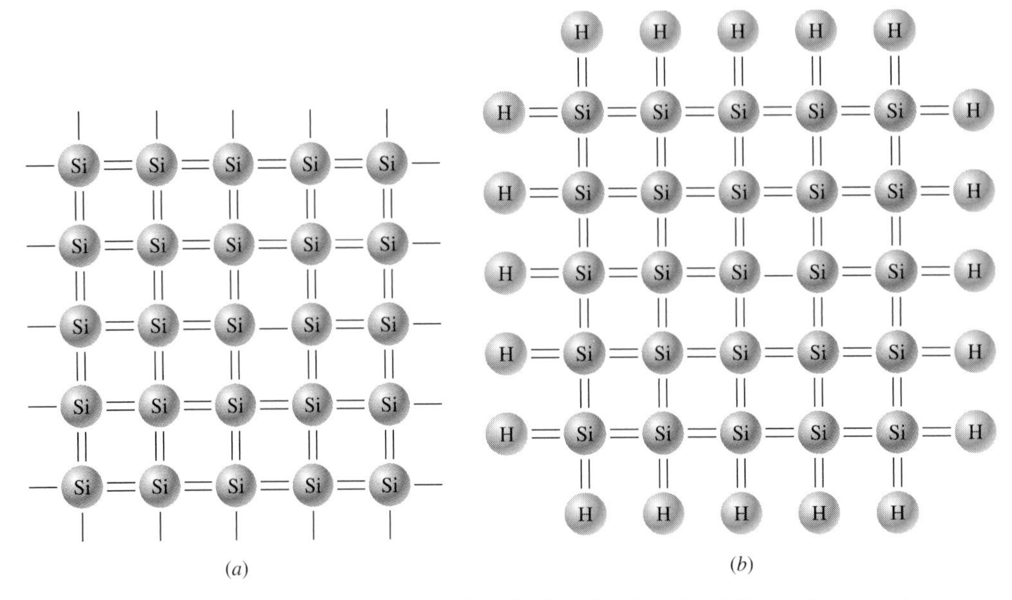

Figure 6.15 Schematics of a crystal showing (a) dangling bonds at the surfaces and (b) passivation of the bonds by atomic hydrogen.

example, a clean Si surface exposed to atomic hydrogen. Each H atom has one electron and one vacant state in its outer shell, and thus can bond to Si as shown in Figure 6.15b. In this case, all of the dangling bonds are saturated (filled) and the surface states have disappeared. Unfortunately, atomic hydrogen is not easily obtainable and is quite mobile on the Si surface. The H atoms are also small and diffuse very easily, so that surface coverage varies with time.

However, silicon dioxide (SiO_2) can also be used to reduce the surface state density. Silicon dioxide is called a *native oxide* for silicon because it forms spontaneously when silicon is exposed to oxygen. It is useful in integrated circuit fabrication because it is insulating, and thus can be easily grown in places where electrical isolation is needed. It is also more stable than hydrogen for rendering the surface states chemically inactive, a process called *passivating* the surface. Hydrogen, however, is often incorporated in the SiO_2 to more effectively passivate the surface.[6]

What are the energies associated with these surface bonds? The surface states are spread over a range of energies, and many of the states are at energies corresponding to the forbidden band in the bulk. Figure 6.16a shows how these states would appear in the energy band diagram for the case of neutrality. Note that for neutrality, the Fermi level of the surface is not the same as that of the bulk. To achieve equilibrium, electrons from the conduction band of the bulk

[6]Sometimes deuterium (an isotope of hydrogen) is used in place of hydrogen for passivation. It is heavier than H and less mobile.

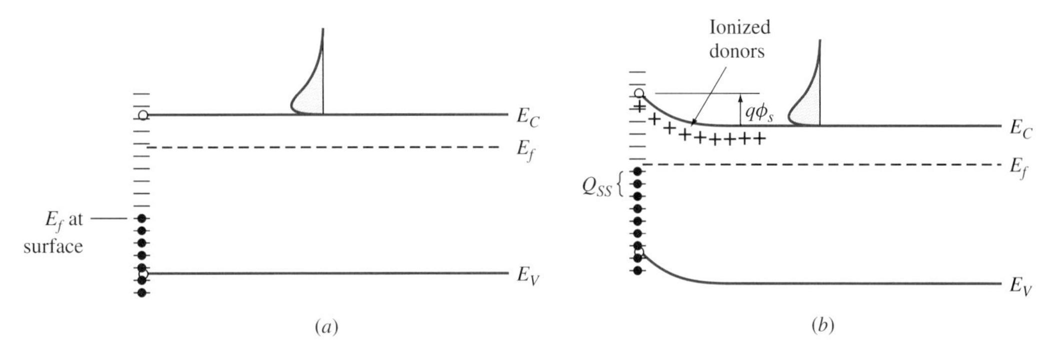

Figure 6.16 Effect of surface states on the energy band diagram. (a) Under the neutrality condition there are empty states at the surface at lower energies than electrons in the n-type semiconductor. (b) At equilibrium the transfer of electrons into the surface states results in a surface potential ϕ_S. The resultant net surface state charge per unit area is designated Q_{SS}.

move in to occupy the empty states at lower energies at the surface. If the surface is not passivated and electrons occupy some of the surface states, a net surface charge builds up (Q_{SS} in Figure 6.16b). A net surface charge requires an equal and opposite charge in the bulk, thus building up an electrostatic field normal to the surface. In this case, the positive charges are ionized donors in the lattice. The bands bend in the semiconductor near the surface to reflect this built-in field, as shown in Figure 6.16b.

We see then that an electrostatic potential difference, or surface potential ϕ_s, exists at the surface of the silicon with respect to the bulk. The value of ϕ_s depends on the density and energies of the surface states and on the doping level of the semiconductor. It is worth pointing out that the polarity of ϕ_s is usually such that minority carriers can be trapped at the surface. In the case of the n-type material in Figure 6.16, holes can be trapped in the valence band at the surface. Note that the band bending is caused by electron capture by the surface states, but the resulting upward band bending creates a potential pocket for holes. The surface state density and surface potential energy induced by the surface states strongly influence many device characteristics. They are controlled by appropriate surface treatments, many of which seem to involve witchcraft.

*6.3.3 EFFECTS OF LATTICE MISMATCH ON HETEROJUNCTIONS

Just as dangling bonds at the surface of a semiconductor affect the energy band profile near the surface, dangling bonds at the interface of two semiconductors with different lattice constants affect the band structure near the interface. Consider the case of a Si:Ge Nn heterojunction, which has a lattice constant mismatch of 4.1 percent. This results in a surface density of dangling bonds on the order of 10^{14} cm^{-2}.[7] [4] The neutrality energy band diagram is indicated in

[7] Lattice mismatch is discussed in more detail in Appendix C.

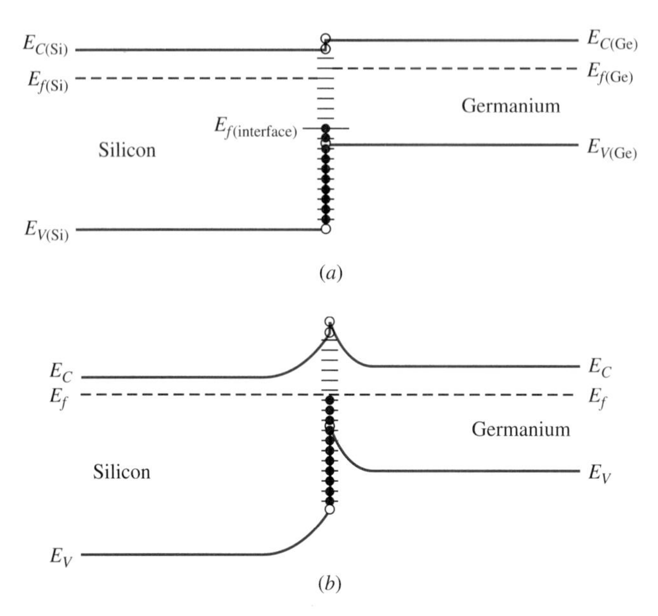

Figure 6.17 Energy band diagram for an Nn heterojunction between silicon and germanium, based on the electron affinity model. The case of neutrality is shown in (a), where the Fermi levels for the Si, Ge, and interface are indicated. The equilibrium case is shown in (b), where the Fermi levels have been aligned.

Figure 6.17a. The Fermi levels for the Si, Ge, and interface are indicated. In this case, the Fermi level for the interface states is at a lower energy than in either semiconductor. To achieve equilibrium, electrons from both materials become trapped in the interface states, which causes a potential energy spike at the interface (Figure 6.17b). (Note that, for simplicity, in this diagram the tunneling-induced dipole effect is neglected.) To cross the interface, electrons must have sufficient energy to surmount this spike. Since few have this much energy, a large interface resistance results.

The interface states above were considered to arise from dangling bands due to the lattice mismatch. This mismatch will also create a mechanical strain in the region near the interface. The strain accumulates with distance and eventually leads to crystal defects. These defects can also produce states within the forbidden bands because of the interruption of the lattice periodicity.

Another source of interface states results from the nonperiodicity of the electron wave function across the junction. Even if the two semiconductor lattices are perfectly matched, the Bloch wave function will change somewhat across the junction, since the atomic sizes on each side are different. This effect, however, is thought to have a relatively small effect on the band lineup at the interface.

The interface states associated with the dangling bonds resulting from the lattice mismatch in heterojunctions is a serious problem. As a result, the semiconductors involved in heterojunctions are lattice-matched as closely as possible.

In compound semiconductor heterojunctions, this is usually accomplished by using ternary or even quaternary semiconductor alloys. This is discussed further in Chapter 11.

6.4 METAL-SEMICONDUCTOR JUNCTIONS

Junctions can also occur between a semiconductor and a metal. When a metal is joined to a semiconductor, for example, there are two types of junctions that can result: rectifying, in which current flows in one direction but not the other, and ohmic, or low-resistance, in which current flows easily in both directions. It is important to be able to control which type of junction will occur. Ohmic contacts are important since every transistor must ultimately make contact to the outside circuitry via a conductor. On the other hand, rectifying metal-semiconductor junctions, called Schottky diodes, also have many applications. In this section, the energy band diagrams and the electrical characteristics of these contacts are considered.

6.4.1 IDEAL METAL-SEMICONDUCTOR JUNCTIONS (ELECTRON AFFINITY MODEL)

We consider ideal metal-semiconductor junctions first, using the electron affinity model. In this ideal model, the effects of tunneling-induced interface dipoles and states in the forbidden region at or near the metal-semiconductor interface are neglected. We will come back and treat the more realistic case in the next section. First, we consider a Schottky contact with a metal and a semiconductor. To be specific, we consider the case of a metal semiconductor between Al and n-type Si. The neutrality energy band diagram is indicated in Figure 6.18a.

Aluminum has a work function of 4.1 eV, and silicon has an electron affinity of 4.05 eV. We choose a donor concentration such that the Fermi level in the silicon is 0.1 eV below E_C. Then we can draw the equilibrium energy band diagram, using the EAM as shown in Figure 6.18b. It can be seen that, for this combination of materials, electrons in the semiconductor do not have a barrier going into the metal. Electrons in the metal see a small barrier of 0.10 eV (measured from the Fermi level in the metal to the conduction band edge of the bulk silicon). The built-in voltage is 0.05 V. Thus we would expect current to flow easily in both directions, and this Al-Si contact would be expected to be low resistance and ohmic. Experimentally, however, this junction is found to be rectifying, with a measured barrier on the order of 0.7 eV. Clearly, the ideal model is inadequate. In the next section, we will improve the model by considering the effects of tunneling-induced interface dipoles and interface states.

6.4.2 INFLUENCE OF INTERFACE-INDUCED DIPOLES

The discrepancy between the predicted and measured barrier in the aluminum-silicon junction is largely attributed to the existence of an electric dipole at the interface. [2]

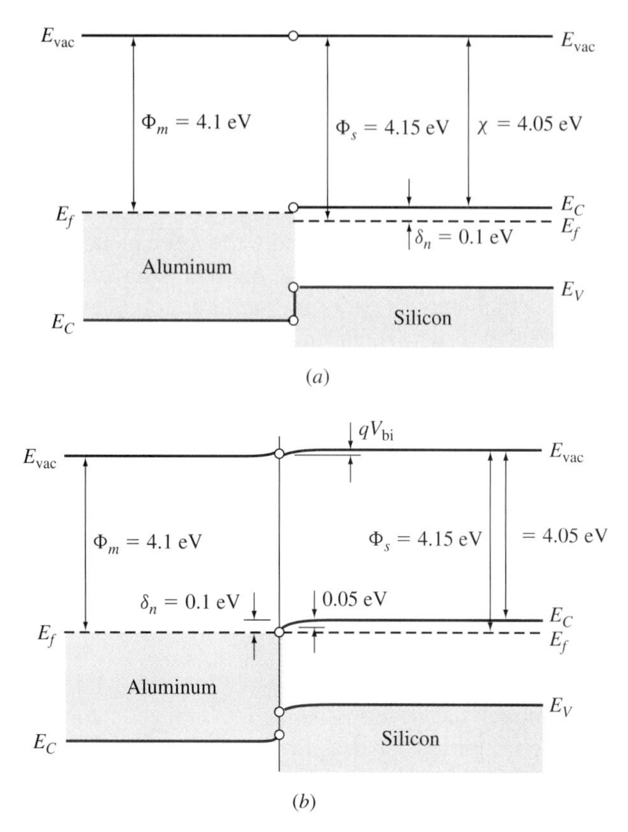

Figure 6.18 Energy band diagram as predicted by the electron affinity model for an Al:n-Si metal semiconductor junction: (a) Neutrality (b) equilibrium. The predicted barrier of 0.10 eV from metal to semiconductor is much less than the experimental value of about 0.7 eV. A more refined model is required.

As in heterojunctions, electrons in the Al conduction band can tunnel into the forbidden band of the Si, thus creating an electric dipole at the interface. As can be seen from Figure 6.18b, it is possible for electrons to tunnel into virtually the entire forbidden band. This represents a large number of electrons or a large amount of charge, and thus the dipole strength is expected to be larger than for the heterojunction case previously considered. Again, the thickness of the dipole region is on the order of a nanometer.

Figure 6.19a shows a close-up of the EAM neutrality case of Figure 6.18, with the effect of the tunneling-induced dipoles added. As the electrons from the metal penetrate a short distance into the semiconductor forbidden band before being reflected back, they carry some negative charge into the silicon, again only for a very short distance (on the order of a nanometer). This bends the bands upward in the Si. Note that there can be no electric field in the interior of a metal

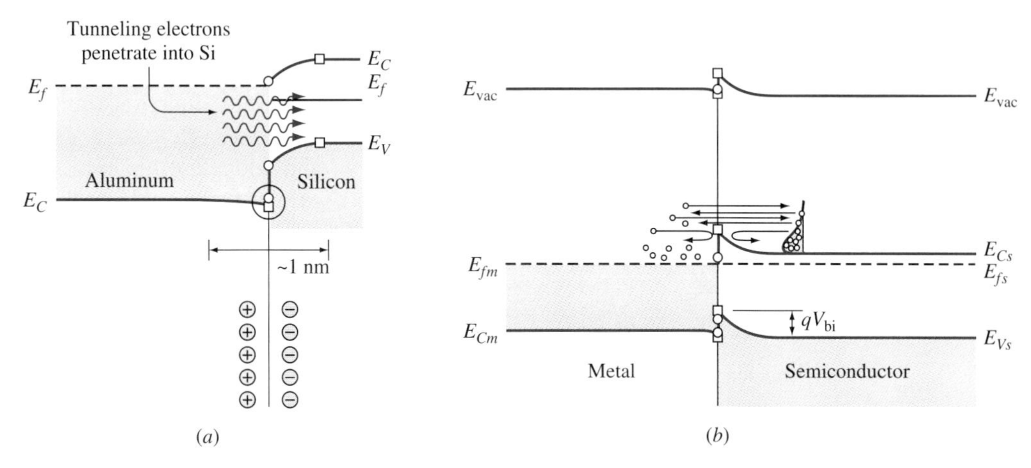

Figure 6.19 (a) The neutrality diagram for the Al:n-Si Schottky barrier diode including the tunneling-induced dipole effect. (b) The equilibrium energy band diagram for an Al:n-Si Schottky barrier diode.

since it is a conductor. There can, however, be charge in the metal at the semiconductor interface, in this case the positive charges of the surface dipoles. The equilibrium energy band diagram, taking into account the effect of the dipole layer, is indicated in Figure 6.19b. Here the horizontal scale is adjusted back to the scale of the depletion region, so that the tunneling-induced band bending appears as abrupt discontinuities. To attain equilibrium in this case, electrons move from the conduction band of the semiconductor to the lower available states in the metal. As the electrons leave the semiconductor, they leave behind ionized donors, which shift the semiconductor energies downward until the Fermi levels line up, as shown. The charge region of the device then consists of the tunneling-induced dipole layer, a depletion region (on the order of a micrometer) in the semiconductor, and a thin region of charge (in this case positive charge) at the surface of the metal. The accumulation region extends into the metal only a short distance—on the order of a fraction of a nanometer, as shown in Figure 6.19a. Because of the relative charge densities in the two materials, virtually the entire voltage drop and space-charge region is within the semiconductor, and usually E_C in the metal is assumed constant as indicated in Figure 6.19b. Also, observe that the barrier for electrons going from the semiconductor to the metal is different from the barrier for electrons going from the metal to the semiconductor. The interfacial dipole creates a near discontinuity in the vacuum level at the interface.

Whereas for lattice-matched heterojunctions the barrier heights for a given semiconductor pair are quite reproducible, there is considerable scatter in the experimentally obtained barrier heights in metal-semiconductor junctions. This is thought to result from interface states caused by adsorbed foreign atoms on the Si surface before the metal is deposited, or from the structure of the Si semiconductor surface. [2]

6.4.3 THE CURRENT-VOLTAGE CHARACTERISTICS OF METAL-SEMICONDUCTOR JUNCTIONS

The electrical characteristics of a metal-semiconductor junction are a strong function of the doping concentration in the semiconductor. Here we consider two limiting cases. First we consider the case of a lightly doped semiconductor. Such a junction is referred to as a *Schottky barrier diode.* Next we consider the case of a heavily (degenerately) doped semiconductor, which results in a low-resistance contact.

Schottky Barrier Diodes (First-Order Model) We have referred to the junction of Figure 6.19 as a *Schottky diode.* In this section we will verify that the electrical behavior is diode-like. At equilibrium the net electron current is zero because the number of electrons having enough energy to surmount the barrier going to the metal from the semiconductor is equal to that going from the semiconductor to the metal, since the Fermi levels are equal. This is indicated in Figure 6.19.

With applied bias, the energy bands are altered. Continuing with the convention that forward bias means reducing the barrier, we apply a forward bias V_a to the Al:n-Si diode as indicated in Figure 6.20a. The potential energy barrier from semiconductor to metal is reduced by qV_a and so the number of electrons in the semiconductor conduction band that have enough energy to surmount the potential barrier has increased (exponentially) with applied voltage. Thus, we expect a large number of electrons to be injected into the metal and a correspondingly high current to flow. It is also possible for electrons in the metal to flow into the holes in the valence band of the semiconductor, but there are few holes, since this is n-type material.

Under reverse bias, the semiconductor-to-metal barrier is increased, as shown in Figure 6.20b. Because of this large barrier, a negligible number of electrons in the semiconductor can flow over the barrier into the metal. There are still a few electrons in the metal that can go over the barrier into the semiconductor (the barrier is smaller in this case than for electrons going in the other direction), contributing a current density of $-J_0$. Since the barrier for electron flow from metal to semiconductor is independent of bias[8] and the barrier to electron flow from semiconductor to metal varies exponentially with bias, the net current density can be expressed as[9]

$$J = J_0[e^{qV_a/kT} - 1] \tag{6.14}$$

[8]The barrier actually decreases slightly with increasing reverse bias because of *image force effects* as discussed in the Supplement to Part 2.

[9]A more accurate expression is $J = J_0(e^{qV_a/nkT} - 1)$, where the diode quality factor n is greater than unity, as discussed in the Supplement to Part 2.

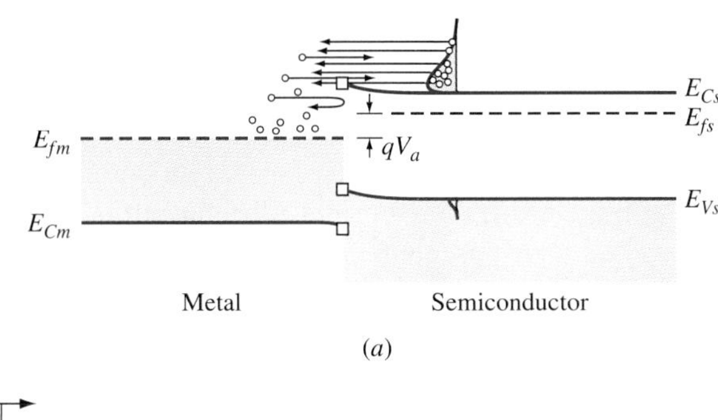

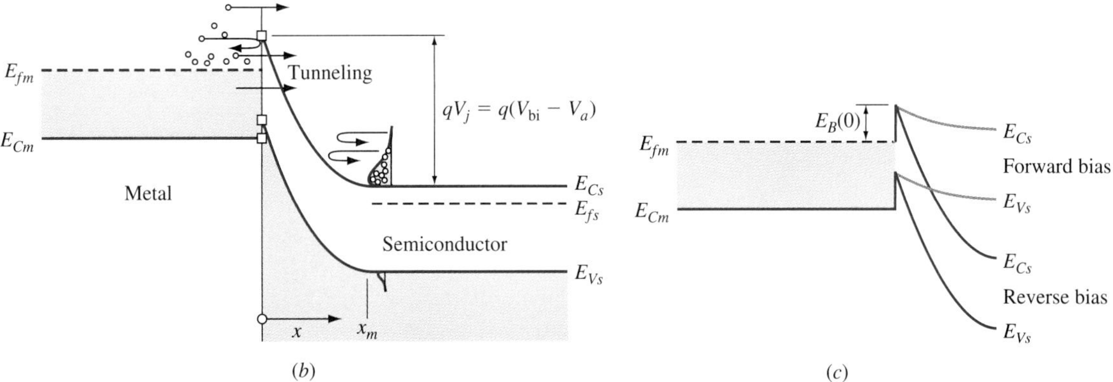

Figure 6.20 Energy band diagrams for a metal:n-semiconductor Schottky barrier. (a) For forward bias, electrons flow from semiconductor to metal. (b) For reverse bias, only a small leakage current flows. (c) For the first-order model, the metal-semiconductor barrier ($E_B(0) = E_C(x = 0) - E_{fm}$) is independent of applied voltage.

The coefficient J_0 has the form

$$J_0 = \frac{qm^*(kT)^2}{2\pi^2\hbar^3} e^{-E_B(0)/kT} \tag{6.15}$$

where m^* is the electron conductivity effective mass and $E_B(0)$ is the barrier height for electrons at the Fermi level in the metal to the conduction band in the semiconductor; i.e., $E_B(0) = (E_{Cs} - E_{fm})$ at the interface.

We see from Equations (6.14) and (6.15) that the current has the same bias dependence as does a pn homojunction [compare with Equation (5.77)]. That is, the current I is exponentially related to the potential barrier $E_B(0)$ from metal to semiconductor for electrons at equilibrium. Equation (6.15) is referred to as the *Richardson-Dushman equation.* It was originally developed for thermionic emission of free electrons (mass m_0) from a metal into a vacuum. Schottky barrier current is often referred to as *thermionic emission current*. Equation (6.15) is often expressed [7]

$$J_0 = \frac{m^*}{m_0} AT^2 e^{-E_g(0)/kT}$$

where

$$A = \frac{qm_0k^2}{2\pi^2\hbar^2} = 120\ \text{A/cm}^2 \cdot \text{K}^2$$

Since electrons are majority carriers on both sides of the interface, they do not contribute to diffusion current. The current is thus limited by the number of electrons that have sufficient x-directed energy to surmount the barrier.

There is also the possibility of tunneling in a Schottky diode under reverse bias. Figure 6.20c shows that for reverse bias the tunneling distance between metal and semiconductor is less than that for a pn junction, and thus the reverse tunnel current is larger in the Schottky barrier diode. This effect is discussed further in the Supplement to Part 2.

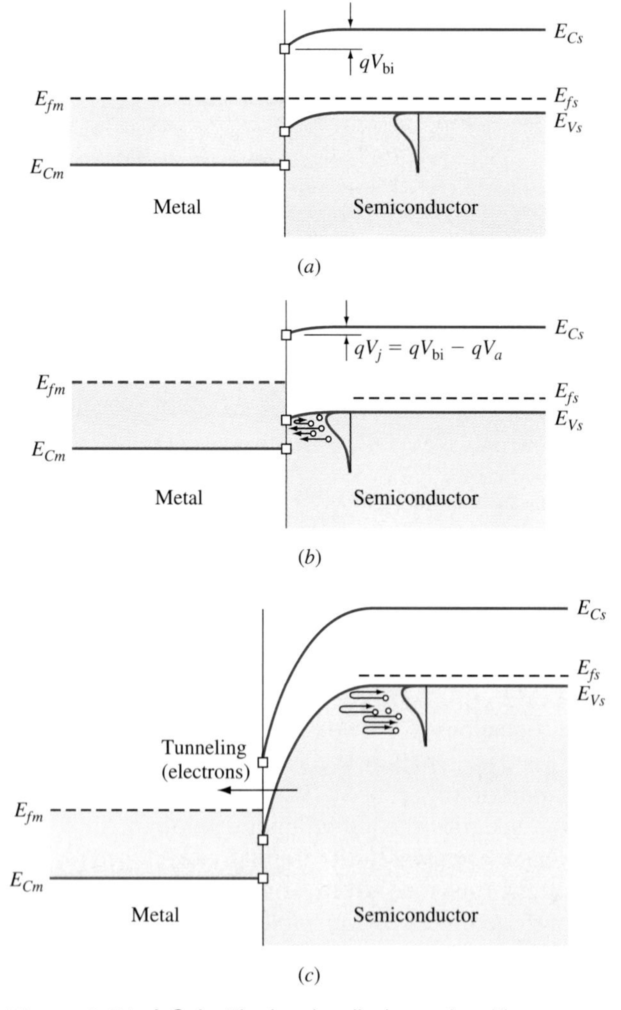

Figure 6.21 A Schottky barrier diode made with a p-type semiconductor. (a) Equilibrium; (b) forward bias; (c) reverse bias.

We have so far considered a metal-semiconductor junction between Al and n-type silicon. It is possible to make a junction with p-type silicon as well. The equilibrium energy band diagram for an Al:p-Si device is shown in Figure 6.21a. The experimentally determined barrier is about 0.6 eV (compared with about 0.9 eV predicted by the EAM. We are now interested in the activity of the holes since they are the majority carriers. The holes in the valence band see a potential barrier. Under forward bias (Figure 6.21b), the barrier for holes is reduced and the hole flow from semiconductor to metal is increased.

Under reverse bias, Figure 6.21c, the barrier to holes is increased, and a negligible number flow into the metal. The electrons in the conduction band of the semiconductor can drift over into the metal, but this is p-type material so there are few electrons and the current is very small. As for an n-type Schottky diode, tunneling contributes to the reverse current.

Rectifying metal-semiconductor contacts are diodes, and they behave qualitatively like pn junctions. The forward current increases exponentially with applied bias, and under reverse bias, there is a small leakage current.

Figure 6.22 compares the I-V_a characteristics of a Schottky diode and a pn junction. The built-in voltage of the Schottky diode is normally less than that of a pn junction, resulting in a larger current at a given forward bias. Under reverse bias, the carrier tunneling discussed above yields a softer reverse characteristic for the Schottky barrier diode.

6.4.4 OHMIC (LOW-RESISTANCE) CONTACTS

We saw in the previous section that metal-semiconductor junctions with low doping in the semiconductor act as diodes. Electrical contacts need to be made to devices, however, and these contacts should be of low resistance, not rectifying.

To make these Schottky barriers low resistance rather than rectifying, the semiconductor surface is degenerately doped before the metal is deposited. Using

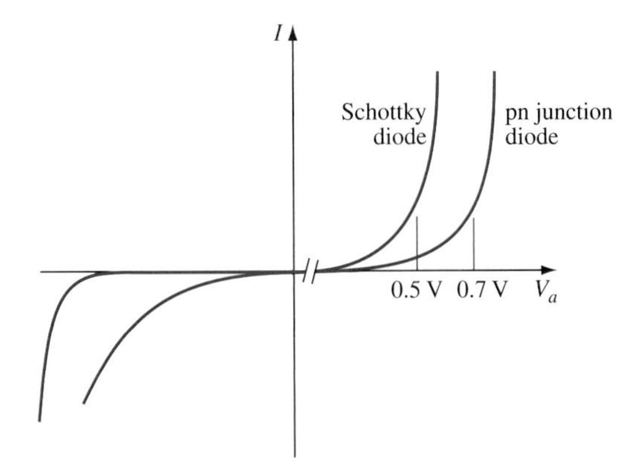

Figure 6.22 Comparison of the I-V_a characteristics of a Schottky diode and a pn junction diode. The scale for the reverse characteristic is compressed compared with the scale for forward bias.

the example of a metallic contact to an n-type semiconductor, this results in a metal-n^+n junction, as shown in Figure 6.23a. Because of the degenerate doping of the semiconductor at the metal interface, the depletion region at the metal-n^+ junction is so thin that the electrons tunnel easily through the barrier, resulting in a low resistance. The low resistance is seen for both directions of current flow. Further, the semiconductor-semiconductor n^+n junction is also of low resistance, and thus the resistance between the metal and the n semiconductor is low. In this case, the contact is ohmic. The p-type case is shown in Figure 6.23b. This tunneling current is discussed further in Supplement to Part 2.

Virtually all low-resistance contacts to semiconductors are made in this manner. This is illustrated for the npn bipolar transistor of Figure 6.24, repeated

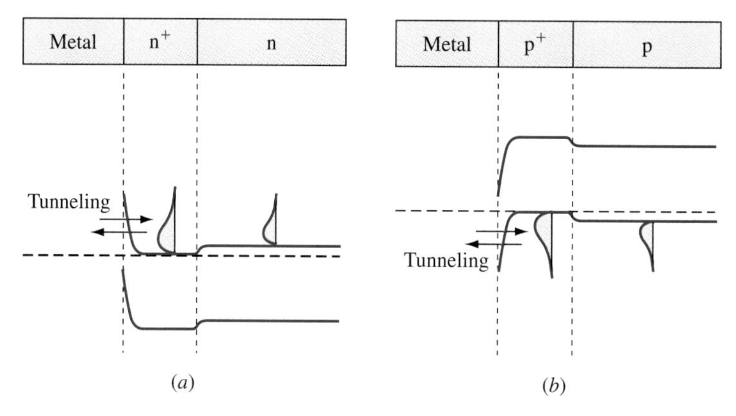

Figure 6.23 Low-resistance metal-semiconductor contacts using degenerate surface layers. Metal-n^+n contact (a) and metal-p^+p contact (b). The Schottky barrier is thin enough to permit tunneling.

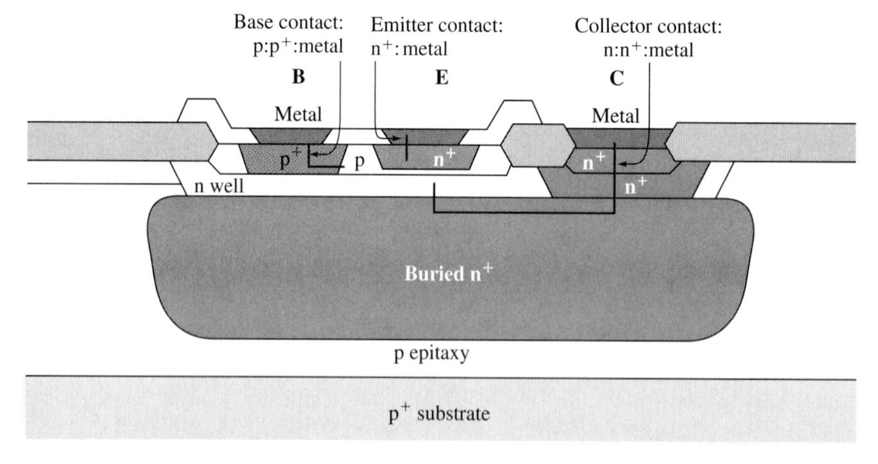

Figure 6.24 Schematic of an npn homojunction transistor indicating the low-resistance contacts. They are the base $p:p^+$:metal contact, the emitter n^+:metal contact, and the collector $n:n^+$:metal contact.

from Figure 6.1. The active transistor is the vertical n^+pn (n^+:p:n well) region under the emitter contact (E). The metal-n^+ emitter contact is a low-resistance contact as discussed. The collector contact (C) is metal-n^+n (metal:n^+:n:buried n^+:n well), as indicated by the dotted line. The base contact (B) is the metal-p^+p structure as indicated by the dashed line. The structure of this transistor is discussed in more detail in Chapter 9.

6.4.5 I-V_a CHARACTERISTICS OF HETEROJUNCTION DIODES

The I-V_a characteristics of heterojunctions depend largely on the band lineup at the interface and the doping type on each side. Further, interface states at the material interfaces cause appreciable generation-recombination current. Because of these complexities, it is difficult to accurately predict the I-V_a characteristics of heterojunctions.

For the lattice-matched Type I heterojunctions considered in this chapter, current across the interface of Np and Pn heterojunctions is primarily by minority carrier diffusion. For Nn and Pp heterojunctions, it results from majority carrier thermionic emission. As in other diodes, the current increases exponentially with applied voltage in the forward direction, and under reverse bias there is a small leakage current.

*6.5 CAPACITANCE IN NONIDEAL JUNCTIONS AND HETEROJUNCTIONS

Capacitance exists in all diodes, but it is generally more difficult to calculate for real diodes than it is for pn prototype homojunctions. In general, however, for a homojunction, the junction capacitance can be expressed as

$$C_j = \frac{\varepsilon A}{w} \qquad \text{homojunctions} \tag{6.16}$$

where the junction width w depends on the doping profile and junction voltage. The same equation applies to a Schottky barrier diode since the junction width w is entirely within the semiconductor.

For Np or Pn Type I heterojunctions, the junction capacitance can be considered as two capacitances in series:

$$C_j = \frac{A\left[\dfrac{\varepsilon_n \varepsilon_p}{w_n w_p}\right]}{\left[\dfrac{\varepsilon_n}{w_n} + \dfrac{\varepsilon_p}{w_p}\right]} \qquad \text{heterojunction} \tag{6.17}$$

where ε_n and ε_p are the permittivities of the p and n regions respectively, while w_n and w_p are the depletion region widths. For N^+n or P^+p heterojunctions and for Schottky barrier junctions, the depletion region is predominantly on one side. In that case, Equation (6.16) is a good approximation.

The stored-charge capacitance is more difficult to calculate. For a homojunction, this is treated in more detail in Part 4, "Bipolar Junction Transistors." For a Schottky barrier device, however, it is interesting to observe that $C_{sc} \approx 0$ because a negligible number of minority carriers are injected.

A final remark about capacitance in heterojunctions: For a heterojunction, the presence of a potential energy notch at the interface can interfere with the reclaiming of injected minority carriers, thus reducing the stored charge capacitance.

6.6 SUMMARY

In this chapter, we extended the results of the prototype homojunction of Chapter 5 to nonprototype homojunctions, heterojunctions, and metal-semiconductor junctions.

The equilibrium energy band diagram at the metallurgical junction was first predicted by the electron affinity model (EAM). This model is adequate for homojunctions where the lattice periodicity is constant across the interface and $\Delta E_V = 0$. A nonzero value of ΔE_V in heterojunctions, however, results in a tunneling-induced dipole within about a nanometer of the interface, which alters the band lineup from that predicted by the electron affinity model. A similar tunneling-induced interface dipole for metal-semiconductor junctions along with the presence of interface states also affects their band lineup.

Although in pn junctions current is limited by the rate at which injected minority carriers can diffuse away from the junction, in metal-semiconductor (Schottky barrier) junctions, minority carrier injection is negligible compared with majority carriers thermionically injected over the barrier. In metal-semiconductor junctions in which the semiconductor is degenerately doped, the depletion region is thin enough that tunneling through the barrier results in low-resistance contacts.

The approach presented here is largely qualitative, because a quantitative analysis requires a knowledge of the doping profiles and band lineups and is not amenable to closed-form solutions. These analyses are better handled with the use of device simulators, which solve the pertinent equations numerically.

6.7 READING LIST

Items 1, 2, 4, 8–12, 17–19, and 26 in Appendix G are recommended.

6.8 REFERENCES

1. S. Tiwari and D. J. Franck, "Empirical Fit to Band Discontinuities and Barrier Heights in III-V Alloy Systems," *Applied Physics Letters,* vol. 60, 1992, pp. 630–632.
2. Winfried Mönch, *Semiconductor Surfaces and Interfaces,* 3rd ed., Chap. 19, Springer, New York, 2001.

3. Federico Capasso and Georgio Margaritando, eds., *Heterojunction Band Discontinuities: Physics and Device Applications,* North Holland, Amsterdam, 1987.
4. G. Margaritondo, *Electronic Structure of Semiconductor Heterojunctions,* Klewer, Dordrecht, 1988.
5. A. G. Milnes and D. L. Feucht, *Heterojunctions and Metal-Semiconductor Junctions,* Chap. 4, Academic Press, New York, 1972.
6. See, for example, A. van der Ziel, *Solid State Physical Electronics,* 2nd ed., Chap. 7, Prentice Hall, Englewood Cliffs, NJ, 1968.
7. C. R. Crowell and V. L. Rideout, "Normalized Thermionic Field Emission in Metal-Semiconductor (Schottky) Barriers," *Solid State Electronics,* vol. 12, 1969, pp. 89–105.

6.9 REVIEW QUESTIONS

1. Outline the steps used to find the charge distribution, electric field, built-in voltage, and shape of the energy band diagram for a given doping profile.
2. Why is it expected, as stated in the text, that a larger grading coefficient produces a narrower depletion region in the linearly graded junction?
3. What is meant by a *hyperabrupt* junction?
4. In a pn junction, the built-in electric field is generated by the ionized donors and acceptors. In the Nn heterojunction of Figure 6.10, what is the source of the electric field on the N side? What is the source on the n side?
5. What is the electron affinity model, and how is it used? Summarize the steps to finding the energy band diagram for a junction using this model.
6. Explain how tunneling dipoles can produce near-discontinuities in the conduction and valence band edges.
7. For a Schottky barrier between aluminum and p-type semiconductor, the hole flow from semiconductor to metal is increased under forward bias and decreased under reverse bias. Since holes are just an artificial concept, explain these same phenomena in terms of the actual electrons in the valence band of the semiconductor.
8. Explain why, for low-resistance contacts to a semiconductor, the contact region of the semiconductor is made degenerate.

6.10 PROBLEMS

6.1 Consider a base-collector junction of a silicon BJT (bipolar junction transistor) like that in Figure 6.1. Assuming a linearly graded junction with $a = 1.2 \times 10^{18}\ \text{cm}^{-3}/\mu\text{m}$, find V_{bi}.

6.2 A pn homojunction has a doping profile as indicated in Figure P6.1.

a. Find the value of the electric field in the bulk on the p side.

b. Find the electric field in the bulk on the n side.

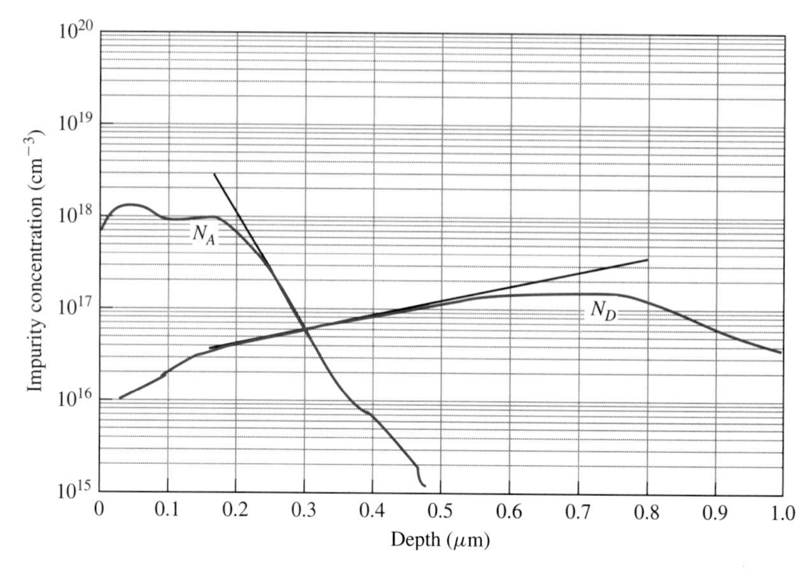

Figure P6.1

c. Plot $N_D - N_A$ as a function of position and find the slope a. [*Hint:* see Equation (4.16)]

d. Find the built-in voltage.

e. Find the junction width at equilibrium.

f. For the width you found in (e), comment on the validity of the linear approximation you used over this distance.

6.3 In Section 6.2.2, it was claimed that hyperabrupt junctions exhibit a large fractional change in junction capacitance with applied voltage. Explain physically why we should expect this to be the case.

6.4 Figure P6.2 shows the equilibrium energy band diagram for a heterojunction between n-type semiconductor A (band gap 1.5 eV) and p-type semiconductor B (1.0 eV).

a. Indicate the directions of:

Electron diffusion

Electron diffusion current $J_{n(\text{diff})}$

Electron drift

Electron drift current $J_{n(\text{drift})}$

Hole diffusion

Hole diffusion current $J_{p(\text{diff})}$

Hole drift

Hole drift current $J_{p(\text{drift})}$

Effective electric field for holes

True electric field

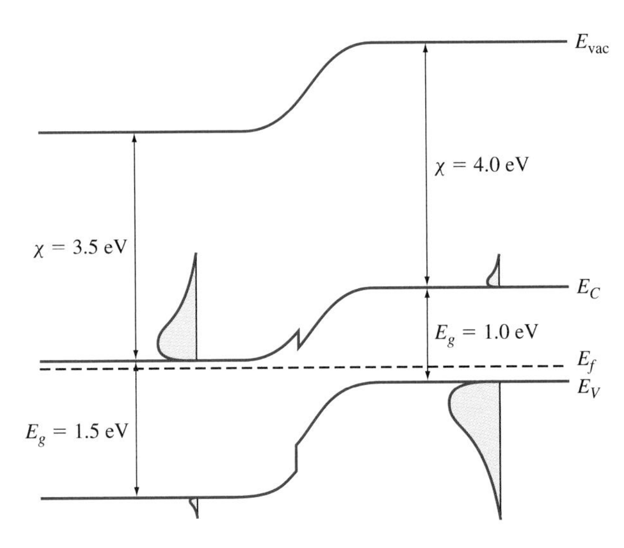

Figure P6.2

b. In the conduction band edge, there is a notch that looks like a quantum well. The quantum well may have one or two discrete states.

  i. Indicate on the drawing where the potential well is for the electrons.

  ii. If an electron is trapped inside the well, list the mechanisms by which it can get out.

6.5 Using the electron affinity model, draw (to scale) the equilibrium energy band diagram for a heterojunction between p-GaAs ($E_g = 1.43$ eV, $\chi = 4.07$ eV), whose $\delta_p = 0.1$ eV, and N-$Al_{0.3}Ga_{0.7}As$ ($E_g = 1.8$ eV, $\chi = 3.74$ eV), whose $\delta_n = 0.15$ eV. Neglect interface states.

6.6 Repeat the previous problem, only now let the GaAs be n type with $\delta_n = 0.1$ eV and let the AlGaAs be p type with $\delta_p = 0.15$ eV.

6.7 Consider the Type I Np heterojunction of Figures 6.8 and 6.9, in which net doping N'_D and N'_A are uniform on the N and p sides respectively. Let ε_n be the permittivity on the n side and ε_p that on the p side. Solve Poisson's equation to find the depletion widths w_n and w_p on the n and p sides and the total depletion width. (*Hint:* At the interface, displacement ($\varepsilon\mathscr{E}$) is continuous.)

6.8 Show that the junction capacitance per unit area for Problem 6.7 can be written as

$$C_j = \left|\frac{dQ_V}{dV_a}\right| = \left|A\sqrt{\frac{q\varepsilon_n\varepsilon_p N'_A N'_D}{2(\varepsilon_n N'_D + \varepsilon_p N'_A)(V_{bi} - V_a)}}\right|$$

6.9 For the Si:Ge Nn junction of Figure 6.17, sketch the energy band diagram that you would expect, using the simple electron affinity model (i.e., ignoring tunneling-induced dipoles and interface states). Discuss the difference in current flow that would result from the two energy band diagrams (electron affinity model and the model that includes the effects of the presence of dipoles). Let $E_C - E_f = 0.1$ eV for silicon and 0.15 eV for Ge.

6.10 We saw that in a heterojunction in which the spike and notch occurred in the conduction band (as in Figure 6.12), it was possible for electrons in the valence band of the narrow-gap material to tunnel a short distance into the forbidden band of the wide-band-gap material. This tunneling induced a dipole that produced a discontinuity in the bands at the junction. Consider a Pn heterojunction such as that in Figure P6.3. Comment on the possibility of tunneling-induced dipoles in this case.

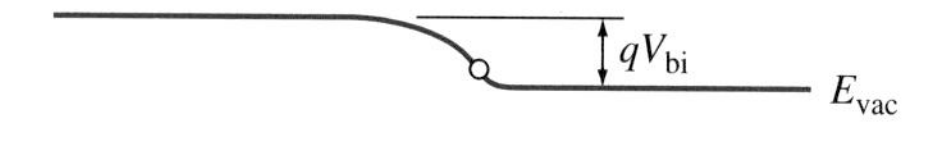

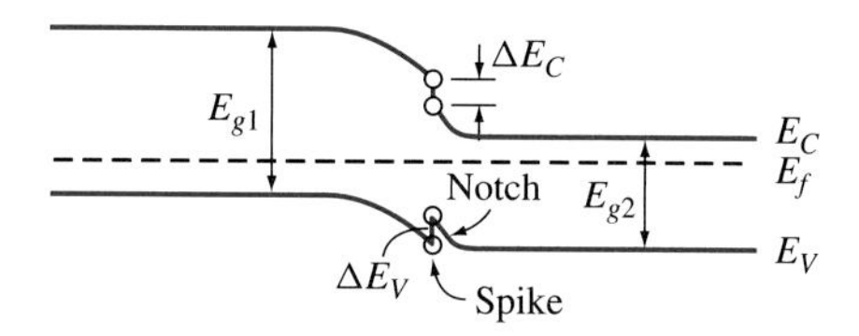

Figure P6.3

6.11 Consider an n-GaAs Schottky barrier diode of area 10 μm^2 and potential barrier ($E_B(0) = 1$ eV).

a. Find the value of I_0 from Equation (6.15).

b. Plot the forward *I-V* characteristic for a diode quality factor of $n = 1.3$.

6.12 Consider the junction of Figure P6.4.

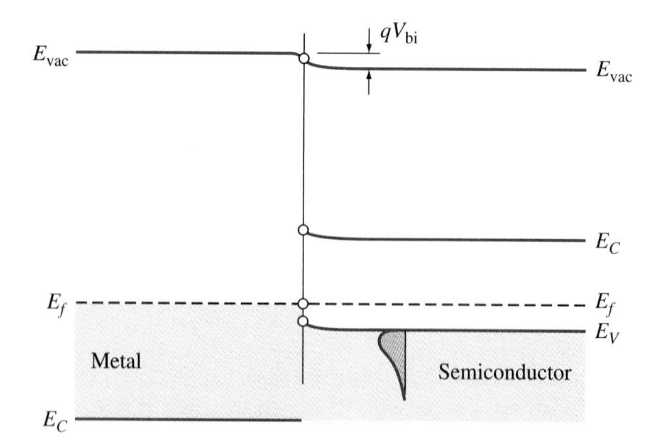

Figure P6.4

a. Draw its energy band diagram under forward and reverse bias.

b. Will this junction be ohmic or rectifying? Why?

6.13 Consider an n-Si:Al Schottky barrier diode for which $N_D' = 10^{17}\ \text{cm}^{-3}$ and the measured built-in voltage is 0.70 V. Let the junction area be $10\ \mu\text{m}^2$.

a. Find the depletion width w. Note that this junction can be treated as a one-sided junction.

b. Find the junction capacitance at $V_a = -5$ V.

6.14 A heterojunction is formed between n-type GaAs of $N_D' = 10^{16}\ \text{cm}^{-3}$ and p-type germanium of $N_A' = 10^{17}$. The measured discontinuities in the band edges are $\Delta E_C = 0.27$ eV and $\Delta E_V = 0.49$ eV.

a. Find the barrier to electrons in electron volts.

b. Find the barrier to holes in electron volts.

c. Find the built-in voltage (measured at E_{vac}) in electron volts.

d. To what is the physical source of the discontinuity in the vacuum level attributed?

Supplement to Part 2: Diodes

S2.1 INTRODUCTION

In this supplement, we discuss some additional points about diodes. First we will look at the physics of the dielectric relaxation time, the time needed to restore charge neutrality to a region when excess carriers of one polarity are suddenly introduced, for example by injection.

Then we explore the capacitance in diodes in somewhat more detail than done earlier. We will show how measurement of the *C-V* (capacitance-voltage) characteristics of a pn junction or a Schottky diode can be used to find doping profiles in junctions. In addition to prototype (step) junctions, nonuniformly doped junctions are considered.

The tunneling current in Schottky diodes and low-resistance metal-semiconductor pn junctions is briefly discussed.

Finally, we introduce the use of the SPICE circuit-analysis program to investigate static and transient diode characteristics.

S2.2 DIELECTRIC RELAXATION TIME

When a pn junction is turned on, and excess carriers are injected across the junction, at the instant of injection there will be an excess charge. For example, if electrons are injected into the p side, there will be extra negative charge there. In this case extra positive charges must be summoned from the external circuit to re-establish charge neutrality in the p-type material. This process takes a small but finite time, called the dielectric relaxation time τ_D. Let us estimate this time, first for majority carriers and then for minority carriers.

S2.2.1 CASE 1: DIELECTRIC RELAXATION TIME FOR INJECTION OF MAJORITY CARRIERS

We take first the case of excess majority carriers. Assume, for example, that Δn excess electrons are injected into n-type material, as might be the case for a forward-biased Schottky barrier junction. At time $t = 0$ there are $\Delta n(0)$ excess electrons in a small region as indicated in Figure S2.1a. This figure shows the carrier concentration as a function of position. This excess negative charge changes the potential locally, causing the band edge to bend as shown in Figure S2.1b. This local band bending creates an electric field $\mathscr{E}$, which tends to conduct the excess electrons away from the region.

We can find this current. For low injection such that $\Delta n \ll n_0$, we can ignore the electron diffusion term because the gradient is small. The electron current is then

$$J_n = \sigma_n \mathscr{E} = q\mu_n n \mathscr{E} \tag{S2.1}$$

The continuity equation for electrons is

$$\frac{\partial n}{\partial t} = -\frac{1}{q}\frac{\partial J_n}{\partial x} \tag{S2.2}$$

We neglect the recombination and generation terms because the excess carriers in this case are majority carriers, and recombination and generation will have a negligible effect on such large concentrations. Taking the derivative of Equation (S2.1), we have

$$\frac{\partial J_n}{\partial x} = q\mu_n \left[n\frac{\partial \mathscr{E}}{\partial x} + \mathscr{E}\frac{\partial n}{\partial x} \right] \tag{S2.3}$$

$$\frac{\partial J_n}{\partial x} = q\mu_n \left[(n_0 + \Delta n)\frac{\partial \mathscr{E}}{\partial x} + \mathscr{E}\frac{\partial (n_0 + \Delta n)}{\partial x} \right] \tag{S2.4}$$

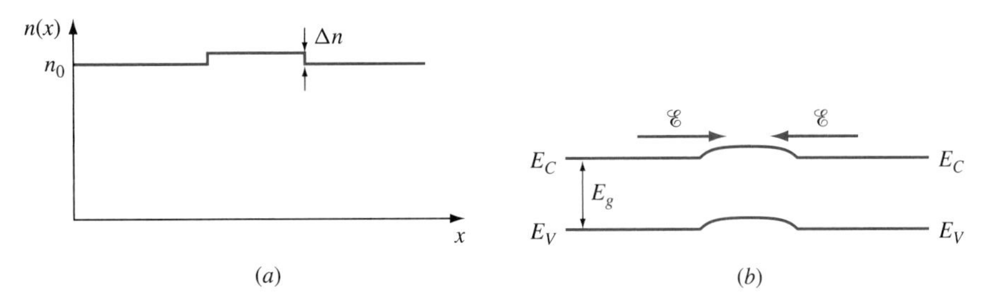

Figure S2.1 Dielectric relaxation time for excess majority carriers. (a) A localized group of excess majority carriers (in this case electrons) is injected into the sample; (b) the charge temporarily distorts the bands. The charge is neutralized with time constant τ_D by holes that come in from the surrounding area (ultimately through the contacts), and the bands return to flat.

The injected excess carriers are small in number because of the assumption of low injection, so we can neglect Δn in comparison with n_0. We also recognize that n_0 is a constant and thus $\partial n_0/\partial x = 0$. This results in

$$\frac{\partial J_n}{\partial x} = q\mu_n n_0 \frac{\partial \mathscr{E}}{\partial x} \tag{S2.5}$$

Since the electric field is given by $\mathscr{E} = -dV/dx$,

$$\frac{\partial \mathscr{E}}{\partial x} = -\frac{\partial^2 V}{\partial x^2} \tag{S2.6}$$

and from Poisson's equation

$$\frac{\partial^2 V}{\partial x^2} = -\frac{q\Delta n}{\varepsilon} \tag{S2.7}$$

Then by combining Equations (S2.5) to (S2.7), we obtain

$$\frac{\partial J_n}{\partial x} = q\mu_n \left[-n_0 q \frac{\Delta n}{\varepsilon}\right] \tag{S2.8}$$

and the continuity equation for electrons becomes

$$\frac{\partial n}{\partial t} = \frac{\partial \Delta n}{\partial t} = -\frac{q\mu_n n_0}{\varepsilon}\Delta n = -\frac{\sigma}{\varepsilon}\Delta n \tag{S2.9}$$

This is a standard differential equation, whose solution is

$$\Delta n = \Delta n(0)e^{-t/\tau_D} \tag{S2.10}$$

where $\Delta n(0)$ is the excess concentration at a point at $t = 0$ and

$$\boxed{\tau_D = \frac{\varepsilon}{\sigma} = \frac{\varepsilon}{qn_0\mu_n}} \tag{S2.11}$$

where μ_n is the majority carrier electron mobility. In p material, a similar equation applies involving the majority carrier hole mobility.

The quantity τ_D is called the *dielectric relaxation time.* It is a measure of the time required to neutralize excess carriers. Referring back to Figure S2.1, the net charge density, and with it the energy bands, decays to the normal (flat) condition with a time constant τ_D.

EXAMPLE S2.1

Find the dielectric relaxation time for Si of 1 Ω-cm (0.01 Ω-m) resistivity.

■ Solution

For this material $\sigma = 100$ S/m. From Equation (S2.11),

$$\tau_D = \frac{\varepsilon}{\sigma} = \frac{8.85 \times 10^{-12}\ \text{F/m} \times 11.8}{100\ \text{S/m}} = 1.04 \times 10^{-12}\ \text{s}$$

The dielectric relaxation time in this example is about an order of magnitude greater than the mean free time $\bar{t}$ between collisions. It is also shorter than the switching speed of semiconductor devices, and for most practical cases the relaxation can be considered instantaneous.

S2.2.2 CASE 2: INJECTION OF MINORITY CARRIERS

The situation is different if excess minority carriers are injected (e.g., in a forward-biased pn junction). In this case the change in minority carrier concentration is locally significant, so the excess carriers will be dispersed by diffusion rather than drift. Diffusion, however, is a relatively slow process.

Suppose holes are injected into an n-type semiconductor. The field built up by the excess holes acts on the electrons. At time $t = 0$, there are excess holes but no excess electrons, Figure S2.2a. The electrons are attracted into this region (Figure S2.2b) by drift because of the field resulting from the excess positive charges, and the negative electrons tend to neutralize those positive charges. This

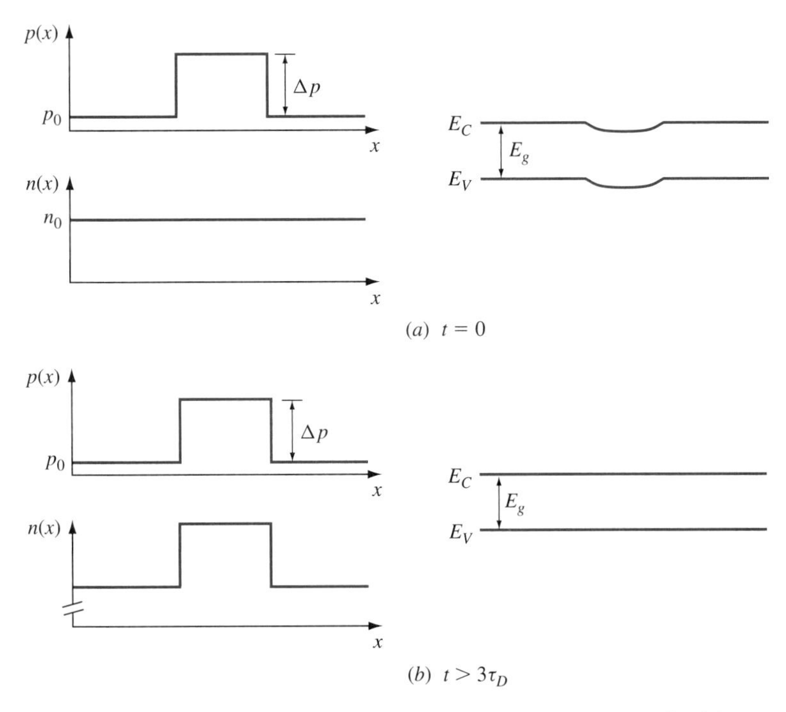

Figure S2.2 Dielectric relaxation for excess minority carriers (in this case holes). (a) Excess holes are injected at time $t = 0$. At that instant the bands bend downward because of the local excess positive charge. (b) During a period amounting to a few dielectric relaxation time constants, excess electrons are drawn in to neutralize the charge and flatten the bands. Note the carrier concentration plots are not to scale.

neutralization occurs in a dielectric relaxation time. While electrical neutrality is quickly established, the excess holes (and excess electrons to ensure neutrality) diffuse slowly while recombining, so there are still excess holes and electrons for some time (the carrier lifetime). The bands are flat during this recombination period because neutrality has been established.

To summarize:

1. If excess charge is injected into a semiconductor, it is electrically neutralized within about a dielectric relaxation time τ_D.
2. If the excess charge consists of majority carriers, the field set up by these carriers conducts them out via the contacts.
3. If the excess charge consists of minority carriers, majority carriers flow in from the contacts to neutralize it. The excess electrons and holes then diffuse and recombine.

S2.3 JUNCTION CAPACITANCE

In Chapter 5 we briefly discussed the capacitance of diodes, both the junction capacitance and the stored-charge capacitance. Here we investigate these two in more detail. We will treat junction capacitance, and show how a measurement of the capacitance versus voltage can be used to determine the doping concentration, the built-in voltage of a junction, and the doping profiles of various structures. After that, we will revisit the stored-charge capacitance, looking at the important case of the short-base diode.

S2.3.1 JUNCTION CAPACITANCE IN A PROTOTYPE (STEP) JUNCTION

Recall that the junction (differential) capacitance is defined as

$$\boxed{C_j \equiv \frac{dQ}{dV_a}} \tag{S2.12}$$

Since dQ is the change in charge at the edges of the depletion region for a change in voltage,[1] the junction capacitance has the form of a parallel-plate capacitor

$$\boxed{C_j = \frac{\varepsilon A}{w}} \tag{S2.13}$$

[1]More accurately, $C_j = dQ/dV_j$, where $V_j = V_a - IR_S$. However, the capacitance is normally measured for reverse bias or small forward bias such that I is small and $V_j \approx V_a$.

While Equations (S2.12) and (S2.13) are general, for the prototype (step) pn junction the result is

$$C_j = A\left[\frac{q\varepsilon N'_A N'_D}{2(N'_D + N'_A)(V_{bi} - V_a)}\right]^{1/2} \quad \text{prototype junction} \tag{S2.14}$$

For a one-sided step junction, the equation becomes

$$C_j = A\left[\frac{q\varepsilon N'}{2(V_{bi} - V_a)}\right]^{1/2} \quad \text{one-sided step junction} \tag{S2.15}$$

where N' is the (uniform) net doping on the lightly doped side. Equation (S2.15) is also valid for a Schottky diode in which the semiconductor is uniformly doped.

Apart from its influence on device response time, experimentally the capacitance yields useful information about the device. The capacitance-voltage characteristic can be used to measure the built-in voltage of a junction. To see this, we square and invert Equation (S2.14) to obtain

$$\frac{1}{C^2} = \frac{2(N'_D + N'_A)(V_{bi} - V_a)}{A^2 q\varepsilon N'_D N'_A} \tag{S2.16}$$

Everything in Equation (S2.16) is constant except V_a. A plot of $1/C^2$ as a function of applied voltage gives a straight line as indicated in Figure S2.3. Extrapolating this line to $(1/C)^2 = 0$ gives $V_a = V_{bi}$, so the built-in voltage may be determined directly from the experimental plot. Further, the slope of the C-V curve is

$$\frac{d\dfrac{1}{C^2}}{dV_a} = \frac{-2(N'_D + N'_A)}{A^2 q\varepsilon N'_D N'_A} \tag{S2.17}$$

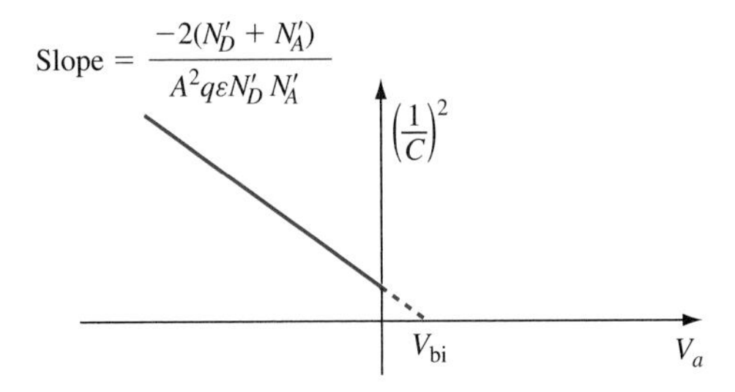

Figure S2.3 The capacitance-voltage characteristic of a prototype (step) pn junction can be used to measure V_{bi} experimentally.

from which

$$\frac{N'_D + N'_A}{N'_D N'_A}$$

can be found, assuming the area of the junction is known. Furthermore, if it is known that $N'_D \gg N'_A$, the slope determines N'_A. If $N'_A \gg N'_D$, it determines N'_D.

S2.3.2 JUNCTION CAPACITANCE IN A NONUNIFORMLY DOPED JUNCTION

The above analysis is valid for a step junction with constant doping on each side, and for a one-sided step junction or a Schottky diode with uniform doping in the semiconductor. The C_j-V_a characteristics are also useful, however, for determining the doping profile for nonuniform doping. Consider the case of Figure S2.4, where a Schottky diode is fabricated on the semiconductor surface, but in which the doping level of the n-type Si varies with distance from the surface. In this case the $1/C^2$-V_a curve will not be a straight line.

Let $N'_D(w)$ be the net donor doping level at the edge of the transition region (of width w) for an applied voltage V_a across the Schottky diode. Since $C = |dQ/dV_a|$ and

$$dQ = qN'_D(w)A\,dw \tag{S2.18}$$

we can write

$$C = \left| qN'_D(w)A\frac{dw}{dV_a} \right| \tag{S2.19}$$

But, since $C = \varepsilon A/w$, we can express $N'_D(w)$ as

$$\frac{1}{N'_D(w)} = \left| q\varepsilon A^2 \frac{1}{C}\frac{d(1/C)}{dV_a} \right| \tag{S2.20}$$

Finally, since

$$\frac{1}{C}\frac{d(1/C)}{dV_a} = \frac{1}{2}\frac{d(1/C^2)}{dV_a} \tag{S2.21}$$

Equation (S2.20) can be expressed as

$$\boxed{\frac{1}{N'_D(w)} = \left| \frac{q\varepsilon A^2}{2}\frac{d(1/C^2)}{dV_a} \right|} \tag{S2.22}$$

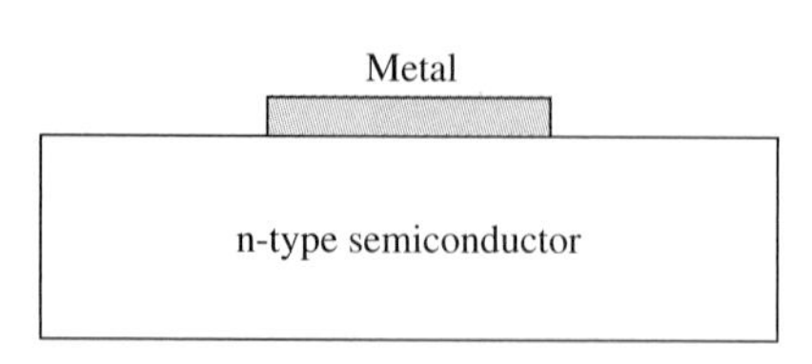

Figure S2.4 A Schottky diode with an n-type semiconductor.

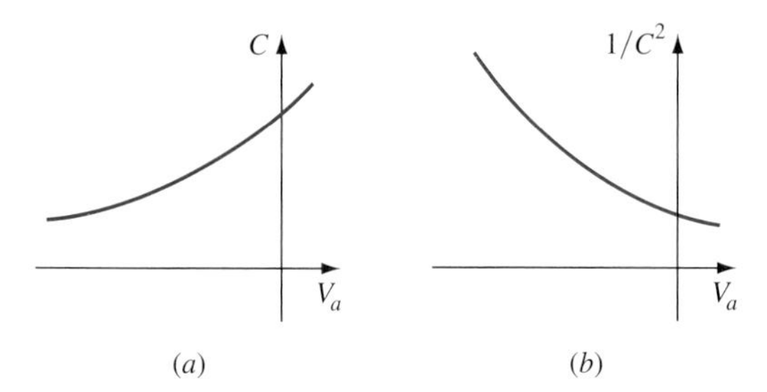

Figure S2.5 The *C-V* characteristics of a Schottky diode. (a) *C* versus V_a; (b) a plot of $(1/C)^2$ as a function of applied voltage.

The doping concentration profile as a function of distance, $N'_D(w)$, can be then found from the C-V_a characteristics and the slope of the experimental $1/C^2$-V_a plot by using Equation (S2.22). This is illustrated in Figure S2.5. In (a) the capacitance is plotted as a function of voltage, and in (b), $1/C^2$-V_a is plotted. To find w at a given voltage, the distance that the depletion region extends into the n-type material can be determined from the capacitance and Equation (S2.13). The doping concentration at that position is determined from the slope of the $1/C^2$-V_a characteristic and Equation (S2.22). The built-in voltage cannot be obtained for this case. This method is still quite useful for measuring the uniformity of semiconductor material however, and automatic systems are available (or easy to make) to plot $1/C^2$ as a function of applied voltage and do the necessary calculation to determine the N'_D-x relation.

S2.3.3 VARACTORS

The variation of junction capacitance with voltage is the basis of a class of devices called *variable capacitance diodes* or *varactors*. They are used for tuning circuits, frequency modulating radio signals, frequency conversion, and parametric amplification. [1]

Consider a one-sided n^+p junction in which the depletion region is predominantly on the p side. Let the doping be represented by some arbitrary functional form

$$N'_A = Bx^m \tag{S2.23}$$

where at the metallurgical junction $x = 0$. Solving Poisson's equation gives

$$\frac{d^2V}{dx^2} = \frac{-qN'_A}{\varepsilon}$$

with the boundary conditions $V(x = 0) = 0$ and $V(x = w) = V_{\text{bi}} - V_a$, where $x = w$ is at the edge of the transition region, gives an expression for the junction width [1, 2]

$$w = \left[\frac{\varepsilon(m+2)(V_{\text{bi}} - V_a)}{qB}\right]^{1/(m+2)} \tag{S2.24}$$

The capacitance is then found to be

$$C = \left[\frac{qB\varepsilon^{m+1}}{(m+2)(V_{\text{bi}} - V_a)}\right]^{1/(m+2)}$$
$$= B^*(V_{\text{bi}} - V_a)^{-1/(m+2)} = B^*(V_{\text{bi}} - V_a)^{-s} \quad \text{(S2.25)}$$

where we have combined the constants into a single constant B^*. We define

$$s = \frac{1}{m+2} \quad \text{(S2.26)}$$

where s is called the *sensitivity.* The larger s (the smaller m), the larger the capacitance variation with applied (reverse) voltage. For linearly graded junctions, $m = 1$ and $s = \frac{1}{3}$. For step (prototype) junctions, $m = 0$ and $s = \frac{1}{2}$.

An interesting case is that for $m = -\frac{3}{2}$. In this situation, $s = 2$ and

$$C \propto \frac{1}{(V_{\text{bi}} - V_a)^2} \quad \text{(S2.27)}$$

If such a capacitance is used with an inductor L in a tuned circuit, the resonant frequency f_r is

$$f_r = \frac{1}{2\pi\sqrt{LC}} \propto (V_{\text{bi}} - V_a) \quad \text{(S2.28)}$$

and varies linearly with applied voltage. This considerably simplifies the design of tunable circuits.

S2.3.4 STORED-CHARGE CAPACITANCE OF SHORT-BASE DIODES

The junction capacitance of short-base diodes is the same as for their long-base counterparts. The stored-charge capacitance, however, is much reduced from that of a long-base diode because the stored minority carrier charge in the short-base region is appreciably less than in the long-base diode.

Consider the n^+p short-base diode of Figure S2.6 where the p region is uniformly doped and $W_B \ll L_n$. The stored charge consists of the total excess charges in the short base. The electron concentration at the junction edge is $n_p(x_p)$ and at the contact end is n_{p0}. Thus, since recombination is negligible, the total stored excess charge under this distribution is

$$Q_s = \frac{qA[n_p(x_p) - n_{p0}]W_B}{2} \quad \text{(S2.29)}$$

and the current is

$$I = \frac{qAD_n[n_p(x_p) - n_{p0}]}{W_B} \quad \text{(S2.30)}$$

Since this is minority carrier diffusion current, D_n is the minority carrier diffusion coefficient.

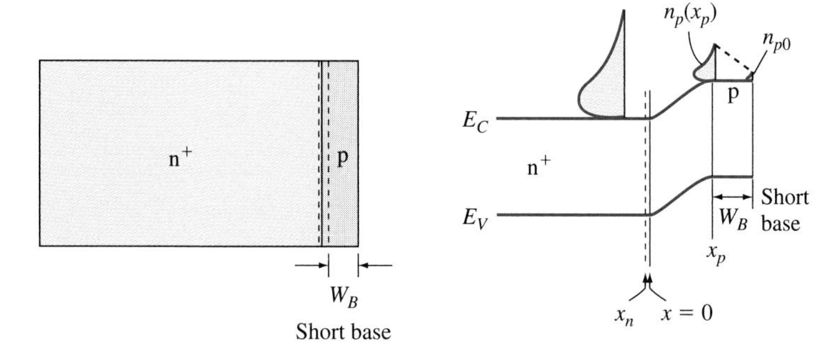

Figure S2.6 A short-base n$^+$p junction showing the charge distribution that creates the stored charge capacitance.

From Equations (S2.29) and (S2.30), we have

$$\frac{Q_s}{I} = \frac{(W_B)^2}{2D_n} \tag{S2.31}$$

or the stored charge is

$$Q_s = I\frac{(W_B)^2}{2D_n} = I\tau_T \tag{S2.32}$$

Here we have defined the quantity

$$\boxed{\tau_T \equiv \frac{Q_s}{I} = \frac{(W_B)^2}{2D_n}} \tag{S2.33}$$

which is called the *minority carrier transit time* or simply *transit time.*

The reclaimable stored charge Q_{sr} is, as in Chapter 5, that fraction of the stored charge that contributes to the actual capacitance:

$$\boxed{Q_{sr} = \delta Q_s = \delta I \tau_T} \tag{S2.34}$$

Therefore the stored-charge capacitance C_{sc} is [3]

$$\boxed{C_{sc} = \frac{dQ_{sr}}{dV_a} = \delta\frac{dQ_s}{dV_a} = \delta I \tau_T} \tag{S2.35}$$

EXAMPLE S2.2

Compare the stored-charge capacitance for long-base and short-base n^+p diodes with $N_A' = N_A = 10^{17}\ \text{cm}^{-3}$, $I = 10$ mA, and a minority carrier lifetime $\tau_n = 2.9 \times 10^{-6}$ s (Figure 3.21). Let $W_B = 0.3\ \mu$m for the short-base diode. The factor δ is $\frac{1}{2}$ for the long-base diode and $\frac{2}{3}$ for the short-base diode. [3]

■ **Solution**

For the long-base diode, from Equation (5.116), we have

$$C_{sc} = \delta I \tau_n = \frac{1}{2} \times 10^{-2} \times 2.9 \times 10^{-6} = 1.45 \times 10^{-8}\,\text{F}$$

For the short-base diode, we use Equations (S2.35) and (S2.33) to write

$$C_{sc} = \delta I \tau_T = \delta I \frac{(W_B)^2}{2D_n} = \frac{2}{3} \times 10^{-2} \frac{(0.3 \times 10^{-4}\,\text{cm})^2}{2 \times 20\,\text{cm}^2/\text{s}} = 1.5 \times 10^{-12}\,\text{F}$$

where we found the minority carrier diffusion length $D_n = 20\,\text{cm}^2/\text{s}$ from Figure 3.11.

For this example, the stored-charge capacitance of the long-base diode is about 10^4 times as large as for the short-base diode, because of its much larger stored charge.

S2.4 SECOND-ORDER EFFECTS IN SCHOTTKY DIODES

In Chapter 6, a first-order model for a Schottky diode was discussed. In this model it was assumed that diode current was thermionic in nature. All carriers with sufficient x-directed energy to surmount the barrier $E_B(0)$ at the metal-semiconductor interface do so. This is illustrated again in Figure S2.7. Part (a) shows the energy band diagram at equilibrium for an n-channel Schottky diode. At equilibrium the net current flow is zero, and the current density J_0 from metal to semiconductor is equal to that flowing from semiconductor to metal as indicated in (a). For forward bias (b), with V_a positive, the barrier from metal to semiconductor, $E_B(0)$, is unchanged while the barrier from semiconductor to metal is reduced by the factor qV_a (V_a positive). In (c) the energy band diagram is shown for reverse bias (V_a negative). Here too, the barrier from metal to semiconductor is unchanged but the barrier from semiconductor to metal is increased. This result is the current-voltage relation given in Equation (6.14), which is repeated here for the first-order model.

$$\boxed{J = J_0(e^{qV_a/kT} - 1)} \tag{S2.36}$$

where J_0 is given by

$$\boxed{J_0 = \frac{qm^*(kT)^2}{2\pi^2\hbar^3}\, e^{-E_B(0)/kT}} \tag{S2.37}$$

There are two important second-order effects, however, that need to be considered in practical Schottky diodes. These are tunneling through the barrier and barrier lowering due to image effects.

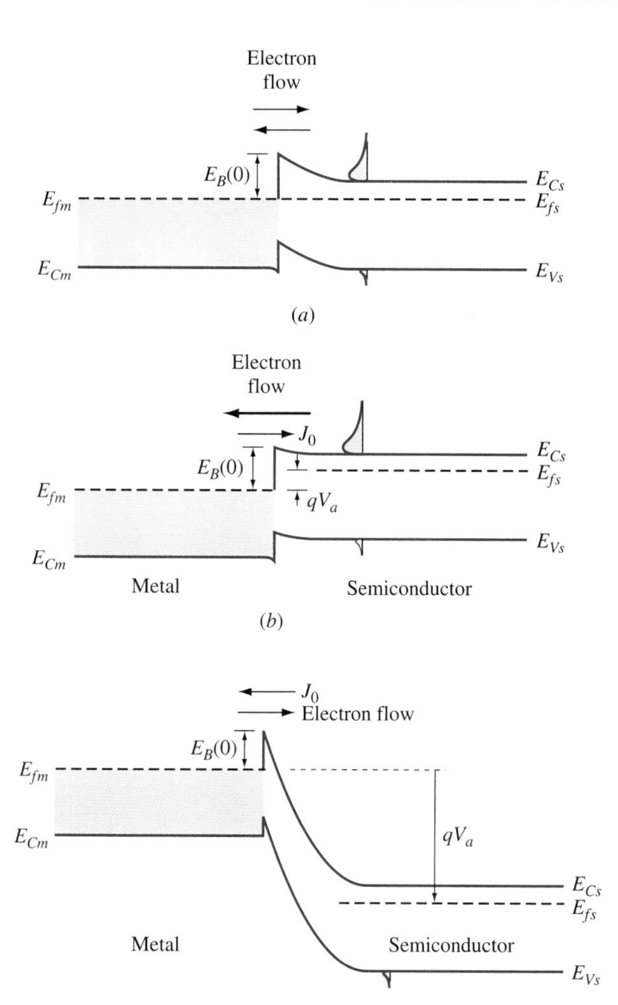

Figure S2.7 First-order current flow in an n-Schottky barrier under (a) equilibrium, where no net current flows; (b) forward bias, where the barrier for electrons is reduced, going from the semiconductor to the metal, but not from the metal to the semiconductor, resulting in a net electron flow from semiconductor to metal; (c) reverse bias. To first order, electrons are assumed to cross the barrier thermionically.

S2.4.1 TUNNELING THROUGH SCHOTTKY BARRIERS

As indicated in Chapter 6, there is a probability that electrons will tunnel through the depletion region in a Schottky barrier. We first treat the case of a rectifier in which the semiconductor is lightly doped (nondegenerate) and then the case of low-resistance metal-semiconductor contacts in which the semiconductor is degenerate.

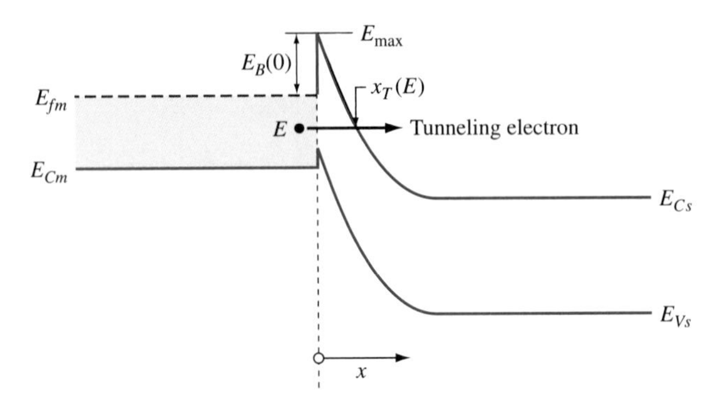

Figure S2.8 In a reverse-biased Schottky diode, electrons can tunnel through the metal-semiconductor barrier. The Schottky barrier is assumed to be close to triangular. The tunneling distance is x_T.

For either case, the probability that an electron at energy E with x-directed velocity tunnels into the semiconductor is

$$T = e^{-2\int_0^{x_T} [(2m^*/\hbar^2)(E_C(x)-E)]^{1/2}\,dx} \qquad 0 \le x \le x_T \tag{S2.38}$$

where x_T is the tunneling distance at energy E and E_C is the conduction band edge in the semiconductor.

Tunneling in Schottky Diodes We first consider the case of a Schottky rectifier. Consider the reverse-biased Schottky diode of Figure S2.8, where x_T is the tunneling distance at energy E, and where the Schottky barrier is assumed triangular. For such a triangular barrier, $x_T = (E_{\max} - E)/q|\mathscr{E}|$ and the tunneling probability becomes

$$T = e^{-4\sqrt{2m^*}(E_{\max}-E)^{3/2}/3q\hbar|\mathscr{E}|} \tag{S2.39}$$

The tunnel current is proportional to the electron concentration in the metal at energy E. For $E > E_{\text{fm}}$, the electron concentration decreases exponentially with E but the tunneling probability decreases exponentially with $(E_{\max} - E/q|\mathscr{E}|)^{3/2}$, or it increases as $n(E)$ decreases. The total tunnel current is dependent on $\mathscr{E}$, which depends on the bias and the doping concentration in the semiconductor. Tunneling through this triangular barrier is referred to as *Fowler-Nordheim tunneling*. Except at high temperatures where the electron concentration is appreciable at higher energies, most of the tunnel current occurs for $E \approx E_{\text{fm}}$ or $E_{\max} - E = E_B(0)$, since the electron concentration in the metal is a maximum in this region. The Fowler-Nordheim tunnel current can then be approximated as [4]

$$J_{FN} \approx \frac{q^2|\mathscr{E}^2|}{16\pi^2\hbar E_B(0)} e^{-4\sqrt{2m^*}(E_B(0))^{3/2}/3\hbar q|\mathscr{E}|} \tag{S2.40}$$

where $\mathscr{E}$ is the field in the semiconductor and m^* is the effective mass of the tunneling electrons.

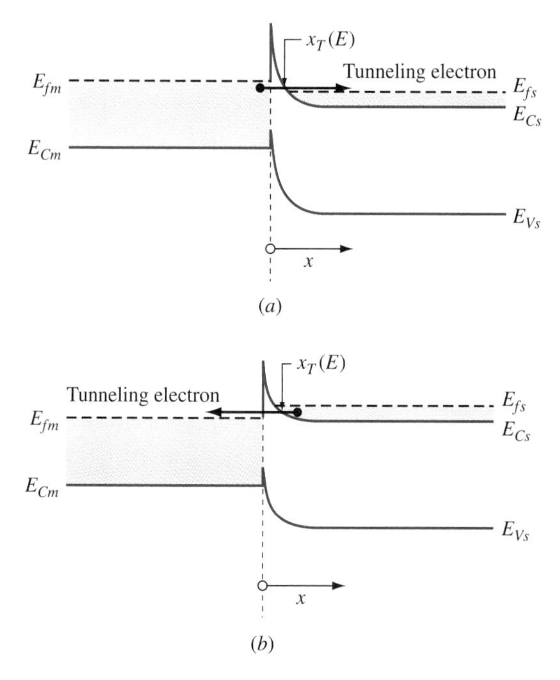

Figure S2.9 For a metal-semiconductor junction in which the semiconductor is degenerate, the tunnel distance is short, resulting in a large current for small reverse (a) or forward (b) bias. This results in a small contact resistance.

Tunneling in Low-Resistance Contacts For a metal-semiconductor contact in which the semiconductor is degenerate, the Fermi level in the semiconductor is within the conduction band and the depletion region at this energy is narrow enough to result in a small x_T and thus a large tunneling probability at any energy. In this case, the current is large for both forward and reverse bias. Figure S2.9 indicates the tunneling for a small reverse bias (a) and for small forward bias (b). In (a), a large current flows for a small reverse bias (in the millivolt range), while in (b) a large current flows for a small forward bias. The heavier the doping, the smaller the tunneling distance and the smaller the junction resistance.

S2.4.2 BARRIER LOWERING IN SCHOTTKY DIODES DUE TO THE IMAGE EFFECT

In addition to the tunneling described above, the current in a Schottky barrier is also modified from the first-order prediction by the image effect, which lowers the barrier. This arises because an electron in the semiconductor near the metal of a Schottky barrier induces a positive charge at the metal surface. Consider an electron of charge $-q$ in the conduction band of an n-type Schottky diode a distance x from the metal interface, Figure S2.10a. There will be an attractive force

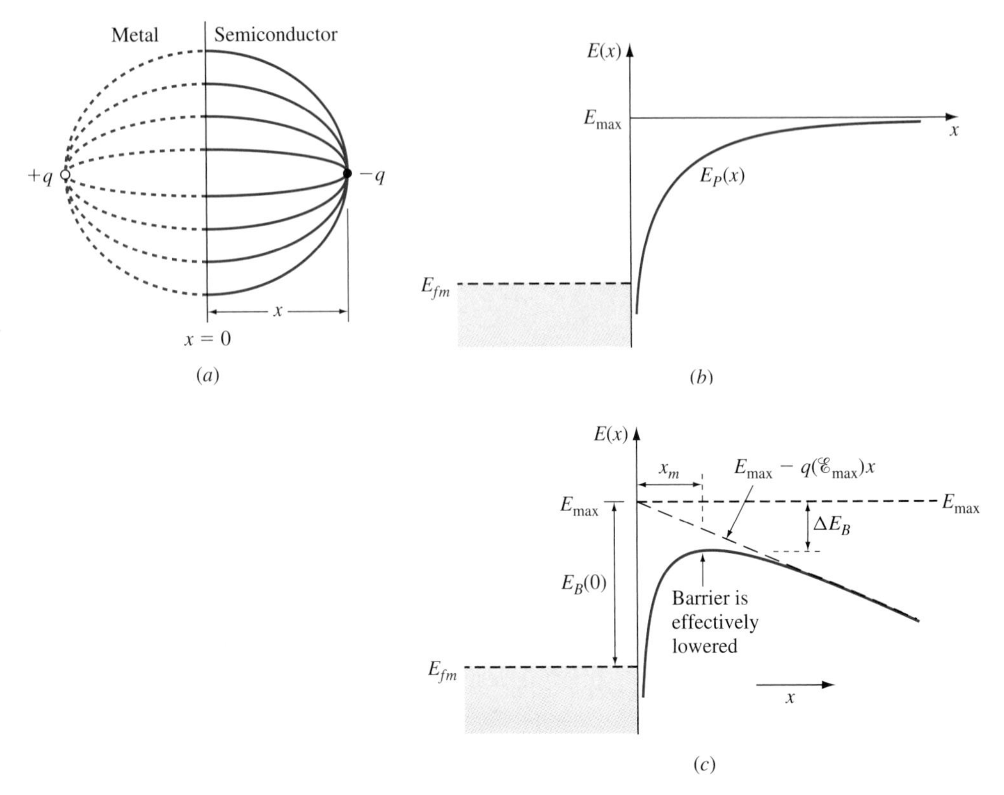

Figure S2.10 An electron in the semiconductor conduction band a distance x from the metal surface creates a positive surface charge in the metal. (a) The field in the semiconductor is equal to that caused by an imaginary $+q$ charge a distance x inside the metal. (b) The resultant potential energy in the semiconductor. (c) This combined with the field in the semiconductor depletion region lowers the metal-semiconductor barrier by an amount ΔE_B.

between the electron and the metal surface. At the interface the resultant field must be perpendicular to the metal, since the metal is assumed to be a perfect conductor. The field lines in the semiconductor then are equivalent to those produced by considering the metal to be a semiconductor with a charge $+q$ a distance x inside the metal to balance the charge $-q$ of the electron. This $+q$ charge is referred to as an *image*. The equivalent field is indicated in the figure.

The force on the electron by its image charge is given by Coulomb's equation

$$F = \frac{-q^2}{4\pi\varepsilon_s(2x)^2} = -\frac{dE_P}{dx} \tag{S2.41}$$

and the electron potential energy due to the image effect is found by integrating E_P from finite x to infinity, which gives

$$E_P(\text{image}) = E_{\text{max}} - \frac{q^2}{16\pi\varepsilon_s x} \tag{S2.42}$$

as indicated in Figure S2.10b. The variation of potential energy with distance produces an electric field. There is also an electric field associated with the depletion region, which varies with distance into the semiconductor. The image effect, however, is concentrated near the interface, so we can assume that the electric field due to the semiconductor depletion region is equal to its maximum value $|\mathscr{E}_{\max}|$, as indicated by the dashed line in Figure S2.10c. The potential energy of an electron at position x becomes

$$E_P(x) = -q|\mathscr{E}_{\max}|x - \frac{q^2}{16\pi\varepsilon_s x} \tag{S2.43}$$

which has a maximum value at $x = x_m$ where

$$x_m = \left[\frac{q}{16\pi\varepsilon_s|\mathscr{E}_{\max}|}\right]^{1/2} \tag{S2.44}$$

and the change in the barrier height due to the image effect is

$$\Delta E_B = q\left[\frac{q|\mathscr{E}_{\max}|}{4\pi\varepsilon_s}\right]^{1/2} \tag{S2.45}$$

For a value of $|\mathscr{E}_{\max}| = 10^5$ V/cm $= 10^7$ V/m and $\varepsilon_s = 11.8\varepsilon_0$ (e.g., Si), $\Delta E_B \approx 0.036$ eV, and occurs at about 7 nm from the interface. Note that this distance is an order of magnitude greater than the influence of the interface dipoles.

Since this image-induced barrier lowering increases the current density for a Schottky diode, the diode equation becomes

$$J_0 = \frac{qm^*(kT)^2}{2\pi^2\hbar^3}e^{-(E_B(0)-\Delta E_B)/kT} \tag{S2.46}$$

The value of $|\mathscr{E}_{\max}|$ (and thus J_0) depends on doping level and applied voltage:

$$|\mathscr{E}_{\max}| = \left[\frac{2qN_D'(V_{\text{bi}} - V_a)}{\varepsilon_s}\right]^{1/2} \tag{S2.47}$$

S2.5 SPICE MODEL FOR DIODES

We have discussed analytical models for the static electrical characteristics of pn junctions. In circuit applications, the diode bias varies with time, and the voltage and current dependencies of the diode parameters influence the circuit response. There are a number of circuit analysis programs that can calculate the circuit response, taking into account the bias (and temperature) dependence of the device characteristics. The most common of these is SPICE. (There are several versions of SPICE.[2]) To utilize SPICE, it is necessary to specify the important SPICE parameters of the devices used in the circuit under consideration.

---

[2]A student version of PSPICE can be obtained from Cadence Design Systems, Inc., Cadence PCB Systems Division, 13221 SW 68th Parkway, Suite 200, Portland OR 97223-8328; http://www.cadence.com/.

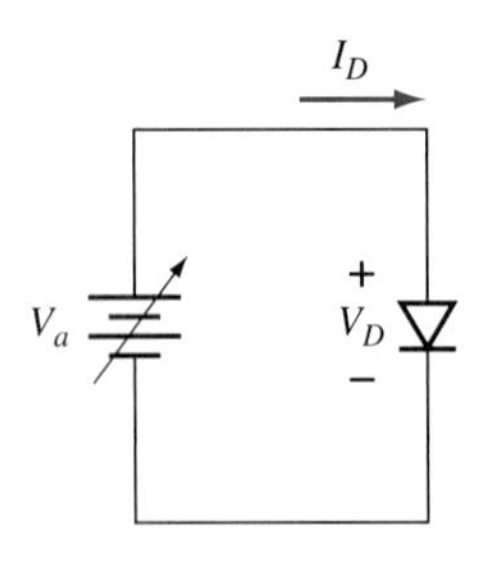

Figure S2.11 Circuit for determining the static *I-V* characteristics of a diode.

We present two examples of the use of SPICE with diodes. One is to simulate the static I-V characteristics of a diode and the second is an example of transient analysis.

S2.5.1 THE USE OF SPICE AS A CURVE TRACER [5]

A curve tracer is an instrument that can measure the static I-V characteristics of semiconductor devices over a range of current and voltage values. The voltage and/or current sources are swept over a range of values and the resultant device currents and voltages are measured and recorded.

To emulate this behavior using SPICE, the DC sweep command is used. The simulated SPICE circuit for a diode is shown in Figure S2.11 where the range of the applied voltage sweep V_a is specified and the resultant diode current I_D is calculated and plotted as a function of V_D. Note that for this case, $V_D = V_a$.

The static I-V characteristics of a diode can be calculated and plotted provided certain diode parameters are specified. The simplest diode model is based on the equation

$$I = I_0(e^{qV_j/nkT} - 1) = I_0(e^{q(V_D - IR_S)/nkT} - 1) \tag{S2.48}$$

where V_j is the junction voltage

$$V_j = (V_D - IR_S)$$

The voltage V_a is the applied voltage ($V_a = V_D$) and R_S is the series resistance of the diode. The parameter n is the diode quality factor, referred to in SPICE as the emission coefficient and takes into account the generation-recombination current as well as the injection current.

In addition, the breakdown voltage V_{BD} and the reverse current I_{BD} at which breakdown occurs can be specified. A partial list of SPICE parameters used for the static characteristics are listed in Table S2.1. Included are the values used for the example whose results are shown in Figure S2.12. In (a) the forward bias characteristics are shown on a linear scale, and in (b) in a semi-log plot.

The reverse and forward I-V characteristics are plotted for this diode in Figure S2.13. The reverse breakdown voltage specified as 10 V is clearly seen.

Table S2.1 Partial list of static SPICE parameters for diodes

Symbol (this book)	Symbol (SPICE)	Model parameter	Units	Values used in example
I_0	IS	Saturation current	A	1×10^{-14}
R_S	RS	Ohmic resistance	Ω	10
n	N	Emission coefficient	—	1.5
V_{BR}	BV	Reverse-bias breakdown voltage	V	10
I_{BD}	IBV	Reverse-bias breakdown current	A	1×10^{-9}

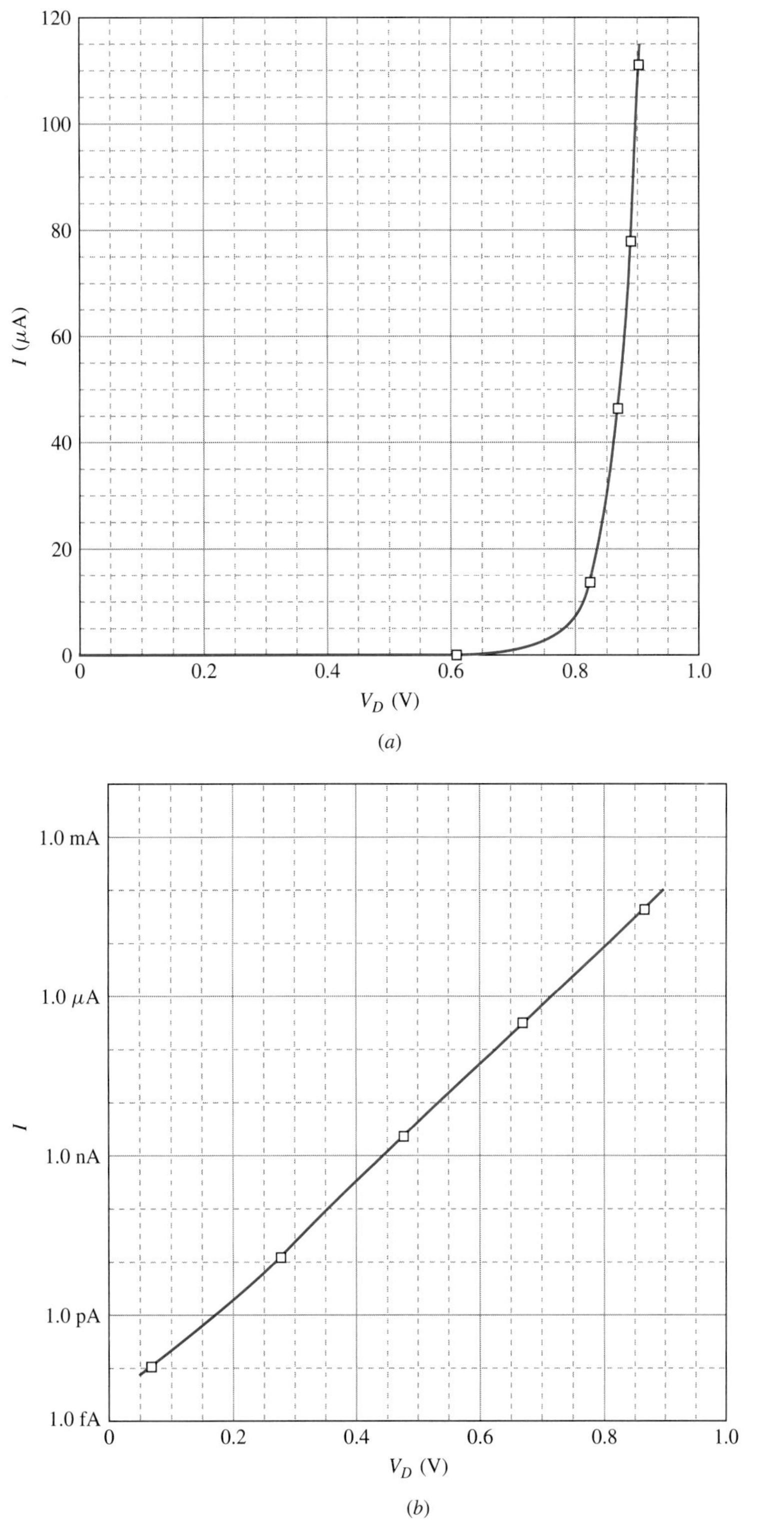

Figure S2.12 SPICE-simulated forward *I-V* characteristics of a diode with the parameters specified in Table S2.1. In (a) the characteristics are indicated on a linear plot; in (b) on a semilogarithmic plot.

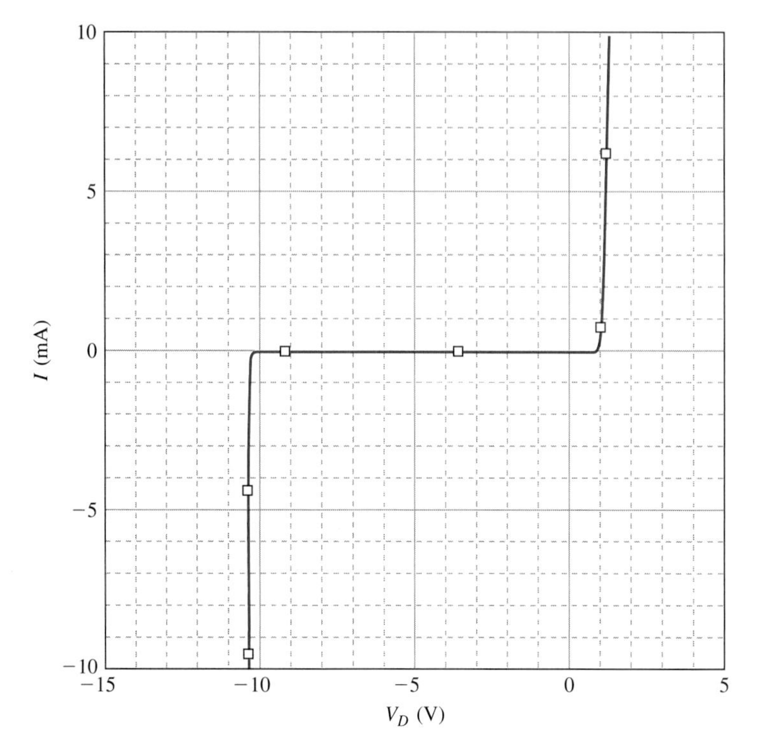

Figure S2.13 SPICE-simulated forward and reverse characteristics for the diode specified in Table S2.1.

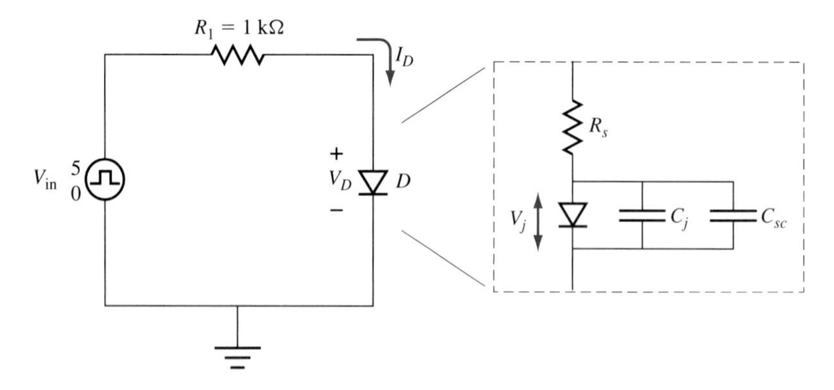

Figure S2.14 Circuit schematic used to simulate the transient response of a diode stimulated by an input pulse. On the right is the model used in SPICE.

S2.5.2 TRANSIENT ANALYSIS

SPICE is a convenient tool for determining the transient characteristics of a circuit. Consider the simple diode current of Figure S2.14, similar to that of Figure 5.34, which consists of a resistor in series with a pn junction. A voltage pulse from 0 to 5 V and of 300 ns duration is applied at $t = 50$ ns. We consider the junction to be a wide-base n^+p diode with an electron lifetime in the wide p region to be 10^{-7} s. The other parameters used in the SPICE transient simulation are given in Table S2.2. The resultant diode current and voltage transients are shown in Figure S2.15a and b respectively.

From part (a) of Figure S2.15 it can be determined that, during turn-on, the current rapidly reaches its limiting value of $(V_{\text{pulse}} - V_D)/R_1 = (V_{\text{pulse}} - V_j)/(R_1 + R_S)$, where V_D is the diode terminal voltage and V_j is the junction voltage or the diode voltage minus IR_S ($V_j = V_D - IR_S$). During the turn-off, the current becomes negative and returns to zero as the diode capacitance discharges.

In (b), during turn-on the diode voltage reaches its limiting value $(V_j + I_F R_S)$ reasonably fast. When the applied pulse decreases to zero, the junction voltage cannot change instantaneously because of its capacitance but drops from $V_j + I_F R_S$ to $V_j + I_R R_S$, where I_F and I_R are indicated in (a). (Note that I_R is negative.)

The transient results depend on the value of V_j as a function of time as the capacitances discharge. While this would be a difficult hand calculation, the calculation routine is built into SPICE.

Table S2.2 Partial list of SPICE parameters for diodes

Symbol (this book)	Symbol (SPICE)	Model parameter	Units	Values used in example
I_0	IS	Saturation current	A	1×10^{-14}
R_S	RS	Ohmic resistance	Ω	12
n	N	Emission coefficient		1
V_{BR}	BV	Reverse-bias breakdown voltage	V	10
I_{BD}	IBV	Reverse-bias breakdown current	A	1×10^{-9}
C_j	CJ0	Zero bias junction capacitance	F	0.1×10^{-15}
m	M	Junction grading coefficient	—	0.5
V_{bi}	VJ	Built-in voltage	V	0.93
t_T	TT	Transit time[3]	s	1×10^{-7}

[3]The term *transit time* evolved from the analysis of bipolar junction transistors, where it represents the average time for minority carriers to cross the base from emitter to collector. It is defined as Q_s/I. For short-base diodes, it is the average time for minority carriers to cross the short-base region ($W_B \ll L_n$ for electrons in a short p region). For long-base prototype diodes, $t_T \approx \tau_n$.

Figure S2.15 Diode transients for the circuit of Figure S2.10 with the diode parameters of Table S2.2. (a) Diode current transient; (b) voltage transient.

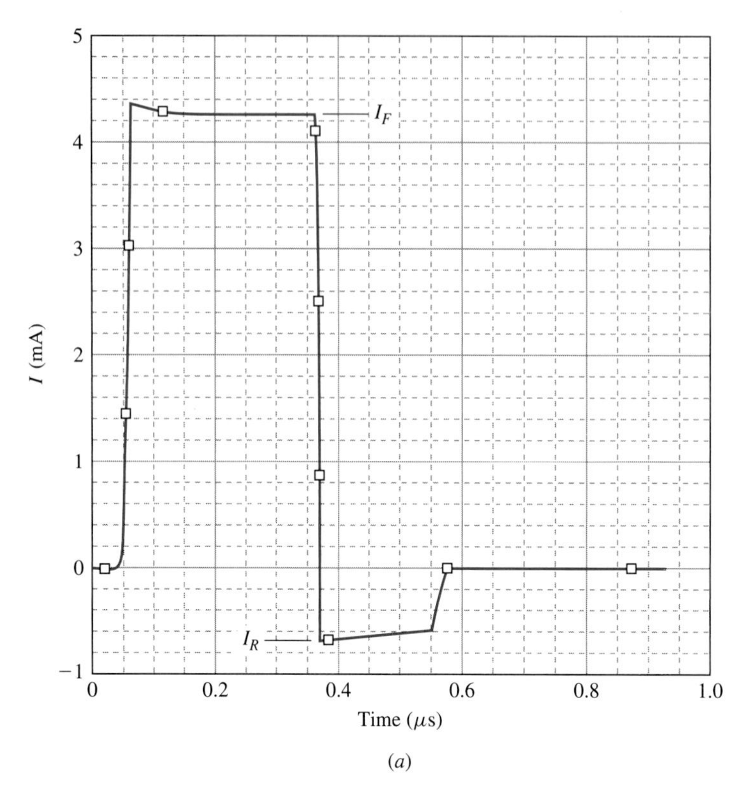

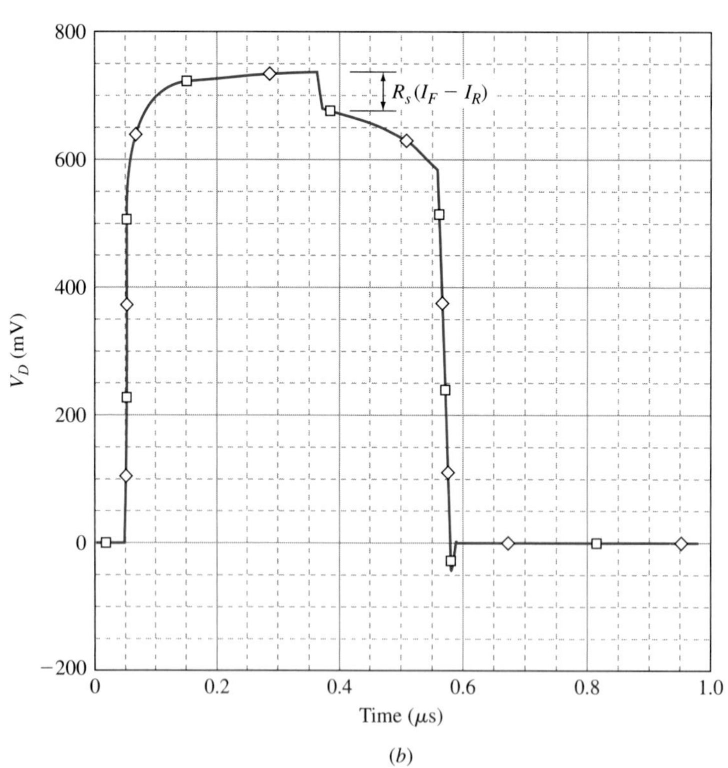

5
4
3
2
1
0
−1
I (mA)
0 0.2 0.4 0.6 0.8 1.0
Time (μs)

(*a*)

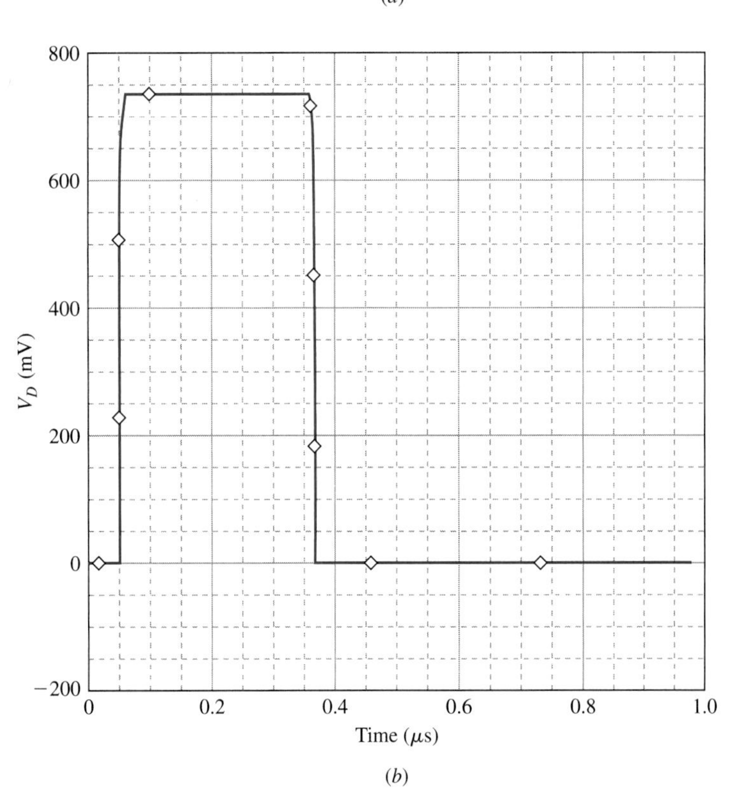

(*b*)

Figure S2.16 SPICE-simulated diode transients for the parameters of Table S2.2, except with zero transit time.

To determine the relative effects of the stored-charge capacitance compared to those of the junction capacitances, the SPICE program is rerun with the same parameters except that the space-charge capacitance is set equal to zero (which is done by setting the transit time to zero, or TT = 0). The results are shown in Figure S2.16. From a comparison of Figures S2.15 and S2.16 it can be seen that the diode turn-on and turn-off times are determined largely by the space-charge capacitance. The diode of Figure S2.16 approximates the case of a Schottky barrier diode, where the minority carrier injection is negligible.

S2.6 SUMMARY

In this supplement to Part 2, "Diodes," we investigated some additional topics in relation to diodes. The dielectric relaxation time is the time required for charge neutrality to be re-established after a sudden change in carrier concentration. An example is when excess carriers are injected across a junction. For injected minority carriers, within a time constant τ_D, majority carriers move into the region to compensate the charge and level out the band edges. For injected majority carriers, τ_D is the time constant associated with ejecting these carriers from the material via the ohmic contacts.

We also reviewed how capacitance measurements can be used to investigate some junction parameters. For a prototype (step) junction, the C-V_a characteristics can be used to experimentally measure the built-in voltage. They can also determine the relationship between the doping on either side of the junction. If the junction is one-sided, the doping on the lightly doped side can be found. In nonuniformly doped, one-sided junctions, the doping profile can be obtained.

The stored-charge capacitance of step short-base diodes (the width of the more lightly doped side much less than a minority carrier diffusion length) was found to be orders of magnitude smaller than that of the prototype long-base diode because of the much reduced minority carrier stored charge. (The stored-charge capacitance for nonuniformly doped semiconductors is treated further in Chapter 9 with reference to bipolar junction transistors.)

Tunnel current in Schottky diodes was briefly discussed along with the current in low resistance metal-semiconductor contacts. We saw that tunneling increases the current, and that the image effect lowers the metal-semiconductor barrier, which also increases the current.

The program SPICE is used extensively for circuit analysis. Because the voltage and current dependences of some of the device parameters, a hand calculation is difficult compared to the use of SPICE, where the calculation routines are built in.

S2.7 READING LIST

Items 6, 8, 11, 12, and 17–19 in Appendix G are recommended.

S2.8 REFERENCES

1. M. H. Norwood and E. Shatz, "Voltage Variable Capacitance Tuning—A Review," *Proc. IEEE,* 56, pp. 788–798, 1968.
2. S. M. Sze, *Physics of Semiconductor Devices,* 2nd ed., Chapter 2, John Wiley & Sons, New York, 1981.
3. Joseph Lindmayer and Charles Y. Wrigley, *Fundamentals of Semiconductor Devices,* Chapter 2, D. Van Nostrand, Princeton, NJ, 1965.
4. M. Lenzlinger and E. H. Snow, "Fowler-Nordheim tunneling into thermally grown SiO_2," *J. Appl. Phys.* 40, pp. 278–283, 1969.
5. Gordon W. Roberts and Adel S. Sedra, *SPICE for Microelectronic Circuits,* 3rd ed., Chapter 3, Saunders College Publishing, New York, 1992.

S2.9 PROBLEMS

S2.1 A prototype (step) junction has $N'_D = 2 \times 10^{16}$ cm^{-3}, $N'_A = 5 \times 10^{15}$ cm^{-3}, and $V_{\text{bi}} = 0.71$ V. Plot the *C-V* characteristic and $1/C^2$ versus V_a for this junction if the area of the junction is 10^{-4} cm^2. Plot for $V_a = -5$ V to -0.5 V.

S2.2 The *C-V* characteristic is measured for a silicon diode that is heavily doped on the n side and lightly doped on the p side. The experimental results are shown in Figure PS2.1. Find the value of the built-in voltage and N'_A. The diode is a square, 10^{-4} cm on a side.

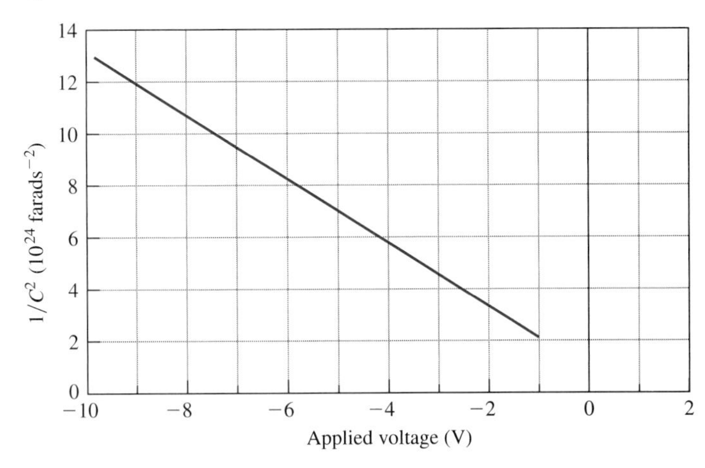

Figure PS2.1

S2.3 The *C-V* measurements for a particular junction are given in Figure PS2.2. Plot the junction width as a function of applied voltage, and plot the doping concentration N'_D as a function of junction width w. Let the junction area be 5×10^{-8} cm^{-3}.

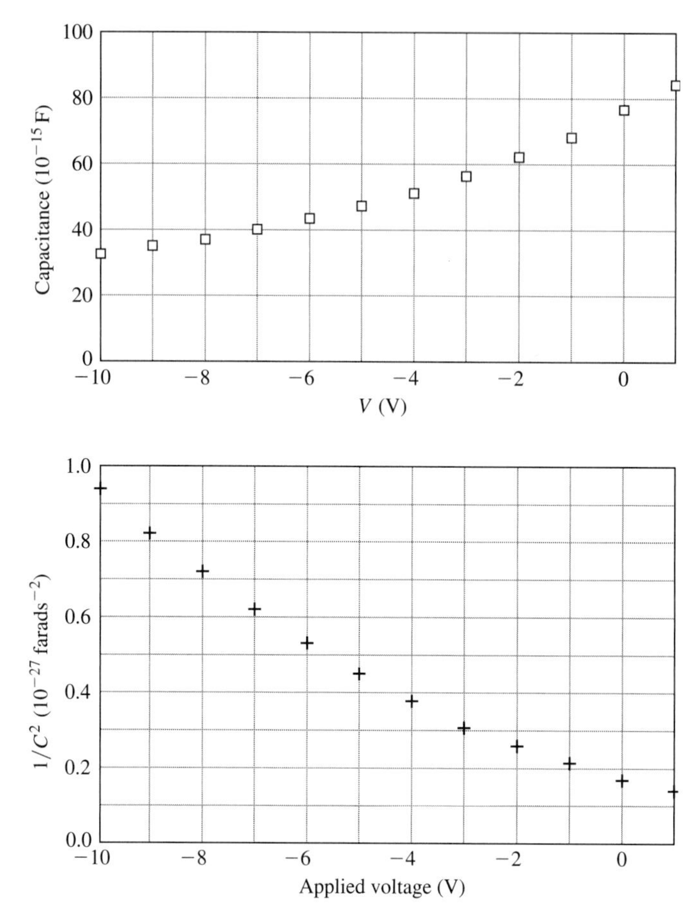

Figure PS2.2

S2.4 In Section S2.3.3, we discussed varactors and mentioned a specific case of interest in which the resonant frequency of a circuit using an inductor with the tunable capacitance of a junction was proportional to the applied voltage. In some circuits using junction capacitance as a tuning element, however, it is convenient to have the capacitance vary linearly with applied voltage. Find the required doping profile (N'_D versus x) of the deposited epitaxial layer.

S2.5 From Figure S2.15b, verify that during turn-off the initial drop in V_D is equal to $R_S(I_F - I_R)$ where $R_S = 12\ \Omega$ and I_F and I_R are obtained from Figure S2.15a.

S2.6 From Figure S2.15a, show that the storage time is in reasonable agreement with the expression $t_S = \tau_n \ln(1 + |I_F/I_R|)$.

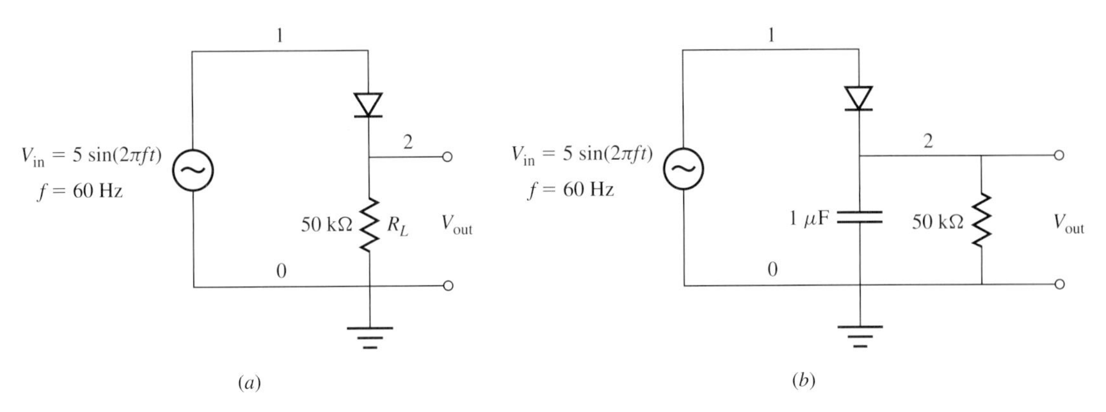

Figure PS2.3

S2.7 Using SPICE, plot the input and output voltages for the circuit of Figure PS2.3 for $0 < t < 50$ ms. The diode parameters are: IS $= 10^{-14}$A and $N = 1$.

S2.8 Repeat Problem S2.7 with a capacitor of 1 μF connected in parallel with R_L as in Figure PS2.3(b).

S2.9 Repeat Problem 5.17a using SPICE, and compare the plots.

PART 3

Field-Effect Transistors

Up to now, we have discussed semiconductor materials and two-terminal devices. By far the most important semiconductor devices, however, are transistors, which are three-terminal devices. Transistors have two very useful modes of operation: they can be amplifiers and they can be switches.

When a transistor is used as an amplifier, as in analog circuits, the current or voltage at one terminal controls the current or voltage between the other two terminals. A small change in the control signal (the electrical equivalent of turning a knob) can produce large changes in the output signal; thus, the small signal is amplified.

In the digital mode, the signal at the control terminal of the transistor controls the state of the switch. A change in the input is the electrical equivalent of throwing a lever—the input controls whether or not current can pass through the transistor.

Any transistor can operate as either an amplifier or a switch; it depends only on the surrounding circuitry. We leave the discussion of circuit design to another course, but in this book, we focus on the physics of operation of the transistors themselves.

There are two major classes of transistors, based on the physics of their operation. These are the field-effect-transistors (FETs), which are discussed in this part of the book, and the bipolar junction transistors (BJTs), which are covered in Part 4. The origin of the names will become clear as we develop an understanding of how these devices work.

Interestingly, the field-effect transistor was invented first, but the bipolar junction transistor was the first be developed into a practical device. For many years, bipolar transistors predominated. More recently, however, FETs have surpassed BJTs in ease of fabrication and low cost, and currently most electronic circuits use FETs as the fundamental circuit elements.

There are many types of field-effect transistors, but they all have in common the element that an electric field across a *gate* structure controls the flow of current between the other two terminals, the *source* and the *drain.* The differences between the various types of FETs are primarily in the structure of the gate and mechanism used to apply the field.

The Generic FET

Before we begin to analyze specific types of field effect transistors, we examine a generic device to understand the basic principles of operation. A simplified perspective view of a field effect transistor is shown in Figure III.1. There are three terminals, called the source (S), drain (D), and gate (G). There is an electrically conducting *channel* extending from the source to the drain. The gate is above but electrically isolated from the channel. As we will see, the voltage on the gate terminal with respect to the source, V_{DS}, is used to control the passage of carriers through the channel by varying the field in the insulating region. The mechanism of this control depends on the particular type of FET, so for now we just show the gate electrode symbolically.

In Figure III.1a the source, channel, and drain are n type and are fabricated in a p-type semiconductor (substrate). This device is referred to as an n-channel field-effect transistor, or NFET. In the p-channel device (PFET) in Figure III.1b the source, channel and drain are p type, and the substrate is n type.

Note that in both cases a pn junction exists between the FET and the substrate. In practice, this junction must never be forward biased (often the substrate is connected to the source). This is because the current should flow between the drain and the source, not into the substrate. In addition, if multiple FETs are put

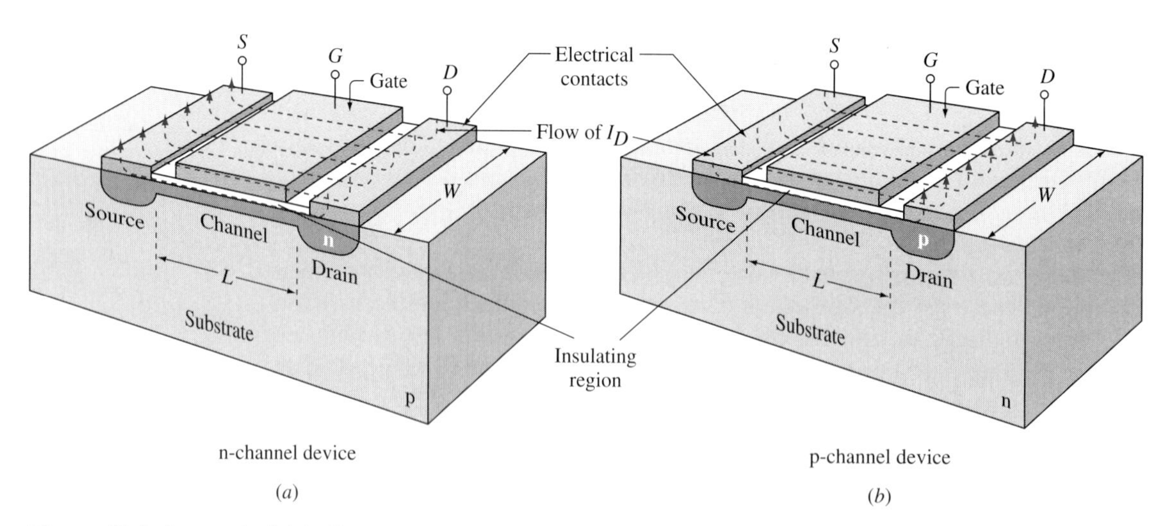

Figure III.1 A generic field-effect transistor contains a source *S*, a drain *D*, and a channel controlled by a gate electrode *G*. Since this is a generic device, the gate is shown only symbolically. Specific gate structures will be discussed later. (a) An n-channel device; (b) p-channel FET. Direction of conventional current flow is in the direction of the dashed arrows. Here *W* is the channel width and *L* is the channel length.

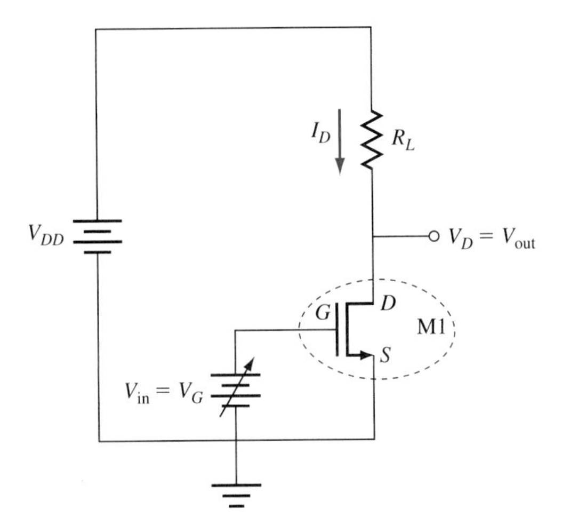

Figure III.2 Simple (inverter) circuit for an NFET. The input voltage V_G controls the channel current I_D and the output voltage V_D.

onto the same substrate, as long as the junctions are not forward biased, then in effect the FETs are all electrically isolated from each other.

We will see later that virtually no dc current flows into or out of the gate. The application of a voltage on the gate produces an electric field that affects the conductance of the channel, but the gate itself does not conduct.

Since the carriers (electrons or holes) cannot flow into the substrate and cannot flow through the gate, they are confined to the channel, and under appropriate bias conditions, these carriers can flow between the source and the drain, producing a current in the channel. We refer to this current as I_D, the current at the drain, but it is also the channel current and the source current.[1] The dashed arrows in Figure III.1 indicate the path of the current flow. In an n-channel FET the channel carriers are electrons and the drain voltage with respect to the source, V_{DS}, is positive. Electrons flow from source to drain, so the direction of current is from drain to source, as shown in the figure.

In a p-channel FET, the carriers are holes and the drain-to-source voltage is negative. Therefore, the holes flow from the source to the drain. The current I_D by convention is defined as positive going from drain to source and is therefore negative.

An example of a FET used in a simple circuit is shown in Figure III.2. The FET, M1, is an n-channel metal-oxide-semiconductor field-effect transistor (MOSFET) discussed in more detail in Chapter 7. The drain supply voltage V_{DD}

[1]Of course, some leakage current does flow between channel and gate and between channel and substrate, but this leakage current is normally small compared with the channel current between drain and source.

is in series with the load resistor R_L and the channel connects the drain D and source S. Recall that the conductivity of a semiconductor is controlled by the concentration of carriers. Since in this illustration, the source is at ground (zero) potential, the source voltage $V_S = 0$, the gate to source voltage $V_{GS} = V_G$, and the drain-to-source voltage $V_{DS} = V_D$. The electron concentration in the channel is controlled by the gate (input) voltage $V_{\text{in}} = V_G$. The output voltage of the circuit is $V_{\text{out}} = V_D$. From Kirchhoff's voltage law,

$$V_{\text{out}} = V_{DS} = V_D = V_{DD} - I_D R_L$$

For V_G such that $I_D = 0$, $V_{\text{out}} = V_{DD}$. For V_G such that I_D is very large, $V_{\text{out}} \approx 0$. (We will see later that when the drain current I_D is very large, the voltage between the drain and the source is small.) For intermediate values of V_G, $0 < V_{\text{out}} < V_{DD}$.

Figure III.3a shows the typical electrical characteristics of an NFET. Here the current through the channel, I_D, is plotted as a function of the voltage across the channel V_{DS}, for various values of the controlling gate voltage V_{GS}. There are three regions of operation: the sublinear region, the saturation region, and the subthreshold region. The subthreshold region is the horizontal line marked $V_{GS} \leq V_T$. We define the threshold voltage V_T as the value of V_{GS} required to initiate a given current flow through the channel, often taken (in the current saturation region) as $I_{D\text{sat}} = 40W/L$ nA. For all gate voltages below this

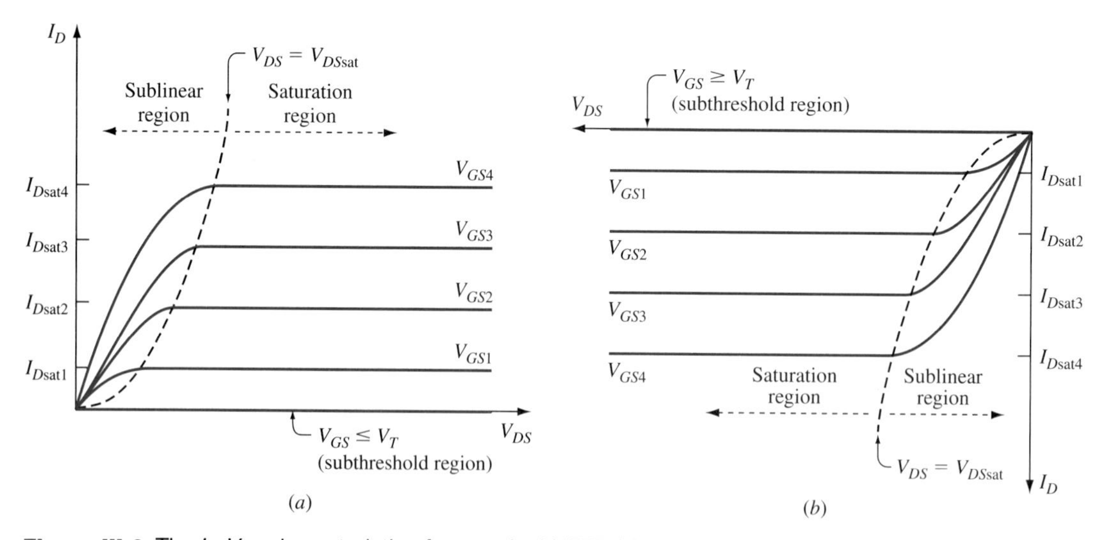

Figure III.3 The I_D-V_{DS} characteristics for a typical NFET (a) and PFET (b) for five values of V_{GS}. The dashed line separates the sublinear and saturation regions at $V_{DS} = V_{DS\text{sat}}$. In the NFET, the drain voltage and drain current are positive, and the gate voltage must be higher than the threshold voltage V_T for appreciable current to flow. In the PFET, the drain voltage and drain current are negative, and the gate voltage must be lower than the threshold for current to flow.

threshold, I_D is small and often considered to be zero, regardless of the value of the drain-source voltage V_{DS}. That means all the I_D-V_{DS} curves for $V_{GS} < V_T$ lie close to the horizontal axis.

When the gate voltage V_{GS} is above threshold, current can flow. In an NFET, for $V_{GS} > V_T$ and V_{DS} positive, electrons flow from source to drain, or current flow is from drain to source.

Next, look at one of the curves on the plot. Notice that for a given V_{GS} above threshold ($V_{GS} > V_T$), as V_{DS} increases, the I_D-V_{DS} relation is sublinear and eventually I_D saturates. The value of saturated I_D is called $I_{D\text{sat}}$, while the value of V_{DS} at which the current saturates is $V_{DS\text{sat}}$. These are both functions of the gate voltage V_{GS} and the properties of the FET.

The region for $V_{DS} > V_{DS\text{sat}}$ is referred to as the *current-saturation region,* or simply the *saturation region.* The region for $V_{DS} < V_{DS\text{sat}}$ is variously called the *linear region* (although it is sublinear), the *triode region,* or simply the *sublinear region.* The dashed line in Figure III.3a indicates the division between these two regions. For higher gate voltages, the value of $I_{D\text{sat}}$ and $V_{DS\text{sat}}$ are both larger.

Typical characteristics for a PFET are shown in Figure III.3b. Here holes flow (and thus current flows) from the source to the drain for negative values of V_{DS}. For the PFET, the gate voltage is less than (more negative than) the threshold voltage for a finite I_D.

In Part 3 of this book, we examine the *I-V* characteristics in some detail. As an introduction here, however, let us consider a geographical analogy to help understand why the curves are shaped as they are. Suppose we have two deep lakes, connected to each other by a shallow canal as illustrated schematically in Figure III.4a. The bottom of the lake system is somewhat analogous to potential energy for electrons as a function of position while the depth of the water represents the electron concentration. The lake on the left (S, representing source) is considered to be at constant depth. Four cases for the lake on the right (D, drain)

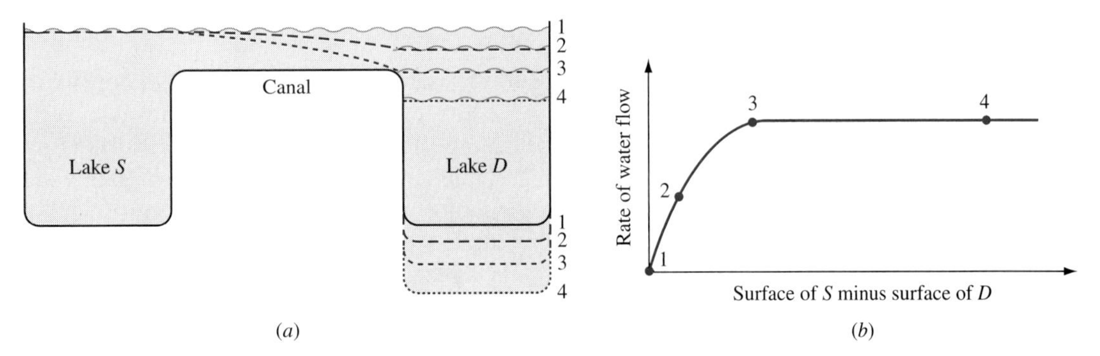

Figure III.4 Lake analogy to FET operation. (a) Two lakes are connected by a canal. The water level of lake S is constant. Four values of water level (and depth) of lake D are shown. (b) The corresponding rates of water flow.

are shown. In case 1, the surface of D is the same height as the surface of S and no water flows between the lakes. This is indicated by position 1 in Figure III.4b, which shows a plot of water flow versus the difference in surface levels of the two lakes. When the surfaces are equal, the flow is zero. In case 2, the surface of D is slightly below that of S and so some water flows from S to D. Case 3 represents the situation in which the surface of D is at the same height as the bottom of the canal. The slope of the water surface in the canal is increased and so is the rate of flow. For the surface of D below the bottom of the canal (case 4), the slope of the surface in the canal is not affected. Since the rate of water flow is determined by the slope of the water surface in the canal, the rate of flow saturated at its value at 3. As the level of surface D continues to drop, the rate of water flow through the canal stays the same.

Transistors in Circuits

Most of Part 3 of this book will be devoted to understanding and deriving the shapes of the I_D-V_{DS} characteristics of FETs. First, though, we will briefly investigate how these transistors can be used as amplifiers and as switches.

We said earlier that in a digital circuit, changing the gate voltage would be like throwing a switch. Figure III.5a shows an NFET inverter circuit with a resistive load. This is the circuit of Figure III.2 repeated. Suppose that a positive voltage supply V_{DD} is applied to the drain through a load resistor R_L. If the input gate voltage V_{GS} is below threshold (logic low), little current can flow through the channel—the switch is open circuited. Since $I_D \approx 0$, the voltage drop across the resistor $I_D R_L$ is also zero and the output voltage is $V_{DS} \approx V_{DD}$ (logic high) as indicated at the "switch open" position in Figure III.5b. If the gate voltage is changed to a value above threshold (for the NFET), then current flows across the channel—the switch is closed. The amount of current that can flow is now determined by the external circuit as well as by the transistor. The circuit determines the *load line* (dashed in the figure). The resistance in the external circuit causes some voltage drop in V_{DS} as I_D increases. The output voltage is then $V_{DS} = V_{DD} - I_D R_L$.

For the analog case, the gate voltage is kept above threshold, but is varied by a small amount within some range, Figure III.5c. As the gate voltage varies, the current I_D and the voltage V_{DS} also vary proportionally. Usually the variation in the gate voltage is quite small (the difference in gate voltage from one curve to the next in the figure may be a fraction of a volt), while the variation in V_{DS} is considerably larger—perhaps about a volt. Hence, the transistor acts as an amplifier, magnifying the small change in V_{GS} to a large change in V_{DS}.

To optimize the transistor design, then, the engineer will want to have a thorough understanding of the I_D-V_{DS} characteristics and how to shape them. Here in this introduction, we will outline the basic approach for analyzing the drain current as a function of the gate and drain voltages. Then, in Chapter 7 we execute and refine that approach.

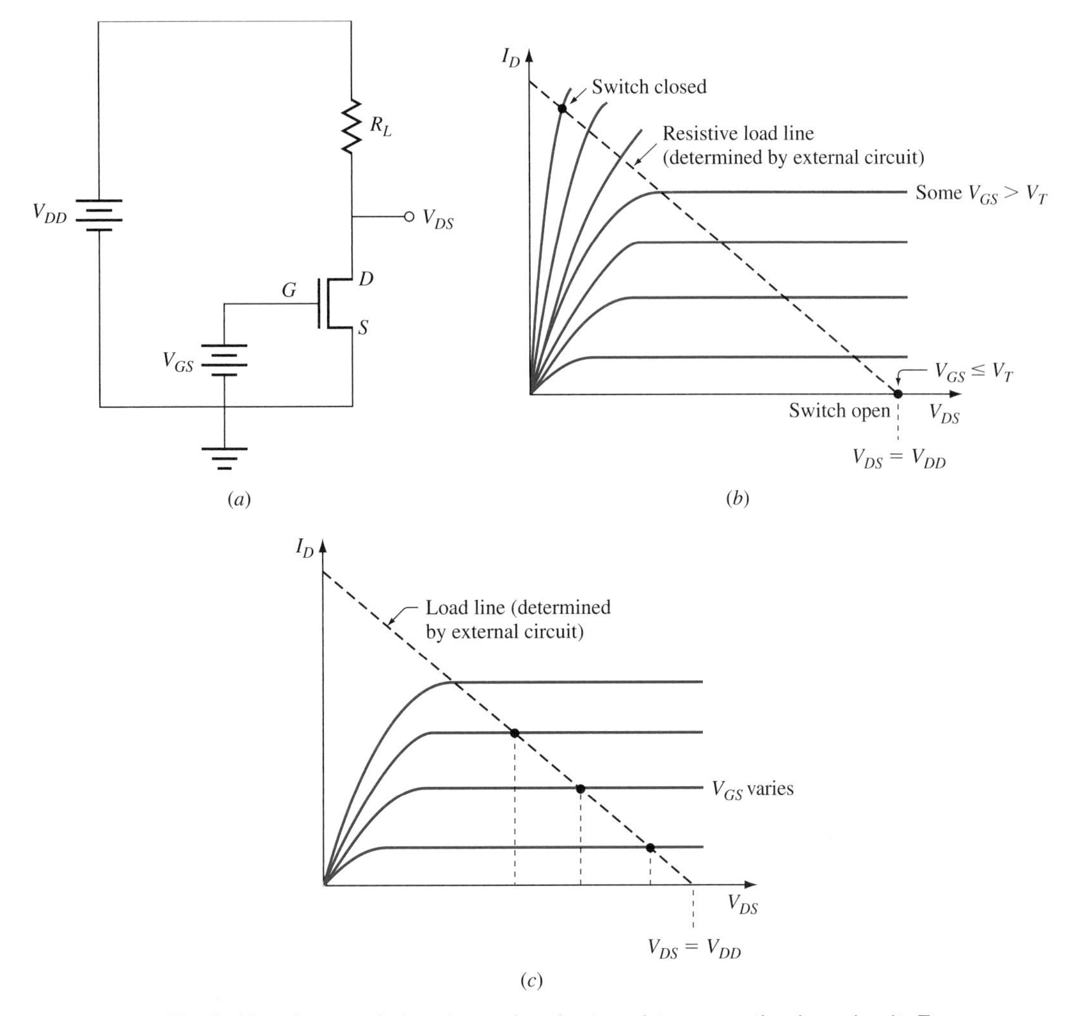

Figure III.5 The I_D-V_{DS} characteristics determine the transistor operation in a circuit. For example, (a) shows a digital circuit, an inverter. In digital operation, (b) the gate voltage is switched between two values, one above threshold and one below. Current is either near zero, or some value determined by both the transistor (solid lines) and the circuit (dashed line). In an analog circuit (c), the gate voltage remains above threshold, but varies. As V_{GS} changes, the current through the transistor also changes, as well as the voltage V_{DS}.

The Basis for Deriving the I_D-V_{DS} Characteristics of a FET

To begin our understanding of the drain current, we recall from elementary physics that current is defined as the amount of charge passing through a given area per unit time. We consider the case of an n-channel FET, Figure III.6. In an NFET, electrons carry the current by moving in the positive y direction. We take the x coordinate to be downward, across the channel.

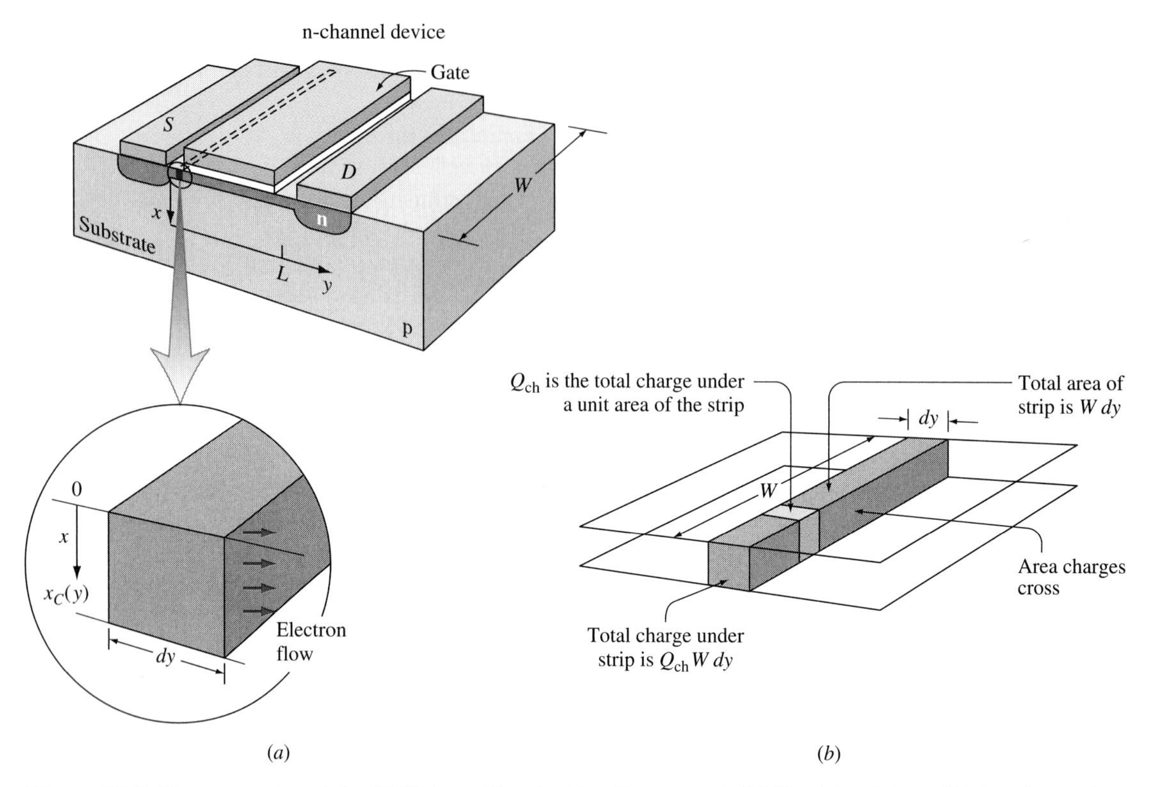

Figure III.6 The geometry of the NFET used for deriving the current. (a) The big picture; (b) the channel charge is the charge per unit area under the strip.

Let us consider the current flowing across the shaded region. Suppose there are n charges per unit volume in the channel where $n = n(x, y)$. The charge per unit area in the channel at position y is

$$Q_{ch}(y) = -q \int_0^{x_C(y)} n(x, y)\, dx \tag{III.1}$$

and the integral is taken across the channel depth, x_C as indicated in Figure III.6a. In an incremental length dy at position y in the channel, the total channel charge is

$$(\text{Channel charge in } dy) = Q_{ch}(y) W\, dy \tag{III.2}$$

Since the current is equal to the charge passing a given area per unit time, the channel current I_D at a given y is

$$I_D = -W Q_{ch}(y) v(y) \tag{III.3}$$

where $v(y)$ is the average channel electron velocity at position y. Note that in Equation (III.3), the velocity v is positive, the charge Q_{ch} is negative, and the negative sign ensures that I_D (from drain to source) is a positive quantity. Note also that I_D is independent of y, so $Q_{\text{ch}}(y)$ and $v(y)$ are inversely proportional.

Equation (III.3) is the fundamental equation for current flow in all FETs and is the starting point for deriving the electrical characteristics for any given structure. What is needed, then, are analytical expressions for $Q_{\text{ch}}(y)$ and $v(y)$ as functions of applied voltages for a particular FET structure. Once these parameters are modeled, the mathematical analysis is similar for all classes of FETs.

Note that the charge per unit area in the channel, Q_{ch}, Figure III.6b, is analogous to the quantity of water per unit area in the canal of Figure III.4a; i.e., the amount of water per unit volume times the depth of the water. The term v is the average velocity of the water flow at position y. From the water analogy, it is clear that the depth of the water varies along the canal, and similarly the amount of charge available for conduction varies with position in a FET.

In general, current in the channel flows by a combination of drift and diffusion. In practice, however, we are primarily concerned with the operation of FETs in which drift current predominates. That occurs when the density of charge in the channel is so great that the drift current is large compared with the diffusion current.

The drift current is driven by the electric field along the channel. We call this the longitudinal field $\mathscr{E}_L$ shown in Figure III.7a. In an n-channel device, the drain is more positive than the source. The longitudinal electric field then, is in the negative y direction. The electrons are accelerated toward the positive terminal, so their velocity is in the opposite direction to the field. Thus for electrons we have

$$v(y) = -\mu(y)\mathscr{E}_L(y) \tag{III.4}$$

where $\mu(y)$ is the average channel mobility (a positive quantity) at position y. Substituting into Equation (III.3), we find the channel current is

$$I_D = WQ_{\text{ch}}(y)\mu(y)\mathscr{E}_L(y) \tag{III.5}$$

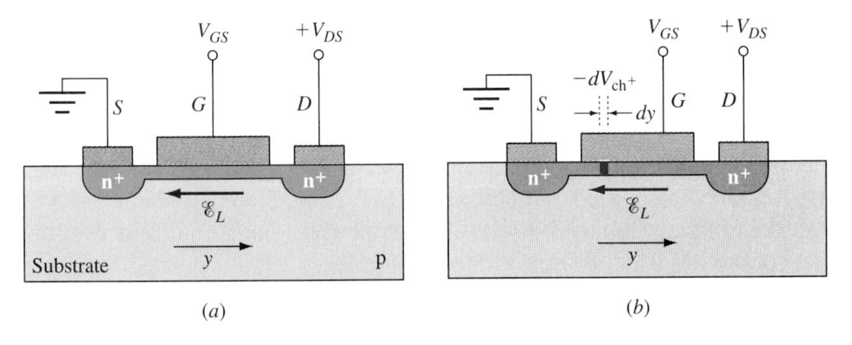

Figure III.7 (a) The longitudinal field $\mathscr{E}_L$ for an NFET. (b) The incremental channel voltage dV_{ch}.

Electric field is, by definition, $\mathscr{E} = -dV/dy$. The incremental voltage across the channel increment dy is dV_{ch}, as shown in Figure III.7b. Then

$$\mathscr{E}_L = -\frac{dV_{\text{ch}}}{dy} \tag{III.6}$$

Equation (III.5) can then be expressed in the form

$$I_D = -WQ_{\text{ch}}(y)\mu(y)\frac{dV_{\text{ch}}(y)}{dy} \quad \text{or} \quad I_D\,dy = -WQ_{\text{ch}}(y)\mu(y)\,dV_{\text{ch}}(y) \tag{III.7}$$

We will use this equation as the starting point for deriving the electrical characteristics of all FETs.

The organization of Part 3 is as follows. Chapter 7 is limited to the dc characteristics of MOSFETs. As we study the static characteristics, we will also come to understand the principles of operation of field-effect transistors. Chapter 8 presents additional material on MOSFET devices, including transient effects and small-signal equivalent circuits. Further, modern devices have very short channels—appreciably less than 1 μm. This makes the devices operate faster, but it affects the physics, too. The models that we develop in Chapter 7 will have to be modified to describe these "short-channel effects." Therefore, a part of Chapter 8 is devoted to examining short-channel effects and their effects on the I_D-V_{DS} characteristics of modern field-effect devices. Three other classes of FETs are also discussed briefly in Chapter 8: the heterojunction field-effect transistor (HFET), the metal-semiconductor field-effect transistor (MESFET), and the junction field-effect transistor (JFET). This material is largely qualitative.

In Chapter 7, the operation of MOSFETS is discussed for an n-channel MOSFET as a model. The results are readily adapted for p-channel MOSFETs. However, most circuits use a combination of NMOS and PMOS devices fabricated in a single substrate. A circuit containing both NMOS and PMOS devices is referred to as a CMOS (complementary MOS) circuit. Figure III.8 illustrates a CMOS inverter, a common structure used in digital circuits. It consists of a p-type substrate in which an NMOS is fabricated, while the PMOS resides in an n-type region within the p substrate.[2] Note that the gates of the two FETs are connected electrically, as are the two drains. The circuit diagram of this inverter is shown in Figure III.9a, while input and output voltages are indicated in Figure III.9b. The input voltage applied to the gates varies between zero volts and V_{DD} (the supply voltage) while the inverted output varies between V_{DD} and zero. Note that in this circuit the source of the NMOS is at ground (zero) potential and the source voltage of the PMOS is V_{DD}. The operation of this switch is discussed further in Chapter 8.

---

[2]This arrangement is called n-well technology.

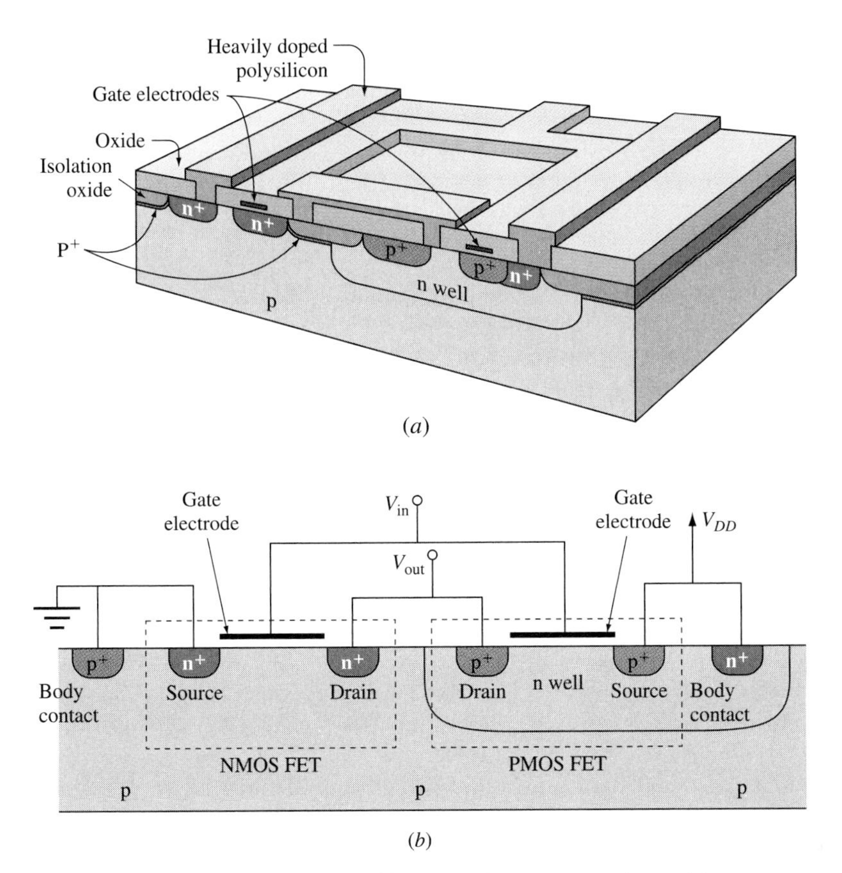

Figure III.8 The CMOS inverter. (a) Physical structure, adapted from C. G. Fonstad, *Microelectronic Devices and Circuits,* McGraw-Hill,1994; (b) cross-sectional diagram, adapted from R. C. Jaeger, *Microelectronic Circuit Design,* McGraw-Hill, 1997.

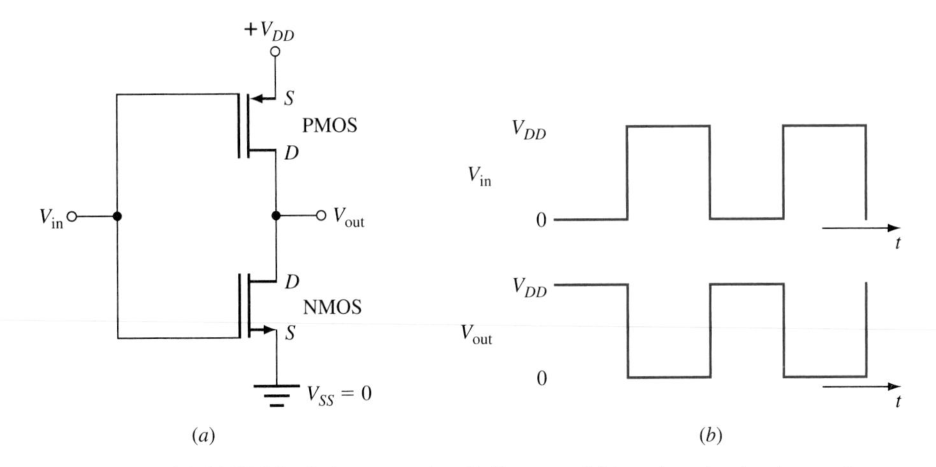

Figure III.9 (a) CMOS logic inverter circuit diagram; (b) input and output waveforms.

Supplemental information on FETs is presented in the Supplement to Part 3. The gate-substrate capacitance is discussed to indicate how it can be used as a fabrication diagnostic tool. Some degradation mechanisms of MOSFETs are briefly discussed. Variations of the MOSFET structures—dynamic random-access memories (DRAMS) and charge-coupled devices (CCDs)—are qualitatively described. Applications of SPICE to MOSFETs are also briefly discussed. ■

CHAPTER 7

The MOSFET

7.1 INTRODUCTION

In this chapter, we discuss the basic operation of the most important class of FETs: The Si-based insulated-gate field-effect transistor (IGFET). The gate material in these devices was originally a metal (aluminum) and the insulator was silicon dioxide (SiO_2). That is the origin of the term metal-oxide-semiconductor field-effect transistor, or MOSFET. For ease of fabrication and reproducibility, however, the current practice is to use degenerately doped polycrystalline Si (poly Si), which is highly conductive, instead of metal for the gate. In n-channel devices, n^+ poly-Si is used for the gates, and p^+ poly-Si is used for p-channel devices. Silicon dioxide is still usually used for the insulator, although nitrogen is sometimes incorporated into the SiO_2 to increase its dielectric constant (and thus device performance). While the term IGFET is a more accurate description, in this book we adopt the practice common in the industry, which is to use the more common term MOSFET to describe this class of devices.

In this chapter, we discuss the MOSFET fundamentals involved in the static electrical characteristics. In Chapter 8, we will cover the short-channel effects important in modern MOSFET devices, the time-dependent characteristics, and in addition, we discuss the other types of field-effect transistors, including the CMOS, HFET, JFET, and MESFET.

7.2 MOSFETs (QUALITATIVE)

In this section, we explore, qualitatively, the basic principles of operation of the MOSFET. To illustrate how the MOSFET channel is formed, a brief qualitative description of MOS capacitors is presented. A more detailed description of MOS capacitors appears in the supplement to FETs.

7.2.1 INTRODUCTION TO MOS CAPACITORS

In this section we consider an ideal MOS capacitor (MOSC) with the structure of Figure 7.1a. This capacitor consists of a degenerate n^+ Si gate, a thin layer of (insulating) SiO_2 separating a gate from a p-type substrate.

Figure 7.1b shows the cross section of the MOSC. The energy band diagram along the cut A-A′ slicing through the gate in the x direction is shown in (c) and (d). We determine the energy band diagrams by following the procedure used in Chapter 6 for heterojunctions. In this case, however, we have two heterojunctions,

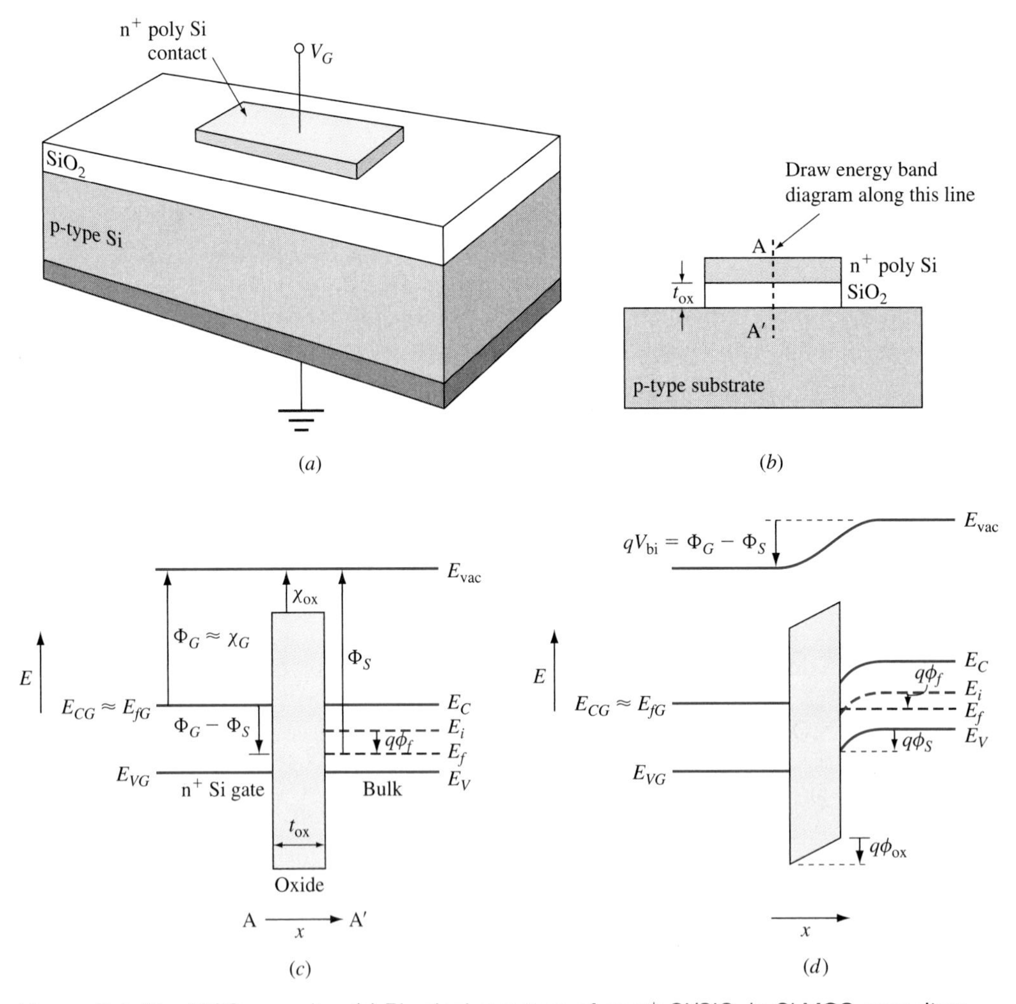

Figure 7.1 The MOS capacitor. (a) Physical structure of an n^+-Si/SiO_2/p-Si MOS capacitor; (b) cross section; (c) the energy band diagram under charge neutrality; (d) the energy band diagram at equilibrium (note that the surface of the p-type substrate near the oxide interface has become weakly inverted).

one between the polysilicon gate and the insulator, and another between the insulator and the semiconductor substrate.

We assume electrical neutrality in every macroscopic region to start, in which case E_{vac} is constant. The result is shown in Figure 7.1c. Here Φ_G and Φ_S are the work functions[1] (in eV) of the gate and the semiconductor respectively.[2] Note that Φ_G is approximated as being equal to χ_G because the gate material is degenerately doped and its Fermi level in the gate is thus near the conduction band edge. The quantity ϕ_f is the potential difference (in volts) between the Fermi level and the intrinsic level. The uppercase Φ's and the χ's are energies, expressed in eV, and the lowercase ϕ's are potentials, expressed in volts. On the energy band diagrams, the ϕ's will always be expressed as $q\phi$, in energy units.

To achieve equilibrium, electrons flow (through an external circuit) from the gate to the semiconductor substrate, causing the Fermi levels to line up. The resulting (equilibrium) energy band diagram is shown in Figure 7.1d.[3] We see that the bands bend, and a built-in voltage results. Some of the voltage is dropped across the oxide and some across the semiconductor. Furthermore, the semiconductor now has a depletion region near its surface—it is depleted of majority carriers, which are holes in this case, since the material is p type.

The total built-in voltage is

$$V_{\text{bi}} = -\frac{1}{q}|\Phi_G - \Phi_S| = \phi_{\text{ox}} + \phi_s \tag{7.1}$$

The voltage drop across the oxide is ϕ_{ox}. The voltage across the Si depletion region, often referred to as the surface potential (i.e., the voltage at the Si surface relative to that in the neutral bulk), is designated as ϕ_s. Notice that the electric field ($\mathscr{E} = (1/q)(dE_{\text{vac}}/dx)$) is discontinuous at both interfaces of the insulator. This is caused by the difference in the permittivities of the materials (Gauss's law, $\mathscr{E}_1\varepsilon_1 = \mathscr{E}_2\varepsilon_2$). The fraction of the built-in voltage appearing across the oxide increases with increasing oxide thickness and with increasing doping level in the Si.

Notice that for this example, at equilibrium (Figure 7.1d), the Fermi level actually crosses the intrinsic level. This means that, while the substrate is *doped* p type, near its surface it is *effectively* n type. The Fermi level near the surface is closer to the conduction band edge than to the valence band edge. At the interface, there is a higher concentration of electrons than holes because of the band

[1]Since the original gate material in a MOSFET was a metal, the term Φ_M is also often used for the work function of the gate.

[2]Note that the gate material is degenerately doped to a degree such that band-gap narrowing (Chapter 4) occurs. This causes the conduction band edge of the gate material, E_{CG}, to be at a slightly lower energy than that of the substrate, E_{Csub}. The gate material band gap is slightly smaller than the substrate band gap. This band-gap shrinkage is, however, normally less than 0.1 eV and we ignore it. For simplicity, we also ignore any charge trapped in the oxide or at the SiO_2/Si interface. These considerations are discussed in Supplement 3.

[3]Here the band lineup at the interface is obtained by the electron affinity model.

bending. The Si surface region is not only depleted, it is *inverted*. We note here that the region consisting of the conductive gate, the insulating oxide, the depletion region and the substrate can be considered to be a capacitor with two different dielectric layers between the electrodes (those of the oxide and of the depletion region), with two different dielectric constants. Note that the built-in voltage is divided between oxide and substrate. If a voltage is applied between gate and substrate, it too will be divided between oxide and substrate and the resultant energy band diagram and charges will be altered.

Figure 7.2 shows the energy band diagrams and charge distributions for various values of gate-substrate voltage V_G for a n^+ Si/SiO_2/p-type semiconductor. In (a), the case for equilibrium is indicated. Because electrons from the gate transfer to the substrate, with the resultant creation of a depletion region in the Si at the Si/SiO_2 interface, the gate charge is positive and the Si depletion region charge is negative as indicated.

If a negative voltage is applied to the gate, this voltage is divided between oxide and semiconductor. The bands in the semiconductor bend up, causing an accumulation of holes in the semiconductor at the oxide interface (b).

If a step voltage (e.g., 2 V) is applied to the gate, a depletion region will be established in the Si within about 3 dielectric relaxation times ($\sim 10^{-12}$ s) resulting in the energy band and charge distribution indicated in (c). However, with time, electrons will be thermally excited into the conduction band where they will be trapped in the potential energy well at the interface. This negative trapped charge (Q_i) raises the energy band in this region until the steady-state condition is reached as indicated in (d), a process that requires on the order of 10 to 100 ms in Si. In steady state, electron generation and recombination rates are equal. Note that in the steady state, the depletion region width is independent of the dc voltage, or Q_B is constant. The charge dependent on voltage resides in the interface charge Q_i.

The capacitance-voltage characteristic of a MOS capacitor (MOSC) is an important diagnostic tool for monitoring the fabrication of MOSFETs. Here we present a brief description of the MOSC *C-V* characteristics. A more detailed description is presented in Supplement 3.

A simple circuit for measuring the *C-V* characteristics of a MOSC is shown in Figure 7.3a. A dc voltage V_{dc} and a small-signal ac voltage v_{ac} are applied between gate and substrate, and the ac current i_{ac} is measured. The ratio $i_{ac}/v_{ac} = 2\pi f C$, and C can be calculated.

A typical *C-V* plot is indicated in (b). Recall that the (differential) capacitance is

$$C = \frac{dQ}{dV} = \frac{\varepsilon A}{W}$$

where dQ is the variation of charge on either side of the oxide with a change dV and W is the spacing between the regions of changing charge. For a reverse dc bias (Figure 7.3b) a small change in dV causes a small change in dQ on either

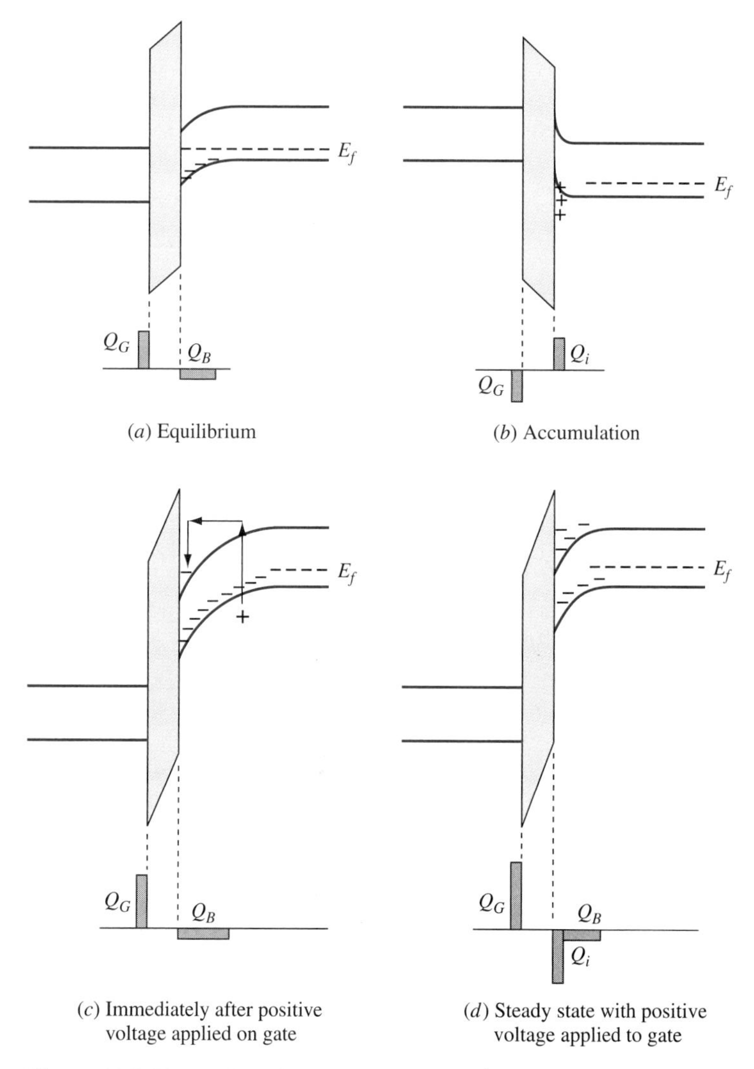

Figure 7.2 Energy band diagrams for the n^+-Si/SiO_2/p-Si capacitor of Figure 7.1 along with the charge distributions for three bias conditions. In (a) the case for equilibrium is indicated. Electrons from the n^+ gate transfer to the p-Si substrate, resulting in a positive gate and a negative depletion region in the substrate. The accumulation condition is indicated in (b). Here a negative voltage is applied to the gate with respect to the substrate such that holes accumulate at the silicon–to–silicon dioxide interface. The situation for a positive 2-V step voltage is shown in (c) immediately after the application of the voltage. With time, electrons generated in the transition region are trapped in the potential well at the interface until steady state is reach as indicated in (d).

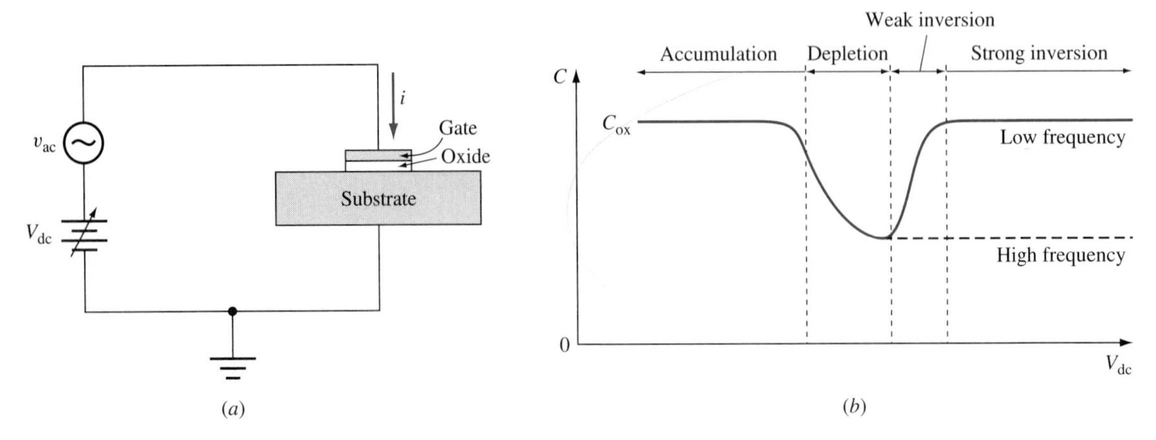

Figure 7.3 (a) Circuit for measuring the capacitance of a MOS capacitor. (b) Capacitance-voltage characteristic for a MOS capacitor at low and high frequencies.

side of the oxide and $C = C_{ox} = \varepsilon A/t_{ox}$. Thus for this condition, with the area of the gate known, the oxide thickness can be determined.

At intermediate voltages, a voltage-dependent depletion region exists in the semiconductor and the MOSC capacitance consists of two capacitors in series.

$$C = \frac{C_{ox} C_s}{C_{ox} + C_s}$$

where C_s is the semiconductor depletion region capacitance and $C < C_{ox}$.

At sufficient positive voltage, an inversion layer exists in the semiconductor and for steady-state dc, the depletion layer width is independent of V_{dc}.

An ac voltage creates charge at the edge of the depletion region. If the frequency is low enough (on the order of 1 to 10 Hz), those created electrons have time to enter the inversion region and the ac influence on the depletion region width is negligible.[4] Thus on the semiconductor side dQ is at the interface $dQ = dQ_i$ and again $C = \varepsilon A/t_{ox}$.

At a high-frequency ac voltage ($\sim$100 MHz), the generated electrons have insufficient time to enter the inversion region before the polarity changes, dQ is at the edge of the depletion region, the capacitance is constant, and $C = C_{ox}C_s/(C_{ox} + C_s)$. Here $C_s = \varepsilon A/w$, where w is the depletion region width and is a measure of the doping level in the substrate.

The MOS capacitance is treated in more detail in Supplement 3.

7.2.2 MOSFETs AT EQUILIBRIUM (QUALITATIVE)

A schematic of an n-channel MOSFET is shown in Figure 7.4. It resembles the MOS capacitor just discussed except that there are source and drain regions of n^+ Si at opposite sides of the gate region. Since electrical connections are made

[4] The low frequency voltage can be thought of as a slowly varying dc voltage which has no effect on the depletion region width.

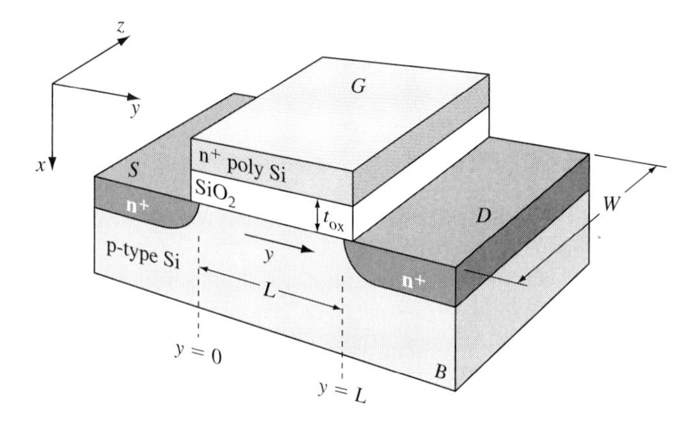

Figure 7.4 Schematic diagram of the structure of an n-channel silicon-based MOSFET. The channel width W, length L, and oxide thickness t_{ox} are shown. The symbols S, G, D, and B represent the source, gate, drain, and substrate (body) respectively.

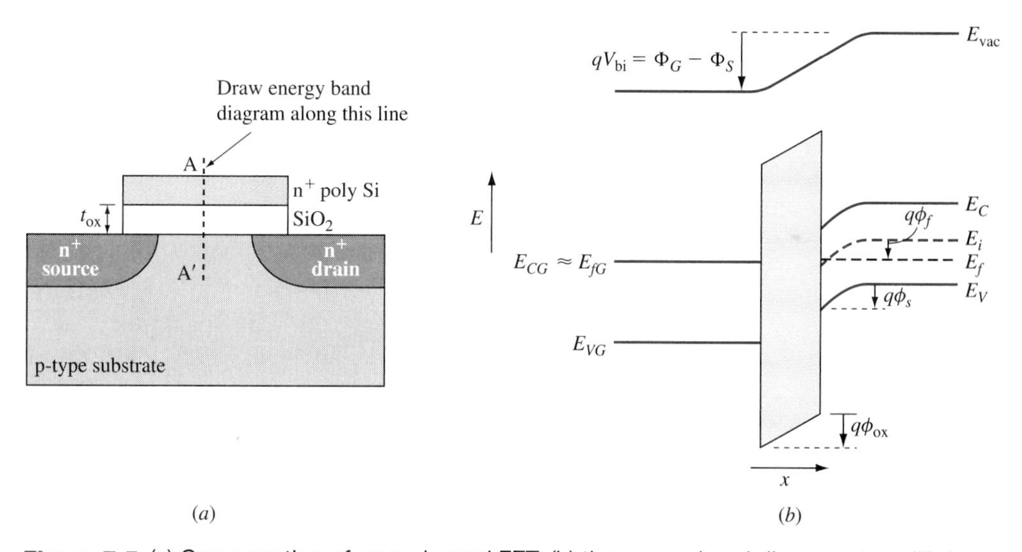

Figure 7.5 (a) Cross section of an n-channel FET; (b) the energy band diagram at equilibrium.

to gate, source, drain, and substrate, a MOSFET is a four-terminal device. Often, however, the substrate is connected to the source, and the MOSFET is then considered to be a three-terminal device. The symbols W and L represent the width and length of the channel.

Figure 7.5a shows the cross section of the MOSFET of Figure 7.4. The equilibrium energy band diagram along the cut A-A′ normal to the gate is shown in (b). From the figure, it appears as though there is no n-type channel from source to drain in this device. From the energy band diagram for this structure,

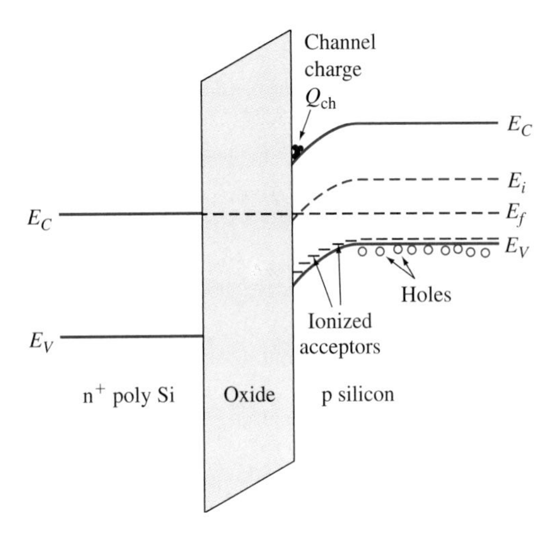

Figure 7.6 The channel charge accumulates in the bulk near the oxide interface. In this case the channel charge consists of electrons.

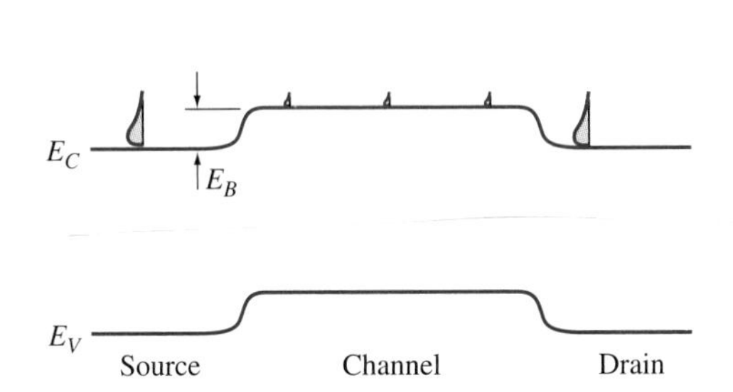

Figure 7.7 The energy band diagram along the channel of the device of Figure 7.4.

we can determine whether a channel in fact exists. For the specific example shown, the band bending in the substrate ($q\phi_s$) is about a half of an electron volt. The equilibrium energy band diagram of (b) is repeated in Figure 7.6, which also indicates the charge in the Si. It can be seen from the figure that the Fermi level is still close to the intrinsic level near the semiconductor-oxide interface. In other words, while a channel does exist, it has a low conductance, since it contains few electrons. Because the electron concentration in the channel is so low, for an applied drain-to-source voltage, only a miniscule current can flow between source and drain. This device is said to be in the *subthreshold* region.

The source and drain are doped n^+, but the induced channel is only weakly n type. The resulting difference in the electron concentration creates a modest (n^+n) potential energy barrier at either end of the channel. Figure 7.7 shows the equilibrium energy band diagram *along* the channel instead of across it. Since the electron density function $n(E)$ decreases exponentially with increasing energy, the barrier E_B is still large enough that only a small electron concentration exists in the channel. Thus, the induced channel is n type but the channel conductivity is negligible. In the next section, we will see how changing the gate voltage affects the barrier height and thus the channel conductance.

7.2.3 MOSFETs NOT AT EQUILIBRIUM (QUALITATIVE)

So far, we have considered the device to be at equilibrium. Now let us examine the physics of MOSFET operation. We will take the substrate to be connected to the source (a common arrangement). There are still two voltages that can be varied, the gate-source voltage V_{GS} and the drain-source voltage V_{DS}.

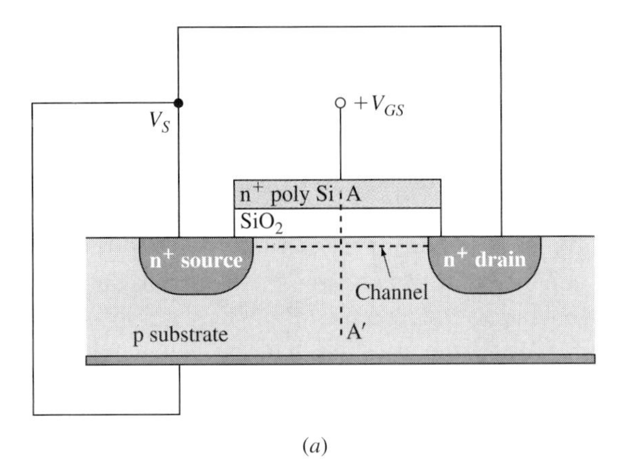

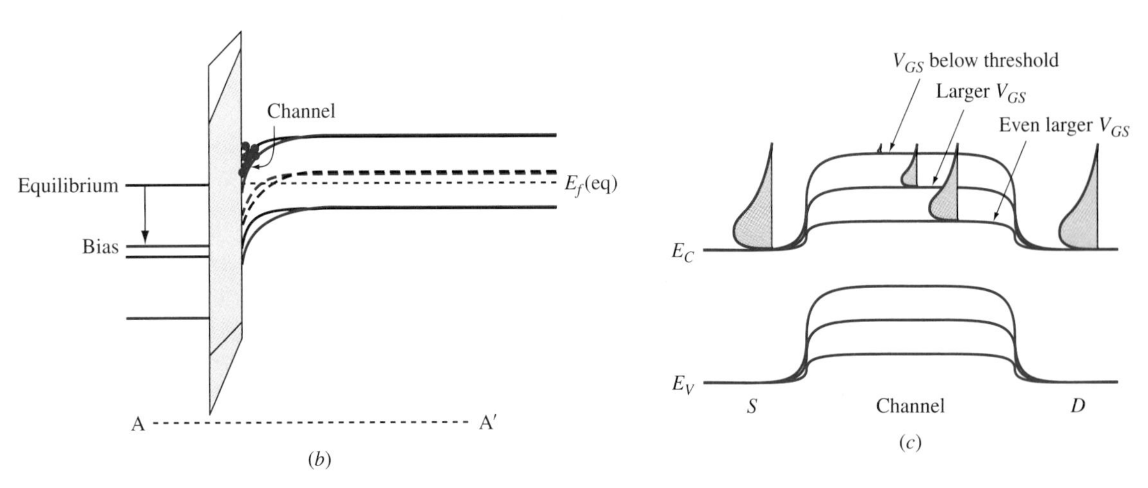

Figure 7.8 A particular MOSFET example, (a) with source, substrate, and drain connected; (b) the energy band diagram along cross section A-A′ at equilibrium (black) and under bias (colored); (c) the energy band diagram along the channel for three values of V_{GS}.

The Case for $V_{DS} = 0$ We begin by connecting the source to the drain electrically, such that those two terminals are at the same potential, as shown in Figure 7.8a. Let us then apply a voltage to the gate with respect to the source. This applied voltage is also divided between oxide and substrate, just as the built-in potential was in Figure 7.5. The effect of applying a positive gate voltage is to lower the channel energy (the conduction band edge), as shown in Figure 7.8b. Here we have drawn the energy band diagram along the line A-A′, perpendicular to the gate. The equilibrium diagram is in black, and the energy band diagram under bias is in color. Under bias, the conduction band edge bends down toward the Fermi level. The surface is now more strongly n type than before, and thus more strongly inverted. There are now more electrons in the channel, increasing its conductance. We say that the channel has been "enhanced."

Another way to look at it is via Figure 7.8c, which shows the energy band diagram along the channel for three different gate voltages. For V_{GS} near threshold, the energy barrier between source and channel at the Si surface, E_B, is fairly high. Few electrons appear in the channel and its conductance is low. As V_{GS} increases above threshold, the barrier decreases. More electrons are able to enter the channel, and thus its conductivity increases.

The channel charge Q_{ch} in the MOSFET channel is analogous to the interface charge Q_i in a capacitor. However, for the capacitor, Q_i is a result of thermal generation and recombination, a relatively slow process. For a MOSFET, the charge is determined by the barriers between source and channel and between drain and channel. The electrons enter and exit the channel by a combination of conduction and diffusion, typically on the order of 10^{-12} to 10^{-11} s, and is normally considered to be instantaneous.

Definition of Threshold We have indicated that, when the gate voltage is below some threshold—i.e., in the subthreshold region—the channel conductance is small. We look at the carrier distributions, and remember that the electron concentration in the semiconductor varies as

$$n = N_C e^{-(E_C - E_f)/kT} \tag{7.2}$$

The electron concentration and thus the conductance of the channel varies exponentially with $E_C - E_f$. This quantity is dependent on the gate-source voltage V_{GS}. An exponential is a smoothly (albeit rapidly) varying function, so it is not clear what value of gate-source voltage should be called *threshold.* A commonly used criterion is that the threshold voltage is the gate-source voltage required to induce an electron concentration at the Si surface that is equal to the hole concentration in the neutral substrate (N'_A).[5] This is equivalent to saying that a channel exists if the Fermi level in the n channel is as far above the intrinsic level as the Fermi level in the bulk is below the intrinsic level. This condition is shown Figure 7.9. Then

$$\phi_s = 2\phi_f \tag{7.3}$$

where ϕ_s is called the *surface potential,* i.e., the voltage at the Si surface relative to that in the bulk.

Enhancement- and Depletion-Type FETs In our earlier example (Figure 7.5), the band bending caused the surface to be n type even with no bias applied. Depending on the doping in the substrate, the surface of the semiconductor may or may not be inverted at equilibrium. If a channel does exist, it may or may not be strong enough to be considered conductive. In the earlier example, the surface was inverted at equilibrium but not enough to be considered a proper channel. With positive gate-source voltage, the channel conductance increased. As mentioned

[5] As indicated earlier, another often-used criterion is that V_T is the value of V_{GS} required to produce a saturation current $I_{D\text{sat}} = 40\, W/L$ nA.

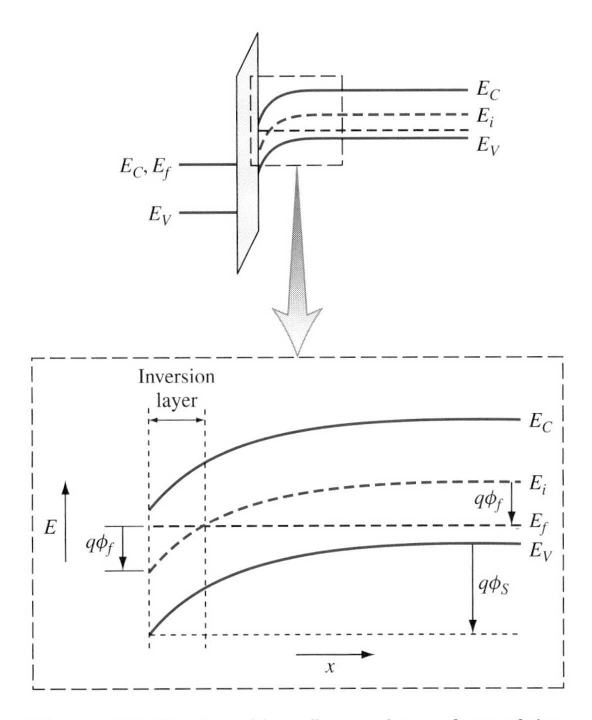

Figure 7.9 The band bending at the surface of the semiconductor. At threshold, the Fermi level is as far above the intrinsic level at the surface (left-hand edge) as it is below the intrinsic level in the bulk.

before, the conductance of the channel is *enhanced* by the application of a positive gate voltage. An *enhancement-type* FET is normally off. It does not conduct appreciably until a channel is created by the application of a gate voltage.

Figure 7.10a and b shows an enhancement-type NFET and an enhancement-type PFET (i.e., n-channel FET and p-channel FET respectively). In the NFET, a positive gate-source voltage must be applied for the transistor to conduct appreciably. For the PFET, the conduction band edge must be bent *upward* to more fully invert the surface such that more holes can enter the channel. This requires a negative voltage on the gate with respect to the source. Thus, for enhancement devices, the threshold voltage of an NFET is positive and the threshold voltage of a PFET is negative. Increasingly negative gate voltage in the PFET causes increasing channel conductance.

It is possible, however, for the device to be fabricated such that a good channel does exist even with no gate-source voltage applied, as shown in Figure 7.10c and d. In other words, a conducting channel exists even at equilibrium. These transistors are normally on. In these devices, one has to apply a gate-source voltage to decrease the band bending, deplete the channel, remove the carriers, and thus turn off the conduction. These are called *depletion devices*.

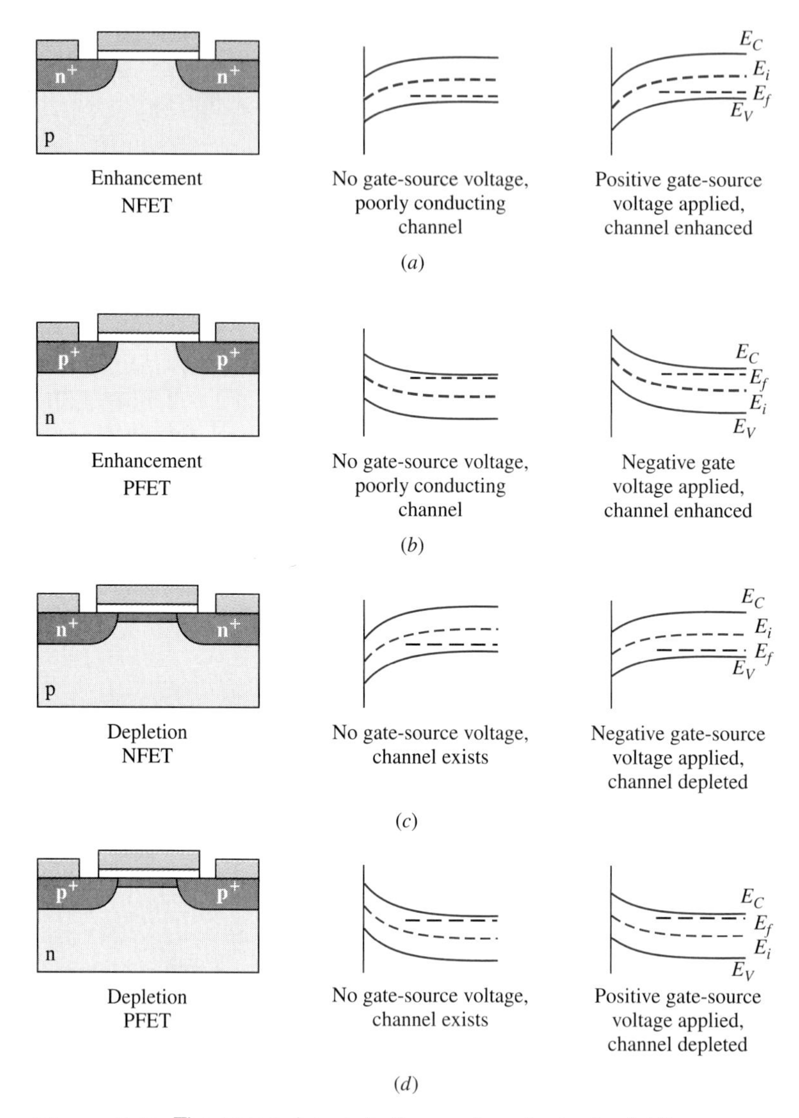

Figure 7.10 The energy bands in the semiconductor for (a) the enhancement NFET; (b) the enhancement PFET; (c) the depletion NFET; (d) the depletion PFET.

Various symbols are used to represent MOSFETs in circuit schematics. Figure 7.11 illustrates some common symbols; M1, M2, and M3 represent n-channel MOSFETs,[6] while M4, M5, and M6 represent p-channel MOSFETs.

The devices M1 and M4 represent NMOS and PMOS enhancement devices respectively. The broken line representing the channel between source and drain

[6]It is common practice to designate MOSFETs by the letter M.

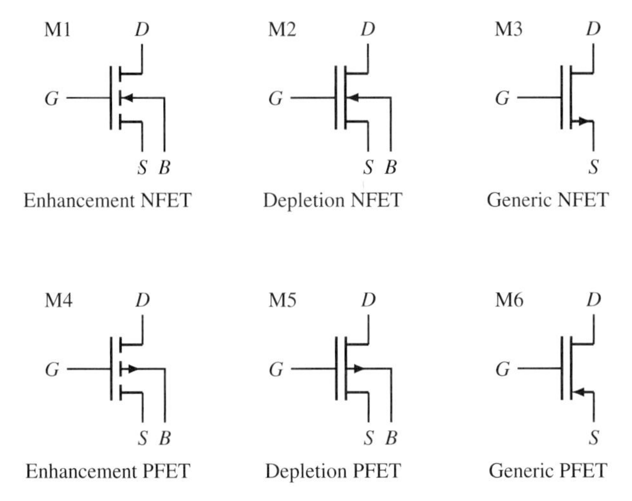

Figure 7.11 Schematic representations (circuit symbols) for MOSFETs. M1 is an enhancement-mode NFET, M2 is a depletion-mode NFET, and M3 can be used for either type. P-channel MOSFETs are represented by M4 (enhancement), M5 (depletion), and M6 (either mode).

indicate that no conducting channel exists for $V_{GS} = 0$. The arrow between substrate (also called the body, B) and channel points in the direction from p to n as for a diode. Depletion NMOS and PMOS devices are indicated by M2 and M5 respectively. The solid (nonbroken) channel indicates that a conducting channel exists for $V_{GS} = 0$.

Devices M3 and M6 represent NMOS and PMOS devices respectively. These symbols are often used where the substrate (body) is connected to the source. The arrow in the source indicates the direction of current flow. These symbols are used for both enhancement and depletion MOSFETs.

More About Threshold Let us look at the threshold conditions more closely. The concentration of electrons at the surface of the semiconductor is

$$n_s = N_C e^{-(E_{C\text{ch}} - E_f)/kT} \tag{7.4}$$

where $E_{C\text{ch}}$ is the energy of the conduction band edge in the channel at the Si surface, Figure 7.12. We see also that the barrier height is equal to $E_B = E_{\text{ch}} - E_f$, since in the heavily doped source $E_f \approx E_C$ (see Figure 7.12b). The concentration of electrons in the channel at the surface can be expressed as

$$n_s = N_C e^{-E_B/kT} \tag{7.5}$$

From this, we can solve for the barrier height:

$$E_B = kT \ln \frac{N_C}{n_s} \tag{7.6}$$

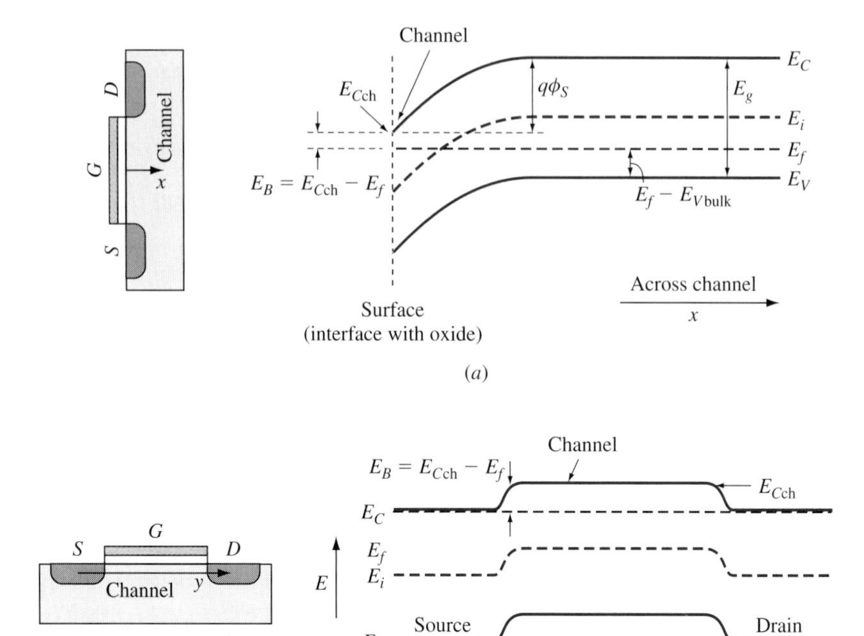

Figure 7.12 The energy band diagram of the NFET across the channel (a) and along the channel (b), with zero voltage between the drain and the source.

From Figure 7.12a, the surface potential ϕ_s can be written

$$\phi_s = \frac{1}{q}[E_g - E_B - \delta_p] \tag{7.7}$$

where δ_p is the energy difference between the Fermi level and the valence band edge in the bulk (neutral) Si.

According to Equation (7.6), E_B varies slowly (logarithmically) with n_s; therefore, from Equation (7.7), ϕ_s also varies slowly. Since ϕ_s is varying slowly, the approximation is often used that above threshold, the band bending ϕ_s remains equal to its threshold value. The gate voltage is dropped partly across the oxide and partly through the band-bending region in the semiconductor (the depletion region). Because the amount of band bending ϕ_s is (roughly) constant, any change in gate voltage above threshold is assumed to be dropped across the oxide. We write this as

$$V_{GS} - V_T = \phi_{\text{ox}} - \phi_{\text{ox}}^{\text{th}} \qquad V_{GS} > V_T \tag{7.8}$$

where $\phi_{\text{ox}}^{\text{th}}$ is the oxide voltage at threshold.

EXAMPLE 7.1

Show that the approximation in Equation (7.8) is valid. That is, how realistic is it that any additional gate voltage above threshold is dropped across the oxide and not the semiconductor?

■ **Solution**

Consider the n-channel MOSFET of Figure 7.4. Let the net substrate doping be $N_A' = 10^{16}\ \text{cm}^{-3}$. We know that threshold occurs when the electron concentration at the surface of the channel, n_s, is equal to N_A', because that is when the surface has the same number of electrons as the bulk has holes.

From Equation (7.6), we have

$$E_{B(\text{threshold})} = kT \ln \frac{N_C}{n_{s(\text{threshold})}} = (0.026\ \text{eV}) \ln \left(\frac{2.86 \times 10^{19}\ \text{cm}^{-3}}{10^{16}\ \text{cm}^{-3}} \right)$$

$$= 7.96kT = 0.207\ \text{eV}$$

The maximum value that n_s can realistically attain in silicon MOSFETs is about $2 \times 10^{18}\ \text{cm}^{-3}$, at which point

$$E_{B(\text{way above threshold})} = kT \ln \frac{N_C}{n_{s(\text{way above threshold})}} = (0.026\ \text{eV}) \ln \left(\frac{2.86 \times 10^{19}\ \text{cm}^{-3}}{2 \times 10^{18}\ \text{cm}^{-3}} \right)$$

$$= 2.66kT = 0.069\ \text{eV}$$

In other words, between threshold and way above threshold, the barrier height E_B and thus $q\phi_s$ vary by only about $5.3kT$, or 138 meV, at room temperature. From Equation (7.7), over this same range the surface potential ϕ_s varies by the value $\Delta E_B/q = 0.138\ \text{V}$.

From Equation (7.3), at threshold, $\phi_{s(\text{threshold})} = 2\phi_f$. Since $\phi_f = (E_i - E_f)/q$ and $E_i - E_f = kT \ln(N_A'/n_i)$, we have

$$\phi_f = \frac{kT}{q} \ln \frac{N_A'}{n_i} = 0.026 \ln \frac{10^{16}}{1.08 \times 10^{10}} = 0.357\ \text{V}$$

Therefore the band bending at threshold is

$$\phi_{s(\text{threshold})} = 2\phi_f = 2(0.357) = 0.714\ \text{V}$$

Now we find the band bending at a surface concentration of $2 \times 10^{18}\ \text{cm}^{-3}$, which is

$$\phi_{s(\text{way above threshold})} = 0.714 + 0.138 = 0.852\ \text{V}$$

or about 20 percent above its value at threshold. The voltage drop across the semiconductor, then, is not exactly constant above threshold but it is changing slowly. Therefore, the approximation that the surface potential ϕ_s is constant above threshold normally is adequate.

The Case for $V_{DS} > 0$ In the previous section, we looked at the effect of the gate-source voltage on the energy band diagram, barrier heights, and carrier concentrations in the channel. The carrier concentration in the channel relates directly to the conductance in the channel. We assumed there that both ends of the channel were at the same voltage.

In this section, we will allow the drain voltage to be different from the source voltage, thus producing a longitudinal electric field in the channel. This field will induce current to flow along the channel. Just as the current that flows through a resistor depends on the voltage across it, the current from the drain to the source depends on the drain-to-source voltage. In a MOSFET, however, the conductance (and thus resistance) of the channel depends on the gate-source voltage. In fact, this is the origin of the name *transistor*. The resistance across (trans) the device is controlled by the gate-source voltage. It is a voltage-controlled resistor.

Figure 7.13a shows an n-channel MOSFET. The drain is at a positive voltage V_{DS} with respect to the source. The gate has some applied voltage above

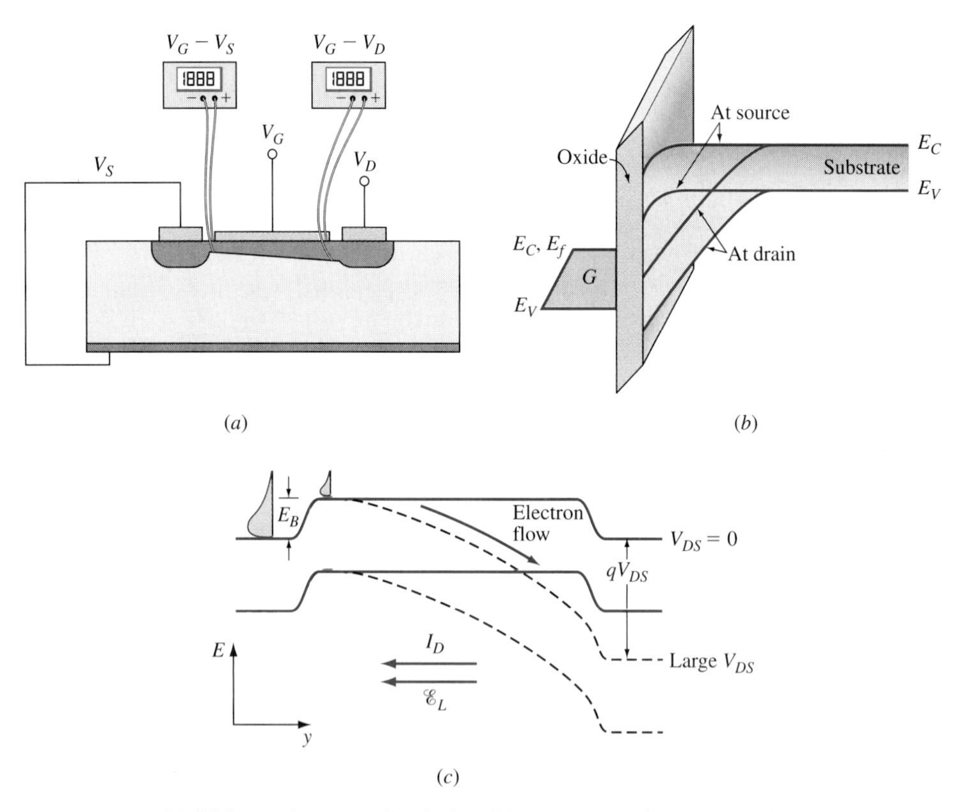

Figure 7.13 (a) With a voltage on the drain with respect to the source, the depletion region width varies along the channel. So does the voltage across the channel at any given point. (b) The energy band diagrams normal to the gate at source and at drain. Here the drain voltage is higher than the source voltage, so the "depth" of the channel varies along its length. (c) The energy band diagram along the channel with no voltage on the drain with respect to the source, and with positive bias applied drain to source.

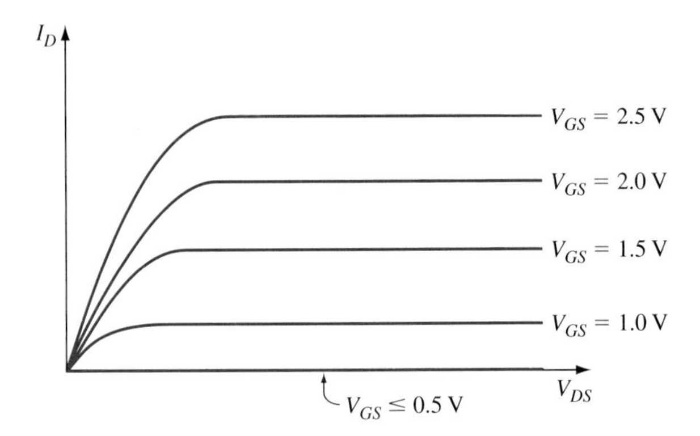

Figure 7.14 The I_D-V_{DS} characteristics of a typical MOSFET. The threshold voltage for this MOSFET is 0.5 V.

threshold, so the channel is conducting. If we could take an imaginary voltmeter and somehow measure the voltage across the oxide at the source end of the channel, it would be $V_G - V_{S'} = V_{GS}$. At the drain end, the voltage across the oxide is $V_G - V_D$. Since the channel voltage varies along the channel and the substrate voltage is constant, this varying channel voltage means also that the depletion region width varies from one end to the other. Thus, the channel is wider at the drain end than at the source end.

Since V_{DS} is positive, electrons entering the potential well from the source end will drift down the channel to the lower electron energy at the drain end. Figure 7.13b shows the energy band diagrams normal to the gate at source and at drain for a positive drain voltage. The channel potential energy (E_C) decreases along the channel from source to drain.

Figure 7.13c shows the energy band diagram along the channel. When no drain-source voltage is applied (solid line), electrons in the channel do not drift, since the longitudinal field $\mathscr{E}_L = 0$. When $V_{DS} > 0$ is applied, the electron energy at the drain end is lowered and electrons are moved toward the drain. Since electrons are negatively charged, the actual current I_D flows from drain to source.

Let us examine the current-voltage characteristics of a typical NFET in Figure 7.14. We expect that when the gate voltage is below threshold ($V_{GS} \leq V_T$), the channel will be weakly conductive and the current I_D will be negligible regardless of the value of V_{DS}. The threshold voltage for the transistor in this example is taken to be $V_T = 0.5$ V, so for any gate-source voltage below 0.5 V the transistor does not conduct appreciably.

As V_{GS} increases above the threshold voltage, the barrier height E_B for electrons entering the channel decreases (recall Figure 7.8c), which results in more electrons entering the channel. The channel conductance increases and thus the current also increases and becomes appreciable for $V_{GS} > V_T$.

There is an interesting feature of Figure 7.14 that bears investigation. We might expect that for a given gate voltage, and thus a given channel conductivity

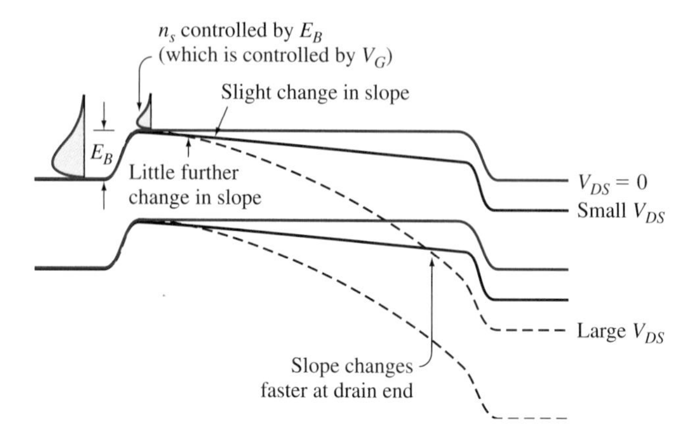

Figure 7.15 The energy band along the channel for three different values of V_{DS}. The current saturates because, as the drain voltage increases, the slope (and thus the electric field) increases faster at the drain end, but at the source end, there is little change. Thus the current is limited by the field at the source end.

(fixed resistor), we would see the current vary linearly with the voltage V_{DS} across the "resistor" (channel). We see this linear behavior in Figure 7.14 for very small values of V_{DS}, but then the current levels off and saturates. We can explain the saturation effect as follows.

Figure 7.15 shows a plot of the energy band diagram along the channel from source to drain for a given value of V_{GS} above threshold. There are three values shown—for $V_{DS} = 0$, for small V_{DS}, and for a larger value of V_{DS}. When the drain voltage is the same as the source voltage ($V_{DS} = 0$), the longitudinal field in the channel is also zero, so there is no slope to the conduction band edge along the channel ($\mathscr{E}_L = (1/q)(dE_C/dy)$). Thus $I_D = 0$ [Equation (III.5) in the Introduction to Part 3]. For small V_{DS}, the energy band diagram tilts slightly. The electrons that enter the channel from the source are accelerated toward the drain by the channel field. At first, as the drain voltage increases, the longitudinal field increases and thus I_D increases (look again at the low V_{DS} end of the I_D-V_{DS} characteristics, Figure 7.14). We will show in the next section that, with increasing V_{DS}, the longitudinal field $\mathscr{E}_L$ increases faster near the drain end than the source end (this is shown in the figure—remember that field is proportional to the slope of the conduction band edge), and most of the incremental drain voltage is dropped near the drain. In other words, with increasing V_{DS} there are increasingly smaller changes in the longitudinal field at the source end, as shown in the figure. Eventually, at some value of V_{DS}, the field at the source reaches a limiting value. Since the current at one end of the channel must be the same as the current at the other end, the current I_D is limited to what can be supported by this field at the source end. The drain current $I_D \propto n_s \mathscr{E}_{L(\text{source})}$, and n_s (the carrier concentration in the channel at the source) is controlled by the barrier height E_B and thus by V_{GS}.

When the field $\mathscr{E}_L(y = 0)$ saturates, so does I_D. (Compare with the analogy with the two lakes and the canal discussed in the Introduction to Part 3.)

Note also that E_B, the barrier from source to channel depends on the gate-source voltage V_{GS}. Therefore, the number of electrons in the channel n_s, which depends on the barrier height, also depends on V_{GS} but not on V_{DS}.[7]

To summarize our qualitative discussion of MOSFET behavior, a MOSFET is a voltage-controlled resistor. The resistor is between the source and the drain. We can control the conductance of the channel between source and drain by controlling the number of channel carriers available for conduction. In the MOSFET, that control results from adjusting the gate voltage, which in turn controls the band bending in the semiconductor. The gate voltage forces the conduction band edge to bend closer to or farther away from the Fermi level.

Current does not flow into the gate terminal, because there is an insulating oxide layer between the gate and the source, channel and drain. The gate voltage induces an electric field in the oxide, which in turn influences the energy bands in the semiconductor. The electric field from which the FET gets its name is the field induced by the gate voltage. For a MOSFET, it is the field across the oxide.

In the next section, we will apply our physical understanding of these processes to derive expressions for the I_D-V_{DS} characteristics of the transistors.

7.3 MOSFETs (QUANTITATIVE)

Now that we have a physical understanding of how MOSFETs work, we can be more quantitative. We will derive expressions for the I_D-V_{DS} characteristics of an NFET. These derivations of the electrical characteristics of MOSFETs are presented in three steps. First, we consider a formulation, in which the carrier mobility is assumed constant along the channel. [1] This is the simplest model, called the *long-channel* model, and it predicts the general form of the I_D-V_{DS} characteristics. It is useful for obtaining insight into the general behavior of MOSFETs but does not closely reproduce the results for modern devices. Therefore, we will then modify the simple model to account for variation in mobility. The mobility is affected by two things: the transverse and the longitudinal electric fields in the channel. Accounting for these are the second and third steps in our development.

We consider the enhancement-type NFET device of Figure 7.4, which is repeated as Figure 7.16. The figure indicates channel width W, the channel length, L, and the oxide thickness t_{ox}. The direction of the longitudinal field $\mathscr{E}_L$ is shown, along with the field component perpendicular to the channel, called the transverse field, $\mathscr{E}_T$. Recall from Figure III.7 that the channel voltage V_{ch} is the voltage at a given point along the channel with respect to the source, and is a function of position along the channel. At the drain end of the channel the channel voltage V_{ch} is equal to the drain-source voltage V_{DS}.

---

[7]As we will discuss in Chapter 8, for very short (submicrometer) channel lengths, V_{DS} does, in fact, affect the value of E_B. In this chapter, however, this effect is ignored.

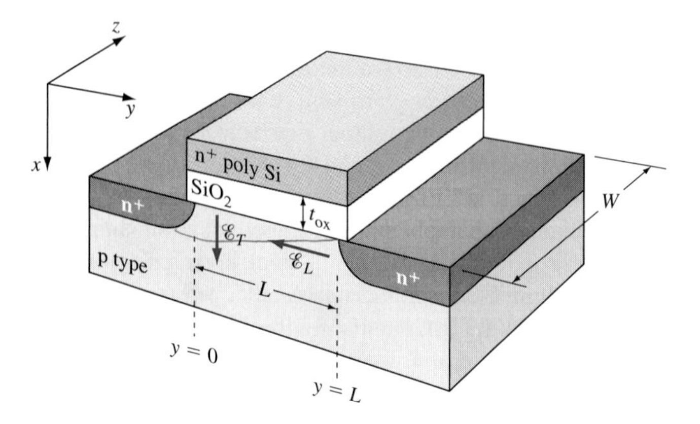

Figure 7.16 The NFET. Longitudinal and transverse electric field directions are indicated.

Our starting point will be Equations (III.3) to (III.7), which we repeat here for convenience:

$$I_D = -W Q_{\text{ch}}(y)v(y) \tag{III.3}$$

$$v(y) = -\mu(y)\mathscr{E}_L(y) \tag{III.4}$$

$$I_D = W Q_{\text{ch}}(y)\mu(y)\mathscr{E}_L(y) \tag{III.5}$$

$$\mathscr{E}_L = -\frac{dV_{\text{ch}}}{dy} \tag{III.6}$$

$$I_D\, dy = -W Q_{\text{ch}}(y)\mu(y)\, dV_{\text{ch}}(y) \tag{III.7}$$

Since our goal is to obtain an expression for I_D, we see from Equation (III.7) that we need the following: an analytical expression for μ, the carrier mobility in the channel; an expression for Q_{ch}, the charge per unit area in the channel (this is the mobile charge, in this case electrons in the conduction band); and an expression for V_{ch} as a function of position y. We will derive these first for the simple long-channel model, to illustrate the physics. Later we will add more realism (and complication) to the equations.

7.3.1 LONG-CHANNEL MOSFET MODEL WITH CONSTANT MOBILITY

There are several models used to describe μ and Q_{ch}. We will start with an oversimplified model for these quantities. While this model gives realistic results only for the I_D-V_{DS} characteristics for MOSFETs with very *long channels* ($L > 5$ to $10\,\mu$m), it is mathematically simple and does illustrate the general principles of operation. This formulation is employed in the SPICE Level 1 model.

Channel Charge Density Since the channel current depends on the channel charge density, our first task will be to analyze the charge in the channel. The charge will depend on the bias conditions. When the gate voltage is below threshold, the conductance is small because the number of electrons available for conduction is small. For simplicity we approximate:

$$Q_{\text{ch}} \approx 0 \qquad \text{and} \qquad I_D \approx 0 \qquad V_{GS} \leq V_T \tag{7.9}$$

For $V_{GS} > V_T$, of course, Q_{ch} is nonzero. We can find how much charge is present by recognizing that the region under the gate acts as a capacitor. Two conductive plates (the heavily doped polysilicon gate electrode and the conductive channel) are separated by an insulator (the oxide). The capacitance of a parallel plate capacitor is given by

$$C = \varepsilon \frac{A}{t} \tag{7.10}$$

where ε is the permittivity of the insulator, A is the area of the plate, and t is the thickness of the dielectric layer. In our device, we use the oxide thickness t_{ox}. The area of the gate electrode is $W \times L$.

On a given integrated circuit, different transistors may have different widths and lengths. The oxide thickness, on the other hand, is usually a constant for a given process and therefore common to all devices on the chip. It is therefore useful to define an *oxide capacitance per unit area* C'_{ox}:

$$C'_{\text{ox}} = \frac{\varepsilon_{\text{ox}}}{t_{\text{ox}}} \tag{7.11}$$

where ε_{ox} is the permittivity of the oxide $\varepsilon_{\text{ox}} = \varepsilon_r \varepsilon_0$, and ε_r is the relative permittivity (dielectric constant) of the dielectric. The dielectric constant is $\varepsilon_r = 3.9$ for SiO_2.

We will need to know the voltage across the capacitor, which is the voltage across the oxide. We don't know what the voltage across the oxide is, exactly, but we do recall from Example 7.1 that to reasonable approximation, for $V_{GS} > V_T$ any *change* in gate voltage appears across the oxide. Since capacitance is $C = |dQ/dV|$, then

$$C'_{\text{ox}} = -\frac{dQ_{\text{ch}}}{dV_{GS}} = -\frac{\Delta Q_{\text{ch}}}{\Delta V_{GS}} = -\frac{Q_{\text{ch}}(V_{GS}) - Q_{\text{ch}}(V_T)}{V_{GS} - V_T} = -\frac{Q_{\text{ch}}(V_{GS}) - 0}{V_{GS} - V_T} \tag{7.12}$$

where we have used the information that Q_{ch} is negative and for $V_{GS} = V_T$, $Q_{\text{ch}} = 0$ [Equation (7.9)]. Letting $Q_{\text{ch}} = Q_{\text{ch}}(V_{GS})$, we have

$$Q_{\text{ch}} = -C'_{\text{ox}}(V_{GS} - V_T) \qquad V_{DS} = 0 \tag{7.13}$$

When V_{DS} is no longer zero but is positive, the voltage on the lower plate of the capacitor between the channel and ground, V_{ch}, is a function of position y along the channel. The voltage dropped across the oxide will thus vary along y and affect Q_{ch}:

$$Q_{\text{ch}}(y) = -C'_{\text{ox}}(V_{GS} - V_T - V_{\text{ch}}(y)) \qquad V_{GS} - V_T > V_{\text{ch}}(y) \tag{7.14}$$

At the source end of the channel, where $V_{\text{ch}} = 0$ since V_{ch} is the channel voltage with respect to the source, this reduces to Equation (7.13).

We now have expressions for the channel charge—but there is a problem. Equation (7.14) is valid only for $(V_{GS} - V_T) > V_{\text{ch}}(y)$. This will always be true at the source end of the channel, provided the gate voltage is above threshold, since $V_{\text{ch}} = 0$. At the drain end, however, $V_{\text{ch}}(y = L) = V_{DS}$. Thus, if $V_{DS} > (V_{GS} - V_T)$, then at some position y in the channel, the channel voltage must be equal to $V_{\text{ch}} = (V_{GS} - V_T)$. At that point along the channel, Equation (7.14) implies that $Q_{\text{ch}} = 0$. However, since V_{GS} is above threshold and V_{DS} is not zero, we know that a current I_D is flowing through the channel. Then from Equation (III.3), since $I_D > 0$, this implies that the electron velocity is infinite. Furthermore, at y greater than this value, $V_{\text{ch}} > (V_{GS} - V_T)$, implying that the channel charge is positive. That would mean that the channel current is carried by holes. From physical arguments, however, we reject both of these scenarios. We know that the maximum possible velocity is the saturation velocity v_{sat}, and from Figure 7.12a holes clearly cannot enter the channel. Thus, since I_D must be a constant at every position in the channel, from Equation (III.3) when the velocity has a maximum the charge has a minimum. The minimum value of Q_{ch} is

$$Q_{\text{ch min}} = -\frac{I_{D\text{sat}}}{W v_{\text{sat}}} \tag{7.15}$$

where $I_{D\text{sat}}$ is the saturation current, which is indicated in Figure 7.17 for several values of V_{GS}. According to this model, the current cannot exceed this amount for a given value of gate voltage V_{GS}. The saturation current is reached when there is a position in the channel for which $V_{\text{ch}} = V_{GS} - V_T$. This happens first at the drain end, when $V_{DS} = V_{GS} - V_T$. We call this the drain saturation voltage $V_{DS\text{sat}}$.

Above this saturation point, Equation (7.14) no longer applies. We will take this point up again later. In the region where it does apply, though, we now have a model for the channel charge.

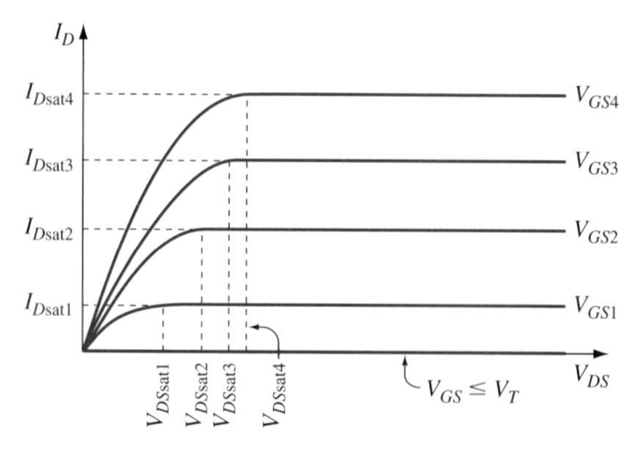

Figure 7.17 The saturation current and saturation voltage are defined.

Channel Mobility Next, we examine the channel mobility μ. In this long-channel simple model, we will take the mobility to be constant. In reality, the mobility depends on the longitudinal electric field $\mathscr{E}_L$ (and thus on V_{ch}). For example, we saw in Chapter 3 that, under high fields, the velocity saturates. Further, the transverse field $\mathscr{E}_T$ will have an effect. We will handle these dependencies of mobility on the fields explicitly later, however. Here we will consider the simplest model, the low-field case, in which the mobility is constant.

Long-Channel Model, Constant Mobility In this model, for mathematical simplicity we assume μ to be constant, and of value somewhere in the neighborhood of one-half to one-third of its bulk value. We use the term *bulk* to distinguish the mobility in a large crystal from the mobility experienced by a carrier in a thin layer such as the channel of a MOSFET. The reason for assuming one-half to one-third of the bulk value is that the small thickness of the channel will tend to slow the carriers down from the bulk value, as we will see later.

To obtain the I_D-V_{DS} characteristics for this model, we integrate both sides of Equation (III.7):

$$\int_0^L I_D\,dy = \int_0^{V_{DS}} -WQ_{\mathrm{ch}}(y)\mu(y)\,dV_{\mathrm{ch}} \tag{7.16}$$

Note the limits of integration. Over the length of the channel, the channel voltage varies from the source voltage $V_{\mathrm{ch}} = 0$ to the drain-source voltage V_{DS}.

As long as the drain voltage is less than $(V_{GS} - V_T)$, the channel voltage V_{ch} will also satisfy $V_{\mathrm{ch}} < (V_{GS} - V_T)$, so we can use Equation (7.14) for the channel charge. Since the mobility μ is constant in this model, it comes out of the integral. The current I_D cannot vary with position along the channel, so it is also a constant and it also comes out of its integral. Thus Equation (7.16) becomes

$$\begin{aligned} I_D\int_0^L dy &= -W\mu\int_0^{V_{DS}} Q_{\mathrm{ch}}\,dV_{\mathrm{ch}} \\ &= -W\mu\int_0^{V_{DS}} [-C'_{\mathrm{ox}}(V_{GS} - V_T - V_{\mathrm{ch}})]\,dV_{\mathrm{ch}} \end{aligned} \tag{7.17}$$

Integrating, the result is, for $V_{DS} \le (V_{GS} - V_T)$,

$$I_D = \frac{WC'_{\mathrm{ox}}\mu}{L}\int_0^{V_{DS}} (V_{GS} - V_T - V_{\mathrm{ch}})\,dV_{\mathrm{ch}} \qquad V_{DS} \le (V_{GS} - V_T) \tag{7.18}$$

or

$$I_D = \frac{WC'_{\mathrm{ox}}\mu}{L}\left[(V_{GS} - V_T)V_{DS} - \frac{V_{DS}^2}{2}\right] \qquad V_{DS} \le (V_{GS} - V_T) \tag{7.19}$$

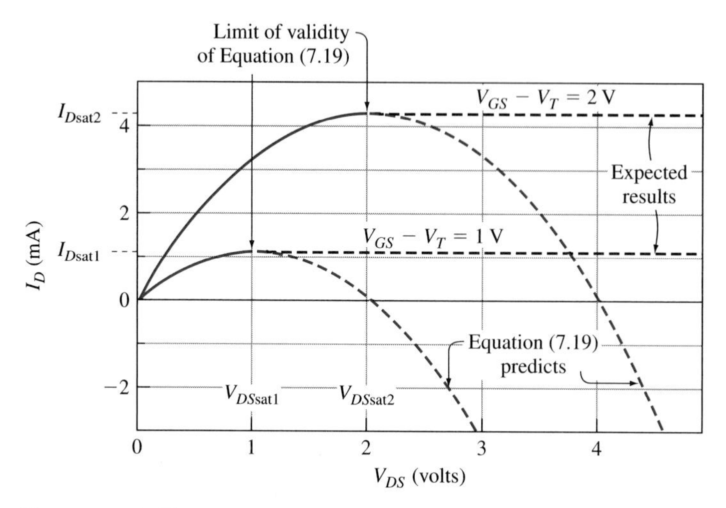

Figure 7.18 The current predicted using Equation (7.19) (solid lines) is only valid up to the point where $V_{DS} = V_{GS} - V_T$. After that the current saturates (black dashed lines), whereas Equation (7.19) would predict a decrease and eventually a sign reversal in the current (colored dashed lines).

There are two independent variables in this equation, the drain voltage and the gate voltage. Figure 7.18 shows the drain current calculated from Equation (7.19) versus the drain voltage for two different values of gate-source voltage V_{GS}. We chose the parameters of the NFET to be $t_{ox} = 4$ nm, $W/L = 5$, and $\mu = 500\,\text{cm}^2/\text{V}\cdot\text{s}$.

We see from the plot that the current reaches a maximum ($dI_D/dV_{DS} = 0$) for $V_{DS} = (V_{GS} - V_T)$. This is also the limit of validity of Equation (7.19). For $V_{DS} > (V_{GS} - V_T)$, from Equation (7.19), I_D would be expected to decrease as indicated by the colored dashed line of Figure 7.18. We will show later that the simple model predicts that once the curve in Figure 7.18 reaches its peak, the current remains (essentially) constant for larger V_{DS}. Thus, the current saturates at some value of I_{Dsat} as indicated.

The value of V_{DSsat} at which I_D saturates is found from taking $\partial I_D/\partial V_{DS}$ in Equation (7.19) and setting it to zero:

$$\frac{\partial I_D}{\partial V_{DS}} = 0 = \frac{WC'_{ox}\mu}{L}(V_{GS} - V_T - V_{DSsat}) \tag{7.20}$$

or

$$V_{DSsat} = (V_G - V_T) \tag{7.21}$$

Above this value, I_D remains constant.

We can use this result in Equation (7.17). By setting the limit of integration to V_{DSsat}, we obtain an expression for the saturation current:

$$I_D = I_{Dsat} = \frac{WC'_{ox}\mu}{L}\left[\left(V_{GS} - V_T - \frac{V_{DSsat}}{2}\right)V_{DSsat}\right] \tag{7.22}$$

But, since in this model $V_{DS\text{sat}} = V_{GS} - V_T$, we can write:

$$I_{D\text{sat}} = \frac{WC'_{\text{ox}}\mu}{2L}(V_{GS} - V_T)^2 = \frac{WC'_{\text{ox}}\mu}{2L}V_{DS\text{sat}}^2 \qquad (7.23)$$

Since $I_{D\text{sat}}$ is proportional to $V_{DS\text{sat}}^2$, this model is sometimes referred to as the *square law model.* It results from the simple long-channel model, assuming constant mobility, and uses Equation (7.14) to represent the channel charge in the region below threshold.

Thus, we can describe the I_D-V_{DS} characteristics for this model with the following three equations:

$$I_D = \frac{WC'_{\text{ox}}\mu}{L}\left[(V_{GS} - V_T)V_{DS} - \frac{V_{DS}^2}{2}\right] \qquad V_{DS} \leq V_{DS\text{sat}}, V_{GS} \geq V_T \qquad (7.24)$$

$$I_{D\text{sat}} = \frac{WC'_{\text{ox}}\mu}{L}\left[\left(V_{GS} - V_T - \frac{V_{DS\text{sat}}}{2}\right)V_{DS\text{sat}}\right] = \frac{WC'_{\text{ox}}\mu}{2L}(V_{GS} - V_T)^2$$

$$V_{DS} \geq V_{DS\text{sat}}, V_{GS} \geq V_T \qquad (7.25)$$

$$V_{DS\text{sat}} = (V_{GS} - V_T) \qquad (7.26)$$

EXAMPLE 7.2

Using the simple long-channel model, assuming constant mobility, plot the I_D-V_{DS} characteristics for an NFET with $W/L = 5$ and $t_{\text{ox}} = 4$ nm. Take the constant mobility for electrons in the channel to be 500 cm²/V · s. Plot for $V_{GS} - V_T = 1, 2, 3$, and 4 V, and V_{DS} from 0 to 5 V.

■ **Solution**

From Equation (7.11), we have

$$C'_{\text{ox}} = \frac{\varepsilon_{\text{ox}}}{t_{\text{ox}}} = \frac{\varepsilon_{r(\text{ox})}\varepsilon_0}{t_{\text{ox}}} = \frac{3.9(8.85 \times 10^{-14}\ \text{F/cm})}{4 \times 10^{-7}\ \text{cm}} = 8.6 \times 10^{-7}\ \text{F/cm}^2$$

For each value of $V_{GS} - V_T$, we must find the saturation point to know whether to use Equation (7.24) or (7.25). For example, from Equation (7.26), we have for $V_{GS} - V_T = 1$ V, $V_{DS\text{sat}} = V_{GS} - V_T = 1$ V. For V_{DS} between 0 and 1 V, then, we use Equation (7.24)

$$I_D = \frac{W}{L}C'_{\text{ox}}\mu\left[(V_{GS} - V_T)V_{DS} - \frac{V_{DS}^2}{2}\right]$$

$$= (5)(8.6 \times 10^{-7}\ \text{F/cm}^2)(500\ \text{cm}^2/\text{V} \cdot \text{s})\left[(1) \cdot V_{DS} - \frac{V_{DS}^2}{2}\right]$$

$$= 2.15 \times 10^{-3}\left[V_{DS} - \left(\frac{V_{DS}}{2}\right)^2\right]$$

At $V_{DS} = V_{DS\text{sat}} = 1$ V, I_D reaches its saturation value of [Equation (7.25)]

$$I_{D\text{sat}} = \frac{WC'_{\text{ox}}\mu}{2L}(V_{GS} - V_T)^2 = (5)\frac{(8.6 \times 10^{-7})(500)}{2}(1)^2 = 1.07\,\text{mA}$$

A similar procedure is used for the other values of $V_{GS} - V_T$. The results are plotted in Figure 7.19. The dashed line indicates the boundary between the sublinear and the saturation region, i.e., for $V_{DS} = V_{DS\text{sat}} = (V_{GS} - V_T)$.

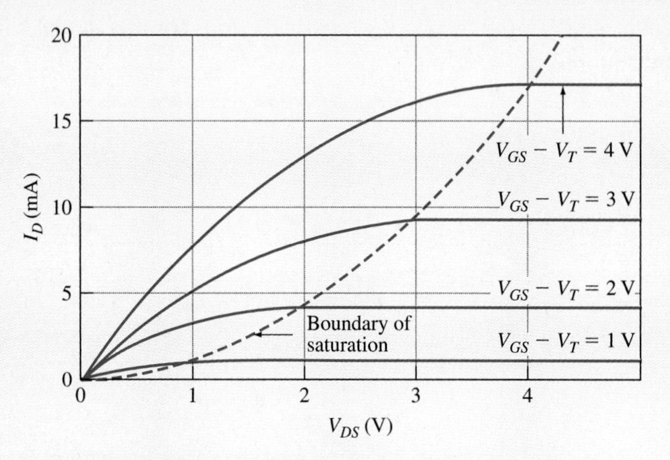

Figure 7.19 The I_D-V_{DS} characteristics of the NFET of Example 7.2: results from the simple model. For this device $W/L = 5$, $t_{\text{ox}} = 4$ nm, $C'_{\text{ox}} = 8.63 \times 10^{-3}$ F/m², and $\mu_n = 500$ cm²/V · s.

Current Saturation Revisited Earlier we claimed that the simple long-channel model (constant μ) predicts that once the drain voltage reaches $V_{DS\text{sat}} = V_{GS} - V_T$, the current saturates and remains constant for all higher drain voltages. This saturation was discussed qualitatively by analogy with water flow between two lakes via a canal as discussed in Part 3, Field-Effect Transistors. Here we discuss current saturation more quantitatively.

When we discussed Figure 7.14, we indicated that for low values of V_{DS}, the longitudinal field $\mathscr{E}_L$ is constant along the channel. At larger drain voltages, the field increases appreciably at the drain end, but not much at the source end, as was shown in Figure 7.15. Let us consider the effect of this point analytically.

For a given V_{GS} and V_{DS} with constant μ, Equation (III.5) becomes

$$I_D = W\mu Q_{\text{ch}}(y)\mathscr{E}_L(y)$$

The current is constant and proportional to the $Q_{\text{ch}}\mathscr{E}_L$ product at any value of y. It is convenient to determine this product and thus the I_D-V_{DS} relation near the source end of the channel ($y = 0$). We can find $\mathscr{E}_L$ from Equation (III.6) ($\mathscr{E}_L = -dV_{\text{ch}}/dy$) if we have an expression for V_{ch} as a function of y. We can find $V_{\text{ch}}(y)$ by integrating Equation (III.7) from 0 to y. From Equations (III.7) and (7.14), then

$$I_D \int_0^y dy = WC'_{\text{ox}}\mu \int_0^{V_{\text{ch}}(y)} (V_{GS} - V_T - V_{\text{ch}})\, dV_{\text{ch}} \tag{7.27}$$

or

$$I_D = \frac{WC'_{\text{ox}}\mu}{y}\left(V_{GS} - V_T - \frac{V_{\text{ch}}(y)}{2}\right) V_{\text{ch}}(y) \tag{7.28}$$

Solving for $V_{\text{ch}}(y)$

$$V_{\text{ch}}(y) = (V_{GS} - V_T) - \sqrt{(V_{GS} - V_T)^2 - \frac{2I_D y}{WC'_{\text{ox}}\mu}} \tag{7.29}$$

where the negative sign associated with the square root is used, since for $y = 0$, $V_{\text{ch}}(0) = 0$.

This gives us an expression for the channel voltage as a function of distance along the channel. The potential energy (E_C) has the same shape as the potential, but inverted ($dE_C/dy = -q\,dV_{\text{ch}}/dy$), which means the conduction band edge has the shape

$$E_C(y) = E_C(0) - qV_{\text{ch}}(y) \tag{7.30}$$

Therefore, we can substitute Equation (7.29) into Equation (7.30) to obtain

$$E_C(y) = E_C(0) - q\left[(V_{GS} - V_T) - \sqrt{(V_{GS} - V_T)^2 - \frac{2I_D y}{WC'_{\text{ox}}\mu}}\right] \tag{7.31}$$

The conduction band edge along the channel is plotted in Figure 7.20a for several values of V_{DS}, with $W/L = 10$, $\mu = 500$ cm^2/V · s, $(V_{GS} - V_T) = 2$ V and $C'_{\text{ox}} = 6.9 \times 10^{-7}$ F/cm^2 ($t_{\text{ox}} = 5$ nm), and with I_D obtained from Equation (7.19). For $V_{DS} = 0$, $I_D = 0$, and there is no voltage drop along the channel and E_C is flat. As V_{DS} increases, the band bends increasingly as seen in the figure. We note that the magnitude of the slope of the E_C-y plot at the source ($y = 0$) increases with increasing V_{DS} and tends toward saturation as V_{DS} approaches $V_{GS} - V_T$ (2 V).

The electric field is proportional to the slope of E_C. The longitudinal field at some point y is

$$\mathscr{E}_L(y) = -\frac{dV_{\text{ch}}}{dy} = \frac{1}{q}\frac{dE_C}{dy} = -\frac{\dfrac{I_D}{WC'_{\text{ox}}\mu}}{\sqrt{(V_{GS} - V_T)^2 - \dfrac{2I_D y}{WC'_{\text{ox}}\mu}}} \tag{7.32}$$

The field $\mathscr{E}_L(y)$ is plotted in Figure 7.20b for the same device, for various values of drain voltage. Notice that near the source end, the magnitude of the

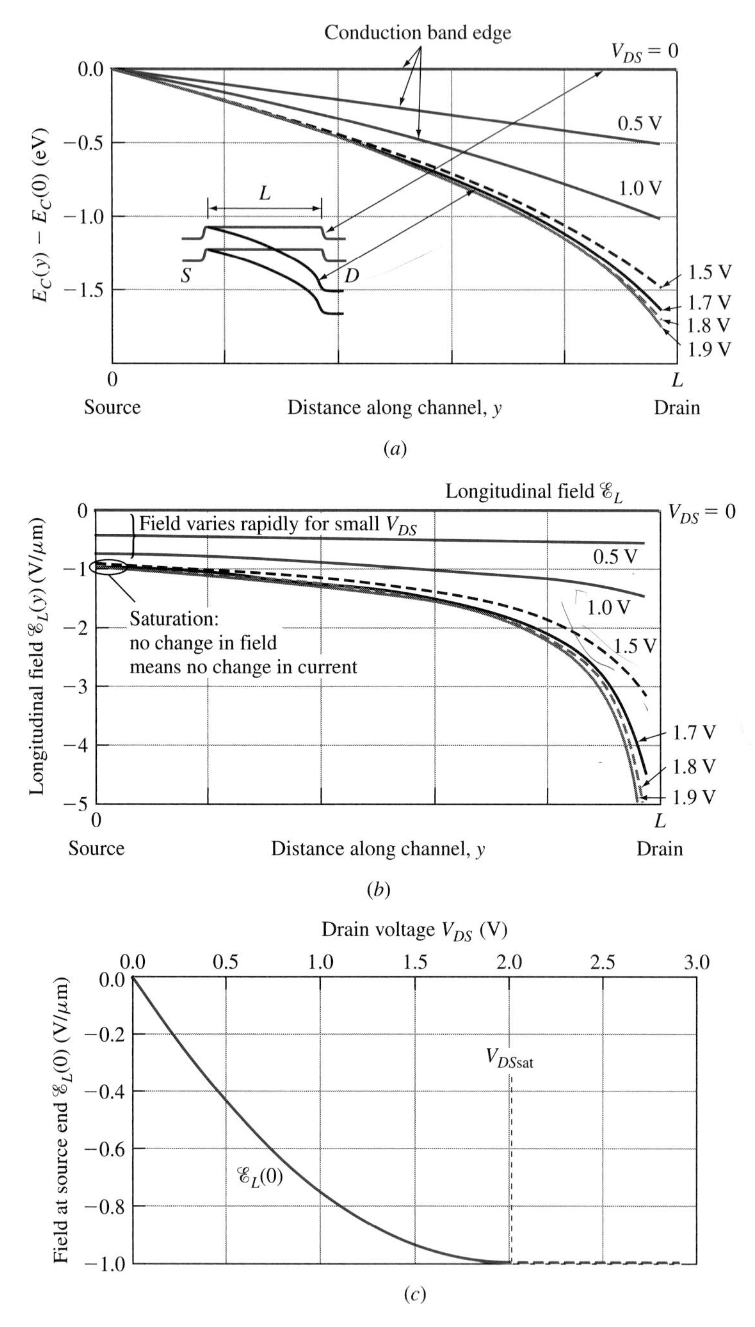

Figure 7.20 Illustration for current saturation. (a) The conduction band edge along the channel bends more at the drain end than at the source end for large drain voltage V_{DS}. (b) Since the longitudinal field is proportional to the slope of E_C, the field changes rapidly at the drain end for increasing values of V_{DS} but not at the source end. (c) The field at the source end is constant as V_{DS} increases beyond a certain point; thus the current is constant as well.

field $\mathscr{E}_L(0)$ varies rapidly with V_{DS} for small values of V_{DS}, but approaches a constant (saturates) as V_{DS} approaches $V_{DS\text{sat}} = V_{GS} - V_T$.

Combining Equations (III.5) and (7.13) at $y = 0$, we have as our final result

$$I_D = -WC'_{\text{ox}}(V_{GS} - V_T)\mu\mathscr{E}_L(0) \tag{7.33}$$

Again, the current is proportional to the electric field at the source end. Since the field is saturating, the current also saturates.

Finally, let us examine the rate at which the last term, the field at the source end $\mathscr{E}_L(0)$, varies with V_{DS}. We use Equation (7.32) with $y = 0$:

$$\mathscr{E}_L(0) = -\frac{\dfrac{I_D}{WC'_{\text{ox}}\mu}}{(V_{GS} - V_T)} = -\frac{I_D}{WC'_{\text{ox}}\mu(V_{GS} - V_T)} \tag{7.34}$$

Substituting for I_D from Equation (7.19) yields

$$\mathscr{E}_L(0) = -\frac{\left(V_{GS} - V_T - \dfrac{V_{DS}}{2}\right)V_{DS}}{L(V_{GS} - V_T)} = -\frac{1}{L}\left[1 - \frac{V_{DS}}{2(V_{GS} - V_T)}\right]V_{DS} \tag{7.35}$$

For $V_{GS} - V_T = 2$ V, we have

$$\mathscr{E}_L(0) = -\frac{\left(1 - \dfrac{V_{DS}}{4}\right)V_{DS}}{L} \tag{7.36}$$

For a channel length of $L = 1\ \mu$m, that produces a field of

$$\mathscr{E}_L(0) = -\left(1 - \frac{V_{DS}}{4}\right)V_{DS} \quad \frac{\text{V}}{\mu\text{m}} \tag{7.37}$$

This is plotted in Figure 7.20c. We observe that the magnitude of the field increases with V_{DS} and then levels off. For saturation, we know that the slope is zero, or

$$\frac{d\mathscr{E}_L(0)}{dV_{DS}} = 0 = -\left(1 - \frac{V_{DS}}{2}\right) \tag{7.38}$$

which occurs for $V_{DS} = (V_{GS} - V_T) = 2$ V, as seen in the figure.

EXAMPLE 7.3

Show that for constant μ, evaluating Equation (III.5) at $y = 0$ gives Equation (7.19) for I_D.

■ Solution

Equation (III.5) is repeated here:

$$I_D = WQ_{\text{ch}}(y)\mu(y)\mathscr{E}_L(y)$$

At $y = 0$, we have $V_{\text{ch}} = 0$. From Equation (7.14), then, $Q_{\text{ch}}(0) = C'_{\text{ox}}(V_{GS} - V_T)$. Combining these with the expression for $\mathscr{E}_L(0)$ from Equation (7.35) into Equation (III.5), we obtain

$$I_D = W\mu Q_{\text{ch}}(0)\mathscr{E}_L(0) = W\mu[-C'_{\text{ox}}(V_{GS} - V_T)]\left[-\frac{\left(V_{GS} - V_T - \frac{V_{DS}}{2}\right)V_{DS}}{L(V_{GS} - V_T)}\right]$$

Canceling $(V_{GS} - V_T)$ gives

$$I_D = \frac{W\mu C'_{\text{ox}}}{L}\left(V_{GS} - V_T - \frac{V_{DS}}{2}\right)V_{DS}$$

which is Equation (7.19).

Channel Length Modulation In the above long-channel MOSFET model, in the sublinear region the drain current is given by Equation (7.19):

$$I_D = \frac{WC'_{\text{ox}}\mu}{L}\left[(V_{GS} - V_T)V_{DS} - \frac{V_{DS}^2}{2}\right] \qquad V_{DS} \leq V_{DS\text{sat}} \qquad (7.19)$$

and in saturation combining Equations (7.25) and (7.26) gives

$$I_D = I_{D\text{sat}} = \frac{WC'_{\text{ox}}\mu}{2L}(V_{GS} - V_T)^2 = \frac{WC'_{\text{ox}}\mu}{2L}V_{DS\text{sat}}^2 \qquad V_{DS} \geq V_{DS\text{sat}} \qquad (7.39)$$

which is constant. In real devices, however, as the drain voltage increases above the saturation point $V_{DS} > V_{GS} - V_T = V_{DS\text{sat}}$, I_D continues to increase slowly with increasing V_{DS}, as indicated in Figure 7.21. This figure represents the experimental results for an n-channel MOSFET with $t_{\text{ox}} = 4.7$ nm, $L = 0.27\ \mu$m,

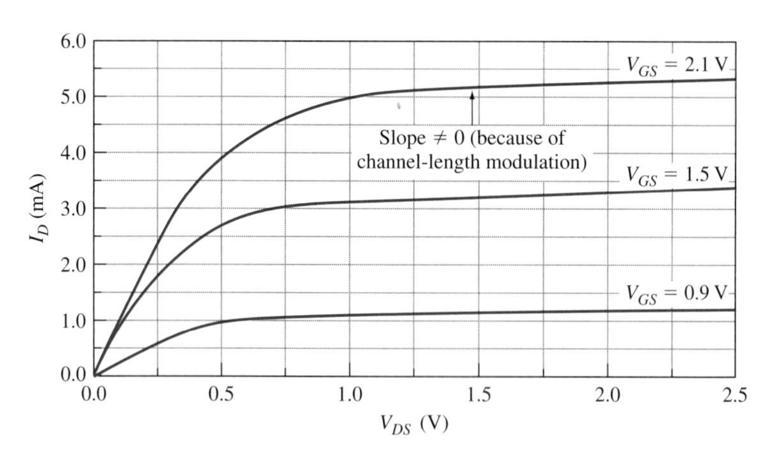

Figure 7.21 Experimental I_D-V_{DS} characteristics for an n-channel MOSFET for three values of gate voltage. The current actually increases with increasing V_D in the "current saturation" region because of channel-length modulation. For this device, $t_{\text{ox}} = 4.7$ nm, $L = 0.27\ \mu$m, $W = 8.6\ \mu$m, and $V_T = 0.3$ V.

and $V_T = 0.3$ V. There are two physical reasons for this increase in drain current that were not taken into account in the simple model: (1) increasing drain voltage V_{DS} reduces the *effective* channel length L, which we will discuss next, and (2) increasing V_{DS} reduces the value of the threshold voltage. The second of these effects is important for very short channels and is discussed in Chapter 8, where short-channel effects are handled.

Let us examine qualitatively why the channel length is effectively shortened as V_{DS} increases above saturation. Figure 7.22a to d indicates the channel energy as a function of y for a given value of $V_{GS} > V_T$ and for four values of V_{DS}. Figure 7.22e shows the corresponding currents. In (a), where $V_{DS} = 0$, the channel charge Q_{ch} is constant in y and no current flows. In (b), for small V_{DS}, Q_{ch} decreases with increasing y and current flows as indicated. For $V_{DS} \geq (V_{GS} - V_T) = V_{DS\text{sat}}$, (c) the current saturates as discussed earlier. For $V_{DS} > V_{DS\text{sat}}$, the channel voltage V_{ch} reaches $V_{DS\text{sat}}$ somewhere before the end of the channel [part (d) of the figure]. The effective channel length then is shorter than

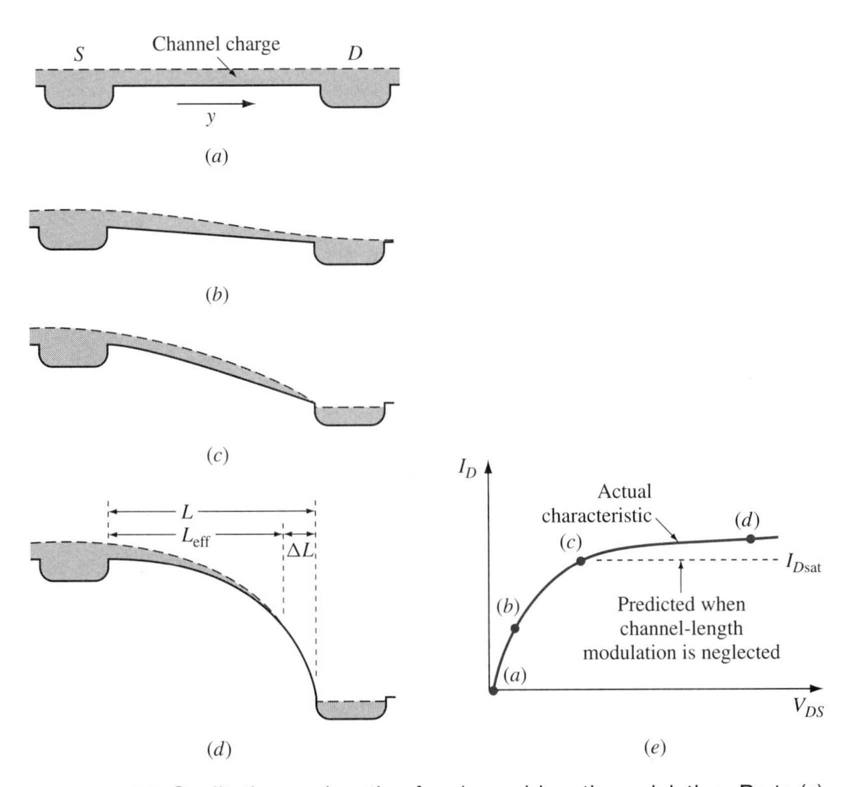

Figure 7.22 Qualitative explanation for channel-length modulation. Parts (a) to (c) repeat the explanation of the simple long-channel model. In (d), as the drain voltage continues to increase, the point at which the channel charge approaches 0 (shaded region), or the point at which $V_{\text{ch}} = V_{GS} - V_T$, moves along the channel toward the source. The channel becomes effectively shorter. (e) The corresponding points on the I_D-V_{DS} characteristics. From point (c) on, the simple model predicts constant current (dashed line).

the physical channel by some amount ΔL:

$$L_{\text{eff}} = L - \Delta L \tag{7.40}$$

Here we will treat ΔL as an empirical quantity (established from measurements).

Now, substituting Equation (7.40) into Equation (7.39), we get for the drain current

$$I_D = \frac{WC'_{\text{ox}}\mu}{2(L - \Delta L)} V_{DS\text{sat}}^2 = \frac{WC'_{\text{ox}}\mu}{2L\left(1 - \dfrac{\Delta L}{L}\right)} V_{DS\text{sat}}^2 = \frac{I_{D\text{sat}}}{\left(1 - \dfrac{\Delta L}{L}\right)} \tag{7.41}$$

where ΔL is a function of V_{DS}. For small $\Delta L/L$,

$$\frac{1}{\left(1 - \dfrac{\Delta L}{L}\right)} \approx \left(1 + \frac{\Delta L}{L}\right)$$

and to first approximation, for $V_{DS} > V_{D\text{sat}}$ the fractional change in channel length is proportional to $V_{DS} - V_{DS\text{sat}}$:

$$\frac{\Delta L}{L} = \lambda(V_{DS} - V_{DS\text{sat}}) \tag{7.42}$$

The quantity λ is known as the *channel length modulation parameter* (a SPICE parameter).[8] For $V_{DS} > V_{DS\text{sat}}$ then

$$I_D \approx I_{D\text{sat}}[1 + \lambda(V_{DS} - V_{DS\text{sat}})] \tag{7.43}$$

The slope $\partial I_D/\partial V_{DS}$ in the I_D-V_{DS} characteristic in saturation is the differential output conductance. The output conductance is thus proportional to λ. From Equation (7.42) it is seen that λ increases with decreasing L.

Extrapolation of the I_D-V_{DS} plots to $I_D = 0$ occurs at a voltage V_A, often referred to as the "Early voltage" analogous to a similar effect in the electrical characteristics in bipolar transistors discussed in Part 4.

EXAMPLE 7.4

Find the SPICE parameter λ for the device of Figure 7.21.

■ Solution

From Equation (7.43), for $V_{DS} > V_{DS\text{sat}}$,

$$I_D = I_{D\text{sat}}[1 + \lambda(V_{DS} - V_{DS\text{sat}})]$$

Then

$$\lambda = \frac{1}{I_{D\text{sat}}} \frac{\partial I_D}{\partial V_{DS}} = \frac{1}{I_{D\text{sat}}} \frac{\Delta I_D}{\Delta V_{DS}}$$

---

[8]In SPICE Level 1, the expression used is $I_D = I_{D\text{sat}}(1 + \lambda V_{DS})$.

For this device, we find

$$C'_{ox} = \frac{\varepsilon_{ox}}{t_{ox}} = \frac{3.9(8.85 \times 10^{-14}\,\text{F/cm})}{4.7 \times 10^{-7}\,\text{cm}} = 7.3 \times 10^{-7}\,\text{F/cm}^2$$

For $V_{GS} = 2.1$ V, from Figure 7.21 the saturation voltage is $V_{DS\text{sat}} \approx 1.2$ V and $I_{D\text{sat}} \approx 5.1$ mA. To find the slope, we extrapolate the straight-line portion of the I_D-V_{DS} curve from 2.5 V to 0 V, as shown in Figure 7.23. We obtain

$$\frac{\Delta I_D}{\Delta V_{DS}} = \frac{5.36 - 4.9}{2.5 - 0} = \frac{0.46\,\text{mA}}{2.5\,\text{V}}$$

Then

$$\lambda = \left(\frac{1}{5.1\,\text{mA}}\right)\left(\frac{0.46\,\text{mA}}{2.5\,\text{V}}\right) = 0.036\,\text{V}^{-1}$$

corresponding to an early voltage (V_A) of 28 V.

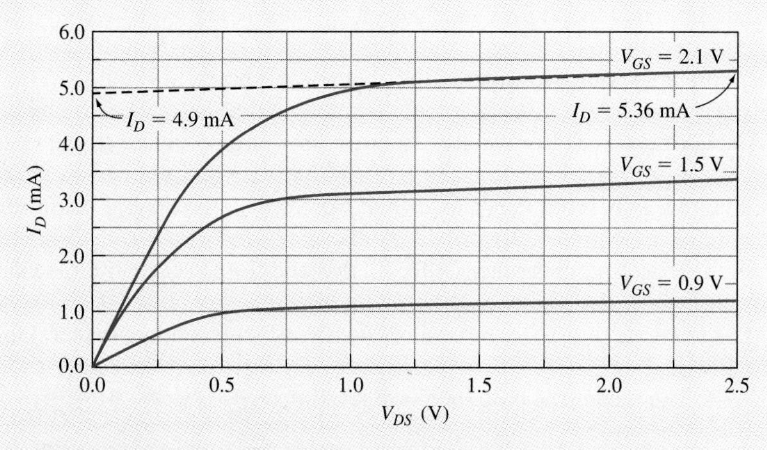

Figure 7.23 Finding the channel-length modulation parameter requires finding the slope of the I_D-V_{DS} characteristic in the "saturation" region (Example 7.4.). This is the same device as in Figure 7.21.

7.3.2 MORE REALISTIC LONG-CHANNEL MODELS: EFFECT OF FIELDS ON THE MOBILITY

In the above long-channel model for finding the I-V_{DS} characteristics of a MOSFET, we assumed that the mobility was a constant. In reality, the carrier mobility is dependent on the transverse and longitudinal fields $\mathscr{E}_T$ and $\mathscr{E}_L$. Thus, the electrical characteristics of FETs will depend on the strengths of these channel fields. For example, for large longitudinal field $\mathscr{E}_L$, the carrier velocities saturate, which limits the current actually obtainable to something lower than predicted by the simple model. At low longitudinal fields, the mobility is independent of $\mathscr{E}_L$ but it does depend on $\mathscr{E}_T$. We will consider the effects of these two fields separately.

When the longitudinal field (along the channel) is small enough that the velocity is proportional to $\mathscr{E}_L$, the carriers have what is called their *low-field mobility,* μ_{lf}. The value of this low-field mobility, however, is influenced by the transverse field (across the channel) $\mathscr{E}_T$.

We first consider the effects of $\mathscr{E}_T$ on the low-field mobility, and then examine how that affects the electrical characteristics of an FET. In the next section, we repeat this to account for the effects of $\mathscr{E}_L$.

Effect of the Transverse Field $\mathscr{E}_T$ on the Low-Field Mobility In this section, we assume that the longitudinal field is low enough that the carrier velocity is small compared with the saturation velocity for electrons moving along the channel. Thus, the electrons have their low-field mobility μ_{lf}. The transverse field $\mathscr{E}_T$, however, influences the value of the low-field mobility, [2] which earlier we took to be constant ($\mu = \mu_{\mathrm{lf}}$). Let us examine the origin of this effect.

In addition to the scattering mechanisms in bulk semiconductors, e.g., lattice and impurity scattering, the electron in the channel of a FET is additionally scattered by collisions with the walls of the channel, as shown in Figure 7.24a. This reduces the mean free time between collisions $\bar{t}$, and thus μ, from the bulk values. As electrons travel from source to drain, they are restricted to the channel region by the potential barrier at the Si/SiO_2 interface and the barrier in E_C in the Si, Figure 7.24b. Note that most of the electrons in the channel are near the bottom of the potential well formed by these barriers, where the channel is extremely narrow. In practical MOSFETs, this additional scattering mechanism reduces the low-field mobility μ_{lf} by a factor of about 2 to 3 from the bulk value.

Let us examine this effect more analytically. We showed in Chapter 3 that in bulk Si the mean free time between collisions for electrons was on the order of 2×10^{-13} s. The mean free path then is approximately

$$\bar{l} \approx \bar{v}\bar{t}$$

where $\bar{v}$ is the average thermal speed. At room temperature, this speed is on the order of 10^7 cm/s $= 10^5$ m/s. Thus for bulk Si, the mean free path is about $\bar{l} \approx 2 \times 10^{-13} \times 10^5 = 2 \times 10^{-8}$ m $= 20$ nm, and is independent of direction.

In a MOSFET, the mean free path $\bar{l}$ and the mean free time between collisions $\bar{t}$ for electrons traveling along the channel (in the y and z directions with no x-directed or transverse velocity) should be about the same as for bulk Si. For electrons with a v_x component, however, there is the additional scattering from collisions with the channel walls just discussed. This additional scattering reduces $\bar{t}$ and the mobility μ, as seen in the following example.

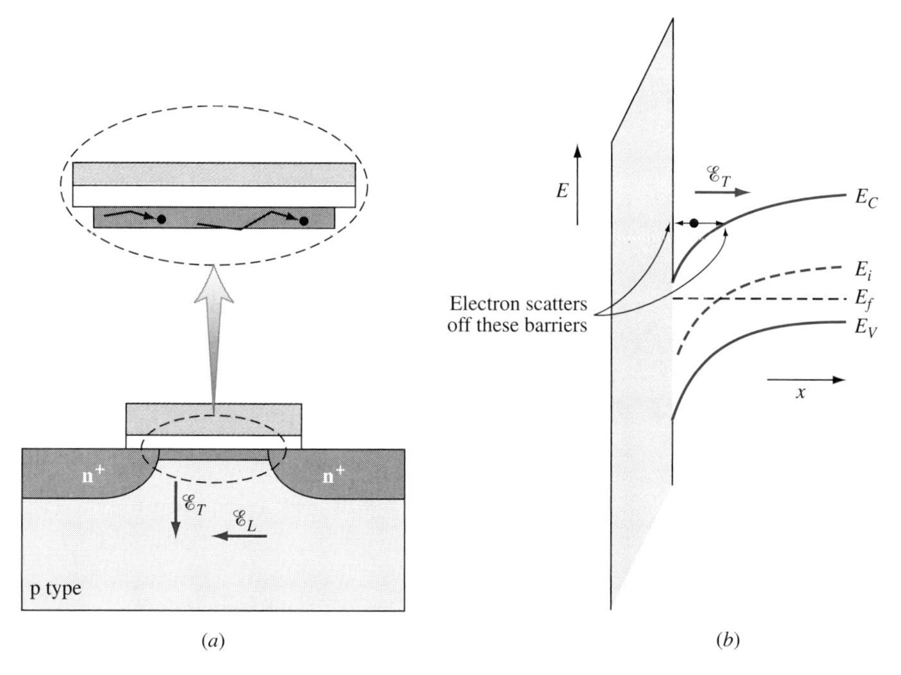

Figure 7.24 The effect of the transverse electric field on the mobility. (a) The electrons in the channel collide with the "walls" of the channel. (b) The energy band diagram shows that the walls are potential barriers at the oxide interface and the barrier of the depletion region in the semiconductor is sloped.

EXAMPLE 7.5

Estimate the time $\bar{t}_x$ between collisions for a channel electron traveling in the x direction, perpendicular to the gate. Compare this with the mean free time $\bar{t}$ of an electron in bulk silicon.

■ **Solution**

Consider an electron with energy $\frac{3}{2}kT$ above the channel floor, and suppose the transverse field is 10^5 V/cm $= 10^7$ V/m and assumed constant with x. Consider the electron to have just made a collision at the Si/SiO$_2$ interface at $t = 0$, where its kinetic energy is $\frac{3}{2}kT = m^* v_{\max}^2/2$. This is its maximum velocity because, as the electron goes across the channel in Figure 7.25, its total energy is constant but the potential energy is increasing so the kinetic energy (and thus the electron velocity) is decreasing. The force on the electron is $F = -q\mathscr{E}_T = m^* dv/dt$ and tends to decelerate it.

We can write

$$-q\mathscr{E}_T \int_0^{\bar{t}_x} dt = m^* \int_{v_{\max}}^{0} dv$$

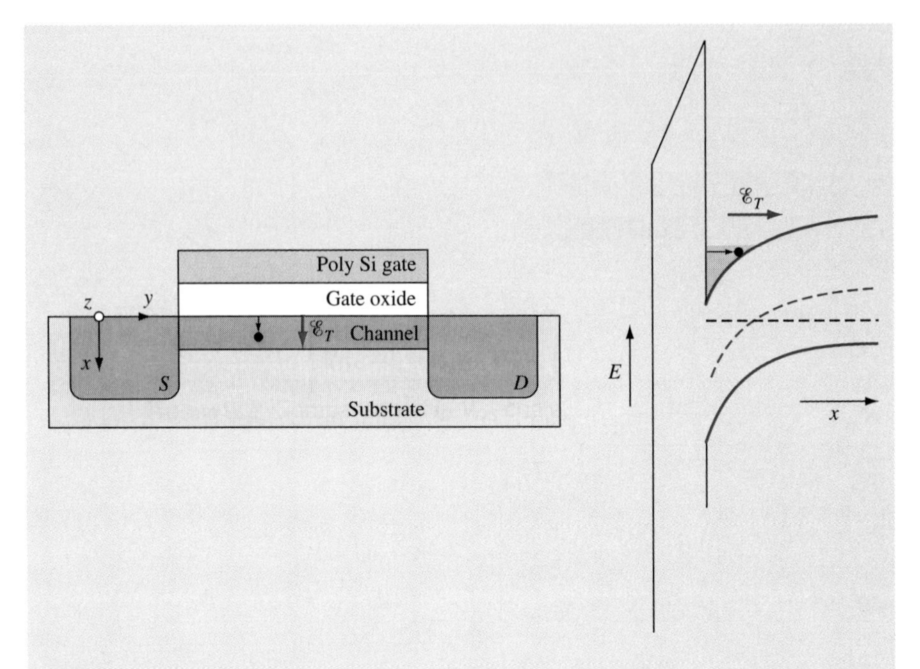

Figure 7.25 Geometry for Example 7.5. We consider only the transverse component of the electron's motion.

and

$$\bar{t}_x = \frac{m^* v_{\max}}{q\mathscr{E}_T}$$

But since

$$\frac{m^* v_{\max}^2}{2} = \frac{3}{2}kT$$

we have for $v_{\max}$:

$$v_{\max} = \sqrt{\frac{3kT}{m^*}}$$

Thus,

$$\bar{t}_x = \frac{m^*}{q\mathscr{E}_T}\sqrt{\frac{3kT}{m^*}} = \frac{\sqrt{3m^*kT}}{q\mathscr{E}_T}$$

Expressing kT in eV gives

$$\bar{t}_x = \frac{1}{\mathscr{E}_T}\sqrt{\frac{3m^*kT\,(\text{eV})}{q}}$$

Letting m^* be the conductivity effective mass, $m^* = 0.26m_0$, we have

$$\bar{t}_x = \frac{1}{10^7}\sqrt{\frac{3 \times 0.26 \times 9.11 \times 10^{-31} \times 0.026}{1.6 \times 10^{-19}}} = 0.34 \times 10^{-13}\,\text{s}$$

which is approximately a factor of 6 smaller than the value of $\bar{t}$ in bulk Si. As a result, sidewall scattering is the predominant scattering mechanism for the x-directed electrons.

EXAMPLE 7.6

For Example 7.5, find the distance $\bar{l}_x$ between electron collisions.

■ **Solution**

For constant $\mathscr{E}_T$, $E_K(\text{eV}) = \mathscr{E}_T \bar{l}_x$

$$\bar{l}_x = \frac{E_K(\text{eV})}{\mathscr{E}_T} = \frac{\frac{3}{2} \times 0.26}{10^7} = 3.9\,\text{nm}$$

This is appreciably less than the 20 nm for the mean free path $\bar{l}$ in bulk silicon.

Note that electrons near the bottom of the channel have smaller kinetic energies and that the transverse field is higher there, reducing their time between collisions even further. Since electrons in the channel have a range of v_x components, however, some average $\bar{t}_x$ must be used, and it must also be averaged with the mean free times in the y and z directions to obtain an overall $\bar{t}$. The point is, $\bar{t}$ and μ are reduced from their bulk values.

The electron mean free path and thus $\bar{t}$ are therefore dependent on $\mathscr{E}_T$, which can be seen another way in Figure 7.26. There we have plotted part of the energy

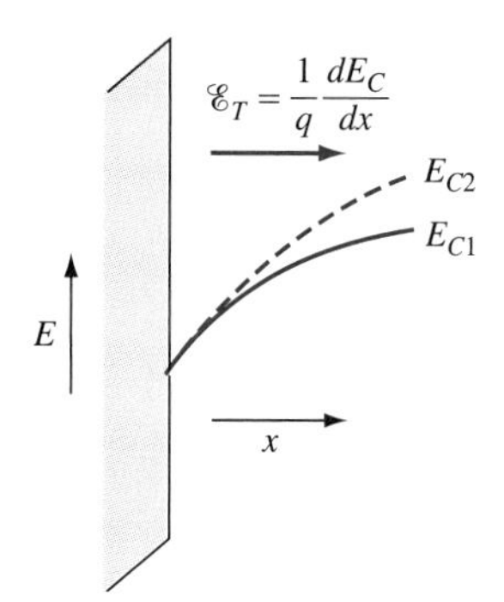

Figure 7.26 With increased band bending, the transverse field $\mathscr{E}_T$ increases. This in turn reduces $\bar{l}$, $\bar{t}$, and μ_{lf}.

band diagram in the region of the channel. The oxide interface is on the left, and two possible conduction band edges are shown on the right. The steeper the slope, the higher the transverse electric field and thus the smaller the mean free time and the mobility.

The value of $\mathscr{E}_T$ in the channel depends on the slope of the conduction band edge. This is a function of two things. First, there is the charge per unit area, Q_B, in the Si depletion region adjacent to the channel. In an n-channel device, these are the fixed, negatively charged ionized acceptors in the p-type substrate, as discussed in the Supplement to Part 3. Second, there is the mobile charge in the channel, Q_{ch}. In an NFET the channel charges are electrons, which tend to raise the potential energy of the channel, affecting the slope. Therefore, the transverse field strength varies with doping, bias conditions, and depth in the channel.

The effect of the transverse field on the low-field mobility is discussed further in the Supplement to Part 3. However, experimentally the low-field mobility can be expressed as

$$\mu_{\text{lf}} = \frac{\mu_0}{1 + \theta(V_{GS} - V_T - V_{\text{ch}})} \tag{7.44}$$

where μ_0 is the channel mobility at the source ($V_{\text{ch}} = 0$) at threshold ($V_{GS} = V_T$). Here the quantity θ is a measured empirical parameter, on the order of 0.03 to 0.2 V^{-1}, and depends on the processing parameters, including the substrate doping, substrate bias, and oxide thickness.

An experimental plot [3] of μ_{lf} versus ($V_{GS} - V_T - V_{\text{ch}}$) is shown in Figure 7.27. We see there that μ_{lf} varies about 30 percent over about a 2-V variation in gate voltage.

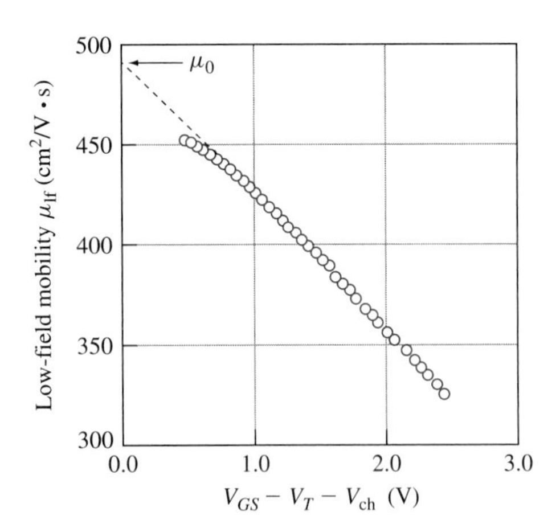

Figure 7.27 Variation of the low-field mobility as a function of $V_{GS} - V_T - V_{\text{ch}}$ for an n-channel MOSFET. The low-field mobility can be expressed as $\mu_{\text{lf}} = \mu_0[1 - \theta(V_{GS} - V_T - V_{\text{ch}})]$.

We saw earlier that the low-field mobility in the channel, μ_{lf}, increases from source to drain. It can be seen in the figure that μ_{lf} decreases approximately linearly with $(V_{GS} - V_T - V_{\text{ch}})$:

$$\mu_{\text{lf}} = \mu_0[1 - \theta(V_{GS} - V_T - V_{\text{ch}})]$$

where μ_0 is the zero-voltage extrapolation and θ the negative of the slope. Note that the maximum mobility is less than 500 $\text{cm}^2/\text{V} \cdot \text{s}$, considerably less than the bulk value of about 1000 (Figure 3.4).

Since $\theta(V_G - V_T - V_{\text{ch}})$ is small compared with unity, and since for $x \ll 1$, $1 - x \approx 1/(1+x)$, μ_{lf} can be expressed in the more customary form of Equation (7.44).

EXAMPLE 7.7

Find the value for θ for the device of Figure 7.27.

■ Solution

Since $\mu_{\text{lf}} = \mu_0[1 - \theta(V_{GS} - V_T - V_{\text{ch}})]$, we can find a formula for θ by taking the derivative of this expression. Using points on the graph gives

$$\theta = -\frac{d\dfrac{\mu_{\text{lf}}}{\mu_0}}{d(V_{GS} - V_T - V_{\text{ch}})} = -\frac{1}{\mu_0}\frac{d\mu_{\text{lf}}}{d(V_{GS} - V_T - V_{\text{ch}})} = -\frac{1}{480}\frac{(355 - 480)}{(2 - 0)} = 0.13\ \text{V}^{-1}$$

Effect of $\mathscr{E}_T$ on the I_D-V_{DS} Characteristics Now that we have considered the effect of $\mathscr{E}_T$ on μ_{lf}, the next step is to see how that affects the I_D-V_{DS} characteristics compared with the constant-mobility model. We correct Equation (7.17) to include the variation of the mobility with V_{ch} [Equation (7.44)]. The expression becomes

$$I_D \int_0^L dy = WC'_{\text{ox}}\mu_0 \int_0^{V_{DS}} \frac{V_{GS} - V_T - V_{\text{ch}}}{1 + \theta(V_{GS} - V_T - V_{\text{ch}})}\, dV_{\text{ch}} \tag{7.45}$$

which yields

$$I_D = \frac{WC'_{\text{ox}}\mu_0}{\theta^2 L}\left\{\theta V_{DS} + \ln\left[\frac{1 + \theta(V_{GS} - V_T - V_{DS})}{1 + \theta(V_{GS} - V_T)}\right]\right\} \tag{7.46}$$

Equation (7.46) is, however, somewhat unwieldy and is not amenable to physical interpretation. Consequently, Equation (7.44) is normally approximated as

$$\mu_{\text{lf}} = \frac{\mu_0}{1 + \theta(V_{GS} - V_T)} \tag{7.47}$$

In this case the dependence on V_{ch} has been conveniently removed, so that $\mu = \mu_{\text{lf}}$ can be moved back outside the integral of Equation (7.16) with the result

$$I_D = \frac{WC'_{ox}\mu_0}{L[1+\theta(V_{GS}-V_T)]}\left(V_{GS}-V_T-\frac{V_{DS}}{2}\right)V_{DS}$$

$$= \frac{WC'_{ox}\mu_{lf}}{L}\left(V_{GS}-V_T-\frac{V_{DS}}{2}\right)V_{DS} \qquad V_{DS} \le V_{DS\text{sat}} \qquad (7.48)$$

$$I_{D\text{sat}} = \frac{WC'_{ox}\mu_{lf}}{L}\left(V_{GS}-V_T-\frac{V_{DS\text{sat}}}{2}\right)V_{DS\text{sat}} \qquad V_{DS} \ge V_{DS\text{sat}} \qquad (7.49)$$

These equations are similar to Equations (7.24) and (7.25) with constant μ replaced by μ_{lf}. The difference is that here the mobility is a function of V_{GS}.

EXAMPLE 7.8

Compare the I_D-V_{DS} characteristics for a MOSFET using the constant mobility model, and then taking the transverse field into account. Let $W/L = 5$, $\theta = 0.13\ \text{V}^{-1}$, $t_{ox} = 5$ nm, $V_T = 1$ V and $\mu_0 = 480\ \text{cm}^2/\text{V}\cdot\text{s}$.

■ Solution

We use Equations (7.48) and (7.49) where $\mu_{lf} = \mu_0$ in the constant mobility model and $\mu_{lf} = \mu_0/[1+\theta(V_{GS}-V_T)]$ to include the effect of $\mathscr{E}_T$ on μ_{lf}.

The oxide capacitance per unit area is

$$C'_{ox} = \frac{\varepsilon_{ox}}{t_{ox}} = \frac{3.9(8.85\times10^{-12})}{5\times10^{-9}} = 6.9\times10^{-3}\ \text{F/m}^2 = 6.9\times10^{-7}\ \text{F/cm}^2$$

The low-field mobility depends on $V_{GS}-V_T$. For $V_{GS}-V_T = 0$ V,

$$\mu_{lf} = \frac{\mu_0}{1+0} = \mu_0 = 480\ \text{cm}^2/\text{V}\cdot\text{s}$$

For $V_{GS}-V_T = 1$ V,

$$\mu_{lf} = \frac{\mu_0}{1+\theta(V_{GS}-V_T)} = \frac{480}{1+(0.13)(1)} = 425\ \text{cm}^2/\text{V}\cdot\text{s}$$

For $V_{GS}-V_T = 2$ V,

$$\mu_{lf} = \frac{\mu_0}{1+\theta(V_{GS}-V_T)} = \frac{480}{1+(0.13)(2)} = 380\ \text{cm}^2/\text{V}\cdot\text{s}$$

For $V_{GS}-V_T = 3$ V,

$$\mu_{lf} = \frac{\mu_0}{1+\theta(V_{GS}-V_T)} = \frac{480}{1+(0.13)(3)} = 345\ \text{cm}^2/\text{V}\cdot\text{s}$$

For $V_{GS}-V_T = 4$ V,

$$\mu_{lf} = \frac{\mu_0}{1+(0.13)(4)} = 316\ \text{cm}^2/\text{V}\cdot\text{s}$$

Figure 7.28 compares the constant mobility curves with the curves obtained by considering the effect of the transverse field. Notice that the drain currents are

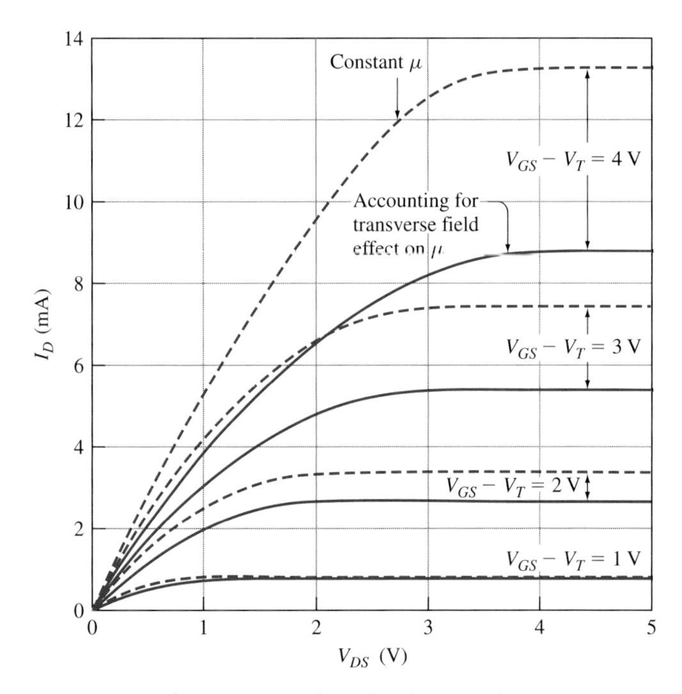

Figure 7.28 Comparison of I_D-V_{DS} characteristics computed by using the constant mobility model ($\theta = 0$, dashed lines) and taking into account the effect of the transverse field (solid line) for $\theta = 0.13\ \text{V}^{-1}$. The transverse field tends to reduce the currents.

noticeably smaller when the transverse field is accounted for. Note, however, that neglecting the effect of V_{ch}, i.e., using Equation (7.47) instead of (7.44), overestimates the effect of V_{GS} on the low-field mobility.

Effect of the Longitudinal Field $\mathscr{E}_L$ on Channel Mobility We have seen that the transverse field has an effect on the mobility and thus affects the values of the saturation current for a given gate voltage. In this section we examine the effect of the longitudinal field $\mathscr{E}_L$ on the mobility and thus on the I_D-V_{DS} curves.

We saw in Chapter 3 that in semiconductors, carrier velocities increase with increasing electric field and eventually saturate. This velocity saturation effect can be significant for carriers in the channel of a FET. In modern devices, the gate lengths are very small (a fraction of a micrometer), resulting in very high fields over a significant fraction of the channel length.

For many semiconductors, including Si, the carrier velocity in the channel of a FET can be empirically expressed as

$$|v| = \frac{\mu_{\text{lf}}|\mathscr{E}_L|}{1 + \dfrac{\mu_{\text{lf}}|\mathscr{E}_L|}{v_{\text{sat}}}} \tag{7.50}$$

where μ_{lf} is the low-field mobility (channel carrier mobility at low $\mathscr{E}_L$) and v_{sat} is the carrier saturation velocity in the channel. Equation (7.50) is used for both electrons in Si n-channel FETs and for holes in p-channel devices.

Since we know that

$$|v| = \mu|\mathscr{E}_L| \tag{7.51}$$

the mobility can be expressed as

$$\mu = \frac{\mu_{\text{lf}}}{1 + \dfrac{\mu_{\text{lf}}|\mathscr{E}_L|}{v_{\text{sat}}}} \tag{7.52}$$

From Equation (7.52) we see that with increasing field $|\mathscr{E}_L|$ along the channel, μ decreases. As we saw in Chapter 3, this is a result of the reduction in the mean free time between collisions due to optical phonon scattering. Figure 7.29 shows experimental data of electron mobility as a function of $\mathscr{E}_L$ in an n-channel MOSFET. The data are matched to Equations (7.51) and (7.52) (solid lines). From the figure, we see that the saturation velocity is $v_{\text{sat}} \approx 4 \times 10^6$ cm/s for this device. Although v_{sat} depends somewhat on μ_{lf}—which depends on temperature, transverse field, and substrate doping concentration—for carriers in the channel of a Si MOSFET, we will use $v_{\text{sat}} = 4 \times 10^6$ cm/s for both electrons and holes,

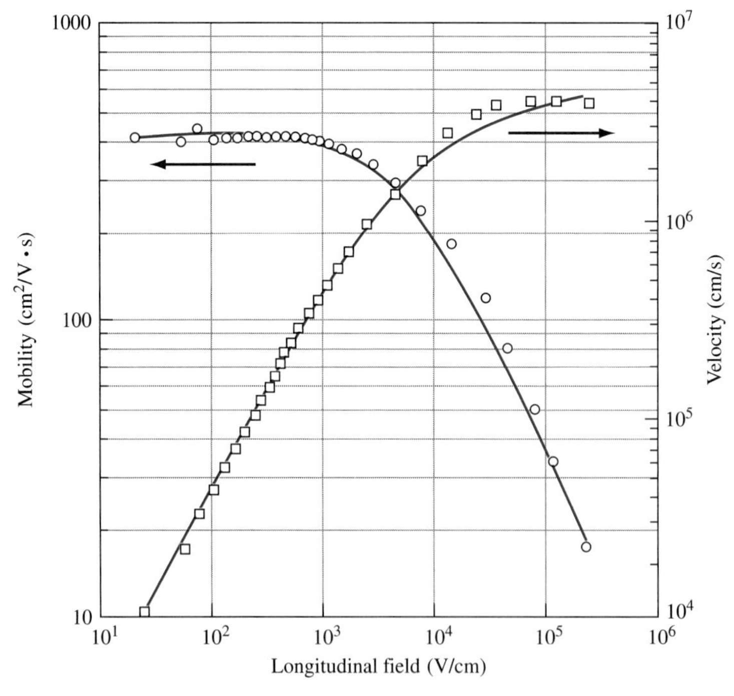

Figure 7.29 Channel electron mobility and velocity ($v = \mu\mathscr{E}$) as a function of lateral field for $V_{GS} = 1.42$ V.

unless otherwise indicated. [3] These values are somewhat smaller than the 10^7 cm/s found for bulk Si.

Effect of $\mathscr{E}_L$ on the I_D-V_{DS} Characteristics of MOSFETs We have seen that μ varies with longitudinal electric field, so now we will revise the simple long-channel model to account for this influence of the carrier mobility on the I_D versus V_{DS} characteristics. If we substitute Equation (7.52) into Equation (III.7), we can write

$$I_D\,dy = \frac{-WQ_{\mathrm{ch}}\mu_{\mathrm{lf}}\,dV_{\mathrm{ch}}}{1+\dfrac{\mu_{\mathrm{lf}}|\mathscr{E}_L|}{v_{\mathrm{sat}}}} \tag{7.53}$$

The electric field can be expressed as $|\mathscr{E}_L| = dV_{\mathrm{ch}}/dy$, so, substituting into the denominator, we have

$$I_D\,dy = -\frac{WQ_{\mathrm{ch}}\mu_{\mathrm{lf}}\,dV_{\mathrm{ch}}}{1+\dfrac{\mu_{\mathrm{lf}}}{v_{\mathrm{sat}}}\dfrac{dV_{\mathrm{ch}}}{dy}} \tag{7.54}$$

Multiplying both sides of Equation (7.54) by the denominator and rearranging gives

$$I_D\,dy + \frac{I_D\mu_{\mathrm{lf}}}{v_{\mathrm{sat}}}\,dV_{\mathrm{ch}} = -WQ_{\mathrm{ch}}\mu_{\mathrm{lf}}\,dV_{\mathrm{ch}} \tag{7.55}$$

Now we can integrate both sides. For the sublinear region, we write

$$I_D\int_0^L dy + I_D\int_0^{V_{DS}}\frac{\mu_{\mathrm{lf}}}{v_{\mathrm{sat}}}\,dV_{\mathrm{ch}} = -W\int_0^{V_{DS}} Q_{\mathrm{ch}}\mu_{\mathrm{lf}}\,dV_{\mathrm{ch}} \qquad V_{DS}\le V_{DS\mathrm{sat}} \tag{7.56}$$

and for the saturation region we have

$$I_{D\mathrm{sat}}\int_0^L dy + I_D\int_0^{V_{DS\mathrm{sat}}}\frac{\mu_{\mathrm{lf}}}{v_{\mathrm{sat}}}\,dV_{\mathrm{ch}} = -W\int_0^{V_{DS\mathrm{sat}}} Q_{\mathrm{ch}}\mu_{\mathrm{lf}}\,dV_{\mathrm{ch}} \qquad V_{DS}\ge V_{DS\mathrm{sat}} \tag{7.57}$$

Integrating, the results are

$$I_D = -\frac{W\mu_{\mathrm{lf}}\displaystyle\int_0^{V_{DS}} Q_{\mathrm{ch}}\,dV_{\mathrm{ch}}}{L+\dfrac{\mu_{\mathrm{lf}}V_{DS}}{v_{\mathrm{sat}}}} \qquad 0\le V_{DS}\le V_{DS\mathrm{sat}} \tag{7.58}$$

$$I_{D\mathrm{sat}} = -\frac{W\mu_{\mathrm{lf}}\displaystyle\int_0^{V_{DS\mathrm{sat}}} Q_{\mathrm{ch}}\,dV_{\mathrm{ch}}}{L+\dfrac{\mu_{\mathrm{lf}}V_{DS\mathrm{sat}}}{v_{\mathrm{sat}}}} \qquad V_{DS}\ge V_{DS\mathrm{sat}} \tag{7.59}$$

Using the same model for the channel charge Q_{ch} as in the long-channel model [Equation (7.14)], we obtain [4]

$$I_D = \frac{WC'_{ox}\mu_{lf}}{L + \dfrac{\mu_{lf}V_{DS}}{v_{sat}}}\left[(V_{GS} - V_T)V_{DS} - \frac{V_{DS}^2}{2}\right] \qquad V_{DS} \leq V_{DSsat} \qquad (7.60)$$

$$I_{Dsat} = \frac{WC'_{ox}\mu_{lf}}{L + \dfrac{\mu_{lf}V_{DSsat}}{v_{sat}}}\left[(V_{GS} - V_T)V_{DSsat} - \frac{V_{DSsat}^2}{2}\right] \qquad V_{DS} \geq V_{DSsat} \qquad (7.61)$$

These are similar to the expressions from the previous model of Equations (7.48) and (7.49), except that L is replaced by $L + \mu_{lf}V_{DS}/v_{sat}$ for $V_{DS} \leq V_{DSsat}$ and by $L + \mu_{lf}V_{DSsat}/v_{sat}$ for $V_{DS} \geq V_{DSsat}$. In effect, the inclusion of the longitudinal field $\mathscr{E}_L$ on the mobility causes the channel length to appear longer by the amount $\mu_{lf}V_{DS}/v_{sat}$ (or $\mu_{lf}V_{DSsat}/v_{sat}$) than for the simpler model. To reflect this effect, Equations (7.58) and (7.59) are normally written

$$I_D = \frac{-W\mu_{lf}\displaystyle\int_0^{V_{DS}} Q_{ch}\,dV_{ch}}{L\left(1 + \dfrac{\mu_{lf}V_{DS}}{Lv_{sat}}\right)} \qquad V_{DS} \leq V_{DSsat} \qquad (7.62)$$

$$I_{Dsat} = \frac{-W\mu_{lf}\displaystyle\int_0^{V_{DSsat}} Q_{ch}\,dV_{ch}}{L\left(1 + \dfrac{\mu_{lf}V_{DSsat}}{Lv_{sat}}\right)} \qquad V_{DS} \geq V_{DSsat} \qquad (7.63)$$

As we shall see shortly, $V_{DSsat} \neq (V_{GS} - V_T)$, unlike the case for the simple long-channel model.[9]

The quantity in the parentheses in the denominator of Equation (7.63), then, represents the influence of the velocity saturation effect on the I_D-V_{DS} characteristics, [4] and

$$I_D = \frac{I_D(\text{no velocity saturation model})}{1 + \dfrac{\mu_{lf}V_{DS}}{Lv_{sat}}} \qquad V_{DS} \leq V_{DSsat} \qquad (7.64)$$

$$I_{Dsat} = \frac{I_{Dsat}(\text{no velocity saturation model})}{1 + \dfrac{\mu_{lf}V_{DSsat}}{Lv_{sat}}} \qquad V_{DS} \geq V_{DSsat} \qquad (7.65)$$

[9]The effect of v_{sat} is often considered to be a short-channel effect since its influence on I_D increases with decreasing L. Here, however, since it is important for L in the low-micrometer range, we will not consider it to be a short-(submicrometer) channel effect. The true short-channel effects are discussed in Chapter 8.

Table 7.1 Some parameters for typical Si MOSFETs

Parameter	n-channel MOSFET	p-channel MOSFET
μ_{lf} (low-field mobility)	500 cm^2/V · s	200 cm^2/V · s
v_{sat} (carrier saturation velocity)	4×10^6 cm/s	4×10^6 cm/s
t_{ox} (gate oxide thickness)	4 nm	4 nm
$C'_{\text{ox}} = \dfrac{\varepsilon_{\text{ox}}}{t_{\text{ox}}}$ (oxide capacitance per unit area)	8.6×10^{-7} F/cm^2	8.6×10^{-7} F/cm^2

where $I_{D\text{sat}}$, without accounting for velocity saturation, is given by Equation (7.49), but $V_{DS\text{sat}} \neq (V_{GS} - V_T)$.

From the above equation, it is seen that the reduction of current from that given by the long-channel model is greater as the channel lengths get shorter. This results from $\mathscr{E}_L$ being large, causing the saturation velocity effects to be important over a larger fraction of the channel.

Since we will present some numerical illustrations of n-channel and p-channel MOSFETs, we present in Table 7.1 values for some parameters for room temperature operation of typical MOSFETs. We will also use these values in the following example.

EXAMPLE 7.9

Find the value of L for an n-channel Si MOSFET for which the velocity saturation effect reduces the subsaturation current by a factor of 2 for a drain-source voltage $V_{DS} = 2\,\text{V}$.

■ Solution

We recognize from Equation (7.64) that we want to set

$$\left(1 + \frac{\mu_{\text{lf}} V_{DS}}{L v_{\text{sat}}}\right) = 2$$

From Table 7.1, for $\mu_{\text{lf}} = 500\,\text{cm}^2/\text{V}\cdot\text{s}$ and $v_{\text{sat}} = 4 \times 10^6\,\text{cm/s}$, solving for L, we find a channel length of

$$L = \frac{\mu_{\text{lf}} V_{DS}}{v_{\text{sat}}} = \frac{(500\,\text{cm}^2/\text{V}\cdot\text{s}) \times (2\,\text{V})}{4 \times 10^6\,\text{cm/s}} = 250 \times 10^{-6}\,\text{cm} = 2.5\,\mu\text{m}$$

This channel length would be typical of the technology in the mid-1980s. Now, let us suppose the (physical) channel length is made even smaller than this value. Taking $L = 0.18\,\mu\text{m}$, a typical value in the late 1990s, we find that the current obtained using the long-channel model is off by a factor of

$$\frac{I_D\ \text{(long-channel model)}}{I_D} = 1 + \frac{\mu_{\text{lf}} V_{DS}}{L v_{\text{sat}}} = 1 + \frac{(500\,\text{cm}^2/\text{V}\cdot\text{s}) \times (2\,\text{V})}{(18 \times 10^{-6}\,\text{cm}) \times (4 \times 10^6\,\text{cm/s})} = 13.9$$

Similarly, for the more recent $0.13\,\mu\text{m}$ technology, the ratio is 19. Clearly, the effect of velocity saturation must be taken into account in realistic FETs.

Continuing the derivation of the I_D-V_{DS} characteristics of the FET, we rewrite Equations (7.64) and (7.65) in the standard form:

$$I_D = \frac{WC'_{ox}\mu_{lf}\left(V_{GS} - V_T - \dfrac{V_{DS}}{2}\right)V_{DS}}{L\left(1 + \dfrac{\mu_{lf}V_{DS}}{Lv_{sat}}\right)} \qquad V_{DS} \le V_{DSsat} \qquad (7.66)$$

$$I_{Dsat} = \frac{WC'_{ox}\mu_{lf}\left(V_{GS} - V_T - \dfrac{V_{DSsat}}{2}\right)V_{DSsat}}{L\left(1 + \dfrac{\mu_{lf}V_{DSsat}}{Lv_{sat}}\right)} \qquad V_{DS} \ge V_{DSsat} \qquad (7.67)$$

We need to find an expression for V_{DSsat}. We can use the same approach we took in Section 7.3.1. By setting $\partial I_D/\partial V_{DS} = 0$ in Equation (7.66):

$$V_{DSsat} = \frac{v_{sat}}{\mu_{lf}}L\left[\left(1 + \frac{2\mu_{lf}(V_{GS} - V_T)}{v_{sat}L}\right)^{1/2} - 1\right] \qquad (7.68)$$

Substituting (7.68) into (7.67) and solving for I_{Dsat}, we obtain

$$I_{Dsat} = WC'_{ox}v_{sat}(V_{GS} - V_T - V_{DSsat}) \qquad (7.69)$$

Let us compare the $I_D = V_{DS}$ characteristics as obtained from the simple model and the model that includes velocity saturation. For this example we choose the parameters of Table 7.1 with $V_{GS} - V_T = 2.6$ V and a channel length of $L = 0.5\,\mu$m. First, we plot the results from the long-channel model using constant mobility in Figure 7.30 (dashed line).

Before we continue, we notice that both the PFET and the NFET have the same result for the earlier model. Since the low-field mobilities for electrons and holes are different ($\mu_{lfn} \approx 2.5\mu_{lfp}$), the currents in the two FETs would be significantly different if the devices were otherwise identical. In many circuits using both NFETs and PFETs, it is desirable to have equal saturation currents of both devices. To achieve this for long-channel devices, the W/L ratios of PFETs are made 2.5 times that of NFETs. This is what has been done here. We choose $W = 10\,\mu$m and 25 μm respectively for the NFET and the PFET. This will equate the characteristics for the two devices as predicted from the long-channel model, i.e., Equation (7.48).

The characteristics from the model that considers velocity saturation are also shown in the figure. From Equations (7.66) and (7.67), it is evident that even with the width-to-length ratio corrected as above, the currents in the velocity saturation model are not the same for the NFET and PFET. We can see from Figure 7.30 that at small V_{DS}, both models predict the same slope. In this region, the longitudinal

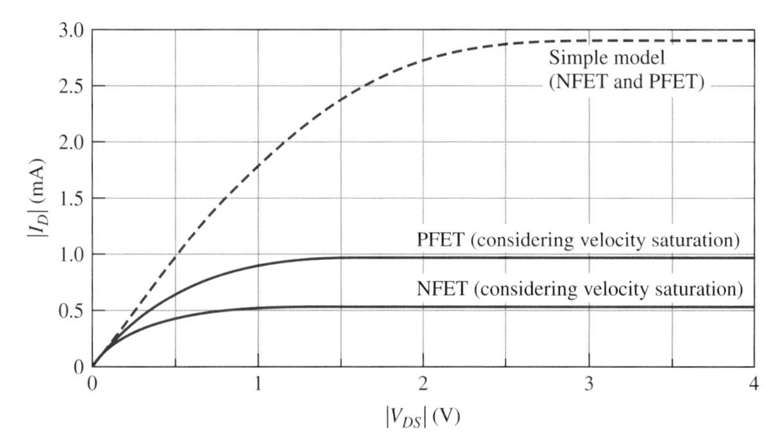

Figure 7.30 The calculated I-V_{DS} characteristics for the simple model and for NMOS and PMOS with carrier velocity saturation accounted for. The case of $V_{GS} - V_T = 2.6$ V and $L = 0.5$ μm is considered. The W/L ratios of the NFET and PFET have been scaled to produce the same I-V_{DS} curve as predicted from the simple model. The differing mobilities cause the scaled devices to be different.

field is not yet large enough for velocity saturation effects to be important. However, the inclusion of velocity saturation effects causes the saturation current to decrease and the voltage at which current saturates ($V_{DS\text{sat}}$) to decrease also. Since μ_{lf} is larger for the NFET than for the PFET, $I_{D\text{sat}}$ and $V_{DS\text{sat}}$ are smaller for the NFET. As we will see in Chapter 8, this means that the velocity saturation effect reduces the performance of the field-effect transistors.

We indicated earlier that, when velocity saturation is accounted for, the saturation voltage is no longer equal to $V_{GS} - V_T$. As the channel length gets shorter, from Equation (7.68) the saturation voltage $V_{DS\text{sat}}$ also decreases. For example, Figure 7.31 shows that for a channel length of $L = 2\,\mu\text{m}$ and $(V_{GS} - V_T) = 2.6\,\text{V}$, the saturation voltages are $V_{DS\text{sat}} = 1.7$ and 2.1 V respectively for the NFET and the PFET. For a shorter channel device, e.g., $L = 0.1\,\mu\text{m}$, these values reduce to 0.57 and 0.84 V respectively. These values are appreciably smaller than the $V_{DS\text{sat}} = (V_{GS} - V_T) = 2.6\,\text{V}$ we would get from the constant mobility long-channel model. Again, this is because the simple long-channel model does not consider velocity saturation.

While the saturation voltage $V_{DS\text{sat}}$ depends on the channel length L (but not the channel width W), the saturation current $I_{D\text{sat}}$ depends on both W and L. Figure 7.32 shows $I_{D\text{sat}}$ as a function of L for a constant W/L ratio of 10. The figure shows results for both n- and p-channel MOSFETs. As expected, for a given W/L ratio, $I_{D\text{sat}}$ decreases with decreasing channel length, again because of velocity saturation.

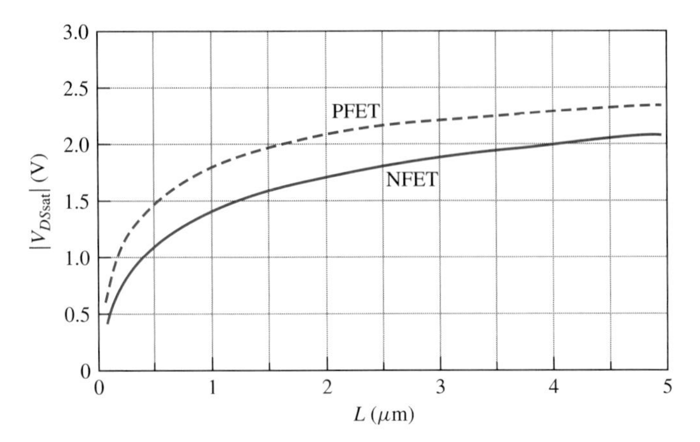

Figure 7.31 The saturation voltage as a function of channel length for $|(V_{GS} - V_T)| = 2.6$ V.

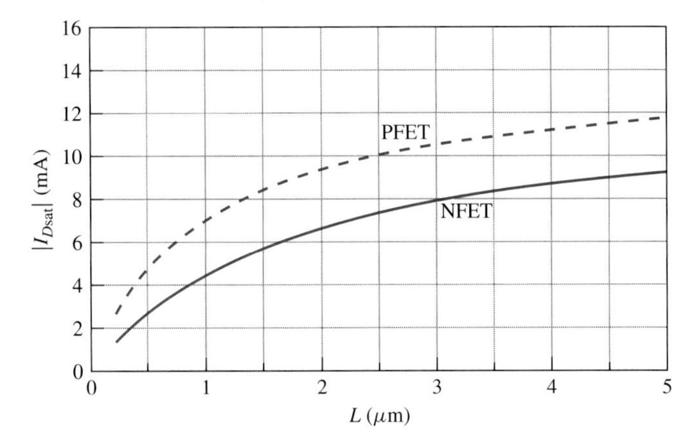

Figure 7.32 Saturation current $|I_{Dsat}|$ as a function of L for n- and p-channel silicon MOSFETs. Here $|(V_{GS} - V_T)| = 2.6$ V, the width-to-length ratio is $W/L = 10$, and $W_p = 2.5\ W_n$.

*7.3.3 SERIES RESISTANCE

Next, we look at the resistance of a FET, which is the resistance between the source and the drain. This includes the resistance along the channel itself, plus the resistances between the source and drain contacts and the channel, as shown in Figure 7.33 for an n-channel MOSFET. The total resistance R_{tot} in a FET is

$$R_{\text{tot}} = \frac{V_{DS}}{I_D} = R_S + R_{\text{ch}} + R_D \tag{7.70}$$

where R_S is the resistance from the source contact to the source end of the

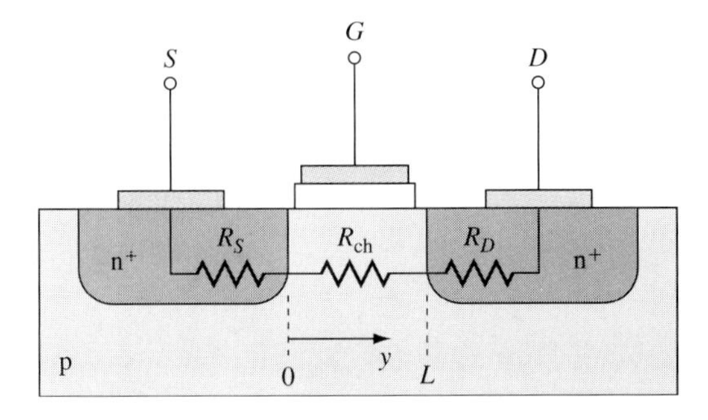

Figure 7.33 Schematic of an NMOS indicating the channel resistance R_{ch}, the source resistance R_S, and the drain resistance R_D.

channel, R_{ch} is the resistance of the channel, and R_D is the resistance from the drain end of the channel to the drain contact.

The region $0 \leq y \leq L$ can be treated as an embedded transistor with external resistances R_S and R_D connected in series to the source and drain contacts respectively. The voltage between the drain and source of the "intrinsic" transistor is then given by the drain-source voltage less the voltage drop in the "external" resistors, $V_{DS} - [I_D(R_S + R_D)]$. At the same time, the voltage between the gate and source of the intrinsic transistor is given by $V_{GS} - I_D R_S$. Then Equation (7.66) becomes, below saturation

$$I_D = \frac{WC'_{ox}\mu_{lf}\left[V_{GS} - I_D(R_S) - V_T - \dfrac{V_{DS} - I_D(R_S + R_D)}{2}\right][V_{DS} - I_D(R_S + R_D)]}{L\left\{1 + \dfrac{\mu_{lf}[V_{DS} - I_D(R_S + R_D)]}{Lv_{sat}}\right\}} \qquad V_{DS} \leq V_{DSsat} \tag{7.71}$$

Because of the symmetry of the MOSFET, the series resistances are normally equal, so that $R_S \approx R_D$ and Equation (7.71) becomes

$$I_D = \frac{WC'_{ox}\mu_{lf}\left(V_{GS} - V_T - \dfrac{V_{DS}}{2}\right)(V_{DS} - 2I_D R_S)}{L\left[1 + \dfrac{\mu_{lf}(V_{DS} - 2I_D R_S)}{Lv_{sat}}\right]} \qquad V_{DS} \leq V_{DSsat} \tag{7.72}$$

In saturation, Equation (7.67) is adapted in a similar manner to obtain

$$I_{Dsat} = \frac{WC'_{ox}\mu_{lf}\left(V_{GS} - V_T - \dfrac{V_{DSsat}}{2}\right)(V_{DSsat} - 2I_{Dsat} R_S)}{L\left[1 + \dfrac{\mu_{lf}(V_{DSsat} - 2I_{Dsat} R_S)}{Lv_{sat}}\right]} \qquad V_{DS} \geq V_{DSsat} \tag{7.73}$$

Equations (7.72) and (7.73) can be solved for the current, algebraically or iteratively.

We can neglect the series resistances for devices in which the channel resistance far exceeds the series resistances, or $R_{ch} \gg 2R_S$. The series resistance depends on the processing details as well as on the channel width W. Typical values for R_S and R_D are about 10 to 100 Ω.

7.4 COMPARISON OF MODELS WITH EXPERIMENT

We have examined several long-channel models for the I_D-V_{DS} characteristics of MOSFETs. These are the constant-mobility model, the model in which transverse field is included, and the model in which velocity saturation is included. We also modified these to account for series resistance. We now wish to explore the question: How good are these models?

Let us consider an actual n-channel MOSFET designed to operate with a supply voltage of 1.8 V. The measured parameters are:

$$L = 0.25\ \mu\text{m}$$

$$W = 9.9\ \mu\text{m}$$

$$t_{ox} = 4.7\ \text{nm}$$

$$\mu_{lf} = 400\ \text{cm}^2/\text{V} \cdot \text{s}$$

$$R_S = R_S = 19.9\ \Omega$$

$$V_T = 0.30\ \text{V}$$

The gate-source voltage is taken equal to the supply voltage, 1.8 V.

The top line in Figure 7.34 gives the results calculated from the simple long-channel model [Equations (7.48) and (7.49)], that is, the model that assumes constant mobility. Since we are considering only a single gate voltage, the appropriate value of low-field mobility has been selected and we do not need to consider the effect of the transverse field variation with gate voltage. Also plotted are the results obtained by using the same device parameters but considering the effects of the longitudinal field by accounting for velocity saturation [Equations (7.66) and (7.67)]. Also indicated in Figure 7.34 are the results of the model taking into account the source and drain series resistances (R_s and R_D) in addition to velocity saturation. [5] The saturation velocity was not measured for this device but was assumed to be $v_{sat} = 7 \times 10^6$ cm/s. Finally, the actual measured data are plotted.

It can be seen from Figure 7.34 that the current predicted by the velocity saturation model is appreciably smaller than that predicted by the constant mobility long-channel model, and is in reasonable agreement with experiment. Taking

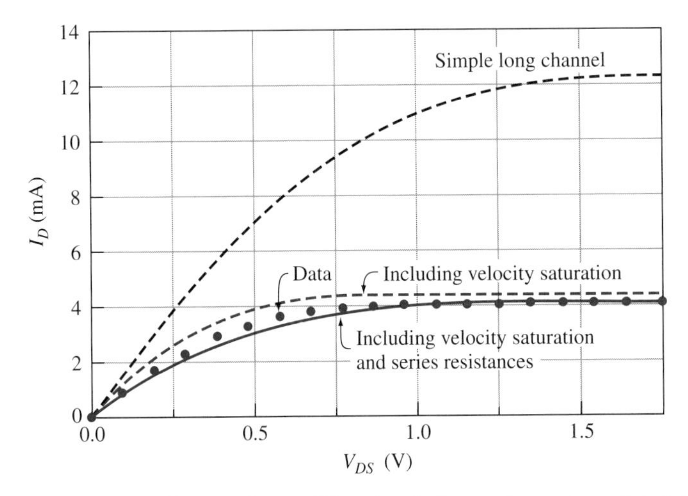

Figure 7.34 Comparison of the simple long-channel model, the model including velocity saturation, the model including both velocity saturation and the series resistances R_S and R_D, and the actual measured data. For this NFET device, $L = 0.25\ \mu$m, $W = 9.9\ \mu$m, and $V_T = 0.3$ V. The gate-source voltage is 1.8 V.

drain and source resistance into account results in quite good agreement between theory and experiment.

7.5 SUMMARY

In this chapter, we discussed the physical principles of operation of the Si-based MOSFET. These transistors have three terminals. The voltage on the gate electrode is used to control the resistance along the channel, between the drain and the source. These are called *field-effect* transistors because the gate controls the channel conductance via an electric field. This field appears across the oxide. Note that only displacement current flows into the gate, because the oxide is an insulator. The voltage applied to the gate creates the electric field that in turn bends the bands in the substrate, inverting the channel (enhancement MOSFETs) or uninverting it (depletion MOSFETs).

We began by considering a simple (long-channel) model for the current-voltage characteristics in a MOSFET, assuming constant channel mobility. We saw that when the gate voltage is greater than some threshold, current can flow in the channel. The greater the voltage across the channel (between the drain and the source), the more current should flow. We found that this is true, but only up to a point. At some V_{DS}, the longitudinal field $\mathscr{E}_L$ is large enough that it sweeps the carriers along the channel from source to drain as fast as they can be supplied by the source, and the current saturates. The saturation voltage $V_{DS\text{sat}}$ and the saturation current $I_{D\text{sat}}$ both depend on the gate voltage.

The equations for the simple long-channel model are:

SIMPLE LONG-CHANNEL MODEL

$$I_D \approx 0 \qquad V_{GS} < V_T \qquad (7.9)$$

$$I_D = \frac{WC'_{ox}\mu}{L}\left[(V_{GS} - V_T)V_{DS} - \frac{V_{DS}^2}{2}\right] \qquad V_{DS} \le V_{DS\text{sat}},\ V_{GS} \ge V_T \qquad (7.24)$$

$$I_{D\text{sat}} = \frac{WC'_{ox}\mu}{L}\left[\left(V_{GS} - V_T - \frac{V_{DS\text{sat}}}{2}\right)V_{DS\text{sat}}\right] \qquad V_{DS} \ge V_{DS\text{sat}},\ V_{GS} \ge V_T \qquad (7.25)$$

$$V_{DS\text{sat}} = V_{GS} - V_T \qquad (7.26)$$

The mobility is taken to be constant for all bias conditions, and is equal to the low-field mobility in the channel. This mobility is considerably less than the bulk value.

Next, we considered the effects of channel length modulation, and saw that under large drain voltages the voltage in the channel reaches saturation somewhere before the drain, producing a nonzero slope in the saturation characteristics. That slope was modeled empirically by

CHANNEL MODULATION EFFECT (SATURATION ONLY)

$$I_D \approx I_{D\text{sat}}(1 + \lambda(V_{DS} - V_{DS\text{sat}})) \qquad (7.43)$$

with

$$\frac{\Delta L}{L} = \lambda(V_{DS} - V_{DS\text{sat}}) \qquad (7.42)$$

where the channel is effectively shortened by ΔL.

Then we considered the effects of the transverse field. The channel is quite thin, and carriers will reflect off the potential barriers at the interface with the oxide and the band bending in the substrate. As a result, carrier velocities are reduced even more than accounted for earlier, with the result that the currents are smaller as well. The carrier low-field mobility can be modeled in terms of the gate voltage:

EFFECT OF TRANSVERSE FIELD (EMPIRICAL)

$$\mu_{\text{lf}} = \frac{\mu_0}{1 + \theta(V_{GS} - V_T)} \qquad (7.47)$$

where the parameters μ_0 (the low-field mobility at the source end at threshold) and θ are determined experimentally. The influence on the I_D-V_{DS} characteristics is to replace constant mobility μ in the simple long-channel model by the low-field mobility μ_{lf}, which is dependent on gate voltage.

These simple long-channel models ignore velocity saturation (i.e., it is assumed that the carriers could attain an arbitrary high velocity such that the $Q_{\text{ch}}v$ product is constant). However, velocity saturation does occur in real devices. The mobility is reduced by the high longitudinal field that results from the short channels. When we consider the reduced mobility, we saw that in a FET, the current could be expressed as

LONG-CHANNEL MODEL WITH VELOCITY SATURATION

$$I_D = \frac{I_D\ \text{(neglecting velocity saturation)}}{\left(1 + \dfrac{\mu_{\text{lf}} V_{DS}}{L v_{\text{sat}}}\right)} \qquad V_{DS} \leq V_{DS\text{sat}} \tag{7.64}$$

$$I_{D\text{sat}} = \frac{I_{D\text{sat}}\ \text{(neglecting velocity saturation)}}{\left(1 + \dfrac{\mu_{\text{lf}} V_{DS\text{sat}}}{L v_{\text{sat}}}\right)} \qquad V_{DS} \geq V_{DS\text{sat}} \tag{7.65}$$

The drain-source voltage at which the current saturates is

$$V_{DS\text{sat}} = \frac{v_{\text{sat}}}{\mu_{\text{lf}}} L \left[\left(1 + \frac{2\mu_{\text{lf}}(V_{GS} - V_T)}{v_{\text{sat}} L}\right)^{1/2} - 1\right] \tag{7.68}$$

The saturation current can be expressed in terms of $V_{DS\text{sat}}$

$$I_{D\text{sat}} = C'_{\text{ox}} W v_{\text{sat}} (V_{GS} - V_T - V_{DS\text{sat}}) \tag{7.69}$$

It is found that the model involving velocity saturation effects agrees reasonably well with experiment.

Finally, we considered the effect of the source and drain series resistances on the I_D-V_{DS} characteristics. The results are, including velocity saturation and assuming $R_S = R_D$,

VELOCITY SATURATION AND SERIES RESISTANCE

$$I_D = \frac{W C'_{\text{ox}} \mu_{\text{lf}} \left(V_{GS} - V_T - \dfrac{V_{DS}}{2}\right)(V_{DS} - 2I_D R_S)}{L\left[1 + \dfrac{\mu_{\text{lf}}(V_{DS} - 2I_D R_S)}{L v_{\text{sat}}}\right]} \qquad V_{DS} \leq V_{DS\text{sat}} \tag{7.72}$$

$$I_{D\text{sat}} = \frac{W C'_{\text{ox}} \mu_{\text{lf}} \left(V_{GS} - V_T - \dfrac{V_{DS\text{sat}}}{2}\right)(V_{DS\text{sat}} - 2I_{D\text{sat}} R_S)}{L\left[1 + \dfrac{\mu_{\text{lf}}(V_{DS\text{sat}} - 2I_{D\text{sat}} R_S)}{L v_{\text{sat}}}\right]} \qquad V_{DS} \geq V_{DS\text{sat}} \tag{7.73}$$

In all of these models, one can include transverse field effects if the value of μ_{lf} is varied with V_{GS} according to Equation (7.47), and one can account for channel length modulation by allowing $I_{D\text{sat}}$ to increase with V_{DS} according to Equation (7.41). In the next chapter, we will explore the MOSFET further. We'll see that as devices become progressively smaller, there are more physical effects that come into play and change the I_D-V_{DS} characteristics. In addition, we have restricted ourselves to static operation up to now; in Chapter 8, we will look at dynamic operation as well.

7.6 READING LIST

Items 1, 2, 4, 8 to 12, 17 to 19, and 27 to 30 in Appendix G are recommended.

7.7 REFERENCES

1. C. Y. Sah, "Evolution of the MOS transistor-from concept to VLSI," *Proc. IEEE,* 76, pp. 1280–1326, 1988.
2. G. Baccarani and M. R. Wordman, "Transconductance degradation in thin-oxide MOSFETs," *IEEE Trans. Electron Devices,* ED-30, pp. 1295–1304, 1983.
3. Dennis Hoyniak, Edward Nowak, and Richard L. Anderson, "Channel electron mobility dependence on lateral electric field in field-effect transistors," *J. Appl. Phys.,* 87, pp. 876–881, 2000.
4. B. T. Murphy, "Unified field-effect transistor theory including velocity saturation," *IEEE J. Solid-State Circuits,* SC-15, pp. 325–327, 1980.
5. Dac C. Pham, "Selective device-temperature scaling for optimum power-delay product in MOSFET circuit design," dissertation, University of Vermont, 1998. Unpublished.

7.8 REVIEW QUESTIONS

1. Explain why virtually no current flows into the gate of a MOSFET. If no current flows into this electrode, how can a signal on the gate have an effect on the operation of the rest of the transistor?
2. Explain in your own words why the current (rate of flow) between source and drain saturates as the potential difference between the drain and the source increases.
3. Summarize in words the steps used to derive the I_D-V_{DS} characteristics of a FET using the long-channel model as an example.
4. Explain in words how applying a voltage to the gate can, in effect, change the material at the surface of the channel from p type to n type.
5. Is the device in Figure 7.5 an enhancement or a depletion FET?

7.9 PROBLEMS

7.1 In any circuit, the transistor operation can be understood by examining the superposition of the transistor characteristics and the conditions imposed by the circuit (the load line). What differentiates, then, a digital transistor circuit from an analog one?

7.2 In modern FETs, the gate is usually degenerately doped silicon, whose Fermi level is essentially at the bottom of the conduction band (for an NFET) or at the top of the valence band (for a PFET).

a. Draw an energy band diagram similar to Figure 7.2c, except making the transistor a PFET.

b. Suppose the device is an NFET, but the gate is made of metal (as was done in the early days). Draw the energy band diagram. Take $\Phi_M < \Phi_S$.

7.3 For the transistor of Figure P7.1, by how much should the gate voltage be changed to produce inversion? Threshold? Assume half the applied voltage appears across the oxide and half across the semiconductor. If the device in the figure is at equilibrium, is this an enhancement or a depletion FET?

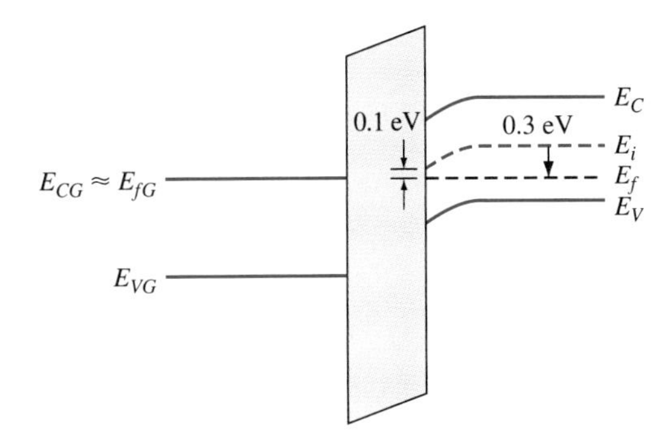

Figure P7.1

7.4 For each of the transistors of Figure 7.10,

a. What is the polarity of the threshold voltage V_T?

b. What is the polarity of V_{DS} that should be used?

c. When $V_{GS} = 0$ (equilibrium), is the transistor on or off?

d. Is the current I_D carried by electrons or holes?

e. For $V_{GS} > V_T$, is the current I_D carried primarily by drift or diffusion?

7.5 An NFET is fabricated with a degenerately doped n-type gate. What doping concentration in the p-type substrate is needed for there to be a channel even with no voltage applied? Assume half the built-in voltage is dropped across the oxide and half across the silicon. Let the definition of "a channel exists" be that the surface of the silicon is inverted such that the electron concentration at the Si/SiO_2 interface is equal to the hole concentration in the bulk is p type. Is the doping you found a minimum or a maximum required to create a depletion-type device?

7.6 Consider two silicon MOSFETs, one n channel and the other p channel, with substrate dopings of 10^{16} cm^{-3}. The NMOS has an n$^+$ gate and the PMOS has a p$^+$ gate, both doped to 10^{19} cm^{-3}. Find the built-in voltage V_{bi} for each, and draw the energy band diagram. Neglect the band-gap narrowing effects discussed in Chapter 2.

7.7 In the MOS process, structures like the gate of a transistor are used to make capacitors as well. If the oxide thickness is 4 nm, what area is needed to achieve a capacitance of 1 pF? The permittivity of silicon diode is $3.9\varepsilon_0$.

7.8 In MOS processing, the W/L ratio is often intentionally made different from transistor to transistor. Plot the I_D-V_{DS} characteristics for the device of Figure 7.19 with the W/L ratio changed to 10 instead of 5. Compare this with the results for $W/L = 5$.

7.9 Explain why the electron affinity model can be used to good approximation to determine the band lineup normal to the gate in a MOSFET. That is, why is the tunneling-induced dipole effect negligible?

7.10 Plot the I_D-V_{DS} characteristics for an NFET, using the long-channel model, for which $W = 10\,\mu$m, $L = 1\,\mu$m, $t_{ox} = 4$ nm, $V_T = 0.25$ V, and the channel length modulation parameter is $\lambda = 0.04\,\text{V}^{-1}$. Use $V_{GS} = 1$ V, 2 V, 3 V, and 4 V. Find the output conductance in saturation for $V_{GS} = 3$ V.

7.11 An enhancement NFET with the characteristics in Table 7.1 has a threshold voltage of $V_T = 1$ V, a channel length of 1 μm, and a width of 5 μm. Considering velocity saturation, with $v_{sat} = 5 \times 10^6$ cm/s, find the current I_D for

a. $V_{GS} = 0\,\text{V},\ V_{DS} = 1\,\text{V}$

b. $V_{GS} = 2\,\text{V},\ V_{DS} = 1\,\text{V}$

c. $V_{GS} = 3\,\text{V},\ V_{DS} = 1\,\text{V}$

7.12 An NFET is made with $t_{ox} = 4$ nm, $L = 1\,\mu$m, $W = 10\,\mu$m, $V_T = 1$ V, and $\mu_{lf} = 500\,\text{cm}^2/\text{V}\cdot\text{s}$.

If the simple model is used, what should the width of the PFET be to get the same saturation current (apart from polarity). Let the low-field mobility for holes be 200 cm^2/V · s.

7.13 An NFET and a PFET are made on the same chip, using the same process. The NFET has $C'_{ox} = 8.6 \times 10^{-7}$ F/cm^2, $t_{ox} = 4$ nm, $L = 0.2\,\mu$m, $W = 15\,\mu$m, $V_T = 1.5$ V, and $\mu_{lf} = 500\,\text{cm}^2/\text{V}\cdot\text{s}$. If the PFET is identical except for its mobility (200 cm^2/V · s) and its width W,

a. What should W be for the PFET to make the characteristics the same as for the NFET, as predicted by the simple model?

b. Find V_{DSsat} and I_{Dsat} for $V_{GS} - V_T = 1$ V.

c. If velocity saturation is considered, how different are V_{DSsat} and I_{Dsat} for the NFET and the PFET compared with the simple model results? Express your result as a ratio (e.g., $I_{DsatNFET}/I_{DsatPFET}$ and $V_{DSsatNFET}/V_{DSsatPFET}$). Assume $v_{sat} = 4 \times 10^6$ cm/s.

7.14 A good way to check the validity of a derivation is to verify that it reduces to the expected result for a particular known case. For example, in the simple model, we neglected velocity saturation, and in the later model we considered it. Since the high field that causes velocity saturation occurs in short-channel devices, we would expect that the later model would reduce to the simple model for long-channel devices.

a. Show that in the limit of large L, V_{DSsat} as given by Equation (7.68) reduces to $V_{DSsat} = (V_{GS} - V_T)$ as given by Equation (7.26) for the simple (long-channel) model.

b. Show that in the limit of large L, I_{Dsat} as given by Equation (7.61) reduces to Equation (7.25) for the simple model.

You may find the following information useful:

For $x < 1$

$$(1 \pm x)^n = 1 \pm nx + \frac{n(n-1)x^2}{2!} \pm \cdots$$

$$(1 \pm x)^{-n} = 1 \mp nx + \frac{n(n+1)x^2}{2!} \mp \cdots$$

7.15 Find an approximate closed-form expression for the drain current, accounting for series resistance and velocity saturation. That is, solve for I_D in Equation (7.72). Neglect the term in I_D^2.

8 CHAPTER

Additional Considerations for FETs

8.1 INTRODUCTION

In Chapter 7, the basic static characteristics of MOSFETs were discussed. In this chapter, we look at some additional considerations for MOSFETs and discuss other types of FETs.

First, we will develop our understanding of the parameters that control MOSFET behavior. Specifically, up to now the threshold voltage and the low-field mobility were treated as known (measured) quantities. In this chapter, we will show how these parameters can be measured for particular MOSFETs.

After that, we will look at MOSFETs in action. We begin with a simple CMOS device. CMOS means complementary **MOS,** circuitry in which both n-channel and p-channel devices are used. We choose the inverter logic circuit as a typical circuit and investigate its operation, power dissipation, and delays in switching (propagation delays).

Although the inverter is a classic example of a digital circuit, MOSFETs are also used frequently in analog circuitry and in fact, CMOS can be used as a high-gain, highly linear amplifier. Thus, we briefly investigate the small-signal equivalent circuits for MOSFETs.

Then we discuss the ways that channel length influences the electrical properties of MOSFETs. As devices become progressively smaller, the models of Chapter 7 must be adapted to include the effects of very short channels ($L < 1\ \mu$m).

Other types of field-effect transistors (MESFETs, JFETs, etc.) are briefly discussed. These devices, while important, currently are used less frequently than MOSFETs.

8.2 MEASUREMENT OF THRESHOLD VOLTAGE AND LOW-FIELD MOBILITY

In Chapter 7, we took the value of the threshold voltage V_T to be known. It is difficult to predict accurately the threshold voltage of a MOSFET during the design phase. That is because it depends on some process characteristics (for example, charges trapped in the oxide and defects at the silicon-oxide interface). Thus in practice, the threshold voltage is measured for a sample device produced in a particular fabrication process.

The threshold voltage can be measured by comparing the experimental current-voltage characteristics with those predicted by theory. To understand these measurements, we begin by recalling from Chapter 7 that above threshold but below saturation, the drain current has the form

$$I_D = \frac{WC'_{\text{ox}}\mu_{\text{lf}}\left(V_{GS} - V_T - \dfrac{V_{DS}}{2}\right)V_{DS}}{L\left(1 + \dfrac{\mu_{\text{lf}}V_{DS}}{Lv_{\text{sat}}}\right)} \tag{8.1}$$

where, from Equation (7.47),

$$\mu_{\text{lf}} = \frac{\mu_0}{1 + \theta(V_{GS} - V_T)} \tag{8.2}$$

Equation (8.1) is a model including the effect of the transverse field (in μ_{lf}) and longitudinal field (in velocity saturation). For V_{DS} small enough that the saturation velocity (i.e., the second term in the denominator) can be neglected, or $(\mu_{\text{lf}}V_{DS}/Lv_{\text{sat}} \ll 1)$ Equation (8.1) simplifies to

$$I_D = \frac{WC'_{\text{ox}}\mu_{\text{lf}}\left(V_{GS} - V_T - \dfrac{V_{DS}}{2}\right)V_{DS}}{L} \tag{8.3}$$

Then substituting Equation (8.2) into Equation (8.3) gives

$$I_D = \frac{WC'_{\text{ox}}\mu_0\left(V_{GS} - V_T - \dfrac{V_{DS}}{2}\right)V_{DS}}{L[1 + \theta(V_{GS} - V_T)]} \tag{8.4}$$

For small $(V_{GS} - V_T)$ this can be approximated

$$I_D = \frac{WC'_{\text{ox}}\mu_0\left(V_{GS} - V_T - \dfrac{V_{DS}}{2}\right)V_{DS}}{L} \tag{8.5}$$

This is the equation of a straight line of I_D against V_{GS}. The intercept ($I_D = 0$) is at

$$V_{GS}(0) = V_T + \frac{V_{DS}}{2} \tag{8.6}$$

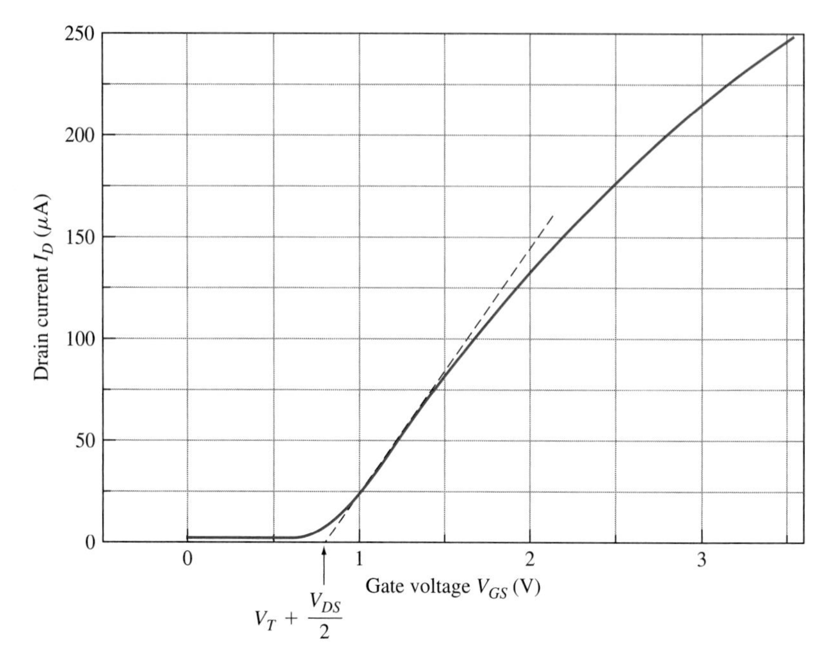

Figure 8.1 MOSFET I_D-V_{GS} characteristics measured experimentally at $V_{DS} = 0.1$ V. The data are plotted on a linear plot. The dashed line shows the extrapolation of the curve at its point of highest slope to the horizontal axis. The extrapolation intersects the V_{GS} axis at the value $V_T + V_{DS}/2$. The slope can be used to find μ_0 and the deviation of the data from the straight line can be used to find θ.

and the slope is

$$\frac{dI_D}{dV_{GS}} = \frac{WC'_{ox}\mu_0 V_{DS}}{L} \tag{8.7}$$

Figure 8.1 shows how the threshold can be determined from an experimental plot of I_D versus V_{GS}. The dashed line has its intercept given by Equation (8.6). From the extrapolated intercept at $I_D = 0$,

$$V_T = V_{GS}(0) - \frac{V_{DS}}{2} \tag{8.8}$$

Further, the slope of the dashed line can be used to find the value of μ_0. Since the values of W, C'_{ox}, ($C'_{ox} = \varepsilon_{ox}/t_{ox}$) are known from the fabrication process, then the low-field mobility at threshold, μ_0, can be determined by using Equation (8.7). [1]

We see, however, that the experimental I_D-V_{GS} curve is not a straight line as predicted from Equation (8.3). This is because μ_{lf} is also dependent on V_{GS}.

Above, the threshold voltage V_T was obtained by extrapolating the tangent to the curve at the point of largest slope[1] to the axis $I_D = 0$. At greater V_{GS}, however, the $\theta(V_{GS} - V_T)$ term in Equation (8.4) results in a deviation of the I_D-V_{DS} characteristic from the dashed line.

EXAMPLE 8.1

Find the values of V_T from the I_D-V_{GS} characteristics of Figure 8.1. The drain voltage is $V_{DS} = 0.1$ V.

■ **Solution**

From Figure 8.1 the straight-line $V_{GS}(0)$ intercept is at $V_{GS} = 0.78$ V. Since $V_{DS} = 0.1$ V, from Equation (8.8),

$$V_T = 0.78 - \frac{0.1}{2} = 0.73 \text{ V}$$

The value of θ can be determined from the deviation of the I_D-V_{GS} characteristics from a straight line as discussed in the Supplement to Part 3.

8.3 SUBTHRESHOLD LEAKAGE CURRENT

In Chapter 7, we developed a formulation for the I_D-V_{DS} characteristics. There we assumed that below threshold ($V_{GS} < V_T$), the drain current was negligible. From Figure 8.1, the current appears to be small below threshold. Figure 8.2, however, plots the same data as in Figure 8.1, but this time on a semilogarithmic scale. The [log I_D-V_{GS}] characteristic approximates a straight line below threshold. This means that I_D decreases approximately exponentially with decreasing V_{GS} below threshold.

We can't neglect this subthreshold leakage current for the following reason. In switching circuits, a device is on when it is operating well above threshold ($V_{GS} \approx V_{DD}$, where V_{DD} is the supply voltage). The device is in the off state when it is operating well below threshold ($V_{GS} \approx 0$).[2] Even in the off state, a small current does flow. In an integrated circuit with hundreds of thousands or millions of transistors, this off-state current can result in considerable power dissipation, and a resulting temperature rise. Thus, it is worthwhile to evaluate the value of the subthreshold leakage current.

We know that the electron concentration in the channel at the source varies exponentially with source-channel barrier height E_B. Therefore the current, which is determined by the number of carriers available, has the same dependence, or

$$I_D = I_0 e^{q(V_{GS}-V_T)/nkT} \qquad (V_{GS} < V_T) \tag{8.9}$$

[1]This approximates the region where $\theta(V_{GS} - V_T) \ll 1$.

[2]We are discussing enhancement devices, such as are common in digital circuits.

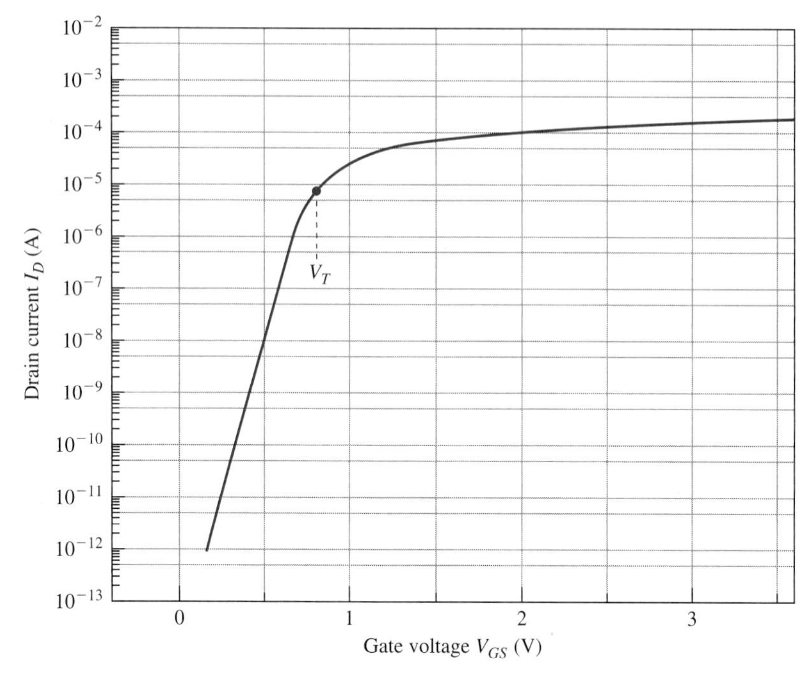

Figure 8.2 The same I_D-V_{GS} characteristics from Figure 8.1, but this time presented on a semilog plot. The current varies approximately exponentially with V_{GS} below threshold.

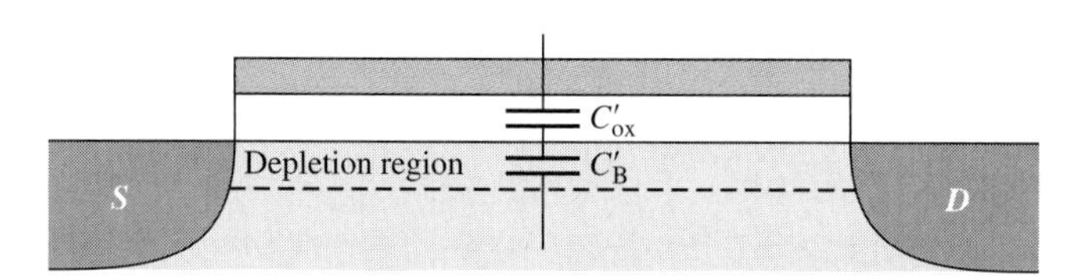

Figure 8.3 The capaciticances of the MOSFET gate oxide and substrate. The substrate capacitance C'_B is the depletion layer capacitance.

where I_0 is the current at threshold ($V_{GS} = V_T$) and $1/n$ is the fraction of $(V_{GS} - V_T)$ that affects the source-channel barrier. As the gate voltage is varied, some of the change in voltage is dropped across the oxide and some is dropped across the semiconductor. This can be written as

$$\frac{1}{n} = \frac{\Delta V_{\text{ch}}(y = 0)}{\Delta V_{GS}} \tag{8.10}$$

where $\Delta V_{\text{ch}}(y = 0)$ is the channel voltage with respect to the source at the source end of the channel.

We can find n as follows. Figure 8.3 shows that the gate structure can be viewed as two capacitors in series. These are C'_{ox} and C'_B, where C'_{ox} is the oxide capacitance per unit area and C'_B is the substrate (bulk) capacitance per unit area, caused by the depletion layer at the semiconductor surface. Below threshold, neglecting the small charge in the channel, C'_{ox} and C'_B act as a voltage divider between gate and substrate. That means that if the gate voltage is changed by ΔV_{GS}, that change is also divided between C'_{ox} and C'_B. Then

$$\frac{\Delta V_{ch}}{\Delta V_{GS}} = \frac{1}{n} = \frac{\dfrac{1}{C'_B}}{\dfrac{1}{C'_B} + \dfrac{1}{C'_{ox}}} = \frac{C'_{ox}}{C'_{ox} + C'_B} \tag{8.11}$$

or the constant n in Equation (8.10) is

$$n \approx \frac{C'_{ox} + C'_B}{C'_{ox}} = 1 + \frac{C'_B}{C'_{ox}} \tag{8.12}$$

Figure 8.2 showed the variation of I_D with V_{GS} for a constant $V_{DS} = 0.1$ V with the value of V_T indicated. For gate voltages well below threshold, because of the exponential dependence, the current is in fact small enough not to be a problem. However, as the gate-source voltage gets close to V_T, the current becomes appreciable. To minimize off-state leakage current, when the transistors are off, they should be strongly off. On the other hand, when the transistor is on the gate voltage must be changed enough to produce a strong change in I_D. We therefore ask, "What variation in gate voltage would produce a factor of 10 change in the drain current?" To answer this, we define the gate voltage variation per decade of current in the straight-line region as the "swing," S. It is found from

$$S = \frac{\partial V_{GS}}{\partial \log I_D} \tag{8.13}$$

Taking this derivative using the expression for V_{GS} as a function of I_D in Equation (8.9),

$$S = \frac{2.3kTn}{q} \tag{8.14}$$

The factor 2.3 comes from converting from the natural log to log base 10.

For the device shown, $S = 84$ mV/decade. We note from Equation (8.12) that the parameter n depends on the ratio of C'_B/C'_{ox}. Ideally, we would like to have the voltage below threshold vary as little as possible and still get large current change. For small S, we want to have a small C'_B (created by a large transition region width or, equivalently, low substrate doping) and a large C'_{ox} (thin oxide, or small t_{ox}). The value of n approaches its minimum value of unity as C'_{ox} approaches infinity or t_{ox} approaches zero. To minimize S, the value of n should be minimized. The minimum value for S is then, for unity n at room temperature,

$$S_{min} = 2.3\frac{kT}{q} \approx 60 \text{ mV/decade} \tag{8.15}$$

For logic circuits, the threshold voltage is generally chosen to be approximately 20 percent of the supply voltage, or $V_T \approx 0.2V_{DD}$. Minimizing S can minimize the required threshold voltage V_T and thus the supply voltage for the chip, V_{DD}. Decreasing V_{DD} lowers the power dissipation during switching as discussed in Section 8.5.

EXERCISE 8.2

Consider a chip with 2 million transistors, half of which are off at a given time. It is desired to keep the total off-state subthreshold current for the chip below 10 μA. For a single device, the current at threshold is 1 μA and the subthreshold slope S is 80 mV per decade of current. Estimate the minimum power supply voltage V_{DD}. The input gate voltage varies between 0 and V_{DD}.

■ Solution

The total current at V_{GS} (below threshold, the off state) for the chip is 10 μA, so the maximum leakage current allowable per transistor is $I_D = 10^{-5}$ A/10^6 transistor = 10^{-11} A/transistor. At threshold the current is $I_D = 10^{-6}$ A/transistor, so between $V_{GS} = 0$ and $V_{GS} = V_T$, there are five decades of current. At 80 mV/decade, that results in a threshold of $V_T = 80 \times 5$ mV $= 0.4$ V. Since normally it is chosen that $V_{DD} \approx 5V_T$, then $V_{DD} = 2$ V.

The power dissipation associated with the leakage current in the previous example is 2 V $\times$ 10^{-5} A $= 2$ μW/chip. This is not the total power dissipation, however. As will be discussed in Section 8.5, the power dissipation associated with switching is normally much greater than that associated with subthreshold leakage.

8.4 COMPLEMENTARY MOSFETs (CMOS)

So far, we have discussed n-channel MOSFETs and p-channel MOSFETs independently. Currently, most integrated circuits use both n-channel and p-channel devices, hence the term complementary MOSFETs, or CMOS. Figure 8.4a shows the schematic cutaway view of a CMOS inverter using the so-called n-well technology. The n well is ion-implanted into a p-type substrate. The p-channel device is fabricated in the n well while the n-channel FET is fabricated directly into the p substrate.

A cross-sectional schematic view of the inverter is shown in Figure 8.4b. Note that the n^+ source is connected to the substrate (body). There is a pn junction between the source of the n-channel device and the substrate. It has zero bias across it, however, preventing current from flowing between the source and the substrate. The p-channel source is connected to the n well. This ensures that current cannot flow out of the p source into the n well. Note that the n-well–p-substrate junction is reverse biased so that the well-substrate current is small. Note also that both gates (inputs) are connected to each other, and the two drains (outputs) are connected to each other.

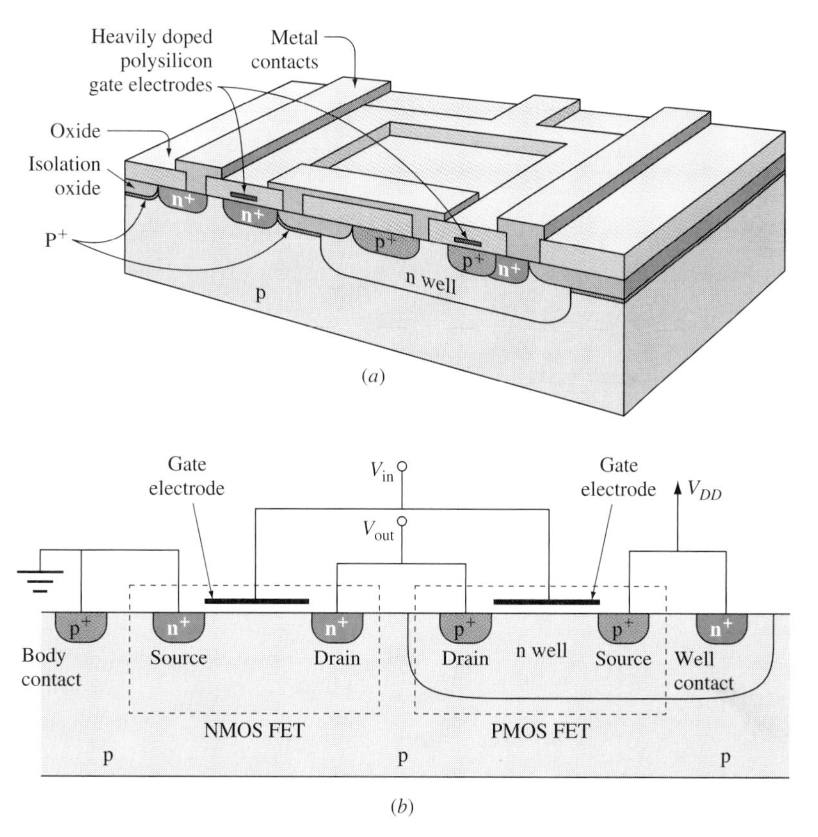

Figure 8.4 The CMOS inverter. (a) Physical structure, adapted from C. G. Fonstad, *Microelectronic Devices and Circuits,* McGraw-Hill,1994; (b) cross-sectional diagram, adapted from R. C. Jaeger, *Microelectronic Circuit Design,* McGraw-Hill, 1997.

8.4.1 OPERATION OF THE INVERTER

A simplified circuit diagram of a digital CMOS inverter is shown in Figure 8.5a. Both channel regions are indicated by broken lines (the lines parallel to the gate electrode) to indicate that these are enhancement devices; i.e., for zero volts between gate and source, they are off. Let us assume a power supply voltage of V_{DD} and an input square wave of 0 to V_{DD} as indicated in Figure 8.5b. When the input voltage is 0 V, the gate-to-source voltage of the NFET is zero and the device is off. Its channel does not conduct. That transistor acts as an open circuit. In the PFET, however, the gate is negative with respect to the source and so this device is on. The channel acts as a conductor between the power supply V_{DD} and the output port. As a result, the output voltage is also V_{DD}, or at a logic 1 state.

When, however, the gate voltage $V_{\text{in}} = V_{DD}$, the NFET is on and the PFET is off. Thus, the output port is effectively tied to ground, or is at logic 0. The circuit inverts the input signal. Note that when the NMOS gate voltage is $V_{GSn} = 0$,

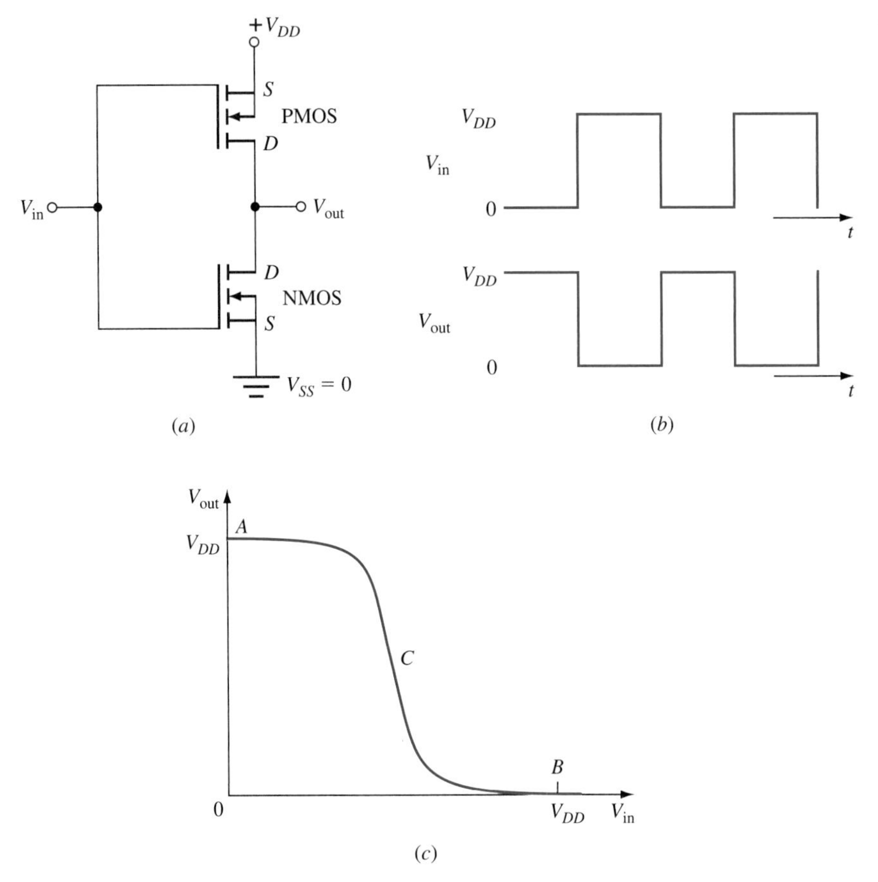

Figure 8.5 (a) CMOS logic inverter circuit diagram; (b) input and output waveforms; (c) the transfer characteristics of a CMOS inverter.

the PMOS gate-to-source voltage is $V_{GSp} = -V_{DD}$, and when $V_{GSn} = V_{DD}$ then $V_{GSp} = 0$. In either state, one device is off, and so in the steady state no current can flow from V_{DD} to ground.[3]

The transfer characteristic (V_{out} versus V_{in}) is shown in Figure 8.5c. For $V_{in} = 0$, the device is operating in region *A*. Here the NMOS is off while the PMOS is on. In region *B*, for $V_{in} = V_{DD}$, the NMOS is on and the PMOS is off. In region *C*, where $V_{in} \approx V_{DD}/2$, both NMOS and PMOS are operating in their current saturation regions. The steepness in region *C* is a measure of the switching speed or, in analog circuits, the voltage gain.

*8.4.2 MATCHING OF CMOS DEVICES

In CMOS logic inverters, for maximum switching speed, it is usually desirable to match the I_D-V_{DS} characteristics of the NMOS and PMOS devices as closely

[3]Actually, of course, the subthreshold leakage current of the off device does flow.

as possible. Since the I_D-V_{DS} characteristics depend in part on the mobilities [see, for example, Equations (7.66) and (7.67)], and the mobilities of electrons and holes are different, the characteristics would be different for otherwise identical devices. The NFETs and PFETs must therefore be "matched." This is typically done by adjusting the widths of the transistor gates to match their saturation currents. From Equation (7.69),

$$I_{D\text{sat}} = C'_{\text{ox}} W v_{\text{sat}} (V_{GS} - V_T - V_{DS\text{sat}}) \tag{8.16}$$

If we assume that the channel lengths are the same for both the NFETs and PFETs, and that $v_{\text{satn}} = v_{\text{satp}}$ and $C'_{\text{oxn}} = C'_{\text{oxp}}$, then to make $I_{D\text{satn}} = I_{D\text{satp}}$, we obtain

$$\frac{W_p}{W_n} = \left| \frac{V_{GS} - V_T - V_{DS\text{satn}}}{V_{GS} - V_T - V_{DS\text{satp}}} \right| \tag{8.17}$$

From Equation (7.68), we have

$$V_{DS\text{sat}} = \frac{v_{\text{sat}}}{\mu_{\text{lf}}} L \left[\left(1 + \frac{2\mu_{\text{lf}} |V_{GS} - V_T|}{v_{\text{sat}} L} \right)^{1/2} - 1 \right] \tag{8.18}$$

All quantities in Equation (8.18) are the same for the n- and p-channel devices except for the low-field mobilities μ_{lf}. Recall from Table 7.1 that typical values are $\mu_{\text{lfn}} = 500\ \text{cm}^2/\text{V}\cdot\text{s}$ and $\mu_{\text{lfp}} = 200\ \text{cm}^2/\text{V}\cdot\text{s}$. Thus, for a given $V_{GS} - V_T$, the ratio W_p/W_n depends on the difference of $V_{DS\text{satn}}$ and $V_{DS\text{satp}}$. Figure 8.6 plots the ratio (W_p/W_n) required to match the transistor currents as a function of channel length L for $(V_{GS} - V_T) = 2.6$ V (with $V_{DD} = 3.3$ V). The widths of the NFETs and PFETs vary from $W_p/W_n = 1.15$ for a channel length of $L = 0.1\ \mu\text{m}$ to $W_p/W_n = 2.2$ for $L = 18\ \mu\text{m}$. This latter compares with $W_p/W_n = 2.5$ predicted by the long-channel model, neglecting velocity saturation effects.

Note, however, that although the saturation currents have been matched, the voltage at which saturation begins, $V_{DS\text{sat}}$, is independent of W. That means that

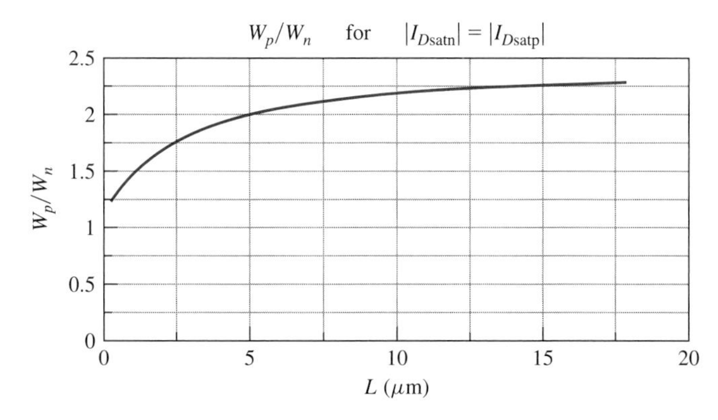

Figure 8.6 The ratio of W_p/W_n needed to match saturation currents ($I_{D\text{satn}} = I_{D\text{satp}}$), as a function of channel length, $L_{\text{nFET}} = L_{\text{pFET}}$.

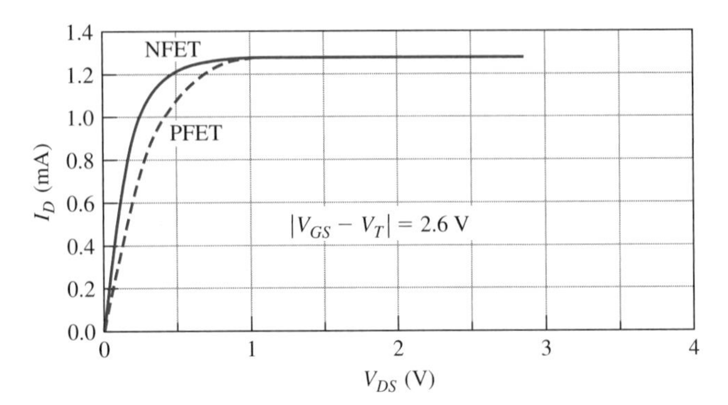

Figure 8.7 Comparison of I_D-V_{DS} characteristics for NMOS and PMOS with matched saturation current. $L_{\text{nFET}} = L_{\text{pFET}} = 0.2\ \mu\text{m}$; $W/L = 10$ for the NMOS and $W/L = 12.2$ for the PMOS.

below saturation, the I_D-V_{DS} characteristics for the NMOS and PMOS still do not match. This is illustrated in Figure 8.7 for a channel length of 0.2 μm, where from Figure 8.6, $W_p/W_n = 1.22$. The width-to-length ratios for the NMOS and PMOS are $W/L = 10$ and 12.2 respectively. The device characteristics can be approximated more closely below threshold by making the channel lengths $L_p < L_n$ and readjusting W_p/W_n.

8.5 SWITCHING IN CMOS INVERTER CIRCUITS

Up to now, we have discussed the steady-state aspects of a CMOS logic inverter switch. In this section, we consider the transient switching effects.

8.5.1 EFFECT OF LOAD CAPACITANCE

Consider the circuit of Figure 8.8a. This is the inverter circuit shown earlier in Figure 8.5, but we have added a load capacitance. The capacitor C_L includes stray wiring capacitance, the output capacitance of the circuit, and the capacitance of the input to the next stage. Unfortunately, it takes some current to charge or discharge the capacitance when the circuit switches from one logic state to the other. This charging and discharging introduces a time delay. In addition, a current flows to ground during the transition, creating additional power dissipation.

When the capacitance is being charged (output goes low to high), current flows from the power supply through the PMOS to the capacitance. During discharging, the current flows from the capacitance through the NMOS to ground. The net result is the transfer of some amount of charge to ground from the power supply, which increases the overall power consumption. The output voltage takes a finite time to change state as indicated in Figure 8.8b. The current, indicated in (c), flows only during switching. (The current waveform and the fall and rise time delays, t_{df} and t_{dr} are discussed in the next section.)

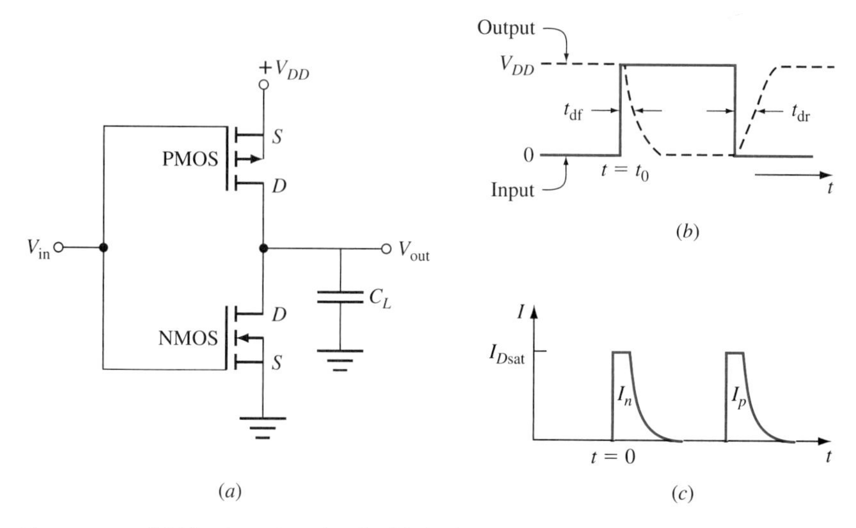

Figure 8.8 (a) The inverter circuit; (b) the input and output signals. The solid line is the input waveform; the dashed line is the output. The output cannot change instantaneously because it requires some time for the load capacitance to charge and discharge. (c) The charging and discharging current that flows during a cycle. One pulse flows through the n-channel device, the other pulse flows through the p-channel device.

During each cycle, then, a charge $Q = C_L V_{DD}$ flows from the power supply to ground. The energy transferred per cycle is:

$$\text{Energy/cycle} = QV_{DD} = C_L V_{DD}^2 \tag{8.19}$$

The power dissipated is the energy per cycle times the number of cycles per second, or frequency. The dynamic power dissipation or the power associated with charging and discharging the load capacitance during switching is then[4]

$$P_{\text{dynamic}} = C_L V_{DD}^2 f \tag{8.20}$$

For high-speed circuits the frequency f is large. To minimize the power dissipation, and thus the temperature of the chip, it is required to minimize C_L and V_{DD}. The minimum supply voltage, however, is dictated by the need to have a small leakage current in the off state as discussed in Section 8.3. This leakage current contributes to static power dissipation.

EXAMPLE 8.3

(a) Find the dynamic power dissipation for a CMOS inverter operating with $V_{DD} = 2.5$ V at 100-MHz frequency. Assume that the load capacitance is 0.1 pF. (b) Repeat for a future CMOS circuit with 20 million CMOS pairs operating at 15 GHz with $V_{DD} = 1.1$ V.

[4]For a more elegant derivation, see Reference 2.

■ **Solution**

a. From Equation (8.20),

$$P_{\text{dynamic}} = C_L V_{DD}^2 f = 10^{-13} \times (2.5)^2 \times 10^8 = 6.25 \times 10^{-5}\ \text{W}$$

b. $P_{\text{dynamic}} = C_L V_{DD}^2 f \times \text{number of CMOS pairs} = 10^{-13} \times (1.1)^2 \times (15 \times 10^9) \times (20 \times 10^6) = 36.3\ \text{kW}$

The dissipation of this much power illustrates a problem in future circuits.

8.5.2 PROPAGATION (GATE) DELAY IN SWITCHING CIRCUITS

The speed of operation of a switching circuit depends on the time required for a change in the input to be reflected in the output. In other words, once a change in signal is applied at the input, it takes some time for that new information to propagate "through" the circuit. One measure of the time delay between input and output is the propagation delay time t_d, sometimes referred to as gate delay. It is defined as the time it takes from the input being at 50 percent of its voltage swing to the output being at 50 percent of its voltage swing. As an example, we consider again the CMOS inverter of Figure 8.8.

For our purposes, an estimate of the delay time is adequate. To get this estimate, we assume a voltage step function series of pulse between zero and V_{DD} is applied to the input at $t = t_0 = 0$ as in Figure 8.8b.

The input and output voltages range between zero and V_{DD}, the power supply voltage. The propagation delay times (there are two, one for rising and one for falling) are the times needed to switch between $V_{\text{in}} = V_{DD}/2$ and $V_{\text{out}} = V_{DD}/2$; these are indicated as t_{dr} for rising and t_{df} for falling in Figure 8.8b. Since these times may differ, the propagation delay is defined as the average of the two:

$$t_d = \frac{t_{\text{df}} + t_{\text{dr}}}{2} \tag{8.21}$$

On the load capacitor, the charge is $Q = C_L V_{\text{out}}$. When the output changes from V_{DD} to $V_{DD}/2$, the charge on the capacitor changes by ΔQ, where

$$\Delta Q = -C_L \frac{V_{DD}}{2} = -\int_0^{t_{\text{df}}} I_{Dn}\, dt \tag{8.22}$$

The current flows through the NFET during this transition from high toward low at the output, hence I_{Dn}.

Consider an integrated CMOS circuit for which $V_{GS} - V_T = 2.6$ V, corresponding to $V_{DD} = 3.3$ V. Knowing I_{Dn} as a function of time, t_{df} can be calculated from Equation (8.22). The value of t_{df} can be estimated with the aid of Figure 8.9. Here we have plotted the current I_{Dn} in the NMOS as the input voltage is switched from logic 1 to logic 0. Initially the NMOS is off and the PMOS

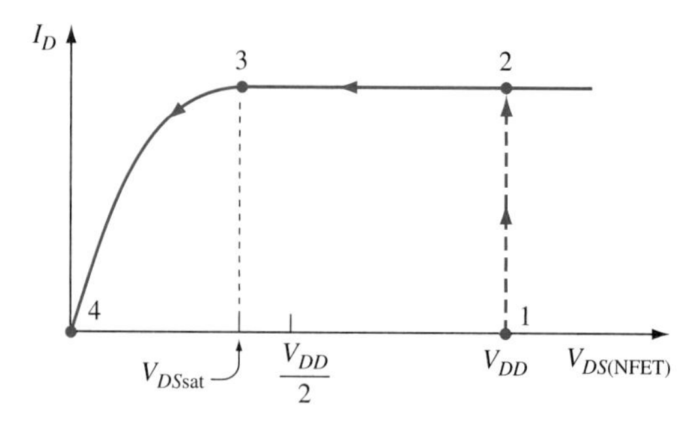

Figure 8.9 The current in the NFET as the input is switched from a logic high to a logic low. Initially the NFET output (drain) is high (point 1). When the input is switched, the NFET turns on and current can flow (point 2). The flowing current I_{Dn} discharges the load capacitance (point 3), sending the output voltage to zero (point 4).

is on, so the output voltage (also the voltage $V_{D(\text{NFET})}$) is high. At $t = 0$, the NMOS is turned on and the PMOS is turned off by a step voltage input as indicated in Figure 8.8a. This corresponds to I_{Dn} going from zero (point 1 in Figure 8.9) to $I_{D\text{sat}}$ (point 2). As the capacitor discharges through the NMOS at a rate proportional to I_{Dn}, V_{DSn} decreases with time from V_{DD} (point 2) to $V_{DS\text{sat}}$ (point 3) and to zero (point 4). For $V_{DS\text{sat}} \le V_{DD}/2$ as indicated, the current is constant at $I_{D\text{sat}}$ during the time t_{df} required for the output to decrease from V_{DD} to $V_{DD}/2$. For this case, from Equation (8.22),

$$t_{\text{df}} \approx \frac{C_L V_{DD}}{2I_{D\text{satn}}} \tag{8.23}$$

In this analysis, we are considering the propagation delay for an inverter discharging its load capacitance through the NFET. Therefore, the values for the parameters (L, W, μ_{lf}, $V_{DS\text{sat}}$, v_{sat}, C'_{ox}) used to calculate $I_{D\text{sat}}$ in Equation (8.23) are those of the NFET. The load capacitance is *charged,* however, through the PFET when the output goes low to high. Thus t_{dr} is

$$t_{\text{dr}} \approx \frac{C_L V_{DD}}{2I_{D\text{satp}}} \tag{8.24}$$

For this, the parameters used to find $I_{D\text{sat}}$ pertain to the PFET. From Equations (8.21), (8.23), and (8.24) we can write

$$t_d = \frac{1}{2}\left[\frac{C_L V_{DD}}{2I_{D\text{satn}}} + \frac{C_L V_{DD}}{2I_{D\text{satp}}}\right] \tag{8.25}$$

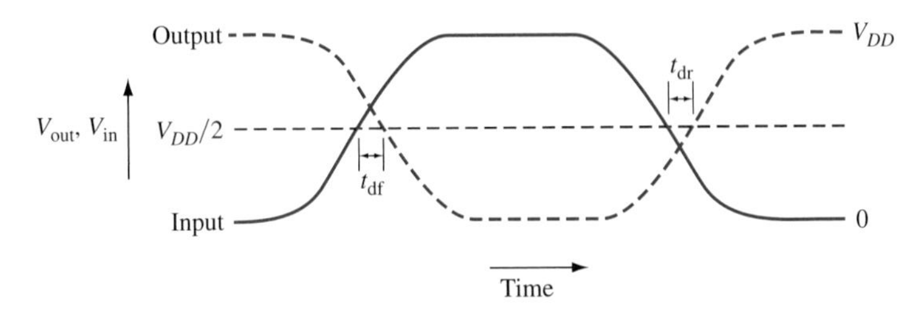

Figure 8.10 The definition of rise and fall delay times when the input is not a perfect step function.

We now investigate the validity of Equations (8.23) and (8.24), which rely on the condition that $|V_{DSsat}| \leq V_{DD}/2$. In the previous chapter, Figure 7.31 indicated the values of $|V_{DSsat}|$ as functions of channel length for NFETs and PFETs with the value of $V_{GS} - V_T = 2.6$ V. The value for the required saturation voltage is $|V_{DSsat}| \leq V_{DD}/2 = 3.3/2 = 1.65$ V. This value of V_{DSsat} can be obtained by making the channel length $L < 1.8\ \mu$m for the NFET, and $L < 0.8\ \mu$m for the PFET. For L in the submicrometer range then, a reasonable approximation is to let $I_D = I_{Dsat}$.

Ideally, to minimize gate delay we would want the rise and fall propagation delay times to be equal. To equate them, we make I_{Dsatn} and I_{Dsatp} approximately equal in Equation (8.25). We do this by appropriately adjusting the (W_p/W_n) ratios of the transistors as discussed earlier.

In the above discussion of propagation delay, it was assumed that a step voltage function was applied to the input. In reality, of course, the input voltage takes some time to change state, since it is coming from the output of a previous circuit. For this case, t_{df} and t_{dr} are indicated in Figure 8.10 as the time between the input and the output reaching their $V_{DD}/2$ values. They are the times between the input crossing $V_{DD}/2$ and the output crossing $V_{DD}/2$ in the falling and rising cases, respectively.

EXAMPLE 8.4

Determine the propagation delay time for the CMOS inverter of Figure 8.8. Assume $I_{Dsatn} = I_{Dsatp} = 1$ mA, $C_L = 0.1$ pF, and $V_{DD} = 2.5$ V.

■ Solution

From Equation (8.25), with $I_{Dsatn} = I_{Dsatp} = I_{Dsat}$,

$$t_d = \frac{C_L V_{DD}}{2I_{Dsat}} = \frac{10^{-13}\ \text{F} \times 2.5\ \text{V}}{2 \times 10^{-3}\ \text{A}} = 1.25 \times 10^{-10}\ \text{s}$$

or $t_d = 125$ ps.

In short-channel MOSFETs (discussed in Section 8.7), the expressions for the currents are more complex than discussed here, and the load capacitance

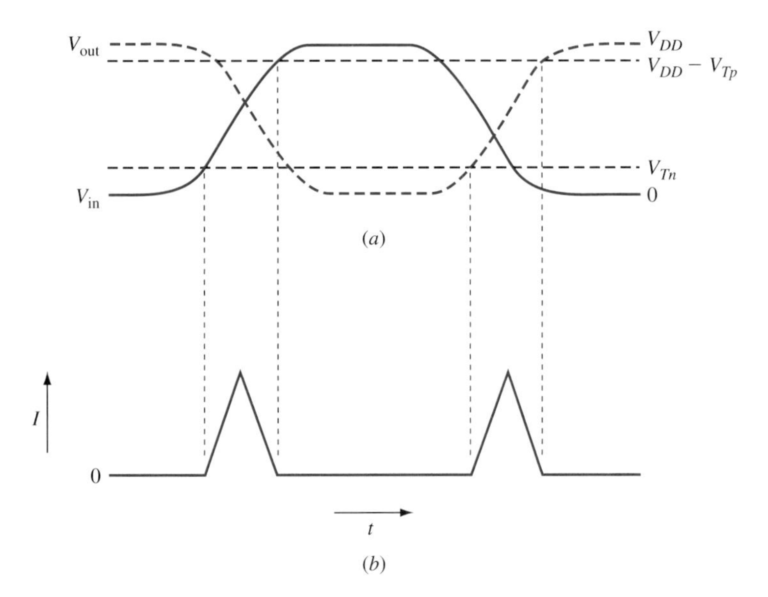

Figure 8.11 During the time that both devices are on, current flows from V_{DD} to ground. (a) The voltage waveforms; (b) the current waveform.

depends on V_{DS}. In such cases, for accurate results, t_d is normally calculated numerically by using a circuit simulator.

8.5.3 PASS-THROUGH CURRENT IN CMOS SWITCHING

The switching waveforms from Figure 8.10 are repeated in Figure 8.11a with some details added. The threshold voltages for the NMOS and PMOS are V_{Tn} and V_{Tp} respectively. We can see that when the input is switching, and passing through the range $V_{\text{Tn}} < V_{\text{in}} < (V_{DD} - |V_{\text{Tp}}|)$, both transistors are briefly on at the same time. During this interval, current flows from the power supply through the transistors to ground as indicated in part (b) of the figure. This results in dynamic power dissipation. The average value of this current decreases as the transition time of the input voltage is reduced. The power dissipation associated with this *pass-through* current can approach 50 percent of that given by Equation (8.20).

8.6 MOSFET EQUIVALENT CIRCUIT

In addition to the fundamental principles of MOSFET operation that have already been discussed in this and the preceding chapter, there are a number of parasitic elements that must be taken into account. These are indicated in Figure 8.12.

The source and drain resistances R_S and R_D are the resistances from the source and drain contacts to the edges of the channel that were discussed in

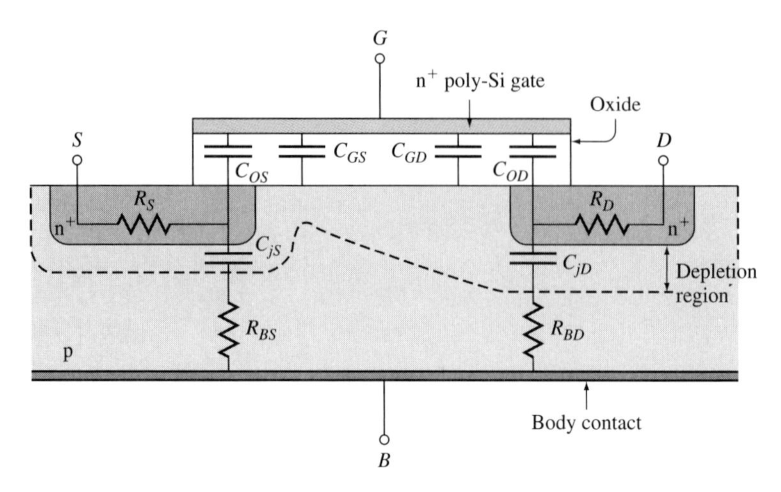

Figure 8.12 Schematic diagram of an n-channel MOSFET showing the various resistances and capacitances.

Chapter 7. The source-to-substrate and drain-to-substrate resistances, R_{BS} and R_{BD}, are the resistances from the edges of the source-substrate and drain-substrate depletion regions to the substrate (body) contact.

The gate oxide prevents direct current from flowing into the gate. There is, however, some capacitance between the gate and the n^+ source, C_{OS}, and between the gate and the n^+ drain, C_{OD}. These are called *overlap capacitances* since the gate overlaps the source and drain regions. In addition, there is capacitance between the gate and the channel. While strictly speaking this is a distributed capacitance, it is normally modeled as two lumped capacitances—between gate and source, C_{GS}, and between gate and drain, C_{GD}. There are also junction capacitances, C_{jS} and C_{jD}, between the n^+ source and drain and the p-type substrate. Note that although the C'_{ox} that we have used in the past is a capacitance per unit area, the capacitances C_{OS}, C_{OD}, C_{GS}, C_{GD}, C_{jS}, and C_{jD} are just capacitances (not per unit area).

The overlap capacitance can be reduced by the use of a self-aligned gate structure in which the gate is used as a mask for the source and drain implantation process. Since the overlap capacitances can therefore be made small, we will neglect them in the next section, where we cover the small-signal equivalent circuit of a MOSFET.

8.6.1 SMALL-SIGNAL EQUIVALENT CIRCUIT

We have discussed the CMOS inverter as a representative example of a digital circuit. MOSFETs are also used in analog circuits, and to analyze such circuits it is useful to have a device characterized by its small-signal equivalent circuit. The small-signal equivalent circuit of Figure 8.12 is shown in Figure 8.13a. For simplicity, in (b) the resistances have been neglected; however, a current

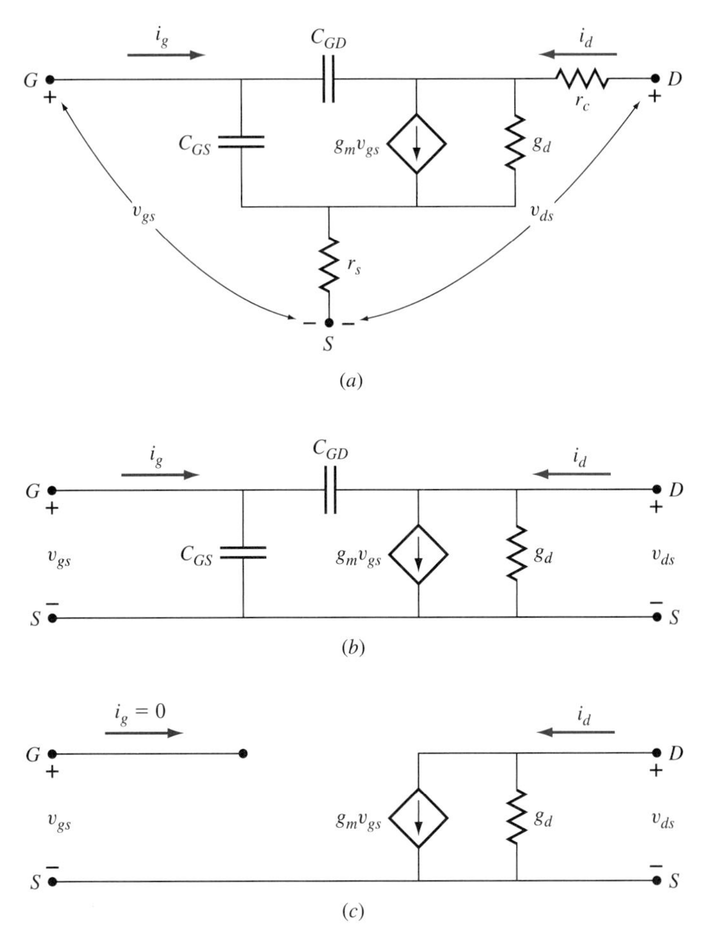

Figure 8.13 Small-signal equivalent circuit for a MOSFET (a) at high frequencies; (b) neglecting the source and drain resistances; (c) at low frequencies.

generator, $g_m v_{gs}$ has been added, where g_m is the transconductance, as will be discussed. An output conductance g_d has also been added. The input small signal voltage[5] is v_{gs}, and the output small signal voltage is v_{ds}. The input and output small signal currents are i_g and i_d.

Figure 8.13c shows an equivalent circuit for a MOSFET at low frequencies. By low frequencies, we mean signals that are varying slowly enough that the capacitances have negligible effect on the signals.

[5]We use the conventional notation that capital letters are used for steady-state or dc values (V_{GS}, V_{DS}, I_D, etc.) and small variations around those steady-state points are represented with small letters (v_{gs}, v_{ds}, i_d, etc.).

At both high and low frequencies, the output current is characterized by a current generator $g_m v_{gs}$ in parallel with an output conductance g_d. It is important to notice that the input signal is a voltage, but the output quantity that we generally consider with these transistors is a current. The transfer function $i_{\text{out}}/v_{\text{in}}$ has units of conductance. The transconductance g_m of the device, then, is

$$g_m \equiv \frac{i_d}{v_{gs}} = \frac{\partial I_D}{\partial V_{GS}} \tag{8.26}$$

where V_{DS} is considered constant, or v_{ds} is zero.

The output conductance results from the slope of the I_D-V_{DS} characteristics with V_{GS} constant.

$$g_d = \frac{i_d}{v_{ds}} = \frac{\partial I_D}{\partial V_{DS}} \tag{8.27}$$

EXAMPLE 8.5

For the device of Figure 7.18 ($t_{\text{ox}} = 4.7\,\text{nm}$, $L = 0.27\,\mu\text{m}$, $W = 8.6\,\mu\text{m}$ and $V_T = 0.3\,\text{V}$), whose I_D-V_{DS} characteristics are shown in Figure 8.14, find the transconductance in saturation, $g_{m\text{sat}}$, at $V_{DS} = 1.5\,\text{V}$, and the output conductance in saturation $g_{d\text{sat}}$ at $V_{GS} = 1.8\,\text{V}$.

■ Solution

To find the transconductance at $V_{GS} = 1.5\,\text{V}$, we choose the points A and B at a constant $V_{DS} = 1.5\,\text{V}$. From Equation (8.26), we have, in saturation,

$$g_{m\text{sat}} = \frac{\partial I_D}{\partial V_{GS}} \approx \frac{\Delta I_D}{\Delta V_{GS}} = \frac{4.3\,\text{mA} - 2.22\,\text{mA}}{1.8\,\text{V} - 1.2\,\text{V}} = 3.5\,\frac{\text{mA}}{\text{V}} = 3.5\,\text{mS}$$

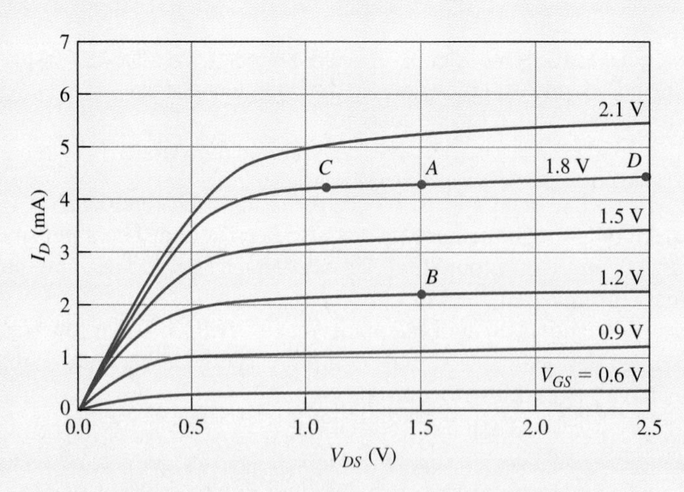

Figure 8.14 How to find g_m and g_d for a MOSFET from electrical measurements (Example 8.5).

The output conductance is found from points C and D, using Equation (8.27), and applying it in the saturation region:

$$g_{d\text{sat}} = \left.\frac{\partial I_D}{\partial V_{DS}}\right|_{\text{sat}} \approx \frac{4.50\,\text{mA} - 4.25\,\text{mA}}{2.5\,\text{V} - 1.2\,\text{V}} = 0.178\,\text{mS}$$

This is equivalent to $(0.178 \times 10^{-3})^{-1} = 5.3\,\text{k}\Omega$ of output resistance.

We will now present some approximate equations for the transconductance g_m and the output conductance g_d. To do this, some parasitic effects are ignored. These are: the source and drain resistances and the influence of the transverse field $\mathscr{E}_T$ on the mobility μ_{lf}, both covered in Chapter 7. We will also ignore the influence of V_{DS} on V_T and on the effective channel length L; these effects are discussed in Section 8.8. We will however, consider velocity saturation here since it has a major impact on the I_D-V_{DS} characteristics, and we will use the velocity saturation model for the I_D-V_{DS} characteristics as the starting point for the derivation.

We see from Equations (8.26) and (8.27) that the two conductances can be found from the I_D-V_{DS} characteristics. The I_D-V_{DS} relation was given by Equation (7.66) and is repeated here:

$$I_D = \frac{WC'_{\text{ox}}\mu_{\text{lf}}\left(V_{GS} - V_T - \dfrac{V_{DS}}{2}\right)V_{DS}}{L\left(1 + \dfrac{\mu_{\text{lf}}V_{DS}}{Lv_{\text{sat}}}\right)} \qquad V_{DS} \le V_{D S\text{sat}} \tag{8.28}$$

This is from the model taking velocity saturation into account. The corresponding saturation current is [Equation (7.67)]

$$I_D = I_{D\text{sat}} = \frac{WC'_{\text{ox}}\mu_{\text{lf}}\left(V_{GS} - V_T - \dfrac{V_{DS\text{sat}}}{2}\right)V_{DS\text{sat}}}{L\left(1 + \dfrac{\mu_{\text{lf}}V_{DS\text{sat}}}{Lv_{\text{sat}}}\right)} \qquad V_{DS} \ge V_{DS\text{sat}} \tag{8.29}$$

We start with the transconductance. Taking the partial derivative as prescribed by Equation (8.26), we obtain for $V_D \le V_{D\text{sat}}$

$$g_m = \frac{WC'_{\text{ox}}\mu_{\text{lf}}V_{DS}}{L\left(1 + \dfrac{\mu_{\text{lf}}V_{DS}}{Lv_{\text{sat}}}\right)} \qquad V_{DS} \le V_{DS\text{sat}} \tag{8.30}$$

This is the result below saturation. Above saturation, the transconductance $g_{m\text{sat}}$ can be determined from Equations (8.26) and (8.29). This is more complicated than it looks, however, because $V_{DS\text{sat}}$ is a function of V_{GS}. Let us do as was done before in Chapter 7. There we used Equation (7.66) [the same as Equation (8.28)], and found the value of $I_{D\text{sat}}$ by setting the slope $\partial I_D/\partial V_{DS}$ to zero. We repeat the result here:

$$I_{D\text{sat}} = Wv_{\text{sat}}C'_{\text{ox}}(V_{GS} - V_T - V_{DS\text{sat}}) \tag{8.31}$$

Inserting this into Equation (8.26) and taking the partial derivative with respect to V_{GS} gives

$$g_{m\text{sat}} = W v_{\text{sat}} C'_{\text{ox}} \left(1 - \frac{\partial V_{DS\text{sat}}}{\partial V_{GS}}\right) \tag{8.32}$$

Using $V_{DS\text{sat}}$ as expressed by Equation (7.68), we obtain

$$g_{m\text{sat}} = W v_{\text{sat}} C'_{\text{ox}} \left\{1 - \left[1 + \frac{2\mu_{\text{lf}}(V_{GS} - V_T)}{v_{\text{sat}} L}\right]^{-1/2}\right\} \tag{8.33}$$

Now let us examine the output conductance g_d. We use the definition of g_d [Equation (8.27)] as the starting point. Differentiating Equations (8.28) and (8.29) below and above saturation, we obtain

$$g_d = \frac{W C'_{\text{ox}} \mu_{\text{lf}}}{L} \left(\frac{(V_{GS} - V_T - V_{DS}) - \dfrac{\mu_{\text{lf}} V_{DS}^2}{2L v_{\text{sat}}}}{\left(1 + \dfrac{\mu_{\text{lf}} V_{DS}}{L v_{\text{sat}}}\right)^2} \right) \qquad V_{DS} \le V_{DS\text{sat}} \tag{8.34}$$

and

$$g_{d\text{sat}} = 0 \qquad V_{DS} \ge V_{DS\text{sat}} \tag{8.35}$$

Of course, the saturation output conductance is not really zero because of some of the effects we neglected in this derivation. One of these is channel-length modulation, discussed in Chapter 7, in which the effective channel length varies with drain-source voltage V_{DS}. We will revisit this effect in Section 8.8.1. The second effect is the dependence of the threshold voltage on V_{DS}, which we will take up in Section 8.8.2.

It is useful to have a graphical representation of $g_{m\text{sat}}$ per unit channel width as a function of channel length for NFETs and PFETs. Figure 8.15 shows such a plot for $0.1 \le L \le 5\ \mu\text{m}$. The parameters used to calculate these plots are those from Table 7.1, with $(V_{GS} - V_T) = 2.6$ V.

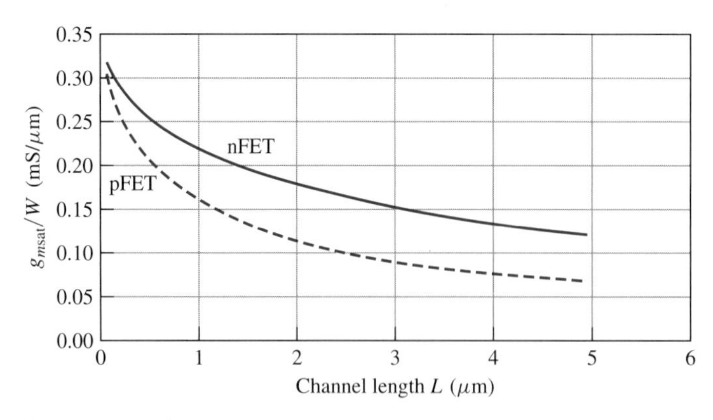

Figure 8.15 Saturation transconductance per unit channel width as a function of channel length for n-channel and p-channel MOSFETs.

From the figure, it can be seen that shorter channel lengths result in higher transconductance. For example, at $V_{GS} - V_T = 2.6\,\text{V}$, the saturation transconductance per unit channel width for the NFET decreases from $0.30\,\text{mS}/\mu\text{m}$ for a $0.1\,\mu\text{m}$ channel to $0.12\,\text{mS}/\mu\text{m}$ for a $5\text{-}\mu\text{m}$ channel. For the PFET, this decrease is from 0.28 to 0.065 $\text{mS}/\mu\text{m}$. In the next section, we will see that shorter channels also result in higher frequency of operation.

8.6.2 CMOS AMPLIFIERS

CMOS inverters have been discussed in relation to digital circuits. They also are useful as small signal amplifiers. Consider the CMOS inverter of Figure 8.5a. The transfer characteristics (V_{out}-V_{in}) of such a device are shown in Figure 8.5c. For $V_{\text{in}} = V_{DD}/2 + v_{\text{ac}}$, where v_{ac} is a small-signal voltage, it can be seen that for a small change in V_{in} (e.g., v_{ac}) the output voltage changes by a large amount or the ac voltage gain is large.

8.7 UNITY CURRENT GAIN CUTOFF FREQUENCY f_T

At low frequencies, the gate capacitance blocks any gate current (excluding a very small leakage through the oxide), so the input current is zero. Since the current gain of the transistor is i_d/i_g, then at low frequencies the current gain is infinite.

At high frequencies, however, there will be (displacement) current flowing in the gate oxide, which will tend to decrease the current gain. The current gain cutoff frequency f_T is defined as the frequency at which the magnitude of the current gain is reduced to unity, with the ac output of the circuit of Figure 8.13 short circuited.[6] In the high-frequency model of Figure 8.13a, there are two capacitances shown explicitly. For the case when the output is short circuited, these two capacitors are in parallel, so their capacitances add. Ignoring the overlap capacitances, the input current is related to the input voltage v_g by the admittance:

$$i_g = j2\pi f(C_{GS} + C_{GD})v_{gs} \approx j2\pi f(C'_{\text{ox}}WL)v_{gs} \tag{8.36}$$

where $(C_{GS} + C_{GD}) \approx C'_{\text{ox}}WL$. (Recall that C_{GS} and C_{GD} are capacitances while C'_{ox} is capacitance per unit area.)

At the drain, the short-circuit output current is

$$i_d = g_m v_{gs} \tag{8.37}$$

Setting the current gain magnitude to unity, then, we make $|i_g| = |i_d|$. Solving for $f = f_T$, we find the current gain cutoff frequency from Equations (8.36) and (8.37):

$$f_T = \frac{g_m}{2\pi C'_{\text{ox}}WL} \qquad \text{or} \qquad \omega_T = 2\pi f_T = \frac{g_m}{C_{\text{gate}}} \tag{8.38}$$

which is a figure of merit for a MOSFET.

[6]The ac output is short circuited to ground via the load capacitance, for example, as shown in Figure 8.8a. The dc output is not short circuited, of course, since to operate, $V_{DS} > 0$.

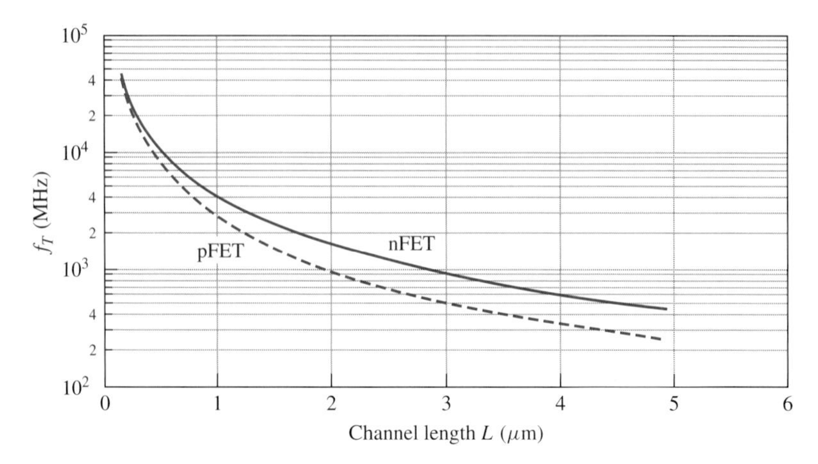

Figure 8.16 Current cutoff frequency as a function of channel length for n-channel and p-channel MOSFETs.

If the MOSFET is operating in the current saturation region, we substitute the expression for $g_{m\text{sat}}$ from Equation (8.33) to obtain

$$f_T = \frac{v_{\text{sat}}}{2\pi L}\left\{1 - \left[1 + \left[\frac{2\mu_{\text{lf}}(V_{GS} - V_T)}{v_{\text{sat}}L}\right]^{-1/2}\right]\right\} \tag{8.39}$$

Figure 8.16 shows a plot of the current cutoff frequency f_T as a function of channel length for the same transistor as in Figure 8.15 [i.e., $v_{\text{sat}} = 4 \times 10^6$ cm/s, $\mu_{\text{lf}} = 500$ cm^2/V $\cdot$ s, and $(V_{GS} - V_T) = 2.6$ V]. Here we see that f_T decreases from 56 GHz for the short-channel NFET ($L = 0.1\ \mu$m) to 430 MHz for a longer-channel NFET ($L = 5\ \mu$m), and from 50 GHz to 234 MHz for the equivalent PFET. Clearly short channels make for significantly higher frequency devices.

In these plots (Figures 8.15 and 8.16), we considered only the velocity saturation effect on the given parameters. These are optimistic results. Series resistance and short-channel effects, presented in the next section, reduce g_m and f_T from the values presented here.

*8.8 SHORT-CHANNEL EFFECTS

While the discussed electrical characteristics of MOSFETs depend on channel length, in this section we discuss some additional parasitic effects that become important for short channels, particularly in the submicrometer range.

8.8.1 DEPENDENCE OF EFFECTIVE CHANNEL LENGTH ON V_{DS}

In Chapter 7, we discussed the effect of channel-length modulation in saturation. Figure 7.19 showed that once saturation was reached, any additional drain voltage effectively moved the point along the channel where saturation occurred,

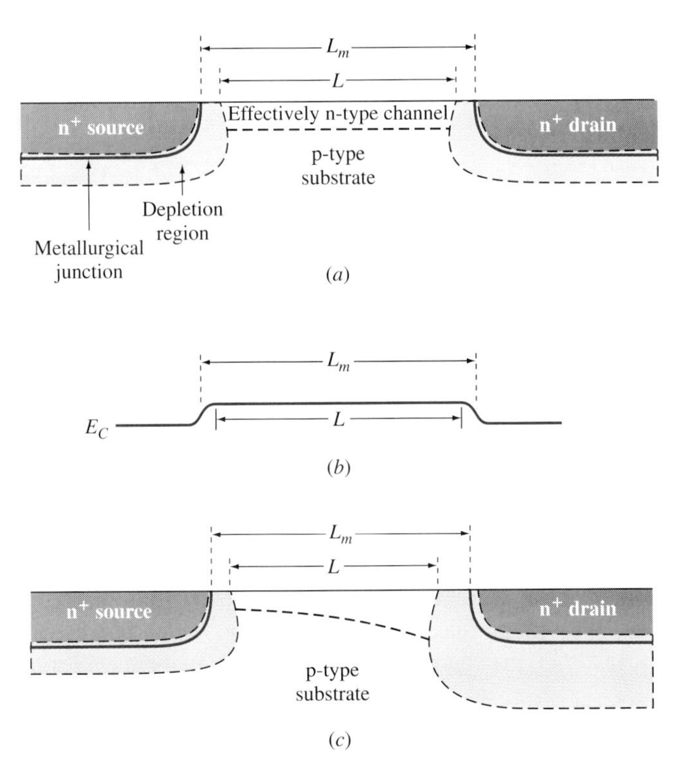

Figure 8.17 Illustration of the depletion regions between the source and the substrate, and between the drain and the substrate (a) for $V_{DS} = 0$. The effective channel length L is reduced from that of the metallurgical channel length L_m. (b) The $E_C - y$ diagram along the channel for $V_{DS} = 0$. (c) With increasing V_{DS}, the effective channel length L decreases.

effectively shortening the channel. You will recall that this caused a nonzero slope in the I_D-V_{DS} characteristics in the saturation region (the channel-length modulation effect). For very short (submicrometer) channels, however, the drain voltage can modulate the channel length even for $V_{DS} < V_{DS\text{sat}}$.

For long-channel MOSFETs, we took the channel length L to be the distance L_m between the metallurgical junctions of source-channel and drain-channel, as shown in Figure 8.17a for an NMOS transistor.[7] The source and drain are much more heavily doped than the substrate, however, such that the depletion regions between n^+ source and drain and p-type substrate extend into the substrate. This is shown in the figure for $V_{DS} = 0$. The depletion region thickness decreases near the channel, because the source-substrate and drain-substrate voltages are reduced there, as a result of the positive surface potential. Remember, the channel

[7]We define the metallurgical junction as the channel edge of the degenerate source or drain.

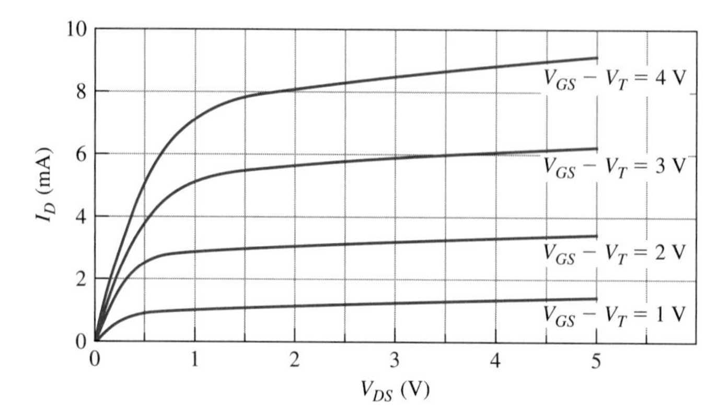

Figure 8.18 For short-channel devices, the reduction of the effective channel length with increasing V_{DS} results in an increase in I_D.

can be effectively n type, even under no bias, because of the band bending. In this region, then, the junctions are n^+n, and the transition regions are primarily in the n region. The energy band diagram along the channel for this case is shown in Figure 8.17b. In effect, the channel length L is less than the metallurgical channel length L_m.

If now a positive drain voltage is applied, the drain-channel depletion region width will increase as shown in Figure 8.17c. Thus, the effective channel length L decreases with increasing V_{DS}. Recall, however, that the channel current increases as the effective channel length gets smaller [as we expect from Equations (8.28) and (8.29), recalling that the *effective* channel length is the entire quantity in the denominator]. Thus, we expect that as V_{DS} increases, the channel length decreases, with a resultant increase in the current, both above and below saturation. This is shown in the I_D-V_{DS} characteristics of Figure 8.18. The long-channel (simple) model, and the velocity saturation model (Figure 7.30), showed the I_D-V_{DS} characteristic as flat in the saturation region.

Why do we call this a short-channel effect? It does occur in long-channel devices too. The reason is that for long-channel devices, the depletion region thickness is small compared with the overall channel length, and $L \approx L_m$. The fractional change in L with V_{DS} is small by comparison. In short-channel devices, the shortening becomes significant, and the slope of the I_D-V_{DS} characteristics is affected in both the sublinear and saturation regions. The channel-length modulation discussed in Chapter 7 happens in both long- and short-channel devices, but only in saturation.

8.8.2 DEPENDENCE OF THRESHOLD VOLTAGE ON THE DRAIN VOLTAGE

The second short-channel effect we discuss is that the value of the threshold voltage can be affected by the drain voltage. [3] Recall that in Chapter 7, we drew the energy band diagram along the channel. This is shown again in Figure 8.19a.

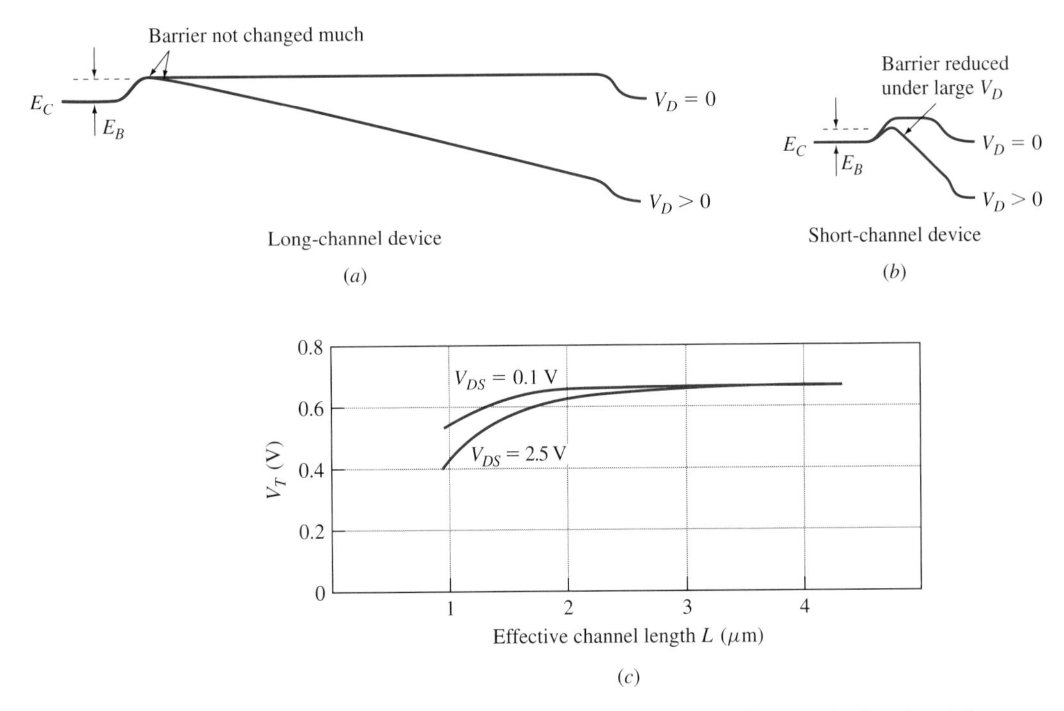

Figure 8.19 (a) For long channels the drain voltage has negligible effect on the barrier at the source-channel interface. (b) For short-channel devices the drain voltage tends to reduce the barrier at the source end. (c) The result is that the threshold voltage is decreased. The effect is more pronounced as the channels get shorter.

There is a barrier at the source-channel interface and another barrier at the drain-channel interface. These two barriers were treated as independent in the long-channel devices. For short-channel devices, however, the drain voltage can influence the barrier height at the source end of the channel, as illustrated in Figure 8.19b. If the drain voltage is large enough and the channel short enough, the barrier E_B from the source to the channel is reduced. This effect then reduces the threshold voltage. The threshold voltage is plotted as a function of effective channel length in Figure 8.19c for two values of drain voltage. Note that as the channel gets shorter, the effect gets more pronounced. This effect is referred to as the *drain-induced barrier lowering* or *DIBL* effect. The DIBL effect is a major cause of the finite slope in the I_D-V_{DS} characteristics (g_{dsat}) of submicrometer devices (e.g., the device of Figure 8.14 with $L = 0.27\ \mu$m).

8.9 MOSFET SCALING

Although having short-channel devices complicates the device models, there are advantages to making short-channel FETs. For one thing, very short channels in MOSFETs make for very fast devices. Furthermore, as the transistor area is reduced, one can put more transistors on a given chip, and/or more chips on a wafer. That increases the yield during manufacturing. An errant piece of dust that ruins one chip becomes less important as the number of chips per wafer increases. Thus, there are powerful incentives to reduce the device size.

We have seen, however, that short channels can have some deleterious effects on the I_D-V_{DS} characteristics of MOSFETs. It is desirable to minimize these effects. We do this by adjusting the device dimensions, voltages, and dopings in the short-channel device.

We start by asking what is the minimum channel length $L_{\min}$ such that short-channel effects do not seriously affect the MOSFET characteristics? There are various approaches. In one of them, $L_{\min}$ is given by the relation [4]

$$L_{\min} = 0.9[t_{\text{ox}} x_j (w_S + w_D)^2]^{1/3} = 0.9\gamma^{1/3} \tag{8.40}$$

where the oxide thickness t_{ox} is expressed in nanometers and the junction depth x_j, the channel length $L_{\min}$, and the depletion widths w_S and w_D at the source and drain respectively are in micrometers. These quantities are indicated in Figure 8.20.

As an example, suppose it is desired to fabricate a MOSFET with a channel length of $L = 0.2\,\mu\text{m}$. To avoid having to consider each short-channel effect separately, the designer instead chooses the quantity in the brackets, γ, to be less than $(0.2/0.9)^3 \approx 0.011$. Then t_{ox}, x_j, w_S, and w_D can be chosen accordingly. Figure 8.21 indicates schematically the variations that can be made in the structure of a device (a) that result in similar behavior in a device with shorter channel length (b). The minimum value of t_{ox} is limited by carrier tunneling through the oxide, and is thought to be on the order of 2 nm with SiO_2 as the gate insulator. Increasing substrate doping and using shallow implants can reduce the value of x_j. Increasing the substrate doping also reduces the depletion region widths w_S and w_D. In addition, reducing the drain voltage further reduces w_D.

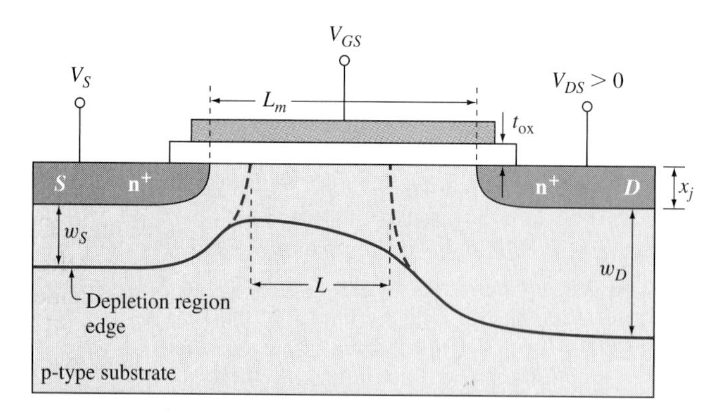

Figure 8.20 The difference between long- and short-channel behavior depends on the oxide thickness t_{ox}, the source and drain junction depth x_j, and the depletion region thickness at the source, w_S, and drain, w_D.

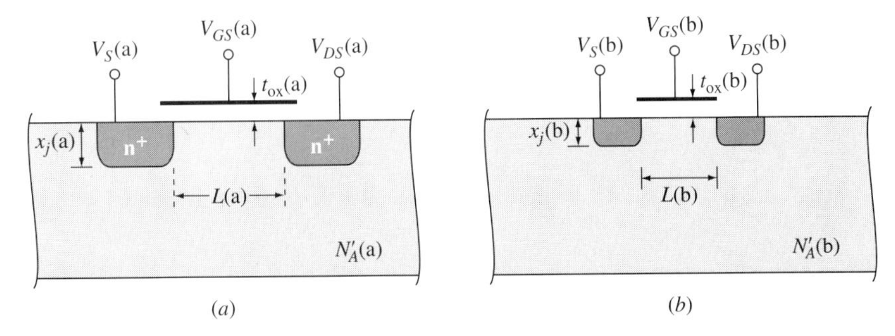

Figure 8.21 Illustration of MOSFET scaling. To reduce short-channel effects, when the channel length is reduced, the dimensions and doping levels of the device (a) are adjusted to reduce the oxide thickness and junction depth (b).

There are trade-offs, however. The increase in substrate doping increases the junction capacitances. Further, it increases the transverse electric field $\mathscr{E}_T$, which reduces the low-field mobility, and that in turn decreases the saturation velocity. The result is reduced transconductance. The resultant decrease in channel thickness can also result in quantum-mechanical effects that can affect the electrical characteristics. [5]

We mention here that other scaling procedures have been proposed in which the device fields are kept constant with changes in dimensions. [6]

8.10 SILICON ON INSULATOR (SOI)

A variation of the MOSFET structure previously discussed is the so-called silicon on Insulator (SOI) device. [6] One method to fabricate these devices is the SIMOX process (separation by implanted oxygen). For an n-channel MOSFET, oxygen is ion implanted beneath the surface of a p-type substrate. The wafer is then annealed at a high temperature, so a layer of SiO_2 is formed that effectively isolates the surface crystalline Si layer from that of the substrate (Figure 8.22a). The devices are then formed in the surface Si by standard means. Finally, the surface Si is selectively etched to isolate the devices as shown in Figure 8.22b, which shows NMOS and PMOS devices for a CMOS technology. With further processing, including SiO_2 deposition, the resultant CMOS structure is indicated in (c). The "metal" contacts are shown as refractory metal silicides (e.g., $TiSi_2$). The oxide spacers serve to confine the silicide contacts to source and drain regions. Such SOI devices can be manufactured with higher speeds than standard MOSFETs, because the isolation of the devices results in reduced parasitic capacitance. [7]

The original SOI devices were made by epitaxially growing Si onto insulating sapphire. Such devices were called *silicon on sapphire* (SOS) [(d) in the figure]. The sapphire substrate must be cut such that its atomic spacing is a close match to that of Si to ensure epitaxial growth.

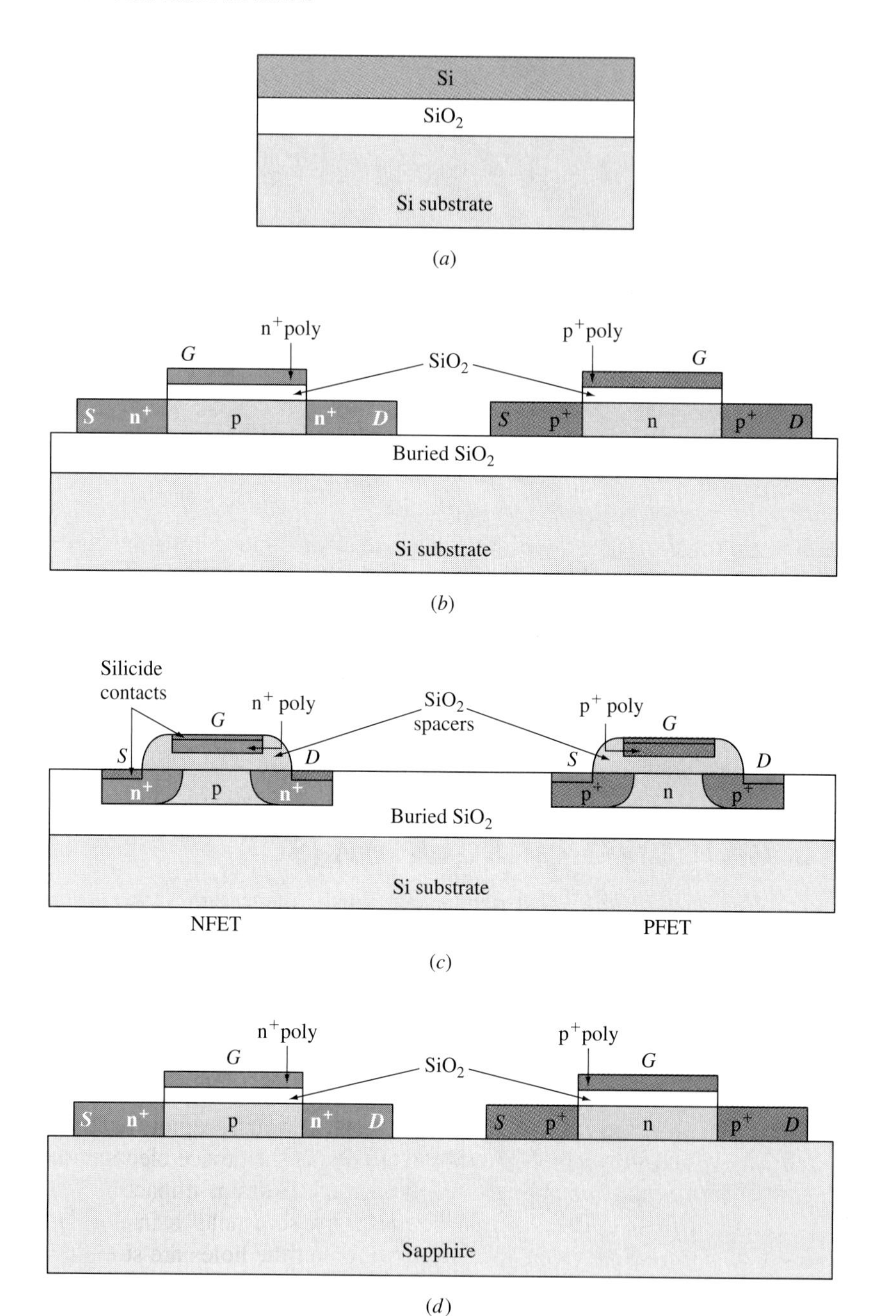

Figure 8.22 Cross section of (a) silicon on insulator; (b) an SOI NMOS and PMOS transistor; (c) a CMOS SOI structure. In (c) the gate, drain, and source contacts are metal silicides. The parasitic junction capacitances and substrate leakage currents are reduced from those of conventional MOSFETs. In (d) another form is shown. The silicon is deposited onto a sapphire substrate.

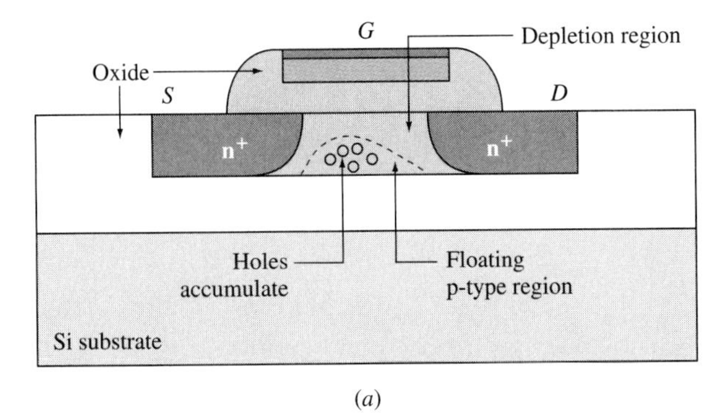

(a)

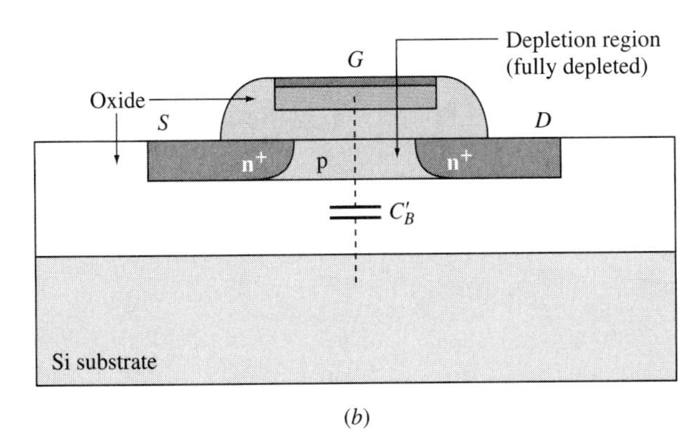

(b)

Figure 8.23 Two styles of SOI (an NMOS is shown). (a) Partially depleted, with a floating semiconductor region; (b) fully depleted.

There are two versions of SOI, partially depleted (PD SOI) and fully depleted (FD SOI). In the partially depleted SOI, the Si thickness under the gate is greater than the Si depletion region adjacent to the channel, Figure 8.23a. This results in an electrically "floating" neutral Si region between the oxide and the channel. This floating region can affect the device electrical characteristics. In the high-field depleted region near the drain, impact ionization can occur. In an n-channel device the depletion region field is in a direction such that the electrons are collected by the drain, but the holes are stored in the floating body, charging it positively. This in turn reduces the source-channel barrier and increases the channel current. This effect is often referred to as the *floating body* or *kink* effect. In high-speed devices, the time to charge the floating body is longer than the input transition time, resulting in an increased transient current.

In a fully depleted SOI device, Figure 8.23b, the Si layer is thin enough that the depletion region extends from the channel to the SiO_2 region, eliminating the floating body effect. This has the effect of reducing the subthreshold swing. From Equation (8.14),

$$S = 2.3\frac{kT}{q}n \tag{8.14}$$

where from Equation (8.12),

$$n = 1 + \frac{C'_B}{C'_{ox}} \tag{8.12}$$

where C'_B is the substrate capacitance per unit area. In the FD SOI the substrate capacitance C'_B extends from the channel across the p-type body and the SiO_2 layer to the Si substrate, resulting in a small value for C'_B. Thus $n \approx 1$ and the subthreshold voltage swing S is very near its ideal value of 60 mV/decade at room temperature. This reduced S permits a lower threshold voltage V_T, and thus a lower supply voltage for a given value of off current. In very short channel devices, however, the oxide effectively increases the field penetration in the region below the gate, and increases the effect of the drain voltage on the threshold voltage (the DIBL effect).

It is interesting to compare the operation of a fully depleted SOI CMOS and a bulk CMOS inverter. For matched devices such that $I_{D\text{satn}} = I_{D\text{satp}} = I_{D\text{sat}}$, from Equation (8.25) the gate delay for either case is

$$t_d = \frac{V_{DD}}{2I_{D\text{sat}}}C_L \tag{8.41}$$

where C_L is the load capacitance. In practice, a CMOS inverter is normally used to drive one or more following circuits. In this case the load capacitance is

$$C_L = C_{\text{out}} + C_w + \text{FO} \times C_{\text{in}} \tag{8.42}$$

where C_{out} = output capacitance of the driving circuit
C_w = stray wiring capacitance
FO = fan-out, or the number of following circuits being driven in parallel
C_{in} = input capacitance of one of the driven circuits, assuming identical driven circuits.

The output capacitance C_{out} is the capacitance at the output of the driving stage and consists of the drain junction capacitances, and the drain-to-gate capacitances including the overlap capacitance. The single-stage input capacitance C_{in} includes the gate-to-source, gate-to-drain, and gate-to-substrate (gate-to-channel + channel-to-substrate) capacitances.

In a bulk CMOS inverter, a large fraction of C_{out} is due to the drain junction capacitances, which are negligible in SOI CMOS. The wiring capacitances C_w and input capacitances C_{in} in SOI CMOS are comparable to those in bulk CMOS. Because of its decreased C_{out}, then, the SOI CMOS is inherently faster than bulk CMOS. However, this speed advantage decreases with increased fan-out.

Another advantage of SOI devices is their reduced susceptibility to *soft errors* relative to their bulk counterparts. Soft errors are a result of high-energy particles striking a device, creating a number of electron-hole pairs in the Si body that can be collected and change a stored logic state. Because of the reduced volume of the bulk Si in SOI devices, the number of electron-hole pairs created is low compared with the case for bulk MOS devices.

8.11 OTHER FETs

Until now, we have concentrated on Si-based MOSFETs. There are, however, other types of FETs that have important applications. In this section, some specific FETs are briefly discussed.

*8.11.1 HETEROJUNCTION FIELD-EFFECT TRANSISTORS (HFETs)

We begin with heterojunction field-effect transistors (HFETs). These have structures similar to that of MOSFETs, except that the gate oxide is replaced by a semiconductor with a higher band gap than the rest of the device.

Figure 8.24a shows the cross section of an HFET device[8] made in the AlGaAs-GaAs system. Figure 8.24b shows the energy band diagram perpendicular to the gate. When the heterojunction is formed, a discontinuity in the conduction band edge E_C results. This is analogous to the barrier between the semiconductor and the oxide in a MOSFET. A channel is thus created at the interface between the two different semiconductors.

To fabricate the device, a semi-insulating GaAs substrate is made by appropriate doping such that the Fermi level lies near midgap. Because the Fermi level is near the center of the gap, there are few free electrons or holes and the substrate is a semi-insulator.

Next, a lightly doped p layer of GaAs is grown epitaxially onto the semi-insulating substrate. This serves as the bulk semiconductor for the device, and the channel at its surface is manipulated by the application of an electric field at the gate (field-effect transistor). Over this bulk layer, a thin layer of nominally undoped AlGaAs is grown, and then a thicker layer of degenerately doped n^+ AlGaAs. The AlGaAs layers serve as the gate structure.

Finally, a layer of n^+ GaAs is deposited as a passivation layer for the AlGaAs. A passivator is required because the Al in the AlGaAs reacts chemically with oxygen in the atmosphere. This GaAs "cap" isolates the AlGaAs from the atmosphere. This layer is later etched in the region of the gate to make the gate contact directly to the wide-band-gap layer.

To create the source and drain, donor impurities are then implanted through the layers to make n^+ regions that extend from the surface to inside the p GaAs. A metal is then deposited on the n^+ GaAs surface to form thin ohmic (tunneling) Schottky barriers. Because the depletion region in the n^+ GaAs is thin, the tunnel current is large and the contact is low resistance, or ohmic.

---

[8]This device is also referred to as a modulation doped field-effect transistor (MODFET) or a high electron mobility transistor (HEMT).

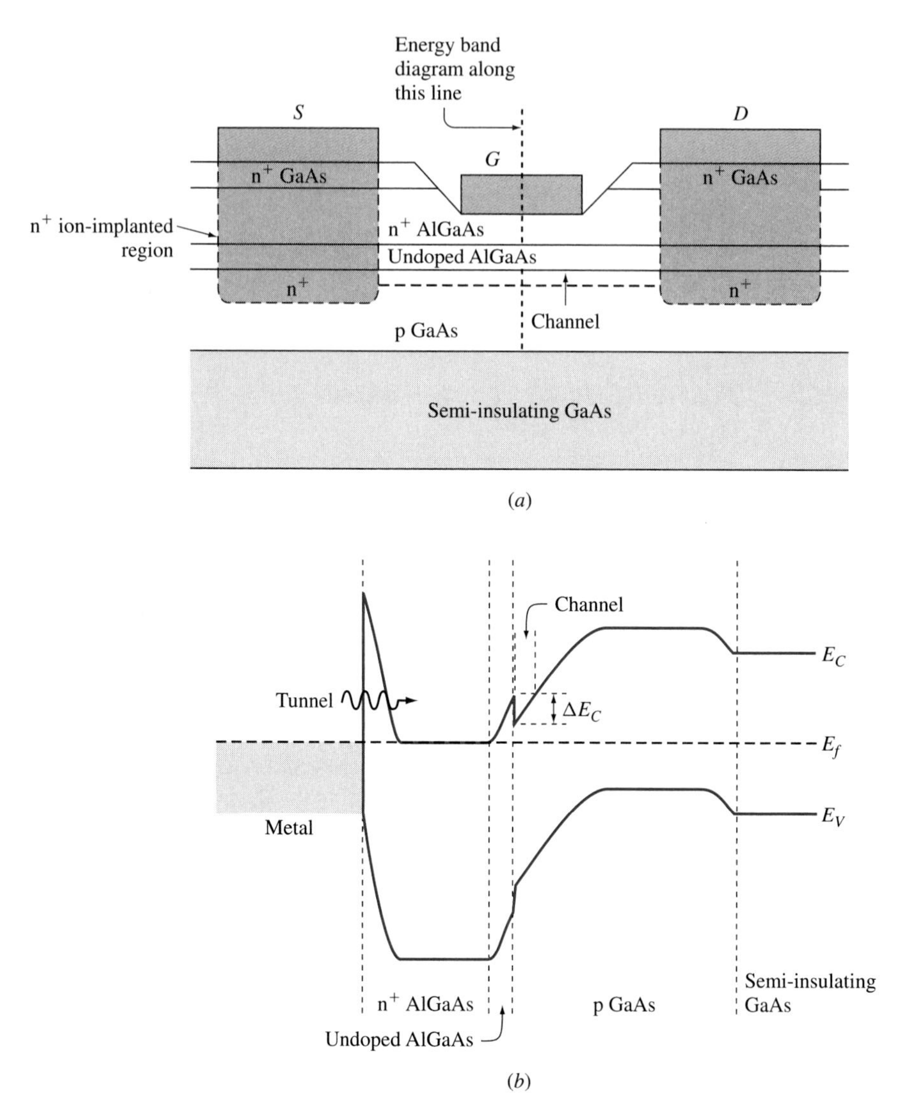

Figure 8.24 (a) The cross-sectional schematic of a GaAs-based HFET; (b) the energy band diagram normal to the gate. The Schottky barriers at the metal-AlGaAs and metal-GaAs interfaces are thin enough to be of low resistance because of tunneling.

Because of the difference in conduction band edges E_C, a channel exists in the p GaAs at its interface with the AlGaAs. Depositing a metal on the AlGaAs makes the gate contact. Since the AlGaAs is degenerately doped, the resultant Schottky barrier is thin enough to permit tunneling, giving it low resistance, just as in the source and drain. The n^+ AlGaAs is also conductive, and in some sense is a continuation of the gate contact. The undoped AlGaAs layer is effectively insulating, and replaces the oxide of a MOSFET in function.

EXAMPLE 8.6

If the undoped layer is insulating in the channel, explain why there is a low resistance from source contact to the internal n^+ source region.

■ Solution

This is best illustrated with the aid of the energy band diagram normal to the source contact, Figure 8.25. Because the source region is heavily implanted with donors, the total region between the source contact and the p region is degenerate n type. The barrier between the metal contact and the n^+ GaAs is thin enough to be a low-resistance tunneling junction. The n^+ GaAs region is heavily doped, so it also has low resistance. The next barrier, between the n^+ GaAs and the n^+ AlGaAs, is also thin, again allowing tunneling, while the deeper n^+ AlGaAs layer between them is heavily doped and thus has low resistance. Finally, the undoped AlGaAs layer, which created the channel in the region under the gate, is degenerate because of the implanted ions and here causes a thin barrier that allows tunneling. Similarly, the drain contact–intrinsic drain junction also has low resistance.

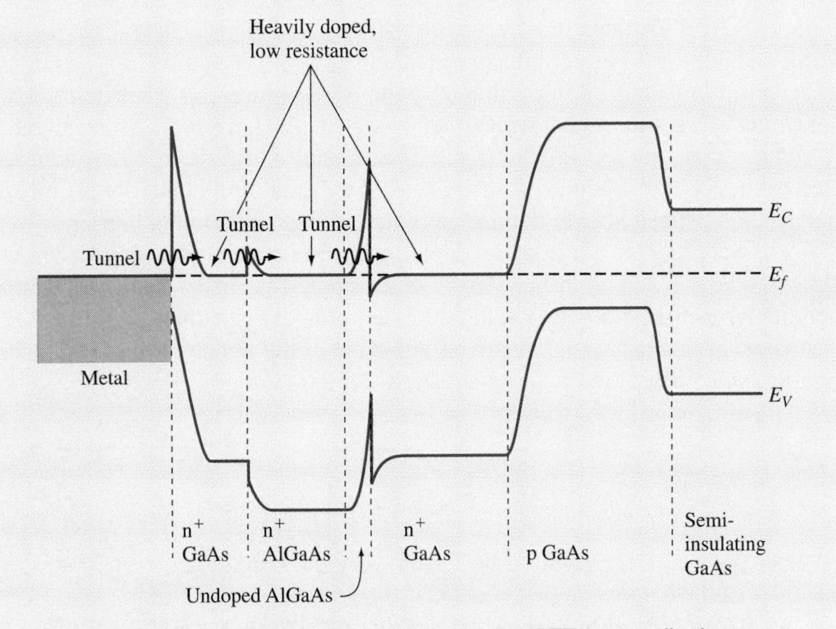

Figure 8.25 The energy band diagram for an HFET perpendicular to the source.

From the energy band diagram of Figure 8.24b, we can see that the built-in voltage between gate and substrate is dropped in part across the undoped AlGaAs and in part across the p GaAs. The energy at the bottom tip of the channel, relative to the Fermi level, is determined by the thickness of the undoped AlGaAs. The AlGaAs must be thick enough that the built-in voltage it supports is greater than $\Delta E_C/q$. The thickness of this layer therefore controls the HFET threshold voltage, an effect similar to the influence of the oxide thickness on V_T in a MOSFET.

The operation of an HFET is similar to that of a MOSFET. For the device in the figure, the channel is above the Fermi level at equilibrium and does not contain many electrons. Thus, the channel does not conduct appreciably. When a positive voltage is applied to the gate, the channel side of the energy band diagram moves down. The bottom of the channel will approach the Fermi level, allowing the channel to fill with electrons and conduct. A more negative gate voltage will deplete the channel.

The advantage of using an HFET structure is high speed. The undoped AlGaAs and the lightly p-doped GaAs provide little scattering of channel electrons by ionized impurities. Since the electron scattering is reduced, their mobility is increased. Thus, high-speed devices can result.

8.11.2 MESFETs

Figure 8.26 illustrates the structure of an n-channel GaAs-based metal-semiconductor field-effect transistor (MESFET). A thin film of n-type GaAs is either ion implanted into or epitaxially deposited onto a semi-insulating GaAs substrate. The source and drain are then formed by heavily doping (degenerately) n^+-type regions by ion implantation. The implant is followed by an annealing step to reduce the structural damage caused by the ion implantation. A metallic gate is then deposited over the n-type region to form a Schottky barrier.

The Schottky barrier causes a depletion region that extends into the GaAs as shown in Figure 8.27a for the case of equilibrium. The energy band diagram perpendicular to the gate (zero drain and gate voltages) is shown in Figure 8.27b. The channel thickness t extends from the edge of the Schottky barrier depletion region to the edge of the n GaAs-substrate depletion region.

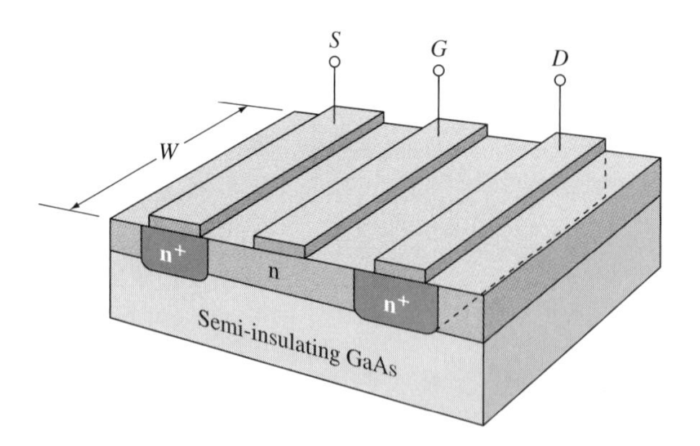

Figure 8.26 Schematic showing the structure of a GaAs MESFET.

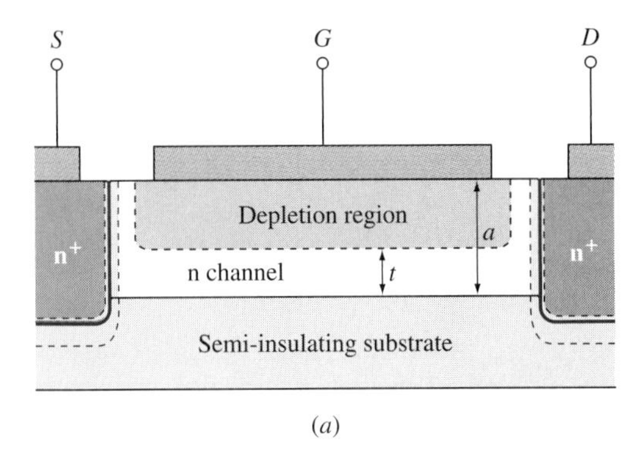

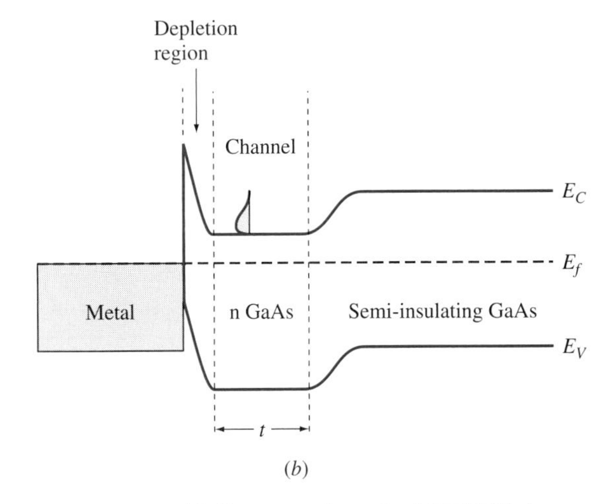

Figure 8.27 (a) Cross section of a MESFET at equilibrium indicating the depletion regions; (b) the energy band diagram perpendicular to the gate. The channel thickness is *t*.

This device is different from the FETs discussed earlier, in that in this case the channel does not form directly at the semiconductor surface, but somewhere below it. The depletion region induced by the Schottky barrier is devoid of carriers, and is thus insulating. The depletion region behaves, then, somewhat like the insulating gate in the MOSFET.

We recognize that since the source, channel, and drain are all n type, a channel exists even when no voltage is applied. Thus, the MESFET shown in Figure 8.27 is a depletion-mode MESFET.

Next, we recall that the channel charge per unit area in a field-effect transistor depends on the barrier height from source to channel, E_B. This barrier height is equal to the built-in voltage between the n^+ source and the n channel, Figure 8.28a. The electron concentration n in the channel is equal to the net doping in the channel, N_D'. The value of the channel charge density (charge per unit area)

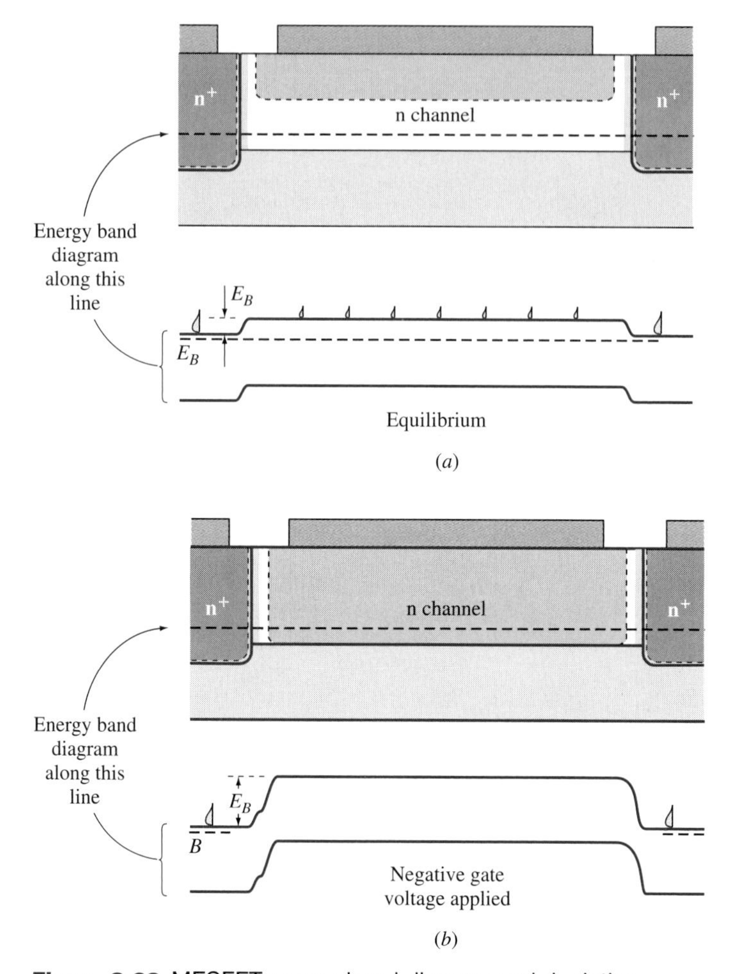

Figure 8.28 MESFET energy band diagram and depletion region (a) at equilibrium and (b) for an applied gate voltage that depletes the channel. In the second case, the channel is still n type but it is empty of carriers because of the increased barrier height.

Q_{ch} at the source end is then proportional to the channel thickness t (from $Q_{\text{ch}} = -qN_D't$, where it is assumed that N_D' is uniform in the channel), and t is controlled by the gate voltage.

We also observe that there are two depletion regions. One is the depletion region caused by the Schottky junction and the other is the depletion region between the channel and the semi-insulating substrate. If a gate voltage V_G is applied such that these two depletion regions overlap, the source-channel barrier increases. This is shown in Figure 8.28b. Increasing this barrier reduces the mobile channel charge Q_{ch} to (almost) zero, which turns off the device. The value

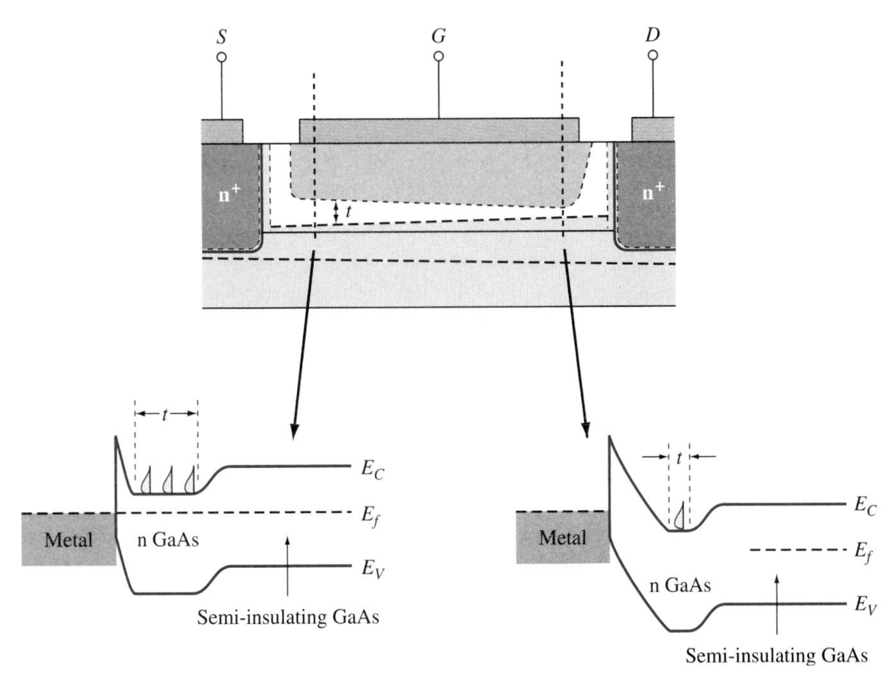

Figure 8.29 Cross section of a MESFET under small V_{DS} bias and the corresponding energy band diagrams at the source end and drain end of the gate.

of gate voltage required to achieve this condition is called the *threshold voltage,* and in this case is negative.[9]

An enhancement-mode MESFET exists if the depletion regions at the source end in Figure 8.28a were to overlap with no gate voltage applied. In that case, the source-to-channel barrier is such that a negligible conducting channel exists at equilibrium. Application of a positive gate voltage then reduces the gate-channel depletion region and forms a channel, which enhances the current. In an enhancement-mode FET, the gate-channel Schottky barrier is actually forward biased to reduce the depletion region width. The gate voltage has to be limited then to about $\frac{3}{4}$ V to avoid excessive gate-channel current. In an enhancement-mode MESFET, the threshold voltage is positive.

The cross-sectional view of a MESFET with a positive drain voltage is shown schematically in Figure 8.29 along with the energy band diagrams perpendicular to the gate near the source and near the drain. At the drain end, the reverse bias across the Schottky barrier is greater than at the source, so the depletion region is wider. As a result, the channel thickness decreases from source to drain.

The I_D-V_{DS} characteristics of the MESFET are qualitatively similar to those of the MOSFET, although the physics is slightly different. Let us consider the case below saturation first.

[9]This voltage is sometimes referred to as the *pinch-off voltage*.

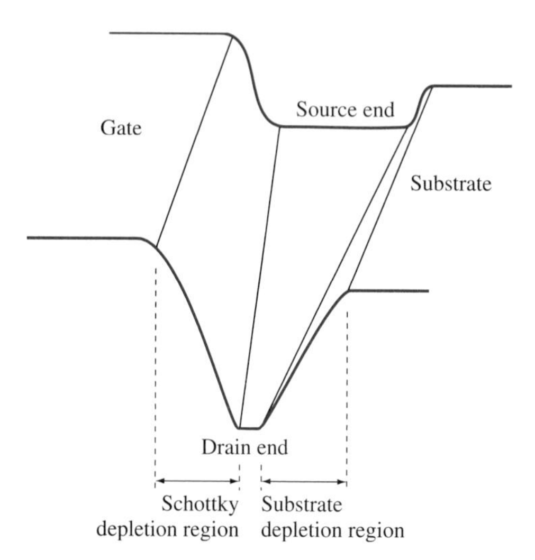

Figure 8.30 The electron potential energy, (E_C) along the channel of a MESFET for $0 < V_{DS} < V_{DS\text{sat}}$. The channel thickness decreases with increasing distance along the channel.

Recall that the current through the channel is proportional to the charge per unit area, Q_{ch}, which in turn is proportional to the thickness t of the channel. Thus, as the drain voltage increases, the resistance also increases. The current therefore increases sublinearly with V_{DS} below saturation. This can be also be seen with the aid of Figure 8.30, which shows the variation of E_C with x and y for $0 < V_{DS} < V_{DS\text{sat}}$ (i.e., operating in the sublinear region) at a given gate voltage V_{GS}. In the channel, the electron concentration is $n_{\text{ch}} = N'_D$, or the net doping concentration in the n-type channel region. The gate voltage controls the thickness of the channel near the source, but this also controls the channel charge per unit area, since $Q_{\text{ch}} = -qN'_D t$. Toward the drain end of the channel, the thickness t decreases, with a corresponding decrease in Q_{ch}. Since the current along the channel is constant, from $I_D = WQ_{\text{ch}}v$ [Equation (III.3)] we can see that if the channel charge decreases, the velocity increases.

We expect the drain current to saturate at some value of V_{DS}. Let us examine how that occurs. With sufficiently high drain voltage, the depletion regions in the channel overlap near the drain as indicated in Figure 8.31, which shows the energy band diagram perpendicular to the junction at the source end and drain end under current saturation. It may seem that current should stop flowing because usually a depletion region has no carriers. In this case, however, although the depletion regions have met, the energy barriers of their walls will form a potential trough, similar to that in Figure 8.30, funneling current from the source into the drain. Thus, current continues to flow.

The current saturates, however, for the same reason as it does in a MOSFET. The number of carriers entering the channel is controlled by the barrier at the

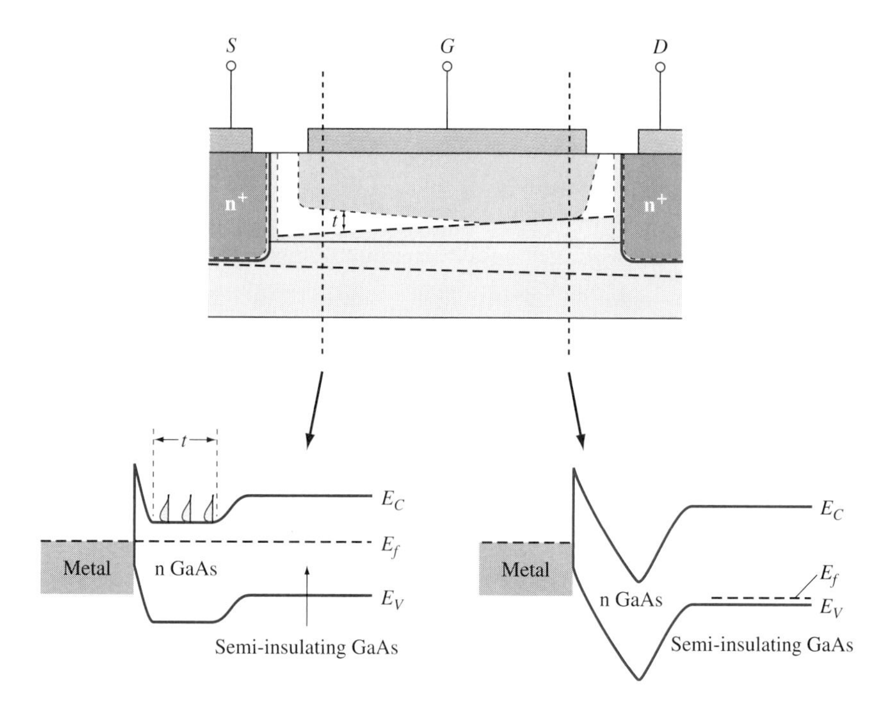

Figure 8.31 The MESFET of Figure 8.29 with $V_{DS} > V_{DSsat}$. At the source, the diagram is the same as for Figure 8.29. At the drain, however, the two depletion regions overlap.

source end, as was shown in Figure 7.15, and the current is proportional to channel charge and channel field, $I_D = W Q_{ch}(0)\mu_{lf}\mathscr{E}_L(0)$. The application of a small drain voltage changes the field even at the source end, but as V_{DS} increases, it bends the band increasingly at the drain end but with little additional effect at the source end.

8.11.3 JUNCTION FIELD-EFFECT TRANSISTORS (JFETs)

The final type of field-effect transistor discussed here is the JFET. A JFET is similar to a MESFET except that a p^+n junction gate replaces the Schottky barrier gate, and the semi-insulating substrate is replaced by a p-type substrate (for an NFET). This is shown schematically in Figure 8.32. As with a MESFET, the control field is applied via a depletion region associated with the reverse-biased gate junction. In the MESFET, that junction is a Schottky barrier; in the JFET, it is a p^+n junction. The channel of thickness t extends from the edge of the gate channel depletion region to the substrate channel depletion region as indicated Figure 8.32a for zero drain-source and gate-source voltages. The corresponding energy band diagram is shown in Figure 8.32b. (Only E_C is shown here.) Figure 8.32c shows the depletion regions for a nonzero drain voltage that is still below saturation. The metallurgical channel thickness is a, the depletion width is w, and the effective channel thickness is t, where $t = (a - w)$.

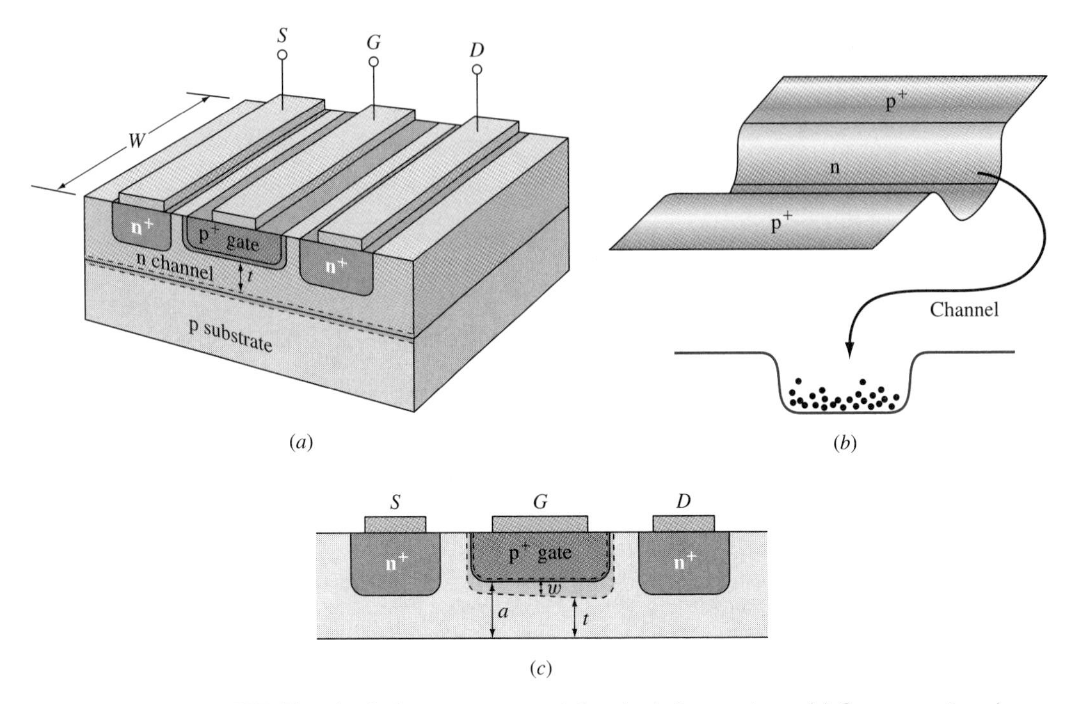

Figure 8.32 The JFET. The shaded areas represent the depletion regions. (a) Cross-sectional diagram; (b) the energy band diagram, (E_C), for $V_{DS} = 0$; (c) when $V_{DS} > 0$ (but not yet in saturation) the depletion region at the drain end increases, narrowing the channel and increasing the channel resistance.

The JFET physics of operation and electrical characteristics are similar to those for a MESFET. A reverse-biased gate-channel voltage reduces the current.

8.11.4 BULK CHANNEL FETs: QUANTITATIVE

We saw that the physics of operation of the MESFET and JFET are similar. Here we will be more quantitative. As with all FETs, the mathematical treatment of MESFETs and JFETs begins with Equation (III.7). For a JFET or a MESFET where $Q_{\text{ch}} = -qn(y)t(y)$,

$$I_D = qW\mu(y)n(y)t(y)\frac{dV_{\text{ch}}(y)}{dy} \tag{8.43}$$

In principle, then, provided we have models for μ, n, and t, Equation (8.43) can be solved for $V_{\text{ch}}(y)$. We can obtain the I_D-V_{DS} characteristics by setting $V_{\text{ch}}(L) = V_{DS}$. We will illustrate the solution for a JFET, but the MESFET solution is similar.

Simple Model for the JFET As we did with the MOSFET, we begin with a simple model for a JFET in which velocity saturation effects are ignored. We start with the I_D-V_{DS} characteristics below saturation ($0 < V_{DS} < V_{DS\text{sat}}$). For this long-channel model, we make the following approximations:

1. The net channel doping is uniform, and $n(y) \approx N'_D$.
2. The electron mobility is the low-field mobility ($\mu = \mu_{\text{lf}}$), and independent of $\mathscr{E}_L$.
3. The gate-channel depletion width $w(y)$ is entirely on the channel side. (This is reasonable since the p^+ gate is heavily doped.)
4. The depletion width in the n-type channel region adjacent to the substrate is negligible. This implies that $t(y) = a - w(y)$, where a is the thickness of the n-type region. (See Figure 8.32c.)
5. The channel is long enough ($L \gg a$) such that $\mathscr{E}_L \ll \mathscr{E}_T$ over most of the channel. We also assume that the gate depletion region thickness w can be considered independent of $\mathscr{E}_L(y)$. This is referred to as the *gradual channel approximation.*

We should point out that approximation 4 is not realistic even for this simple model. The actual depletion region in the channel adjacent to the substrate depends on the net doping in the channel compared with that of the substrate, which must be known for an accurate analysis.

With these assumptions, Equation (8.43) becomes

$$I_D\,dy = qW\mu N'_D a\left[1 - \frac{w(y)}{a}\right]dV_{\text{ch}} \tag{8.44}$$

For a given N'_D, the depletion width depends only on the gate-to-channel voltage $V_G - V_{\text{ch}}(y)$. Adapting Equation (5.39), the expression for the depletion width in an n^+p junction, we have

$$w(y) = \left[\frac{2\varepsilon_s}{qN'_D}(V_{\text{bi}} - V_{GS} + V_{\text{ch}}(y))\right]^{1/2} \tag{8.45}$$

where we have expressed the junction voltage as $V_j = V_{\text{bi}} - V_{GS} + V_{\text{ch}}(y)$. Here the applied voltage V_{GS} is negative to ensure that the junction is reverse biased.

It is convenient to express I_D in terms of its threshold voltage (a readily measurable quantity). At threshold, $V_{GS} = V_T$ and at the source end, $w(0) = a$ and $V_{\text{ch}} = 0$, so we can rewrite Equation (8.45) at $y = 0$ as

$$w(0) = a = \left[\frac{2\varepsilon_s}{qN'_D}(V_{\text{bi}} - V_T)\right]^{1/2} \tag{8.46}$$

Solving for the threshold voltage yields:

$$V_T = V_{\text{bi}} - \frac{a^2 q N'_D}{2\varepsilon_s} \tag{8.47}$$

The threshold therefore depends on doping and the thickness of the n-type layer. Note that V_T is positive for an enhancement JFET and negative for a depletion JFET.

Substituting the expressions for $w(y)$ and a from Equations (8.45) and (8.46) into the brackets of Equation (8.44) gives

$$I_D\,dy = qW\mu N_D' a\left[1-\left(\frac{V_{\text{bi}}-V_{GS}+V_{\text{ch}}}{V_{\text{bi}}-V_T}\right)^{1/2}\right]dV_{\text{ch}} \tag{8.48}$$

Now we invoke assumption 2, which is that $\mu = \mu_{\text{lf}}$ and is independent of V_{ch}. Integrating Equation (8.48) from source to drain, after some algebra we obtain, below saturation,

$$I_D = \frac{qW\mu_{\text{lf}}N_D' a}{L}\left\{V_{DS}-\frac{2}{3}(V_{\text{bi}}-V_T)\left[\left(\frac{V_{DS}+V_{\text{bi}}-V_{GS}}{V_{\text{bi}}-V_T}\right)^{3/2} - \left(\frac{V_{\text{bi}}-V_{GS}}{V_{\text{bi}}-V_T}\right)^{3/2}\right]\right\} \qquad V_{DS} \le V_{DS\text{sat}} \tag{8.49}$$

In current saturation, we know the slope $\partial I_D/\partial V_{DS} = 0$, the same as for a MOSFET. The onset of saturation occurs when $V_{DS} = V_{DS\text{sat}}$. Taking the derivative of Equation (8.49) and setting it to zero gives

$$V_{DS} = V_{DS\text{sat}} = V_{GS} - V_T \tag{8.50}$$

Substituting this back into Equation (8.49), at (and above) the point of saturation we find

$$I_{D\text{sat}} = \frac{qW\mu_{\text{lf}}N_D' a}{L}\left\{V_{GS}-V_T-\frac{2}{3}(V_{\text{bi}}-V_T)\left[1-\left(\frac{V_{\text{bi}}-V_{GS}}{V_{\text{bi}}-V_T}\right)^{3/2}\right]\right\} \qquad V_{DS} \ge V_{DS\text{sat}} \tag{8.51}$$

Velocity Saturation Model for the JFET For the MOSFET, the simple long-channel model was poor for short-channel devices, and the same is true for the JFET. We now correct this JFET model to include velocity saturation. As for a MOSFET, the mobility μ decreases at high longitudinal field $\mathscr{E}_L$. Again using the relation

$$\mu = \frac{\mu_{\text{lf}}}{1+\dfrac{\mu_{\text{lf}}\mathscr{E}_L}{v_{\text{sat}}}} \tag{8.52}$$

and using the relation $|\mathscr{E}_L| = dV_{\text{ch}}/dy$, we substitute into Equation (8.52). Integrating Equation (8.48), we obtain for $V_D \le V_{D\text{sat}}$

$$I_D = \frac{qW\mu_{\text{lf}}N_D' a}{L\left(1+\dfrac{\mu_{\text{lf}}V_{DS}}{v_{\text{sat}}L}\right)}\left\{V_{DS}-\frac{2}{3}(V_{\text{bi}}-V_T)\left[\left(\frac{V_{DS}+V_{\text{bi}}-V_{GS}}{V_{\text{bi}}-V_T}\right)^{3/2} - \left(\frac{V_{\text{bi}}-V_{GS}}{V_{\text{bi}}-V_T}\right)^{3/2}\right]\right\} \qquad V_{DS} \le V_{DS\text{sat}} \tag{8.53}$$

and in saturation:

$$I_{Dsat} = \frac{qW\mu_{lf}N'_D a}{L\left(1 + \dfrac{\mu_{lf}V_{DSsat}}{v_{sat}L}\right)} \left\{ V_{DSsat} - \frac{2}{3}(V_{bi} - V_T)\left[\left(\frac{V_{DSsat} + V_{bi} - V_{GS}}{V_{bi} - V_T}\right)^{3/2} - \left(\frac{V_{bi} - V_{GS}}{V_{bi} - V_T}\right)^{3/2}\right]\right\} \quad V_{DS} \geq V_{DSsat} \tag{8.54}$$

We observe that the above equations can be expressed as for a MOSFET:

$$I_D = \frac{I_D \text{ (model neglecting velocity saturation)}}{1 + \dfrac{\mu_{lf}V_{DS}}{Lv_{sat}}} \tag{8.55}$$

$$I_{Dsat} = \frac{I_{Dsat} \text{ (model neglecting velocity saturation)}}{1 + \dfrac{\mu_{lf}V_{DSsat}}{Lv_{sat}}} \tag{8.56}$$

As for a MOSFET, for short channels, the velocity saturation effect reduces the current from that predicted with velocity saturation neglected.

An expression for V_{DSsat} can be found by setting $\partial I_D/\partial V_{DS} = 0$ as before, but in this case the resulting expressions are difficult to solve analytically.

***I_D-V_{DS}* Characteristics of MESFETs** The I_D-V_{DS} characteristics for Si-based MESFETs are the same as those for JFETs. However, since many MESFETs are made from GaAs rather than silicon, we should point out that for n-channel GaAs-based FETs, Equation (8.52) is a poor approximation for electron mobility and the dependence of mobility on $\mathscr{E}_L$ is appreciably more complicated than considered here. This is because of velocity overshoot, as discussed in Chapter 3 (Figure 3.9).

8.12 SUMMARY

In this chapter, we took a closer look at the effects of the device dimensions and doping on the electrical behavior of MOSFETs.

In Chapter 7 the threshold voltage and low-field mobility were taken to be measurable parameters. In this chapter, their measurement techniques were discussed, including the gate voltage dependence of mobility. This is discussed further in the Supplement to Part 3.

We also examined the power consumption of MOSFETs. Although we usually assume that $I_D \approx 0$ for $V_{GS} < V_T$, in reality some current does flow. We must choose V_T large enough such that in the off state, when $V_{GS} = 0$, negligible current flows. This value of V_T fixes the supply voltage ($V_{DD} \approx 5V_T$), which in turn influences the dynamic power dissipation associated with switching. For a simple CMOS inverter,

$$P_{\text{dynamic}} = C_L V^2{}_{DD} f$$

For a CMOS inverter, expressions for propagation delay t_d in switching, the saturation transconductance, $g_{m\text{sat}}$ and the current cutoff frequency f_t are

$$t_d = \frac{1}{2}\left[\frac{C_L V_{DD}}{2I_{D\text{satn}}} + \frac{C_L V_{DD}}{2I_{D\text{satp}}}\right]$$

$$g_{m\text{sat}} = W v_{\text{sat}} C'_{\text{ox}} \left\{1 - \left[1 + \frac{2\mu_{\text{lf}}(V_{GS} - V_T)}{v_{\text{sat}} L}\right]^{-1/2}\right\}$$

$$f_t = \frac{g_{m\text{sat}}}{2\pi C'_{\text{ox}} L} = \frac{v_{\text{sat}}}{2\pi L}\left\{1 - \left[1 + \frac{2\mu_{\text{lf}}(V_{GS} - V_T)}{v_{\text{sat}} L}\right]^{-1/2}\right\}$$

neglecting parasitic capacitances and resistances.

Shorter channel lengths lead to faster devices but introduce some second-order deleterious effects, which degrade the operating characteristics. One method for scaling MOSFETs to reduce channel length while minimizing these short-channel effects was described. For short-channel devices, the current above saturation increases gradually with increasing V_{DS}. Furthermore, as V_{DS} increases, the threshold voltage decreases.

These undesirable short-channel effects can be compensated for in the design of the transistors: by using shallow source and drain implants, increased substrate doping, thin oxides, and small power supply voltages.

In normal (bulk) MOSFETs the junction capacitance between source and substrate and drain and substrate imposes a limit on the frequency response. This capacitance can be reduced by using a silicon-on-insulator (SOI) technology resulting in faster devices.

A variety of other field-effect devices were briefly described. These include heterojunction field-effect transistors (HFETs), metal-semiconductor field-effect transistors (MESFETs), and junction field-effect transistors (JFETs). These behave similarly to MOSFETs but each has unique physics of operation and unique advantages. For example, HFETs and MESFETs can operate at higher speed.

8.13 READING LIST

Items 1 to 4, 8 to 12, 18, 19, 27 to 30, 31, 32 in Appendix G are recommended.

8.14 REFERENCES

1. See for example, Y. Taur, G. J. Hu, R. H. Dennard, L. M. Terman, C. Y. Ting, and K. E. Petrillo, "A self-aligned 1 μm channel CMOS technology with retrograde n well and thin epitaxy," *IEEE Trans. Electron Devices,* ED-32, pp. 203–209, 1985.
2. Richard C. Jaeger, *Microelectronic Circuit Design,* Section 7.7, McGraw-Hill, New York, 1997.
3. R. R. Troutman, "VLSI limitations from drain-induced barrier lowering," *IEEE Journal of Solid State Circuits,* SC-14, pp. 389–391, 1979.

4. J. R. Brews, W. Fitchner, E. H. Nicollian, and S. M. Sze, "Generalized guide for MOSFET miniaturization," *IEEE Electron Device Lett.*, EDL-1, pp. 2–4, 1980.
5. S. Takagi, J. Koga, and A. Toriumi, "Subband structure engineering for performance enhancement in Si MOSFETs, *IEDM Tech. Dig.*, pp. 219–222, 1997.
6. B. Davari, R. Dennard, and G. Shadidi, "CMOS scaling for high performance and low power—The next ten years," *Proc. IEEE,* 93, pp. 596–606, 1995.
7. M. Yoshimi, H. Hazama, M. Takahashi, S. Kambayashi, J. Wada, and H. Tango, "Two-dimensional simulations and measurement of high-performance MOSFETs made on very thin SOI films," *IEEE Trans. Electron Devices,* ED-36, pp. 493–503, 1989.

8.15 REVIEW QUESTIONS

1. Explain why the I_D-V_{GS} characteristic is not a straight line for large V_{GS}.
2. Why is it important to reduce the subthreshold leakage current? What must be traded off against this in the design process?
3. Explain the operation of the inverter circuit of Figure 8.5.
4. Why should NFETs and PFETs have different dimensions when used in CMOS circuits?
5. What is pass-through current?
6. What is meant by transconductance?
7. What is meant by current gain cutoff frequency f_T of a transistor?
8. List the short-channel effects and explain the physics of each and its effect on the I_D-V_{DS} characteristics of a FET.
9. Comparing Figure 8.22c and Figure 8.4, explain how a silicon-on-insulator design decreases the junction capacitances C_{jS} and C_{jD} below those of conventional CMOS circuits.
10. Considering the MOSFET, HFET, MESFET, and JFET, identify for each the region in which the controlling field occurs. For each, what is the mechanism used to control it (e.g., field across an oxide, changing the voltage across a depletion region, etc.)

8.16 PROBLEMS

8.1 Figure P8.1 shows the I_D-V_{GS} characteristics for an NMOS with $V_{DS} = 50\,\text{mV}$. It is known for this device that $W = 10\,\mu\text{m}$, $L = 0.5\,\mu\text{m}$, and $t_{\text{ox}} = 5\,\text{nm}$.

a. Find the threshold voltage.

b. Find μ_0, the electron channel mobility at threshold.

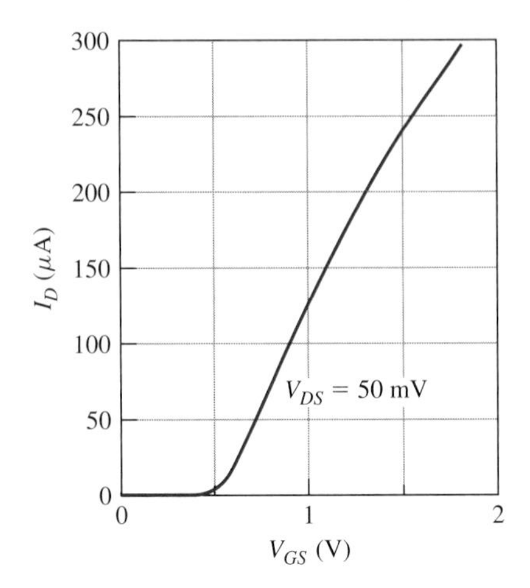

Figure P8.1

8.2 A particular MOSFET process produces $C'_B = 10^{-7}$ F/cm^2 and $I_0 = 4 \times 10^{-20}$ A, and a threshold voltage of $V_T = 0.5$ V. For gate oxide thicknesses of 6.5 nm and 4 nm, find n and S in the subthreshold region. Which device is better, and why?

8.3 a. Find W_p/W_n needed to match $I_{D\text{sat}}$ for CMOS transistors if $\mu_{\text{lfn}} = 500$ cm^2/V · s, $\mu_{\text{lfp}} = 200$ cm^2/V · s, $L = 0.5\ \mu$m, and $V_G - V_T = 2.6$ V. Assume $v_{\text{sat}} = 4 \times 10^6$ cm/s.

b. Find $V_{DS\text{sat}}$ for the NMOS and the PMOS.

c. Adjust the length of the NFET to equalize the $V_{DS\text{sat}}$'s. What should the new W_p/W_n be to keep the $I_{DS\text{sat}}$'s equal?

8.4 A CMOS inverter drives a load that consists of the gate of another FET. The gate area is $0.2 \times 5\ \mu$m. Find the dynamic power dissipation if the clock frequency is 350 MHz, the supply voltage is $V_{DD} = 2.5$ V, and the oxide thickness is 5 nm. If a medium-scale-integration (MSI) circuit has 1000 transistors, what is the power consumption due just to switching? (Neglect feedthrough current.)

8.5 A CMOS inverter has $W_p/W_n = 1.5$. The channel lengths are $L = 1\ \mu$m, and $W_n = 10\ \mu$m. Find the propagation delay time if the load capacitance is 1 pF. Let $t_{\text{ox}} = 4$ nm, $V_{DD} = 2.5$ V, and $V_{GS} - V_T = 2$ V.

8.6 For the transistor whose I_D-V_{DS} characteristics are shown in Figure P8.2, find the small signal parameters:

a. g_d for $V_{GS} = 1.0$ V, $V_{DS} < V_{DS\text{sat}}$

b. g_d for $V_{DS} > V_{DS\text{sat}}$, $V_{GS} = 1.1$ V

c. g_m for $V_{DS} = 0.2$ V, $V_{GS} = 1.1$ V

d. $g_{m\text{sat}}$ for $V_{DS} > V_{DS\text{sat}}$, $V_{GS} = 1.1$ V

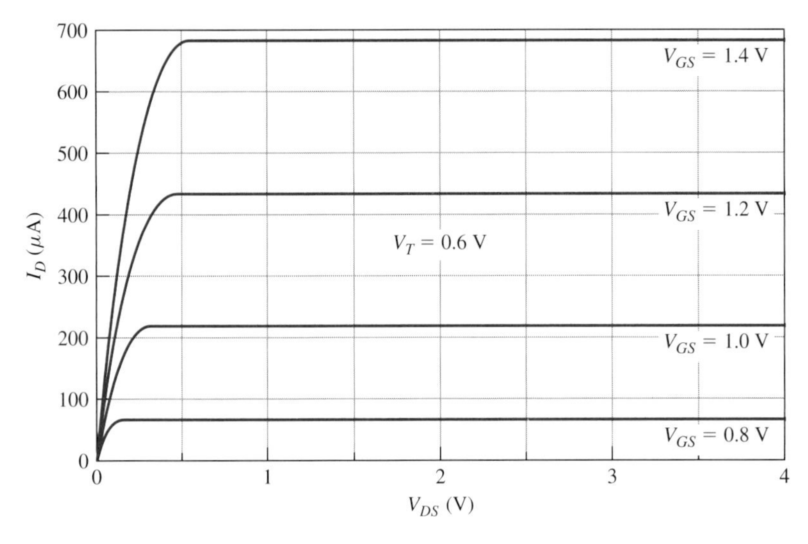

Figure P8.2

8.7 A CMOS circuit must operate at 500 MHz. What is the maximum channel length that can be tolerated, neglecting short-channel effects?

8.8 Practice using circuit models. Suppose a FET was found to have a finite input resistance R_{in}. How would you include this in the model of Figure 8.13a? How would you include the channel resistance?

8.9 A designer wishes to design a device for a 0.3 μm process (meaning that the FETs have $L = 0.3\ \mu\text{m}$). Furthermore, the current fabrication line is limited to $t_{\text{ox}} = 4\ \text{nm}$. What should the junction depth x_j be such that the short-channel effects can be neglected? Let $V_{DD} = 3.3\ \text{V}$, and let N_A' in the channel be 10^{17}. If the channel doping is reduced, what is the effect on the allowable junction depth? If the supply voltage is reduced, what is the effect?

8.10 SOI devices are more rad-hard (insensitive to radiation environments such as those in outer space or nuclear reactors) than conventional CMOS as discussed in the text. Examples of ionizing radiation are bombardment with high kinetic energy neutrons, electrons, protons, or gamma rays (which are photons with energies of a few MeV). The high-energy particles can disrupt the silicon crystal in conventional CMOS, while in SOI the amount of crystalline Si in the device is much smaller. Furthermore, charges that accumulate in the buried SiO_2 layer from the bombardment will generally not affect device operation. Explain why SOI devices will also be less affected by the absorption of gamma rays.

8.11 a. Draw the energy band diagram for the HFET, perpendicular to the gate, for the case when the transistor is strongly on. Indicate the polarity of the gate voltage.

b. Repeat for the depletion case (no channel). Now what is the polarity of the gate voltage?

c. Draw the energy band diagram along the channel for the equilibrium case and when the channel is conducting.

8.12 There are two primary reasons that GaAs-based HFETs are faster than silicon MOSFETs. One reason, given in the text, is that the electrons are traveling in a lightly doped channel. Explain how this increases their velocity. What do you think is the other reason?

8.13 The channel in the HFET is often referred to as supporting a two-dimensional electron gas. The electron gas part of this name refers to the sea of electrons that exists in the channel when the transistor is conducting. Why is it called two-dimensional?

8.14 Draw the energy band diagram at the source and the channel ends of the JFET under saturation (e.g., the same type of figure as Figure 8.31 for the MESFET). Point out the similarities and differences in operation of the two different types of device.

8.15 An n-channel JFET has a channel doping of 10^{16} cm^{-3}. Let the thickness of the n layer be 1 μm, the length of the channel be 4 μm, and the width be 40 μm. Neglect the depletion region adjacent to the substrate and neglect velocity saturation effects.

a. What is the threshold voltage?

b. If the gate voltage is -2 V, at what drain voltage does the current saturate?

c. What is the saturation current under these conditions, using the simple long-channel model?

Supplement to Part 3

S3.1 INTRODUCTION

Chapters 7 and 8 were dedicated to describing the basic principles of FET operation. In this supplement, some of the more specialized topics associated with MOSFETs and MOS-based capacitance structures are briefly described.

S3.2 COMMENTS ON THE FORMULATION FOR THE CHANNEL CHARGE Q_{ch}

In Chapters 7 and 8, various models were used to describe the I_D-V_{DS} characteristics for MOSFETs. In all of those models, the depletion region width in the Si was considered constant along the channel, which is not strictly true. In addition, we neglected any effect of velocity saturation on the channel charge. In this section, we refine the earlier models to account for these two effects.

S3.2.1 EFFECT OF VARYING DEPLETION WIDTH ON THE CHANNEL CHARGE

In Chapter 7, as we developed various models for the current in MOSFETs, we implicitly assumed that the ionized acceptor (or donor) charge in the depletion region to be constant along the channel, since the charge distribution in the depletion region is uniformly distributed. In reality, however, we know that the channel voltage V_{ch} varies with position y along the channel, and thus the depletion width also varies. Technically, this should be taken into account in the determination of the channel charge Q_{ch}. For modern MOSFETs, which have thin gate oxides, this refinement makes little difference in the electrical characteristics. Still, the greater accuracy is sometimes needed, so we will investigate an improved model for the channel charge here.

We start with an expression for Q_{ch} that includes a term dependent on y. For an NMOS, we have

$$Q_{ch}(y) = -C'_{ox}[V_G - V_T - V_{ch}(y)] + qN'_A[w(y) - w_T] \tag{S3.1}$$

where $w(y)$ is the depletion region width at position y,[1] and w_T is its width at the source, at and above threshold.[2] The last term in Equation (S3.1) represents the *body effect.* That is, the depletion width varies along the channel because of the difference in potential between the channel and the body (substrate), which is tied to the source.

Since the voltage across the Si depletion region is the surface potential, and $\phi_s(y) = 2\phi_f + V_{\text{ch}}(y)$, then from the discussion of Chapter 5 [Equation (5.39), for a one-sided junction],

$$w(y) = \left[\frac{2\varepsilon_{\text{Si}}}{qN'_A}(2\phi_f + V_{\text{ch}}(y))\right]^{1/2} \tag{S3.2}$$

and Q_{ch} becomes

$$Q_{\text{ch}}(y) = -C'_{\text{ox}}\left[V_{GS} - V_T - \frac{qN'_A w_T t_{\text{ox}}}{\varepsilon_{\text{ox}}}\left(\sqrt{1 + \frac{V_{\text{ch}}}{2\phi_f}} - 1\right)\right] \tag{S3.3}$$

We can use this model for the channel charge Q_{ch} to find the drain current. Using the same approach as in Chapter 7, we start with Equation (III.7), repeated here:

$$I_D(y) = -WQ_{\text{ch}}(y)\mu(y)\frac{dV_{\text{ch}}(y)}{dy} \tag{III.7}$$

Using Equation (S3.3) in Equation (III.7) and integrating gives, for $V_{DS} \leq V_{DS\text{sat}}$,

$$I_D = \frac{WC'_{\text{ox}}\mu_{\text{lf}}}{L\left(\dfrac{\mu_{\text{lf}}V_{DS}}{v_{\text{sat}}L}\right)}\left\{\left(V_{GS} - V_T - \frac{V_{DS}}{2}\right)V_{DS} - \frac{8}{3}\frac{\phi_f{}^{3/2}t_{\text{ox}}}{\varepsilon_{\text{ox}}}(q\varepsilon_{\text{Si}}N'_A)^{1/2}\right.$$
$$\left.\times\left[\left(1 + \frac{V_{DS}}{2\phi_f}\right)^{3/2} - \left(1 + \frac{3V_{DS}}{4\phi_f}\right)\right]\right\} \qquad V_{DS} \leq V_{DS\text{sat}} \tag{S3.4}$$

Here we have also included the effect of the longitudinal field on $\mu(y)$ (recall "Effect of $\mathscr{E}_L$ on the I_D-V_{DS} Characteristics of MOSFETs" in Section 7.3.2).

In saturation, the current $I_{D\text{sat}}$ is that current obtained in Equation (S3.4) by setting $V_{DS} = V_{DS\text{sat}}$.

For thin oxides (small t_{ox}, large C'_{ox}) characteristic of modern MOSFETs, the last term in Equation (S3.4) can be neglected, and Q_{ch} becomes

$$Q_{\text{ch}} \approx -C'_{\text{ox}}[V_{GS} - V_T - V_{\text{ch}}] \tag{S3.5}$$

---

[1] As in Chapter 8, we will use lower case w for depletion width and upper case W for channel width.

[2] The charge in the depletion region at the source end of the channel is $-qN'_A w_T$, but this value is included in the value for the threshold voltage in the first term. The magnitude of Q_{ch} at y is then reduced by the excess depletion region charge, $qN'_A(w(y) - w_T)$.

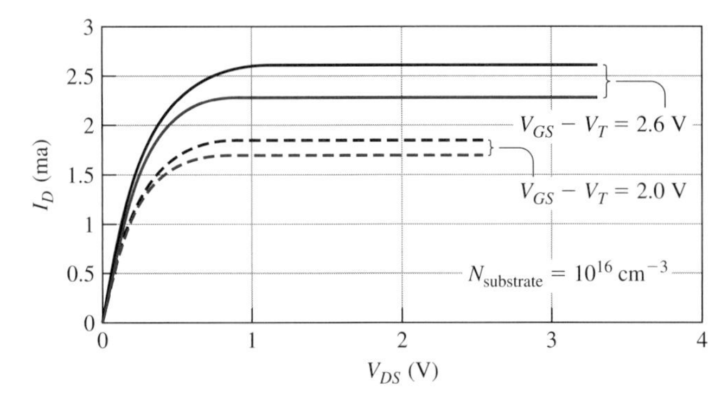

Figure S3.1 Effect of variable depletion width (variable bulk charge) on the I_D-V_{DS} characteristics. Black: simple model. Color: effect of variable depletion width. The effect is small and can be neglected for thin-oxide devices.

In this case, Equation (S3.4) simplifies to

$$I_D \approx \frac{W\mu_{\text{lf}}C'_{\text{ox}}}{L\left(1+\dfrac{\mu_{\text{lf}}V_{DS}}{v_{\text{sat}}L}\right)}\left(V_{GS}-V_T-\frac{V_{DS}}{2}\right)V_{DS} \tag{S3.6}$$

which is identical to the results of Chapter 7, obtained with the simpler model for Q_{ch} [Equation (7.66)]. Thus, for thin-oxide devices, the change in channel charge along the channel can usually be neglected.

A comparison of the above two models is presented in Figure S3.1. Here current is plotted against V_{DS} for a typical device with the MOSFET parameters of Table 7.1. Two gate voltages are shown, $V_{GS} - V_T = 2.6$ V and 2.0 V, corresponding to supply voltages of 3.3 V and 2.5 V respectively. The substrate doping is taken as $N'_A = 10^{16}$ cm^{-3}, with $W = 5\,\mu$m and $L = 0.5\,\mu$m. The gate oxide is 4 nm thick. At small V_{DS} the results are identical. Taking the variation of bulk charge (body effect) into account, for this set of parameters $I_{D\text{sat}}$ is about 9 percent less than for the model in which this effect is neglected. If the oxide were infinitely thin, the results would be identical. Considering the uncertainties in knowing N'_A, μ_{lf}, v_{sat} and t_{ox}, 9 percent is a small error, and the use of the simpler model is often justified.

S3.2.2 DEPENDENCE OF THE CHANNEL CHARGE Q_{ch} ON THE LONGITUDINAL FIELD $\mathscr{E}_L$

In the previous model (Chapter 7) for the I_D-V_{DS} characteristics of a MOSFET, we let the channel charge Q_{ch} approach zero when the device was in saturation. Recall that in that model, the carrier velocity v approached infinity such that the $Q_{\text{ch}}v$ product was finite and thus I_D remained at its constant value of $I_{D\text{sat}}$. For high longitudinal fields, however, the carrier velocity saturates, meaning the channel charge Q_{ch} will reach some minimum value to keep the $Q_{\text{ch}}v$ product constant. In this section, the Q_{ch} dependence on v_{sat} is discussed.

Recall from Equation (III.3) that $I_D = -WQ_{ch}v$. Thus,

$$Q_{ch}(y) = -\frac{I_D}{Wv(y)} \tag{S3.7}$$

where Q_{ch} and the carrier velocity v are evaluated at a given position y. At small longitudinal field strengths $\mathscr{E}_L$, such that $v \ll v_{sat}$, we have

$$Q_{ch} = Q_{chlf} = -C'_{ox}[V_{GS} - V_T - V_{ch}(y)] \tag{S3.8}$$

where Q_{chlf} is the channel charge at small longitudinal fields and is independent of $\mathscr{E}_L$.

Since I_D is constant everywhere in the channel, the product $Q_{ch}v$ is constant. When the longitudinal field is large enough that the carrier velocity saturates, the channel charge reaches some minimum value Q_{chmin}:

$$Q_{chmin} = -\frac{I_D}{Wv_{sat}} \tag{S3.9}$$

We define a critical longitudinal field $\mathscr{E}_{Lc}$ to be that field at which Q_{chlf}, given by Equation (S3.8), is equal to Q_{chmin}, given by Equation (S3.9). Then, for $\mathscr{E}_L \geq \mathscr{E}_{Lc}$

$$Q_{chlf} = Q_{chmin} \qquad \mathscr{E}_L \geq \mathscr{E}_{Lc} \tag{S3.10}$$

Figure S3.2 shows a plot of Q_{ch} for this model (solid line) as a function of field strength $\mathscr{E}_L$ for an n-channel MOSFET with $V_{GS} = 2$ V and assuming

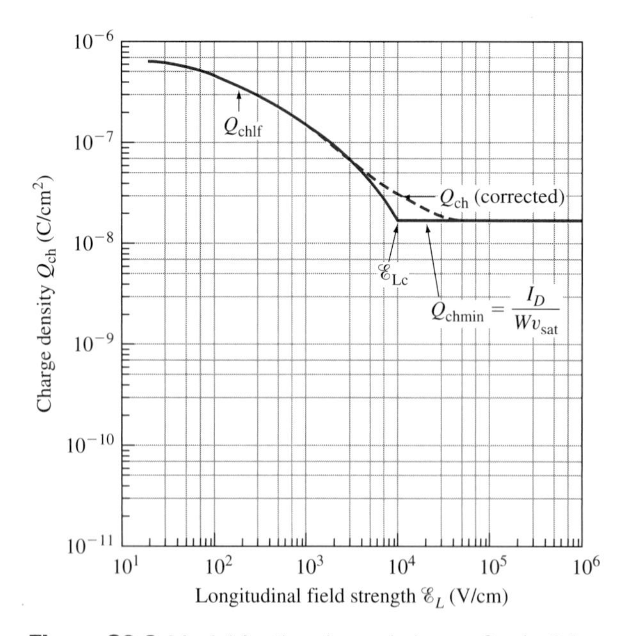

Figure S3.2 Model for the channel charge Q_{ch} (solid line) for $Q_{ch} = Q_{chlf}$ below the critical field $\mathscr{E}_{Lc}$ and $Q_{ch} = Q_{chmin} = I_D/Wv_{sat}$ for $\mathscr{E}_L \geq \mathscr{E}_{Lc}$. Also shown (dashed line) is a correction to Q_{ch} in the vicinity of $\mathscr{E}_{Lc}$.

$v_{sat} = 4 \times 10^6$ cm/s. [1] This model assumes that the field dependence of the carrier velocity changes abruptly at $\mathscr{E}_{Lc}$ from its low field value $v = \mu_{lf}\mathscr{E}_L$ to v_{sat} at the critical field $\mathscr{E}_{Lc}$. This underestimates Q_{ch} where $\mathscr{E}_L$ is close to $\mathscr{E}_{Lc}$.

An empirical model for Q_{ch} has been proposed to eliminate this discrepancy [1]:

$$Q_{ch} = Q_{lf} + \frac{I_D}{W v_{sat}}\left(1 - \frac{I_D}{W Q_{lf}\mu_{lf}\mathscr{E}_L}\right) \tag{S3.11}$$

While this model is somewhat arbitrary, it does reduce to Q_{lf} at small $\mathscr{E}_L$ as required, and at large $\mathscr{E}_L$ where Q_{lf} is small, it reduces to Equation (S3.9). The result from Equation (S3.11) is shown by the dashed line in Figure S3.2.

S3.3 THRESHOLD VOLTAGE FOR MOSFETs

In Chapter 7, we explored the basic physics of the MOSFET, and saw how the application of a gate voltage could enable (or disable) the conducting channel. We considered the threshold voltage V_T to be the gate voltage that would be required to just form a good conducting channel. While this parameter can be determined experimentally once the device is made, it is important to know its dependence on the various material properties of the device so that its value can be controlled during fabrication.

As discussed in "Definition of Threshold" in Section 7.2.3, the threshold voltage is defined as the gate voltage required to produce a surface potential of $\phi_s = 2\phi_f$ at the source end of the channel. It is the gate voltage that produces a carrier concentration at the surface at the source end of the channel equal to, but opposite in sign to, that in the substrate. This surface potential is dependent on the work function difference between the gate and the substrate, and the oxide thickness (recall Figure 7.2). In addition to these previously discussed causes of the surface potential, there can be charges in the material below the gate, as shown in Figure S3.3.

The figure shows an n-channel enhancement MOSFET; the substrate is p type. There is charge at the oxide-substrate interface, along with the usual bulk (depletion) charge in the substrate adjacent to the channel. (Note that this is an enhancement mode device, so at equilibrium there is no conducting channel and thus no mobile charge carriers are shown in the channel.)

The concentrations of these charges, as well as their locations, influence the threshold voltage. These charges need to be controlled during the fabrication process, or, if they cannot be controlled, they should be accounted for and compensated for in the design of the device.

The various types of charges (coulombs per unit area) that influence the value of the threshold voltage are:

Q_{ot}, oxide trapped charges, which are fixed in position and distributed throughout the oxide.

Q_f, fixed oxide charge, the charge associated with the positive Si^+ ions in the SiO_2 near (within about 2 nm) the SiO_2/Si interface. These also are fixed in position.

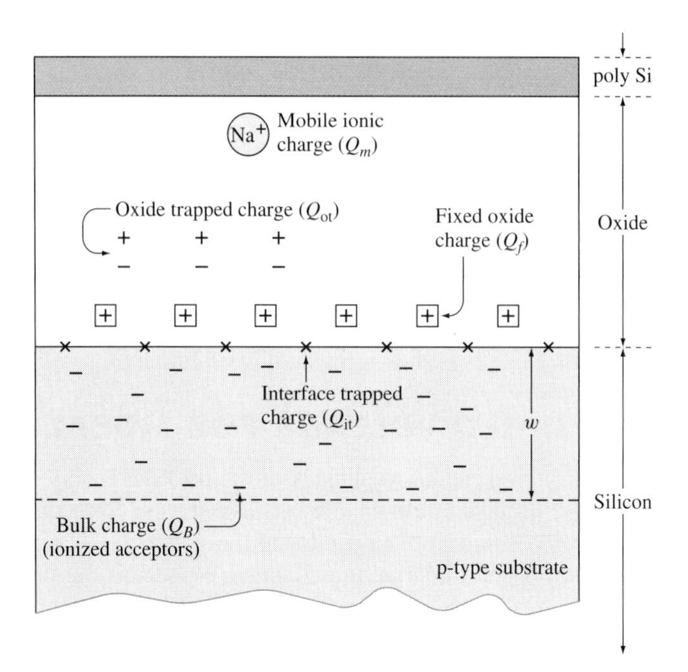

Figure S3.3 The charges in a MOSFET affect the threshold voltage. These include the trapped charge Q_{ot} distributed through the oxide, fixed charge Q_f in the oxide near the Si/SiO_2 interface, bulk charge Q_B in the substrate depletion region (w), mobile ionic charge Q_m in the oxide, and charge Q_{it} trapped in the interface states at the Si/SiO_2 interface. Not to scale.

Q_m, mobile ion charge in the oxide (e.g., Na^+ ions), which are quite mobile in the SiO_2 and can move under the influence of the electric field in the oxide.

Q_{it}, interface trapped charge, the charge associated with interface states at the SiO_2/Si interface (these states are due to crystal defects at the interface, e.g., dangling bonds). This charge is dependent on the surface potential, which is dependent on the gate voltage.

Q_B, depletion region or bulk charge, the charge associated with the ionized acceptors and donors in the Si depletion region at the source end of the channel.

In addition, if a channel exists:

Q_{ch}, the mobile electron charge that is responsible for channel current (i.e., the electrons in the channel)

The charges Q_{it} and Q_B depend on the surface potential, and for $V_{DS} \neq 0$ they depend on position y in the channel since the surface potential is dependent on y. Since the threshold voltage depends on their values at the source end of the channel, however, they are evaluated at $y = 0$.

Of these, the concentrations of oxide trapped charge (Q_{ot}) and the oxide mobile ion charge (Q_m) have been problems in the past, but modern processing methods for silicon have reduced them to levels small enough that they have a small effect on the device characteristics. As a result, we will not consider them further. The other charges remain important and we will consider their effects next.

S3.3.1 FIXED CHARGE

The fixed oxide charge Q_f in the SiO_2 near the interface is a natural result of the thermal oxidation process used to produce the SiO_2 insulating layer. During thermal oxidation, oxygen diffuses through the existing SiO_2 layer and reacts with the Si in the substrate, thus increasing the oxide thickness (and eating into the substrate). This oxidation process is not uniform, however, and there exists a region about 2 nm thick between the Si and the SiO_2 that contains some partially oxidized Si, which is positively charged. As the oxidizing process continues, these Si ions become oxidized (bond with oxygen), but some positive Si ions are incorporated into the newly formed oxide. With current processing methods the concentration of these Si ions can be reduced to the low level of about $10^{10}\ cm^{-2}$.

S3.3.2 INTERFACE TRAPPED CHARGE

The interface trapped charge Q_{it} results from dangling bonds at the Si/SiO_2 interface. These are the same dangling bonds discussed for a Si surface in Chapter 6. There, the silicon surface was next to vacuum, and now it is next to silicon dioxide, but the principle is the same. While the bonding of the SiO_2 to the Si surface reduces the number of dangling bonds and thus the number of interface states, some danglers remain. One method for reducing the number of remaining interface states is to introduce some hydrogen into the silicon dioxide. Since hydrogen, being a small atom, is relatively mobile in SiO_2, at elevated temperatures during processing some of the H atoms diffuse to the SiO_2/Si interface, where they attach to the dangling bonds. The dangling bonds are said to be *passivated,* meaning rendered inert. However, since hydrogen can diffuse in the SiO_2, even at room temperature, the degree of passivation can change with time, with a resultant change in the electrical characteristics of the transistor. Alternatively, deuterium, an isotope of hydrogen, can be used instead of hydrogen for passivation. Deuterium, being larger than hydrogen, has a smaller diffusion coefficient and results in a more stable device. [2]

The surface potential ϕ_s, and therefore the threshold voltage, depends on the charge in these interface states. Figure S3.4 shows the energy band diagram at equilibrium. Note that the interface trapped charge Q_{it} is the charge in *interface* states. In principle Q_{it} can be either positive or negative. For Si MOSFETs, however, the net Q_{it} is normally positive at equilibrium.

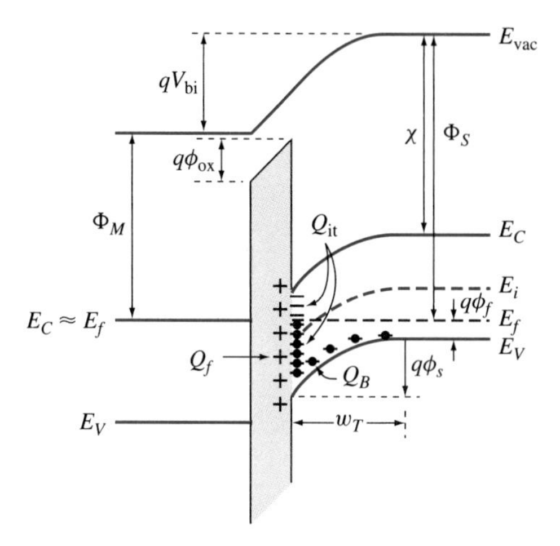

Figure S3.4 Equilibrium energy band diagram normal to the gate. The fixed charge Q_f, the bulk charge Q_B, and the interface trapped charge Q_{it} are indicated. The interface states are largely filled below the Fermi level and empty above it.

S3.3.3 BULK CHARGE

The threshold voltage also depends on the bulk charge Q_B, which consists of fixed ionized acceptors (in p-channel devices they would be ionized donors) in the depletion region at the source end of the channel. The bulk charge is a function of the net doping level in the substrate and also depends on the width $w(y)$ of the depletion region.

$$Q_B(y) = -qN'_A w(y) \tag{S3.12}$$

where Q_B is the bulk charge per unit area, and

$$w(y) = \left[\frac{2\varepsilon_s}{qN'_A}\phi_s(y)\right]^{1/2} \tag{S3.13}$$

Since the threshold voltage depends on the source-channel barrier (at $y = 0$), and since at threshold $\phi_S(0) = 2\phi_f$, then at threshold at the source end of the channel

$$Q_{B(\text{threshold})}(y = 0) = -[2qN'_A\varepsilon_s(2\phi_f)]^{1/2} \tag{S3.14}$$

S3.3.4 EFFECT OF CHARGES ON THE THRESHOLD VOLTAGE

Consider a MOSFET with source, substrate (body), drain, and gate connected together (equilibrium). Since we are using the source as a reference, $V_{GS} = V_{DS} = V_{BS} = 0$, where V_{BS} is the voltage between substrate and source. The built-in potential energy qV_{bi} between gate and substrate is the difference between the work functions Φ_M for the gate and Φ_S for the semiconductor, or $V_{bi} = |\Phi_{MS}|$ where $\Phi_{MS} = (\Phi_M - \Phi_S)$ (see Figure S3.4).[3] The built-in potential is dropped partly across the oxide and partly across the substrate as indicated in Figure S3.4. If now a voltage $V_{GB} = V_T$ is applied to the gate, the electrostatic potential between gate and substrate is at threshold

$$V_{GB} + V_{bi} = V_T - \frac{\Phi_{MS}}{q} = \phi_{ox}(V_{GB}) + \phi_S(V_{GB}) \qquad \text{(S3.15)}$$

where $\phi_{ox}(V_{GB})$ and $\phi_S(V_{GB})$ represent respectively the electrostatic potential across the oxide and the surface potential for a gate-substrate voltage V_{GB}. The threshold voltage is then

$$V_T = \frac{\Phi_{MS}}{q} + \phi_{ox}(V_T) + \phi_s(V_T) \qquad \text{(S3.16)}$$

At threshold, $\phi_s = \phi_s(V_T) = 2\phi_f$ and

$$\phi_{ox}(V_T) = -\left[\frac{Q_f + Q_{it}(2\phi_f) + Q_B(2\phi_f)}{C'_{ox}}\right] \qquad \text{(S3.17)}$$

where Q_{it} and Q_B are evaluated at $\phi_s = 2\phi_f$.

The threshold voltage can then be expressed as

$$V_T = \frac{\Phi_{MS}}{q} + 2\phi_f - \frac{Q_f}{C'_{ox}} - \frac{Q_{it}(2\phi_f)}{C'_{ox}} - \frac{Q_B(2\phi_f)}{C'_{ox}} \qquad \text{(S3.18)}$$

S3.3.5 FLAT BAND VOLTAGE

As the gate voltage is varied, the band bending in Figure S3.4 also varies. It is possible for the applied voltage to cause the energy bands in the semiconductor to bend up instead of down. Figure S3.5a shows an enhancement-type NMOS at equilibrium. In this case, applying a negative gate voltage will reduce the band bending. At some value of applied gate voltage the bands in the Si are flat, as shown in Figure S3.5b. The value of V_{GS} required for the substrate band edges to be flat, or $\phi_s = 0$, is referred to as the flat band voltage V_{FB}:

$$V_{FB} = \frac{\Phi_{MS}}{q} + \phi_{ox}(0) = \frac{\Phi_{MS}}{q} - \frac{(Q_f + Q_{it}(0))}{C'_{ox}} \qquad \text{(S3.19)}$$

where $Q_{it}(0)$ and $\phi_{ox}(0)$ are evaluated at the flat band condition ($\phi_s = 0$). Figure S3.5c illustrates that a more negative gate voltage can reverse the direction of band bending.

[3]The subscript M is for "metal," although in modern FETs, the gate is usually heavily doped polysilicon.

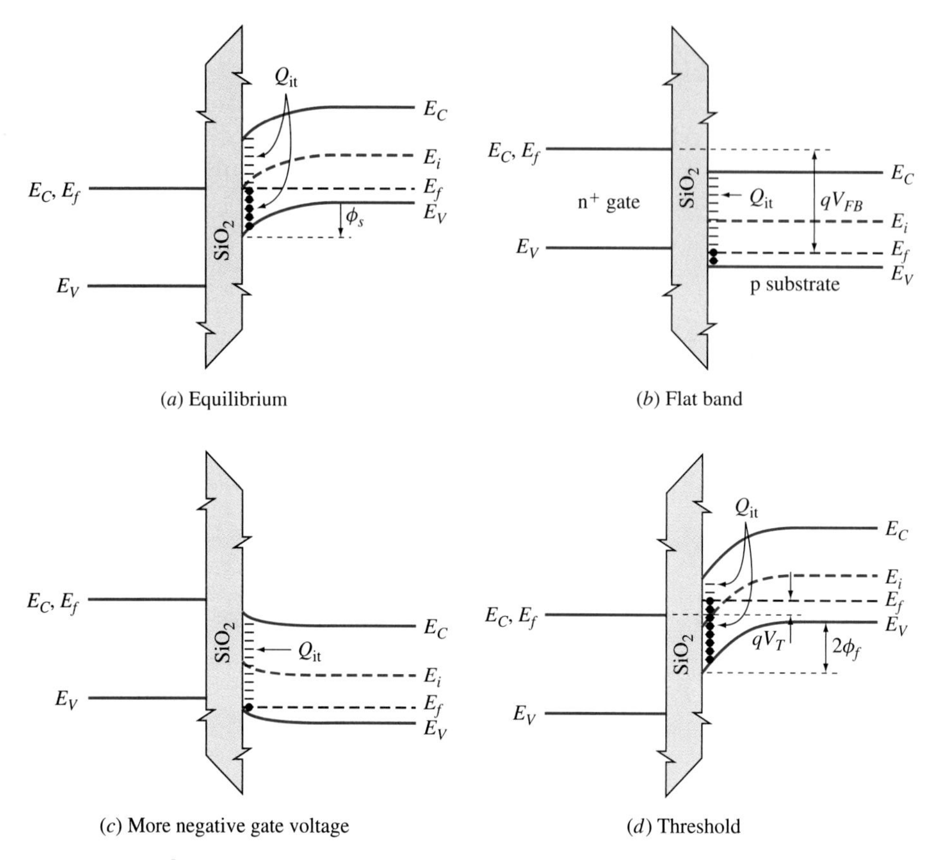

Figure S3.5 Occupation of surface states (traps). (a) Equilibrium. (b) Flat band condition. (c) Under increasingly negative gate voltage, the bands can actually bend up. (d) Threshold. For donor traps, those below the Fermi level are occupied and thus neutral. Traps above the Fermi level are vacant and thus positive. It can be seen that Q_{it} is greater (more positive) at flat band than at threshold.

The energy band diagram at threshold is shown in Figure S3.5d. Here the applied gate voltage is positive. Note that the amount of charge trapped at the interface is different from the flat band condition, or $Q_{it}(0) \neq Q_{it}(2\phi_f)$.

The threshold voltage can be expressed in terms of the flat band voltage,

$$V_T = V_{FB} + 2\phi_f - \frac{Q_B(2\phi_f)}{C'_{ox}} - \frac{Q_{it}(2\phi_f) - Q_{it}(0)}{C'_{ox}} \tag{S3.20}$$

It is often assumed that

$$Q_{it}(2\phi_f) \approx Q_{it}(0) \tag{S3.21}$$

and the last term in Equation (S3.20) is neglected.

EXAMPLE S3.1

Find the threshold voltage for an n-channel MOSFET with an n$^+$ poly Si gate with $N'_A = 10^{16}\,\text{cm}^{-3}$, and $Q_f = Q_{\text{it}}(2\phi_f) = q \times 10^{10}\,\text{C/cm}^2$, an oxide thickness of 4 nm, and a poly Si gate doped to $3 \times 10^{19}\,\text{cm}^{-3}$.

■ Solution

We evaluate the contribution of each term in Equation (S3.18). First, to find Φ_{MS}/q, we use

$$\Phi_{MS} = \Phi_M - \Phi_S$$

The polysilicon is degenerately doped. From Figure 2.25 we see that for $N'_D = 3 \times 10^{19}\,\text{cm}^{-3}$, E_f is 0.07 eV below the unperturbed conduction band edge (at the apparent conduction band edge). Since at equilibrium $E_{fM} = E_{fS}$,

$$\Phi_M = \chi_S + 0.07\,\text{eV}$$

and

$$\Phi_S = \chi_S + E_G - (E_f - E_V)$$

Then

$$\Phi_{MS} = 0.07 - [E_G - (E_f - E_V)]$$

To locate the Fermi level $(E_f - E_V)$ in the substrate under the gate, we use the relation (since the p region is not degenerately doped)

$$p_0 = N'_A = N_V e^{-(E_f - E_V)/kT}$$

or

$$E_f - E_V = kT \ln \frac{N_V}{N'_A} = 0.026 \ln \frac{3.1 \times 10^{19}}{10^{16}} = 0.21\,\text{eV}$$

Since the band gap $E_g = 1.12$ eV, we combine these values and get

$$\phi_{MS} = 0.07 - 1.12 + 0.21 = -0.84\,\text{eV}$$

The next term in Equation (S3.18) is found from the relation

$$p_0 = N'_A = n_i e^{q\phi_f/kT}$$

$$\phi_f = \frac{kT}{q} \ln \frac{N'_A}{n_i} = 0.026 \left(\ln \frac{10^{16}}{1.08 \times 10^{10}} \right) = 0.36\,\text{V}$$

$$2\phi_f = 0.72\,\text{V}$$

We will need the oxide capacitance per unit area for the last three terms:

$$C'_{\text{ox}} = \frac{\varepsilon_{\text{ox}}}{t_0} = \frac{3.9 \times 8.85 \times 10^{-14}\,\text{F/cm}}{4 \times 10^{-7}\,\text{cm}} = 8.63 \times 10^{-7}\,\text{F/cm}^2$$

Now it remains to find the charges. It is given that

$$Q_f = Q_{\text{it}} = q \times 10^{10}\,\text{C/cm}^2 = (1.6 \times 10^{-19}\,\text{C})(10^{10}\,\text{charges/cm}^2)$$

from which we find the quantity:

$$\frac{Q_f}{C'_{\text{ox}}} = \frac{Q_{\text{it}}(2\phi_f)}{C'_{\text{ox}}} = 0.0018 \text{ V}$$

For the bulk charge term:

$$\begin{aligned} Q_B &= -\sqrt{2\varepsilon_s q N'_A(2\phi_f)} \\ &= -\sqrt{2 \cdot (11.8 \times 8.85 \times 10^{-14})(1.6 \times 10^{-19})(10^{16})(2)(0.36)} \\ &= -4.9 \times 10^{-8} \text{ C/cm}^2 \\ \frac{Q_B}{C'_{\text{ox}}} &= -0.057 \text{ V} \end{aligned}$$

Substituting all of these into Equation (S3.18), we find the value of the threshold voltage to be, for this n-channel device,

$$V_T = -0.84 + 0.72 - 0.0018 - 0.0018 + 0.057 = -0.07 \text{ V}$$

It is seen in the above example that the contribution of the interface charge is small and Equation (S3.21) is a reasonable approximation. Sometimes, nitrogen is introduced into the gate oxide to increase the oxide capacitance. However, the introduction of nitrogen increases the interface state density and in that case, Equation (S3.21) is not necessarily a good approximation.

In the above calculation, we considered the impurity-induced band-gap narrowing for the degenerate gate. If we had neglected this effect and taken the Fermi level to coincide with the unperturbed conduction band edge (as is often done for degenerate semiconductors), the calculated value of V_T would be 0.07 eV less than calculated above; i.e., $V_T \approx -0.14$ V.

EXAMPLE S3.2

Find the flat band voltage associated with the device of Example S3.1. Assume $Q_{\text{it}}(2\phi_f) = Q_{\text{it}}(0)$.

■ **Solution**

From Equation (S3.19),

$$V_{FB} = \frac{\Phi_{MS}}{q} - \frac{(Q_f + Q_{\text{it}})}{C'_{\text{ox}}} = -0.84 - \frac{1.6 \times 10^{-19}(10^{10} + 10^{10})}{8.63 \times 10^{-7}} = -1.03 \text{ V}$$

S3.3.6 THRESHOLD VOLTAGE CONTROL

In Example S3.1, a threshold voltage of $V_T = -0.07$ V was obtained for the n-channel MOSFET. This makes it a depletion mode device—a voltage more positive than this (including $V_G = 0$) results in a conducting channel. In digital circuits, however, enhancement MOSFETs are normally preferred, such that a voltage is required to turn them on rather than to turn them off. A typical design rule is to aim for threshold voltages of about 20 percent of the supply voltage.

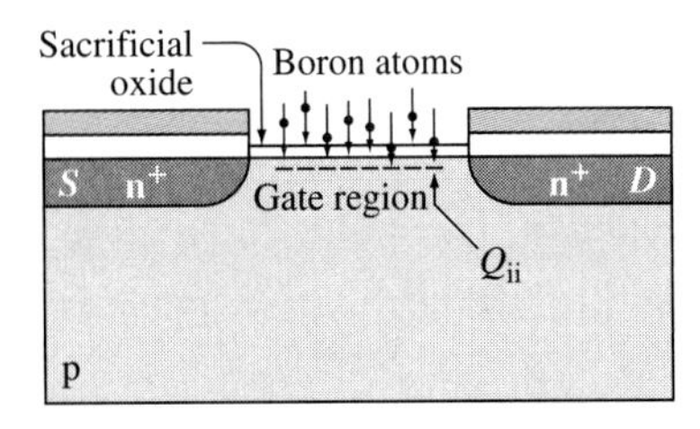

Figure S3.6 Ion implantation of (in this case) boron atoms leaves excess negative charges near the silicon surface. The gate oxide is grown later. The excess charges are used to adjust the threshold voltage.

Thus, for a device expected to operate with a supply voltage of +2.5 V, the threshold voltage should be $V_T \approx +0.5$ V.

This threshold voltage adjustment can be accomplished by ion implanting a p-type impurity (e.g., boron) into the Si substrate in the channel region. Figure S3.6 shows how this implant is done. In ion implantation, discussed in more detail in Appendix C, dopant atoms are ionized and accelerated electrostatically toward the sample, where they strike with sufficient force that they become implanted. The implant is done through a thin sacrificial oxide layer to prevent the ions from going too deep. After the implant, the oxide is removed, and the gate oxide is grown in a later processing step.

If the implant is very shallow, the approximation can be made that all of the impurities are at the SiO_2/Si interface. Since these acceptors are ionized, they contribute an extra charge layer (in this case negative charge), assumed to be at the interface. This localization of the impurities then adjusts the threshold voltage by the value

$$\Delta V_T = -\frac{Q_{ii}}{C'_{ox}} = +\frac{qN_{ii}}{C'_{ox}} \tag{S3.22}$$

where Q_{ii} is the implanted charge per unit area and N_{ii} is the number of implanted boron atoms per unit area. In this case, the change in the threshold voltage is positive since the added charges are negative.

EXAMPLE S3.3

To change the NMOSFET of the example in the previous section to an enhancement device following the design guidelines, we would want to correct the fabrication process to raise the threshold voltage by 0.57 V.

■ **Solution**

Using the ion-implantation technique just described, we would have to implant:

$$N_{ii} = \frac{8.63 \times 10^{-7} \times 0.57}{1.6 \times 10^{-19}} = 3.1 \times 10^{12} \text{ boron atoms/cm}^2$$

Ion implantation is a technique for adjusting the threshold voltage during device fabrication. There is a way, however, to adjust the threshold voltage of a MOSFET after it has already been made—that is, to change the bias on the substrate with respect to the source. The voltage applied to the substrate changes the width of the depletion region between the channel and the bulk substrate below it, and thus changes the bulk charge Q_B at the source end of the channel to the value

$$Q_B = -[2\varepsilon_S q N_A'(2\phi_S - V_{BS})]^{1/2} \tag{S3.23}$$

where V_{BS} is the substrate-to-source voltage.

If we use this adjusted value of the bulk charge in Equation (S3.18), we have an adjustment to the threshold voltage of

$$\Delta V_T = \frac{\sqrt{2\varepsilon_S q N_A'}}{C_{\text{ox}}'}\left[(2\phi_f - V_{BS})^{1/2} - (2\phi_f)^{1/2}\right] \tag{S3.24}$$

or, to express it more tidily,

$$\Delta V_T = \frac{\sqrt{4\varepsilon_S q N_A'\phi_f}}{C_{\text{ox}}'}\left[\left(1 - \frac{V_{BS}}{2\phi_f}\right)^{1/2} - 1\right] \tag{S3.25}$$

Typically, the bulk voltage is chosen such that $|V_{BS}| \gg 2\phi_f$, in which case the threshold adjustment is

$$\Delta V_T \approx \frac{\sqrt{2\varepsilon_S q N_A'\phi_f}}{C_{\text{ox}}'} = \frac{t_{\text{ox}}\sqrt{2\varepsilon_S q N_A'\phi_f}}{\varepsilon_{\text{ox}}}\sqrt{V_{BS}} \tag{S3.26}$$

or

$$V_{BS} = \frac{-1}{2\varepsilon_S q N_A'\phi_f}\left(\frac{\varepsilon_{\text{ox}}\Delta V_T}{t_{\text{ox}}}\right)^2 \tag{S3.27}$$

Note that for devices with thin oxides, since V_{BS} varies as $(1/t_{\text{ox}})^2$, V_{BS} in Equation (S3.27) becomes large. For example, the device in Example S3.1 had an oxide thickness of 4 nm. If we wished to adjust its threshold voltage by $\Delta V_T = 0.57\,\text{V}$, we would need to apply a bulk voltage of

$$V_{BS} \approx 200\,\text{V}$$

which is much higher than the breakdown voltage. Clearly, this method is impractical for devices with the thin oxides typical of modern devices.

*S3.3.7 CHANNEL QUANTUM EFFECTS

Up to now, we have considered the channel in a MOSFET classically. That is, the density of states function $S(E)$ has the value given in Chapter 2:

$$S(E) = \frac{1}{2\pi^2}\left(\frac{2m_{ds}^*}{\hbar^2}\right)^{3/2}\sqrt{E - E_C}$$

This is a continuous function, and an electron could be at any energy in the band.

In the channel of a MOSFET, however, the carriers are confined in the narrow potential well at the Si/SiO_2 interface. At high electric fields ($\mathscr{E}_T > 10^5$ V/cm), the width of the well is small enough that the carriers must be treated quantum mechanically, as was illustrated in Supplement A to Part 1. You will recall from the one-dimensional potential well problem that the electron will be restricted to a finite number of discrete states—not every energy is allowed. This is shown in Figure S3.7a on the left. There the potential energy is constant in the well. In a MOSFET channel, however, shown on the right, the potential varies with position in the well. The principles are the same, in that there are discrete states and the number and actual energies vary with well depth and width, but the quantitative results are somewhat different since the well shape is different. The potential well in a MOSFET, however, is three-dimensional and is narrow only in the x direction. For this case each discrete state in the figure is actually the lowest energy of a "miniband." The bottom of the lowest energy band is somewhat above the classical band minimum at the interface. [3]

The existence of the quantization of states perpendicular to the gate has two effects. We consider an n-channel MOSFET.

1. Because the lowest energy states are above the classical bottom of the well at the surface, it requires a larger gate voltage to bend the bands sufficiently to permit electrons to enter the channel. This then increases the threshold voltage. Figure S3.7b indicates this for a particular case of 10^{12} electrons/cm^2 in the well. [4] The lowest allowed state is 40 meV above the bottom of the well, increasing the potential barrier E_B for electrons entering the channel, Figure S3.7c.
2. Recall that the electron in a potential well has a wave function that oscillates in the well and goes to zero outside the well. This forms a standing wave whose maxima are in the interior of the well. Because of the standing electron wave in the x direction in the well, $\psi^*\psi$, and thus the maximum charge concentration for a given electron, is removed from the Si/SiO_2 interface as indicated in Figure S3.7d for the device of (a). [4] Figure S3.7d shows a plot of charge density as a function of position for the classical case and for the actual (quantum-mechanical) case. Here it can be seen that, from classical considerations, the channel charge should be concentrated near the interface, but when the quantization is considered, the maximum charge is on the order of 1 or 2 nm from the interface. Since the charge is located deeper into the semiconductor, this effect is analogous to having an increased oxide thickness. This results in a reduced transconductance.

For this case the threshold voltage has increased somewhat more than the 40 meV difference between the lowest occupied state and the classical bottom of the well. Because the oxide is in effect thicker, and because any applied gate voltage is dropped partly across the "oxide" and partly in the semiconductor, to produce the required extra band bending of 40 meV, approximately 60 meV must be applied to the gate.

Since these quantum mechanical effects increase with increasing transverse field, they are more pronounced for heavier substrate doping and for increased channel charge.

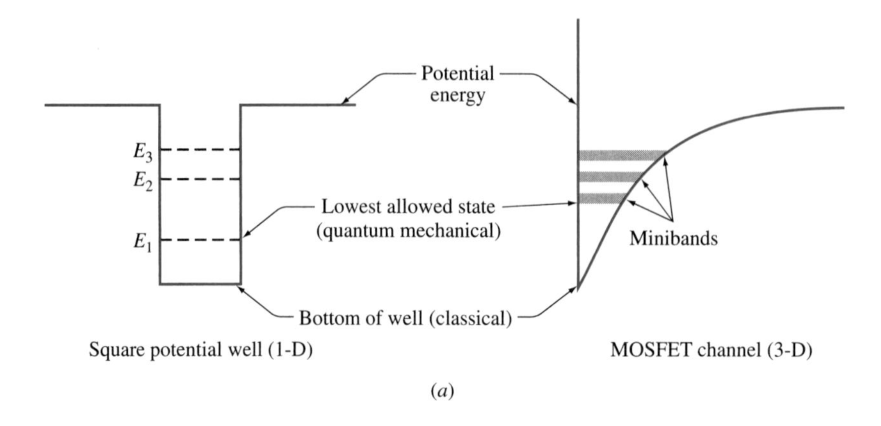

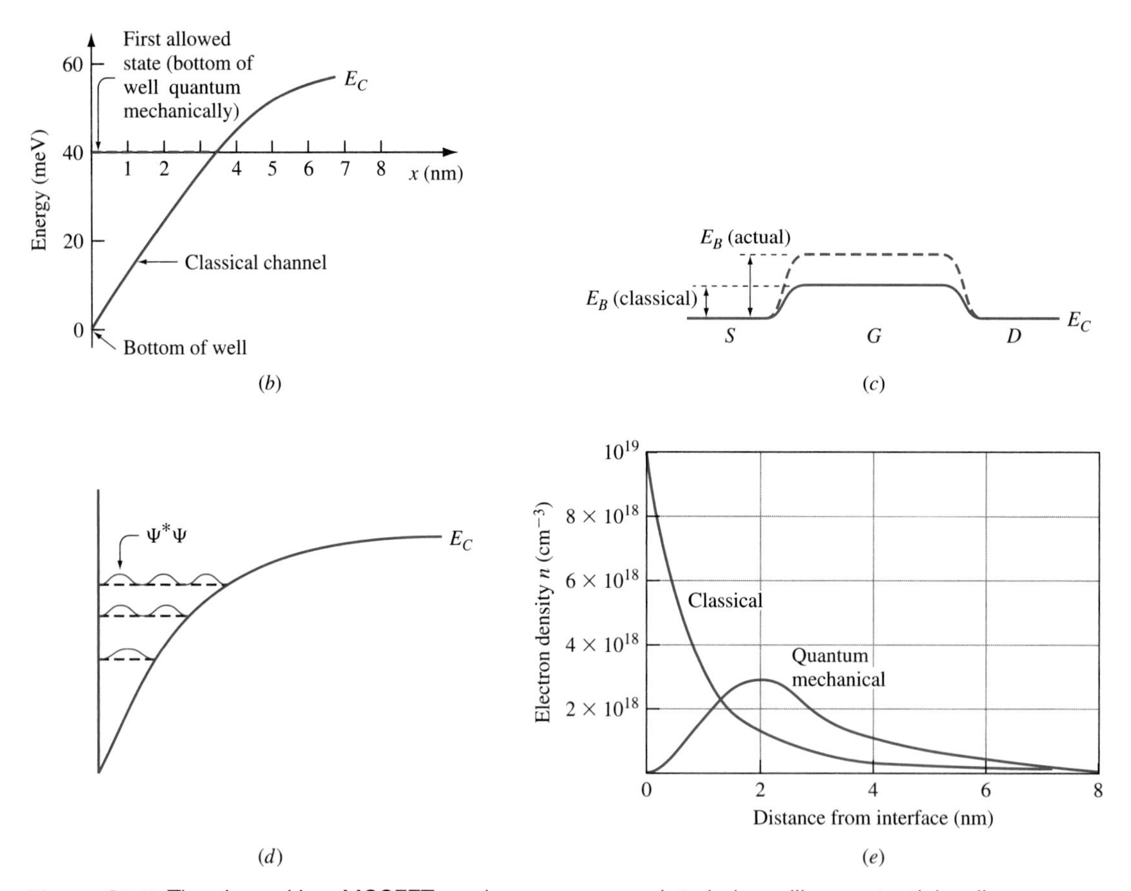

Figure S3.7 The channel in a MOSFET can be narrow enough to behave like a potential well. (a) Comparison of square well from Supplement A to Part 1 to MOSFET channel well; (b) comparison of classical channel depth to quantum-mechanical lowest energy level; (c) the minimum state above the classical bottom of the well results in a higher potential barrier for electrons entering the well; (d) the wave functions indicate that the charge is concentrated somewhere in the interior of the well rather than at the interface; (e) electron concentration normal to the gate for the classical case and the quantum case. In both cases, $Q_{ch}/q = 10^{12}$ cm^{-3}. Adapted from References 3 and 4.

S3.4 UNIVERSAL RELATIONS FOR LOW-FIELD MOBILITY

Experimentally, it has been found that a universal relation exists between the low-field mobility μ_{lf} and an *effective transverse field* $\mathscr{E}_{Teff}$, independent of substrate doping, substrate bias, and oxide thickness. Let us give the relationship first, then discuss the effective transverse field.

For electrons, the relationship between the low-field mobility and $\mathscr{E}_{Teff}$ is [5]

$$\mu_{lf} = 32500 \mathscr{E}_{Teff}^{-1/3} \qquad \text{for electrons} \tag{S3.28}$$

where μ_{lf} is expressed in $cm^2/V \cdot s$ and $\mathscr{E}_{Teff}$ is in V/cm.

The effective transverse field is an artificial idea that arose from experimental results. It works out, however, that the value of $\mathscr{E}_{Teff}$ corresponds to the field value calculated at a position x_0 such that half the channel charge Q_{ch} is on either side of x_0. This is illustrated in Figure S3.8. The field at this point, $\mathscr{E}_T(x_0) = \mathscr{E}_{Teff}$, is evaluated from Gauss's law at $x = x_0$:

$$\mathscr{E}_{Teff} = \frac{\left| Q_B + \dfrac{Q_{ch}}{2} \right|}{\varepsilon_{Si}} \tag{S3.29}$$

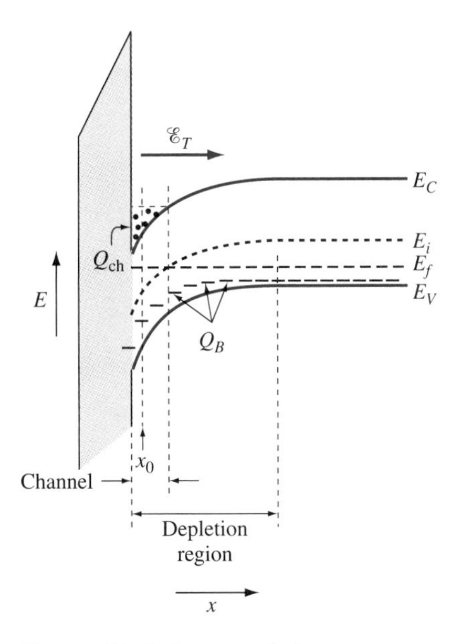

Figure S3.8 Source of charges contributing to transverse field. There are ionized acceptors distributed throughout the depletion region, and there are mobile electrons in the channel.

where the depletion charge is found from the doping and band bending in the depletion region:

$$|Q_B| = \sqrt{2\varepsilon_{\text{Si}} q N'_A \phi_s} = \sqrt{2\varepsilon_{\text{Si}} q N'_A (2\phi_f + V_{\text{ch}})} \tag{S3.30}$$

and the channel charge is found from

$$|Q_{\text{ch}}| = C'_{\text{ox}}(V_{GS} - V_T - V_{\text{ch}}) \tag{S3.31}$$

EXAMPLE S3.4

For an NFET with net substrate doping of $N'_A = 10^{16}\ \text{cm}^{-3}$, find the channel low-field mobility at the source and drain ends of the channel for $t_{\text{ox}} = 5$ nm and $V_{DS} = V_{GS} - V_T = 1$ V.

■ **Solution**

We must determine C'_{ox} and ϕ_f, from which we can find ϕ_s, Q_D, $\mathscr{E}_{T\text{eff}}$, and μ_{lf} at both source and drain.

$$C'_{\text{ox}} = \frac{\varepsilon_{\text{ox}}}{t_{\text{ox}}} = \frac{3.9 \times 8.85 \times 10^{-12}}{5 \times 10^{-9}} = 6.9 \times 10^{-3}\ \text{F/m}^2 = 6.9 \times 10^{-7}\ \text{F/cm}^2$$

$$\phi_f = \frac{kT}{q} \ln\left[\frac{N'_A}{n_i}\right] = (0.026\ \text{V}) \ln\left(\frac{10^{16}}{1.08 \times 10^{10}}\right) = 0.357\ \text{V}$$

At the source end, $\phi_s = 2\phi_f = 2(0.357) = 0.714$ V. The channel voltage is $V_{\text{ch}} = 0$ and the channel charge

$$|Q_{\text{ch}}| = C'_{\text{ox}}(V_{GS} - V_T - V_{\text{ch}}) = (6.9 \times 10^{-7})(1)\ \text{C/cm}^2 = 6.9 \times 10^{-7}\ \text{C/cm}^2$$

The depletion charge at the source end is

$$\begin{aligned} |Q_B(0)| &= \sqrt{2\varepsilon_{\text{Si}} q N'_A (2\phi_f)} \\ &= \sqrt{2(11.8)(8.85 \times 10^{-14}\ \text{F/cm})(1.6 \times 10^{-19}\ \text{C})(10^{16}\ \text{cm}^{-3}) \times (2 \times 0.357\ \text{V})} \\ &= 4.88 \times 10^{-8}\ \text{C/cm}^2 \end{aligned}$$

The effective transverse field is, from Equation (S3.29):

$$\mathscr{E}_{T\text{eff}}(0) = \frac{\left|Q_B(0) + \dfrac{Q_{\text{ch}}(0)}{2}\right|}{\varepsilon_{\text{Si}}} = \frac{4.88 \times 10^{-8} + \dfrac{6.9 \times 10^{-7}}{2}}{(11.8)(8.85 \times 10^{-14})} = 3.8 \times 10^5\ \text{V/cm}$$

and the low-field mobility at the source end is found from Equation (S3.28):

$$\mu_{\text{lf}}(0) = 32500 \mathscr{E}_{T\text{eff}}{}^{-1/3} = 455\ \text{cm}^2/\text{V} \cdot \text{s}$$

We repeat the process for the drain end. The channel charge there is essentially zero, $Q_{ch} \approx 0$. The drain voltage is $V_{DS} = 1\,\text{V}$, and $\phi_s = (2\phi_f + V_{DS})$. The depletion charge is

$$|Q_B(L)| = \sqrt{2\varepsilon_S q N_A'(2\phi_f + V_{DS})}$$
$$= \sqrt{2(11.8 \times 8.85 \times 10^{-14}\,\text{F/cm})(1.6 \times 10^{-19}\,\text{C})(10^{16}\,\text{cm}^{-3})(2 \times 0.357 + 1\,\text{V})}$$
$$= 7.6 \times 10^{-8}\,\text{C/cm}^2$$

The effective transverse field is

$$\mathscr{E}_{Teff}(L) = \frac{\left|Q_B(L) + \dfrac{Q_{ch}(L)}{2}\right|}{\varepsilon_{Si}} = \frac{7.57 \times 10^{-8} + 0}{11.8(8.85 \times 10^{-14})} = 7.3 \times 10^4\,\text{V/cm}$$

(about one-fifth the value of 3.8×10^5 V/cm at the source end), and the low-field mobility is

$$\mu_{lf}(L) = 780\,\text{cm}^2/\text{V} \cdot \text{s}$$

The electrons have higher low-field mobility at the drain end of the channel.

From the above example, it can be seen that the field $\mathscr{E}_{Teff}$ decreases and thus the mobility μ_{lf} increases with y going from source to drain.

The previous discussion was for electrons in n-channel devices. In p-channel MOSFETs, the low-field mobility for holes can be fit to a universal relation [6]

$$\mu_{lf} \approx 50000 \mathscr{E}_{Teff}^{-1/2} \qquad \text{for holes} \tag{S3.32}$$

where

$$\mathscr{E}_{Teff} = \frac{\left(Q_B + \dfrac{Q_{ch}}{3}\right)}{\varepsilon_{Si}} \qquad \text{for holes} \tag{S3.33}$$

The above formulations are somewhat unwieldy. There is an alternative relation that is more amenable to modeling. It is also approximate and empirical. It expresses the dependence of μ_{lf} on the gate to channel voltage as discussed in Chapter 7:

$$\mu_{lf} \approx \frac{\mu_0}{1 + \theta(V_{GS} - V_T - V_{ch})} \approx \frac{\mu_0}{1 + \theta(V_{GS} - V_T)}$$

S3.5 MEASUREMENT OF V_T

In Chapter 8 a method was indicated for the measurement of threshold voltage V_T and the low-field mobility at threshold μ_0. This technique was based on analysis of the I_D-V_{GS} characteristics at a small drain voltage such that velocity saturation effects are negligible; e.g., where

$$\frac{\mu_{lf} V_{DS}}{L v_{sat}} \ll 1 \tag{S3.34}$$

and in that case

$$\mu_{\text{lf}} \approx \frac{\mu_0}{1 + \theta(V_{GS} - V_T)} \tag{S3.35}$$

Since θ is on the order of 0.1 V^{-1}, for small $V_{GS} - V_T$,

$$\mu_{\text{lf}} \approx 1 - \theta(V_{GS} - V_T) \tag{S3.36}$$

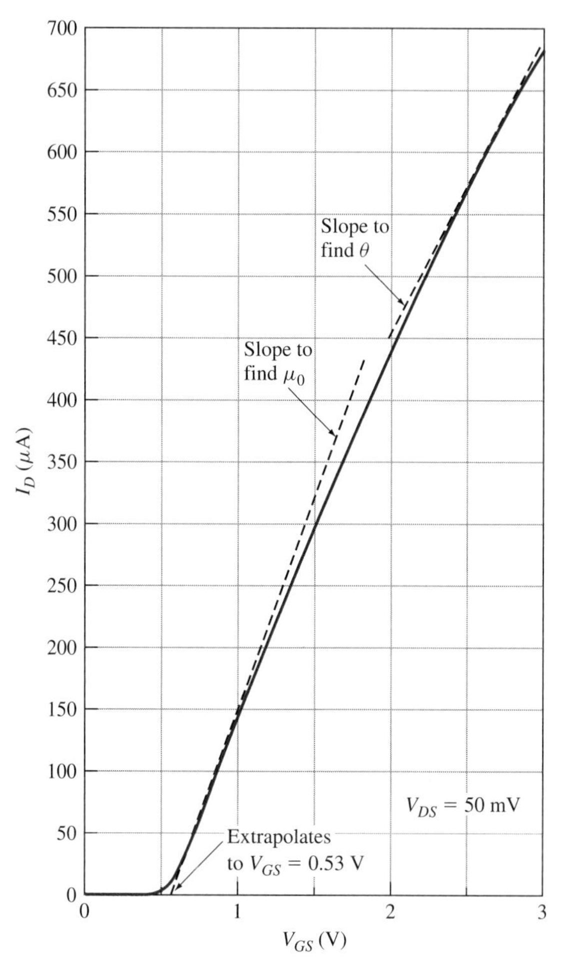

Figure S3.9 Plot of I_D-V_{GS} characteristics for $V_{DS} =$ 50 mV (solid line) for a MOSFET with $W = 10\ \mu$m, $L = 0.5\ \mu$m, and $t_{\text{ox}} = 5$ nm. The lower dashed line is the tangent to the maximum slope region of the curve. The upper line is tangent at a chosen value of V_{GS}. From the plot, the threshold voltage V_T, mobility at the source at threshold μ_0, and the gate voltage induced-mobility modulation factor θ can be determined.

and

$$I_D \approx \frac{WC'_{ox}V_{DS}\mu_0[1-\theta(V_{GS}-V_T)]\left(V_{GS}-V_T-\frac{V_{DS}}{2}\right)}{L} \qquad \text{(S3.37)}$$

Figure S3.9 shows the I_D-V_{GS} characteristics for a MOSFET device with $V_{DS} = 50\,\text{mV}$, $W = 10\,\mu\text{m}$, $L = 0.5\,\mu\text{m}$, and $t_{ox} = 5\,\text{nm}$. In Chapter 8 we discussed how to find V_T and μ_0 from the line tangent to the curve at maximum slope. The value of θ can be deduced from the deviation of the actual characteristics from the straight line approximation.

For $(V_{GS} - V_T)$ large enough that $(V_{GS} - V_T) \gg V_{DS}/2$, the slope of the I_D-V_{GS} characteristic from Equation (S3.37) is

$$\frac{dI_D}{dV_{GS}} = \frac{WC'_{ox}\mu_0 V_{DS}}{L}[1-2\theta(V_{GS}-V_T)] \qquad \text{(S3.38)}$$

The utility of this is that the parameter θ can be determined by comparing the slope of the I_D-V_{GS} characteristic at a given $(V_{GS} - V_T)$ with the maximum slope.

EXAMPLE S3.5

Find the values of V_T, μ_0, and θ from the I_D-V_{GS} characteristics of Figure S3.9. The drain-source voltage is $V_D = 0.05$ V.

■ **Solution**

From Figure S3.9 the straight line $V_{GS}(0)$ intercept is at $V_{GS} = 0.53\,\text{V}$. Since $V_{DS} = 0.05\,\text{V}$, from Equation (8.8),

$$V_T = V_{GS}(0) - \frac{V_{DS}}{2} = 0.53 - \frac{0.05}{2} \approx 0.5\,\text{V}$$

The value of θ can be found by comparing the slope of the I_D-V_{GS} characteristic at threshold with that at some arbitrary point. Suppose we choose $V_{GS} = 2.5\,\text{V}$. From Equation (8.7),

$$\left.\frac{dI_D}{dV_{GS}}\right|_{\text{threshold}} = \frac{WC'_{ox}\mu_0 V_{DS}}{L}$$

and from Equation (S3.38) we have

$$\left.\frac{dI_D}{dV_{GS}}\right|_{V_{GS}=2.5\,\text{V}} = \frac{WC'_{ox}\mu_0 V_{DS}}{L}[1-2\theta(V_{GS}-V_T)]$$

Thus

$$\frac{\text{Slope of data at } V_{GS} = 2.5\,\text{V}}{\text{Slope of straight line}} = 1 - 2\theta(V_{GS}-V_T)$$

The slope of the data at $V_{GS} = 2.5$ V is 236 μA/V and that of the straight line section is 333 μA/V. Thus

$$\theta = \frac{1 - \dfrac{236}{333}}{2(2.5 - 0.5)} = \frac{0.29}{4} = 0.073\ \text{V}^{-1}$$

From Equation (8.7),

$$\mu_0 = \left.\frac{dI_D}{dV_{GS}}\right|_{\text{threshold}} \cdot \frac{L}{WC'_{\text{ox}}V_{DS}} = 483\ \text{cm/V}\cdot\text{s}$$

We remind the reader that this development is valid only for small V_{DS}.

EXAMPLE S3.6

Find the maximum value of V_{DS} such that the above method yields valid results.

■ **Solution**

We used the condition of small V_{DS} to approximate Equation (S3.6) by

$$I_D = \frac{W\mu_{\text{lf}}C'_{\text{ox}}}{L}\left(V_{GS} - V_T - \frac{V_{DS}}{2}\right)V_{DS}$$

saying in effect that

$$\frac{\mu_{\text{lf}}V_{DS}}{Lv_{\text{sat}}} \ll 1$$

Choosing a condition of "maximum acceptable V_{DS}" to be such that the error is 10 percent, we write

$$\frac{\mu_{\text{lf}}V_{DS}}{Lv_{\text{sat}}} < 0.1$$

Rearranging gives us our result:

$$V_{DS\text{max}} \leq \frac{0.1v_{\text{sat}}}{\mu_{\text{lf}}}L$$

For $v_{\text{sat}} = 4 \times 10^6$ cm/s and $\mu_{\text{lf}} = 400\ \text{cm}^2/\text{V}\cdot\text{s}$, the maximum drain voltage for which this graphical method of finding the threshold voltage is valid is:

$$V_{DS\text{max}} \leq 10^3 L\ \text{cm} = 0.1L\ \mu\text{m}$$

For a channel length of 1 μm, that gives

$$V_{DS\text{max}} \leq 0.1\ \text{V}$$

This voltage decreases with decreasing L.

*S3.6 ALTERNATIVE METHOD TO DETERMINE V_T AND μ_{lf} APPLICABLE TO LONG-CHANNEL MOSFETs

There is an alternative method for finding the threshold voltage of a MOSFET, provided the MOSFET is of the long-channel type. Unlike the previous method, where the device was measured below saturation, this method relies on the $I_{D\text{sat}}$-V_{GS} characteristics for the device being in the current saturation region. The saturation current $I_{D\text{sat}}$ can be expressed as

$$I_{D\text{sat}} = \frac{WC'_{\text{ox}}\mu_{\text{lf}}\left(V_{GS} - V_T - \dfrac{V_{DS\text{sat}}}{2}\right)V_{DS\text{sat}}}{L\left(1 + \dfrac{\mu_{\text{lf}}V_{DS\text{sat}}}{v_{\text{sat}}L}\right)} \tag{S3.39}$$

Recall from Chapter 7 that for long-channel devices such that $\mu_{\text{lf}}V_{DS\text{sat}}/v_{\text{sat}}L \ll 1$, the saturation voltage is $V_{DS\text{sat}} = (V_{GS} - V_T)$. Thus we can simplify Equation (S3.39) to

$$I_{D\text{sat}} = \frac{WC'_{\text{ox}}\mu_{\text{lf}}}{L}\left(V_{GS} - V_T - \frac{V_{DS\text{sat}}}{2}\right)V_{DS\text{sat}}$$

$$= \frac{WC'_{\text{ox}}\mu_{\text{lf}}}{2L}(V_{GS} - V_T)^2 \tag{S3.40}$$

We take the square root of both sides to obtain:

$$\sqrt{I_{D\text{sat}}} = (V_{GS} - V_T)\sqrt{\frac{WC'_{\text{ox}}\mu_{\text{lf}}}{2L}} \tag{S3.41}$$

or

$$\mu_{\text{lf}} = \frac{2L}{WC'_{\text{ox}}}\left(\frac{d\sqrt{I_{D\text{sat}}}}{dV_{GS}}\right)^2 \tag{S3.42}$$

Then a plot of $\sqrt{I_{D\text{sat}}}$ versus V_{GS} intersects the V_{GS} axis at $(V_{GS} = V_T)$. For μ_{lf} constant, the plot is a straight line of slope $\sqrt{WC'_{\text{ox}}\mu_{\text{lf}}/2L}$. Since W, L, and C'_{ox} are known for a given process, this measurement gives a result for μ_{lf}.

However, this method is not applicable to short-channel devices. Let us use the following example to demonstrate this.

EXAMPLE S3.7

Find the maximum value of V_{GS} that can be used to determine V_T by this alternative method.

■ Solution

This method involves the saturation current and thus relies on the condition that $\mu_{\text{lf}}V_{DS\text{sat}}/Lv_{\text{sat}} \ll 1$ (i.e., velocity saturation can be neglected). For a device in which this is true, it is also true that

$$V_{DS\text{sat}} = V_{GS} - V_T$$

Choosing the criterion that "much less than is smaller than 10 percent of," we write

$$\frac{\mu_{\text{lf}} V_{D\text{Ssat}}}{v_{\text{sat}} L} = \frac{\mu_{\text{lf}}(V_{GS} - V_T)}{v_{\text{sat}} L} \leq 0.1$$

We can solve for the maximum value of $V_{GS} - V_T$:

$$\begin{aligned}(V_{GS} - V_T) \leq 0.1 \frac{v_{\text{sat}} L}{\mu_{\text{lf}}} &= 0.1 \left[\frac{(4 \times 10^6\ \text{cm/s})(L)}{500\ \text{cm}^2/\text{V} \cdot \text{s}} \right] \approx (0.1\ \text{V/cm}) \left(\frac{10^6}{10^2} \right) L \\ &= (0.1\ \text{V/cm})(10^4\ \text{cm})(L\ \text{cm}) \left(\frac{1\ \text{cm}}{10^4\ \mu\text{m}} \right) \\ &= (0.1\ \text{V}/\mu\text{m})(L\ \mu\text{m}) = 0.1L\end{aligned}$$

In the last line, the factor 0.1 has units of V/μm, and the value of L can be used directly in μm.

The second condition for this alternative method is that the device must be in saturation. The gate voltage must also be above threshold for saturation to occur, yielding a range of

$$0 \leq (V_{GS} - V_T) \leq 0.1L\ \mu\text{m}$$

This is equivalent to saying that Equation (S3.41) will accurately represent the actual measurements over this range of $V_{GS} - V_T$, and a plot of $\sqrt{I_{D\text{sat}}}$ versus $V_{GS} - V_T$ will be linear. For a long-channel device of, say, $L = 10\ \mu$m, this results in a range of $0 \leq V_{GS} - V_T \leq 1$ V.

For a shorter-channel device, for example of $L = 1\ \mu$m, the valid range is only 0.1 V wide, and for $L = 0.25\ \mu$m, the range is 25 mV wide. To extrapolate a straight line requires choosing two points and the smaller the distance between the points, the less accurate the extrapolation. Thus for short-channel devices it is difficult to obtain reliable results using the $\sqrt{I_{D\text{sat}}}$ versus $(V_{GS} - V_T)$ method.

S3.7 MOS CAPACITORS

We noted in Chapter 7 that the gate structure of a field-effect transistor looks like a capacitor. In fact, the same fabrication processes can be used to put capacitors as circuit elements into integrated circuits. Further, the electrical characteritics of a MOS capacitor can be used as a diagnostic tool to determine various material properties resulting from the fabrication processes. Thus, we will examine MOS capacitors in some detail in this section.

The structure of a MOS capacitor is shown in Figure S3.10. It is similar to a MOSFET except that there are no source or drain regions. As is the case for MOSFETs, the "metal" gate is often degenerately doped Si. We will consider the case of a metal electrode, so we have a metal/SiO_2/Si MOS capacitor, and we'll take the Si to be p type.

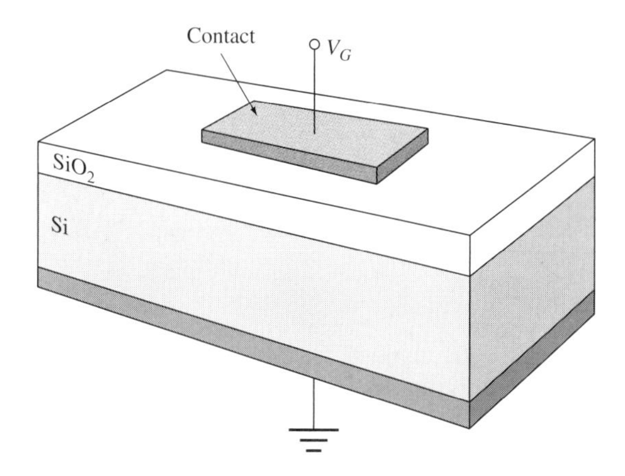

Figure S3.10 Schematic of a MOS capacitor.

S3.7.1 IDEAL MOS CAPACITANCE

The capacitance between the gate and the substrate depends on the applied voltage, the metal-semiconductor work function difference, and all the charges involved in the determination of MOSFET threshold voltage. In this section we idealize the problem by assuming that Φ_{MS}, Q_f, and Q_{it} are all equal to zero. (We will include these terms in the following section.) We will consider only the influence of the depletion region width in the bulk substrate on the bulk charge Q_B, and the mobile charge in the Si at the Si/SiO_2 interface, which we call Q_i. Note that in MOSFETs, this mobile charge was referred to as Q_{ch}, the channel charge per unit area. There is no channel in the capacitor since there is no source or drain.

Figure S3.11 shows the energy band diagram for various values of gate-substrate voltage V_G for a metal/SiO_2/p-type semiconductor and for a metal/n-type semiconductor. In (a), a gate voltage of the polarity shown is applied to bend the bands such that majority carriers accumulate near the surface of the semiconductor. Since in the simplified ideal case being considered, $\Phi_{MS} = 0$, the flat band case (b) is also the equilibrium case.

In (c), a gate voltage is applied, which creates a depletion region in the Si, and in (d), the gate voltage is large enough that the Si is inverted near the interface.

Let us consider the metal/p-type semiconductor. The charges associated with each of the bias conditions of Figure S3.11 are indicated schematically in Figure S3.12.

- In accumulation (a), holes in the valence band exist at the interface, creating a positive mobile interface charge Q_i. An equal and opposite electron charge exists on the metal gate.
- In the flat band condition (b), neutrality exists everywhere.

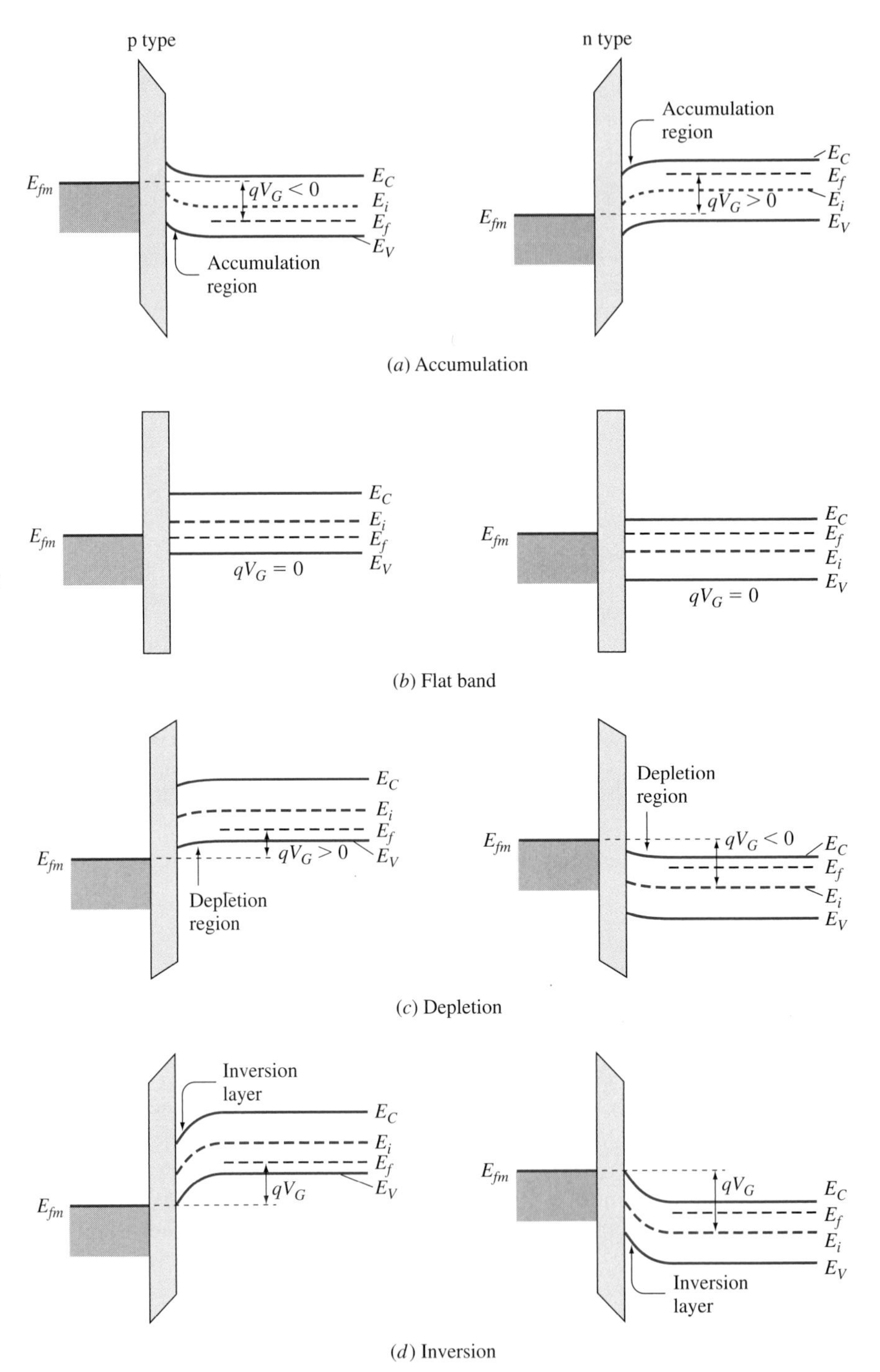

Figure S3.11 Energy band diagram for ideal MOS structures using a metal gate and p-type silicon substrate (*left*) and n-type substrate (*right*). In (a) the bands bend in such a way that the surface of the silicon is accumulated (with majority carriers). For this special case where $\Phi_M = \Phi_S$, the flat-band case (b) is also the equilibrium case. In (c) the polarity of V_G is such that a depletion region is established in the silicon. In (d), V_G is sufficient to produce an inversion layer.

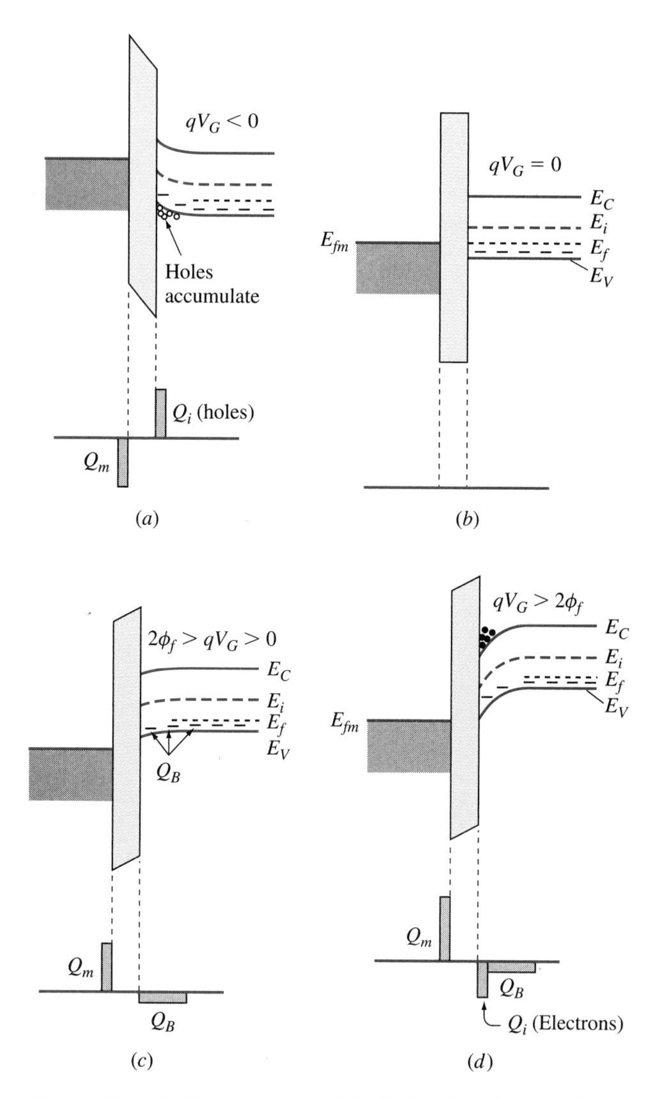

Figure S3.12 Charges associated with the ideal MOS p-Si capacitor of Figure S3.10. In (a) holes accumulate near the surface, attracting an equal and opposite negative charge on the gate. In (b) flat band is also the case for equilibrium (for this case in which $\Phi_M = \Phi_S$). (c) For small positive V_G, a depletion region is formed in the silicon, with negatively ionized acceptors near the surface not neutralized by holes. An equal and opposite positive charge appears on the gate. (d) For V_G such that $\phi_s > 2\phi_f$, the bulk charge Q_B remains, and in addition there are mobile electrons occupying states in the silicon potential well.

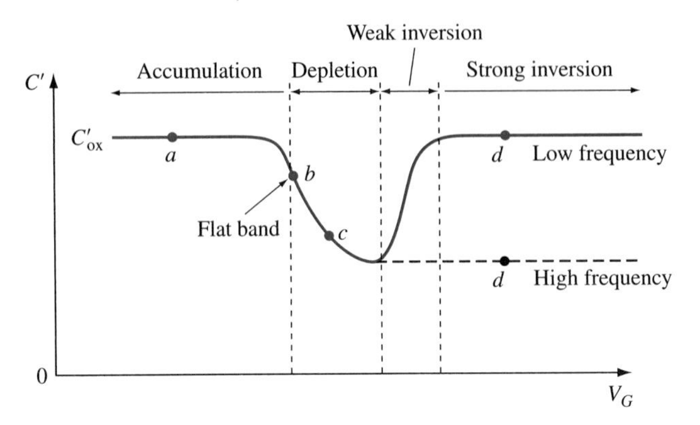

Figure S3.13 Capacitance-voltage characteristic for the ideal MOS capacitor at low and high frequencies. (a) Accumulation; (b) flat band; (c) depletion; (d) strong inversion.

- In depletion (c), the charge in the semiconductor results from the ionized acceptors Q_B in the depletion region. There are no mobile charges, so $Q_i \approx 0$.
- In inversion (d), in addition to the bulk charge Q_B resulting from the ionized acceptors in the depletion region, a negative Q_i exists resulting from electrons at the interface.

The gate-substrate capacitance C' per unit area as a function of V_G is indicated in Figure S3.13 for this ideal case at low frequencies ($f \sim 10$ Hz) and at high frequency ($f \sim 1$ MHz). Here V_G varies slowly, so that above threshold ($\phi_S = 2\phi_f$), the variation of V_G causes a change in Q_i, and Q_B is constant. The C'-V_G plot can be explained as follows:

- For $V_G < 0$ V, the Si surface is in accumulation as indicated in Figure S3.11a. Since $C' = |dQ_m/dV_G|$, when there is a change dV_G, then the charge on the metal changes by $dQ_m = -dQ_i$. This structure looks like and behaves as a parallel plate capacitor, with SiO_2 as the dielectric and with capacitance per unit area of

$$C' = C'_{\text{ox}} \qquad \text{accumulation} \tag{S3.43}$$

Point a on the C-V_G plot indicates accumulation.

- At flat band (point b in Figure S3.13), there is still some mobile charge at the interface. If the gate voltage is varied by a small amount from V_{FB}, the bands will bend up or down slightly. Thus, the effect of dV_G penetrates some small distance into the substrate. This distance is known as the Debye length L_D. This adds, in effect, a second capacitor in series with the oxide capacitor. Thus, at $V_G = V_{FB}$,

$$\frac{1}{C'} = \frac{1'}{C'_{\text{ox}}} + \frac{L_D}{\varepsilon_{\text{Si}}} = \frac{t_{\text{ox}}}{\varepsilon_{\text{SiO}_2}} + \frac{L_D}{\varepsilon_{\text{Si}}} \qquad \text{near flat band} \tag{S3.44}$$

- Note that in accumulation, the number of mobile charges is large and they are all close to the interface, such that the second term in Equation (S3.44) is negligible. Equation (S3.44) then reduces to Equation (S3.43).
- For $V_G > V_{FB}$ (remember $V_{FB} = 0$ for this ideal capacitor), a depletion region forms in the substrate. For point c on the plot, there is no mobile charge, and $Q_i = 0$. The capacitance is a series combination of the oxide capacitance C'_{ox} and the (bulk) junction capacitance C'_B

$$\frac{1}{C'} = \frac{1}{C'_{\text{ox}}} + \frac{1}{C'_B} \tag{S3.45}$$

where C'_B is the capacitance per unit area associated with the depletion region:

$$C'_B = \left(\frac{\varepsilon_{\text{Si}} q N'_A}{2\phi_s}\right)^{1/2} \tag{S3.46}$$

and w_B, the depletion region depth, is

$$w_B = \left(\frac{2\varepsilon_{\text{Si}}\phi_s}{q N'_A}\right)^{1/2} \tag{S3.47}$$

- In the inversion region (point d), in addition to the depletion charge Q_B, there exists an interface charge Q_i. The capacitance in this case depends on whether the measurements are made at high frequencies or at low frequencies.

At high frequencies (typically 1 to 10 MHz), $f \gg 1/\tau$, where τ is the lifetime associated with the carrier generation and recombination in the depletion region. In this case, the change in V_G is rapid enough that negligible generation or recombination can occur during one cycle, meaning that Q_i is not affected. Thus $|dQ_m/dV_G| = |dQ_B/dV_G|$, and the capacitance is the series combination of C'_{ox} and C'_B

$$\frac{1}{C'} = \frac{1}{C'_{\text{ox}}} + \frac{1}{C'_B} \tag{S3.48}$$

This value is independent of the value of V_G, as shown by the dashed line in Figure S3.13.

At low frequencies, however (typically 1 to 100 Hz), for $f \ll 1/\tau$, the generation and recombination of minority carriers in the depletion region can follow the change in V_G. Thus, Q_i can vary in response to the changing gate voltage. Recall from "More About Threshold" in Section 7.2.3, however, that above threshold, ϕ_s is nearly constant and that any additional change in gate voltage is dropped across the oxide. That implies that Q_B is constant. Thus, in this case $|dQ_m/dV_G| = |dQ_i/dV_G|$. As a result, in this regime C' is essentially equal to C'_{ox} and is independent of the dc value of V_G.

S3.7.2 THE C-V_G CHARACTERISTICS OF REAL MOS CAPACITORS

In the above, Φ_{MS}, Q_{it}, and Q_f were assumed to equal zero. In a real MOS capacitor, however, they must be considered. We will take the effect of each in turn.

The Effect of Φ_{MS} In the ideal MOS capacitor, flat band occurred at $V_G = 0$. As discussed in Section S3.3.5, and expressed in Equation (S3.19), for a real MOS capacitor the flat band voltage becomes

$$V_{FB} = \frac{\Phi_{MS}}{q} - \frac{[Q_f + Q_{\text{it}}(0)]}{C'_{\text{ox}}} \tag{S3.49}$$

The effect, then, is to shift the C-V plot of Figure S3.13 to the right or left on the voltage scale, depending on the sign of Φ_{MS}.

Interface States It was indicated earlier that the charge Q_{it} trapped in interface states depends on the band bending in the semiconductor ϕ_S. At a given value of ϕ_S, the interface trap states below the Fermi level in the semiconductor are predominantly occupied while those above the Fermi level are mostly vacant. As V_G and thus the band bending changes, the value of Q_{it} also changes. The interface states, also referred to as *fast interface states,* however, respond to V_G at frequencies below about 1 kHz but not at frequencies on the order of 1 MHz. Thus, their concentration can be determined from a comparison of high-frequency capacitance with the low-frequency capacitance (Figure S3.14).

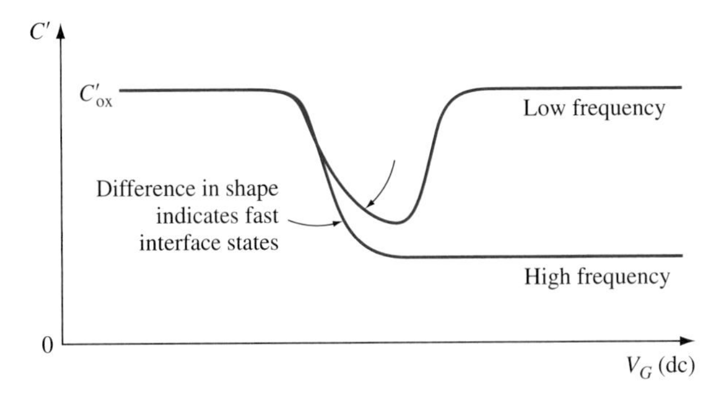

Figure S3.14 C-V characteristics of a real MOS capacitor at low and high frequency. The difference in the general shape from that of the ideal MOS capacitor in Figure S3.13 results from the presence of fast interface states. Since the occupancy of these states can follow the low-frequency variation in V_G but not the high-frequency variation, the difference between the two plots is a measure of the concentration of fast states.

S3.7.3 PARAMETER ANALYSES FROM C-V_G MEASUREMENTS

We have seen how some of the physical parameters of the MOSFET can affect the C'-V_G curves. Conversely, examination and interpretation of these curves can reveal the values of some parameters, including the oxide thickness, the bulk or depletion capacitance C'_B, the depletion region width w_B, and the substrate doping level.

In the accumulation mode, and at low frequencies in the inversion mode (see Figure S3.13), the oxide thickness can be determined if the oxide permittivity is known, from

$$C' = C'_{\text{ox}} = \frac{\varepsilon_{\text{SiO}_2}}{t_{\text{ox}}} \tag{S3.50}$$

From the high-frequency C-V_G characteristics in the inversion mode, and using the measured value of C'_{ox}, one can find the bulk or depletion region capacitance:

$$C'_{\text{inversion}} = \frac{C'_{\text{ox}} C'_B}{C'_{\text{ox}} + C'_B} = \frac{C'_{\text{ox}}}{\dfrac{C'_{\text{ox}}}{C'_B} + 1} \tag{S3.51}$$

This can be solved for C'_B. Furthermore, since

$$C'_B = \frac{\varepsilon_{\text{Si}}}{w_B}$$

The maximum depletion width w_B is now known. From

$$w_B = \left[\frac{2\varepsilon_{\text{Si}}(2\phi_f)}{qN'_A} \right]^{1/2} \tag{S3.52}$$

we can find the doping concentration in the substrate using

$$\phi_f = \frac{kT}{q} \ln \frac{N'_A}{n_i} \tag{S3.53}$$

As discussed in Supplement 2, N'_A can also be determined from the C-V characteristics of a Schottky barrier obtained by depositing a metal onto the Si surface with the oxide removed.

*S3.8 MOS CAPACITOR HYBRID DIAGRAMS

In devices incorporating MOS capacitors, it is often convenient to use "hybrid" diagrams. A hybrid diagram is a combination of the physical diagram and the diagrams. A hybrid diagram is a combination of the physical diagram and the energy band diagram as indicated in Figure S3.15 for a well partially filled with electrons. While the energy band diagram plots electron energy versus position x, the direction normal to the gate, and the physical diagram plots the lateral structure y versus x, the hybrid diagram, Figure S3.15a, plots lateral position

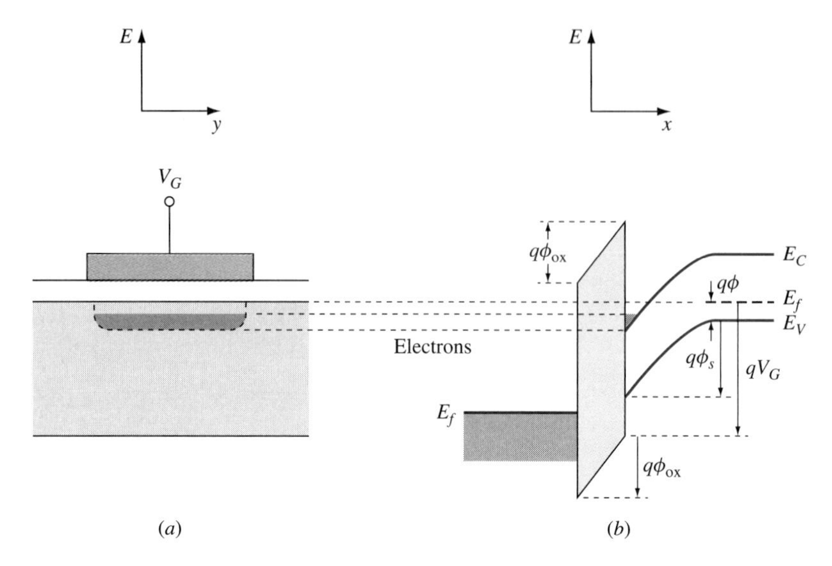

Figure S3.15 (a) Hybrid diagram and (b) corresponding energy band diagram for a MOS capacitor. The shaded areas represent the electron charges in the potential well.

y versus electron energy. The regular energy band diagram is shown in Figure S3.15b.

To construct a hybrid diagram:

- The Fermi level in the bulk semiconductor of the energy band diagram (Figure S3.15b) is aligned with the SiO_2/Si interface of the physical diagram (a) as indicated.
- The energy at the bottom of the well in (b) is indicated in (a) (the dashed line is the shape of the well).
- The concentration of electrons in the well is indicated in the shaded region of the hybrid diagram.[4]

We use a hybrid diagram to illustrate how the MOS capacitor charges with time. In Figure S3.16a, the hybrid diagram and energy band diagram are shown for a p-type MOS capacitor with no gate voltage applied. At some time $t = 0$, a positive step function voltage V_G is applied to the gate of the MOS capacitor (b). The total voltage across the device then is

$$V_j = V_{\text{bi}} + V_G = \phi_{\text{ox}} + \phi_s \tag{S3.54}$$

Immediately after the application of V_G, the potential well is empty, as the capacitor has not yet charged. However, electrons thermally generated in and near the Si depletion region become trapped in the well. The accumulating negative charge causes the energy band diagram near the interface (where the

[4]Remember, however, in the physical diagram the electrons are concentrated at the SiO_2/Si interface.

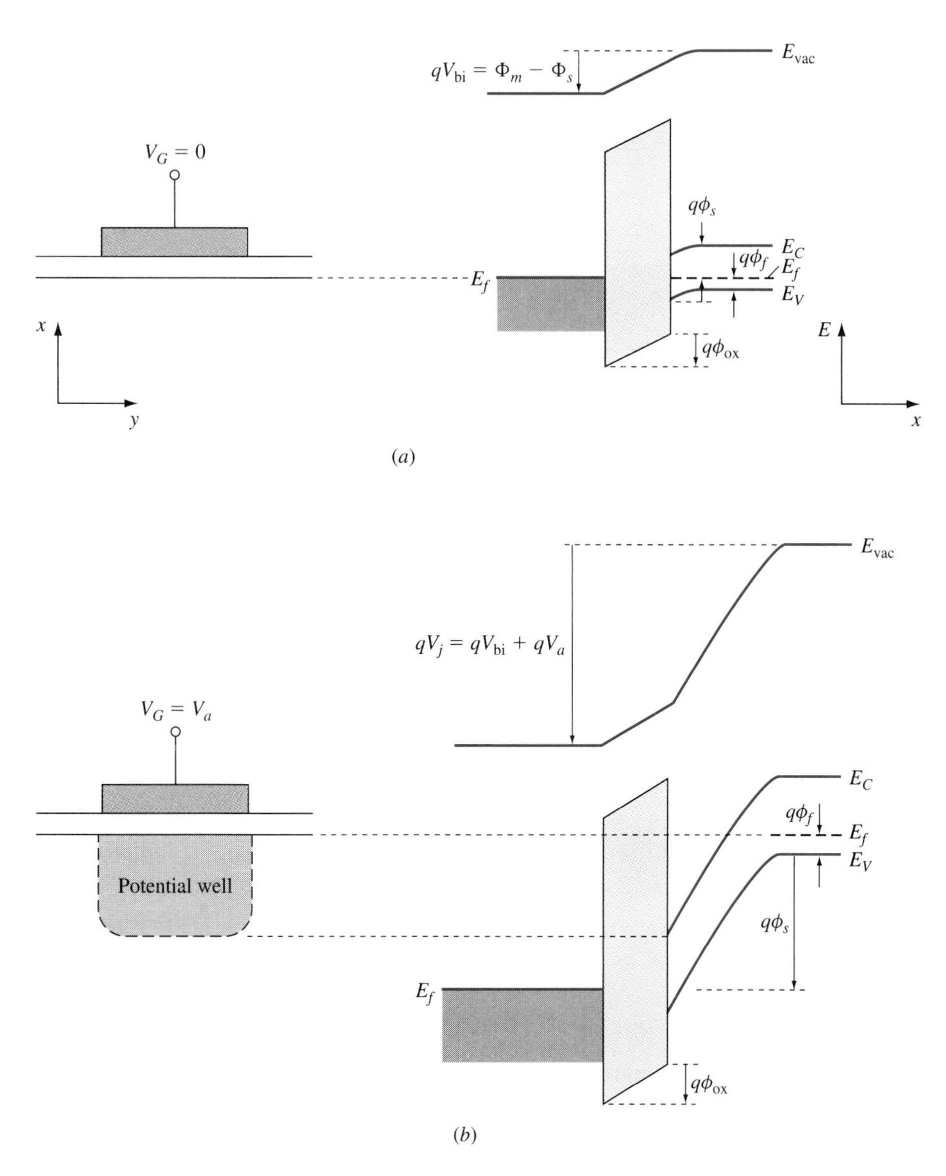

Figure S3.16a–b Hybrid diagram and corresponding energy band diagram for a MOS capacitor with a step voltage applied to the gate. (a) Equilibrium. (b) A positive voltage V_G is applied to the gate. Some of the voltage is dropped across the oxide and some across the semiconductor. At $t = 0$, the capacitor has not yet charged and the potential well is empty.

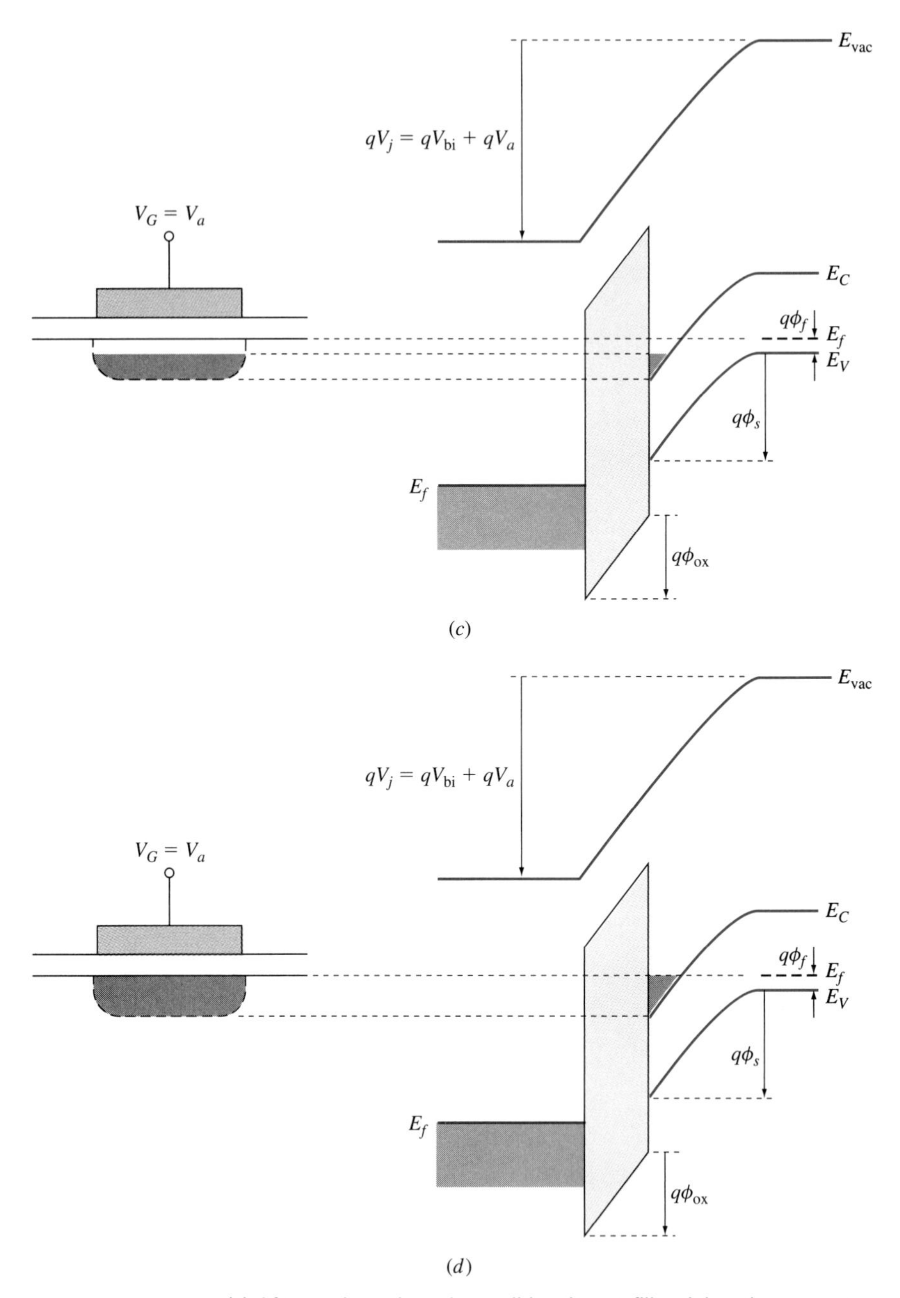

Figure S3.16c–d (c) After a short time the well begins to fill, raising the energy band diagram locally at the semiconductor-oxide interface. Thus ϕ_{ox} increases and ϕ_s decreases. (d) The well is completely full and the capacitor is charged.

negative charges are) to move upward (higher electron energies). This movement both decreases the surface potential ϕ_s and increases the oxide potential ϕ_{ox} as indicated in Figure S3.16c. When the capacitor is fully charged, the well is filled as pictured in (d). Here the barrier has been reduced such that the rate of electrons entering the well (by thermal generation) is equal to that of those escaping (by recombination).

In a MOSFET transistor, when the positive step function is applied to the gate, electrons enter the potential well (channel) almost instantaneously from the source to decrease the surface potential to $2\phi_f$. In an isolated MOS capacitor, however, electrons enter the well by thermal generation, a process that in Si can take several milliseconds at room temperature.

*S3.8.1 DYNAMIC RANDOM-ACCESS MEMORIES (DRAMs)

MOS technology is also used to make memory cells, for example random-access memory (RAM). [8] An attractive feature of a RAM is that each memory cell can be written to and read from individually. In RAM memory, the information stored in each cell remains there until the cell is rewritten or until the power to the circuit is turned off (volatile memory).

The most common type of RAM is the dynamic random-access memory (DRAM). Each cell is composed of a MOS capacitor and an n^+p junction separated by a transfer gate. A cross section of one such "one-transistor memory cell" is shown schematically in Figure S3.17a. The capacitor is connected to a constant voltage, 2.5 V in this illustration. The transfer gate of the DRAM cell, when it is at zero volts, isolates the capacitor from the diode as in any other enhancement field-effect device. When $V_G = 2.5$ V, a conducting channel connects capacitor well and diode.

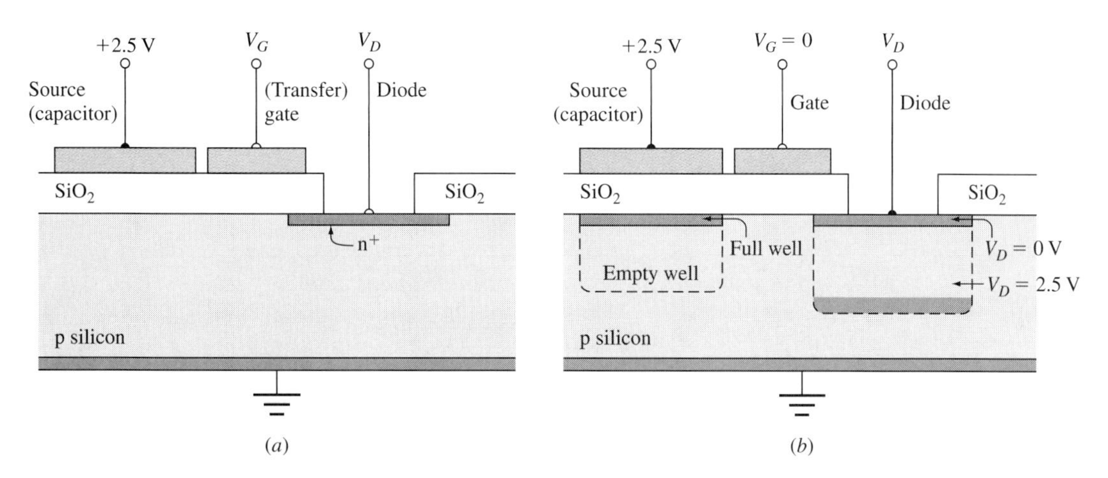

Figure S3.17 (a) The cross section of a DRAM and (b) the hybrid diagram. This one-transistor memory cell consists of a diode and a MOS capacitor, separated by a transfer gate. When $V_G = 0$, the source and drain are isolated.

Figure S3.17b shows the hybrid diagram for this memory cell for $V_G = 0$, when the capacitor and diode are isolated from each other. We define a zero (0) to be stored in the capacitor well if it is empty and a one (1) to be stored if it is filled. The dashed lines represent the case for the source well empty of charge (logical 0) and for the diode biased at 2.5 V. The solid lines represent the state for a logical 1, in which case the potential well under the capacitor contains charge. Note that the state stored depends only on the charge in the capacitor well, not on the diode voltage V_D. We illustrate two different values of V_D, however, because we will use both conditions in what follows.

We now examine the process of writing a 1 or a 0 into the capacitor well. For the write operation, the capacitor well is initially empty (logical zero). To write a 1, V_D is set to zero volts at the same time that the gate is pulsed on. While $V_G = 2.5$ V, a channel exists. Charge then flows from diode to capacitor as indicated in Figure S3.18a. For this operation, the diode plays the role of the source of a MOSFET, while the capacitor acts as a drain.

Note that charge flow is by drift as well as diffusion. There is more charge at the right-hand end of each well during the transfer process, since the charge starts from the diode and diffuses to the left. Because the depth of the wells under the capacitor and the gate depends on the electron concentration at any position, the increased charge on the right of each well creates an electric field

$$\left(\mathscr{E} = \frac{1}{q}\frac{dE_C}{dy}\right)$$

which accelerates the electrons to the left. The drift and diffusion continue until the capacitance well is filled or contains a 1. The gate pulse is removed, the channel disappears, and the charge remains in the capacitor potential well.

To write a 0, we set $V_D = 2.5$ V and pulse the gate on. Although a channel exists (the channel is not shown in the figure), we can see from Figure S3.18b

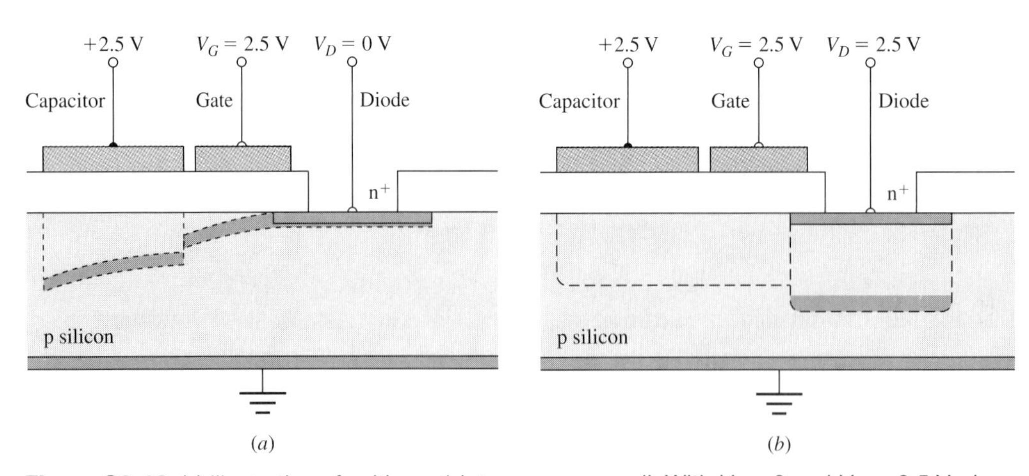

Figure S3.18 (a) Illustration of writing a 1 into a memory cell. With $V_D = 0$ and $V_G = 2.5$ V, charge flows from diode to capacitor to fill the potential well of the source capacitor. (b) To write a 0, V_D is made positive when the gate is opened. In this case no charge flows from diode to capacitor.

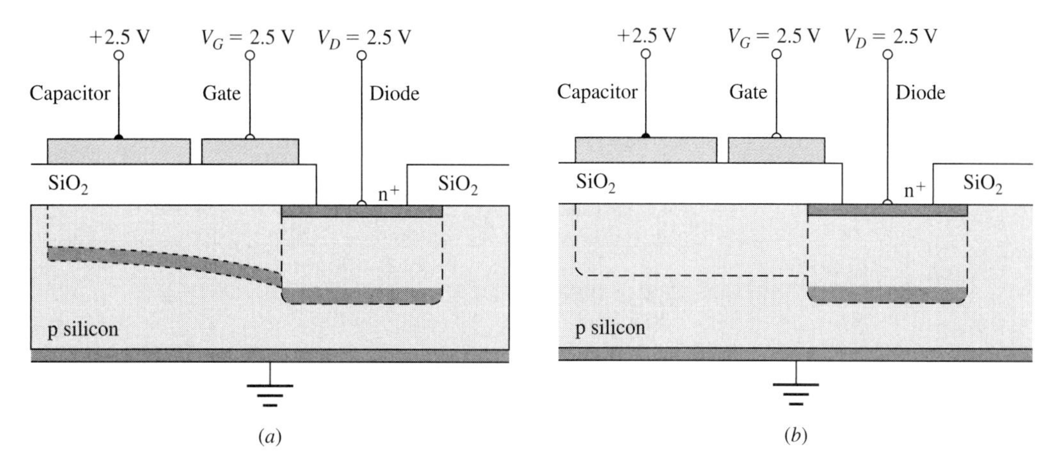

Figure S3.19 To read the contents of a DRAM memory cell, V_D is made positive when the gate is opened ($V_G = 2.5$ V). (a) If the cell contains a 1, charge flows from capacitor to diode. (b) If the cell contains a 0, no charge flows.

that no charge transfers because in this case the diode is at the same potential as the capacitor. When the gate is turned off ($V_G = 0$) the source well is still empty, i.e., uncharged.

Thus, to write a 1, we set the diode voltage to 0 V and pulse the gate on. This charges the capacitor. To write a 0, we set the diode voltage to 2.5 V and pulse the gate on. In either case, the capacitance voltage remains at 2.5 V.

To read what is stored in the well, V_D is set to 2.5 V and the gate turned on. If the capacitor contains a 1, the charge is transferred to the diode as indicated in Figure S3.19a. The current resulting from this charge transfer is interpreted as having read a stored 1. For this read operation, the capacitor acts as a MOSFET source while the diode acts as a MOSFET drain. If the capacitor well is empty (uncharged), no charge flows during the read cycle and the absence of diode current is interpreted as reading a stored 0, as indicated in Figure S3.19b.

Thermal generation in the region of the capacitor well, however, will trap electrons in the well. With time, an empty well will fill; that is, a 0 will become a 1. For this reason the stored charge must be "refreshed" periodically (every few milliseconds), i.e., read out and rewritten.

*S3.8.2 CHARGE-COUPLED DEVICES (CCDs)

A charge-coupled device (CCD) is essentially an analog shift register. It consists of a linear array of MOS capacitors as indicated in Figure S3.20. The basic principle of charge-coupled devices involves the movement of charge from one physical location to another in a controlled manner by the use of sequenced clock pulses. [9] While CCDs are capable of performing numerous electronic functions such as storing data, signal processing, and logic operations, the most important application is in optical image sensing (e.g., in digital cameras). This application will be discussed further in Chapter 11. Here we simply indicate one method of charge transfer.

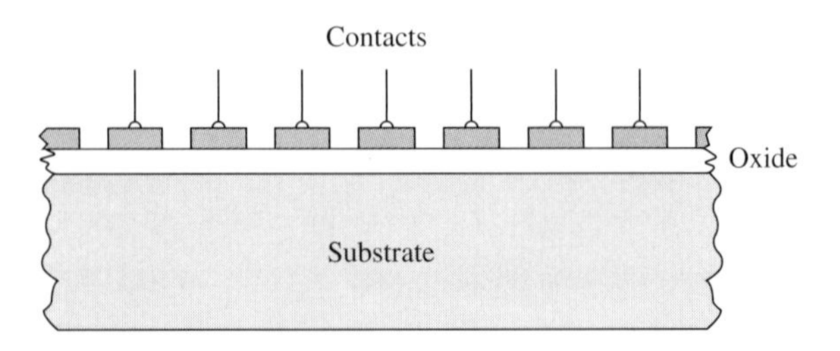

Figure S3.20 A linear array of adjacent MOS capacitors form a CCD shift register.

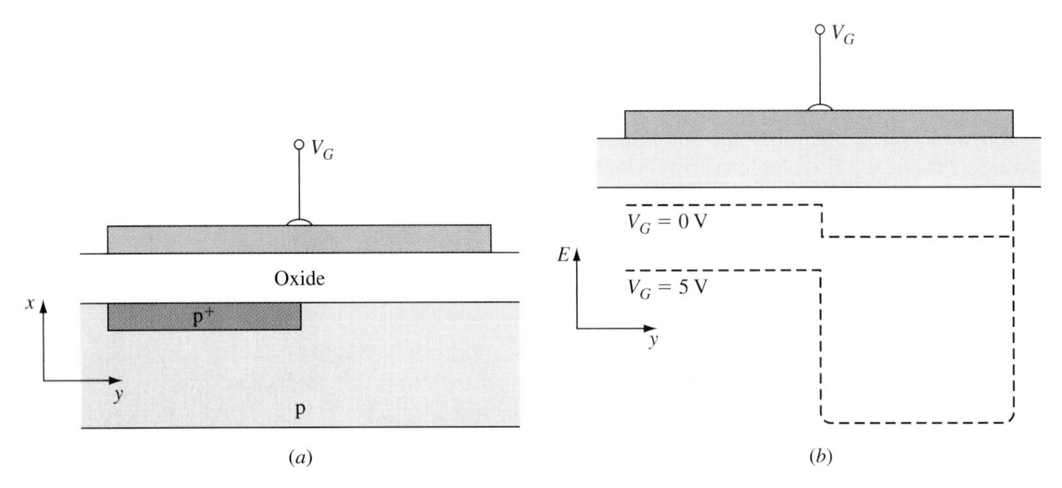

Figure S3.21 (a) Physical diagram of a MOS capacitor for use in a CCD and (b) the corresponding hybrid diagram for empty wells. The required asymmetry in the potential well is achieved by ion implanting acceptors to increase the doping concentration and thus reduce the depth of the potential well in one section of the capacitor.

We take the case of a two-phase CCD. In these devices, the potential well under each electrode has two different depths—shallow under a portion of the electrode and less shallow under the other portion. One method to build such a structure is to selectively ion-implant acceptors into the p substrate as indicated in Figure S3.21a. In the region of increased acceptor concentration, the surface potential is reduced (and ϕ_{ox} increased), as shown in the hybrid diagram of Figure S3.21b for the two cases of $V_G = 0$ V and 5 V.

We illustrate the operation of this two-phase shift register by assuming the gates connected to the timing signal of phase 1 to be at 5 V, and those connected to phase 2 to be at 0 V as shown in Figure S3.22. These two voltages, combined with the stepped doping, create four different well depths in the array.

We also assume that there are various numbers of electrons in the wells under the gates of phase 1, as in Figure S3.22a. For example, these could have been optically generated, resulting from the light of an optical image being focused on the CCD. To transfer electrons one-half step to the right, the voltage on phase 1 is made 0 V and that of phase 2 is 5 V. This increases the well energy

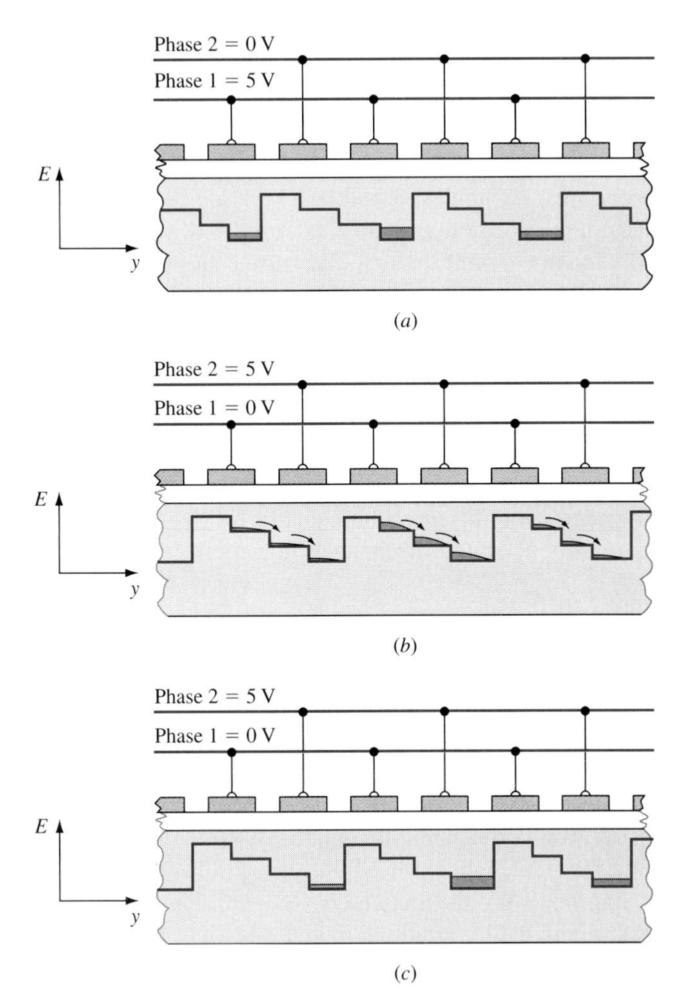

Figure S3.22 Hybrid diagram of a section of a two-phase CCD. In (a) the electrodes connected to phase 1 are at 5 V and those connected to phase 2 are at 0 V. Any mobile charge is confined to the wells under Phase 1. In (b), Phase 1 is made 0 while phase 2 is at 5 V. Charge flows from the phase 1 capacitors to those of phase 2. In (c) the charge shifting process is completed; the charges now reside in a well one-half step to the right.

under the capacitors of phase 1 and decreases it for those of phase 2. The electrons then transfer to the phase 2 gates as shown in Figure S3.22b. The hybrid diagram after the transfer is completed is indicated in (c). Next, the two voltages are cycled again and the charges transfer another half-step to the right. As the voltages cycle on the two phases, the electron "signals" keep being transferred to the right. These signals are then detected by the use of a MOSFET at the output of the shift register. A two-dimensional image sensor is discussed in Chapter 11.

*S3.9 DEVICE DEGRADATION

In this section, we examine a number of failure mechanisms in MOSFETs. Some of these are thermally activated, meaning that the degradation mechanism operates on thermal energy. That in turn implies that degradation will be faster at higher temperatures. Examples would be (1) oxide breakdown, in which thermally excited charges lodge in the oxide and eventually cause it to be conductive; (2) ionic diffusion in the oxide, which can cause changes in the threshold voltage over time; and (3) electromigration.

Electromigration occurs in the metal lines often used to interconnect the various transistors and devices on the chip. As electrons flow through these lines, they can collide with the metal atoms, giving them enough momentum to move. With time, this atomic movement can cause voids in the line or shorts between adjacent lines. Electromigration is thermally activated because at higher temperatures, the atoms have more vibrational kinetic energy and thus are more easily dislodged from their equilibrium positions.

For a thermally activated mechanism, the mean time to failure (MTTF) is given by

$$\text{MTTF} \propto e^{E_a/kT} \tag{S3.55}$$

where E_a is the activation energy. Typical values for E_a are 0.3 eV for oxide breakdown, 0.7 eV for ionic diffusion in the oxide, and 0.7 eV for electromigration. These mechanisms, however, can be reduced by proper device design.[5]

Another failure mechanism, known as *hot-carrier-induced degradation,* is not thermally activated but does depend on the kinetic energy of the channel carriers. It can cause severe degradation. To illustrate the origin of this phenomenon, we consider an n-channel MOSFET biased in the current saturation region. The energy diagram along the channel is shown in Figure S3.23a. In the region near the drain where the longitudinal field $\mathscr{E}_L$ is high, the electrons can attain a high kinetic energy between collisions. They are called *hot carriers* by analogy to electrons that have high kinetic energy due to high temperatures.

There is also a transverse field $\mathscr{E}_T$, which accelerates these fast-moving carriers toward the oxide as indicated in Figure S3.23b. Those electrons that strike the oxide with sufficient kinetic energy can create interface states that can trap electrons, or can penetrate a short distance into the oxide and become trapped. This trapped negative charge shifts the conduction band edge upward in the region adjacent to the drain since $\mathscr{E}_L$ is maximum there.

Figure S3.24 shows the energy band diagram for $V_{DS} = 0$ for a virgin device (a) and for a device that has been subjected to hot-carrier stress (b). For the case shown, the potential energy "hump" prevents the conducting channel from connecting source to drain.

---

[5]For example, electromigration can be reduced by the use of copper conducting lines as shown on the cover of this book rather than the much lighter aluminum lines.

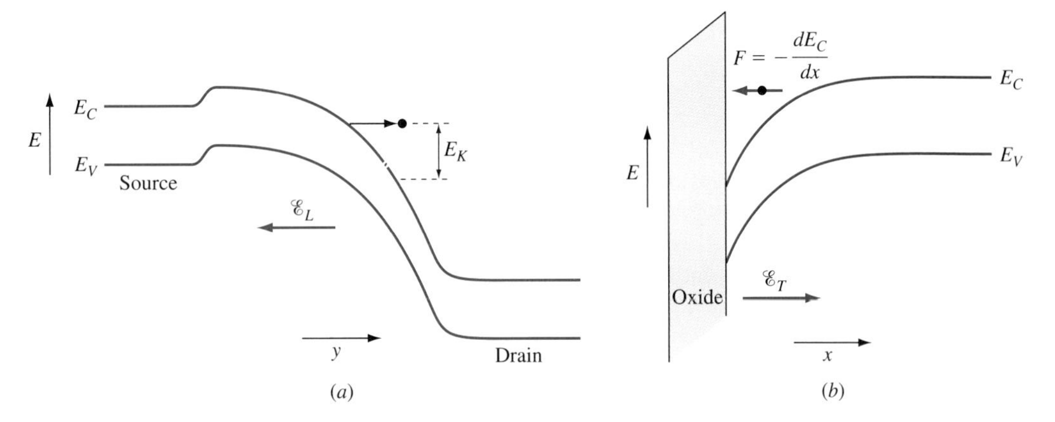

Figure S3.23 (a) The energy band diagram along the channel. The high lateral field means that between collisions, the electron can gain appreciable kinetic energy. (b) Because of the transverse field, the hot electrons are accelerated toward the oxide, where they can create interface states or become embedded in the oxide, thus causing a negative charge buildup.

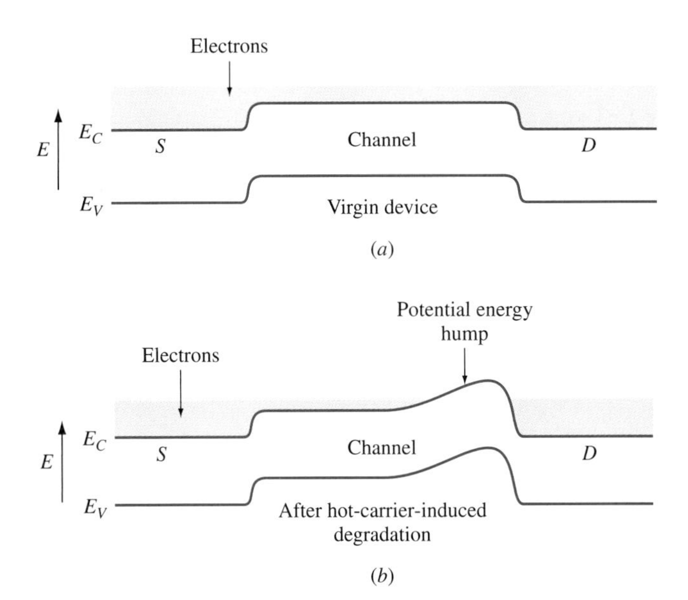

Figure S3.24 Effect of hot-carrier stress on a MOSFET energy band diagram along the channel of a MOSFET for $V_{DS} = 0$ and V_{GS} large enough to create a channel. (a) The virgin device just off the shelf, and (b) the device after long operation in the forward mode. For the stressed device, no channel exists near the drain.

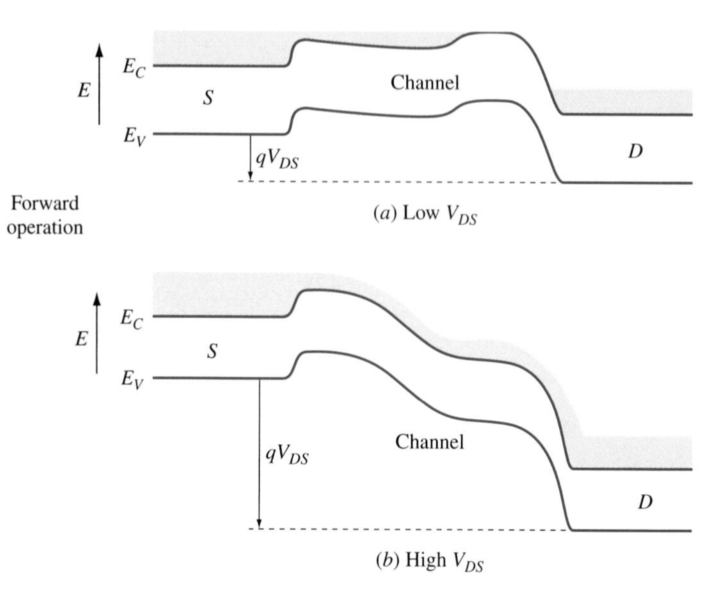

Figure S3.25 Energy band diagrams for stressed device operating in the forward mode. (a) For small V_{DS}, I_D is reduced from that of a virgin MOSFET. (b) At high V_{DS}, the hot-carrier-induced hump has little effect on current flow.

Figure S3.25 indicates what happens in the stressed device as the drain voltage V_{DS} is varied. The energy band diagram is drawn for a small value of V_{DS} (a) and for a larger V_{DS} (b). At small V_{DS} the hump prevents current flow. In (b), the drain voltage is large enough that the hump has a negligible effect on the current, so that at large V_{DS} there is not a noticeable aging effect. For intermediate values of V_{DS}, however, the presence of the hump reduces the current from that of a virgin device.

Recall that MOSFETs are normally fabricated such that the source and drain are interchangeable. On a virgin device, therefore, the electrical characteristics are the same for forward and inverse operation. The hot-carrier-induced potential hump in the channel near the drain, however, causes an asymmetry in the I_D-V_{DS} characteristics. The energy band diagram for inverse operation is shown in Figure S3.26. Here the polarity of the drain voltage is reversed from the forward operation of Figure S3.25. Because the hump is near the acting source (*D*), the positive voltage on the acting drain (*S*) has little effect on the size of the barrier. This means that the threshold voltage for inverse operation is increased.

Figure S3.27 shows the I_D-V_{DS} characteristics for forward and inverse operation of a MOSFET in its virgin state and after stress. The forward characteristics are plotted in the first quadrant while the inverse characteristics are in the third quadrant. The solid lines indicate the characteristics of the virgin device, while the dashed lines show the characteristics after prolonged operation in the

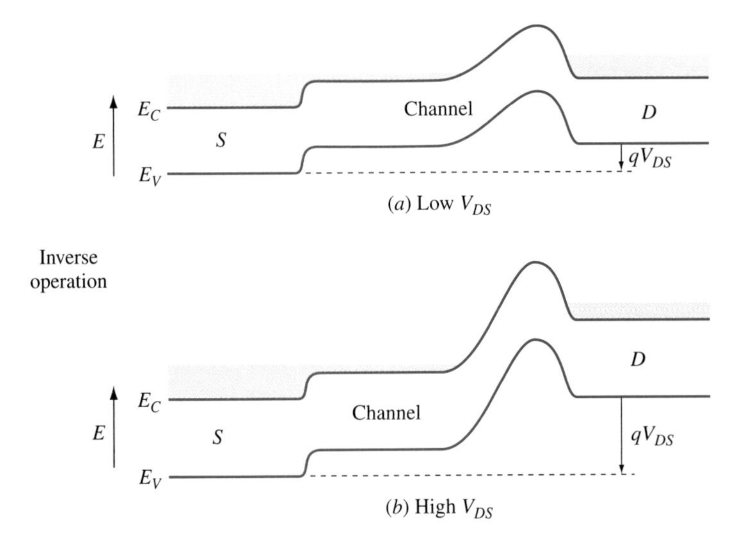

Figure S3.26 Energy band diagrams for inverse operation (drain voltage polarity is inverted) for a stressed device. In this mode of operation the actual drain is effectively the source and the actual source is effectively the drain. Since the potential energy hump is near the acting source, the voltage on the acting drain is not enough to create a conducting channel. In effect, the threshold voltage V_T has increased.

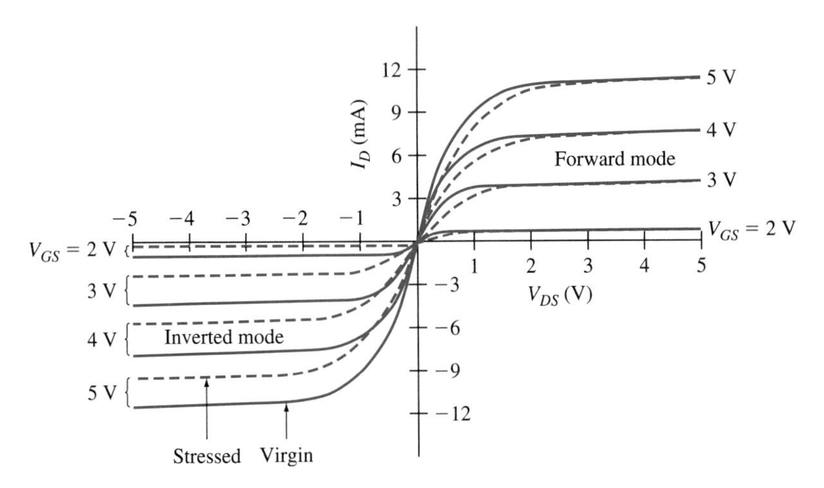

Figure S3.27 The I_D-V_{DS} characteristics of a virgin device (solid lines) and a hot-carrier-stressed device (dashed lines) for forward and inverse operation. The reduction in V_T and thus I_D is apparent for inverse operation.

forward mode. The I_D-V_{DS} characteristics for the virgin device are symmetric in forward and inverse operation, but that is not true for the stressed device.

The characteristics change for both forward and inverse operation. For forward operation with small V_{DS}, the drain current I_D is reduced somewhat from its virgin value. At large enough V_{DS}, in the saturation region, the current is equal to that before stress, and the device appears to operate normally.

Next, compare the I_D-V_{DS} characteristic at $V_{GS} = 2$ V in Figure S3.27. In the forward mode, the saturation current is the same for the virgin and stressed device. In the inverse mode, however, even with a large V_{DS}, no conducting channel exists. To initiate conduction a larger V_{GS} is required. Put another way, V_T is increased, which results in reduced current at all drain-to-source voltages.

There are some circuits in which the transistors operate in both forward and inverse modes. These circuits (e.g., bilateral transmission gates) are more susceptible to the effects of hot-carrier-induced degradation.

Hot-carrier-induced degradation is increasingly pronounced with decreasing L. This is because, in a short channel, the high longitudinal field region at the drain end reaches closer to the source. The potential energy hump extends over a larger fraction of the channel.

As expected, hot-carrier-induced degradation is reduced by the use of smaller drain voltages.

*LIGHTLY DOPED DRAIN (LDD) MOSFETs

The hot-carrier degradation in MOSFETS discussed in the previous section can be reduced by reducing the maximum longitudinal field $\mathscr{E}_L$. One way to reduce the field is to use smaller drain voltages, as mentioned earlier. Another method is to use a structure known as a lightly doped drain (LDD).

In the LDD MOSFET, there is a more lightly doped n region between the n^+ drain and the channel, as shown schematically in Figure S3.28. The more lightly doped region is also more resistive, and part of the drain voltage is dropped across it. This reduces the field in the channel near the drain. Because it is easier from a fabrication standpoint to make the transistors symmetric, a similar lightly doped region is present between the n^+ source and channel. This region, however, serves no useful purpose.

There is a trade-off. These lightly doped regions introduce extra series resistance in the source and drain, which degrades the I_D-V_{DS} characteristics as discussed in Section 7.3.3.

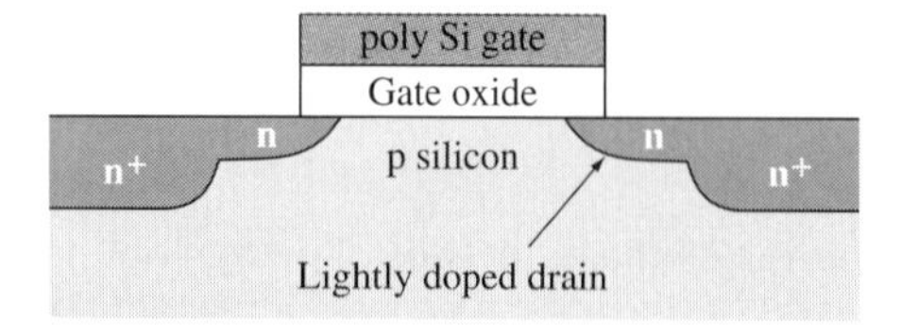

Figure S3.28 Cross section of the lightly doped drain (LDD) MOSFET.

*S3.10 LOW-TEMPERATURE OPERATION OF MOSFETs

Most electronic equipment is operated at room temperature (300 K), or somewhat higher because of self-heating from the power dissipation in the circuits. There are, however, some advantages to operating at reduced temperatures. The most cost-effective way to achieve low-temperature operation is to immerse the operating circuits in liquid nitrogen (77 K). Thus, we compare room temperature electrical characteristics with those at 77 K.

Figure S3.29 shows the experimental I_D-V_{DS} characteristics at 300 K and at 77 K for NFETs from the same chip with three different channel lengths: 20, 2, and 0.7 μm. These devices were designed with $W = 50\,\mu$m for each case. The characteristics are plotted for $(V_{GS} - V_T) = 1.0, 2.0, 3.0,$ and 4.0 V.

There are several things to notice in the figures. One is that, for a given gate voltage, the drain current is higher at lower temperatures. This means the transconductance, $g_m = \partial I_D / \partial V_{GS}$, is increased. Secondly, at both temperatures, the saturation current increases with decreasing channel length, consistent with the results of Chapter 7. Finally, as the channels get shorter, the effect of the low temperature on I_D is not as strong.

These results can be explained with the aid of the expression for the I_D-V_{DS} characteristics in saturation. We use Equations (7.60) and (7.61), repeated here:

$$I_D = \frac{W C'_{ox} \mu_{lf} \left(V_{GS} - V_T - \dfrac{V_{DS}}{2} \right) V_{DS}}{L \left(1 + \dfrac{\mu_{lf} V_{DS}}{L v_{sat}} \right)} \qquad V_{DS} \le V_{DSsat} \qquad \text{(S3.56)}$$

$$I_{Dsat} = \frac{W C'_{ox} \mu_{lf} \left(V_{GS} - V_T - \dfrac{V_{DSsat}}{2} \right) V_{DSsat}}{L \left(1 + \dfrac{\mu_{lf} V_{DSsat}}{L v_{sat}} \right)} \qquad V_{DS} \ge V_{GSsat} \qquad \text{(S3.57)}$$

Of the terms in these equations, the one that is particularly temperature sensitive is the low-field mobility, which is a factor of 4 to 8 times greater at 77 K than at 300 K. This compares with an increase by 10 to 20 percent at 77 K for the saturation velocity. As a result, at reduced temperatures, I_D is increased over its room temperature value for a given $V_{GS} - V_T$.

Neglecting the small temperature dependence of V_{DSsat}, it can be seen that for long-channel devices, where the term $\mu_{lf} V_{DSsat} / L v_{sat}$ is small, I_{Dsat} is approximately proportional to μ_{lf}. With decreasing channel length L, this term becomes appreciable and the relative increase in I_{Dsat} is reduced. For submicrometer channel lengths of modern devices, the improvement in I_{Dsat} obtained by going to lower temperatures is small.

Increased current at low temperatures results from higher transconductance, but larger currents would tend to imply increased power consumption. We recall

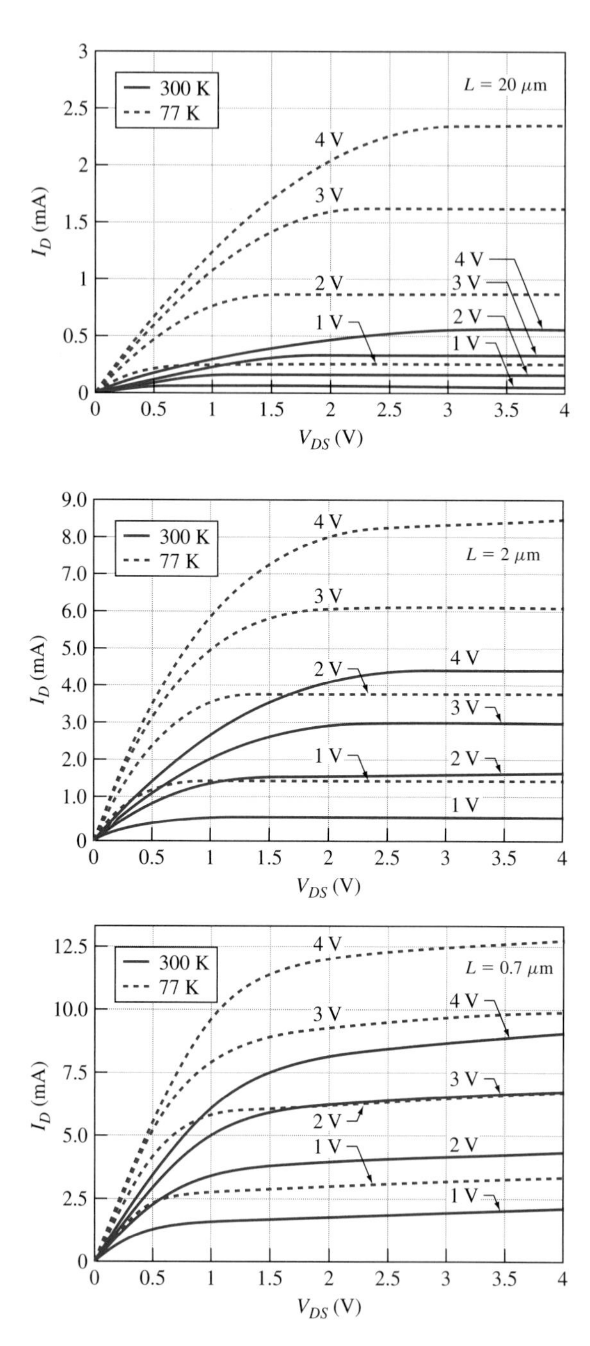

Figure S3.29 I_D-V_{DS} characteristics for n-channel MOSFETs at 77 K (dashed lines) compared with those at room temperature (solid lines) for three different channel lengths and four values of $V_{GS} - V_T$. The differences at the two temperatures are less pronounced for shorter channels.

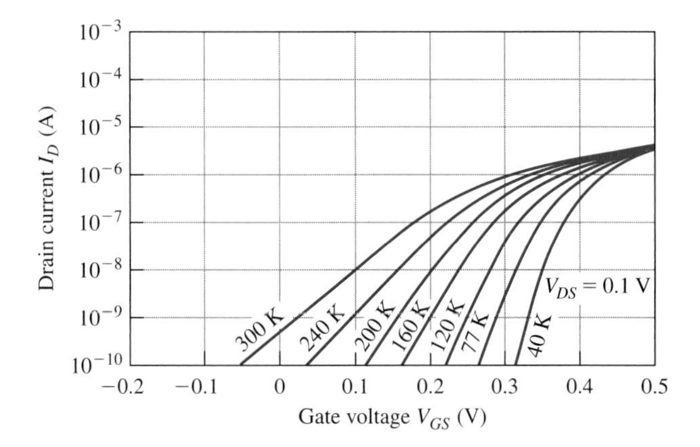

Figure S3.30 Subthreshold current dependence on gate voltage for a range of temperatures.

from Chapter 8 that in a switching circuit the dynamic power and static power are

$$\begin{aligned} P_{\text{ac}} &= C_L V_{DD}^2 f \\ P_{\text{dc}} &= I_{\text{leakage}} V_{DD} \end{aligned} \tag{S3.58}$$

We concluded that one could reduce dynamic power consumption by reducing the supply voltage V_{DD}. A reduction in V_{DD}, however, requires a design to reduce threshold voltage. As we saw in Chapter 8, at a given temperature, a lower V_T causes an increase in I_{leakage} in the off state.

Figure S3.30 shows the I_D-V_{GS} characteristics below threshold for an NFET at $V_{DS} = 0.1$ V. The curves are shown for seven different temperatures ranging from 40 to 300 K. Observe that, at lower temperatures, a smaller change in gate voltage is required to produce a given change in drain current. Recall that the subthreshold swing [Equations (8.14) and (8.15)] is given by

$$S = \frac{\partial V_{GS}}{\partial \log I_D} \tag{S3.59}$$

The swing decreases with decreasing temperature and has approximately one-quarter the value at 77 K as it has at 300 K. The reduced swing implies that for the same leakage current at $V_{GS} = 0$, the threshold voltage, and thus the supply voltage, can be reduced by a factor of 4. This in turn reduces the dynamic power dissipation by a factor of 16.

Operating at low temperatures can have other benefits. For example, the packing density (number of devices per unit area) on a chip is limited largely by the temperature rise caused by power dissipation. Assuming that (for a digital circuit) the power dissipation is primarily dynamic, the packing density at 77 K can be increased by a factor of 16 without overheating.

Furthermore, the heat transfer rate from the channel, or the rate at which heat can be extracted from the device to keep it cool, is determined largely by the thermal conductivity of Si. This is a factor of 6 larger at 77 K than at 300 K. Combining the improved heat extraction with the reduced power dissipation, the temperature-limited packing density is increased by a factor of nearly 100. Of course, there are other limitations on packing density such as minimum device size and spacing between devices.

Semiconductor devices are also expected to be appreciably more reliable at low temperatures. Those failure mechanisms that are temperature activated are reduced at low temperature.

We indicated earlier that the activation energy for the gate oxide breakdown is about $E_a \approx 0.3\,\text{eV}$. Therefore, from Equation (S3.55), the MTTF for devices that fail because of oxide breakdown is expected to be increased by a factor of about 10^{15}. We also discussed hot-carrier-induced degradation, which we indicated is not thermally activated. At a given supply voltage the hot-carrier-induced degradation would actually be increased slightly at lower temperatures, because the mean free time between collisions increases and electrons have more time to accelerate and attain higher kinetic energies. At lower temperatures, however, the devices can operate at lower voltages, resulting in smaller values of the longitudinal field $\mathscr{E}_L$. The net result is that hot-carrier-induced degradation can be reduced in MOSFETs designed for and operated at low temperatures.

While in principle there are advantages for MOSFET operation at cryogenic temperatures, the improvement in the electrical characteristics in modern submicrometer-channel-length devices is small and currently the cost and complexity of the required refrigeration systems does not justify this technology.

*S3.11 APPLICATIONS OF SPICE TO MOSFETs

There are many models used to simulate the operation of MOSFETs. A popular class are those used in SPICE. [10] In SPICE, there are three basic models referred to as Level 1, Level 2 and Level 3.[6] Each of these levels has many variations, depending in part on the vendor.

The simplest MOSFET model is Level 1. It is based on the equations for the simple long-channel MOSFET of Chapter 7. The equations for the simple long-channel model are:

$$I_D \approx 0 \qquad V_{GS} \le V_T$$

$$I_D = \frac{WC'_{\text{ox}}\mu}{L}\left[\left(V_{GS} - V_T - \frac{V_{DS}}{2}\right)\right]V_{DS} \qquad V_{DS} \le V_{DS\text{sat}},\ V_{GS} \ge V_T$$

$$I_{D\text{sat}} = \frac{WC'_{\text{ox}}\mu}{L}\left[\left(V_{GS} - V_T - \frac{V_{DS\text{sat}}}{2}\right)\right]V_{DS\text{sat}} \qquad V_{DS} \ge V_{DS\text{sat}},\ V_{GS} \ge V_T$$

$$V_{DS\text{sat}} = V_{GS} - V_T$$

[6] A more recent model for use in simulating short-channel MOSFETs is the Berkeley short-channel IGFET model (BSIM).

In this model, μ is assumed constant.[7] SPICE Level 1 can be used for a first-order evaluation for long-channel MOSFETs ($L > 20\,\mu$m), but is inadequate for modern devices with submicrometer channel lengths.

For more realistic calculations, SPICE Level 3 models are used. There are several parameters incorporated in these models. Here we discuss only those parameters associated with a simple static n-channel MOSFET model.

SPICE Level 3 is based on Equations (7.72) and (7.73), with the low-field mobility, from Equation (7.47),

$$\mu_{\mathrm{lf}} = \frac{\mu_0}{1 + \theta(V_{GS} - V_T)}$$

SPICE expresses the I_D-V_{DS} characteristics somewhat differently from those presented here, to consider second- (and third-) order terms due to inclusion of parasitic effects.

Another difference is that rather than use the expression for $V_{D\mathrm{sat}}$ of Equation (7.68)

$$V_{D\mathrm{sat}} = \frac{v_{\mathrm{sat}}}{\mu_{\mathrm{lf}}} L \left[\left(1 + \frac{2\mu_{\mathrm{lf}}(V_{GS} - V_T)}{v_{\mathrm{sat}} L} \right)^{1/2} - 1 \right]$$

SPICE instead uses the expression

$$V_{D\mathrm{sat}} = \frac{(V_{GS} - V_T)}{(1 + F_B)} + \frac{v_{\mathrm{sat}} L}{\mu_{\mathrm{lf}}} - \sqrt{\left(\frac{V_{GS} - V_T}{(1 + F_B)} \right)^2 + \left(\frac{v_{\mathrm{sat}} L}{\mu_{\mathrm{lf}}} \right)^2} \tag{S3.60}$$

where the factor F_B accounts for the influence on Q_{ch} of the variation of depletion region charge along the channel. That is,

$$F_B = \frac{\gamma}{2\sqrt{2\phi_f}} \tag{S3.61}$$

where in SPICE the GAMMA function γ is

$$\gamma = \frac{\sqrt{2\varepsilon_S q N'_A}}{C'_{\mathrm{ox}}} \tag{S3.62}$$

Table S3.1 shows some of the major static SPICE Level 3 parameters for MOSFETs.

S3.11.1 EXAMPLES OF THE USE OF SPICE WITH MOSFETs

Here we give three examples of using SPICE for MOSFETs. We use SPICE to emulate a curve tracer to obtain the transfer characteristics and the output

[7] In SPICE, the product $\mu C'_{\mathrm{ox}}$ is often replaced by the transconductance coefficient, KP.

Table S3.1 Some parameters used for SPICE Level 3

Symbol	SPICE keyword	Parameter name	Typical value	Units
L	L	Channel length	1E-6	m
W	W	Channel width	5E-6	m
μ_0	UO	Low-field mobility	500	$cm^2/V \cdot s$
t_{ox}	TOX	Gate oxide thickness	5E-9	m
V_T	VTO	Threshold voltage	0.5	V
$2\phi_f$	PHI	Surface potential	0.72	V
θ	THETA	Mobility modulation factor	0.10	V^{-1}
v_{sat}	VMAX	Maximum drift velocity	5E4	m/s
λ	LAMBDA	Channel length modulation factor	0.05	V^{-1}
N_A, N_D	NSUB	Substrate doping	1E16	cm^{-3}
I_s	IS	Bulk junction saturation current	1E-16	A
R_D	RD	Drain ohmic resistance	20	Ω
R_S	RS	Source ohmic resistance	20	Ω

characteristics. Then the transient response of a simple CMOS inverter is investigated.

In Section S3.5 the I_D-V_{DS} characteristics with small V_{DS} were obtained from experimental data to determine threshold voltage V_T, low-field channel mobility at threshold μ_0, and the mobility modulation factor θ. If the values of these parameters are known, SPICE can be used to simulate the I_D-V_{GS} characteristics.

EXAMPLE S3.8

Find the I_D-V_{GS} characteristics using SPICE for an n-channel MOSFET having the following parameters.

$$L = 0.3\,\mu m$$

$$W = 3\,\mu m$$

$$VTO = 0.5\ V$$

$$PHI = 0.75\ V$$

$$TOX = 5\ nm$$

$$VMAX = 5 \times 10^6\ cm/s$$

$$THETA = 0.1\ V^{-1}$$

■ **Solution**

To obtain these characteristics we use the SPICE curve tracer arrangement of Figure S3.31a. The drain-source voltage is set at 40 mV to ensure that the operation is in the linear region and the current can be given by

$$I_D = \frac{WC'_{ox}}{L} \cdot \frac{\mu_0}{1 + \theta(V_{GS} - V_T)} \cdot \left(V_{GS} - V_T - \frac{V_{DS}}{2}\right) V_{DS} \qquad \text{(S3.63)}$$

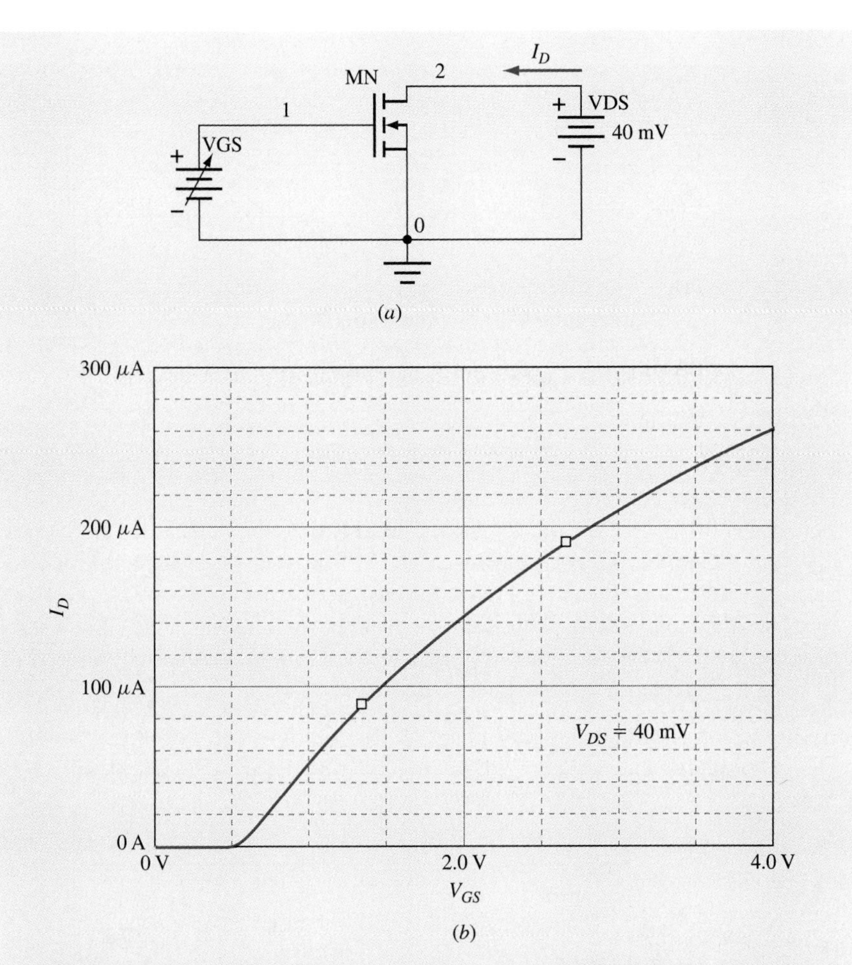

Figure S3.31 (a) Circuit diagram used to simulate the I_D-V_{GS} characteristics with constant V_{DS} small enough such that the mobility is independent of $\mathscr{E}_L$. (b) The SPICE simulation results. The deviation from a straight line results from the finite value of θ.

The gate voltage is then swept from $0 \le V_{GS} \le 4$ V and the resultant drain current is plotted as a function of V_{GS}. The results are shown in Figure S3.31b. The derivation from a straight line is the result of a finite value of θ in Equation (S3.63) as discussed in Section S3.5.

EXAMPLE S3.9

Plot the output characteristics for the NMOS of Example S3.8. For this we use the SPICE circuit of Figure S3.32.

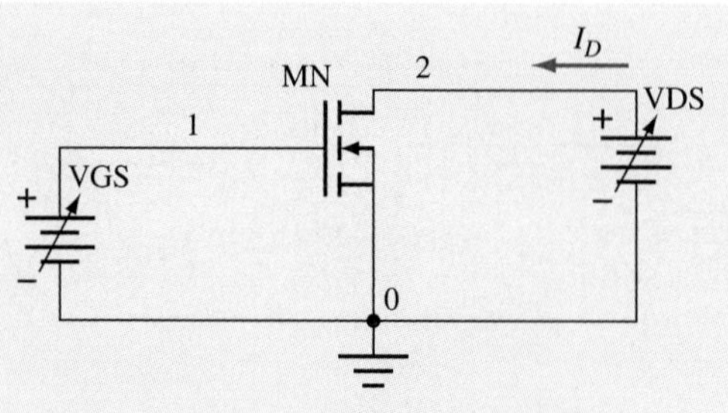

Figure S3.32 Circuit diagram used to simulate the I_D-V_{DS} characteristics. The value of V_{DS} is swept for different values of V_{GS}.

■ Solution

The drain voltage is swept over the range $0 \leq V_{DS} \leq 5$ V with the gate-source voltage incremented in 1-V steps from 0 to 4 V. The SPICE I_D-V_{DS} plots for this case are shown in Figure S3.33a. For comparison, the SPICE results neglecting the velocity saturation effect are shown in (b). (Note the change in scale on the vertical axis.) The effect of saturation velocity on the saturation current $I_{D\text{sat}}$, at which the current saturates, is clearly visible. At $V_{GS} = 4$ V, the saturation current is reduced by a factor of 13 when the velocity saturation effect is considered.

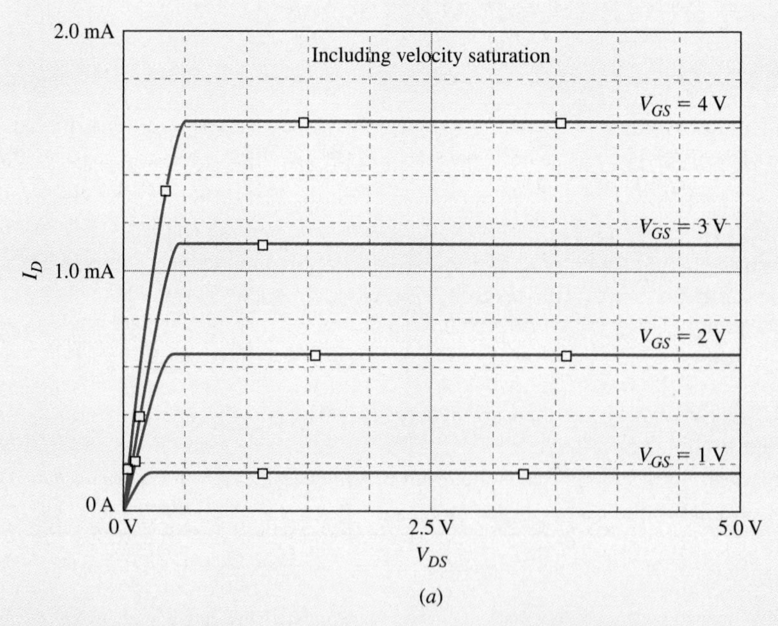

Figure S3.33 SPICE simulation for the device of Example S3.8 considering the effect of velocity saturation (a) and neglecting this effect (b).

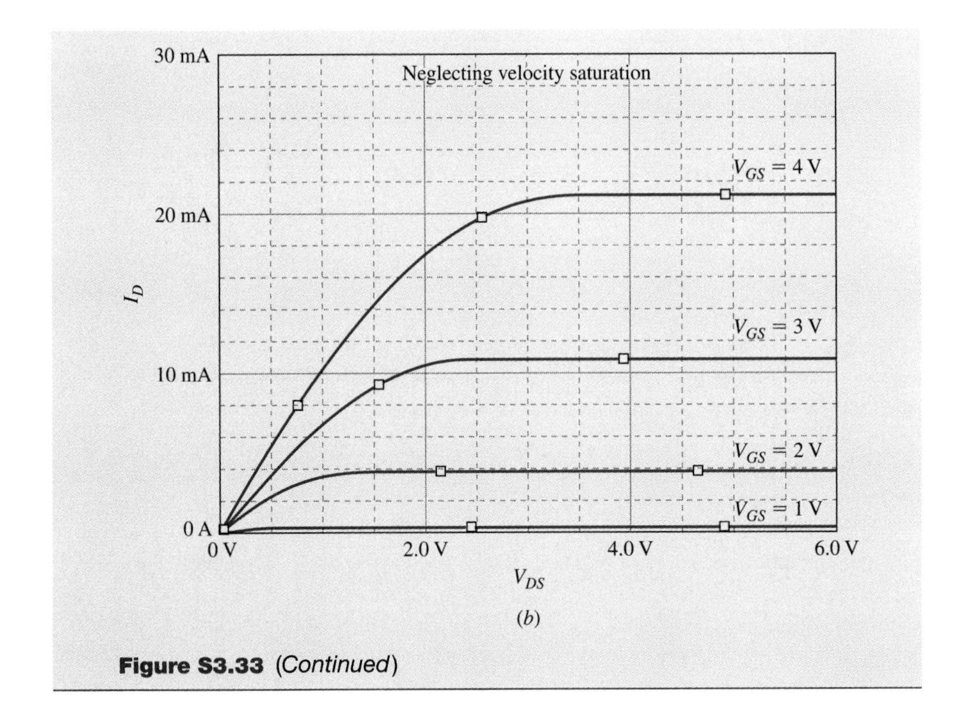

Figure S3.33 *(Continued)*

S3.11.2 DETERMINING THE TRANSIENT CHARACTERISTICS OF A CMOS DIGITAL INVERTER

As indicated earlier, CMOS is an important technology in modern digital systems. Here we present an example of using SPICE to simulate the transient response to a simple CMOS digital inverter designed to operate at a supply voltage of 2.5 V. For simplicity we use the static MOSFET characteristics and consider a capacitance load of 0.2 pF as indicated in Figure S3.34. The input is a

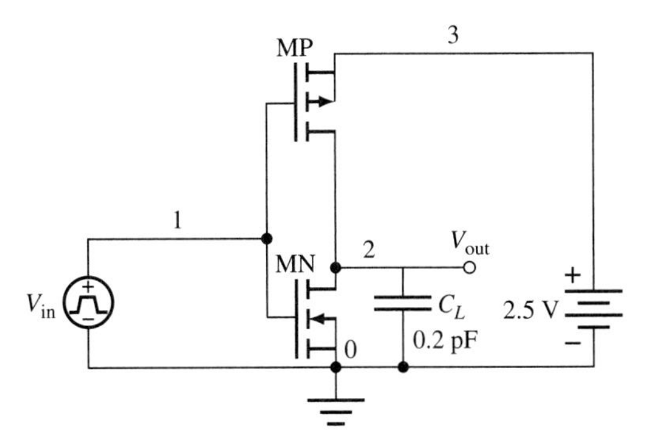

Figure S3.34 Circuit for SPICE simulation of the response of a CMOS digital inverter with a capacitive load.

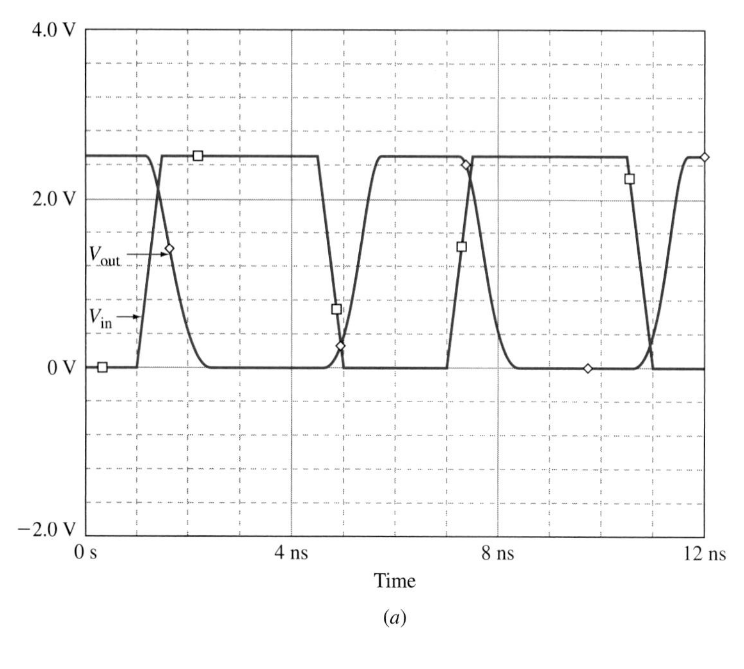

(*a*)

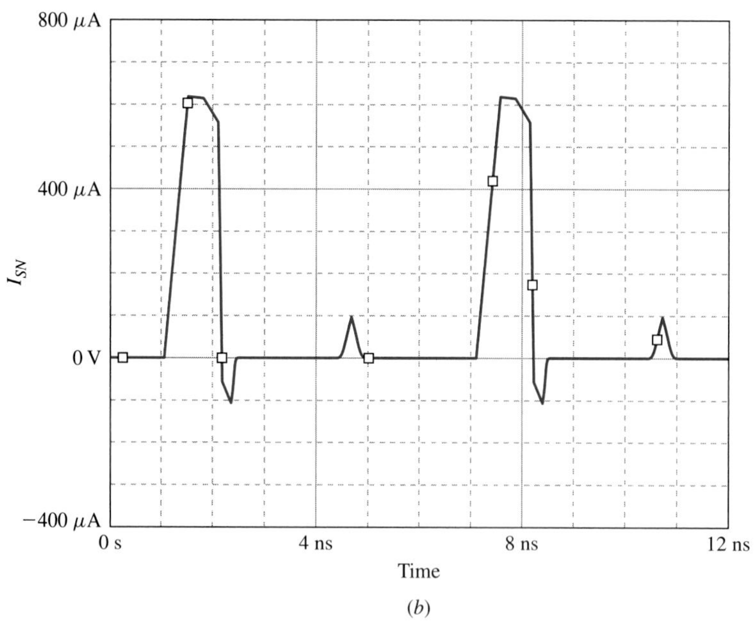

(*b*)

Figure S3.35 (a) Spice simulated input and output voltage waveforms and (b) current waveform for the NMOS in the circuit of Figure S3.34 and the parameters of Section S3.11.2.

repetitive voltage pulse of 2.5 V amplitude of 3-ns duration and 6-ns period. The rise and fall times of the input are each taken to be 0.5 ns. The parameters for the FETs are taken to be:

NMOS	PMOS
L = 0.2 μm	L = 0.2 μm
W = 2 μm	W = 2.44 μm
VTO = 0.5 V	VTO = −0.5 V
TOX = 5 nm	TOX = 5 nm
UO = 500 cm^2/V · s	UO = 200 cm^2/V · s
VMAX = 5 × 10^6 cm/s	VMAX = 5 × 10^6 cm/s

The width of the PMOS is taken to be 1.22 times that of the NMOS, as discussed in Section 8.4.2.

The input and output voltage waveforms produced by SPICE are shown in Figure S3.35a. It can be seen that the rise and fall times of the output are greater than those of the input because of the effect of the capacitance load.

The current flow transients through the NMOS are shown in Figure S3.35b. A large current flows through the NMOS when it is on and discharging the load capacitance. A small current spike is observed when the NFET is switching off and the PFET is switching on. This is the passthrough current that flows when both devices are partially on as discussed in Section 8.5.3.

S3.12 SUMMARY

This supplement amplifies some of the topics presented in Chapters 7 and 8. In those earlier chapters, we presented various models from which to find the current in a MOSFET. In the Supplement to Part 3, we refined the formulation for the channel charge Q_{ch} to include the effects of the variations of depletion region charge along the channel and the dependence of Q_{ch} on the longitudinal field $\mathscr{E}_L$.

In the earlier chapters, the threshold voltage was treated as an experimentally measured quantity whose value for a given process was assumed to be known. In this supplement, the threshold voltage was related to physical parameters. Its dependence on the work function difference between the gate and the substrate, Φ_{MS}, was investigated, as well as the effects of various charges, including charges in the oxide, charges at the oxide semiconductor interface, and charges within the semiconductor depletion region. Adjustment of the threshold voltage to its desired value is achieved by ion implantation of impurities in the semiconductor near its interface with the oxide.

Universal relations for low-field mobility for Si NMOS and PMOS transistors in terms of an effective transverse field were presented. The technique to measure threshold voltage, the low-field mobility at threshold, and the dependence on gate voltage was discussed.

Next, we examined the MOS capacitor, a useful circuit element and valuable diagnostic tool. The capacitance-voltage characteristics in particular are useful to measure material properties resulting from the fabrication processes, including the concentration of interface states, the depletion width, the oxide thickness, and the substrate doping. In the category of capacitance-related devices, two classes of devices utilizing MOS capacitance were briefly described: dynamic random-access memories (DRAMs) and charge-coupled devices (CCDs).

Mechanisms responsible for device degradation were also briefly examined. These include thermally activated mechanisms such as oxide breakdown and electromigration, as well as hot-carrier-induced degradation resulting from operation at high lateral channel fields.

The operation of MOSFETs at 77 K was briefly described. While this showed promise for improving device performance for long-channel devices, the performance improvement is minimal in modern short-channel (submicrometer) devices.

Finally, Level 1 and Level 3 models of SPICE for MOSFETs were briefly described. Although the simple Level 1 model is adequate for a first approximation in long-channel MOSFETs, more complex models (e.g., Level 3) are required for short-channel devices. Examples of the use of SPICE with MOSFETs were presented.

S3.13 READING LIST

Items 2, 4, 8, 10, 12, 17 to 19, 29, 32, and 33 in Appendix G are recommended.

S3.14 REFERENCES

1. Dennis Hoyniak, Edward Nowak, and Richard L. Anderson, "Channel electron mobility dependence on lateral electric field in field-effect transistors," *J. Appl. Phys.,* 87, pp. 876–881, 2000.
2. Kangguo Cheng, Jinju Lee, Karl Hess, Joseph W. Lyding, Young-Kwang Kim, Young-Wug Kim, and Kwang-Pyuk Suh, "Improved Hot-Carrier Reliability of SOI Transistors by Deuterium Passivation of Defects at Oxide/Silicon Interfaces," *IEEE Trans. Electron Devices,* ED-49, pp. 529–531, 2002.
3. F. Stern and W. E. Howard, "Properties of semiconductor surface inversion layers in the quantum limit," *Phys. Rev.,* 163, p. 816, 1967.
4. F. Stern, "Quantum properties of surface space charge layers," *CRC Crit. Rev. Solid-State Sci.,* 4, p. 499, 1974.
5. G. Baccarani and M. R. Wordeman, "Transconductance degradation in thin-oxide MOSFETs," *IEEE Trans. Electron Devices,* ED-30, pp. 1295, 1983.
6. S. Takagi, M. Iwase, and A. Toriumi, "On universality of inversion layer mobility in n- and p-channel MOSFETs," *IEDM Technical Digest,* pp. 398–401, 1988.
7. Dieter K. Schroder, *Semiconductor Material and Device Characterization,* John Wiley and Sons, New York, Section 4.8, 1990.

8. R. H. Dennard, "Scaling Challenges for DRAM and Microprocessors in the 21st Century," *Electrochemical Society Proceedings,* Vol. 97–3, pp. 519–532, 1997.
9. Carlo H. Sequin and Michael F. Tomsett, *Charge Transfer Devices,* Academic Press, New York, 1975.
10. G. Massobrio and P. Antognetti, *Semiconductor Device Modeling with SPICE,* 2nd ed., McGraw-Hill, New York, 1993.

S3.15 REVIEW QUESTIONS

1. Why should we realistically expect the channel charge to vary with position along the channel? When this is taken into account, what effect does it have on the current-voltage characteristics of a MOSFET?
2. Explain how the presence of fixed oxide charge can influence the value of the threshold voltage. What other charges must also be considered?
3. What is meant by flat band voltage?
4. How is the threshold voltage of MOSFETs controlled after fabrication?
5. When the channel is very narrow, quantum mechanical effects come into play. What is the result on the allowed energies for electrons in the channel?
6. Explain qualitatively how the I_D-V_{GS} relationship can be used to find the low-field mobility and its dependence on the gate voltage.
7. Explain how *C-V* measurements can be used to experimentally find various parameters of the process. Which parameters can be obtained with this technique?
8. Explain the operation of a DRAM cell.
9. What is meant by hot-carrier-induced degradation? What effect does it have on the MOSFET operation? How can it be mitigated?
10. What is the purpose of a lightly doped drain structure? How does it work?
11. What are the advantages of low-temperature operation of CMOS circuits? What is the engineering trade-off?

S3.16 PROBLEMS

S3.1 A Si NMOS with $t_{ox} = 5$ nm has a threshold voltage of 0.2 V. What is the required implant dose (atoms/cm^2) to increase the threshold voltage to 0.5 V?

S3.2 Adjacent MOSFETs are isolated from each other by a thick *field oxide* region as indicated in Figure PS3.1. The gate oxide thickness is 5 nm and the field oxide region is 0.5 μm thick. Both gate and interconnect are of the same metal with $\Phi_{MS} = -0.4$ eV. The Si net doping concentration is 10^{16} cm^{-3}. The threshold voltage for the NMOS is 0.5 V. What is the threshold voltage for the parasitic NMOS between the active devices?

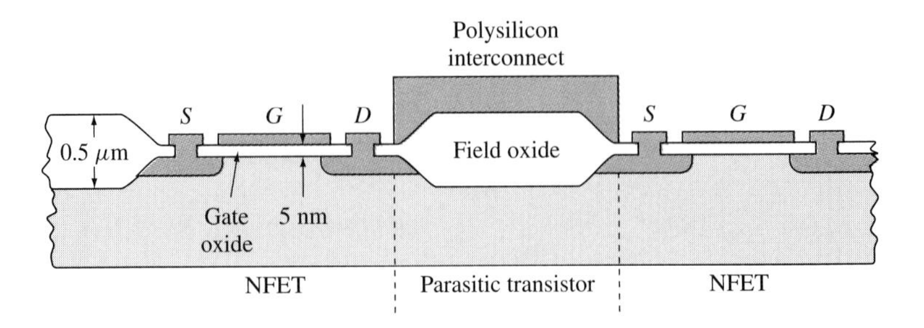

Figure PS3.1

S3.3 If the amount of positive charge Q_f in Figure 8.2 is increased, what will be the effect on V_T? Explain the physics of your answer.

S3.4 Find V_T for a silicon n-channel FET with the following parameters:

$N_{D\text{gate}} = 10^{20}\ \text{cm}^{-3}$

$N_A = 10^{17}\ \text{cm}^{-3}$ in the channel

$t_{\text{ox}} = 4\ \text{nm}$

$Q_f = q \times 5 \times 10^{10}\ \text{C/cm}^2$

$Q_I = q \times 10^{10}\ \text{C/cm}^2$

a. Is this an enhancement or depletion device?

b. What is the result if bandgap narrowing is neglected?

S3.5 For the device of Example S3.1, what is the minimum value the combined fixed charge and interface charge can be and still ensure an enhancement device? What is the corresponding density of charges?

S3.6 A MOSFET process produces

$\Phi_{MS} = -0.5\ \text{eV}$

$C'_{\text{ox}} = 6.9 \times 10^{-7}\ \text{F/cm}^2$

$Q_f = q \times 10^{10}\ \text{C/cm}^2$

$Q_{\text{it}}(2\phi_f) = q \times 10^{10}\ \text{C/cm}^2$

$Q_B(2\phi_f) = -5 \times 10^{-8}\ \text{C/cm}^2$

$\phi_f = 0.4\ \text{eV}$

What should the ion implantation be to make this an enhancement mode device intended to operate in a 3.3 V operating circuit?

S3.7 For an NFET with channel doping $N'_A = 5 \times 10^{16}\ \text{cm}^{-3}$, what is the value of the effective transverse field at threshold? What is the low-field mobility for this case? Let $t_{\text{ox}} = 5\ \text{nm}$.

S3.8 For the device whose I_D-V_{GS} characteristics are shown in Figure PS3.2, find the value of μ_0, V_T, and θ. The measurements were taken at $V_{DS} = 50\ \text{mV}$ and $W = 10\ \mu\text{m}$, $L = 0.5\ \mu\text{m}$, and $t_{\text{ox}} = 5\ \text{nm}$.

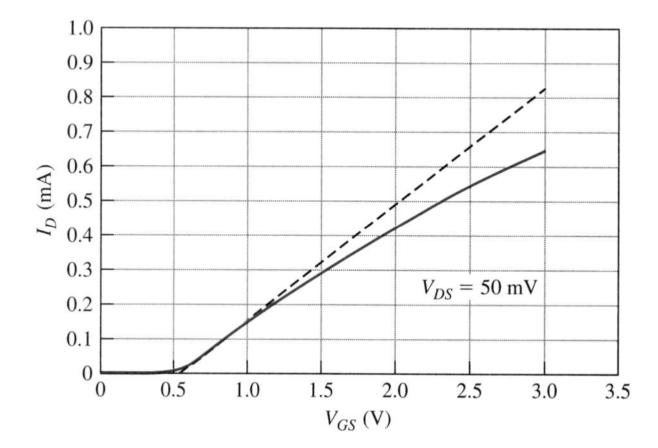

Figure PS3.2

S3.9 Figure PS3.3 shows a plot of $\sqrt{I_{D\text{sat}}}$ versus V_{GS} for a particular FET of channel length 0.5 mm. Use the alternative method of Section S3.6 to estimate the value of the threshold voltage and the low-field mobility. Over what range of gate voltages is a straight line extrapolation valid? Estimate the accuracy to which you can find μ_{lf} using this plot and this technique.

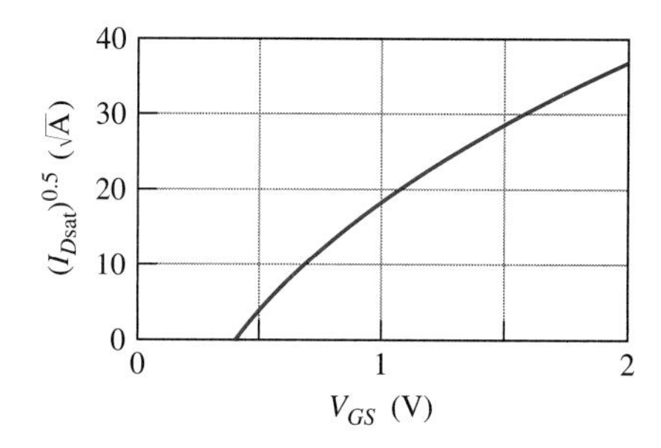

Figure PS3.3

S3.10 Consider a MOS capacitor where the gate is degenerate n^+ Si and the substrate is n-type Si with $N'_A = 10^{16}\ \text{cm}^{-3}$. Except for the difference in work functions of gate and substrate, the capacitor can be considered to be ideal. Neglect band-gap narrowing in the degenerate gate.

a. What is the built-in voltage of the device?
b. Sketch the equilibrium energy band normal to the gate.
c. What is the flat band voltage?
d. Sketch the energy band diagram at flat band.

e. Sketch the low-frequency and the high-frequency C-V_G characteristics of the device.

f. Sketch the charge distribution as a function of position for $V_G = 0$, $V_G = +5$ V and $V_G = -5$ V.

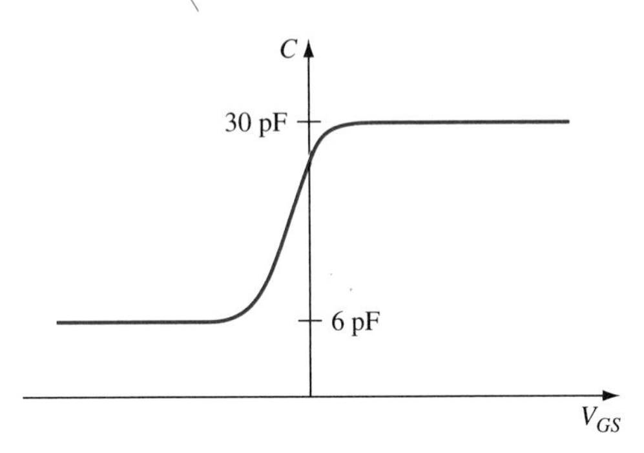

Figure PS3.4

S3.11 The C-V_G characteristics of an (ideal) MOS capacitor of area 10^{-3} cm is shown in Figure PS3.4.

a. Is the measurement made at high frequency or low frequency?

b. Is the semiconductor n type or p type?

c. What is the oxide thickness?

d. What is the net doping concentration of the semiconductor?

S3.12 Explain how a logic 1 and logic 0 can be written into a DRAM cell if the initial state of the cell is a 1 instead of 0 as assumed in the text.

S3.13 For the two-phase CCD discussed in Section S3.8.2, explain the operation if no p^+ implant were made.

S3.14 From the plot of Figure S3.31b, verify that $V_T = 0.5$ V, $\mu_0 = 500\,\text{cm}^2/\text{V}\cdot\text{s}$, *and* $\theta = 0.1\,\text{V}^{-1}$.

S3.15 Determine the gate delay for the CMOS inverter whose waveforms are shown in Figure S3.35a.

S3.16 For the circuit of Section S3.11.2 (Figure S3.34), using SPICE, plot the voltage waveforms and the NMOS current transients if the rise and fall times are increased to 1 ns.

S3.17 a. For the above problem with C_L replaced with a resistor of 50 kΩ, plot the drain current of the NFET (this is the passthrough current).

b. Repeat with V_{out} open circuited (no capacitor and no resistor). What is the cause of the problem?

S3.18 For the circuit of Section S3.11.2, using SPICE, plot the current waveform of the load capacitance.

PART 4

Bipolar Junction Transistors

The preceding chapters dealt with field-effect transistors, which are unipolar devices, i.e., devices in which only one type of carrier, electrons or holes, contributes to current flow in the device. In this section, we discuss bipolar devices, in which both electrons and holes must be considered.

Like a FET, a bipolar junction transistor (BJT) is a two-junction, three-terminal device in which one terminal controls the current flow between the other two terminals. The terminology is different, however. In Table IV.1, the analogous regions of a BJT and a FET are compared.

In a FET, the source acts as a source of carriers, which are emitted into the channel where they flow into the drain. In a BJT, the emitter acts as the carrier source and emits carriers into the base where they flow to the collector and are collected. In both cases, the voltage at the control electrode (gate or base), determines the number of carriers available to flow from source to drain or from emitter to collector. The physics of the control mechanism differs between a FET and a BJT, however. In a FET, the control (gate) voltage is capacitively coupled to the channel. In a BJT, a direct connection is made to the base. Another distinction is that in a FET, carrier flow from source to drain is primarily by drift. In a BJT, carrier flow from emitter to collector is at least in part by diffusion.

Figure IV.1 indicates the (planar) structure of an integrated circuit (IC) npn (emitter-base-collector) BJT in which all contacts, the emitter (E), base (B) and collector (C) are made from the top.[1] In (a) the top view is indicated and in (b) the cross-sectional schematic of the device is shown. The structure of Figure IV.1 is representative of BJTs of the mid 1970s. Because of its relative simplicity, it will be used to develop the basic operating principles of bipolar junction transistors. The structure and operation of more modern devices is discussed later. Note that this integrated circuit transistor exists in an n-type *well* in a p-type substrate. The

[1] This is in contrast to a discrete BJT in which the collector contact is usually made from the bottom.

Table IV.1 Comparison of terminology of a BJT with that of a FET

FET	BJT
Source	Emitter
Drain	Collector
Channel (gate electrode)	Base

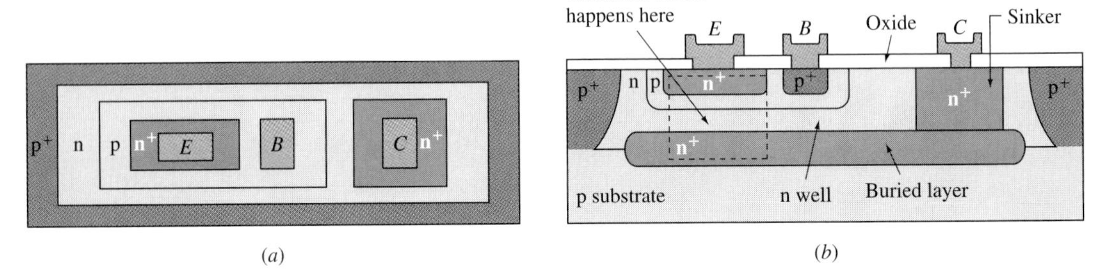

Figure IV.1 Schematic diagram of an npn bipolar transistor used in integrated circuits. (a) Top view and (b) cross-sectional view.

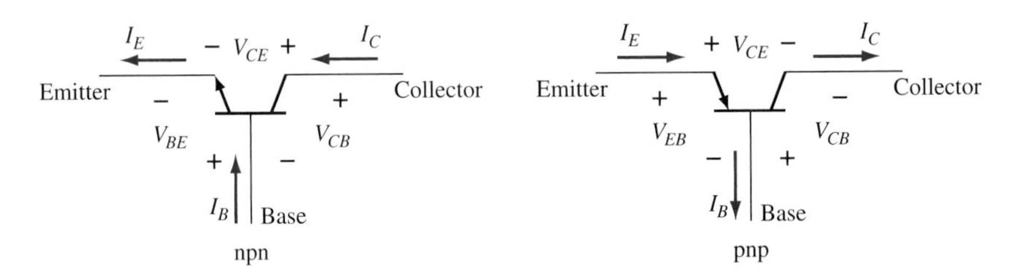

Figure IV.2 Circuit symbols for npn and pnp BJTs. The arrow in the emitter indicates the direction of current flow. The polarities of the voltages are those for bias in the active mode.

p-type substrate is connected to the most negative voltage of the circuit such that the junctions between the n wells and the p substrate are never forward biased. In this way, the various transistors on a chip are isolated from each other. The transistor action, which consists of electrons being injected from the emitter into the base and ending up in the collector, occurs in the region indicated in the figure. The buried n^+ layer provides a low-resistance path from the collector region underneath the emitter to the collector contact on the right.

The circuit symbols for npn and pnp devices are indicated in Figure IV.2, where the various voltage differences and currents are indicated. From Kirchhoff's voltage law,

$$V_{CE} = V_{CB} + V_{BE} \tag{IV.1}$$

and from Kirchhoff's current law,

$$I_E = I_C + I_B \tag{IV.2}$$

The arrow in the emitter of the circuit symbol represents the direction of current flow in the active mode, thus indicating whether it is npn or pnp.

Figure IV.3 indicates two common circuit configurations for an npn BJT, along with the input and output voltages and currents. In (a), the base is common to both input and output and so is referred to as the *common base configuration*. In (b), the *common emitter configuration,* the emitter is common to both input and output.

Figure IV.4a and b show the I_C-V_{CE} characteristics for the npn and pnp transistors operating in the common emitter configuration. Qualitatively, they are similar to those of the FET. One key difference, however, is that the different curves correspond to different values of base *current,* whereas in the FET the varying parameter was the gate *voltage*. We will see why later.

With two junctions, each of which can be forward or reverse biased, there are four possible modes of operation, as indicated in Table IV.2. The analogous bias regimes for the FET are shown in the last column. The forward and saturation bias regimes are also shown in Figure IV.4.

These biasing modes are used in circuits in various ways, as shown in Figure IV.4c and d. In digital circuits, for example, a BJT is operated in two regions:

Table IV.2 Biasing Modes for a BJT

	Biasing polarity and typical values for Si BJTs		
Biasing mode	***E-B* junction**	***C-B* junction**	**FET analog**
Active	Forward (0.7 V)	Reverse (5 V)	Current saturation
Voltage saturation	Forward (0.7 V)	Forward (0.7 V)	Sublinear
Inverted	Reverse (5 V)	Forward (0.7 V)	Inverted
Cutoff	Reverse, zero, or weakly forward (0 V)	Reverse (5 V)	Cutoff

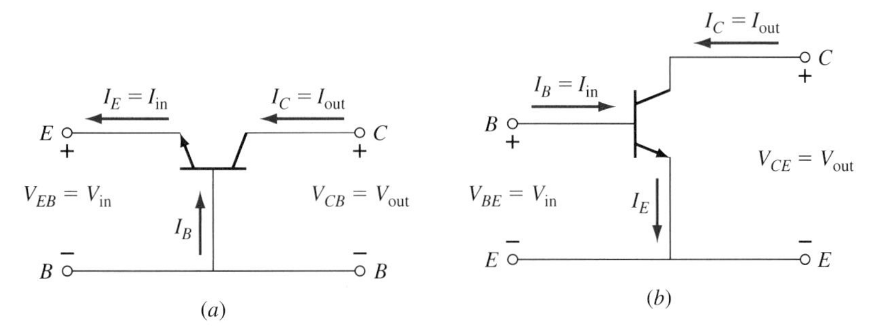

Figure IV.3 (a) Common base and (b) common emitter configurations for an npn BJT.

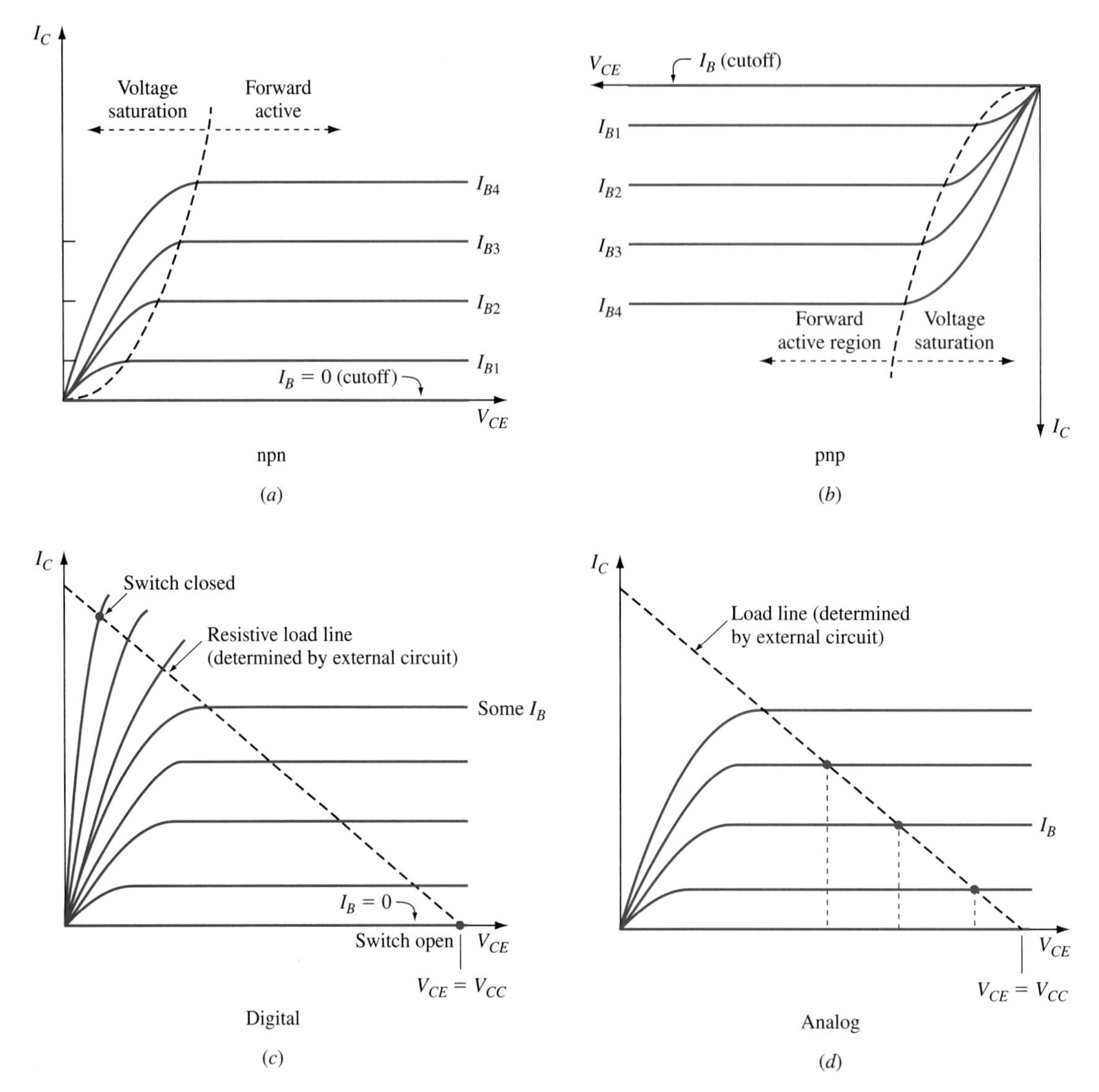

Figure IV.4 Typical I_C-V_{CE} curves for (a) an npn transistor and (b) a pnp transistor in the common emitter configuration. An npn BJT can be used in a digital circuit (c) or an analog circuit (d).

the low-voltage, high-current *saturation* mode, corresponding to the on state, and in the *cutoff* mode, which is a low-current, high-voltage off state. In the on state, both the emitter-base and base-collector junctions are forward biased, and in cutoff they are both reverse biased. The switching behavior of BJTs is discussed in Chapter 10. In analog circuits, the BJT is operated in the *forward active* mode. In this mode, the emitter-base junction is forward biased and the base-collector junction is reverse biased.

The electrical characteristics of a BJT are obtained by solving the continuity equations with appropriate boundary and initial conditions in each region of the device. These equations are given in Equations (3.63) and (3.64) and repeated here for zero optical generation (i.e., in the dark).

$$\frac{\partial n}{\partial t} = \frac{\partial \Delta n}{\partial t} = \frac{1}{q}\frac{\partial J_n}{\partial x} - \frac{\Delta n}{\tau_n} \tag{IV.3}$$

$$\frac{\partial p}{\partial t} = \frac{\partial \Delta p}{\partial t} = -\frac{1}{q}\frac{\partial J_p}{\partial x} - \frac{\Delta p}{\tau_p} \tag{IV.4}$$

where, in general, J_n or J_p is the sum of the respective drift and diffusion current densities.

$$J_n = J_{n\text{drift}} + J_{n\text{diff}}$$
$$J_p = J_{p\text{drift}} + J_{p\text{diff}}$$

To obtain the steady-state (dc) characteristics, the time-dependent terms are set to zero. For transient analysis they must be considered.

In Chapter 9, the dc model of BJT operation is developed. First, the ideal (prototype) BJT is considered, in which the doping is uniform in each region—emitter, base, and collector. Later, the effects of nonideal parameters are discussed, and the graded-doping base transistor is introduced. In Chapter 10, we examine the ac operation of a BJT. We introduce the Ebers-Moll model and the hybrid pi small-signal model. We examine BJT capacitance and transient behavior. Chapter 10 also describes some specific BJTs, including the double-poly (polysilicon) self-aligned bipolar junction transistor, BJT switching transistors, and BiMOS, a hybrid technology that involves both BJTs and FETs. This technology combines the advantages of FETs with those of BJTs. The Supplement to Part 4 examines various types of heterojunction bipolar transistors (HBTs), including uniformly doped, graded-base, and graded-composition HBTs. In addition to Si-based devices, compound semiconductor BJTs are considered, since some such materials provide increased performance. Four-layer switching devices are also briefly discussed. The application of SPICE to simple BJT circuits is briefly covered. ■

CHAPTER 9

Bipolar Junction Devices: Statics

9.1 INTRODUCTION

In the introduction to Part 4, we described the basic bipolar junction transistor (BJT). We saw that the current-voltage characteristics are similar to those of a field-effect transistor, in that the current flowing between two terminals is controlled by a control signal applied to the third terminal. In a FET the control signal is the voltage applied to the gate, but in a BJT the control signal is the current applied to the base. The physics of operation of the two types of transistors is completely different.

Recall that in a pn junction under forward bias, minority carriers are injected across the junction, and the number of carriers injected depends on the bias voltage. In an npn BJT in the active mode, the (forward-biased) base-emitter voltage controls the electron flux F_n injected from the emitter into the base. This flux is indicated in Figure 9.1. The electron flux is the number of electrons crossing an area per unit time. Current is equal to the product of the carrier charge and the flux. The electron current flows in the direction opposite to the electron flux, but hole current flows in the same direction as hole flux.

In realistic transistors, the base is currently made quite thin (tenths of a micrometer), while the emitter area dimensions of 1970-era BJTs were on the order of micrometers. A thin, or short, base is essential to the efficient operation of a BJT, as will be seen later.

Recall from Figure IV.1 that the transistor action occurs between the emitter, base, and collector under the emitter region. The actual transistor is much larger to permit electrical contact to the base and collector regions. For clarity in discussing the ideal transistor, the "box" diagram is used as indicated in Figure 9.2a, where the base thickness is exaggerated for clarity.

The primary current flow in an npn BJT results from electrons from the emitter contact flowing through the emitter, then through the base, then into the n and

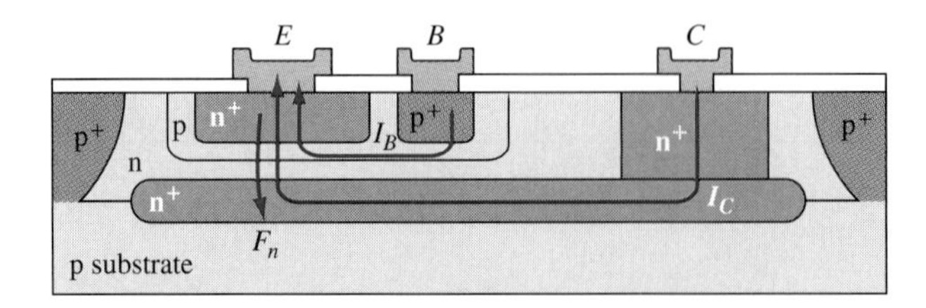

Figure 9.1 Illustration of electron flux F_n and currents of an npn BJT.

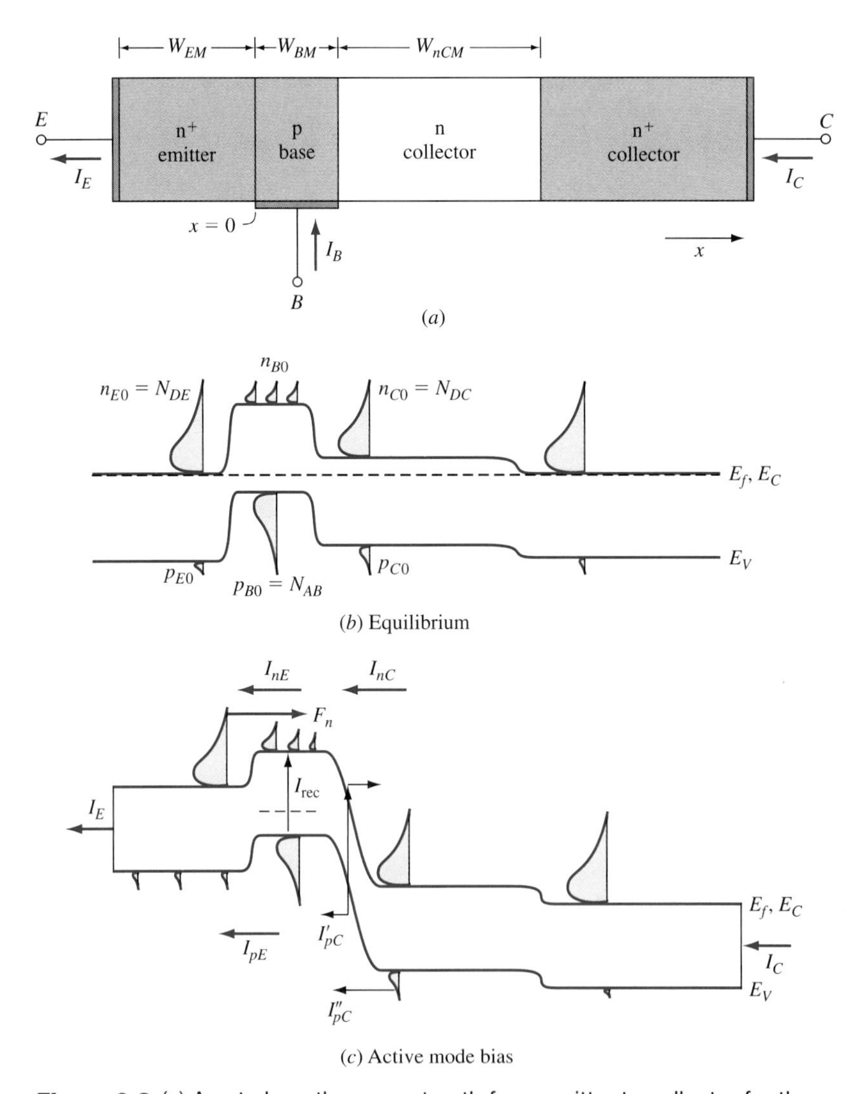

Figure 9.2 (a) A cut along the current path from emitter to collector for the npn device of Figure 9.1. (b) The equilibrium energy band diagram for a BJT with uniform doping in each region. (c) The energy band diagram for operation in the active mode.

n^+ regions of the collector and out the collector contact. The width of the emitter region (between the metallurgical junctions) is W_{EM}, that of the base region is W_{BM}, and that of the n collector region is W_{nCM}.

The best way to understand the physics of any semiconductor device is by use of the energy band diagram. The equilibrium energy band diagram is shown in Figure 9.2b. To construct this energy band diagram, we make the following assumptions:

1. The doping in the n^+ emitter is degenerate and the emitter is uniformly doped.
2. The doping in the p-type base is nondegenerate and uniform (except for the p^+ region under the base contact, which is not in this energy band diagram).
3. The n collector region is nondegenerate and uniformly doped.
4. The degenerate n^+ collector region is uniformly doped.
5. Impurity-induced band-gap narrowing is neglected.
6. The Fermi level coincides with the conduction band edge in n^+ emitter and n^+ collector.

We refer to a BJT with uniform doping in each region as a *prototype* transistor. The energy band diagram of this prototype BJT under normal (analog) operating conditions (active mode) is shown in Figure 9.2c. The emitter-base junction is forward biased, which permits electrons to be injected from the emitter into the base and holes to be injected from the base into the emitter. The base-collector junction is reverse biased. For n_E electrons injected into the base from the emitter, a small fraction recombine in the base and contribute to base current I_B. Most of the electrons injected from emitter into the base reach the collector-base junction and are collected and contribute to collector current I_C. Since $I_C < I_E$, the current gain $\alpha = I_C/I_E$ from emitter to collector is less than unity. However, most circuits use the common emitter configuration in which the input current is I_B and the output current is I_C. The current gain then is $\beta = I_C/I_B$, which is typically on the order of 100. Note that the electrons, which are minority carriers, flow across the base only by diffusion since there is no electric field in the base, or

$$\mathscr{E} = -\frac{1}{q}\frac{dE_{\text{vac}}}{dx} = -\frac{1}{q}\frac{dE_C}{dx} = 0$$

In Figure 9.2c the emitter current I_E consists of the current resulting from electron injection from emitter to base, I_{nE}, plus the hole current injected from base to emitter, I_{pE}:

$$I_E = I_{nE} + I_{pE} \tag{9.1}$$

Note that for this example, the recombination current within the E-B junction is neglected. This recombination current will be considered later.

The collector current I_C is composed of an electron component and a hole component. The electron current I_{nC} is the electron current resulting from the electrons that cross the base from the emitter and reach the collector. The hole current I_{pC} has two components: the hole current, I''_{pC} extracted from the collector into the base, and I'_{pC}, the current due to the electron-hole generation in the reverse-biased B-C junction. Thus

$$I_C = I_{nC} + I_{pC} \tag{9.2}$$

where $I_{pC} = I'_{pC} + I''_{pC}$.

The collector electron current I_{nC} is the emitter electron current injected into the base minus the current lost to recombination in the base:

$$I_{nC} = I_{nE} - I_{\text{rec}} \tag{9.3}$$

where I_{rec} is the current due to electron-hole recombination within the base.

The current into the base, I_B, is then

$$I_B = I_{pE} + I_{\text{rec}} - I_{pC} \tag{9.4}$$

Some general observations concerning the relative magnitudes of the various current components can be qualitatively determined.

Since the emitter is much more heavily doped than the base, from Chapter 5 [it can be seen, for example, from Equation (5.80)] we know that

$$I_{nE} \gg I_{pE}$$

Further, recalling that the base is made thin enough that its width is much less than an electron diffusion length, the probability of electrons recombining within the base is small, and

$$I_{\text{rec}} \ll I_{nE}$$

Thus

$$I_{nC} \approx I_{nE}$$

In the active mode, I_C is on the order of milliamperes, but from Chapter 5 the leakage current for a reverse-biased pn junction is several orders of magnitude smaller than this. Thus, I_{pC} is small and

$$I_C \approx I_{nC}$$

The relative magnitudes of the above currents are shown in Figure 9.3, where I_{pE}, I_{pC} and I_{rec} are exaggerated for clarity.

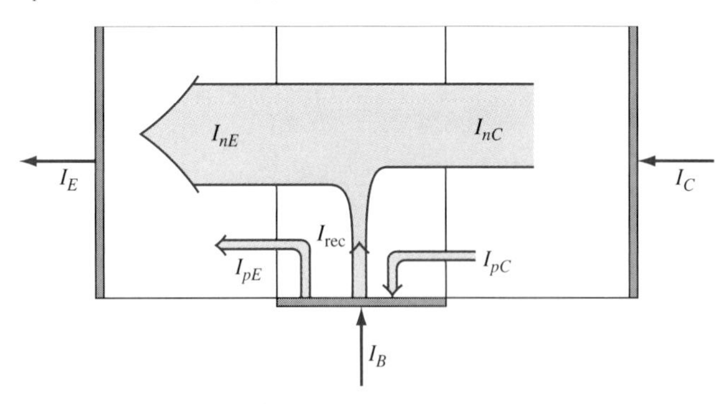

Figure 9.3 Schematic indicating the relative magnitudes (not to scale) of the various current components in an npn BJT operating in the active mode.

9.2 OUTPUT CHARACTERISTICS (QUALITATIVE)

A typical (idealized) family of curves for the output characteristics of an npn BJT operated in the *common base configuration* with input (emitter) current as a parameter is shown in Figure 9.4a, and for the *common emitter configuration* in Figure 9.4b with the input (base) current as a parameter. Figure 9.4c shows the output characteristics for the common emitter configuration with input voltage (V_{BE}), as a parameter.

The output characteristics for the common base configuration of Figure 9.4a are easily explained. First take the case of $I_E = 0$. With no emitter current, $I_C = -I_B$, and the current from the collector flows through the collector-base junction and out the base contact. This is then a simple diode and the I_C-V_{CB} characteristic of an np junction is obtained. The diode curve in this figure looks inverted compared with the way it is usually drawn. When the collector-base junction is forward biased, V_{CB} and I_C are negative and when it is under reverse bias, they are positive, according to convention.

Now let us consider what happens when the emitter-base junction is forward biased so that I_E is positive. Since the emitter is n type in this case, electrons are injected from the emitter into the base. Note that while electrons are majority carriers in the emitter, once they reach the base they become minority carriers. Normally those electrons would recombine in the p-type base region in an average time (lifetime) τ_n, or within a diffusion length L_n. The minority carrier lifetime is on the order of $\tau_n = 1\ \mu$s in Si, corresponding to L_n on the order of tens of micrometers. The base region is thin, however, and before the injected carriers recombine, most of them diffuse to the collector edge of the base junction. If the C-B junction is reverse biased, the field of this junction accelerates them into the collector. Thus, most of the emitter current contributes to collector current. The effect on the I_C-V_{CE} characteristics of Figure 9.4a is that the entire diode curve translates upward by an amount αI_E, where α is the fraction of I_E that contributes to I_C. This fraction is typically on the order of 99 percent.

For the common emitter configuration, the physics is the same but the I-V characteristics look different because now I_C is plotted versus V_{CE} with the varying parameter I_B. If the p base is made positive with respect to the emitter, thus forward biasing the base-emitter junction, electrons will be injected from the emitter into the base and holes from the base into the emitter. The injected holes recombine in the emitter or at the emitter contact and flow out the common emitter contact, but a small fraction of the electrons injected into the base will recombine and flow out the base contact. Most of the electrons diffuse across the thin base to be collected by the collector.

As can be seen from Figure 9.3, the base input current I_B (that is supplied by the base) has three components. It consists of hole current injected into the emitter I_{pE}, plus the electron-hole recombination current I_{rec}, minus the collector-base leakage current I_{pC}. Each of these is small. Thus, I_B is on the order of 1 percent of the injected electron current (supplied through the emitter contact). Most of the emitter electron current ends up at the collector, and we take the collector

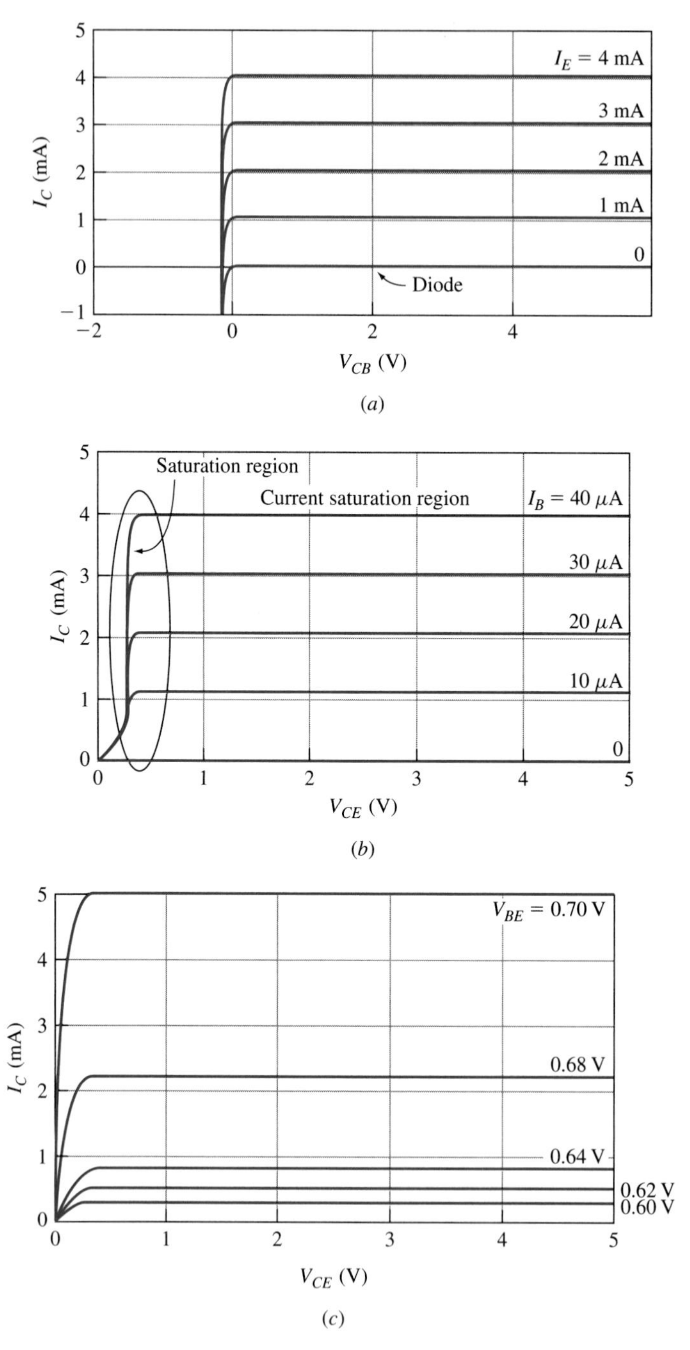

Figure 9.4 Idealized output characteristics of an npn transistor with input current as a parameter operating in (a) the common base configuration and (b) the common emitter configuration. (c) The output characteristics of the common emitter configuration with input voltage as a parameter.

current to be the output. The output current I_C is thus on the order of 100 times the input current I_B. This is shown in Figure 9.4b.

From Equation (IV.1), we see that for a given collector-emitter voltage (V_{CE}), applying a base-emitter voltage (V_{BE}) tends to reduce the collector-base voltage (V_{CB}), since their sum is a constant. At low V_{CE}, the base-emitter voltage may be larger than V_{CE}, which would make V_{CB} negative. This means that both base-emitter and base-collector junctions are forward biased and the device is said to operate in the *saturation region*; i.e., the collector-emitter voltage (V_{CE}) is saturated at a small value (about 0.2 to 0.3 V). At larger V_{CE}, the *current* saturates for a given value of I_B as indicated. This current saturation corresponds to the saturation region for FETs, but in BJTs the term *saturation* refers to the region circled in Figure 9.4b. In Figure 9.4a and b, where input current is the varying parameter between curves, the output currents are approximately proportional to the input currents. In (c) however, the output current increases exponentially with input voltage. Because of this nonlinearity, this latter representation is seldom used and will not be considered further here.

9.3 CURRENT GAIN

As indicated earlier, two common modes of operation of a BJT are the common base and common emitter configurations shown in Figure IV.3. In the common base circuit, the emitter current is taken as the input and the collector current is the output. The current gain for common base operation is

$$\alpha = \frac{I_{\text{out}}}{I_{\text{in}}} = \frac{I_C}{I_E} \tag{9.5}$$

This ratio is close to unity. A small fraction of the electrons injected into the base from the emitter are lost by recombination, and a small hole current is injected from the base to the emitter. In addition, as will be discussed later, the hole current from collector to base is negligible. The collector current is close to but smaller than the emitter current, so $I_C < I_E$ and $\alpha < 1$, typically about 0.99.

In the common emitter circuit, the base current is the input and the output is the collector current, so the common emitter current gain is

$$\beta = \frac{I_{\text{out}}}{I_{\text{in}}} = \frac{I_C}{I_B} \tag{9.6}$$

Since $I_E = I_C + I_B$, the common emitter current gain β can be expressed as

$$\beta = \frac{I_C}{I_B} = \frac{I_C}{I_E - I_C} = \frac{\dfrac{I_C}{I_E}}{1 - \dfrac{I_C}{I_E}} = \frac{\alpha}{1 - \alpha} \tag{9.7}$$

or

$$\alpha = \frac{\beta}{\beta + 1} \tag{9.8}$$

However, since $\alpha < 1$, for large β it is desired to have α approach unity.

To improve our understanding of these current gains, we can break up the various currents into their electron and hole components. First, let us consider α. Multiplying and dividing Equation (9.5) by I_{nE} and I_{nC} and rearranging gives

$$\alpha = \frac{I_{nE}}{I_E}\frac{I_{nC}}{I_{nE}}\frac{I_C}{I_{nC}} \tag{9.9}$$

where I_{nE} is the electron current produced by the electron flux injected from emitter into the base and I_{nC} is the current that results from electrons reaching the collector, as shown in Figure 9.2c.

We can look at each of the terms in Equation (9.9) individually. The first term is called the injection efficiency γ, which is defined as the fraction of the emitter current that is due to electron injection into the base.

$$\gamma = \frac{I_{nE}}{I_E} = \frac{I_{nE}}{I_{nE} + I_{pE}} = \frac{1}{1 + \dfrac{I_{pE}}{I_{nE}}} \tag{9.10}$$

where I_{pE} is the hole current injected from base to emitter.

Ideally, γ is close to unity, which is accomplished by making $I_{pE} \ll I_{nE}$.

The transport efficiency α_T (often called the current transport factor) is the fraction of the electrons injected from the emitter into the base that reach the collector.

$$\alpha_T = \frac{I_{nC}}{I_{nE}} = \frac{I_{nE} - I_{\text{rec}}}{I_{nE}} = 1 - \frac{I_{\text{rec}}}{I_{nE}} \tag{9.11}$$

where I_{rec} is the electron recombination current in the base. The value of α_T approaches unity for $I_{\text{rec}} \ll I_{nE}$.

The last term is the collection efficiency M:

$$M = \frac{I_C}{I_{nC}} = \frac{I_{nC} + I_{pC}}{I_{nC}} = 1 + \frac{I_{pC}}{I_{nC}} \tag{9.12}$$

where I_{pC} is the hole leakage current from collector to base in the reverse-biased B-C junction. This collection efficiency is often called the *collection multiplication factor*. From Equation (9.9) then,

$$\alpha = \gamma\alpha_T M \tag{9.13}$$

To determine values for α and β then, we must relate the quantities γ, α_T, and M to the currents I_{pE}, I_{nE}, I_{rec}, I_{nC}, and I_{pC}. To do this analytically requires a model for the device.

9.4 MODEL OF A PROTOTYPE BJT

Now we will go through the operation of the BJT in the active mode more quantitatively, starting with a simple but not entirely realistic model. Once we understand the basic operating principles, we will add refinements.

We take as our example the Si npn BJT considered before in Figure IV.1 and in Figure 9.1. It consists of a degenerately doped n^+ emitter (E), a nondegenerate p base (B), and a collector region (C) that includes both nondegenerate n and degenerate n^+ regions. In this prototype transistor, all regions are assumed to be uniformly doped.

Notice that the metal terminal contacts are made to degenerate n^+ emitter and n^+ collector regions and to a degenerate p^+ region in the base. As discussed in Chapter 6, these metal-degenerate semiconductor contacts form Schottky barriers thin enough to permit tunneling and thus ensure low-resistance contacts.

The n collector region is relatively lightly doped to reduce the collector-base capacitance and to ensure that the breakdown voltage of the collector-base junction is high enough that breakdown doesn't occur (recall that the base-collector junction is reverse biased in forward active bias). The n^+ collector region (buried layer) is used to provide a low-resistance path in the collector. The n^+ *sinker* between the collector contact and the n^+ collector region also reduces the collector resistance.

To obtain quantitative results for the currents in each region, the continuity equations must be solved with the appropriate boundary conditions. To obtain the steady-state (dc) currents, the continuity equations [Equations (IV.3) and (IV.4)] become

$$\frac{\partial J_n}{\partial x} = q\frac{\Delta n}{\tau_n} \tag{9.14}$$

$$\frac{\partial J_p}{\partial x} = -q\frac{\Delta p}{\tau_p} \tag{9.15}$$

For the prototype transistor being considered, since the emitter, base, and collector are each uniformly doped, at equilibrium there are no electric fields in these regions. Under forward bias only miniscule fields exist and minority carrier flow is primarily by diffusion. In the p-type base, the influx of minority carriers, electrons in this case, increases the minority carrier concentration near the junction. To maintain neutrality, an equal number of holes are drawn into the base via the base contact. For the low-injection condition considered here, the excess hole concentration is negligible compared with the equilibrium hole concentration resulting from ionized acceptors. Similarly, the holes injected into the emitter have negligible effect on the electron concentration there. The currents of primary interest, then, as discussed in Chapter 5, are the minority carrier diffusion currents. In general, to find these currents we must solve the continuity equation in each region using the appropriate boundary conditions, not always an easy task. However, since the doping in each region is uniform, $\mathscr{E} \approx 0$, only diffusion current needs to be considered for minority carriers.

Figure 9.5 repeats Figure 9.2a with the emitter-base and base-collector depletion regions exaggerated. Because, for an npn BJT, positive current is from collector to emitter, in the negative x direction, the currents obtained from solutions of the continuity equations will result in negative quantities, i.e., in the

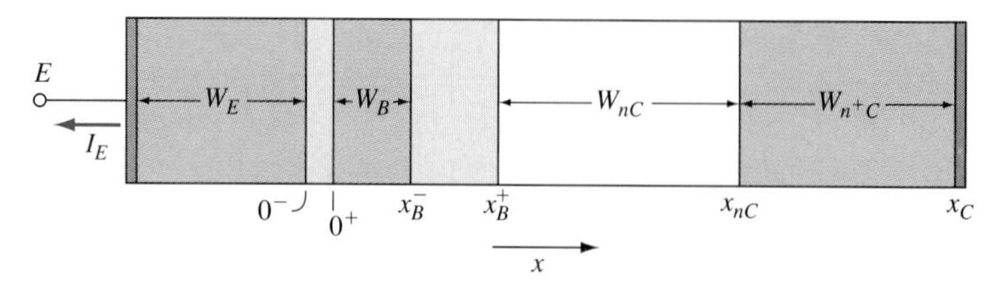

Figure 9.5 Widths and coordinates for simple prototype model. The depletion region widths have been exaggerated.

negative x direction. The emitter edge of the E-B junction is designated $x = 0^-$ and the base edge is 0^+. Similarly the base edge of the B-C junction is designated x_B^- and the collector edge is x_B^+. The width of the quasi-neutral region in the emitter is W_E; of the base, W_B, and of the n collector, W_{nC}.

In the emitter the diffusion current density is

$$J_{pE} = -qD_{pE}\frac{d\Delta p_E}{dx} \tag{9.16}$$

and Equation (9.15) becomes

$$D_{pE}\frac{d^2\Delta p_E}{dx^2} = \frac{\Delta p_E}{\tau_{pE}} \tag{9.17}$$

with the boundary conditions

$$\Delta p_E(x = 0^-) = p_{E0}(e^{qV_{BE}/kT} - 1)$$
$$\Delta p_E(x = x_E) = 0$$

where p_{E0} is the equilibrium hole concentration in the emitter. In the base, the electron diffusion current density is

$$J_{nB} = qD_{nB}\frac{d\Delta n_B}{dx} \tag{9.18}$$

and Equation (9.14) becomes

$$D_{nB}\frac{d^2\Delta n_B}{dx^2} = -\frac{\Delta n_B}{\tau_{nB}} \tag{9.19}$$

with the boundary conditions

$$\Delta n_B(0^-) = n_{B0}(e^{qV_{BE}/kT} - 1)$$
$$\Delta n_B(x_B^-) = n_{B0}(e^{qV_{BC}/kT} - 1)$$

In the collector, the hole diffusion current density is

$$J_{pC} = -qD_{pC}\frac{d\Delta p_C}{dx} \tag{9.20}$$

and

$$D_{pC}\frac{d^2\Delta p_C}{dx^2} = \frac{\Delta p_C}{\tau_{pC}} \tag{9.21}$$

with the boundary conditions

$$\Delta p(x_B^+) = p_{C0}(e^{qV_{BC}/kT} - 1)$$
$$\Delta p(x_{nC}) \approx 0$$

The last boundary condition assumes that $W_C \gg L_{pC}$, and therefore all the excess minority carriers have recombined by the time they reach the end of the collector region.

To calculate the diffusion current in each region, Δp_E, Δn_B, and Δp_C are solved with appropriate boundary conditions and used in Equations (9.16), (9.18), and (9.20) to determine the minority carrier diffusion currents in each region. The results in general involve hyperbolic functions. [1, 2] The task can be simplified, however, with the above assumptions and for operation in the active mode.

To find the current gain parameters $\alpha = \gamma\alpha_T M$ and $\beta = \alpha/(1-\alpha)$ we must solve for I_{nB}, I_{pE}, I_{nC}, I_{rec}, and I_{pC}. To obtain a quantitative result, we make the following additional assumptions appropriate for most modern BJTs:

1. The width of the base region W_B, measured between the edges of the depletion regions, is much smaller than the diffusion length L_{nB} for electrons in the base, or $W_B \ll L_{nB}$.
2. The width of the emitter region $W_E \ll L_{pE}$, where L_{pE} is the diffusion length for holes in the emitter.
3. The concentration of electrons injected into the base is small enough that, everywhere in the base, the electron concentration is always much less than the hole concentration, or $n_B \ll p_B$. This is the "low-injection" condition. We will consider high injection in a later model.

The energy band diagram of Figure 9.2c for operation in the active mode is redrawn in Figure 9.6, where the E-B and B-C transition regions are indicated. For simplicity, the n$^+$ collector region is not shown. Note that the minority carrier concentration in the base goes to zero at the collector edge (where the minority carriers are extracted because of the reverse bias on that junction), and the excess minority carrier concentration injected into the emitter decays to zero at the emitter edge because of the presence of the contact (not shown).

9.4.1 COLLECTION EFFICIENCY *M*

From Equation (9.12), the collection efficiency M is

$$M = \frac{I_{nC} + I_{pC}}{I_{nC}} = 1 + \frac{I_{pC}}{I_{nC}} \tag{9.22}$$

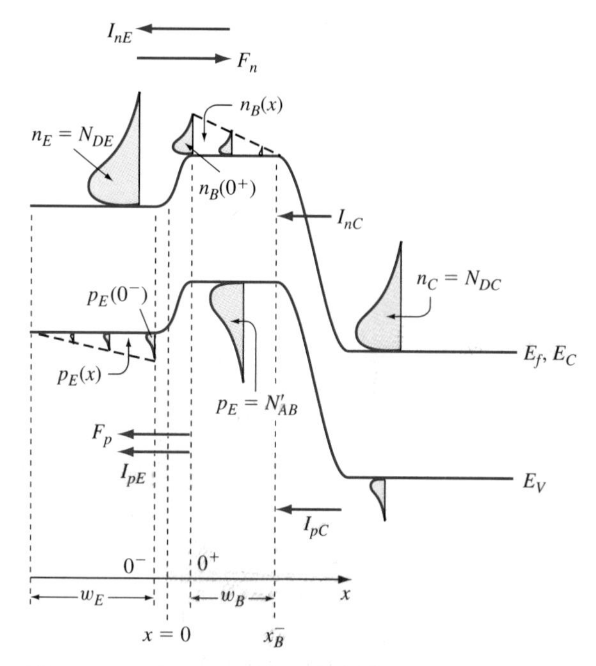

Figure 9.6 Active mode energy band diagram of a prototype npn BJT showing currents used to determine transport properties.

where I_{pC} consists of the hole current diffusing from collector to base plus the hole generation current in the collector-base junction. Electron-hole pairs are continually being generated thermally in the junction, and the electric field separates them and drives them away from the junction. Except when the reverse bias is large enough that significant carrier multiplication occurs (approaching breakdown, a situation to be avoided), I_{pC} is several orders of magnitude less than I_{nC}. Thus, for normal operation

$$M \approx 1 \tag{9.23}$$

and

$$\alpha \approx \gamma \alpha_T \tag{9.24}$$

Finally, recall that in Chapter 5 we found that for a reverse-biased junction, the generation current was much greater than the diffusion current. Thus, for the base-collector junction, Equation (9.20) need not be solved.

9.4.2 INJECTION EFFICIENCY γ

From Equation (9.10)

$$\gamma = \frac{1}{1 + \dfrac{I_{pE}}{I_{nE}}} \tag{9.10}$$

However, I_{nE} is the injected electron current across the forward-biased E-B junction, and thus can be expressed as

$$I_{nE} = qA_E D_{nB} \left.\frac{dn_B}{dx}\right|_{x=0^+} \tag{9.25}$$

where A_E is the area of the emitter-base junction, D_{nB} is the diffusion constant for electrons in the base, and $x = 0^+$ represents the position of the base edge of the emitter-base transition region.

To find an expression for dn_B/dx, we note that excess carriers are injected into the base at the base-emitter junction and $n_B(0^+)$ is the electron concentration at the emitter edge of the base ($x = 0^+$), but at the collector end of the base, the carriers are extracted. Since the base is much shorter than a minority carrier diffusion length ($W_B \ll L_{nB}$), the electron-hole recombination in the base is small, and under these conditions, to first approximation, the excess carrier concentration in the base varies linearly from $n_B(0^+)$ to $n_B(x_B^-) \approx 0$. Thus

$$n_B(x) = n_B(0^+)\left[1 - \frac{x - 0^+}{W_B}\right]$$

as indicated by the straight dashed line in the base region conduction band in Figure 9.6. Therefore

$$\left.\frac{dn_B}{dx}\right|_{x=0^+} = -\frac{n_B(0^+)}{W_B}$$

and Equation (9.25) becomes

$$I_{nE} = -\frac{qA_E D_{nB} n_B(0^+)}{W_B} \tag{9.26}$$

Similarly, since the emitter is also thin, such that $W_E \ll L_{pE}$, we can write at the emitter end of the E-B junction

$$I_{pE} = I_{pB} = -qA_E D_{pE} \left.\frac{dp}{dx}\right|_{x=0^+} \approx -\frac{qA_E D_{pE} p_E(0^-)}{W_E} \tag{9.27}$$

where D_{pE} is the diffusion constant for holes in the emitter. Then substituting these expressions into Equation (9.10) and simplifying gives

$$\gamma \approx \frac{1}{1 + \dfrac{p_E(0^-)}{n_B(0^+)}\dfrac{D_{pE}}{D_{nB}}\dfrac{W_B}{W_E}} \tag{9.28}$$

To maximize the injection efficiency, from Equation (9.28) one should minimize the second term in the denominator. One has little control of D_{pE} or D_{nB}, but one can minimize $p_E(0^-)/n_B(0^+)$, the ratio of the minority carrier concentrations. In other words, one should dope the emitter more heavily than the base. In addition, W_B/W_E should also be made small, or the base should be made thinner than the emitter.

9.4.3 BASE TRANSPORT EFFICIENCY α_T

The electron collector current I_{nC} is equal to the emitter electron injection current, less any that is lost to recombination in the base. Recalling that the recombination current is the ratio of the electron charge in the base $qA_E\langle n_B\rangle W_B$ to the minority carrier lifetime τ_{nB} [refer to Equation (3.59)], we have

$$I_{\text{rec}} = \frac{qA_E\,\langle n_B\rangle\,W_B}{\tau_{nB}} \tag{9.29}$$

where $\langle n_B\rangle$ is the average electron concentration in the base. Then

$$\langle n_B\rangle \approx \frac{n_B(0^+)}{2} \tag{9.30}$$

With the aid of Equations (9.26), (9.29), and (9.30), Equation (9.11) becomes

$$\alpha_T \approx 1 - \frac{W_B^2}{2D_{nB}\tau_{nB}} = 1 - \frac{W_B^2}{2L_{nB}^2} \tag{9.31}$$

where we have used $D_{nB}\tau_{nB} = L_{nB}^2$. Therefore, to maximize the current transport factor α_T, the base width W_B should be small. Note that Equation (9.31) is for an npn transistor, and that L_{nB} is the minority carrier diffusion length *in the base*. For a pnp transistor, the equation is the same except that L_{pB} would be used.

It is of interest to calculate β for an npn transistor. We will do two examples below. In Example 9.1, a nondegenerate emitter will be considered. In Example 9.2, these results will be adapted to that of a somewhat more realistic BJT with a degenerately doped emitter. In both cases it is assumed that each region is uniformly doped such that the diffusion current predominates and relations for γ [Equation (9.28)] can be used.

EXAMPLE 9.1

Find β for a bipolar junction transistor with a nondegenerate emitter. Assume that the emitter, base, and collector are noncompensated and that

$$N'_{DE} = N_{DE} = 2\times 10^{18}\ \text{cm}^{-3}$$
$$N'_{AB} = N_{AB} = 10^{16}\ \text{cm}^{-3}$$
$$N'_{DC} = N_{DC} = 10^{15}\ \text{cm}^{-3}$$
$$W_E = 0.2\ \mu\text{m}$$
$$W_B = 0.1\ \mu\text{m}$$

■ **Solution**

To find β we must first determine γ, α_T, and α. From Equation (9.28)

$$\gamma \approx \frac{1}{1 + \dfrac{p_E(0^-)}{n_B(0^+)}\dfrac{D_{pE}}{D_{nB}}\dfrac{W_B}{W_E}} \tag{9.28}$$

To find the first factor in the denominator, we recall that for a given value of V_{BE},

$$p_E(0^-) = p_{E0}e^{qV_{BE}/kT} \tag{9.32}$$

$$n_B(0^+) = n_{B0}e^{qV_{BE}/kT} \tag{9.33}$$

where p_{E0} is the equilibrium hole concentration in the emitter and n_{B0} is the equilibrium concentration of electrons in the base. Taking the ratio, we obtain

$$\frac{p_E(0^-)}{n_B(0^+)} = \frac{p_{E0}}{n_{B0}} \tag{9.34}$$

Since both regions are considered to be nondegenerate

$$p_{E0}n_{E0} = n_i^2 \tag{9.35}$$

$$n_{B0}p_{B0} = n_i^2 \tag{9.36}$$

Substituting $n_{E0} = N'_{DE}$ and $p_{B0} = N'_{AB}$, we have

$$p_{E0} = \frac{n_i^2}{N'_{DE}}$$

$$n_{B0} = \frac{n_i^2}{N'_{AB}}$$

and

$$\frac{p_E(0^-)}{n_B(0^+)} = \frac{N'_{AB}}{N'_{DE}} = \frac{10^{16}}{2 \times 10^{18}} = \frac{1}{200}$$

Next, we determine D_{pE}/D_{nB}. From Figure 3.11, for the doping levels given, the minority carrier diffusion coefficients are

$$D_{pE} = 5.5\,\text{cm}^2/\text{s}$$

$$D_{nB} = 32\,\text{cm}^2/\text{s}$$

and thus $D_{pE}/D_{nB} = 0.17$.

The base and emitter widths W_B and W_E are given, so

$$\gamma = \frac{1}{1 + \dfrac{p_E(0^-)}{n_B(0^+)}\dfrac{D_{pE}}{D_{nB}}\dfrac{W_B}{W_E}} = \frac{1}{1 + \dfrac{1}{200} \times 0.17 \times \dfrac{0.1}{0.2}} = 0.9996$$

Next, the transport efficiency α_T can be found from Equation (9.31). For $N_{AB} = 10^{16}\,\text{cm}^{-3}$, from Figure 3.23, $L_{nB} = 400\,\mu\text{m}$ (remember, it is the minority carrier diffusion length in the base) and

$$\alpha_T = 1 - \frac{1}{2}\left(\frac{0.1}{400}\right)^2 = 0.9999999 \approx 1$$

As an aside, we observe that our assumption that the base is much shorter than a minority carrier diffusion length is met, since $W_B = 0.1\,\mu\text{m}$ and $L_{nB} = 400\,\mu\text{m}$. Similarly, from

Figure 3.23, the minority carrier diffusion length in the emitter is $L_{pE} = 10\,\mu$m, and the emitter length of 0.2 μm is short compared with this and so the assumption that $W_E \ll L_{pE}$ is reasonably valid. Continuing with our example, from Equation (9.24), we have for $\alpha = \gamma\alpha_T \approx \gamma$. Then, from Equation (9.7)

$$\beta = \frac{\alpha}{1-\alpha} \approx \frac{\gamma}{1-\gamma} = \frac{0.9996}{1-0.9996} = 2500$$

This result is clearly much larger than typical values of β, which are on the order of 100 to 200 for a Si npn BJT. One reason is that in practical transistors, the emitter and base regions are more heavily doped than in the previous example. In the degenerately doped n-type emitter, the conduction band edge is lowered by the band-gap narrowing effect discussed in Chapter 2. This is shown in Figure 9.7. Thus, electrons being injected into the base from the emitter face a slightly larger energy barrier than was accounted for in Example 9.1. The barrier for holes is essentially unaffected, however, since neither the emitter nor the base have degenerate p-type doping in this example.[1] The increased barrier for electrons reduces the injection of electrons, but the hole injection remains the same. This, in turn, reduces the injection efficiency γ [Equation (9.10)]. Since the injection varies exponentially with the barrier height, the impact is significant. In Example 9.1, the injection efficiency was the most important factor in the value of β. Note that the conduction band edge in the p-type base is unaffected.

Let us now consider a somewhat more realistic example and include the effects of band-gap narrowing.

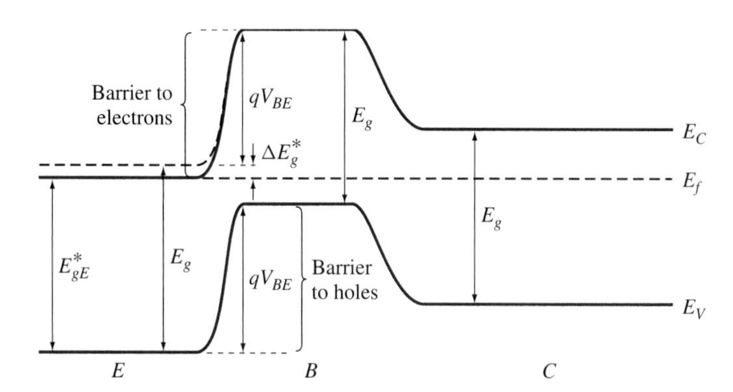

Figure 9.7 The band-gap narrowing in the heavily doped emitter results in a higher barrier to electrons than to the holes at the emitter-base junction.

[1]It is often the case in modern transistors that the base is doped heavily enough that its band gap is also affected.

EXAMPLE 9.2

We consider a degenerate emitter. Assume again an npn prototype transistor of the same geometry as Example 9.1 in which each region is uniformly doped, but now the emitter is degenerate. Let $N'_{DE} = 5 \times 10^{19}\ \text{cm}^{-3}$, and to keep the N'_{DE}/N'_{AB} ratio the same as in the previous example, let $N'_{AB} = 2.5 \times 10^{17}\text{cm}^{-3}$.

■ **Solution**

The injection efficiency is still given by Equation (9.28):

$$\gamma \approx \frac{1}{1 + \dfrac{p_E(0^-)}{n_B(0^+)} \dfrac{D_{pE}}{D_{nB}} \dfrac{W_B}{W_E}}$$

To determine γ we must again find D_{pE}/D_{nB} and $p_E(0^-)/n_B(0^+)$, but in this case, to find the $p_E(0^-)/n_B(0^+)$ ratio we must consider the case of the degenerate emitter.

As in the previous case,

$$p_E(0^-) = p_{E0}e^{qV_{BE}/kT}$$

$$n_B(0^+) = n_{B0}e^{qV_{BE}/kT}$$

$$\frac{p_E(0^-)}{n_B(0^+)} = \frac{p_{E0}}{n_{B0}}$$

and in the nondegenerate base, as before

$$n_{B0} = \frac{n_i^2}{N'_{AB}} \tag{9.37}$$

But, in the degenerate emitter, from Equation (2.111),

$$n_{E0}p_{E0} = \frac{N'_{DE}}{N_C}e^{\Delta E_g^*/kT} \times n_i^2 \tag{9.38}$$

where ΔE_g^* is the impurity-induced apparent band-gap narrowing, and N_C is the effective density of states in the conduction band. Since $n_{E0} = N'_{DE}$,

$$p_{E0} = \frac{e^{\Delta E_g^*/kT}}{N_C} \times n_i^2$$

and

$$\frac{p_E(0^-)}{n_B(0^+)} = \frac{N'_{AB}}{N_C}e^{\Delta E_g^*/kT}$$

The injection efficiency in this case is

$$\gamma \approx \frac{1}{1 + \dfrac{p_E(0^-)}{n_B(0^+)} \dfrac{D_{pE}}{D_{nB}} \dfrac{W_B}{W_E}} = \frac{1}{1 + \dfrac{N'_{AB}}{N_C}e^{\Delta E_g^*/kT} \dfrac{D_{pE}}{D_{nB}} \dfrac{W_B}{W_E}} \quad \text{emitter degenerate} \tag{9.39}$$

Since $N_C = 2.86 \times 10^{19}\ \text{cm}^{-3}$ and, from Figure 2.25, $\Delta E_g^* = 0.08$ eV, we have, using $D_{pE} = 3.3\ \text{cm}^2/\text{s}$ and $D_{nB} = 15\ \text{cm}^2/\text{s}$ from Figure 3.11,

$$\gamma = \frac{1}{1 + \dfrac{2.5 \times 10^{17}}{2.86 \times 10^{19}} e^{0.08/0.026} \times \dfrac{3.3}{15} \times \dfrac{1}{2}} = 0.980$$

The value of α_T can be obtained from Equation (9.31) as before, but for this base doping, $L_n = 80\ \mu\text{m}$ (note that the assumption that $W_B \ll L_{nB}$ is still valid), and

$$\alpha_T = 1 - \frac{1}{2}\left(\frac{0.1}{80}\right)^2 = 0.999999$$

Again $\alpha_T \approx 1$ and

$$\alpha \approx \gamma$$

Then

$$\beta = \frac{\alpha}{1-\alpha} = \frac{\gamma}{1-\gamma} = 49$$

which is on the order of the expected value.[2] Note that, although we changed the doping in this example, we kept the ratio of N'_{DE}/N'_{AB} the same. If band-gap narrowing had been neglected with this new doping, we would have obtained a result similar to that in Example 9.1, modified only by a slightly different value of D_{pE}/D_{nB}.

The main factor in reducing the calculated value of β in Example 9.2 from that calculated in the previous example is the consideration of the impurity-induced apparent band-gap narrowing, ΔE_g^*. This effect cannot be ignored in practical transistors. In the above example, the effect of band-gap narrowing in the base is neglected since the base is nondegenerate and the effect is small. If this effect is taken into account, Equation (9.39) becomes

$$\gamma = \frac{1}{1 + \dfrac{N'_{AE}}{N_C} e^{\Delta E^*_{gBE}/kT} \dfrac{D_{pE}}{D_{nB}} \dfrac{W_B}{W_E}} \qquad \text{base and emitter degenerate} \qquad (9.40)$$

where

$$\Delta E^*_{gBE} = E^*_{gB} - E^*_{bE} \qquad (9.41)$$

and E^*_{gB} is the apparent band gap of the base and E^*_{gE} is that of the emitter. Thus ΔE^*_{gBE} is the difference in the (apparent) band gaps of base and emitter.

EXAMPLE 9.3

Show that for a prototype transistor, β can be approximated by

$$\beta \approx \frac{n_B(0^+)}{p_E(0^-)} \frac{D_{nB}}{D_{pE}} \frac{W_E}{W_B}$$

[2] It is somewhat low compared with realistic values on the order of 100 to 200. Section 9.5 shows how the nonuniform doping in real BJTs tends to raise the value of β.

■ **Solution**

Since $\beta \approx \gamma/(1-\gamma)$, $\gamma = \beta/(1+\beta)$. From Equation (9.28), letting

$$Z = \frac{p_E(0^-)}{n_B(0^+)} \frac{D_{pE}}{D_{nB}} \frac{W_B}{W_E}$$

we can write $\gamma = 1/(1+Z)$, and

$$\beta \approx \frac{\dfrac{1}{1+Z}}{1 - \dfrac{1}{1+Z}} = \frac{\dfrac{1}{1+Z}}{\dfrac{Z}{1+Z}} = \frac{1}{Z}$$

and

$$\beta \approx \frac{n_B(0^+)}{p_E(0^-)} \frac{D_{nB}}{D_{pE}} \frac{W_E}{W_B}$$

Note that, for a nondegenerate emitter,

$$\beta \approx \frac{N'_{DE}}{N'_{AB}} \frac{D_{nB}}{D_{pE}} \frac{W_E}{W_B} \tag{9.42}$$

and, for a degenerate emitter,

$$\beta \approx \frac{N_C}{N'_{AB}} \frac{D_{nB}}{D_{pE}} \frac{W_E}{W_B} e^{-\Delta E^*_{gBE}/kT} \tag{9.43}$$

Note that the apparent bandgap narrowing can be determined from the variation of β with temperature.

It should be emphasized that the model for a prototype BJT is overly simplified. In most BJTs, the doping levels in each region are not constant. As we will see in the next section, the base impurity grading tends to increase the value of β over that calculated from the prototype model in which band-gap narrowing is considered.

9.5 DOPING GRADIENTS IN BJTs

The prototype transistor we considered had uniform doping in each region. We now consider a more realistic bipolar junction transistor. We take as an example an npn BJT fabricated in the *bipolar-CMOS* (BiCMOS) technology.[3] Its doping profile, of early 1980 BJT vintage, and which we discussed in Chapter 4, is shown again in Figure 9.8a. The profile, with x positive from left to right and $x = 0$ at the emitter surface, was measured by a technique known as *secondary-ion mass spectroscopy* (SIMS). The impurity concentration is plotted vertically on a log scale, and the horizontal axis is distance into the material. The base region is doped with boron. The n-type dopants (phosphorus, antimony, and arsenic) are used to dope the emitter and collector. Because the ion implants used

[3]In this technology, n-channel and p-channel MOSFETs and npn and pnp BJTs are fabricated in the same chip.

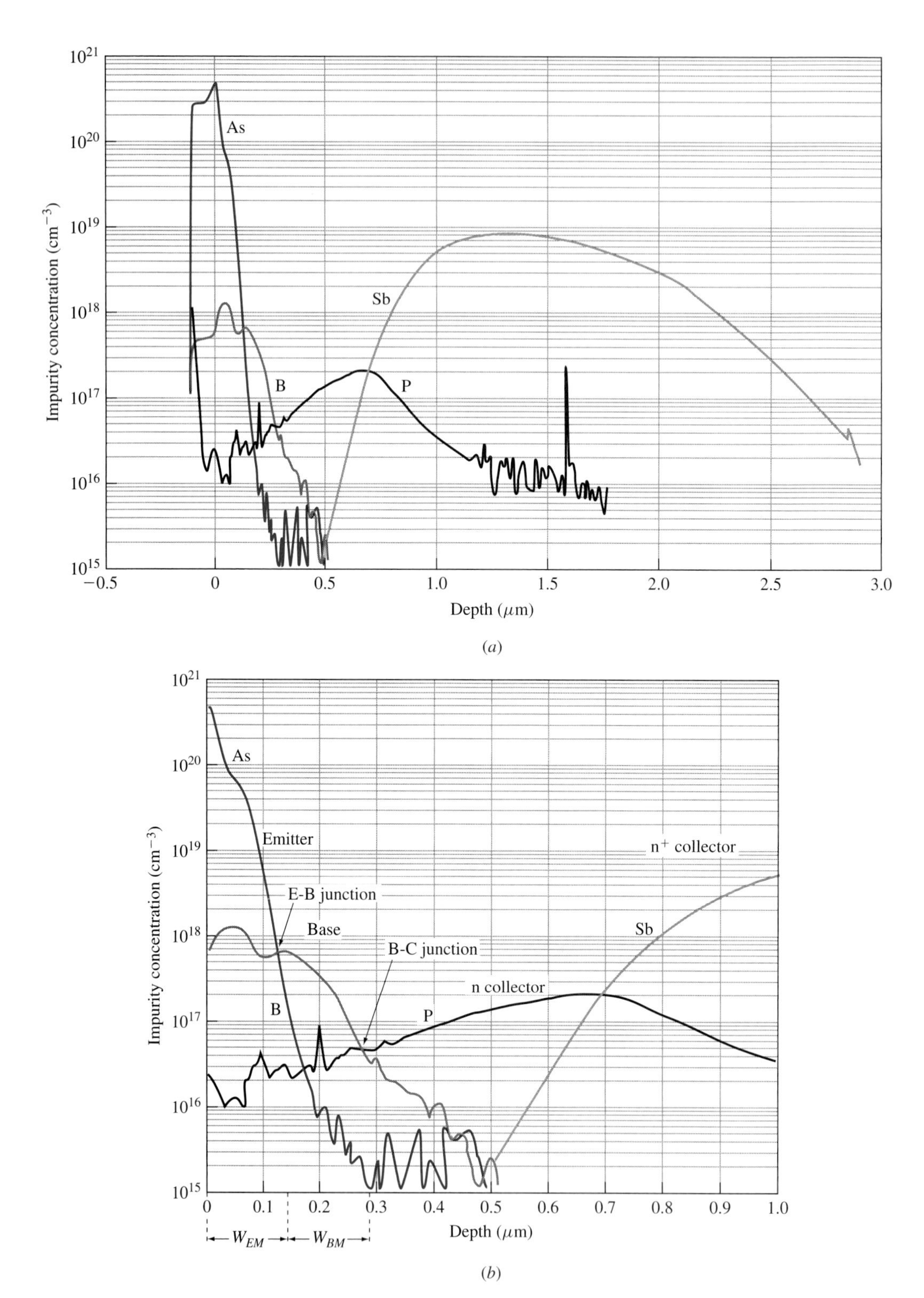

Figure 9.8 SIMS data. (a) Measured impurity concentration profile for an npn BJT. (b) Simplified donor (solid line) and acceptor (dashed line) profiles.

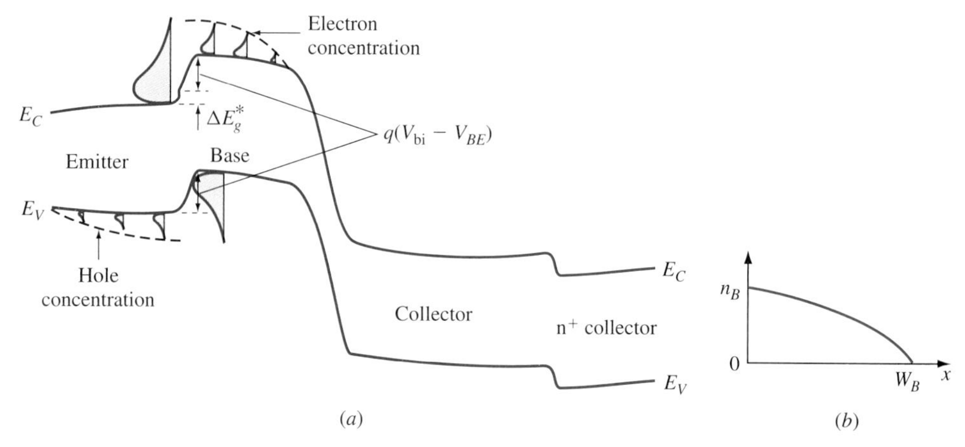

Figure 9.9 (a) Approximate active mode energy band diagram for the BJT of Figure 9.8 considering band-gap shrinkage. (b) The resultant electron distribution in the base.

to create this profile were made through a surface oxide layer, some dopants remain in the oxide region ($x < 0$).

This profile is replotted on an expanded scale in Figure 9.8b for the region of the active transistor. Where the donor concentration exceeds the acceptor concentration, the region is n type. The emitter-base and base-collector metallurgical junctions are at the locations where $N_A = N_D$. The metallurgical emitter width is $W_{EM} = 0.13\,\mu\text{m}$ and the metallurgical base width is $W_{BM} = 0.14\,\mu\text{m}$.

It can be seen from Figure 9.8b that the doping profiles in the emitter, base, and collector are all nonuniform, which complicates the analysis. Because of the impurity gradients, electric fields exist in each region, and the current flow is by a combination of drift and diffusion. These added complications are not amenable to hand calculations, and device simulation programs are required for analysis.[4] We will, however, discuss the physics of effects of the graded doping, and present some results.

The doping gradients in the emitter, base, and collector cause the conduction band and valence band edges to bend, producing internal electric fields as discussed in Chapter 4. These fields must be considered. An energy band diagram of this device is shown in Figure 9.9a, in which the variation in doping is taken into account. This figure is for operation in the active mode. The emitter-base junction is forward biased by an amount V_{BE}. The collector-base junction is reverse biased.

From the energy band diagram we can see:

1. Because of the doping dependence of the band-gap narrowing in the emitter (primarily a decrease in E_C), an effective field exists for electrons, $\mathscr{E}_e^* = (1/q)/(dE_C/dx)$. The conduction band edge bends closer to the valence band edge where the doping is heaviest, near the surface.
2. The effective field for holes in the emitter, $\mathscr{E}_h^* = (1/q)/(dE_V/dx)$, is in the opposite direction as that for electrons.

[4]The nominal common emitter gain for this device is $\beta = 90$.

3. In the base, an electric field exists that accelerates electrons toward the collector.
4. Because of the relatively small doping in the base and the corresponding small band-gap narrowing, here the slopes of the conduction and valence band edges are equal ($dE_C/dx = dE_V/dx$). In other words, the effective fields for electrons and holes (Chapter 4) are equal ($\mathscr{E}_e^* \approx \mathscr{E}_h^*$).
5. The barrier for electrons at the emitter-base junction is greater than that for holes by the amount ΔE_g^*. This is the amount of apparent band-gap narrowing in the heavily n-doped emitter.

This device is referred to as a *graded-base transistor.* Because of the effective field for electrons in the base, the injected electrons are accelerated toward the collector. Thus, the electron concentration gradient is not constant as was the case for uniform base doping. The electron concentration in the base for a graded base BJT is shown in Figure 9.9b. The presence of the field in the base creates a drift component to the current that increases I_{nB} above its value for diffusion. The electron distribution in the base of Figure 9.9b and the way in which the base field increases β are treated in the next section.

9.5.1 THE GRADED-BASE TRANSISTOR

It was seen in the previous section that, by grading the base doping, an electric field could be created in the base region that helps accelerate electrons across the base. This tends to improve the injection efficiency. Now we will look at this more closely.

Consider an npn BJT with some arbitrary base doping profile, $N'_{AB}(x)$. The hole (majority) current density in the base is

$$J_{pB} = q\mu_p p\mathscr{E} - qD_P\frac{dp}{dx} \tag{9.44}$$

From the Einstein relation, $D/\mu = kT/q$, Equation (9.44) can be rewritten as

$$J_{pB} = qD_p\left[\frac{q}{kT}p\mathscr{E} - \frac{dp}{dx}\right] \tag{9.45}$$

At equilibrium, there is no net current, so $J_{pB} = 0$ and

$$\frac{q}{kT}p\mathscr{E} = \frac{dp}{dx} \tag{9.46}$$

Considering that all acceptors are ionized, we have $p(x) = N'_{AB}(x)$, and the field in the base is

$$\mathscr{E} = \frac{kT}{qN'_{AB}(x)}\frac{dN'_{AB}(x)}{dx} \tag{9.47}$$

Equation (9.47) is a general relation for nondegenerate semiconductors.

For the special case where the doping profile in the base varies exponentially with position, a condition that is a reasonable approximation in many BJTs,

$$N'_{AB}(x) = N'_{AB}(x_0)e^{-x/\lambda} \tag{9.48}$$

where $N'_{AB}(x_0)$ is the doping level at some position $x = x_0$ in the base and λ is a characteristic length that is a measure of the doping gradient. Recall from Example 4.1 that an exponential doping gradient results in a constant electric field:

$$\mathscr{E} = -\frac{kT}{q\lambda} \tag{9.49}$$

From Poisson's equation,

$$\frac{d\mathscr{E}}{dx} = \frac{Q_V}{\varepsilon} \tag{9.50}$$

where Q_V is the charge per unit volume. Since $\mathscr{E}$ is constant, however, $Q_V = 0$, meaning the base is electrically neutral. If N_{AB} is not exponential, a small net charge exists and the interior of the base is referred to as a *quasi-neutral* region.

Let us take an example. We consider the impurity concentration of the graded base transistor of Figure 9.10. Here the net doping concentration $|N_D - N_A|$ is plotted against depth from the surface. On a semilog plot, a straight line represents an exponential distribution. Thus, from the graph, over most of

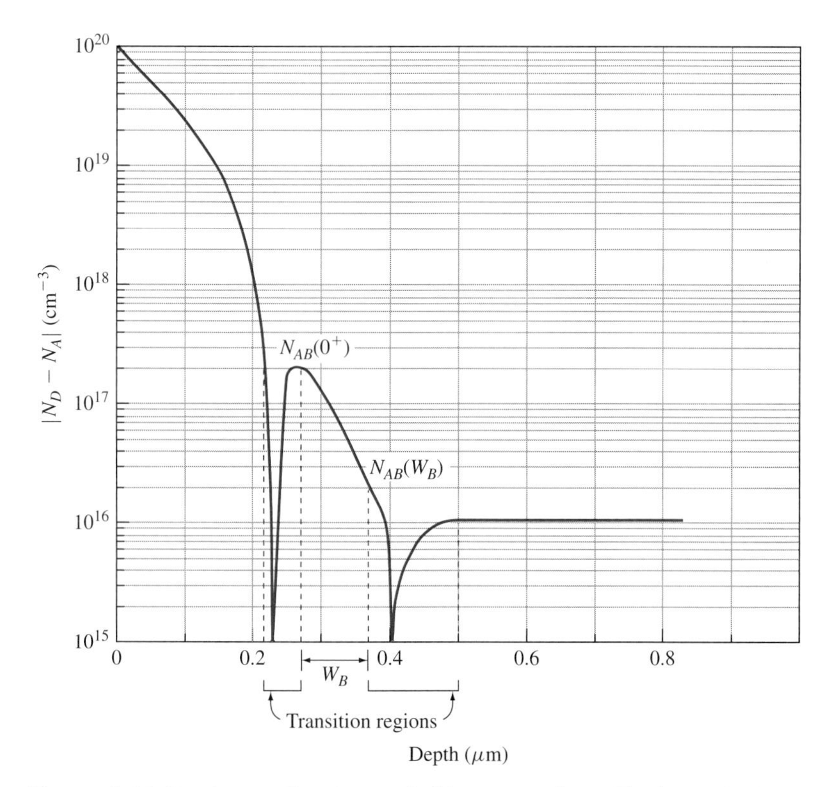

Figure 9.10 Doping profile of a graded-base transistor. The base doping profile as a function of position *x* is approximated as a straight line in this semilog plot, indicating an exponential impurity distribution (and thus a constant electric field) in the base.

the base region we can approximate the net acceptor concentration as in Equation (9.48).

It is convenient to introduce the parameter [3]

$$\eta = \frac{W_B}{\lambda} \tag{9.51}$$

This is, in effect, a measure of the value of the constant electric field in the base, since it varies with the parameter λ, which describes the doping profile. Substituting Equation (9.48) into Equation (9.51) and solving for η gives

$$\eta = \frac{W_B}{x} \ln \frac{N_A(0^+)}{N_A(x)} \tag{9.52}$$

Evaluating this expression at $x = W_B$ results in

$$\eta = \ln \frac{N_A(0^+)}{N_A(W_B)} \tag{9.53}$$

The electric field strength can then be written

$$\mathscr{E} = \frac{kT}{q} \frac{1}{\lambda} = \frac{kT}{q} \frac{\eta}{W_B} \tag{9.54}$$

Let us apply these results.

EXAMPLE 9.4

Find η and $\mathscr{E}$ in the base for the transistor of Figure 9.10.

■ Solution

Since N'_{AB} approximates a straight line on this semilog plot, it can be considered an exponential function, and so we can use the analysis above. From Figure 9.10, the base width is $W_B = 0.10\,\mu\text{m}$. The doping at each end of the base is

$$N_{AB}(0^+) = 2 \times 10^{17}\,\text{cm}^{-3} \qquad N_{AB}(W_B) = 2 \times 10^{16}\,\text{cm}^{-3}$$

From Equation (9.53), we can find the factor

$$\eta = \ln \frac{2 \times 10^{17}}{2 \times 10^{16}} = 2.3$$

From Equation (9.54) the field strength is

$$\mathscr{E} = 0.026 \times \frac{2.3}{0.10} = 0.6\,\text{V}/\mu\text{m} = 6\,\text{kV/cm}$$

Note that for this high a field, electrons in the base are traveling at their saturation velocity.

Also note that this constant built-in field is established by the exponential *majority* carrier (acceptor) distribution. However, this field affects the *minority* carrier electron current in the base (and the rate at which the electrons cross the base to arrive at the collector). We can analyze this situation as follows. The electron current density in the base is the sum of the diffusion and drift currents:

$$J_{nB} = q\mu_n n\mathscr{E} + qD_n \frac{dn_B}{dx} \tag{9.55}$$

Again using the relation $D/\mu = kT/q$, and solving Equation (9.55) for dn_B/dx, we obtain

$$\frac{dn_B}{dx} = \frac{J_{nB}}{qD_n} - \frac{qn_B\mathscr{E}}{kT} \tag{9.56}$$

We can substitute the expression for $\mathscr{E}$ from Equation (9.54) into Equation (9.56). We also set the electron concentration at the collector end of the base to $n(W_B) = 0$, since the collector is reverse biased. This expression for the electron distribution in the base, $n_B(x)$, then becomes

$$n_B(x) = \frac{J_{nB}W_B}{qD_n}\left[\frac{1 - e^{\eta(x/W_B - 1)}}{\eta}\right] \quad \text{graded-doping base} \tag{9.57}$$

where D_n is evaluated at the electric field associated with η. (Recall that at large $\mathscr{E}$, μ and D decrease with increasing $\mathscr{E}$.)

The term outside the brackets is a constant for a given transistor, for a given bias level. The bias level determines J_{nB}. We will therefore plot the normalized electron concentration distribution in the base,

$$n_{B(\text{norm})} = \frac{n_B(x)}{\dfrac{J_{nB}W_B}{qD_n}}$$

as a function of normalized distance x/W_B, Figure 9.11. This normalization is useful because it means that the plot can be applied to any exponentially doped region.[5] The plot shows the term in brackets, with η as a parameter. The case of $\eta = 0$ is the case for uniform base doping. With increasing field (increasing η) the carrier distribution with distance becomes flatter, meaning that the diffusion current ($J_{n(\text{diff})} = qD_n\, dn_B/dx$) becomes increasingly less important. The current becomes more dominated by drift. Note that for a given J_{nB}, the total electron charge in the base decreases with increasing η.

As an example, consider a transistor having the fairly typical value of $\eta = 4$. We see from Figure 9.11 that for $\eta = 4$, the electron concentration is reasonably constant over most of the base. Since the diffusion current is $J_{nB} = qD_n\, dn/dx \approx 0$, the diffusion current is negligible and drift predominates. With increasing x, toward the collector end of the base region, diffusion current becomes more pronounced.

[5]This formulation is most useful for $\mathscr{E}$ small enough that the diffusion coefficient is independent of $\mathscr{E}$, e.g., the low-field diffusion coefficient.

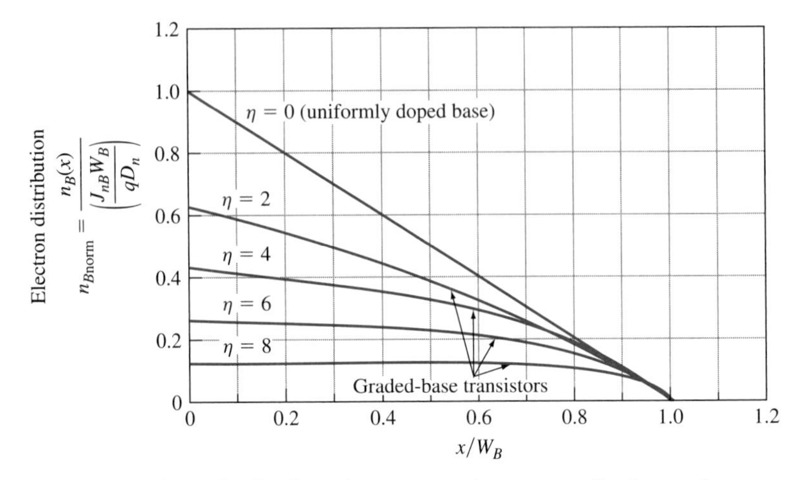

Figure 9.11 The distribution of excess carriers across the base of a graded-base npn transistor for a given I_C. The electron distribution is normalized as indicated, and the horizontal axis is a function of distance across the base. The distributions are plotted for various values of the grading parameter η; a uniformly doped transistor has an η of 0. As η and thus $\mathscr{E}$ increases, the slope of the concentration decreases, particularly at the emitter end, meaning the current is carried less by diffusion and more by drift. This plot assumes that D_n is independent of $\mathscr{E}$ and thus η.

9.5.2 EFFECT OF BASE FIELD ON β

The presence of a field in the base alters the mechanism by which the injected carriers cross the base, and thus will have an effect on β. We recall that

$$\beta = \frac{I_C}{I_B} \approx \frac{J_{nB}}{J_{pE}} \tag{9.58}$$

We use Equation (9.57) for a graded-base transistor and solve for J_{nB} at $x = 0^+$:

$$J_{nB} = \frac{qD_n n_B(0^+)}{W_B}\left(\frac{\eta}{1 - e^{-\eta}}\right) \tag{9.59}$$

For a given doping at the base edge $N'_{AB}(0^+)$ and emitter-base voltage, the minority carrier concentration $n_B(0^+)$ and back injection J_{pE} are constant. Thus, the only thing different between this transistor and a uniformly doped one is the factor in brackets. To see the effect of η on the current gain, we again normalize, this time normalizing the value of β for a graded-doping $\beta(\eta)$ to the uniformly doped transistor's $\beta(0)$ (η is zero). We therefore write for the normalized β

$$\frac{\beta(\eta)}{\beta(0)} \approx \frac{\eta}{1 - e^{-\eta}} \tag{9.60}$$

This function is plotted in Figure 9.12. We can see that β increases with increasing η, or increasing base field. At $\eta = 4$, for example, β is 4 times larger than for

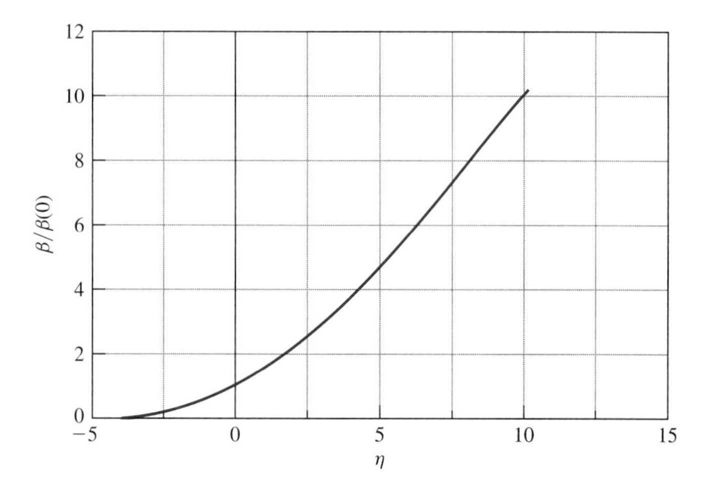

Figure 9.12 The effect of the grading parameter η on the current gain β of a BJT. If the grading goes in the wrong direction ($\eta < 0$), β actually becomes smaller.

the uniform base device.[6] We also note that for negative η (N'_{AB} increasing with x), β is quite small.

9.6 THE BASIC EBERS-MOLL DC MODEL

Until now, we have considered only BJT operation in the active mode, in which the emitter-base junction is forward biased and the base-collector junction is reverse biased. This mode of operation is used for analog or linear circuits. In digital circuits, however, operation in all four of the modes that were listed in Table IV.2 is possible. To model transistors operating in digital switching circuits, the Ebers-Moll model [4] is commonly used.

In the Ebers-Moll model, an npn BJT is represented by two back-to-back diodes with interaction between the two diodes, as shown in Figure 9.13. The figure shows the common base configuration. Each junction is represented by a diode in the model. Each junction also has a dependent current source whose current depends on the current in the *other* junction.

Let I_F be the forward-biased emitter-base current. Because of the thin base region, a fraction of the carriers injected into the base from the forward-biased emitter-base junction will flow into the collector. The factor α_F is the forward common base current gain. Then $\alpha_F I_F$ represents the emitter-base current that is collected by the reverse-biased collector. This current is modeled by the dependent current source belonging to the collector-base junction. Thus far, we can model the active mode of operation.

For operation in the inverse mode, the collector-base junction is forward biased while the base-emitter junction is reverse biased. For this model, the

[6]These results are somewhat optimistic, since the analysis neglects saturation velocity at high base fields.

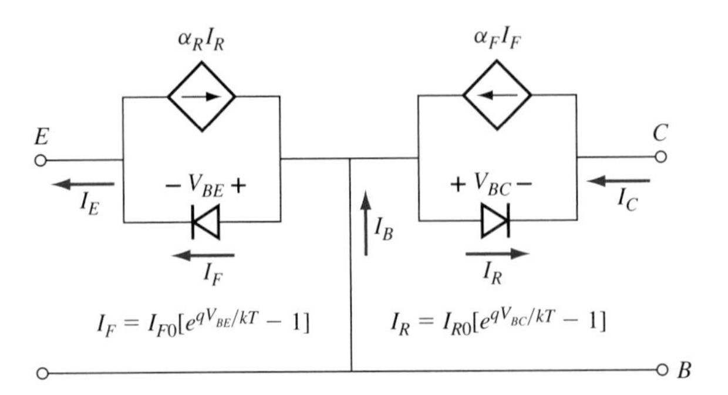

Figure 9.13 The Ebers-Moll model represents a bipolar junction transistor as two interacting back-to-back diodes. This is the common-base configuration.

collector-base current is represented by I_R, while $\alpha_R I_R$ is that current collected by the reverse-biased base-emitter junction.[7] This current is represented by the dependent current source corresponding to the emitter-base junction. Here α_R is called the inverse-mode common-base current gain.

Normally transistors work poorly in the inverse mode, and $\alpha_R \ll \alpha_F$. Recall that the current transfer ratio $\alpha = I_C/I_E \approx \gamma\alpha_T$. We showed in Section 9.4 that to maximize the injection ratio γ, the emitter should be much more heavily doped than the base. In the inverse mode, the collector is acting as an emitter, but because the collector doping is less than the base doping, the injection efficiency γ for inverse operation is small. Further, the emitter area is normally much smaller than the collector area. Thus, few electrons injected from collector to base find their way to the emitter, resulting in a small (inverse-mode) transport efficiency, α_T.

The currents I_F and I_R of the diodes in the model are the usual diode currents, and can be expressed

$$\begin{aligned} I_F &= I_{F0}(e^{qV_{BE}/kT} - 1) \\ I_R &= I_{R0}(e^{qV_{BC}/kT} - 1) \end{aligned} \tag{9.61}$$

where I_{F0} and I_{R0} are the saturation currents for the emitter-base and collector-base diodes respectively. The emitter and collector terminal currents are

$$\begin{aligned} I_E &= I_F - \alpha_R I_R \\ I_C &= \alpha_F I_F - I_R \end{aligned} \tag{9.62}$$

or, from Equation (9.61),

$$\begin{aligned} I_E &= I_{F0}(e^{qV_{BE}/kT} - 1) - \alpha_R I_{R0}(e^{qV_{BC}/kT} - 1) \\ I_C &= \alpha_F I_{F0}(e^{qV_{BE}/kT} - 1) - I_{R0}(e^{qV_{BC}/kT} - 1) \end{aligned} \tag{9.63}$$

[7]The subscripts F and R may be confusing; they refer to the bias of the respective junctions under forward active bias conditions. In the inverse mode, the emitter-base junction is reverse biased and the current I_F is actually a reverse-bias current. The model will still work, however; α_F is still the fraction of emitter-base current arriving at the collector and α_R is the fraction of the base-collector current arriving at the emitter.

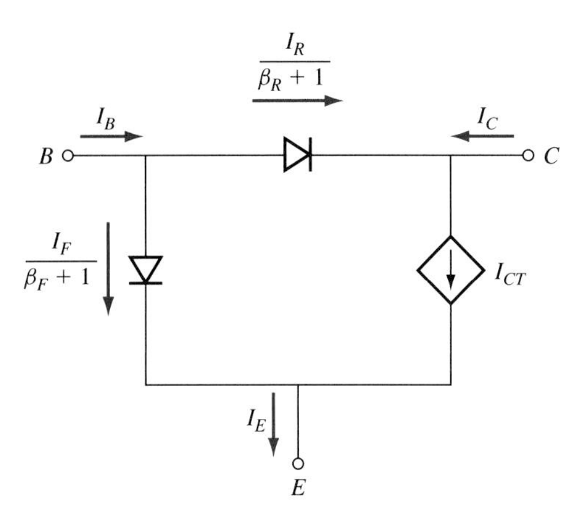

Figure 9.14 The common-emitter representation of the Ebers-Moll model.

The base terminal current is then

$$I_B = I_E - I_C = I_F(1 - \alpha_F) + I_R(1 - \alpha_R) \tag{9.64}$$

The forward and reverse common base current gains are related by the expression [5]

$$\alpha_F I_{F0} = \alpha_R I_{R0} = I_S \tag{9.65}$$

where the term I_S is introduced since it is a parameter used in SPICE.

For circuit analysis, it is normally more convenient to use the common-emitter representation of the Ebers-Moll model shown in Figure 9.14, where

$$\begin{aligned}
\beta_F &= \frac{\alpha_F}{1 - \alpha_F} \qquad \alpha_F = \frac{\beta_F}{\beta_F + 1} \\
\beta_R &= \frac{\alpha_R}{1 - \alpha_R} \qquad \alpha_R = \frac{\beta_R}{\beta_R + 1} \\
I_{CT} &= (\alpha_F I_F - \alpha_R I_R) = \frac{\beta_F I_F}{\beta_F + 1} - \frac{\beta_R I_R}{\beta_R + 1}
\end{aligned} \tag{9.66}$$

The terminal currents are then

$$\begin{aligned}
I_E &= I_{CT} + \frac{I_F}{\beta_F + 1} = \left(1 + \frac{1}{\beta_F}\right) I_S(e^{qV_{BE}/kT} - 1) - I_S(e^{qV_{BC}/kT} - 1) \\
I_C &= I_{CT} + \frac{I_R}{\beta_R + 1} = I_S(e^{qV_{BE}/kT} - 1) - \left(1 + \frac{1}{\beta_R}\right) I_S(e^{qV_{BC}/kT} - 1) \\
I_B &= \frac{I_F}{\beta_F + 1} + \frac{I_R}{\beta_R + 1} = \frac{I_S}{\beta_F}(e^{qV_{BE}/kT} - 1) + \frac{I_R}{\beta_R}(e^{qV_{BC}/kT} - 1)
\end{aligned} \tag{9.67}$$

where the right-hand side of Equation (9.67) uses SPICE variable names. However in SPICE, I_E is defined as positive in the opposite direction to the conventional direction of I_E. Thus in SPICE, I_E will have the opposite sign from that indicated above.

EXAMPLE 9.5

Find the terminal currents for an npn BJT operating in the active mode.

■ Solution

In the active mode $e^{qV_{BE}/kT} \gg 1$ and $e^{qV_{BC}/kT} \ll 1$. Then from Equation (9.67),

$$I_E = \left(1 + \frac{1}{\beta_F}\right) I_S e^{qV_{BE}/kT} = \frac{I_C}{\alpha_F}$$

$$I_C = I_S e^{qV_{BE}/kT}$$

$$I_B = \frac{I_S}{\beta_F} e^{qV_{BE}/kT} = \frac{I_C}{\beta_F}$$

The Ebers-Moll equations, or variations of them (e.g., the Gummel-Poon equations) [6] are often used in circuit analysis programs such as SPICE.[8]

9.7 CURRENT CROWDING AND BASE RESISTANCE IN BJTs

As indicated in Section 9.3, the base region should be thin in a good transistor to increase current gain and frequency response. A thin base does have a drawback, however: a high base resistance that degrades the device performance.

Consider the discrete transistor of Figure 9.15. The emitter length L and width h are indicated. There is a base resistance, which consists of two parts. These are the resistance R'_B from base contact to the emitter edge, due primarily to the resistance of the p-type material, and R_B, the effective resistance under the emitter (often referred to as the *intrinsic base resistance*), which is normally larger because here the base is thin. In this section, we are interested only in R_B. We call R_B an "effective" resistance; it is really a distributed quantity, because I_B decreases with increasing position y, but for circuit analysis it is convenient to consider it a lumped resistance.

The base current I_B flows laterally (parallel to the junction plane) and causes a lateral *IR* drop in the base region. This drop along the emitter edge means that the base-emitter voltage V_{BE} is a function of position. Since the emitter current density is a strong (exponential) function of V_{BE},

$$J_E(y) = J_{E0}(e^{qV_{BE}(y)/kT} - 1) \tag{9.68}$$

a small lateral voltage drop in the base causes a large spatial dependence on emitter current density. The polarity of this lateral voltage drop is such that V_{BE}

---

[8]The Gummel-Poon equations include second-order effects not present in the Ebers-Moll model, e.g., nonconstant β.

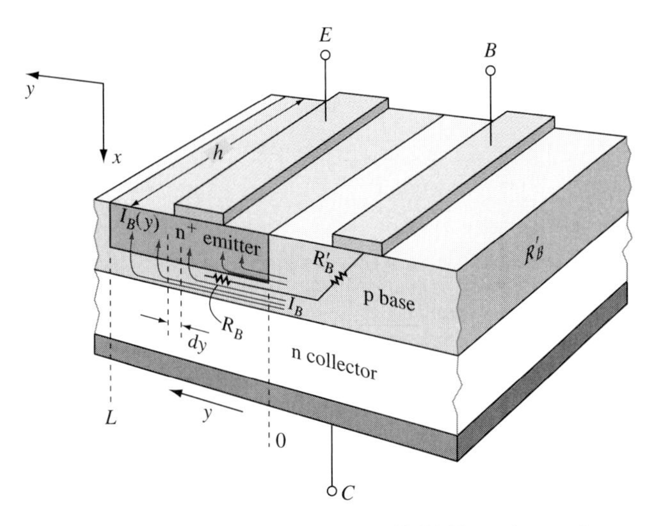

Figure 9.15 Current crowding in a BJT. There is a voltage drop across the base resistance R_B, which varies the emitter-base junction bias along the edge of the emitter. The effect is a greater bias at the end toward the base contact, and thus greater base-emitter current flow at that end.

is largest at the emitter edge nearest to the base contact. The emitter current then "crowds" toward the base contact.

This current crowding (often called emitter crowding) increases with increasing base current, since that increases the effect of the *IR* drop. That means that as I_B increases, the current crowding increases and a larger fraction of the current is flowing into the emitter at the end near the base contact than at lower currents. The implication is that the effective distance that the base current flows is reduced with increasing current, and therefore so is the equivalent lumped base resistance, R_B.

There are several ways to define a lumped resistance. [7] Here we define R_B as the magnitude of the average lateral base voltage divided by the terminal base current

$$R_B \equiv \frac{\langle V_B \rangle}{I_B} \tag{9.69}$$

where the average lateral base voltage is

$$\langle V_B \rangle = \frac{\displaystyle\int_0^L V_B(y) \frac{dV_B(y)}{dy}\, dy}{\displaystyle\int_0^L \frac{dV_B(y)}{dy}\, dy} \tag{9.70}$$

The quantity $V_B(y)$ is the base voltage at position y, and $dV_B(y)/dy$ is the base voltage distribution function with position. It can be written

$$\frac{dV_B(y)}{dy} = -\frac{R_\square}{h} I_B(y) \tag{9.71}$$

where $I_B(y)$ is the base current at position (y), and $R_\square$ is the sheet resistance in the base (ohms per square),

$$R_\square = \frac{1}{q\mu_p \int_0^{W_B} N'_{AB}\, dx} \tag{9.72}$$

where W_B is the base width.

Solving for R_B in general is somewhat involved. However, at small base currents (low injection) such that $V_{BE}(y)$ is nearly constant, $I_B(y)$ varies linearly from I_{B0} at $y = 0$ to zero at $y = L$:

$$I_B(y) = I_B\left(1 - \frac{y}{L}\right) \tag{9.73}$$

where I_B is the base terminal current.

Combining Equations (9.71) and (9.73) and integrating, we find

$$V_B(y) = -\frac{R_\square}{h} I_B \left[y - \frac{y^2}{2L}\right] \tag{9.74}$$

Then solving for R_B with the aid of Equations (9.69), (9.70), (9.71), and (9.74) gives

$$R_B = \frac{R_\square L}{4h} \tag{9.75}$$

Equation (9.75) is valid for small I_B, where current crowding is negligible. With increasing I_B, the emitter current is increasingly concentrated near the emitter edge adjacent to the base contact, which as we saw causes a reduction in R_B. Figure 9.16 shows a plot of the normalized base resistance (normalized taking into account the h/L aspect ratio of the emitter and the sheet resistance) as a function of normalized base current (normalized as indicated on the graph). There is a second line on the graph—it is an alternative normalized base equivalent resistance R_{B1}, arrived at using another definition of equivalency. This alternative R_{B1} is the basis of a homework problem.

While it might appear that the reduction of base resistance due to current crowding would be beneficial, this is not the case for two reasons:

1. Since the base-emitter current is concentrated at the edge of the emitter near the base contact, much of the emitter is inactive. However, the emitter-base junction capacitance associated with this inactive emitter remains, with its adverse effect on frequency response.
2. As discussed in Section 9.10, the high current density at the emitter edge reduces the current gain, β.

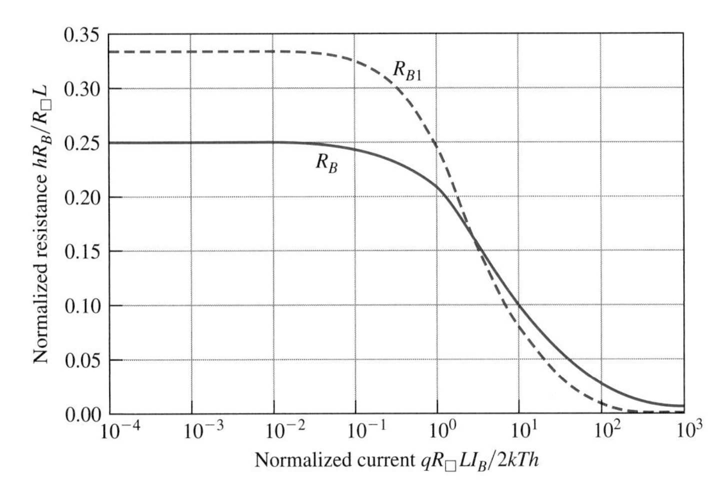

Figure 9.16 The effect of the base current on the effective resistance R_B used to represent the effect of current crowding. Two different models for lumping this distributed resistance are used. The solid line is based on the average voltage drop across the lumped resistance. The dashed line is based on the average power dissipated in the resistor.

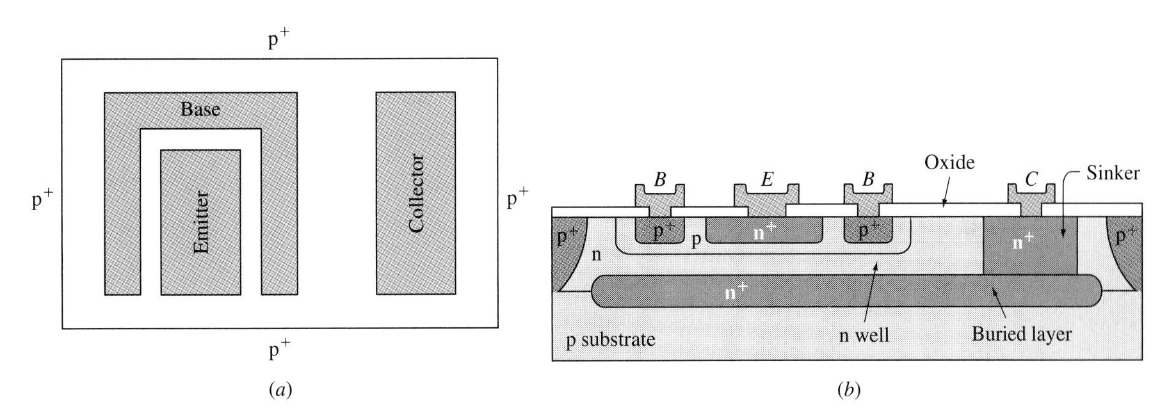

Figure 9.17 A double-base transistor. (a) Top view; (b) cross section.

In *power transistors,* the currents are large enough that the current crowding causes high injection effects that tend to reduce the current gain. In that case, two bases are often used, one on either side of the emitter as indicated in Figure 9.17a and b to reduce the crowding. (The base contact is often made to only one side.) This double base reduces the base resistance by a factor of 4. To reduce the base resistance even further, interdigitated structures, like that shown schematically in Figure 9.18a and b are often used in power transistors.

In modern integrated circuit BJTs, the emitter stripe width L can be made small (in the quarter-micrometer range) and for such cases, current crowding is

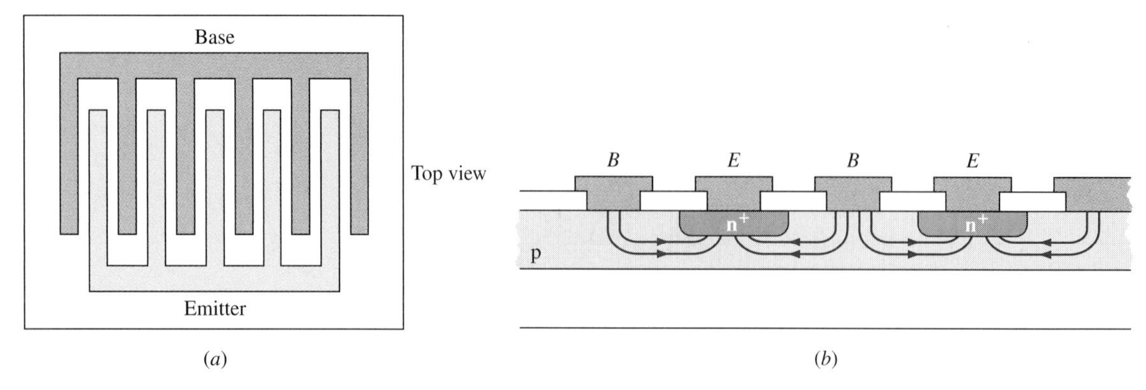

Figure 9.18 The contacts can be interdigitated, as in (a) top view and (b) cross section.

not a major problem. Typical emitter sizes in modern devices are on the order of $0.2 \times 2\,\mu$m.

9.8 BASE WIDTH MODULATION (EARLY EFFECT)

In Section 9.4.2, we obtained an expression for the injection efficiency γ [Equation (9.28)] that contained the quantity W_B/W_E in the denominator. The quantity W_B is the base width measured between the *E-B* depletion region and the *B-C* depletion region edges as was shown in Figure 9.5. As the reverse bias across the collector-base junction is increased, the depletion region gets wider and the effective base width gets narrower. This has the effect of increasing γ and thus increasing the current gain β.

Figure 9.19a shows the electron distribution n_B in the base as a function of distance across the base for different values of V_{CE} for a uniformly doped base transistor, so the distribution is a straight line ($\eta = 0$). Because the reverse-biased C-B junction extracts carriers, n_B goes to zero at W_B, which shrinks with V_{CB} (and V_{CE}, assuming a constant base-emitter bias).

The effect of the shrinking base on the I_C-V_{CE} characteristics is shown in Figure 9.19b. Recall that $\beta = I_C/I_B$. For a given value of I_B, I_C increases with increasing V_{CE}, reflecting the increase in β. Extrapolations of these I_C-V_{CE} curves meet (approximately) at a voltage $-V_A$, where V_A is called the *Early voltage.* [8] Note that this is similar to the influence of drain voltage in a MOSFET on channel length, and thus on drain current. The agreement is different but the effect on the I-V characteristics is the same.

Figure 9.19a is repeated in Figure 9.20 for the case of the graded base transistor. Although a change in V_{CE} does change the effective base width, it has no effect on the field in the base and little effect on dn_B/dx at the emitter edge of the base, and thus it has little influence on saturation current. This in turn increases the magnitude of the Early voltage V_A and results in a more constant value of β. In Chapter 10 we will see that the small-signal output conductance ($g_d = dI_C/dV_{CE}$) is also reduced.

As the collector voltage gets larger, and the depletion region at the base-collector junction gets wider, it can become wide enough such that W_B becomes

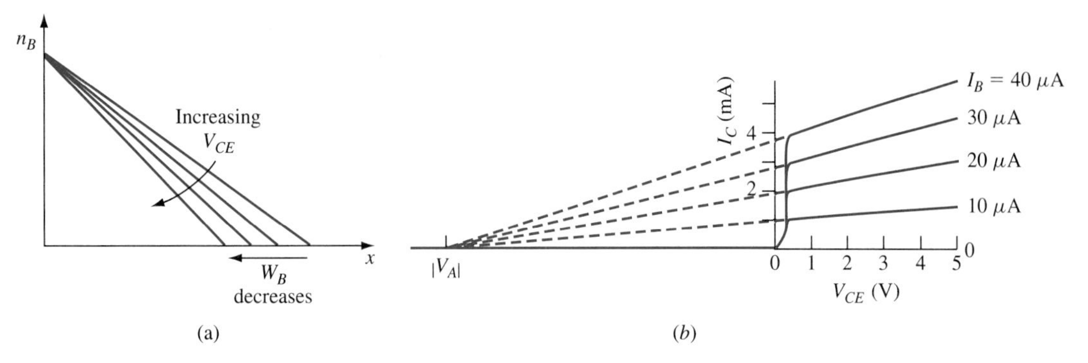

Figure 9.19 Increasing collector voltage causes a decrease in effective base width (a) resulting in increased I_C and β (b) for a uniform base transistor. The Early voltage V_A is indicated.

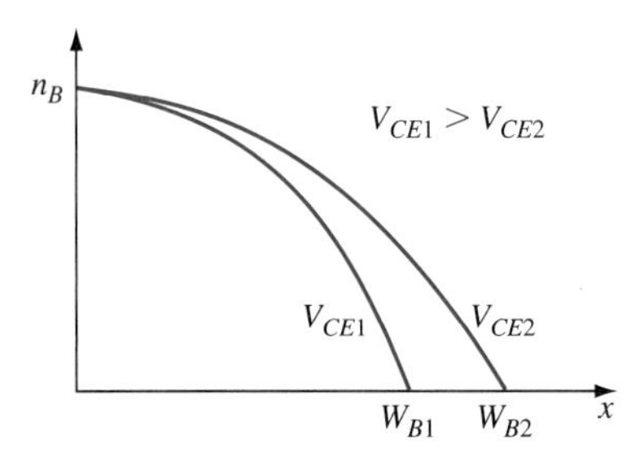

Figure 9.20 For a graded-base transistor, a change in collector voltage has little effect on n_B near the emitter, and thus little effect on I_C or β, resulting in increased V_A.

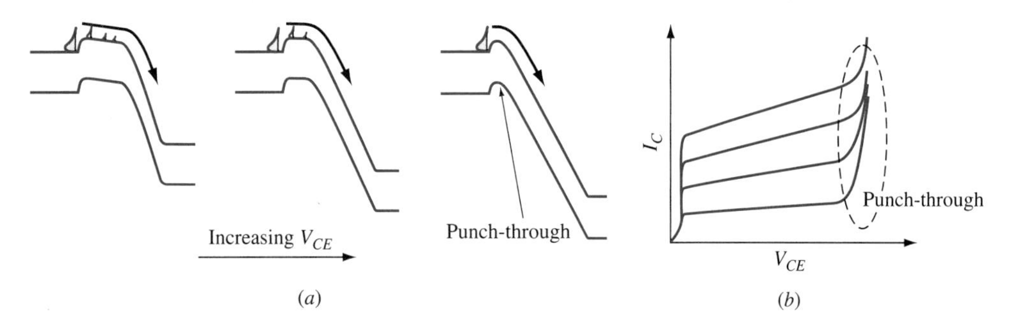

Figure 9.21 (a) As the collector-base voltage increases, the effective base width decreases. At punch-through, the two depletion regions meet, and the emitter electrons are swept directly into the collector. (b) The effect is loss of transistor action.

zero. When this happens, the depletion region of the collector actually meets the depletion region of the emitter, as shown in Figure 9.21a. The current at that point becomes very large. The electrons are swept out of the emitter to the collector and the base loses control over I_C. Transistor action disappears, as can be seen from the I_C-V_{CE} characteristics in Figure 9.21b. This effect is called *punch-through*.

EXAMPLE 9.6

Consider a prototype npn BJT with emitter, base, and collector dopings:

$$N'_{DE} = 9 \times 10^{19}\ \text{cm}^{-3}$$
$$N'_{AB} = 4 \times 10^{17}\ \text{cm}^{-3}$$
$$N'_{DC} = 6 \times 10^{16}\ \text{cm}^{-3}$$

What is the minimum value of W_{BM}, the metallurgical base width, such that the punch-through voltage $V_{CE}(PT)$ is greater than 6 V? Assume $V_{BE} = 0.75$ V.

■ **Solution**

Figure 9.22a shows the depletion region widths in the base for a BJT under normal bias. From the figure, we see that

$$W_B = W_{BM} - w_{pEB} - w_{pCB}$$

where W_{BM} is the metallurgical base width and w_{pEB} and w_{pBC} are the depletion widths on the p (base) sides of the *E-B* and *B-C* junctions.

At punch-through, the *C-B* depletion region meets the *E-B* depletion region, and $W_B = 0$. Then

$$W_{BM} = w_{pEB} + w_{pCB}$$

For the collector-base junction, from Equation (5.36), we have

$$w_{pCB} = \left[\frac{2\varepsilon V_{jCB}}{qN'_A\left(1 + \dfrac{N'_{AB}}{N'_{DC}}\right)}\right]^{1/2} = \left[\frac{2\varepsilon(V_{\text{bi}BC} + V_{CB})}{qN_{AB}\left(1 + \dfrac{N'_{AB}}{N'_{DC}}\right)}\right]^{1/2}$$

where $V_{jCB} = V_{\text{bi}CB} + V_{CB}$, N'_A is the net doping on the p side of the junction and N'_D is that on the n side.

Since the emitter-base junction can be considered to be a one-sided step junction, the depletion region width w_{pEB} is, from Equation (5.39),

$$w_{pEB} = \left[\frac{2\varepsilon(V_{\text{bi}EB} - V_{EB})}{qN'_{AB}}\right]^{1/2}$$

To find the built-in voltages we refer to Figure 9.22b, which shows the neutrality energy band diagram for this transistor. From Figure 2.25, the impurity-induced band-gap narrowing in the emitter is $\Delta E_g^* \approx 0.09\,\text{eV}$. We neglect the small band-gap narrowing in the base and collector.

In the base the hole concentration is $p_{B0} = N'_{AB} = N_V e^{-\delta p/kT}$, so

$$\delta p = kT \ln \frac{N_V}{N'_{AB}} = 0.11\,\text{eV}$$

as indicated in the figure. The emitter-base built-in voltage then is

$$V_{\text{bi}EB} = \frac{1}{q}(E_g - \Delta E^*_{gBE} - \delta p) = 1.12 - 0.09 - 0.11 = 0.92\,\text{V}$$

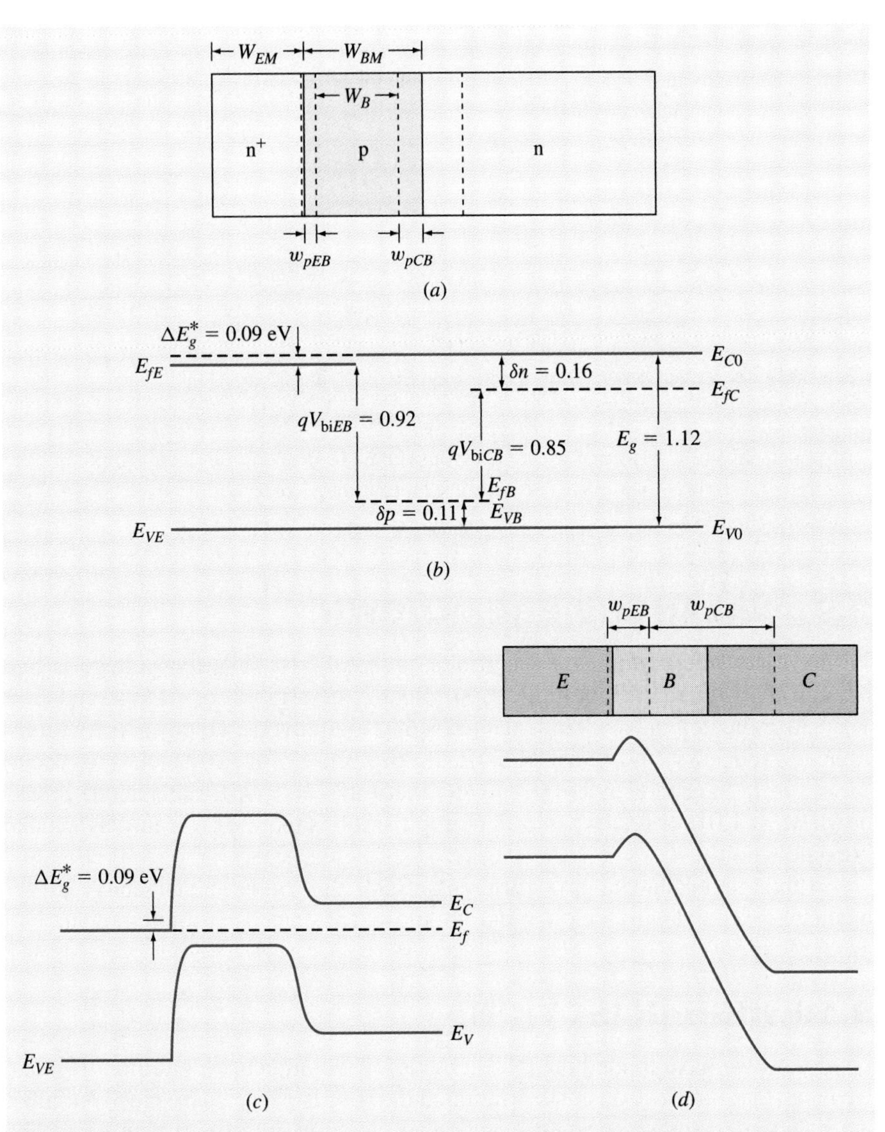

Figure 9.22 (a) The widths of the emitter, base, and the depletion regions relevant to Example 9.4; (b) the energy band diagram under neutrality; (c) the equilibrium energy band diagram; (d) under punch-through.

Similarly in the collector, $\delta n = 0.16\,\text{eV}$ and the base-to-collector built-in voltage is

$$V_{\text{bi}CB} = E_g - \delta n - \delta p = 0.85\,\text{V}$$

Alternatively, from Equation (5.13),

$$V_{\text{bi}CB} = kT \ln \frac{N'_{DC} N'_{AB}}{n_i^2} = 0.85\,\text{V}$$

The equilibrium energy bond diagram is shown in Figure 9.22c, for the specified punch-through voltage of $V_{CE} = 6\,\text{V}$, the collector-base reverse bias at $V_{BE} = 0.75\,\text{V}$ is $V_{CB} = V_{CE} - V_{BE} = 6 - 0.75 = 5.25\,\text{V}$. At this bias, the depletion region width on the base side of the C-B junction is

$$w_{pCB} = \left[\frac{2\varepsilon(V_{\text{bi}CB} + V_{CB})}{qN'_{AB}\left(1 + \dfrac{N'_{AB}}{N'_{DC}}\right)}\right]^{1/2} = \left[\frac{2 \times (11.8 \times 8.85 \times 10^{-12}) \times (0.85 + 5.25)}{1.6 \times 10^{-19} \times 4 \times 10^{23}\left(1 + \dfrac{4 \times 10^{23}}{6 \times 10^{22}}\right)}\right]^{1/2}$$

$$= 0.051\,\mu\text{m}$$

Similarly, the depletion width in the base on the emitter end is

$$w_{pEB} = 0.024\,\mu\text{m}$$

At punch-through, (d), the depletion regions meet. To prevent this from happening at a value of V_{CE} less than 6 V, the minimum metallurgical base width is

$$W_{BM} = w_{pEB} + w_{pCE} = 0.024 + 0.051 = 0.075\,\mu\text{m}$$

9.9 AVALANCHE BREAKDOWN

We have considered the case where the collector-base junction voltage is low enough that the carrier multiplication effect is negligible, or $M = 1$ and $\beta = \gamma\alpha_T$. As indicated in Chapter 5, at sufficiently high reverse voltage, current multiplication occurs and thus $M > 1$. In this case, for avalanche to occur, $M = 1 + 1/\beta$. (See Problem 9.4.) Note that for $\beta = 100$, for avalanche breakdown $M = 1.01$, or relatively little multiplication is required to cause avalanche breakdown.

Although avalanche is a different effect, the I_C-V_{CE} characteristics for avalanche breakdown look similar to those for punch-through indicated in Figure 9.21c.

9.10 HIGH INJECTION

Until now, we have treated the BJT in the low-level injection condition. By this, we mean that the electron concentration in the base is everywhere much less than the hole concentration, despite injection of electrons into the p-type base. The energy band diagram of Figure 9.23a illustrates the low-injection case. Under operation, an excess electron concentration Δn is injected into the base. This excess concentration of negative charges tends to make the base more negative. This, in turn, creates a field that attracts excess holes, Δp, from the base contact. To achieve neutrality, Δp must equal Δn. Under low injection $\Delta p \ll N'_A$; however, this makes almost no difference to the relative hole concentration. Thus, the back injection of holes from the base to the emitter is essentially unaffected.

At large forward bias, such that Δn is not small compared with $p_{B0} = N'_A$, as in Figure 9.23b, it is possible for the number of holes, $p = p_{B0} + \Delta p = N'_A + \Delta n$, to be changed significantly. Since at large forward bias the barrier to hole injection is small, many of these excess holes have enough energy to be

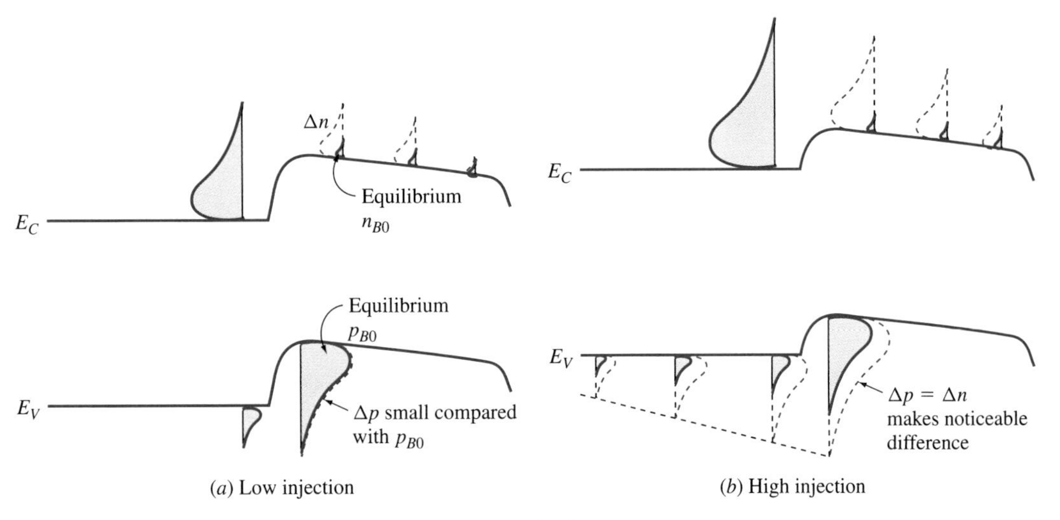

Figure 9.23 To achieve space charge neutrality, electrons injected into the base (Δn) draw an equal number of excess holes (Δp). (a) For low injection, these excess holes have negligible influence on the base-emitter hole current. (b) Under high injection the base-emitter barrier for holes is small enough that many of these excess holes are injected into the emitter.

injected into the emitter, thus increasing both I_{pE} and I_{pB}. The increase in I_{pE} decreases the injection efficiency [Equation (9.10)]. The increase in I_{pB} means an increase in the total base current I_B, which, since $\beta = I_C/I_B$, causes β to decrease from its low injection value.

9.11 BASE PUSH-OUT (KIRK) EFFECT

There is a second reason β is reduced under high injection. At high currents, the effective base width *increases,* with a concurrent *reduction* in β. This effect, often referred to as the *base push-out effect* or the *Kirk effect* [9] is discussed with the aid of Figure 9.24.

When we considered the low-injection condition we assumed the electron concentration at W_B to be zero, and we neglected the small electron concentration inside the base-collector transition region. This is incorrect because the base electron current density, which is equal to the collector current density, J_C is

$$J_C = J_{nB} = -qn_B(x)v(x) \tag{9.76}$$

where $v(x)$ is the velocity of the carriers. Since in the high-field C-B junction the electron velocity is limited by its saturation velocity v_{sat} we know

$$v(x) \le v_{\text{sat}} \tag{9.77}$$

From Equation (9.76), if $v(x)$ has a maximum for a given current, then the electron concentration has a minimum, even near and within the CB transition region:

$$n \ge \left| \frac{J_C}{qv_{\text{sat}}} \right| = n_{\text{min}} \tag{9.78}$$

Thus n_{min} is the minimum electron concentration in the depletion region.

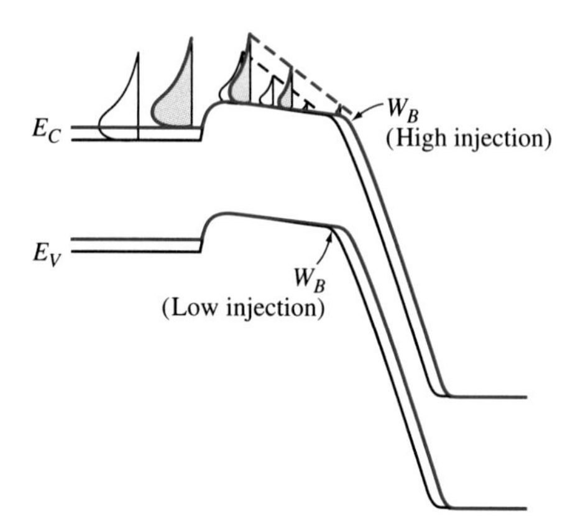

Figure 9.24 At high currents the electron charge density near the collector edge of the base, and for a short distance into the base-collector transition region, must be considered. This results in an increased effective base width and a decreased β.

For the low injection condition (small J_C), this value of n_{min} is small enough to be ignored. For high injection, however, it must be considered. As electrons enter the high-field transition region, they are accelerated toward the collector. Their velocity increases with increasing x until they reach their saturation velocity. The effect of this negative charge is to cause the electron potential energy E_C (and also E_V) to increase within the transition region. This effectively increases the effective base width W_B, with a corresponding decrease in β.

For currents large enough such that $n_{\text{min}} > N_{\text{DC}}$, where N_{DC} is the donor concentration in the n collector, the *apparent* base region extends well into the n collector region and can approach the n^+ collector region, where N_{DC} is much larger. With increasing J_C then, the high-field region shifts into the collector, as indicated in Figure 9.24. To avoid appreciable base push-out, it is desirable to design for

$$J_C \leq 0.3|qN_{\text{DC}}v_{\text{sat}}| \tag{9.79}$$

EXAMPLE 9.7

Estimate the maximum collector current density J_C for an n collector doping of $N_{\text{DC}} = 5 \times 10^{16}\ \text{cm}^{-3}$ to avoid excess base push-out.

■ Solution

From Equation (9.79),

$$\begin{aligned} J_{C\text{max}} &\approx 0.3|qN_{\text{DC}}v_{\text{sat}}| = 0.3(1.6 \times 10^{-19}\ \text{C})(5 \times 10^{16}\ \text{cm}^{-3})(10^7\ \text{cm/s}) \\ &= 2.4 \times 10^4\ \text{A/cm}^2 \\ &= 0.24\ \text{mA}/\mu\text{m}^2 \end{aligned}$$

From the above, it can be seen that typically J_C is less than 1 mA/μm^2.

9.12 RECOMBINATION IN THE EMITTER-BASE JUNCTION

Just as for a pn junction diode, some electron-hole recombination occurs in the transition region of the emitter-base junctions of a BJT. The B-E current resulting from this recombination never reaches the collector and thus does not contribute to I_C. It does, however, increase both I_E and I_B from those values found by the earlier models. Since $\alpha = I_C/I_E$ and $\beta = I_C/I_B$, these increases in I_E and I_B reduce the values for α and β. Figure 9.25 shows plots of I_C and I_B as functions of base-emitter voltage. This is referred to as a *Gummel plot.* The collector current I_C is essentially equal to the emitter electron current I_{nE} that is injected into the base and which is proportional to $e^{qV_{BE}/kT}$. Thus on the semilog scale of Figure 9.25, the I_C-V_{BE} plot is a straight line of slope $q \log e/kT$.

The base current has two components, injection current and recombination current. Injection current results from holes injected into the emitter and varies

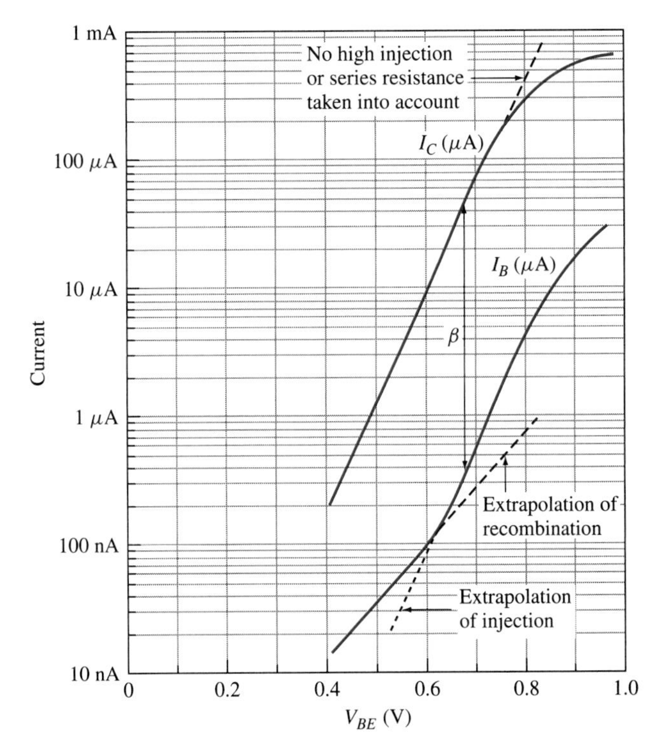

Figure 9.25 Gummel plot for a BJT. Except at high current, the log I_C-V_{BE} plot is a straight line of slope $q \log e/kT$. The I_B-V_{BE} plot is a straight line of slope $q \log e/2kT$ at low currents, changing to $q \log e/kT$ at higher currents. The deviation at the higher currents results from a portion of V_{BE} being dropped across the base series resistance (and for power transistors with small base resistance, high-injection effects).

as $e^{qV_{BE}/kT}$. The recombination current results from the base supplying the holes to support the electron-hole recombination in the emitter-base transition region. The recombination current varies as $e^{qV_{BE}/nkT}$, where n is an empirical parameter and $n \approx 2$. Since the injection current varies more rapidly with V_{BE} than the recombination current, injection predominates at higher base-emitter voltage while the recombination current predominates at lower voltages.

For modern Si BJTs, the midgap impurity concentration responsible for carrier recombination in the emitter-base transition region is low enough such that recombination current is negligible compared with the collector current for $J_B > 0.1\,\text{nA}/\mu\text{m}^2$.

Recall that $\beta = I_C/I_B$. Thus, on this semilog plot, the vertical distance between the I_C and the I_B plots is proportional to β. For analog circuits, BJTs are normally operated in the region of near-constant β.

At high currents, the slope reduces as a result of series resistance, and as discussed in Sections 9.10 and 9.11, β is again reduced because of high injection effects.

9.13 SUMMARY

In this chapter, we have introduced some concepts fundamental to the operation of bipolar junction transistors. In BJTs, carriers are injected into the base from the emitter. Ideally, all of the injected carriers cross the base and become collected. The amount of injection depends strongly on the barrier height at the emitter-base junction, so that small changes in base voltage create large changes in transistor current.

BJTs are characterized by the currents in the various modes of operation. A figure of merit for BJTs operating in the active mode used for analog circuits is

$$\beta = \frac{I_C}{I_B} = \frac{\gamma \alpha_T M}{1 + \gamma \alpha_T M}$$

where, for a prototype nondegenerate emitter with $W_B \ll L_{nB}$ and $W_E \ll L_{pE}$,

$$\gamma = \frac{1}{1 + \dfrac{N'_{AB}}{N'_{DE}} \dfrac{D_{pE}}{D_{nB}} \dfrac{W_B}{W_E}} \qquad \text{uniformly doped npn}$$

$$\gamma = \frac{1}{1 + \dfrac{N'_{DB}}{N'_{AE}} \dfrac{D_{nE}}{D_{pB}} \dfrac{W_B}{W_E}} \qquad \text{uniformly doped pnp}$$

and

$$\alpha_T = 1 - \frac{W_B^2}{2L_{nB}^2} \qquad \text{npn}$$

$$\alpha_T = 1 - \frac{W_B^2}{2L_{pB}^2} \qquad \text{pnp}$$

But since α_T and M are close to unity,

$$\beta \approx \frac{\gamma}{1+\gamma}$$

For both emitter and base nondegenerate (not a common situation),

$$\beta_{npn} \approx \frac{I_{nE}}{I_{pE}} \approx \frac{N'_{DE}}{N'_{AB}} \frac{D_{nB}}{D_{pE}} \frac{W_E}{W_B} \qquad \text{nondegenerate emitter and base}$$

$$\beta_{pnp} \approx \frac{I_{pE}}{I_{nE}} \approx \frac{N'_{AE}}{N'_{DB}} \frac{D_{pB}}{D_{nE}} \frac{W_E}{W_B} \qquad \text{nondegenerate emitter and base}$$

The normal situation is for the emitter to be degenerately doped. The heavy emitter doping causes band-gap narrowing in the emitter, which tends to increase the barrier for electron injection and significantly reduce the current gain β. For degenerate emitter and nondegenerate uniform base,

$$\beta_{npn} \approx \frac{N_C}{N'_{AB}} \frac{D_{nB}}{D_{pE}} \frac{W_E}{W_B} e^{-\Delta E^*_{gBE}/kT} \qquad \text{npn degenerate emitter, uniform base}$$

$$\beta_{pnp} \approx \frac{N_V}{N'_{DB}} \frac{D_{pB}}{D_{nE}} \frac{W_E}{W_B} e^{-\Delta E^*_{gBE}/kT} \qquad \text{pnp degenerate emitter, uniform base}$$

where $\Delta E^*_{gBE} = E^*_g(\text{base}) - E^*_g(\text{emitter})$ is the difference in the apparent band-gap narrowing in base and emitter.

If the impurity gradient and the resultant field in the base are considered, β is increased somewhat from the above values. The field in the base increases the fraction of injected electrons reaching the collector, and thus increases the collector current.

A portion of the transition region of the base-collector junction extends into the base, thus reducing the effective base width (W_B) from its metallurgical width (W_{BM}). With increasing collector voltage, W_B decreases and thus β increases (Early effect).

Under high-injection conditions, such as in power transistors, the electronic charge associated with the high current causes the effective base width to increase. Thus at a given collector voltage, β decreases with increasing current (Kirk effect).

9.14 READING LIST

Items 2, 4, 8 to 12, 17 to 19, 26, and 34 in Appendix G are recommended.

9.15 REFERENCES

1. Yuan Taur and Tak H. Ning, *Fundamentals of Modern VLSI Devices,* Cambridge University Press, Cambridge, U.K., Chap. 6, 1999.

2. Robert F. Pierret, *Semiconductor Device Fundamentals,* Addison Wesley, New York, Chap. 11, 1996.
3. Joseph Lindmayer and Charles Wrigley, *Fundamentals of Semiconductor Devices,* Chap. 4, Van Nostrand, Princeton, 1965.
4. J. J. Ebers and J. L. Moll, "Large-signal behavior of junction transistors," *Proc. IRE,* 42, pp. 1761–1772, 1954.
5. R. S. Muller and T. I. Kamins, *Device Electronics for Integrated Circuits,* John Wiley & Sons, New York, 1977.
6. H. K. Gummel and H. C. Poon, "An integral charge control model for bipolar transistors," *Bell Syst. Tech. J.,* 49, pp. 827–852, 1970.
7. J. L. Lary and R. L. Anderson, "Effective Base resistance of Bipolar Transistors," *IEEE Transactions on Electron Devices,* ED-32, pp. 2503–2505, 1985.
8. J. M. Early, "Effect of space-charge layer widening in junction transistors," *Proc. IRE* 40, pp. 1401–1406, 1952.
9. C. T. Kirk Jr., "A theory of transistor cutoff frequency (f_T) falloff at high current densities," *IEEE Trans. Electron Devices,* ED-9, pp. 164–174, 1962.

9.16 REVIEW QUESTIONS

1. What is the purpose of the n^+ sinker in Figure IV.1?
2. Equation (9.1) describes the prototype BJT under forward active bias. Why is it reasonable to neglect the recombination current in the emitter-base junction?
3. Explain why it is advantageous to make the base thin in a bipolar junction transistor.
4. Explain the reasons for doping the emitter heavily compared with the base, and for doping the collector lightly.
5. Why should we reduce back-injection (from the base back into the emitter)? What can be done to control it?
6. Explain how band-gap narrowing in the degenerate emitter affects the current gain β. What is the physics behind this?
7. From the SIMS plot of Figure 9.8, it is clear that the doping is graded in all three regions—the emitter, base, and collector. Why, then, is this device referred to as a *graded-base transistor?*
8. Explain in words how grading the base doping can increase the common-emitter current gain of a transistor.
9. Explain why, in Figure 9.9, E_C bends downward and E_V bends upward toward the surface of the semiconductor.
10. Explain why an increasing V_{CE} causes the collector current to increase, even in the active region.

9.17 PROBLEMS

9.1 For each of the transistors in Figure P9.1, indicate the mode of operation (forward active, cutoff, saturation, etc.)

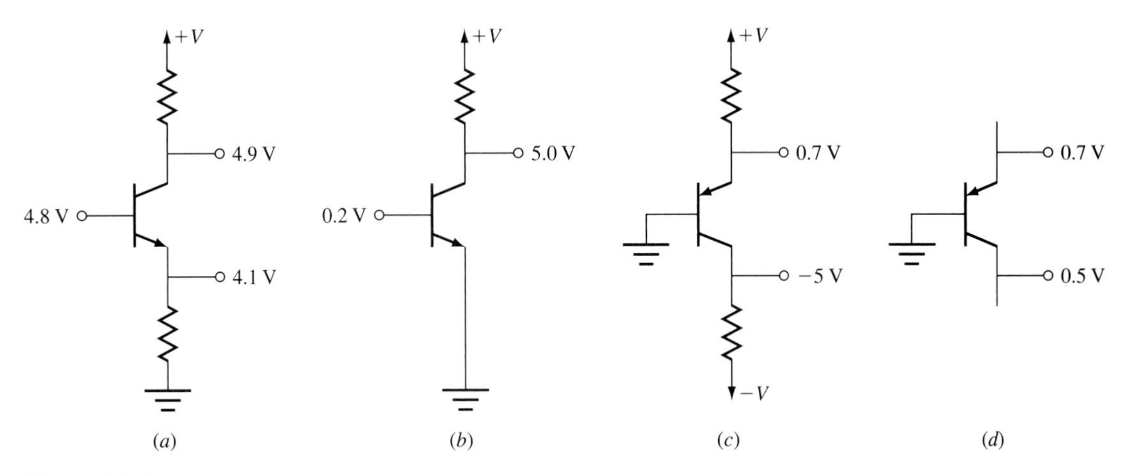

Figure P9.1

9.2 For each area in Figure P9.2, identify how the region should be doped to make a good pnp transistor (n^+, n, intrinsic, p, p^+), and indicate the reason(s). Indicate where the active pnp transistor occurs.

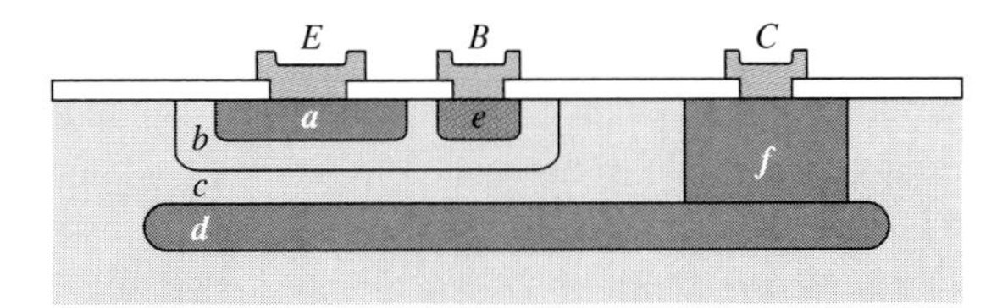

Figure P9.2

9.3 Draw the energy band diagram for a pnp transistor at equilibrium and under forward active bias.

9.4 From Equation (9.13), $\alpha = \gamma \alpha_T M$, where M is the carrier multiplication factor in the base-collector junction. For small base-collector voltage, $M = 1$ and $\alpha = \gamma \alpha_T$ and $\beta = \alpha/(1 - \alpha)$. Show that for avalanche breakdown, $M = 1 + 1/\beta$.

9.5 Using the prototype model (i.e., ignore the apparent emitter band-gap narrowing reduction on the injection efficiency), find α and β for an npn BJT with $N_{DE} = 10^{19}\ \text{cm}^{-3}$, $N_{AB} = 2 \times 10^{17}\ \text{cm}^{-3}$, and $N_{DC} = 10^{17}\ \text{cm}^{-3}$. Indicate clearly all your steps. Let $V_{BE} = 0.8$ V and $V_{CB} = 2.0$ V. The metallurgical widths are $W_{EM} = 0.2\ \mu\text{m}$ and $W_{BM} = 0.2\ \mu\text{m}$.

9.6 For the device of Example 9.1, estimate the emitter-base junction width, the width appearing on the emitter side, and that appearing on the base side. Assume that the junction is forward biased with $V_j = V_{bi} - V_a = 0.2\,\text{V}$. Draw the emitter and base to scale and indicate the locations of the edges of the transition region.

9.7 Derive the simple-model expression for β for a pnp transistor with a nondegenerate emitter [the pnp equivalent of Equation (9.42)].

9.8 A prototype npn transistor has its emitter doped to $N_D = 5 \times 10^{19}\,\text{cm}^{-3}$. The base is doped with $N_A = 2 \times 10^{17}\,\text{cm}^{-3}$. Find ΔE_g^*, accounting for narrowing of the emitter conduction band. Find β if $W_E = 0.12\,\mu\text{m}$ and $W_B = 0.07\,\mu\text{m}$.

9.9 In Example 9.2, we neglected the band-gap narrowing in the base. Find the current gain β, this time taking ΔE_{gB}^* into account. Draw the equilibrium energy band diagram, and indicate the true band edges and apparent band edges in both the emitter and the base. What is the effective barrier height for electrons? Holes?

9.10 Suppose that a degenerate Si layer of $N_D = 10^{20}\,\text{cm}^{-3}$ and $N_A = 0$ is epitaxially deposited onto the emitter of the transistor of Section 9.4. Explain how such a layer could increase β slightly. Both emitter layers together are still much shorter than the diffusion length of either electrons or holes in the emitter. [*Hint:* There is a small amount of band-gap narrowing due to the acceptors in the base (and the compensated acceptors in the emitter)].

9.11 Working engineers must be able to teach themselves new things throughout their careers to keep up to date. Research and learn about SIMS, and write up a short explanation of what it does and how it works.

9.12 If a pnp transistor is made with the same dimensions and doping concentrations as an npn, explain why the pnp will have a lower β. Neglect band-gap narrowing effects.

9.13 If an npn transistor is considered to have all step junctions, with the properties in the table below, find:

$N_{DE} = 5 \times 10^{19}\,\text{cm}^{-3}$	$N_{AB} = 4 \times 10^{17}\,\text{cm}^{-3}$	$N_{DC} = 10^{17}\,\text{cm}^{-3}$
$W_{EM} = 0.13\,\mu\text{m}$	$W_{BM} = 0.20\,\mu\text{m}$	
$V_{BE} = 0.8\text{ V}$ (forward bias)	$V_{CB} = 2\text{ V}$ (reverse bias)	

a. The minority carrier lifetimes
b. The corresponding diffusion coefficients and diffusion lengths
c. The built-in voltage of the collector-base junction
d. The built-in voltage of the emitter-base junction (neglecting band-gap narrowing)
e. W_B

f. γ, neglecting band-gap narrowing

g. α_T

h. α, assuming $M = 1$,

i. β

9.14 Repeat Problem 9.13, but this time account for band-gap narrowing in the emitter.

9.15 Find β for an npn transistor with a degenerately doped emitter at $N'_{DE} = 10^{20}\ \text{cm}^{-3}$ and $N'_{AB} = 5 \times 10^{17}\ \text{cm}^{-3}$. Let $W_E = 0.15\ \mu\text{m}$ and $W_B = 0.1\ \mu\text{m}$.

9.16 Consider the graded-base transistor whose net doping profile is shown in Figure P9.3.

a. Find η.

b. Find the built-in electric field in the base.

c. Estimate β.

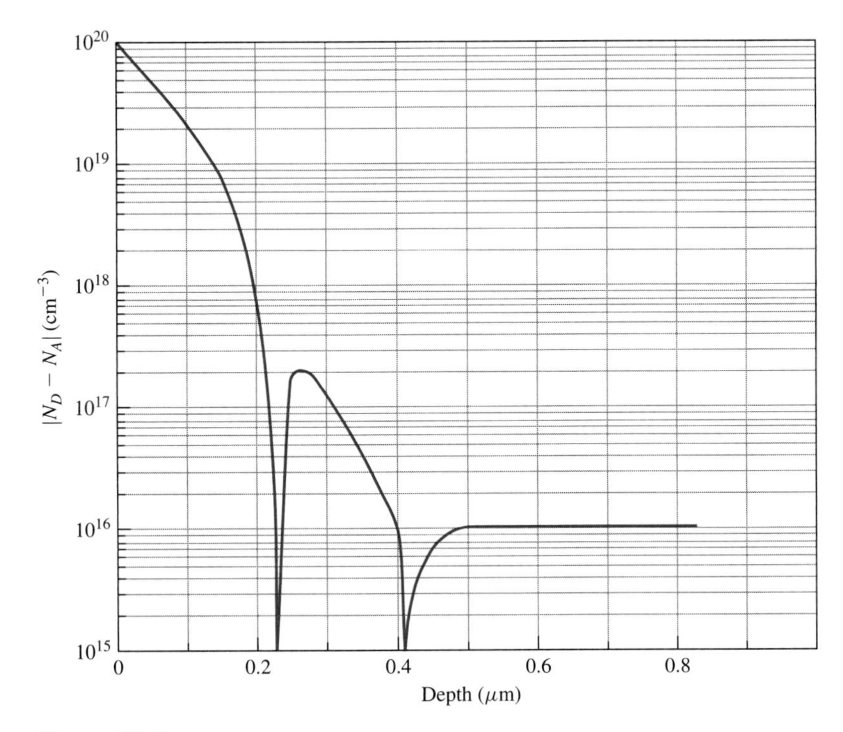

Figure P9.3

9.17 A graded-base transistor has its emitter degenerately doped to $10^{19}\ \text{cm}^{-3}$ and its base has a doping concentration of $5 \times 10^{17}\ \text{cm}^{-3}$ at the emitter edge and a grading parameter $\eta = 3$. The emitter width is $W_E = 0.2\ \mu\text{m}$ and base width is $W_B = 0.15\ \mu\text{m}$. Find β.

9.18 An npn BJT is operating in the forward active region. Assuming the default SPICE values $I_S = 10^{-16}$ A, $\beta_F = 100$, find the terminal currents I_C, I_B, and I_E for $V_{BE} = 0.7$ V.

9.19 Typical forward β's are on the order of 100, while typical reverse β's are on the order of 0.1. Find the corresponding α_F and α_R.

9.20 Fill in the missing steps to verify Equation (9.75).

9.21 We previously chose an equivalent resistance in the base using the average voltage drop across the base. Here we will use the power dissipated in a resistor, $P = I^2R$. We define a new equivalent resistance based on the power P_B that is dissipated by the base current flowing through a lumped resistance $R_B = P_B/I_B^2$, where P_B is the power dissipated in the base. Find an expression for R_B. Assume I_B to be small enough that Equation (9.73) is valid.

9.22 For the transistor whose I_C-V_{CE} curves are shown in Figure P9.4, find the Early voltage.

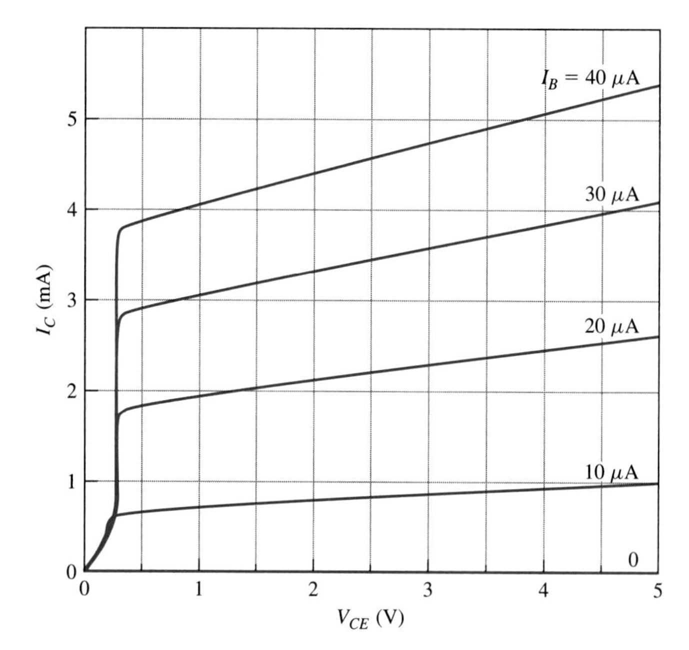

Figure P9.4

9.23 An npn transistor has $N'_{DE} = 10^{20}$, $N'_{AB} = 5 \times 10^{17}$, and $N'_{DC} = 10^{16}$. For $V_{BE} = 0.75$ V, how small can the base width W_{BM} be to keep the punch-through voltage above 12 V?

9.24 As indicated in Section 9.11, to avoid excessive base push-out, J_C is less than about 0.25 mA/μm^2. Consider a BJT with an emitter width (L) of 0.5 μm and a base sheet resistance of 10 k$\Omega/\square$. For $\beta = 100$, find I_B and the lateral voltage drop in the intrinsic base region and discuss the current crowding effect for this device.

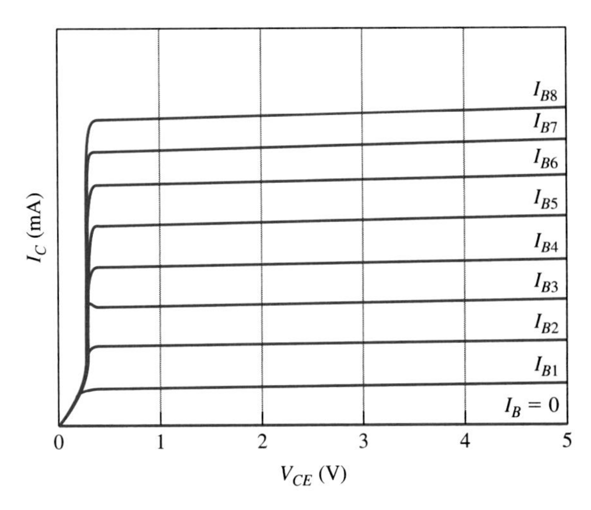

Figure P9.5

9.25 For the transistor whose I_C-V_{CE} curves are shown in Figure P9.5, explain why the lines are closer together as I_B increases.

9.26 Find β for the device whose Gummel plot is shown in Figure P9.6.

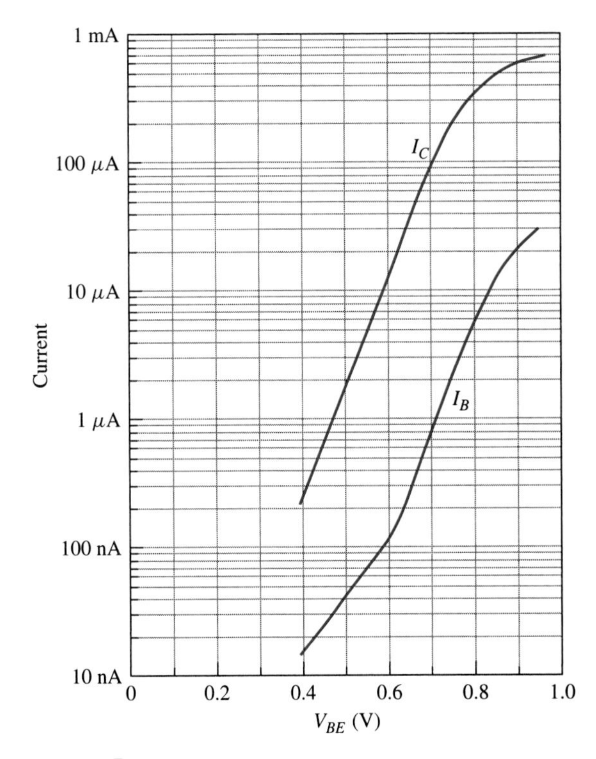

Figure P9.6

9.27 Figure P9.7 shows some experimental data for the current gain (normalized to that at 300 K) of a bipolar junction transistor as a function of temperature. Also included is a straight-line approximation to the linear region. From the data, estimate the value of ΔE_g^*.

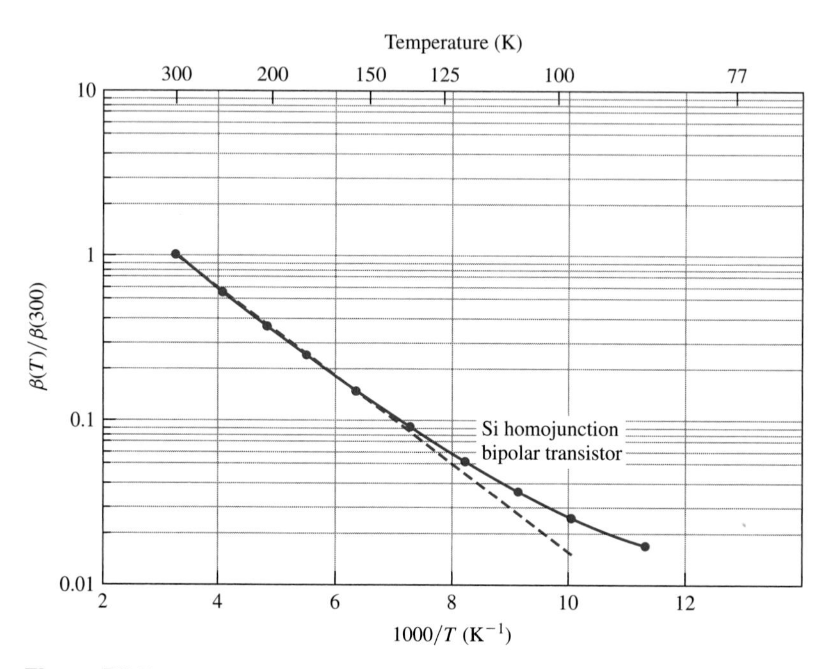

Figure P9.7

CHAPTER 10

Time-Dependent Analysis of BJTs

10.1 INTRODUCTION

In Chapter 9 the dc characteristics of bipolar junction transistors were discussed to illustrate the physics of operation. However, in many systems high frequency operation in analog systems and fast switching speed in digital circuits are of great importance. In this chapter, the time-dependent characteristics of BJTs are examined. First, small-signal ac models are examined for use in analog circuits. Then switching transients are investigated for digital circuits. The advantages and disadvantages of BJTs compared with MOSFETs are examined and the use of both types of transistors on a chip (BiMOS) is discussed.

10.2 EBERS-MOLL AC MODEL

In Chapter 9, Section 9.6, the Ebers-Moll dc model of a BJT was presented. To model the ac behavior, the parasitic resistances and capacitances must be considered. Figure 10.1 shows the Ebers-Moll ac common emitter equivalent circuit for a BJT. It is the dc equivalent circuit of Figure 9.14 with the parasitic capacitances added. Since the capacitances are more important than the parasitic resistances in determining the time-varying behavior, for simplicity, the parasitic resistances are omitted (we will attend to them later). In Figure 10.1, C_{jBE} and C_{jBC} represent the base-emitter and base-collector junction capacitances respectively, while C_{scBE} and C_{scBC} represent the stored-charge capacitances associated with the forward-biased base-emitter and base-collector junctions. The terminal currents from Section 9.6 are repeated here.

$$I_E = I_{CT} + \frac{I_F}{\beta_F + 1}$$

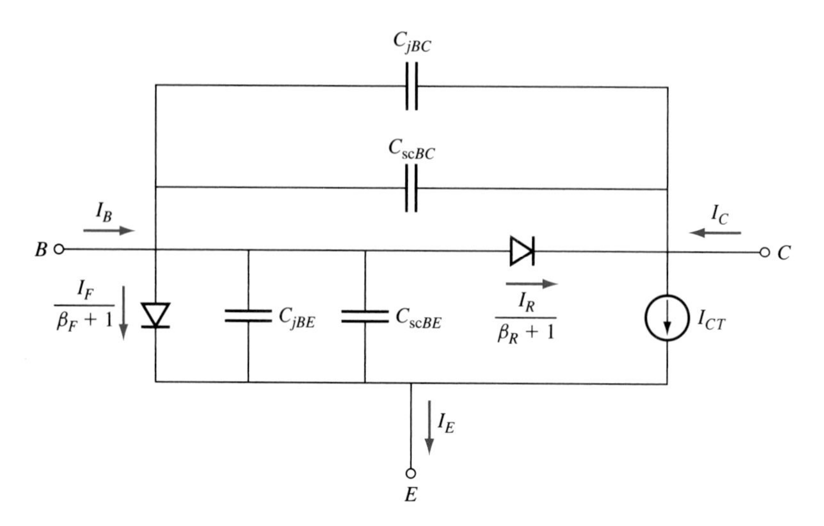

Figure 10.1 Ebers-Moll common emitter ac model for a BJT.

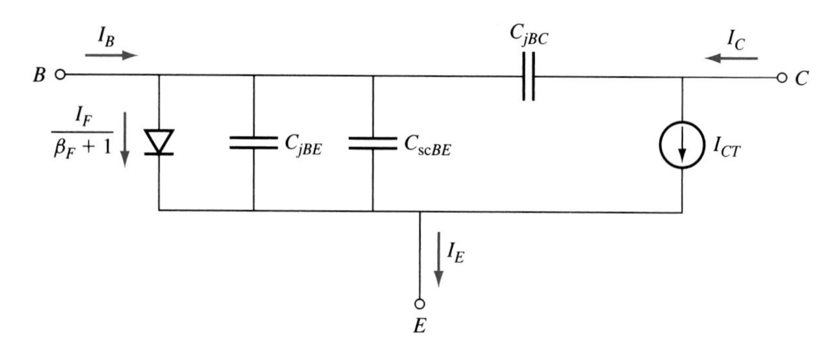

Figure 10.2 Ebers-Moll common emitter model for a BJT operating in the forward active mode. The collector-base junction is reverse biased, so C_{scBC} and I_R are neglected.

$$I_C = I_{CT} + \frac{I_R}{\beta_R + 1}$$

$$I_B = \frac{I_F}{\beta_F + 1} + \frac{I_R}{\beta_R + 1}$$

where

$$I_{CT} = \frac{\beta_F I_F}{\beta_F + 1} - \frac{\beta_R I_R}{\beta_R + 1}$$

The Ebers-Moll equivalent circuit operating in the active mode is shown in Figure 10.2. Since the base-collector junction is reverse biased, I_R and C_{scBC} are neglected.

The Ebers-Moll model is useful for analyzing both switching and analog circuits, and in the next section the use of Ebers-Moll for small-signal analysis is discussed.

10.3 SMALL-SIGNAL EQUIVALENT CIRCUITS

Many analog circuits use BJTs operating in the active mode to amplify small-signal voltages and currents. We will consider an amplifier operating in the common emitter configuration as indicated in Figure 10.3. In (a) the input and output

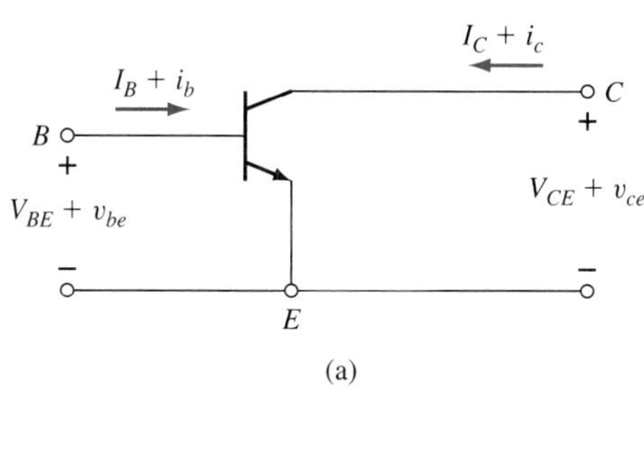

(a)

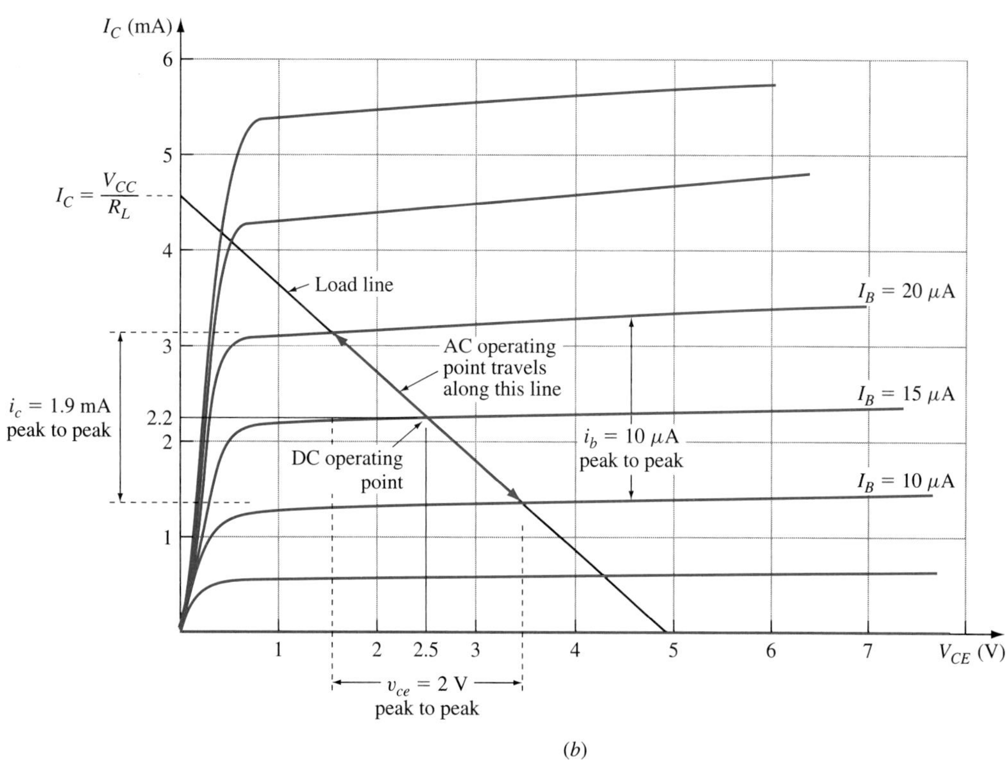

(b)

Figure 10.3a–b (a) The common emitter circuit with the dc (capital letters) and small-signal, or ac, quantities indicated; (b) illustration of operation along the load line. For $I_B = 15\ \mu A$, the dc operating point is at $I_C = 2.2$ mA and $V_{CE} = 2.5$ V.

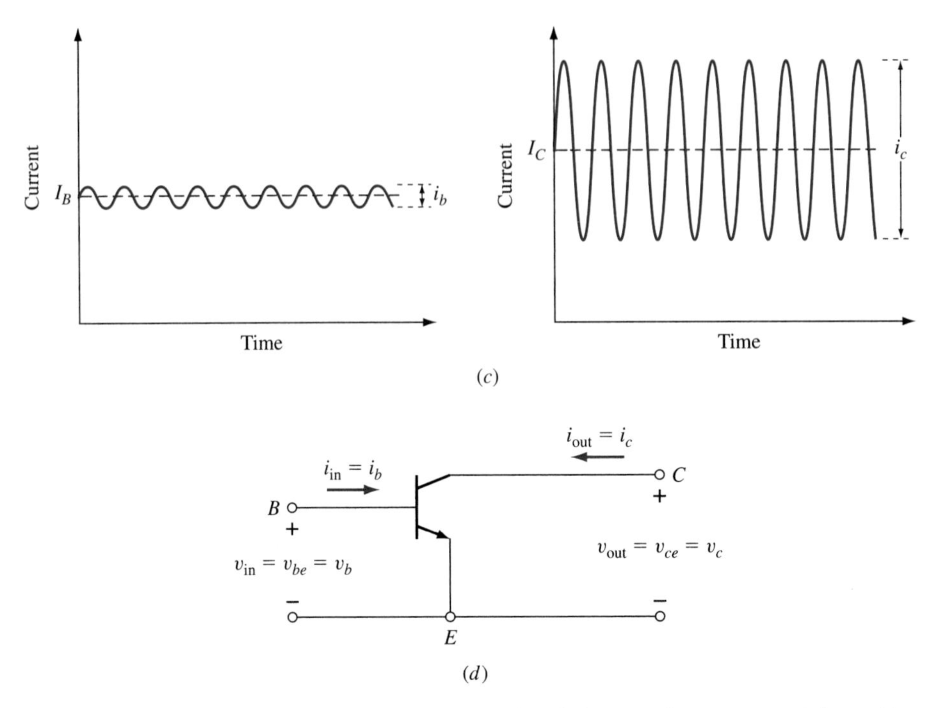

Figure 10.3c–d (c) The total input and output signals have a dc component (I_B and I_C respectively) and an ac component (i_b and i_c respectively); (d) the common emitter circuit, considering only the ac quantities.

voltages and currents are indicated. We use the convention that uppercase symbols such as I_B, I_C, V_{BE}, and V_{CE} represent dc quantities, and lowercase symbols i_b, i_c, v_{be}, and v_{ce} represent ac quantities. Each node will have a dc bias point and a dc current associated with that bias point, and the analog signal is taken to be a small variation around that bias point. For example, in Figure 10.3b, the dc bias point is at $I_B = 15\ \mu\text{A}$, $V_{CE} = 2.5\ \text{V}$, and $I_C = 2.2\ \text{mA}$. The base current varies around the dc point by 10 μA peak-to-peak. For a load resistance R_L and a supply voltage V_{CC}, the dc collector current is

$$I_C = \frac{V_{CC} - V_{CE}}{R_L}$$

which is represented by the load line in (b). Since $I_C = \beta I_B$, the operating point corresponds to the intersection of the load line with the I_C-V_{CE} characteristics for a given value of I_B. The operating point moves along the load line indicated in the figure. If I_B increases by a small amount, I_C increases by β times that amount. The output voltage V_{CE} drops proportionally.

If the dc bias point is fixed and small ac changes are made, then small variations in i_b cause large variations in i_C, as shown in Figure 10.3c. The corresponding circuit for small-signal analysis is shown in Figure 10.3d. Since the

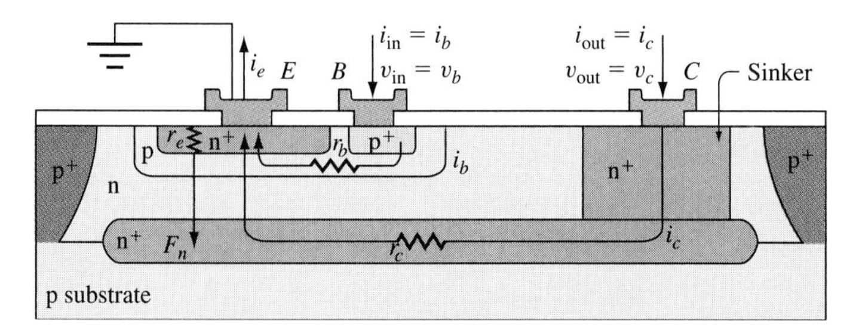

Figure 10.4 An integrated circuit bipolar junction transistor in the common emitter configuration, showing the small-signal currents.

common emitter is used as reference, the ac input current, output current, input voltage, and output voltage are respectively i_b, i_c, v_b, and v_c.

There are a number of possible small-signal equivalent circuits obtainable from Figure 10.3d, but the most common is the so-called hybrid-pi equivalent circuit, so we will consider this next.

10.3.1 HYBRID-PI MODELS

To model the small-signal behavior, we consider the integrated circuit BJT of Figure 10.4 (the BJT of the previous chapter). The small-signal currents and voltages are indicated. Also shown are the series base resistance r_b, the collector resistance r_c, and the emitter resistance r_e. These arise from the finite conductivities of the semiconductor regions. These are the actual, physical resistances in the device, and include the contact resistances.

The equivalent circuit model for an ideal transistor in the common emitter configuration is shown in Figure 10.5a. This is known as the *hybrid-pi model.*[1] The resistances are shown, along with several parasitic capacitances. This model can be analyzed just like an actual circuit, and will yield the same (small signal) behavior as an ideal transistor. In an ideal transistor, the resistances would be zero, and the hybrid-pi model would reduce to the Ebers-Moll model of Figure 10.2.

In the hybrid-pi model, the capacitance C_μ between collector and base is the junction capacitance of the reverse-biased collector-base junction:

$$C_\mu = C_{jBC} \tag{10.1}$$

The capacitance of a reverse-biased diode was discussed in Chapter 5.

The capacitance between the base and the emitter is

$$C_\pi = C_{jBE} + C_{scBE} \tag{10.2}$$

[1]It is called this because the arrangement of the resistors in the model could be imagined to form the Greek letter pi.

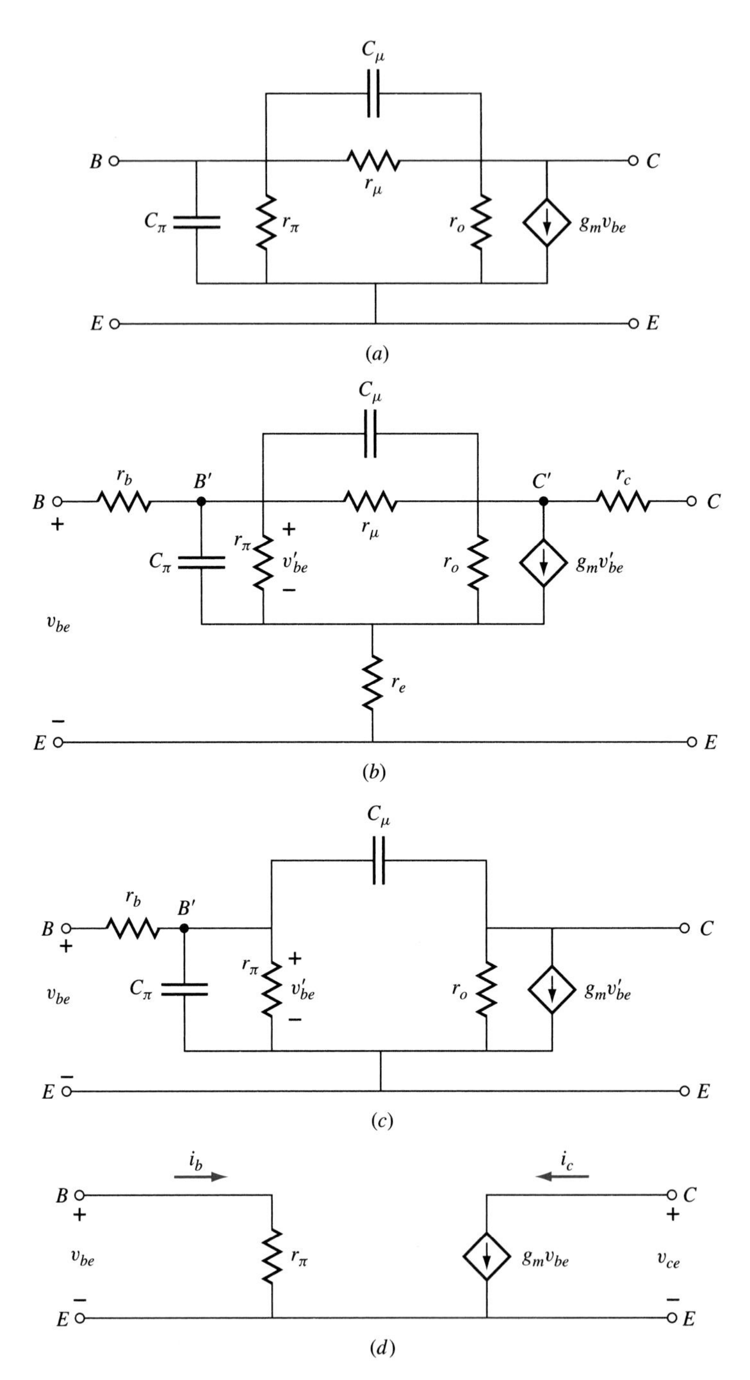

Figure 10.5 Hybrid-pi models: (a) Ideal BJT (all resistances can be neglected). (b) All resistances including contact resistances are included. In practical BJTs, r_e and r_c are often small enough to be neglected, and the feedthrough resistance r_μ is large (the resistance of the reverse-biased *C-B* junction). These approximations simplify the hybrid-pi model to (c). (d) Low-frequency model in which the capacitances are neglected.

This capacitance C_π is the sum of two capacitances, C_{jBE}, the junction capacitance of the forward-biased base-emitter junction, and C_{scBE}, the base-emitter stored-charge capacitance. Charge storage capacitance was discussed in connection with diodes and has the same meaning in BJTs.

The hybrid-pi model shows a differential input resistance r_π, which is a measure of how the base current changes for a differential change in base voltage. Since the *I-V* characteristics are not linear for a diode junction, this depends on the slope of the *I-V* curve at the operating point of interest:

$$r_\pi \approx \frac{1}{\left.\dfrac{\partial I_B}{\partial V_{BE}}\right|_{V_{CE}}} = \frac{v_{be}}{i_b} \tag{10.3}$$

We know that the relationship between I_B and V_{BE} is $I_B \approx I_{BO}(e^{qV_{BE}/kT} - 1)$ so

$$\left.\frac{\partial I_B}{\partial V_{BE}}\right|_{V_{CE}} = \frac{q}{kT} I_{B0} e^{qV_{BE}/kT} \approx \frac{qI_B}{kT} \tag{10.4}$$

Therefore combining Equations (10.3) and (10.4) gives

$$r_\pi \approx \frac{kT}{qI_B} \approx \frac{\beta_{\text{DC}} kT}{qI_C} \tag{10.5}$$

In this last step we also used the relationship $\beta_{\text{DC}} = I_C/I_B$, which represents the dc or low-frequency current gain.

We can express r_π in terms of the device parameters by expressing I_B as

$$I_B = I_{nB} + I_{pB} \approx I_{pB} \tag{10.6}$$

since the recombination current in the base, I_{nB}, is normally much smaller than the base-to-emitter injection current I_{pB}.

Since from Chapter 9, for a nondegenerate prototype transistor, $I_{pB} = I_{pE} = (qA_E D_{pB} p_E(0^-)/W_E)$ [Equation (9.27)][2]

$$I_B = qA_E \left(\frac{D_{pE}}{W_E} p_E(0^-) \right) \tag{10.7}$$

Substituting $p_E(0^-) = (n_i^2/N_{DE}) \exp(qV_{BE}/kT)$ yields

$$I_B = qA_E n_i^2 \left(\frac{D_{pE}}{W_E N_{DE}} \right) e^{qV_{BE}/kT} \tag{10.8}$$

From Equation (10.5),

$$r_\pi = \frac{\dfrac{kT}{q^2 A_E n_i^2} e^{-qV_{BE}/kT}}{\left(\dfrac{D_{pE}}{W_E N_{DE}} \right)} = \frac{kT W_E N_{DE}}{q^2 A_E n_i^2 D_{pE}} e^{-qV_{BE}/kT} \tag{10.9}$$

[2]Here the currents are taken as being positive in the negative x direction.

The feedthrough resistance r_μ between the collector and the base is

$$r_\mu = \frac{1}{\left.\dfrac{\partial I_C}{\partial V_{CB}}\right|_{V_{BE}}} \tag{10.10}$$

This is the differential resistance of the reverse-biased collector-base junction (the reciprocal slope of the common base output characteristics). This is normally large enough that it can be ignored.

The output resistance r_o is

$$r_o \approx \frac{1}{\left.\dfrac{\partial I_C}{\partial V_{CE}}\right|_{V_{CB}}} = \frac{V_A}{I_C} \tag{10.11}$$

which is the reciprocal slope of the common emitter output characteristics. The quantity V_A is the Early voltage.

The hybrid-pi model's dependent current generator produces a current proportional to the base-emitter voltage v_{be}. The value of that current is $g_m v_{be}$, where g_m is the device transconductance,

$$g_m = \left.\frac{\partial I_C}{\partial V_{BE}}\right|_{V_{CE}} = \frac{i_c}{v_{be}} \tag{10.12}$$

The transconductance is a measure of how the input voltage controls the output current in a transistor.

In a transistor operated in the forward active region the collector current is essentially equal to the emitter current, and the emitter current depends on the base-emitter voltage, giving

$$I_C \approx I_{C0} e^{qV_{BE}/kT} \tag{10.13}$$

where I_{C0} is the base-collector junction leakage current (Equation 5.78). Thus, from Equation (10.12),

$$g_m = \frac{qI_C}{kT} \tag{10.14}$$

But, from Equation (10.5)

$$\frac{1}{r_\pi} = \frac{qI_B}{kT}$$

Combining the last two equations gives $g_m/(1/r_\pi) = I_C/I_B = \beta_{DC}$, or

$$g_m = \frac{\beta_{DC}}{r_\pi} \tag{10.15}$$

Now we have expressions for the parameters of the hybrid-pi model, but so far we have discussed this model only for the ideal transistor. If the series resistances are included, the equivalent circuit of Figure 10.5b results. For example, now some voltage is dropped across the input resistor r_b. Here, then, the dependent current generator has the value $g_m v'_{be}$, where v'_{be} is the voltage across r_π; i.e.,

$$v'_{be} = v_{be} - i_b r_b - i_e r_e \tag{10.16}$$

Because the emitter is heavily doped and thin, r_e is small and can often be neglected.[3]

The value of the collector resistance r_c is reduced by the use of a degenerate n^+ collector or buried layer and the n^+ sinker that were indicated in Figure 10.4. In a well-designed BJT, r_c is small enough that it also can often be neglected. Further, the resistance r_μ between the base and collector is very large because that junction is reverse biased. Thus, this feedthrough resistance can be approximated as being infinite. With these approximations, the equivalent circuit of Figure 10.5c results.

Since the parasitic resistances reduce the voltage gain of the circuit, they should be minimized. Designers can decrease the base resistance by making the base region more heavily doped and increasing the base width. There are trade-offs here, however, since increased base doping increases C_{jBE}, while increasing the base thickness decreases β_{DC} and thus g_m.

For a well-designed transistor, then, r_e, r_b, and r_c are often small enough to be neglected. At low frequencies, such that the capacitances can be ignored, one arrives at the simplified equivalent circuit of Figure 10.5d. This circuit, because of its simplicity, is used extensively for first-order analysis.

For higher frequencies, the hybrid-pi model of Figure 10.5b or c is used. In the next section, we examine the effect of the capacitances on the high-speed behavior of BJTs. The model of Figure 10.5b, however, is valid only up to frequencies on the order of a few gigahertz. Above this, the accuracy decreases with increasing frequency because it does not take into account the carrier delay (transit time) from emitter to collector.

10.4 STORED-CHARGE CAPACITANCE IN BJTs

We discussed the stored-charge capacitance of a pn junction in Chapter 5. Here we re-examine it in terms of the BJT with the aid of Figure 10.6. This figure shows a transistor having uniform base and emitter dopings. In part (a) of the figure we show the energy band diagram for an npn BJT operating in the active mode for a given emitter-base voltage. Also shown are the minority carrier distributions in the base and the emitter. Excess minority carriers (electrons) are injected into the base from the emitter and extracted at the base-collector junction. Similarily excess holes are injected from base to emitter and extracted at the emitter contact.

---

[3]In modern BJTs designed for high-frequency operation, r_e is not always negligible. We discuss this further in Section 10.6.

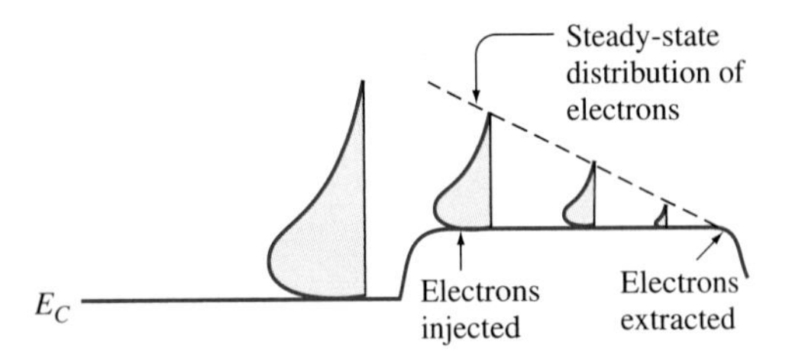

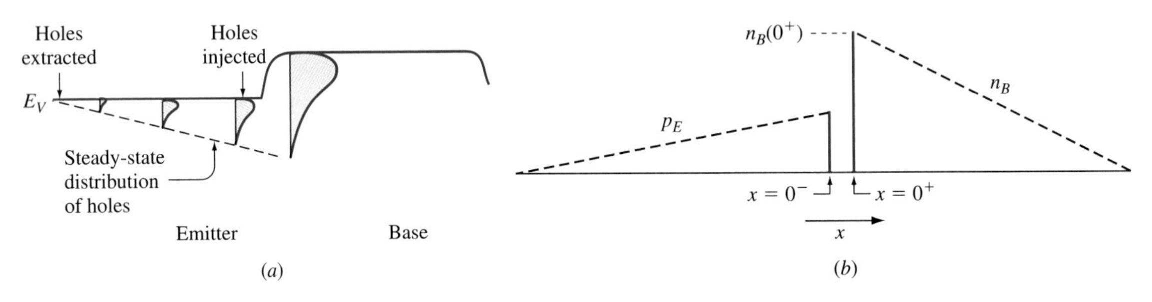

Figure 10.6 The injected charge on each side of the base-emitter junction acts as a capacitance. (a) The energy band diagram; (b) the charge distributions.

The distribution of charge in the base, if uniformly doped, is very close to a straight line, as shown in Figure 10.6b. The total stored charge in the base is

$$Q_B = -\frac{qA_E n_B(0^+)W_B}{2} \tag{10.17}$$

where $n_B(0^+)$ is the electron density at the emitter edge of the base, and A_E is the area of the emitter. But for a uniformly doped base the electron current moves by diffusion, so from Equation (3.40), and using $I = JA$, we have

$$I_n = -qA_E D_n \frac{dn(x)}{dx}$$

where the minus sign arises because current is taken as positive in the negative x direction in our present coordinate system.

The slope of the electron concentration is $-n_B(0^+)/W_B$, and

$$I_{nB} = \frac{qA_E n_B(0^+)D_n}{W_B} \tag{10.18}$$

Thus, combining Equations (10.17) and (10.18) gives

$$Q_B = -\frac{W_B^2 I_{nB}}{2D_n} \tag{10.19}$$

Similarly, the charge stored in the emitter is

$$Q_E = \frac{W_E^2 I_{pE}}{2D_p} \tag{10.20}$$

In a well-designed BJT, however, the hole current in the emitter is much smaller than the electron current in the base ($I_{pE} \ll I_{nB}$), and so we can often neglect the effect of stored carrier charge in the emitter.

Figure 10.7a shows the distribution of carriers in a uniformly doped base at some time $t = 0$. (Compare this with the charge distribution for a long-base

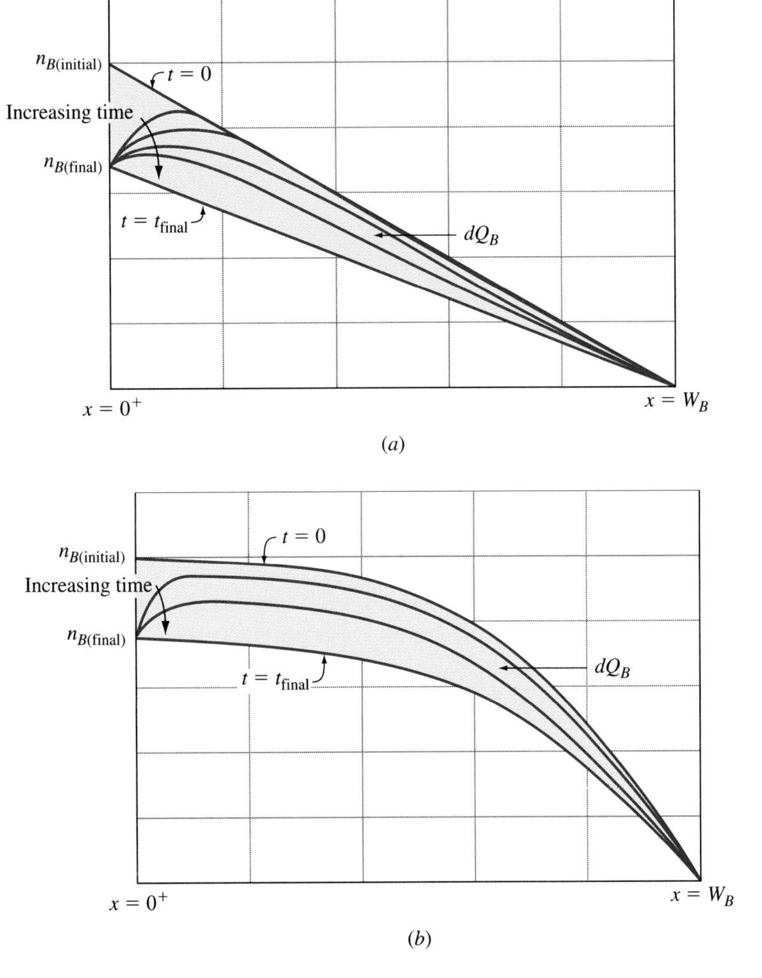

Figure 10.7 The change in the charge distribution when the injection is changed in a uniformly doped junction. It takes time to get rid of the excess charge, which is equivalent to discharging a capacitor. (a) Uniformly doped base; (b) graded base.

diode in Figure 5.33.) When the emitter-base voltage is decreased by an amount dV_{BE}, it reduces the number of carriers being injected across the $x = 0$ plane (in other words, $n_B(0^+)$ changes). The steady-state stored base charge is decreased by an amount equal to the shaded area:

$$dQ_B = \frac{W_B^2}{2D_n} \frac{dI_{nB}}{dV_{BE}} dV_{BE} \tag{10.21}$$

The evolution of the charge distribution in a uniform base is also shown in Figure 10.7a. Of this charge dQ_B that must be gotten rid of, some diffuses back to the emitter and a portion diffuses to the collector, as was the case for the diode of Chapter 5. Only the recoverable fraction δ of dQ_B that is recovered by the emitter flows in the external circuit and contributes to the stored-charge capacitance. Defining the recoverable charge dQ_{Br} by

$$dQ_{Br} = \delta dQ_B \tag{10.22}$$

For the case of the uniformly doped base, δ is about $\frac{2}{3}$, or two-thirds of dQ_B is recovered by the emitter.

For the case of the graded-base transistor of Figure 10.7b, the field in the base accelerates the base electrons to the collector and reduces the fraction that returns to the emitter. That reduces the reclaimable charge, and so correspondingly reduces the stored-charge capacitance.[4] For a typical base gradient, the recovered fraction of dQ_B is on the order of $\delta = 0.2$ to 0.3.

Determination of this reclaimable charge is beyond the scope of this book, so we simply state without proof that the stored charge in the base, Q_B, is given by [1]

$$Q_B = \frac{I_C W_B^2}{\langle D_{nB} \rangle} \left[\frac{\eta - 1 - e^{-\eta}}{\eta^2} \right] \tag{10.23}$$

where η is the doping parameter, $\langle D_{nB} \rangle$ is the average diffusion coefficient in the base, and the charge that is reclaimable, Q_{Br}, is

$$Q_{Br} = \frac{I_C W_B^2}{\langle D_{nB} \rangle} \left[\frac{1 - e^{-\eta}}{\eta^2} \right] \left[\frac{\sinh \eta - \eta}{\cosh \eta - 1} \right] \tag{10.24}$$

Then the fraction δ of reclaimable charge for a change in voltage dV_{BE} is

$$\delta = \frac{dQ_{Br}}{dQ_B} = \frac{(1 - e^{-\eta}) \left(\dfrac{\sinh \eta - \eta}{\cosh \eta - 1} \right)}{\eta - 1 + e^{-\eta}} \tag{10.25}$$

[4]For uniform base doping, diffusion is responsible for the recovered base charge, and the resultant capacitance is referred to as *diffusion capacitance*. Since in most BJTs the recovered charge is influenced by drift as well as diffusion, we refer to the resultant capacitance as *stored-charge capacitance*.

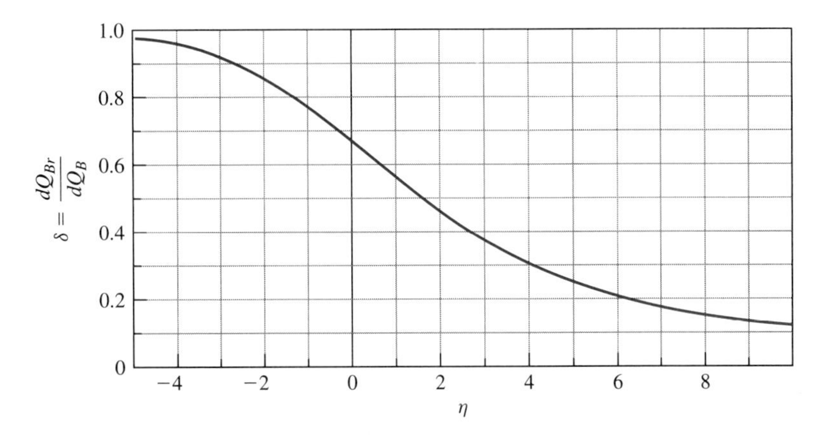

Figure 10.8 Fraction of reclaimable charge in the base as a function of grading parameter η in the base. As the fraction of reclaimable charge decreases, the stored-charge capacitance also decreases, and the device response time decreases (the device operates faster).

which is plotted in Figure 10.8. We can see that the fraction of the reclaimable charge reduces with increasing η, from $\frac{2}{3}$ for $\eta = 0$ (uniformly doped case), to 0.13 for $\eta = 10$. Note that this corresponds to a decrease in stored-charge capacitance, so that the graded-base transistor will have a faster operating speed than a uniformly doped one. Negative values of η indicate a field that accelerates injected charge back toward the emitter. Thus for $\eta = -3$, about 90 percent of the injected charge is reclaimable, which would increase the capacitance considerably and slow the device response.

We define the stored charge capacitance as

$$C_{scBE} \equiv \delta \left| \frac{dQ_B}{\partial V_{BE}} \right| \tag{10.26}$$

Then from Equation (10.21),

$$C_{scBE} = \frac{\delta W_B^2}{2\langle D_{nB}\rangle} \frac{\partial I_{nB}}{\partial V_{BE}} \tag{10.27}$$

Since the collector current consists primarily of the electrons coming from the base, we have

$$\frac{\partial I_{nB}}{\partial V_{BE}} \approx \frac{\partial I_C}{\partial V_{BE}} \approx \beta_{DC} \frac{\partial I_B}{\partial V_{BE}} = \frac{\beta_{DC}}{r_\pi} \tag{10.28}$$

where we have used $I_C = \beta_{DC} I_B$ and Equation (10.3). Combining Equations (10.27) and (10.28) gives

$$C_{scBE} = \frac{\delta W_B^2 \beta_{DC}}{2\langle D_{nB}\rangle r_\pi} = \frac{\delta W_B^2 q I_C}{2\langle D_{nB}\rangle kT} \tag{10.29}$$

For constant base doping, $\delta = \frac{2}{3}$ and

$$C_{scBE} = \frac{W_B^2 \beta_{DC}}{3D_{nB} r_\pi} \qquad \text{uniformly doped base} \tag{10.30}$$

For the graded-base considered above, $\delta \approx 0.2$ and

$$C_{scBE} \approx \frac{W_B^2 \beta_{DC}}{10\,\langle D_{nB} \rangle\, r_\pi} \qquad \text{graded-base with } \eta = 6 \tag{10.31}$$

We can see that the stored-charge capacitance decreases with decreasing base width and it also decreases through the recoverable fraction δ, which decreases with increasing field in the base.

10.5 FREQUENCY RESPONSE

At high frequencies, the capacitance reduces the current gain of a BJT. In this section, we investigate the frequency response of the short-circuit current gain, using a BJT in the common emitter configuration. Figure 10.9 shows the hybrid-pi circuit of Figure 10.5c with r_b and r_c neglected and the output short-circuited. The output current is

$$i_c = (g_m - j\omega C_\mu) v_{BE} \tag{10.32}$$

while the base current is

$$i_b = \left(\frac{1}{r_\pi} + j\omega C_\pi + j\omega C_\mu \right) v_{BE} \tag{10.33}$$

The short-circuit current gain is then

$$\beta(\omega) = \frac{i_c}{i_b} = \frac{g_m - j\omega C_\mu}{\dfrac{1}{r_\pi} + j\omega(C_\pi + C_\mu)} = \frac{r_\pi (g_m - j\omega C_\mu)}{1 + j\omega r_\pi (C_\pi + C_\mu)} \tag{10.34}$$

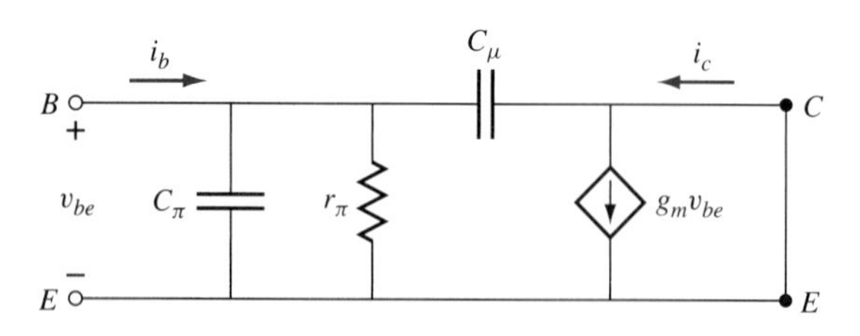

Figure 10.9 The hybrid-pi model for high-frequency short-circuit current gain.

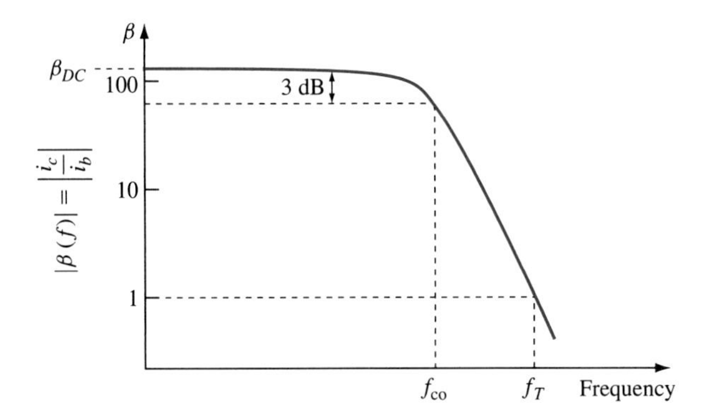

Figure 10.10 The frequency response of a BJT, showing the cutoff frequency f_{co} and the unity gain frequency f_T.

At normal operating frequencies and currents, $g_m \gg \omega C_\mu$, and so

$$\beta(\omega) \approx \frac{r_\pi g_m}{1 + j\omega r_\pi (C_\pi + C_\mu)} = \frac{\beta_{DC}}{1 + j\omega r_\pi (C_\pi + C_\mu)} \tag{10.35}$$

The magnitude of the current gain as a function of frequency is

$$\beta(f) = \frac{\beta_{DC}}{\sqrt{1 + \left(\frac{f}{f_{co}}\right)^2}} \tag{10.36}$$

where the cutoff frequency for the frequency response of β, f_{co} is

$$f_{co} = \frac{1}{2\pi r_\pi (C_\pi + C_\mu)} \tag{10.37}$$

This function is plotted in Figure 10.10.

10.5.1 UNITY CURRENT GAIN FREQUENCY f_T

Figure 10.10 indicates the frequency f_T at which current gain is unity. This figure of merit is known as *the unity current gain frequency* or *unity gain frequency*. From Equation 10.36, $\beta(f) = \beta(f_T) = 1$, or $\beta_{DC} = \sqrt{1 + (f_T/f_{co})^2}$. However, for $\beta = 1$, $f \gg f_{co}$ and

$$f_T = \beta_{DC} f_{co} \tag{10.38}$$

Note that f_T is equal to the current gain-bandwidth product ($\beta_{DC} f_{co}$).

EXAMPLE 10.1

Estimate f_T for a prototype BJT with a base doping of 10^{17} cm^{-3} and a base width of $W_B = 0.1\ \mu$m.

■ **Solution**

To obtain a rough estimate of f_T, we use the equivalent circuit of Figure 10.9, but neglect C_μ since it is usually small. The base current i_b results from the voltage v_{be} across r_π and C_π in parallel:

$$i_b = v_{be}\left(\frac{1}{r_\pi} + j\omega C_\pi\right)$$

The collector current is $i_c = g_m v_{be}$. Then the current gain $\beta(\omega)$ is

$$\beta(f) = \frac{i_c}{i_b} = \frac{g_m v_{be}}{v_{be}\left(\dfrac{1}{r_\pi} + j\omega C_\pi\right)} = \frac{g_m}{\left(\dfrac{1}{r_\pi} + j\omega C_\pi\right)}$$

which can be expressed

$$\beta(f) = \frac{g_m r_\pi}{1 + j\omega r_\pi C_\pi} = \frac{\beta_{\text{DC}}}{1 + j\omega r_\pi C_\pi}$$

This reduces to β_{DC} for $\omega = 0$, as expected.

For $|\beta(f)| = 1$, $\omega r_\pi C_\pi \gg 1$ and

$$|\beta(f)| = 1 = \frac{\beta_{\text{DC}}}{2\pi f_T r_\pi C_\pi} = \frac{\beta_{\text{DC}}}{2\pi f_T r_\pi (C_{jBE} + C_{\text{sc}BE})}$$

Except at low currents, the storage capacitance is appreciably larger than the junction capacitance and thus $C_\pi \approx C_{\text{sc}BE}$. Solving for f_T and using Equation (10.29) gives

$$f_T \approx \frac{\beta_{\text{DC}}}{\dfrac{2\pi\delta r_\pi W_B^2 \beta_{\text{DC}}}{2D_n r_\pi}} = \frac{2D_n}{2\pi\delta W_B^2}$$

For a prototype transistor, $\delta = \frac{2}{3}$, and for a base doping of 10^{17} cm^{-3}, $D_n \approx 20$ cm^2/s, from Figure 3.11. For a base width of $0.1\ \mu\text{m} = 10^{-5}$ cm, we have

$$f_T = \frac{2 \times 20\ \text{cm}^2/\text{s}}{2\pi \times \frac{2}{3} \times (10^{-5}\ \text{cm})^2} = 9.5 \times 10^{-10}\ \text{Hz} = 95\ \text{GHz}$$

This result is approximate, however, since the presence of parasitic resistances will lower this value.

In Example 10.1, we estimated the value of f_T as $f_T = 1/2\pi r_\pi C_{\text{sc}BE}$. Another approach is to estimate f_T by the time delay for carriers to cross the base from emitter to collector, discussed next.

10.5.2 BASE TRANSIT TIME

For an npn transistor, the electron current in the base is

$$I_{nB} = qA_E\, \Delta n(x) v_n(x) \tag{10.39}$$

where $v(x)$ is the average velocity at position x in the base. Neglecting the small recombination current, the overall base current I_B is independent of x, and the $\Delta n(x) v_n(x)$ product is constant. The time an electron requires to traverse a distance dx is

$$dt = \frac{1}{v_n(x)}\, dx \tag{10.40}$$

The time required to cross the base, then, is

$$t_T = \int_0^{t_{tB}} dt = \int_0^{W_B} \frac{1}{v_n(x)}\, dx \tag{10.41}$$

The velocity can be obtained from Equation (10.39) and

$$t_T = \int_0^{W_B} \frac{qA_E\, \Delta n}{I_{nB}}\, dx = \frac{1}{I_{nB}} \int_0^{W_B} qA_E\, \Delta n\, dx \tag{10.42}$$

The integral on the right-hand side (not containing I_{nB}) is the total stored minority carrier charge in the base, Q_B. The base transit time is then

$$t_T = \frac{Q_B}{I_{nB}} = \frac{Q_B}{I_C} \tag{10.43}$$

Equation (10.43) is valid for an arbitrary base doping profile. For uniform doping in the base, the electron distribution is linear, decreasing from emitter to base:

$$\Delta n(x) = \Delta n(0^+) \left(1 - \frac{x}{W_B}\right) \tag{10.44}$$

and the stored charge is

$$Q_B = \frac{qA_E\, \Delta n(0^+) W_B}{2} \tag{10.45}$$

Since the collector current is almost entirely due to electrons arriving from the base,

$$I_C = I_{nB} = qA_E D_n \frac{d\,\Delta n}{dx} \tag{10.46}$$

Thus, with the aid of Equation (10.44),

$$I_C = \frac{qA_E D_n\, \Delta n(0^+)}{W_B} \tag{10.47}$$

and from Equation (10.43),

$$t_T = \frac{W_B^2}{2D_n} \tag{10.48}$$

For this approximation,

$$f_T = \frac{1}{2\pi t_T} \tag{10.49}$$

EXAMPLE 10.2

Find the base transit time for electrons in the npn prototype transistor of Example 10.1 [base doping level of 10^{17} cm^{-3} and base width of $W_B = 0.1\ \mu\text{m}\ (10^{-5}\ \text{cm})$].

■ Solution

From Equation (10.48),

$$t_T = \frac{W_B^2}{2D_n} = \frac{(10^{-5})^2}{2 \times 20} = 2.5 \times 10^{-12}\ \text{s} = 2.5\ \text{ps}$$

where $D_n \approx 20$ cm^2/s was found from Figure 3.11. This result is more than 5 orders of magnitude smaller than the electron lifetime in the base, indicating that on the order of 10^{-5} of the base electrons recombine. That is, only one electron in about 100,000 is lost to recombination in the base, and $\alpha_T \approx 1$.

In this approximation,

$$f_T = \frac{1}{2\pi t_T} = \frac{1}{2\pi \times 2.5 \times 10^{-12}\ \text{s}} = 63.6\ \text{GHz}$$

It is reduced from that of Example 10.1 by the factor $1/\delta$.

In a graded-base transistor, the above value of t_T is reduced and thus f_T is increased because the field in the base accelerates the electrons toward the collector (i.e., the transport is by drift as well as diffusion.) For a base field greater than about 5 kV/cm, the electrons traverse the base at the saturation velocity. In this case,

$$t_T \approx \frac{W_B}{v_{\text{sat}}} = \frac{0.1\ \mu\text{m}}{10^7\ \text{cm/s}} = 1\ \text{ps}$$

for the above example.

10.5.3 BASE-COLLECTOR TRANSIT TIME t_{BC}

In addition to the delay due to the transit time across the base, there is a delay due to the time t_{BC} required for a carrier to traverse the depletion region between base and collector. Because in this region the field is large, the carrier velocity over most of the depletion region is equal to its saturation velocity:

$$t_{BC} = \frac{w_{BC}}{v_{\text{sat}}} \tag{10.50}$$

where w_{BC} is the width of the base-collector depletion region and depends on the doping levels and collector-base voltage.

10.5.4 MAXIMUM OSCILLATION FREQUENCY f_{max}

While the unity current gain frequency f_t is a convenient figure of merit for BJTs, particularly at low current levels, it does not consider the effects of parasitic resistances. Another common figure of merit is the *maximum oscillation frequency* f_{max}, the frequency at which the power gain of the device is unity when the base resistance is considered. This can be expressed [2]

$$f_{max} = \left(\frac{f_T}{8\pi r_b C_{jBC}} \right)^{1/2} \tag{10.51}$$

where r_b is the base resistance and C_{jBC} is the base-collector junction capacitance. Note that, depending on the values of r_b and C_{jBC}, f_{max} may be either larger or smaller than f_T. In modern high-performance transistors, usually $f_{max} > f_T$.

10.6 HIGH-FREQUENCY TRANSISTORS

In Chapter 9, the dc characteristics of conventional transistors were discussed, and so the parasitic capacitances were not important. For high-frequency operation, however, it is necessary to use structures that minimize the parasitic resistances and capacitances. In the next section, we discuss one such structure, the double poly Si self-aligned structure with a polysilicon emitter. Following that we will look at the effects of the base transit time on the high-speed operation of BJTs.

10.6.1 DOUBLE POLY Si SELF-ALIGNED TRANSISTOR

The cross section of the double poly Si self-aligned transistor[5] is shown in Figure 10.11. [3] In this high-speed design, there are two base regions, or more accurately the base region is divided so that half of it appears on either side of the emitter.[6] To form the "extrinsic" base, p^+ polysilicon is deposited in the region indicated, and at elevated temperatures, acceptors diffuse to form a low-resistance p^+ base. A more lightly doped active base is formed earlier by ion implantation. To form the emitter, an n^+ polysilicon layer is deposited, and, under heat treatment, donors diffuse to form a shallow (on the order of 30 nm) n^+ emitter in the single crystalline Si. The emitter of this device consists of an n^+ single-crystal region and an n^+ polysilicon region in series.

The advantages of this structure over that of a conventional BJT are

1. The base-collector junction area can be made smaller, thus reducing the associated junction capacitance.

[5]The term *self-aligned* means that, as each layer is grown, the structures previously created on the layer below are used as the mask, instead of using multiple photomasks, each of which would have to be independently aligned.

[6]In Figure 10.11 it appears that there is only one base contact. However, both base regions are active since they are connected internally, similar to the structure illustrated in Figure 9.17a.

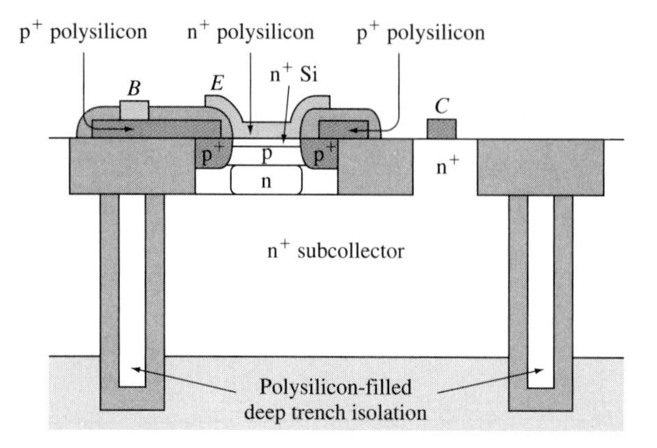

Figure 10.11 Schematic cross section of a double poly self-aligned BJT. The emitter consists of a shallow n^+ single-crystal region and an n^+ polysilicon region. The extrinsic base consists of a p^+ polysilicon region and a p^+ single-crystal region obtained from diffusion from the polysilicon. (Adapted with permission from T. Ning, "History and Future Perspectives of the Modern Silicon Bipolar Transistor," *IEEE Transactions on Electron Devices,* 48, pp. 2485–2491, 2001, © 2001 IEEE.)

2. The extrinsic base regions can be made short and heavily doped, reducing the base resistance.
3. The intrinsic base resistance is reduced by a factor of 4 by the use of a double base.
4. The active emitter area can be made small, reducing its junction capacitance.
5. The polysilicon emitter can extend over an appreciably larger area than the active emitter, reducing the emitter resistance.
6. As a result of the self-alignment process, the device area can be reduced compared with conventional devices, where photolithographic alignment tolerances require larger separation between base and emitter contacts.
7. The presence of the polycrystalline portion of the emitter increases the current gain, as discussed below.

These transistors are designed for high speed, but they also exhibit higher-than-expected β's. The increase in β that comes from the use of the polysilicon emitter structure is not well understood, and several models to explain this effect exist. [4] One of these models is based on the effects of band-gap narrowing being less in polysilicon than in monocrystalline emitters. This is explained with the aid of Figure 10.12a. The base is doped p type, and with $N_{AB} \approx 10^{18}$ cm^{-3},

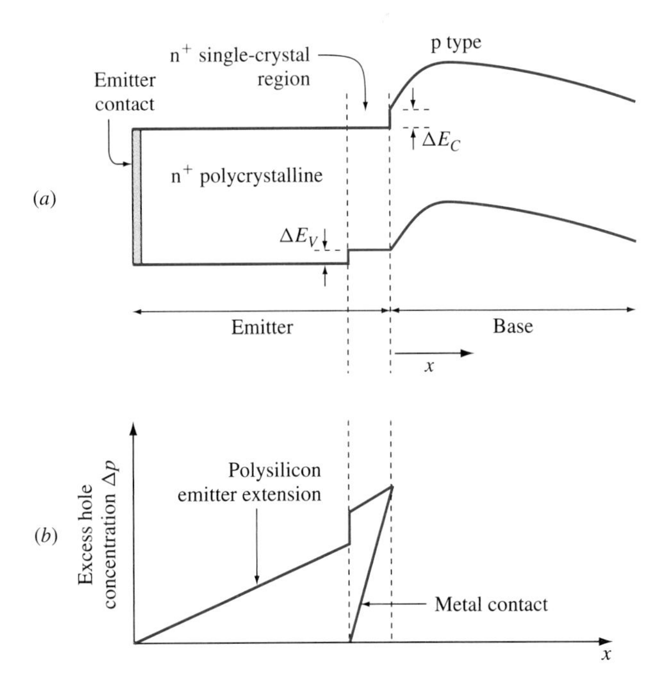

Figure 10.12 The use of a polysilicon n^+ emitter extension layer. (a) The energy band diagram. The discontinuity in the valence band acts as a barrier to holes. (b) The excess hole concentration of the polysilicon emitter compared with that of a metal contact directly on the monocrystalline emitter. The smaller gradient of Δp results in a reduced hole current and an increased β.

there is some band-gap narrowing in the base, raising E_V. The n^+ monocrystalline emitter is diffused into the p base region and thus, in addition to the increase in E_V already present, there is an additional decrease in E_C due to donors in the emitter. The n^+ polysilicon, however, is grown on top of these structures, and so contains a negligible concentration of acceptors. Thus in the n^+ polysilicon, there is only the band-gap narrowing due to the donors and E_V is not affected. As a result, the single-crystal emitter will have its valence band at a higher energy than in the polysilicon emitter. This discontinuity in E_V acts as a barrier to hole flow, reducing the base-emitter hole current and increasing the injection efficiency and thus β.

A comparison of the hole concentration profile in a polysilicon emitter device with that of a metal contact to the single-crystal emitter is shown schematically in Figure 10.12b. The gradient in the hole concentration for the polysilicon emitter extension layer case is smaller than for a metal contact applied directly to the single-crystalline emitter layer, thus reducing the hole diffusion current. There is also a discontinuity in Δp. Those holes "lost" in the discontinuity do not make it to the emitter contact and do not contribute to base-emitter hole current. This reduction of I_{pE} increases the injection efficiency γ [Equation (9.10)], which in turn increases the current gain β.

10.7 BJT SWITCHING TRANSISTOR

In Section 10.4, it was shown that the injected minority charge in a forward-biased junction leads to stored-charge capacitance. Here, we will examine the effect of that capacitance on the switching time.

As mentioned in the discussion on field-effect transistors, the time to switch a circuit between off and on depends on the time required to charge and discharge the circuit capacitances, Section 8.5. This is true for bipolar junction transistors also. In the BJT case, however, both minority carriers and majority carriers are involved. In addition to the majority carriers in the collector and emitter flowing into the circuit to discharge the circuit capacitances, the time associated with injecting and removing minority carrier stored charge further increases the turn-on and turn-off times.

We illustrate the effects of stored minority carriers with the circuit of Figure 10.13, in which R_C is the load resistor and C_L is the load capacitance. The base current is supplied via an input voltage V_{in}, through a base resistor R_B. In the circuit shown, the input voltage is switched from V_R to V_F. For V_R negative, the base-to-emitter junction is reverse biased and the BJT is off. The output voltage $V_{\text{out}} = V_{CC}$, where V_{CC} is the supply voltage. For $V_{\text{in}} = V_F$, the base-emitter is forward biased and the transistor is conducting.

For operation as a switch, V_F, R_B, and R_C are chosen such that for $V_{\text{in}} = V_F$, the BJT is operating in saturation, or the transistor is on, and when $V_{\text{in}} = V_R$, the transistor is in cutoff. These two states are illustrated in Figure 10.14, which illustrates the I_C-V_{CE} characteristics of a BJT with its associated resistive (R_C)

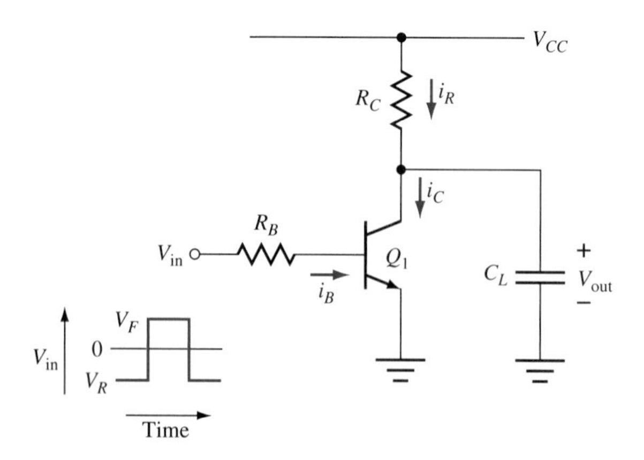

Figure 10.13 A simple bipolar transistor inverter circuit.

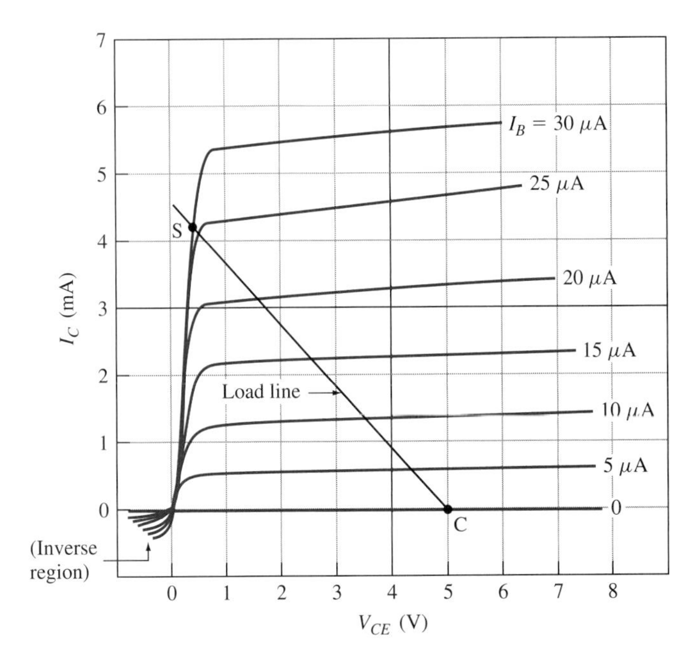

Figure 10.14 The I_C-V_{CE} characteristics of the transistor in Figure 10.13, along with the circuit load line. When the input is high, the *E-B* junction of transistor Q1 is forward biased, and the transistor is in saturation (point S). Current flows through the transistor and thus through R_C such that voltage is dropped across the resistor and the output goes low. When the input is low, the transistor is in cutoff (point C); the *E-B* junction is reverse biased. No current flows through Q1, so no voltage is dropped across R_C, and the voltage output is high.

load line. The collector current I_C is controlled by the base current I_B. For V_{in} negative (or zero), the E-B junction is not forward biased, and $I_E \approx 0$. Therefore $I_C = I_E - I_B \approx 0$. In this case there is no voltage drop across R_C, and thus when V_{in} is negative, $V_{CE} = V_{CC}$ as indicated by point C (cutoff). For most of the region along the load line, $I_B = (V_{\text{in}} - V_{BE})/R_B \approx V_{\text{in}}/R_B$. For $V_{\text{in}} = V_F$, the E-B junction is forward biased and current flows. The voltage V_F is chosen such that a large I_C flows and most of V_{CC} is dropped across R_C, with V_{CE} remaining across the transistor (and thus the output). Thus, I_C saturates at V_{CE}, indicated by point S (saturation).

10.7.1 OUTPUT LOW-TO-HIGH TRANSITION TIME

Because it takes some time for the minority carriers injected into the base to reach the collector, the output high-to-low transition time t_{hl} is increased over the time required just to discharge the junction capacitance. More important is the

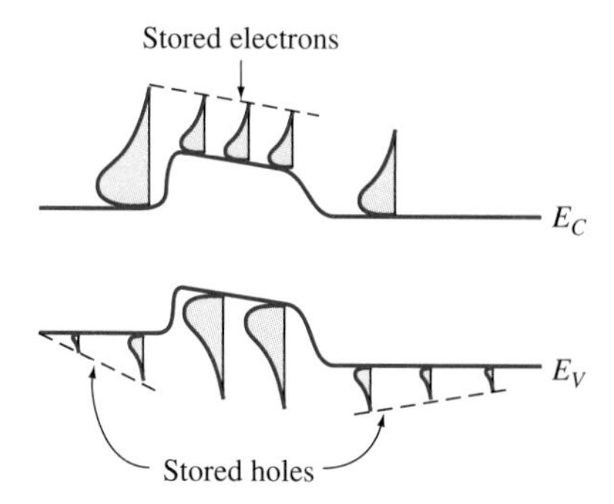

Figure 10.15 The charge "stored" in the excess carrier distributions when the transistor is in saturation. Carriers are constantly injected across both junctions, resulting in excess holes in the emitter and collector and excess electrons in the base. When the transistor goes out of saturation, it takes time for the excess carrier concentration to dissipate, and the response of the transistor is therefore slowed.

effect on the output low-to-high transition time t_{lh}, or the time required to go from S to C in Figure 10.14. In saturation, both the emitter-base and collector-base junctions are forward biased. Thus, there is injection at both junctions, and there are minority carriers stored in the emitter, base, and collector. This can be seen from the energy band diagram of Figure 10.15. When the input V_{in} is switched from V_F to V_R, the new input voltage attempts to reverse-bias the base-emitter junction. However, most of the excess electrons in the base, which were already injected from the emitter before the input was switched, continue on to the collector. This tends to keep the collector current constant for a finite time. The rest of the excess base electrons flow out the base contact. Some of the excess holes in the emitter and collector diffuse to the base and also contribute to base current. This base current is $I_B \approx V_R/R_B$ and is constant until the stored charge is small enough that the base current finally decreases. Only then can the transistor begin to change state from S (low) to C (high). This delay caused by the removal of stored minority carriers is a serious problem associated with the output low-to-high transition time of BJTs. As might be expected, this delay is dependent on the thickness of the emitter and base, and on the built-in fields of the emitter, base, and collector.

Note that t_{lh} depends on the value of the off voltage V_R as indicated in Figure 10.16. With V_R more negative, I_B is larger, and the carriers can be dissipated more quickly and the delay time is reduced.

The value of transition time is not easily calculated even if the device geometry and doping profiles are known. This is because of the dependence of the

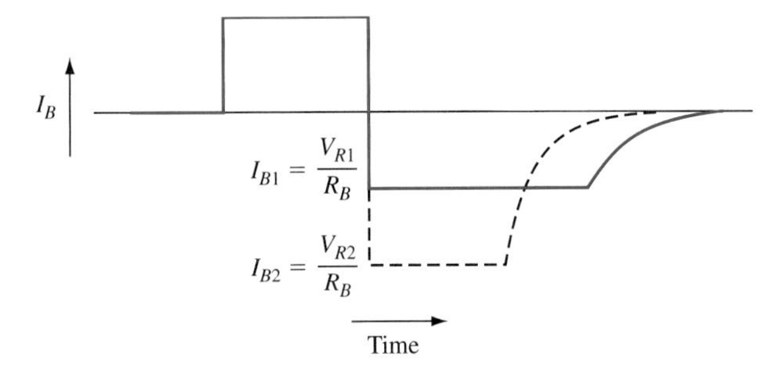

Figure 10.16 The turn-off time decreases as the voltage for the off state (V_R) increases.

junction voltage with stored charge.[7] For the simple circuit shown (Figure 10.13), the transition time can easily approach 1 μs, a value much too long for high-performance logic circuits. Furthermore, the use of V_{in} of two polarities is inconvenient—but using $V_R = 0$ V results in an even larger transition time.

While the circuit of Figure 10.13 is useful to illustrate the effects of minority carrier storage, its switching time is too great to be of practical use in high-speed circuits. In the next section we look at an alternative.

10.7.2 SCHOTTKY-CLAMPED TRANSISTOR

The speed of the output low-to-high transition can be increased appreciably by the use of a Schottky-clamped transistor. Such a transistor is shown schematically in Figure 10.17a. The base metallization extends over the p^+ base region (making an ohmic contact there) and over the n-collector region (creating a Schottky barrier). In effect, the base-collector pn junction and the metal-collector Schottky barrier junction are in parallel, as shown in (b).

In the off state (output high), both of these junctions are reverse biased and the transistor behaves normally. In the on state (output low) both junctions are forward biased, and we would normally expect the transistor to operate in saturation, with a large current flowing through the base-emitter and the base-collector junctions. In the Schottky-clamped transistor, however, the built-in voltage for the Schottky barrier diode from base to collector is less than that for base-collector pn junction. From Figure 10.18 we can see that, for a given forward voltage from base to collector, the current through the Schottky barrier is greater than that through the pn junction. Further, the Schottky current is carried by majority carriers while the pn junction current is carried by minority carriers. This implies that the minority carrier injection across the base-collector junction, and thus the stored charge, is greatly reduced compared with that of a conventional BJT. With reduced stored charge, the time required to remove it is decreased.

[7]This problem, however, can be handled in SPICE as illustrated in the Supplement to Part 4.

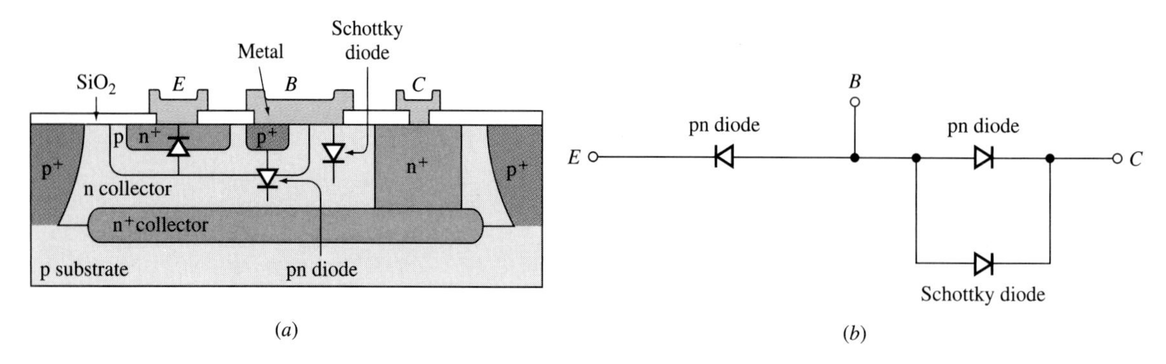

Figure 10.17 The Schottky-clamped transistor. (a) Cross-sectional view; (b) equivalent circuit schematic. The base-collector junction consists of two diodes in parallel.

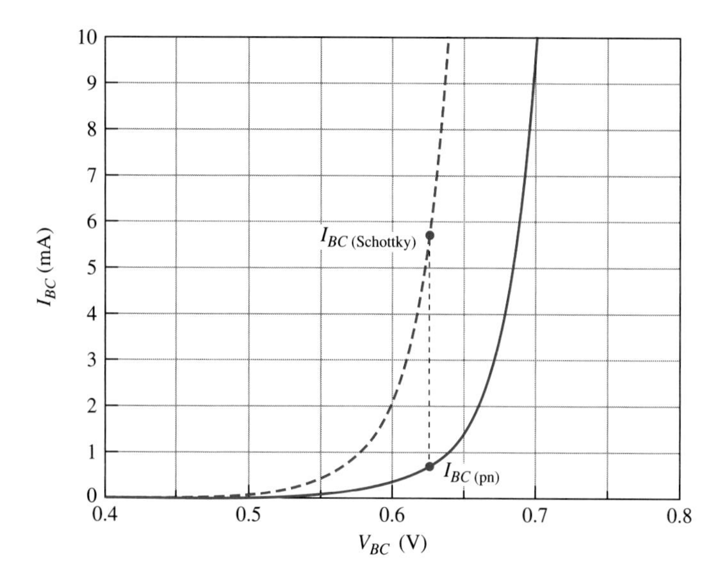

Figure 10.18 Comparison of the forward-biased I-V_{BC} characteristics of the Schottky and pn diodes of the Schottky-clamped transistor. The Schottky diode carries more current than the pn junction. Since the Schottky current consists primarily of majority carriers, little minority carrier injection exists. Thus, for a given I_C, the stored-charge capacitance is less in the Schottky-clamped transistor.

10.7.3 EMITTER-COUPLED LOGIC

Another approach to increasing switching speed is by the use of emitter-coupled logic (ECL), in which the collector-base junction is prevented from being forward biased and so saturation is avoided.

Figure 10.19a shows a schematic diagram of an ECL gate. The dc voltage V_E and the resistor R_E are chosen such that Q5 always operates in the region of

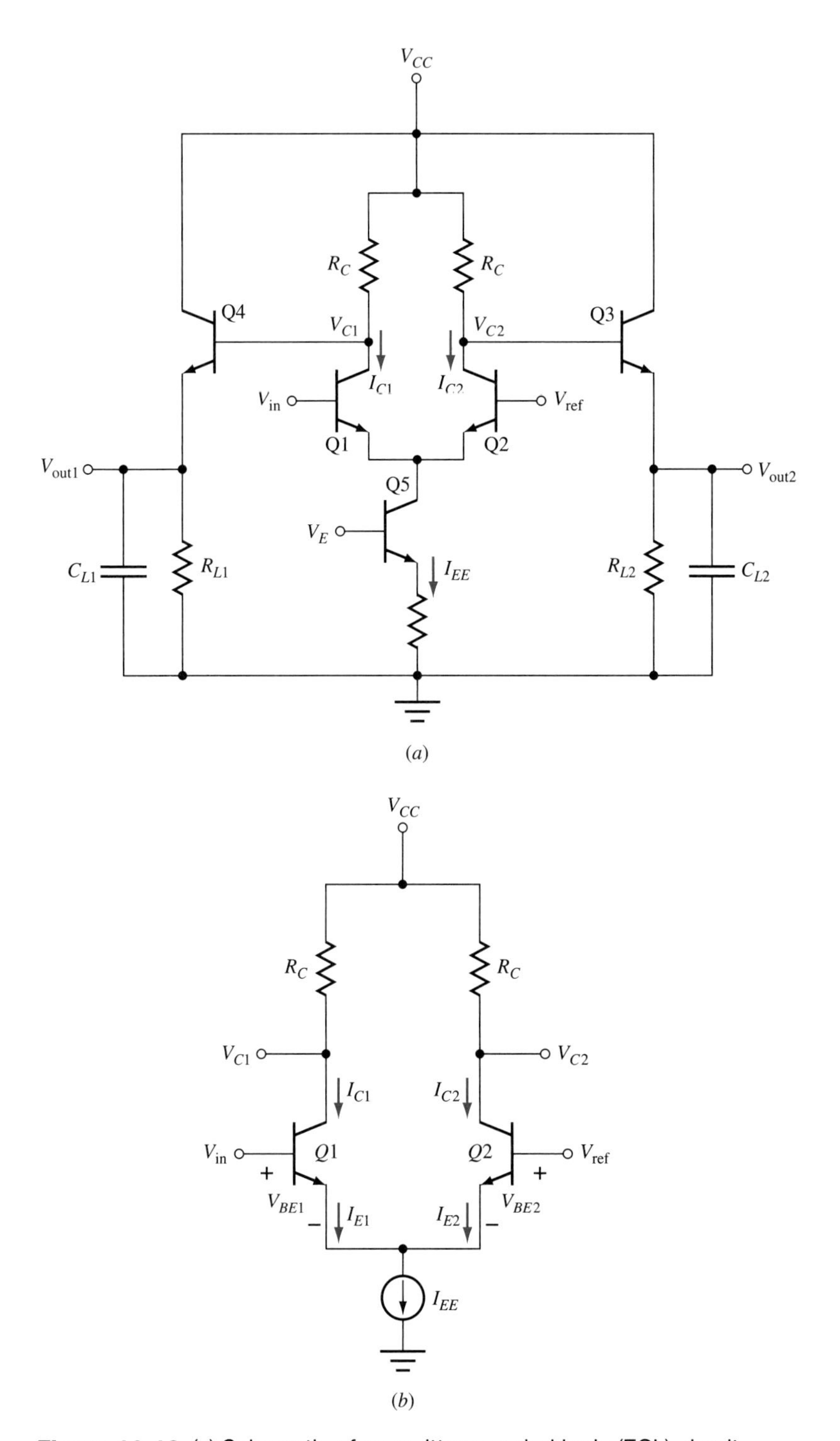

Figure 10.19 (a) Schematic of an emitter-coupled logic (ECL) circuit. (b) The current switching portion. The sign of $V_{in} - V_{ref}$ determines which BJT is conducting the constant current I_{EE}.

current saturation, producing a constant current I_{EE}. This current flows through either Q1 or Q2, or both.

The current switching portion of the ECL gate is shown in Figure 10.19b. It consists of two matched transistors Q1 and Q2, and matched resistance loads R_C. The emitters are connected together and supplied by the constant current I_{EE}. A reference voltage is applied to the base of Q2. The input voltage switches from a few tenths of a volt below V_{ref} to a few tenths of a volt above V_{ref}, which causes the output voltages to switch from zero to some (positive) value. In effect, the current I_{EE} is switched from one transistor to the other. An illustration follows.

EXAMPLE 10.3

Consider the case for $V_{\text{in}} = V_{\text{ref}} + 0.3$ V. Find the ratio of current through Q1 to that through Q2, and the output voltages. Repeat for $V_{\text{in}} = V_{\text{ref}} - 0.3$ V.

■ Solution

Since the emitters are common, $V_{BE1} = V_{BE2} + 0.3$ V. We know

$$I_C = I_{C0} e^{qV_{BE}/kT}$$

so we have

$$\frac{I_{C1}}{I_{C2}} = e^{q(V_{\text{in}} - V_{\text{ref}})/kT} = e^{0.3/0.026} \approx 10^5$$

Or I_{C1} is 5 orders of magnitude greater than I_{C2}. The output voltage is $V_{C1} = V_{CC} - I_{EE}R_C$.

Since I_{EE} is constant, $I_{E1} + I_{E2} = I_{EE}$. And since V_{ref} and R_C are chosen such that the saturation region is avoided, $I_{C1} + I_{C2} \approx I_{EE}$. Since $I_{C1} \approx 10^5(I_{C2})$, $I_{C2} \approx 0$, and thus $V_{C2} \approx V_{CC}$.

Similarly for $V_{\text{in}} < V_{\text{ref}}$,

$$V_{C1} \approx V_{CC}$$
$$V_{C2} \approx V_{CC} - I_{EE}R_C$$

Thus the circuit of Figure 10.19 has an inverting output (V_{out1}) and a noninverting output (V_{out2}).

In Figure 10.19a, the outputs V_{C1} and V_{C2} act as the inputs to the emitter-follower circuits consisting of Q3 and Q4 and their load resistances R_{L1} and R_{L2}. The capacitors C_{L1} and C_{L2} represent the total load capacitances connected to the gates. The purpose of the emitter-follower circuits is to isolate the outputs V_{C1} and V_{C2} from the load.

The speed of ECL circuitry comes from the fact that neither transistor ever enters saturation. Thus the stored charge in the base is smaller (no injection from the collector side as there was in Figure 10.15) and in the collector (no injection into the collector). While ECL circuits are extremely fast, the power dissipation can be quite large since one of the devices is always operating in the active mode.

10.8 BJTs, MOSFETs, AND BiMOS

10.8.1 COMPARISON OF BJTs AND MOSFETs

Both BJTs and MOSFETs are used in electronic circuits. It is instructive to compare the important electrical parameters of these devices.

Input Impedance The dc input resistance is virtually infinite in a MOSFET. At high frequencies,

$$Z_{\text{in}} = \frac{1}{j\omega(C_{GS} + C_{GD})} \qquad \text{FET} \tag{10.52}$$

For a BJT with a forward-biased E-B junction,

$$Z_{\text{in}} = r_\pi + \frac{1}{j\omega(C_{jBE} + C_{\text{sc}BE})} \qquad \text{BJT} \tag{10.53}$$

Because $(C_{jBE} + C_{\text{sc}BE}) \gg (C_{GS} + C_{GD})$, the input impedance is much larger for a MOSFET than for a BJT.

Transconductance Transconductance is defined as the variation of output current with a change in input voltage

$$g_m = \left.\frac{\partial I_{\text{out}}}{\partial V_{\text{in}}}\right|_{V_{\text{out}}} = \frac{i_{\text{out}}}{v_{\text{in}}}$$

For a MOSFET, according to the simple long-channel model, in the current saturation region

$$\begin{aligned} I_{\text{out}} = I_{D\text{sat}} &= \frac{W\mu C'_{\text{ox}}(V_{GS} - V_T)^2}{2L} \\ g_m &= \frac{W\mu C'_{\text{ox}}(V_{GS} - V_T)}{L} = \frac{2I_{D\text{sat}}}{(V_{GS} - V_T)} \approx \frac{2I_{D\text{sat}}}{V_{DD}} \qquad \text{FET} \end{aligned} \tag{10.54}$$

where $V_{GS} = V_{DD} \gg V_T$.

For a BJT

$$\begin{aligned} I_{\text{out}} &= I_C = I_{Cs} e^{qV_{BE}/kT} \\ g_m &= \frac{I_C}{kT/q} \qquad \text{BJT} \end{aligned} \tag{10.55}$$

The ratio of the transconductances is

$$\frac{g_m(\text{BJT})}{g_m(\text{MOSFET})} = \frac{\dfrac{I_C}{kT/q}}{\dfrac{2I_{D\text{sat}}}{V_{DD}}}$$

For equal values of output current, $I_C = I_{D\text{sat}}$, and $V_{DD} = +2.5$ V,

$$\frac{g_m(\text{BJT})}{g_m(\text{MOSFET})} = \frac{2.5}{2 \times 0.026} \approx 50$$

or the transconductance of a BJT is much larger than that of a MOSFET.

The above comparison was for a long-channel MOSFET. In most modern MOSFETs, channel lengths are small enough that velocity saturation must be considered. When velocity saturation is taken into account, the transconductance g_m of the MOSFET is reduced with decreasing channel length [see Section 7.3.2 and Equation (8.33)], and the discrepancy is even more pronounced.

Speed The unity current gain frequencies for unloaded circuits are

$$f_T(\text{MOSFET}) \approx \frac{g_m(\text{MOSFET})}{2\pi(C_{GS} + C_{GD})}$$

$$f_T(\text{BJT}) \approx \frac{g_m(\text{BJT})}{2\pi(C_{jBE} + C_{\text{sc}BE})}$$

Even though $g_m(\text{MOSFET}) \ll g_m(\text{BJT})$, it is also true that $(C_{GS} + C_{GD}) \ll (C_{jBE} + C_{\text{sc}BE})$, which makes the cutoff frequencies for unloaded devices comparable. For a circuit with an appreciable load capacitance C_L, however, the load capacitance predominates. Thus

$$f_T(\text{MOSFET}) \ll f_T(\text{BJT})$$

Power Dissipation The power dissipation in MOSFETs is appreciably less than for fast BJTs (e.g., ECL).

Ease of Manufacture MOSFETs require fewer processing steps than BJTs, and thus can be made at lower cost.

Summary From the above it would appear that MOSFET integrated circuits would be preferable to BJT circuits. This is the case for low- and medium-speed applications. The greatest advantage of the BJT is its much larger transconductance, which permits it to be used where the load capacitance is appreciable. The load capacitance consists of the wiring capacitance and the capacitance due to a large fan-out. The high transconductance of the BJT means a higher-output current, which can charge and discharge the capacitance more quickly.

The advantages of MOSFETs and of BJTs can be combined in a single chip using BiMOS or BiCMOS technologies, as discussed next.

10.8.2 BiMOS

The basic circuit diagram of a simple BiMOS analog amplifier circuit is shown in Figure 10.20. The input is to the gate of an n-channel field-effect transistor, and the output is at the collector of the bipolar transistor. Both devices are fabricated on the same chip. The input to the gate of the MOSFET (M) consists of a

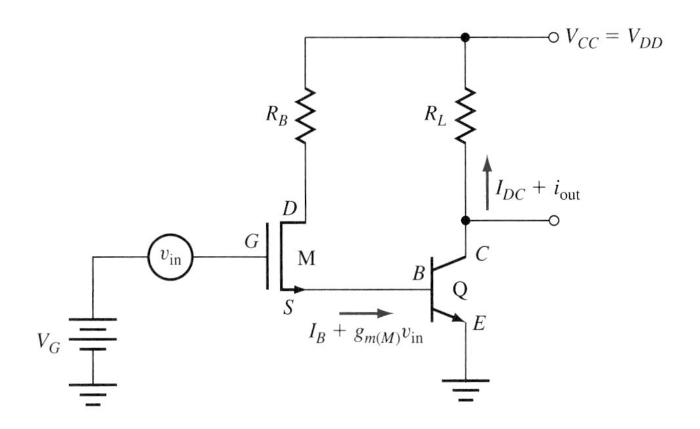

Figure 10.20 Schematic of a simple analog BiMOS amplifier. The input to the circuit is at the gate of the MOSFET (M), while the output is at the collector of the BJT (Q).

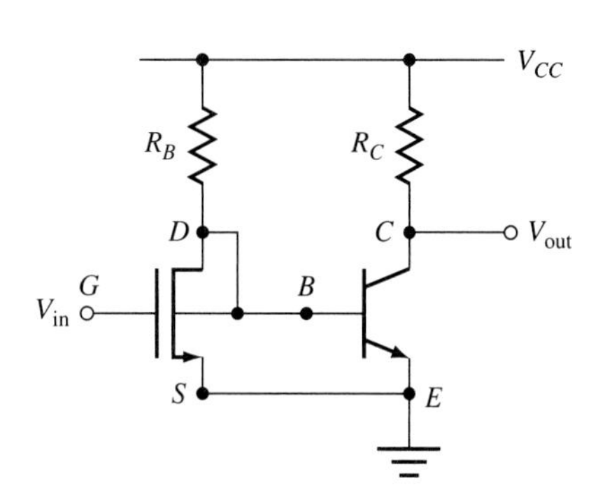

Figure 10.21 Circuit diagram of a BiMOS digital inverter. This circuit has the high input resistance and small input capacitance of the MOSFET and the high transconductance of the BJT.

dc voltage and a small signal ac analog voltage v_{in}. The dc gate voltage biases the MOSFET such that a drain current flows through the source and into the base of the BJT (Q). This current is sufficient to bias the BJT in the active mode. The signal voltage v_{in} produces a drain ac current $i_M = i_B = g_{m(M)}v_{in}$. This signal current is amplified by the BJT whose output signal current is

$$i_{(Q)} = i_{out} = \beta i_B = \beta g_{m(M)} v_{in}$$

The overall transconductance of this circuit is then

$$g_{m\text{total}} = \frac{i_{out}}{v_{in}} = \beta g_{m(M)}$$

This circuit has the advantage of combining the high-input impedance of the MOSFET with a high transconductance.

A simple digital inverter switch is indicated schematically in Figure 10.21. Consider the MOSFET to be an enhancement device. Then when $V_{in} = 0$, the MOSFET is off and current flows from V_{CC} through R_B and into the base of the BJT, turning it on. This causes a large I_C to flow through R_C. Since $V_C = V_{CC} - I_C R_C$, the output voltage V_{out} is low.

For a positive voltage ($V_{in} > V_T$), the MOSFET turns on and diverts the current from R_B through the MOSFET. The base current is then small and the BJT turns off. Now $V_{out} = V_{CC}$. Thus the BiMOS circuit of Figure 10.21 is an inverter. Figure 10.22 shows the cross sectional view of this simple BiMOS circuit. The n^+ source region also serves as emitter. The BJT consists of the left n^+pn device. Note that the drain contacts overlap the p base region.

The above illustrations are quite simple. More advanced circuits also use both CMOS and complementary BJTs (e.g., npn and pnp) and are referred to as BiCMOS.

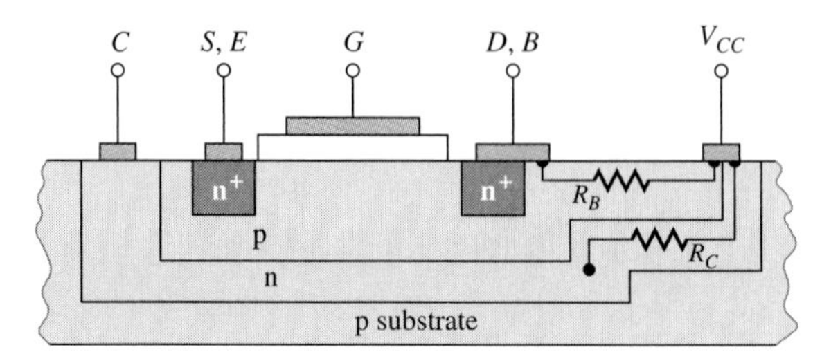

Figure 10.22 Cross-sectional schematic of the BiMOS circuit of Figure 10.20. The source and emitter are common, as are the drain and base.

10.9 SUMMARY

In this chapter, the dc models of Chapter 9 were extended to describe the time-dependent behavior of BJTs. The Ebers-Moll dc model was modified to include base-emitter and base-collector junction and the stored-charge capacitances. The small-signal hybrid-pi model was illustrated for the common emitter configuration for operation in the active mode.

The operation of BJTs as digital switches was discussed. To operate at high switching speeds it is necessary to keep the transistor out of the saturation region to minimize the base-collector stored charge. To accomplish this the base-collector should not be forward biased, or if it is, the bias should be small enough that negligible minority carrier injection exists between base and collector. Two schemes to implement this strategy were discussed: the Schottky-clamped transistor in which a Schottky diode exists in parallel with the base-collector pn junction, and emitter-coupled logic circuits in which a constant current is switched between two BJTs. The ECL mode of operation provides for fast switching, but requires high power dissipation.

Next, the electrical parameters of BJTs were compared with those of MOSFETs, to see the advantages of each technology:

- The input impedance of MOSFETs is much larger than that of BJTs. The input resistance of MOSFETs is nearly infinite for MOSFETs and is in the kilohm range for BJTs. The MOSFET input capacitance is much smaller than that for BJTs.
- The transconductance is much larger for BJTs than for MOSFETs.
- The unity current gain frequencies are comparable for two classes of devices for unloaded circuits. However, for circuits with appreciable load capacitance, the BJT is faster because of its higher transconductance.
- The advantages of each transistor type can be exploited by using MOSFETs for the input stage, and BJTs for the output stage. Two examples of such BiMOS circuits were discussed.

10.10 READING LIST

Items 1 to 4, 8 to 12, 18, and 26 in Appendix G are recommended.

10.11 REFERENCES

1. Joseph Lindmayer and Charles Y. Wrigley, *Fundamentals of Semiconductor Devices,* D. Van Nostrand, Princeton, NJ, Chapter 4, 1965.
2. R. L. Prichard, "High Frequency Power Gain of Junction Transistors," *Proc. IRE,* 43, pp. 1075–1085, 1955.
3. Tak H. Ning, "History and Future Perspective of the Modern Silicon Bipolar Transistor," *IEEE Trans. Electron Devices,* 48, pp. 2485–2491, 2001.
4. A. K. Kapoor and D. J. Roulston, eds., *Polysilicon Emitter Bipolar Transistors,* IEEE Press, New York, 1989.

10.12 REVIEW QUESTIONS

1. Under what conditions can the capacitance C_{scBC} in Figure 10.1 be ignored, and why? Why is the diode representing the base-collector junction not shown in Figure 10.2?
2. Explain how a small variation in I_B produces a large, proportional variation in I_C.
3. How do designers minimize the parasitic base resistance in a BJT?
4. What is meant by unity current gain frequency?
5. In Chapter 9 we saw that making the base region thin increases the β of a transistor. Explain how a thin base also increases the cutoff frequency.
6. How does the use of self-aligned structures allow one to make a higher-frequency BJT?
7. Explain the operation of a Schottky-clamped transistor.
8. How does emitter-coupled logic improve circuit switching time?

10.13 PROBLEMS

10.1 For an npn transistor with $N'_{DE} = 10^{19}$ cm^{-3}, $N'_{AB} = 2 \times 10^{17}$ cm^{-3}, $N'_{DC} = 5 \times 10^{16}$ cm^{-3}, $W_E = 0.13\ \mu$m, and $W_B = 0.15\ \mu$m under the bias conditions of $I_B = 20\ \mu$A and $V_{BC} = -2.5$ V, find

a. β

b. I_C

c. r_π

d. g_m

e. C_{BE} (C_π)

f. C_μ

g. f_{co}

h. f_T

Note that band-gap narrowing should be accounted for, and that both sides of the E-B junction are short. Let the area of the emitter junction be $A_E = 2.5 \times 10^{-7}$ cm^{-2} and the area of the collector junction be $A_C = 8 \times 10^{-7}$ cm^{-2}.

10.2 For the transistor of Problem 10.1, plot $|i_c|/|i_b|$ as a function of frequency. What is the unity current gain frequency?

10.3 An npn BJT with uniformly doped emitter, base, and collector has $\beta = 95$, base width W_B of 0.15 μm, electron diffusion coefficient of 10 cm^2/s, r_π of 1000 Ω, and a C-B junction capacitance of 0.05 pF. What is its cutoff frequency?

10.4 If the base in the above problem is actually graded in doping, with a grading parameter $\eta = 2$,

a. What is the field in the base?

b. For the same value of I_C, what is the value of β?

c. What is the resultant cutoff frequency?

10.5 Two transistors that are otherwise identical differ in that one has a uniform base and the other has a graded base with $\eta = 6$. Assume the field in the base is small enough that the diffusion constant has its low-field value.

a. What is the improvement in β for the graded device?

b. What is the change in the stored charge capacitance if both transistors are operated at the same value of I_C?

10.6 Equation (10.39) indicates that the electron velocity in the base increases as $\Delta n_B(x)$ decreases (I_{nB} constant). Explain the physics of this.

10.7 Find the base transit time t_T for an npn prototype BJT with $W_B = 0.05$ μm and base doping of 5×10^{17} cm^{-3}.

10.8 Repeat the above problem for a pnp prototype BJT with $W_B = 0.05$ μm and base doping of 5×10^{17} cm^{-3}.

10.9 Find the transit time t_{BC} across the base-collector transition region for an npn prototype BJT with base doping of 5×10^{17} cm^{-3}, collector doping of 5×10^{16} cm^{-3}, and collector-base voltage of 2.5 V.

10.10 In a particular prototype transistor fabrication process, the base width is cut in half. What is the effect on the base transit time?

10.11 An npn transistor's base region is doped to 10^{18} cm^{-3}. How thick would the base region need to be for the base transit time to be equal to $1/100$ of the electron lifetime in the base? If such a transistor were manufactured, what would be the effect on β and the operating frequency?

10.12 For high-speed BJTs, the double poly self-aligned transistor is preferred to the conventional transistor discussed in Chapter 9. For similar emitter, base, and collector dopings compare the following parameters and explain your reasoning.

a. Collector-base junction capacitance

b. Forward current gain β_F

c. Reverse current gain β_R

d. Base resistance

e. Early voltage

10.13 For a particular Schottky-clamped transistor, the leakage current density in the Schottky diode is $J_{0SD} = 10^{-5}$ A/cm^2, and the leakage current for the base-collector junction is $J_{0BC} = 10^{-11}$ A/cm^2. For a given bias voltage across both in parallel, what is the ratio of the current flowing through the Schottky contact to the current flowing through the B-C junction? What is the impact on turn-off time?

10.14 From Equations (10.43) and (10.23), find an expression for the base transit time in a graded-base transistor as a function of η. Assume that the area of the pn junction is 20 times that of the Schottky diode.

Supplement to Part 4: Bipolar Devices

S4.1 INTRODUCTION

In the previous two chapters, the operation of homojunction BJTs was discussed. This supplement briefly treats two additional classes of bipolar devices. First, heterojunction BJTs (HBTs) are described. The use of an emitter with a larger band gap than that of the base decreases the ratio of the base-to-emitter minority current to the emitter-to-base injected current, thus increasing injection efficiency and current gain. As for a homojunction BJT, grading the base doping creates a field for carriers that accelerates them across the base, thus reducing transit time and increasing current gain and frequency response. The field for electrons can be further increased by appropriately grading the base composition and thus the band gap.

The use of four-layer devices (npnp or pnpn) are briefly described for use as switches in power circuits. Although not transistors, their operation can be understood by considering them as interacting pn junctions.

The Ebers-Moll model for a BJT was discussed in Chapters 9 and 10. In this supplement the more accurate Gummel-Poon model, which includes second-order parasitic effects, is introduced. The use of SPICE in bipolar transistors based on the Ebers-Moll and Gummel-Poon models is briefly discussed with some simple examples.

S4.2 HETEROJUNCTION BIPOLAR TRANSISTORS (HBTs)

In Chapter 9, we saw that, in a homojunction BJT, to increase β we wanted to maximize the injection efficiency γ, so the emitter was doped much more heavily than the base. The heavy emitter doping, however, produced impurity-induced band-gap narrowing in the emitter, which resulted in the potential energy barrier for electrons being greater than that for holes for an npn transistor, Figure S4.1a.

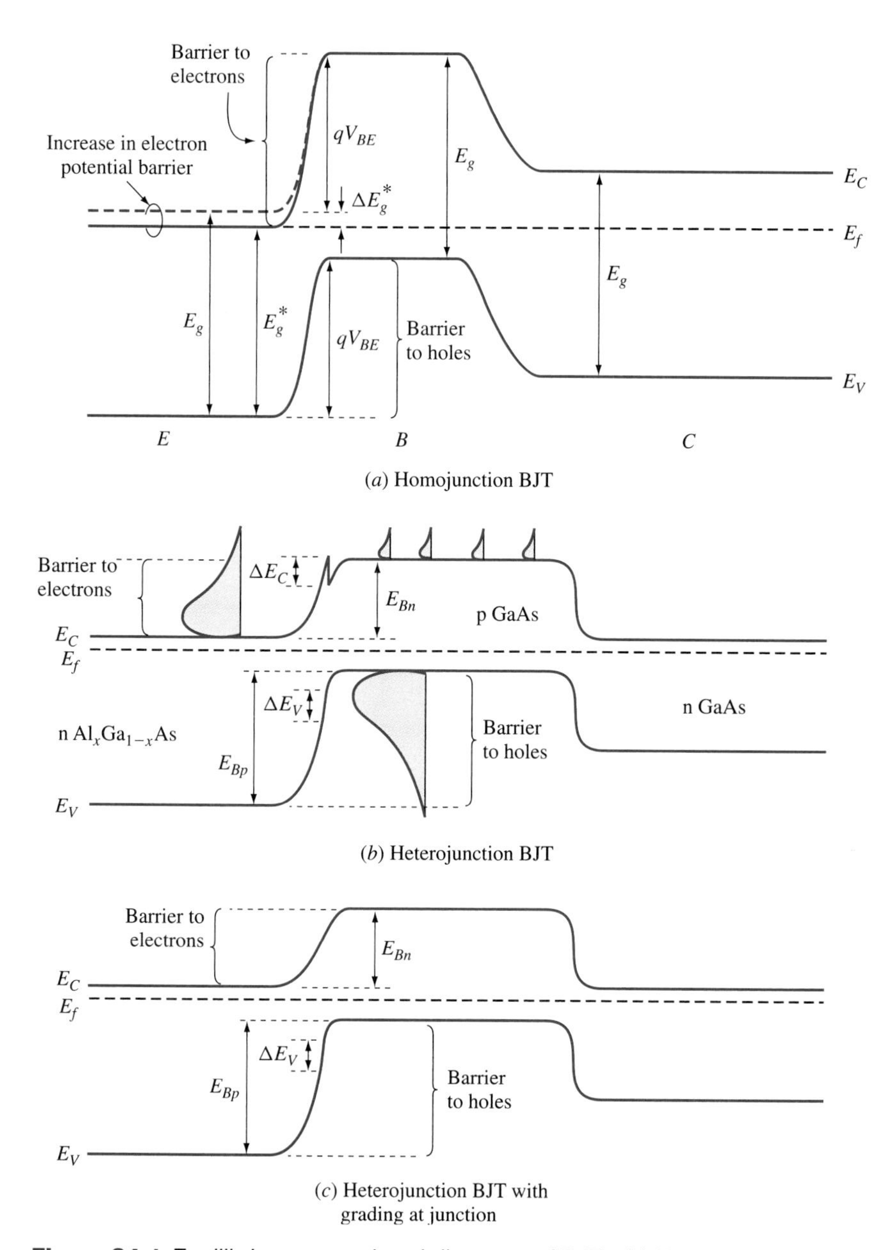

Figure S4.1 Equilibrium energy band diagrams of BJTs. (a) Homojunction npn transistor: The degenerately doped emitter causes band-gap narrowing, which increases the barrier to injection of electrons and reduces β. (b) The heterojunction bipolar transistor (HBT) uses a wide-band-gap emitter to create a large barrier to holes to reduce back-injection. (c) The wide-gap material is graded to eliminate the potential energy spike in the conduction band.

This lowered the electron injection efficiency while the back-injection of holes remained the same, reducing β. Recall from Example 9.2 that the result is that $\beta \approx \gamma/(1-\gamma)$, which is reduced by the factor $e^{-\Delta E^*_{gBE}/kT}$, where

$$\Delta E^*_{gBE} = E^*_g \text{ (base)} - E^*_g \text{ (emitter)} \tag{S4.1}$$

In effect, the heavy doping produced an extra barrier that worked against us. We would like it to be the other way around—we'd like to *increase* the barrier to back-injected holes while we *decrease* the barrier for the injected electrons.

This can be done by choosing an emitter material with a larger band gap than that of the base, Figure S4.1b. Such a structure is referred to as a heterojunction bipolar transistor, or HBT.

S4.2.1 UNIFORMLY DOPED HBT

Consider the uniformly doped $Al_xGa_{1-x}As$:GaAs (AlGaAs:GaAs) HBT, whose equilibrium energy band diagram is shown in Figure S4.1b. Here we assume that each region (emitter, base, and collector) is uniformly doped and nondegenerate. The emitter consists of the wider band-gap material $Al_xGa_{1-x}As$, and the base and collector are made of GaAs. There is a discontinuity ΔE_C in the conduction band edge and a discontinuity ΔE_V in the valence band edge. Normally, during fabrication the Al content of the emitter is graded near the base contact such that the potential energy spike is not present, as indicated in Figure S4.1c. We can see that the barrier for electrons (E_{Bn}) is appreciably smaller than that for holes (E_{Bp}).

Figure S4.2 shows the energy band diagram for this HBT operating in the active mode. The injected electron current is several orders of magnitude greater than the injected hole current. In practical devices, however, unless the emitter and base have virtually identical lattice constants, β is smaller than predicted. This is because if the lattices are not well matched, there is some base-emitter current that flows by recombination via interface states at the heterojunction, which reduces the injection efficiency. Still, for this structure, β can be made appreciably greater than for homojunction transistors.

There are other advantages to using an HBT structure apart from increased current gain. We saw in Chapter 9 that in a homojunction transistor, to get adequate γ and thus β, the emitter must be much more heavily doped than the base. In a homojunction BJT, however, γ is degraded by the ΔE^*_{gBE} resulting from the heavy emitter doping. In a heterojunction, γ is controlled by the differences in the band gaps, removing these doping restrictions. Thus, the base can be heavily doped to reduce the base resistance, while the emitter can be lightly doped to reduce the emitter-base junction capacitance.

A plot of β as a function of collector current I_C for an npn $Al_xGa_{1-x}As$:GaAs HBT is shown in Figure S4.3 for three temperatures. [1] For this device the aluminum fraction is $x = 0.4$, the emitter thickness is $W_E = 1.0\ \mu\text{m}$, the base width $W_B = 0.4\ \mu\text{m}$, and the doping concentrations are $N'_{DE} = 3 \times 10^{17}\ \text{cm}^{-3}$, $N'_{AB} = 10^{18}\ \text{cm}^{-3}$, and $N'_{DC} = 10^{15}\ \text{cm}^{-3}$. [2] The current gain is low at small I_C, since here the parasitic (recombination) currents predominate. The increase

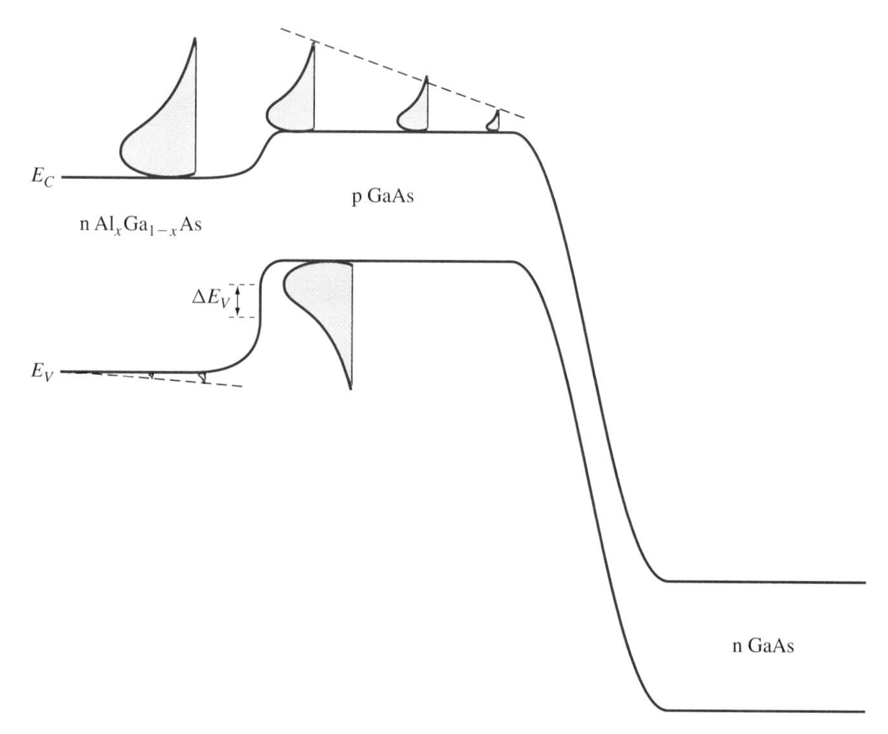

Figure S4.2 The HBT under forward active bias.

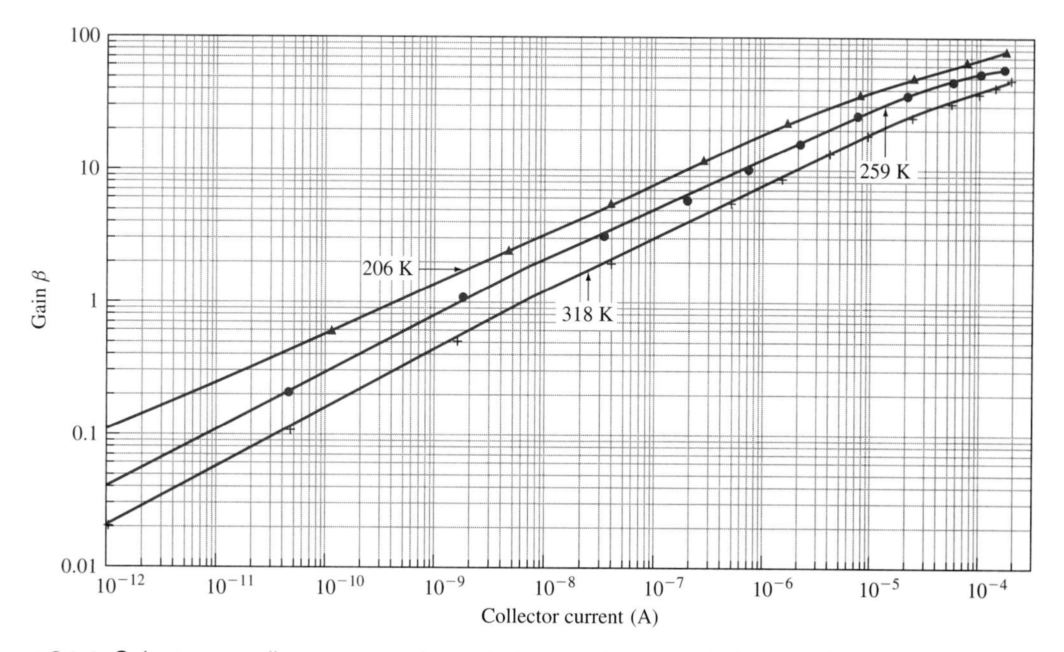

Figure S4.3 Gain versus collector current for an AlGaAs:GaAs:GaAs HBT at three temperatures. For this device the aluminum fraction is $x = 0.4$, the emitter thickness is $W_E = 1.0\ \mu$m, the base width $W_B = 0.4\ \mu$m, and the doping concentrations are $N_{DE}' = 3 \times 10^{17}$ cm^{-3}, $N_{AB}' = 10^{18}$ cm^{-3}, and $N_{DC}' = 10^{15}$ cm.

in β with increasing I_C results from the greater dependence of I_C on V_{BE} ($I_C \propto e^{qV_{BE}/kT}$) compared with the dependence of the recombination current on V_{BE} ($I_B \propto e^{qV_{BE}/nkT}$, where $n \approx 2$).

It is seen in Figure S4.3 that β increases with decreasing temperature. This is the opposite behavior of homojunction transistors, where β decreases with decreasing temperature. This can be explained as follows: For an npn homojunction transistor with degenerate emitter and nondegenerate uniformly doped base, from Equation (9.43),

$$\beta \approx \frac{N_C}{N'_{AB}} \frac{D_{nB}}{D_{pE}} \frac{W_E}{W_B} e^{-\Delta E^*_{gBE}/kT} \tag{S4.2}$$

The effective density of states in the conduction band, N_C, varies as $T^{3/2}$, while D_{nB} and D_{pE} vary slightly with temperature. The temperature dependence of these parameters is, however, small compared with that of the term $e^{-\Delta E^*_{gBE}/kT}$, which is the impurity-induced reduction of the apparent band gap in the emitter. Thus, to first approximation,

$$\beta \propto e^{-\Delta E^*_{gBE}/kT} \tag{S4.3}$$

and β decreases with decreasing temperature. The value of ΔE^*_{gBE} (which is positive) can then be estimated from the slope ($\Delta E^*_{gBE}/k$) of a plot of ln β as a function of $1/T$.

In a heterojunction transistor, however, ΔE^*_{gBE} is normally negative since the emitter band gap is larger than that of the base, and β can increase with decreasing temperature. While the current gain of the Si homojunction transistor decreases appreciably with decreasing temperature, that of the HBT increases. The HBT therefore appears to have promise for operation at cryogenic temperatures.

S4.2.2 GRADED-COMPOSITION HBT

The results of Figure S4.3 are for an HBT with uniform base doping. Just as for a homojunction, performance can be improved by appropriately grading the base doping to create a built-in field in the base, which accelerates the minority carriers to the collector. [3]

Another way to create a built-in field in the base is to grade the composition of the material, so that the band gap changes with position across the base. Such a grading is shown schematically in Figure S4.4 for an npn Si:Si_xGe_{1-x}:Si HBT with a uniformly doped base. [4] Here both the emitter and collector are silicon, and the base is a Si-Ge alloy. The Ge content is increased from zero slightly inside the Si emitter to the final Ge content somewhat inside the collector. The grading continues into the collector to avoid potential energy spikes in the conduction band. Since the band gap E_g decreases with increasing Ge content, the E-B junction is a heterojunction, graded in this case to avoid spikes in the band edges. The decreasing E_g with position induces a built-in field for electrons in the base.

The base field can be further increased by combining a graded composition with graded doping in the base, as shown in the SIMS plot of Figure S4.5. [5]

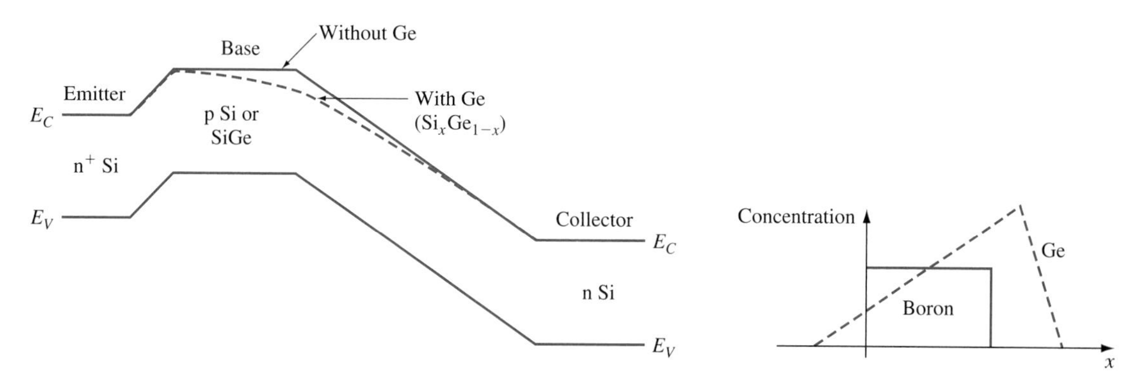

Figure S4.4 The Si:Si_xGe_{1-x}:Si HBT energy band diagram (left) and plot of germanium concentration (right). The addition of Ge starts slightly into the emitter and extends a short distance into the collector. Adapted from J. D. Cressler, "Re-engineering silicon: Si-Ge heterojunction bipolar transistor technology, *IEEE Spectrum,* pp. 49–55, 1995. Copyright IEEE 1995.

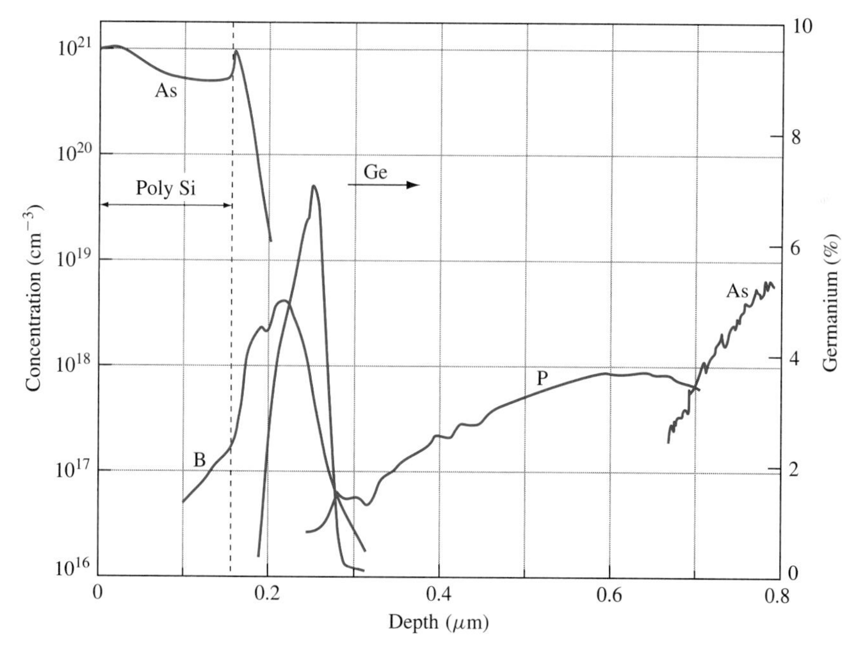

Figure S4.5 SIMS data showing the concentrations of the dopants and the Ge in the Si:Si_xGe_{1-x}:Si heterojunction bipolar transistor. From G. Niu, J. D. Cressler, S. Zhang, W. E. Ansley, C. S. Webster, and D. L. Harame, "A unified approach to RF and microwave noise parameter modeling in bipolar transistors, *IEEE Transactions on Electron Devices*, 48, no. 11, pp. 2568–2574, 2001. Copyright IEEE 2001.

Here the increasing Ge content and the decreasing acceptor doping in the base combine to create a high base field (on the order of 10^3 to 10^4 V/cm). This field reduces the base transit time, which increases the frequency response. Values for the unity gain frequency f_T well in excess of 100 GHz have been measured. [6]

In Si-Ge alloys, however, there is a difference in lattice constant (4 percent) between the Si and the Ge. Therefore we would expect dangling bonds, and thus interface states, to be present at the emitter-base junction and within the base because of the variation of germanium concentration with position. These states would allow recombination at the interface and within the base and degrade the current gain β of the transistor. The lattice-matching defects can be avoided, however, if the Ge content is kept relatively small (less than about 20 percent), and the alloy layer is thin. In this case, the Si_xGe_{1-x} alloy on one side of the interface bonds with the Si on the other side atom for atom, without dangling bonds. Because this produces strain in the alloy, it is referred to as a *strained layer*. If the layer is too thick or the Ge content too high, the strain will be relieved with the creation of interband states. These will trap carriers and thus reduce β.

We make one final point about HBTs. In Section S4.2.1 we saw that β would be expected to increase with decreasing temperature, and this is experimentally observed. Figure S4.6 shows plots of β normalized to its room temperature value $[\beta(T)/\beta(300\text{ K})]$ as a function of inverse temperature $1/T$. The figure compares a Si homojunction transistor to a $Si{:}Si_xGe_{1-x}{:}Si$ heterojunction transistor.

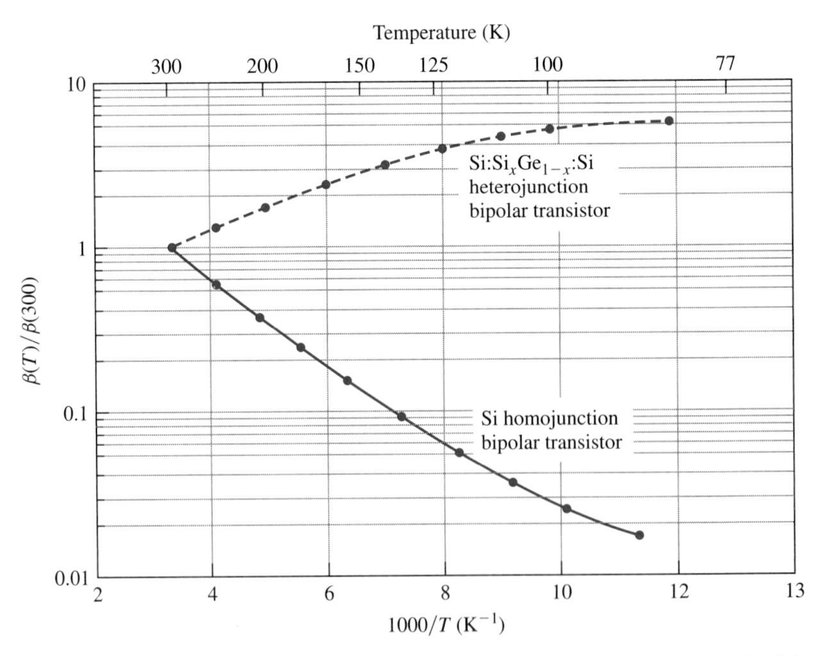

Figure S4.6 Variation of β with temperature for a Si bipolar transistor (solid line) and a $Si{:}Si_xGe_{1-x}{:}Si$ HBT (dashed line).

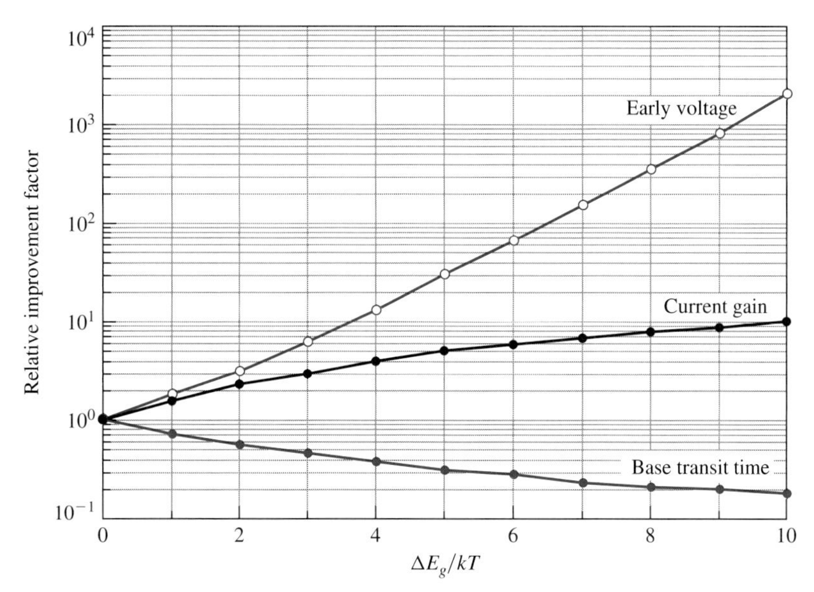

Figure S4.7 Improvement in Early voltage, current gain, and base transit time for a Si:Si_xGe_{1-x}:Si BJT compared with a similar device with a silicon base. (Adapted with permission from T. Ning, "History and Future Perspectives of the Modern Silicon Bipolar Transistor," *IEEE Trans. Electron Devices,* 48, pp. 2485–2491, © IEEE 2001.)

As indicated in Chapters 9 and 10, the graded doping in a BJT increases the Early voltage and reduces the base transit time as well as increasing β. Figure S4.7 shows the relative improvements in these quantities as a function of maximum bandgap difference across the base of a Si:Si_xGe_{1-x}:Si BJT relative to that of a similar device with a Si base. [7]

S4.3 COMPARISON OF Si-BASE, SiGe-BASE, AND GaAs-BASE HBTs

Device parameters of Si (Si-base) BJTs, SiGe-base, and GaAs-base HBTs are presented for devices similar in structure to the double poly self-aligned BJT of Figure 10.11 and at the same collector current density. [7]

- The base transit time of a SiGe-base transistor is less than for a Si-base device since the graded composition in the base creates a quasi electric field for electrons, which accelerates them to the collector.
- The base transit time for a GaAs-base HBT is smaller than for a Si-base BJT because of the higher low-field electron mobility of GaAs.
- The base transit time of a (uniformly doped) GaAs-base HBT is comparable to that of a SiGe-base device.[1]

[1]The base transit time of a GaAs-base HBT can be increased by grading the base doping and/or composition.

- The base collector junction transit times t_{tBC} are comparable because for the high fields in the base-collector depletion region, the electron velocities are at their saturation values, which are comparable (Figure 3.9).
- The emitter-base junction capacitances of the SiGe-base and Si-base transistors are comparable because of the comparable emitter and base dopings. For the GaAs-base HBT, however, the emitter-base junction capacitance is much reduced since the emitter can be lightly doped (on the order of 5×10^{17} cm^{-3} compared with a doping on the order of 10^{20} cm^{-3} for the Si-base and SiGe-base transistors.
- The base resistance of a GaAs-base HBT is about an order of magnitude smaller than that of the Si-base and SiGe-base devices since the GaAs base can be heavily doped (on the order of 10^{19} cm^{-3}) without degrading the current gain.
- The unity current gain frequency f_T is larger for the GaAs-base HBT compared with a comparable SiGe-base transistor. The f_T of a SiGe-base transistor is in turn larger than that of a similar Si-base device.
- A major advantage of the SiGe-base device over the GaAs-base HBT is that SiGe is compatible with Si integrated circuit processing and can be used to fabricate BiCMOS circuits.

S4.4 THYRISTORS (npnp SWITCHING DEVICES)

The BJTs that we have discussed consist of two closely spaced and electrically interacting pn junctions. Here we consider the case of three interacting junctions. These devices are used primarily as switches for high-power circuits. The general name for this class of device is *thyristor*.

S4.4.1 FOUR-LAYER DIODE SWITCH

Figure S4.8a shows the structure of an npnp diode. Here the intermediate p and n layers are relatively thin—appreciably less than a minority carrier diffusion

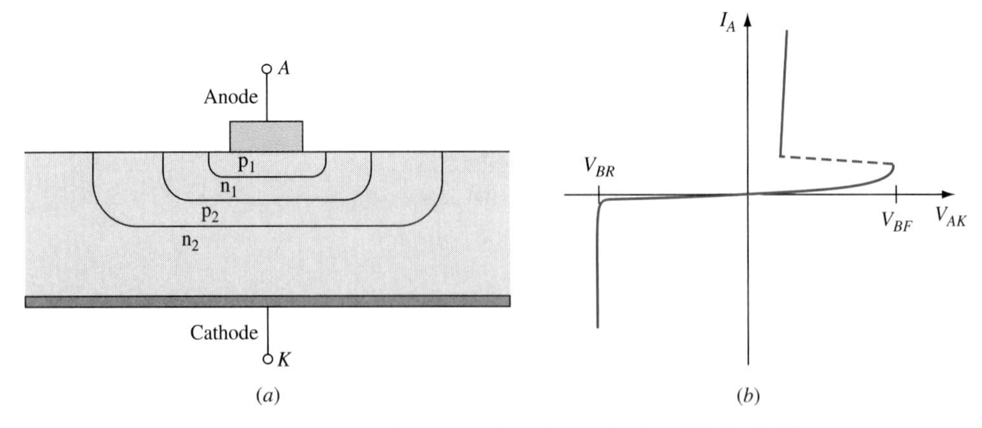

Figure S4.8 (a) The cross section of an npnp four-layer device; (b) the I_A-V_{AK} characteristic.

length. We call the top terminal (the p layer) the anode (A), and the bottom n layer is the cathode (K). The I_A-V_{AK} characteristics of such a diode are indicated in Figure S4.8b. We use I_A for the anode current while V_{AK} is the anode voltage with respect to the cathode. Note this characteristic is distinctly different from that of a two-layer diode. The I-V characteristics look similar in the reverse region, but in the forward region there is a forward breakdown, whose character is different from that of a usual junction breakdown. The forward-bias breakdown voltage V_{BF} and the reverse-bias breakdown voltage V_{BR} are indicated.

Using this I_A-V_{AK} characteristic, we first discuss the circuit behavior of this switch, and then describe the physics of its operation.

Figure S4.9a indicates an npnp diode circuit with an adjustable supply voltage V_S and a load resistance R_L. The I_A-V_{AK} characteristic and its load

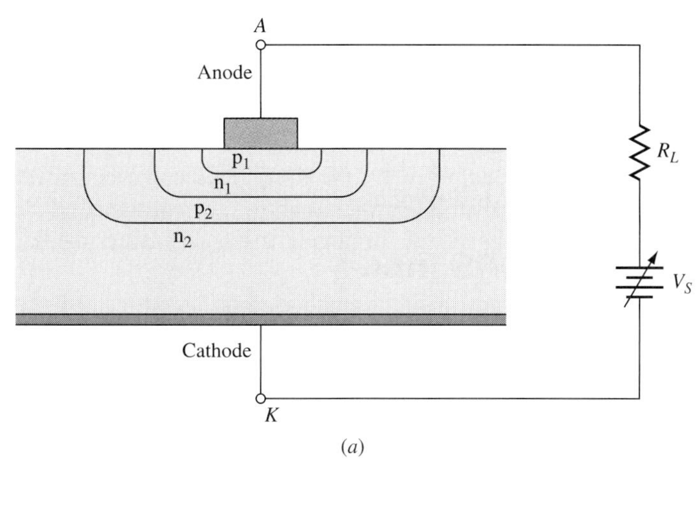

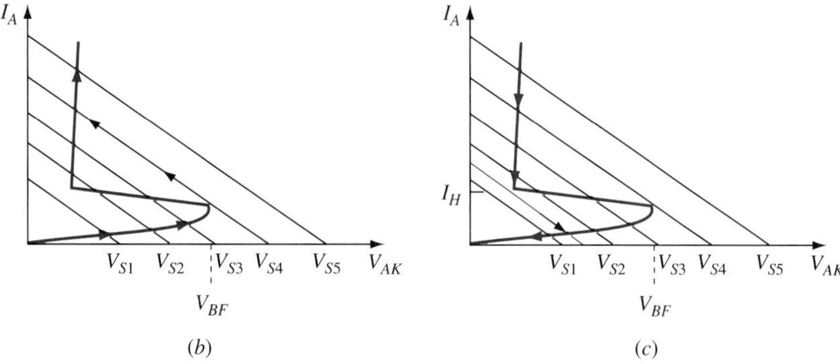

Figure S4.9 The npnp four-layer diode in operation. (a) A varying voltage is applied from anode to cathode. (b) As V_S increases from zero, the voltage across the device increases up to V_{BF}. Then the device switches to a low-voltage, high-current state. (c) When the V_s is decreased, the operating point follows the path shown, remaining in a low-voltage, high-current state until the holding current I_H is reached.

lines are shown in Figure S4.9b for V_S increasing from zero. For $V_S = 0$, $I_A = 0$. As V_S increases, the current is given by the intersection of the I_A-V_{AK} characteristic and the instantaneous load line. Note that for V_{S2} and V_{S3} there are three intersections; however, since V_S is increasing from zero, the operating point is that indicated on the lower branch, since there is no way to change to another current value.

For V_{S4}, however, the load line intersects the I_A-V_{AK} characteristic only at one point—at high current and low voltage. Thus for $V_S = V_{S4}$ the diode switches from a high-voltage, low-current (off) state to a low-voltage, high-current (on) state. For $V_S > V_{S4}$, the device operates on the upper branch.

Once the device is in the on state, as V_S is reduced, the operating point follows the path shown in Figure S4.9c. It will remain on until I_A becomes smaller than the *holding current* I_H, and will then switch to the off state. If V_S is then increased again, the characteristics will follow the lower branch until V_S exceeds V_{S4} as indicated in Figure S4.9b and the device switches again from off to on.

Note that this switching results from a negative differential resistance (negative slope) in the I_A-V_{AK} characteristics. There is no such negative resistance for V_{AK} negative and so, for this polarity of V_S, no switching occurs.

Now let us explain (qualitatively) the npnp device characteristic with the use of energy band diagrams. Figure S4.10a shows the energy band diagram for V_S positive while Figure S4.10c is for negative V_S. We will come back to part (b) of the figure.

For positive V_S there are two forward-biased pn junctions and one reverse-biased junction. Because the n_1 and p_2 layers are thin, holes injected into the narrow n_1 region from the p_1 region are collected in the narrow p_2 region. Since p_2 is electrically floating, it becomes positive from the collected holes. Positive potential on the energy band diagram means lower energy, so the p_2 section moves down on the energy band diagram. This reduces the p_2n_2 barrier, allowing more electrons to be injected from n_2 into p_2. These electrons are then collected by n_1, causing it to become negative, effectively moving it upward on the diagram. That reduces the p_1n_1 barrier, which causes more holes to be injected from p_1 into n_1, and collected by p_2. When the voltage V_S is low, this regenerative effect continues until the injection current in the forward-biased junctions is just equal to the recombination current (device off), Figure S4.10a. Thus, the overall current is limited to some small value. When V_S increases above some amount, the feedback causes the current to increase very rapidly, and the current is limited by the $I_A R_L$ drop in the circuit. This is the situation indicated in part (b). Here all junctions are forward biased.

For V_S negative, Figure 4.10c, there are two reverse-biased junctions and only one that is forward biased. Carriers trapped in n_1 and p_2 do reduce the n_1p_2 barrier, but that has no effect on current injected from cathode or anode.

S4.4.2 TWO-TRANSISTOR MODEL OF AN npnp SWITCH

For V_S positive, the energy band diagram of Figure S4.10a, repeated in Figure S4.11a, appears to be that of a pnp transistor in series with an npn transistor.

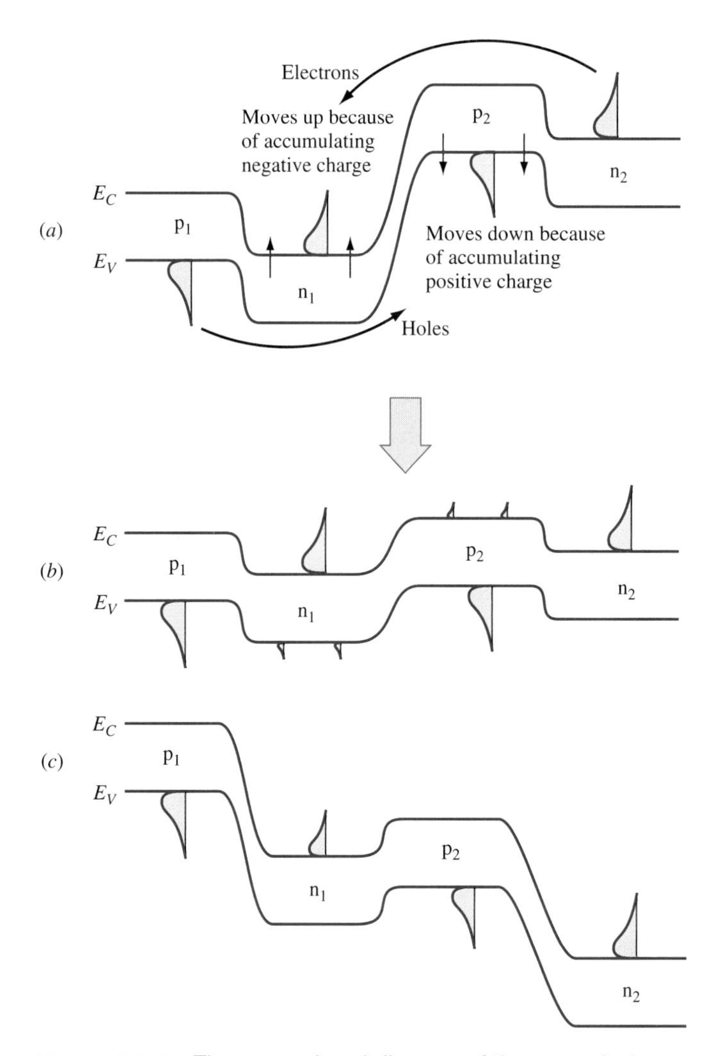

Figure S4.10 The energy band diagram of the npnp device. (a) Under positive V_S, two of the junctions are forward biased but the center one is reverse biased, so little current flows through the device. (b) Injected electrons accumulate in the n_1 region and injected holes accumulate in the p_2 region, causing the reverse bias across the junction between those two areas to decrease, producing the low-voltage, high-current state shown. (c) Under reverse bias, two of the junctions are reversed biased.

The figure indicates the emitters, bases, and collectors of these two transistors. The circuit schematic in Figure S4.11b shows the two transistors explicitly.

In general for a transistor,

$$
\begin{aligned}
I_C &= \alpha I_E + I_{C0} \\
I_B &= (1 - \alpha) I_E - I_{C0}
\end{aligned}
\tag{S4.4}
$$

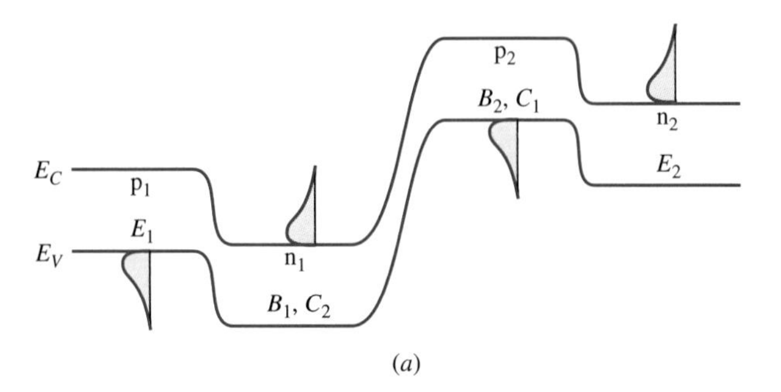

(a)

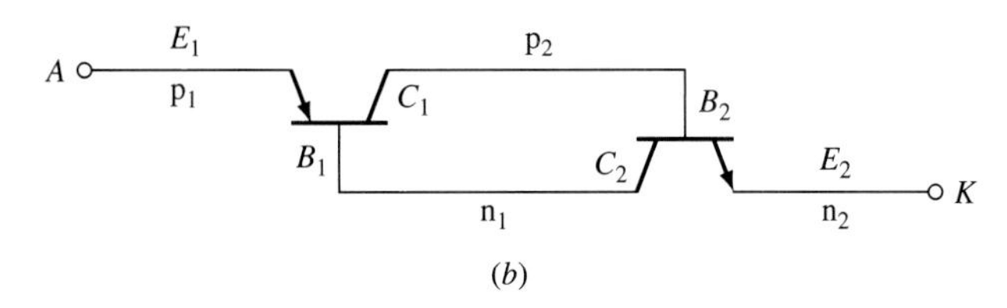

(b)

Figure S4.11 The npnp device can be thought of as an npn transistor connected to a pnp transistor. (a) The energy band diagram; (b) the equivalent circuit.

where I_{C0} is collector-base current with $I_E = 0$. Rearranging each of these, we have

$$\begin{aligned} I_{B1} &= (1 - \alpha_1)I_{E1} - I_{C01} \\ I_{C2} &= \alpha_2 I_{E2} + I_{C02} \end{aligned} \tag{S4.5}$$

But $I_{B1} = I_{C2}$. Also the emitter current I_{E1} is the same as the anode current I_A, and the other emitter current is the same as the cathode current. Furthermore, the anode current has to equal the cathode current, so we have $I_{E1} = I_{E2} = I_A = I_K$, from which we obtain the relation

$$I_A = \frac{I_{C01} + I_{C02}}{1 - (\alpha_1 + \alpha_2)} \tag{S4.6}$$

The current gains α_1 and α_2, however, are functions of I_A. In the forward active region of a regular transistor, they are close to unity, but here the transistors never get to that mode of operation. For small I_A, recombination current in the forward-biased junction transition regions predominates, and the α's are small. This produces a small current. When I_A is large enough that $(\alpha_1 + \alpha_2) = 1$, I_A attempts to become infinite, but is limited by the $I_A R_L$ drop.

S4.5 SILICON CONTROLLED RECTIFIERS (SCRs)

While the four-layer diode discussed is useful for describing the switching mechanism, the device itself is of limited use because the applied voltage must be changed to initiate the off-to-on transition. In other words, the signal to be

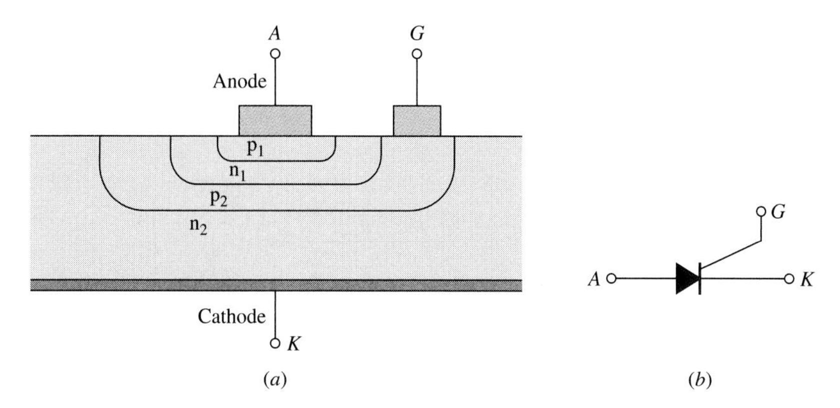

Figure S4.12 (a) The silicon controlled rectifier is an npnp structure with a gate connection at one of the internal layers; (b) the circuit symbol.

switched must itself be changed, instead of being switched by another voltage as in the transistor. A variation of this npnp device is the silicon controlled rectifier[2] or SCR.

The structure of an SCR is indicated schematically in Figure S4.12a. It is similar to that of a four-layer diode except that an additional contact is made to one of the interior regions—in this example, to p_2. This region is called the gate because a current pulse applied to this region is used to initiate the off-to-on transition. Figure S4.12b shows the SCR device circuit symbol.

The energy band diagram for the off state is indicated in Figure S4.13a. For this case $(\alpha_1 + \alpha_2) < 1$. If the gate ($p_2$) is made positive by applying a positive current pulse, the p_2 section will move down on the energy band diagram, and the n_2p_2 junction barrier will be reduced, allowing more electrons to be injected into p_2. These extra electrons are collected by n_1, causing that (floating) region to become negative, which in turn reduces the p_1n_1 barrier and increases the hole current injected from p_1 to n_1. This increase in current results in an increase in α_1 and α_2. Thus, for I_G large enough to cause $(\alpha_1 + \alpha_2) = 1$, the device will switch from off to on. This is illustrated in Figure S4.13b for three values of gate current. For $I_G = 0$ the characteristics are those of a pnpn diode. Increasing I_G decreases the device voltage V_{AK} at which $(\alpha_1 + \alpha_2) = 1$, thus changing the switching point.

Note that once the device is on, it will remain on until V_S is reduced to the point where I_A reaches I_H, the holding current, or the minimum value of I_A such that $(\alpha_1 + \alpha_2) = 1$.

A variation of the SCR is the semiconductor controlled switch (SCS), in which gate contacts are made to both the n_1 and the p_2 regions. The device can be switched from off to on by either a positive current pulse applied to p_2 or a negative current pulse applied to n_1. An SCR can be used to control the current

[2]Sometimes referred to as a *semiconductor controlled rectifier,* but the semiconductor is normally silicon.

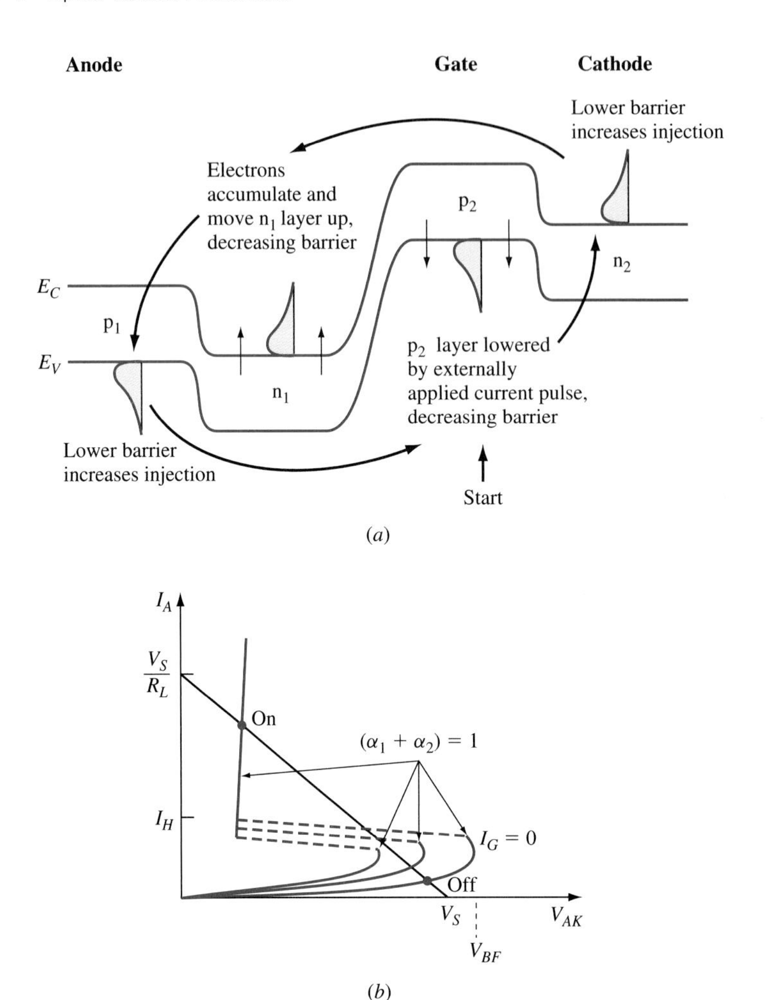

Figure S4.13 (a) The SCR energy band diagram and (b) operating characteristics. In this case, switching is activated by the application of a gate current I_G, which alters the I_A-V_{AK} characteristic so the device can switch.

(and thus the power) in a load by the timing of the gate pulse. Consider the circuit of Figure S4.14. The ac power line feeds a load whose average current is controlled by the SCR. The time delay circuit controls the timing of the input cycle, in which a positive pulse is applied to the gate (at times $t_1, t_3, t_5, \ldots$). These pulses trigger the SCR to turn on, and it conducts for the remaining of the positive cycle until the current reduces to the holding current, I_H ($t_2, t_4, \ldots$). The SCR remains off during the negative of the input cycle and is turned on again by the next gate pulse.

In the SCR the load current flows for less than half of the input cycle. A more useful thyristor is the triac, whose structure is indicated schematically in Figure S4.15a. It consists of two SCRs connected in antiparallel. Its I_A-V_{AK}

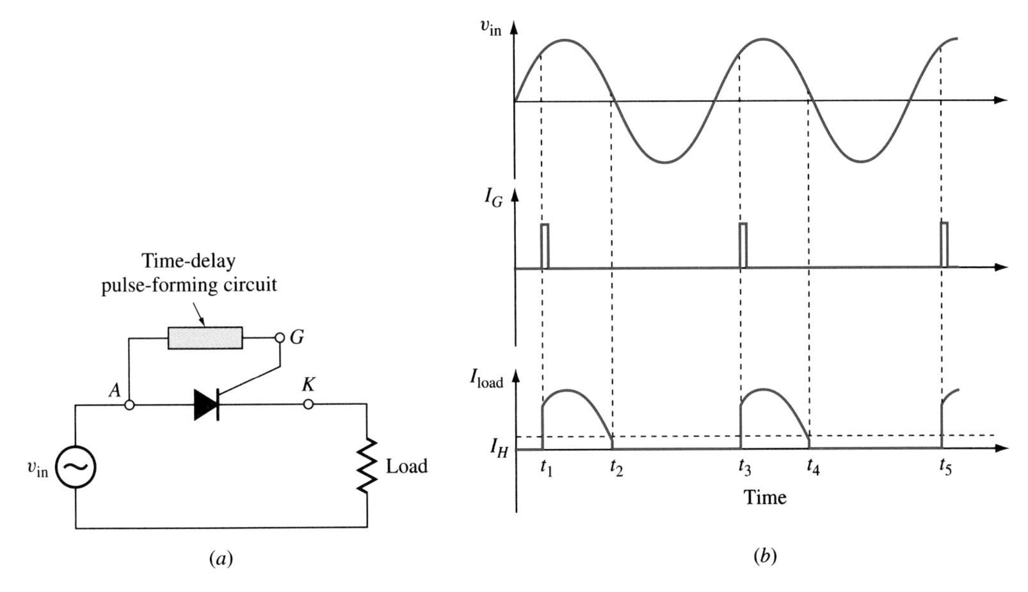

Figure S4.14 (a) Use of an SCR to control the power delivered to a load. (b) Input voltage, gate, and load current waveforms.

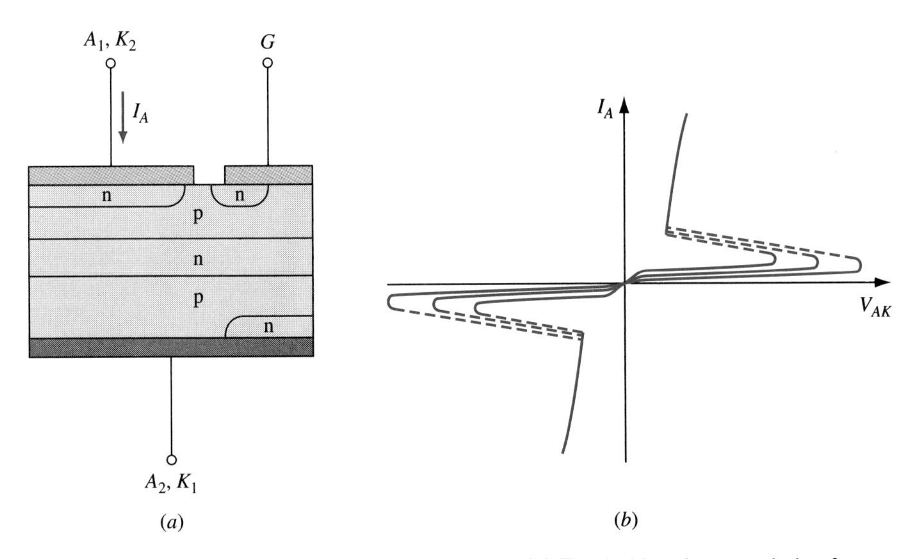

Figure S4.15 (a) Simplified structure of a triac. (b) The I_A-V_{AK} characteristics for gate pulse at three different times in the input cycle.

characteristics are indicated in Figure S4.15b for three values of gate pulse time. The time delay circuit turns the triac on during each half cycle, which permits increased control of the load power.

Triacs are quite useful for ac power control. A common example is a dimmer switch to control the ambient light level.

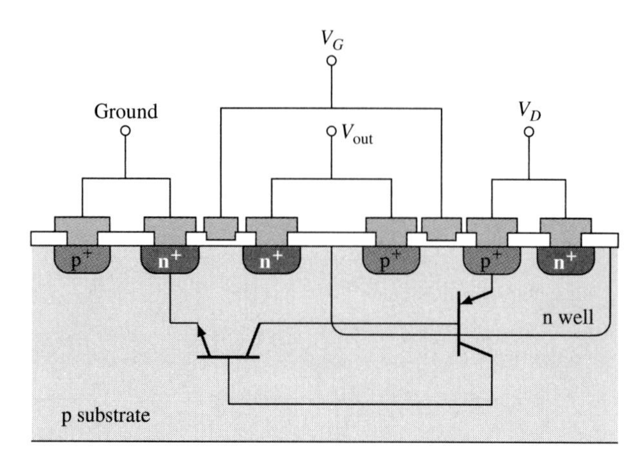

Figure S4.16 In a CMOS integrated circuit, parasitic npnp structures can appear. Unless care is taken in the design, the parasitic npnp can latch-up, preventing the CMOS circuit from operating as intended.

S4.6 PARASITIC pnpn SWITCHING IN CMOS CIRCUITS

We have seen how useful devices can be made by using an npnp structure. Parasitic npnp structures can occur in integrated circuits, however, with deleterious effects. For example, the CMOS inverter discussed in Chapter 8 is shown schematically in Figure S4.16. Here V_D is positive. Note that a pnpn structure is inadvertently created between the source of the p-channel device and the source of the n-channel FET. The two-transistor model of this structure is also indicated. If $(\alpha_1 + \alpha_2) = 1$, the device will attempt to switch to a small value of V_D. If the resistance in the V_D supply is large enough, this parasitic npnp device can "latch up," but then the CMOS circuit ceases to operate as an inverter. If, on the other hand, the V_D supply resistance is not large, the current will be large enough to damage the circuit or the power supply.

Latch-up can be avoided by reducing the transport efficiency and thus α of the two parasitic bipolar transistors. One scheme is to increase the separation of the n-channel and p-channel devices, but this decreases the packing density. Another method is to implant regions (with, for example, oxygen) along the switching path to reduce the carrier lifetime and thus increase recombination and reduce transport efficiency.

S4.7 APPLICATIONS OF SPICE TO BJTs

Just as there are several levels of SPICE for use in MOSFET circuits, there are two SPICE levels for BJTs. These levels are based on the Ebers-Moll and the

Table S4.1 SPICE parameters for basic static Ebers-Moll model

Symbol	SPICE keyword	Parameter	Units
I_s (see note)	IS	Saturation current	A
β_F	BF	Common emitter current gain (normal mode)	
β_R	BR	Common emitter current gain (inverse mode)	
V_{AF}	VA	Early voltage (normal mode)	V
V_{AR}	VB	Early voltage (inverse mode)	V

Note: This is the same quantity as I_0 in previous chapters. Here we adjust our notation to match SPICE's.

Gummel-Poon formulations of the electrical characteristics of BJTs. Unlike MOSFETs, for which the SPICE levels must be specified, for BJTs SPICE is based on the Gummel-Poon formulation. This model reverts to the Ebers-Moll model unless a parameter specific to the Gummel-Poon model is specified (e.g., ISE).

In Chapter 9, the Ebers-Moll static model for BJTs was discussed. In the basic model, the parasitic elements were neglected. The SPICE characteristics for this model are listed in Table S4.1.

The Ebers-Moll dc model introduced in Chapter 9 can be used for first-order calculations, but it treats the current gains β_F and β_R as constant values. These values are current dependent, however. This is not of great importance for β_R, since it is normally small, but it is important for β_F, the current gain in the active mode. Figure S4.17 compares the experimental value of β_F with the constant value assumed in the Ebers-Moll model. It can be seen that β_F increases with collector current, reaches a maximum, and then decreases. While the Ebers-Moll model is useful for analysis of BJT circuits at intermediate current levels, at very low currents, junction recombination currents must be considered. Likewise, at high currents, high-injection effects are important. For these considerations the more complete Gummel-Poon model must be used.

The Gummel-Poon model differs from the Ebers-Moll model in that it takes the current dependence of current gains β_F and β_R into account. The SPICE parameters for the static Gummel-Poon model are listed in Table S4.2.

In the Gummel-Poon model the parameters β_F and β_R (BF and BR in SPICE) represent the maximum common emitter current gains for forward operation and inverse operation respectively. The normal knee current I_{KF} represents the (approximate) value of I_C for maximum β_F. In what follows we discuss the parameters for normal operation. Analogous considerations apply for inverse operation. The base and collector currents and thus β (recall $\beta = I_C/I_B$) are handled differently at low to medium currents (less than those for maximum β) and high currents (above that value). We will discuss these two current regimes shortly, but first let us address parasitic effects.

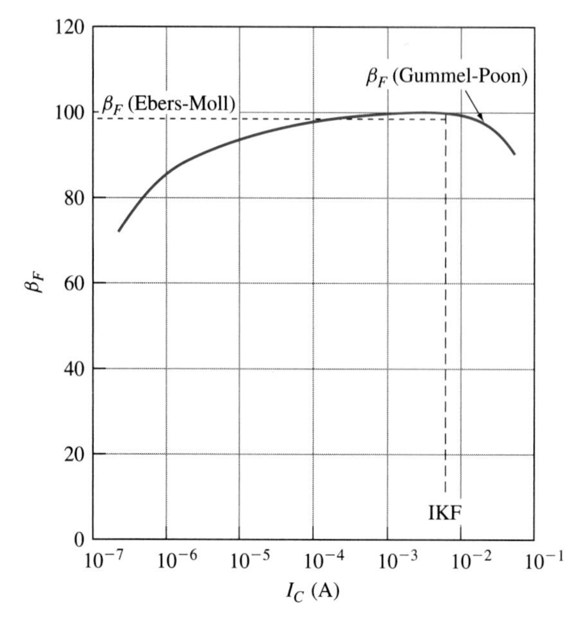

Figure S4.17 Active-mode current gain as a function of collector current. The solid line represents the experimental data while the dashed line is the Ebers-Moll approximation. The maximum value of β_F is that used in the Gummel-Poon model. The collector current at the maximum β_F is approximately the knee current, IKF.

Table S4.2 SPICE parameters for static Gummel-Poon model

Symbol	SPICE keyword	Parameter	Units
I_S	IS	Saturation current	A
β_F	BF	Maximum normal-mode current gain	
β_R	BR	Maximum inverse-mode current gain	
V_{AF}	VA	Early voltage (normal mode)	V
V_{AR}	VB	Early voltage (inverse mode)	V
I_{SBE}	ISE	*B-E* recombination saturation current	A
n_{BE}	NE	*B-E* recombination current emission coefficient	
I_{SBC}	ISC	*B-C* recombination saturation current	A
n_{BC}	NC	*B-C* recombination current emission coefficient	
I_{KF}	IKF	Normal knee current	A
I_{KR}	IKR	Inverse knee current	A

Table S4.3 Some important parasitic elements associated with SPICE for BJTs

Symbol	SPICE keyword	Parameter name	Units
R_B	RB	Base resistance	Ω
R_C	RC	Collector resistance	Ω
R_E	RE	Emitter resistance	Ω
$C_{jBE}(0)$	CJE	Zero bias B-E junction capacitance	F
$V_{bi}(B\text{-}E)$	VJE	B-E built-in voltage	V
$m(B\text{-}E)$	MJE	B-E grading coefficient	—
$C_{jBC}(0)$	CJC	Zero bias B-C junction capacitance	F
$V_{bi}(B\text{-}C)$	VJC	B-C built-in voltage	V
$m(B\text{-}C)$	MJC	B-C grading coefficient	—
t_{TF}	TF	Base transit time (normal mode)	s
t_{TR}	TR	Base transit time (inverse mode)	s

S4.7.1 PARASITIC EFFECTS

For the time-dependent analysis, the parasite elements must be considered. The major parasitic elements are shown in Table S4.3, with their SPICE keywords.

Note that the t_{TF} and t_{TR} in the SPICE formulation are related to the base transit time in the normal mode and in the inverse mode respectively. Thus, as discussed in Section 10.5.2,

$$t_{TF} = \frac{Q_{BF}}{I_{CF}} \tag{S4.7}$$

where Q_{BF} is the minority carrier charge stored in the base in the forward mode, (Equation 10.45) and t_{TF} is the transit time in the forward mode. There is a corresponding relation for t_{TR}.

S4.7.2 LOW TO MEDIUM CURRENTS

Next, let us consider the β's. At low currents, β_F is reduced from its maximum value by the recombination in the base-emitter depletion region.

The collector current varies exponentially with B-E voltage:

$$I_C = I_S e^{qV_{BE}/kT} \tag{S4.8}$$

while the base current has two terms,

$$I_B \approx \frac{I_S}{\beta_F} e^{qV_{BE}/kT} + I_{SBE} e^{qV_{BE}/n_{BE}kT} \tag{S4.9}$$

Here β_F is the current gain in the near-constant β region, I_{SBE} is the saturation current due to the recombination in the B-E junction, and n_{BE} is an empirical

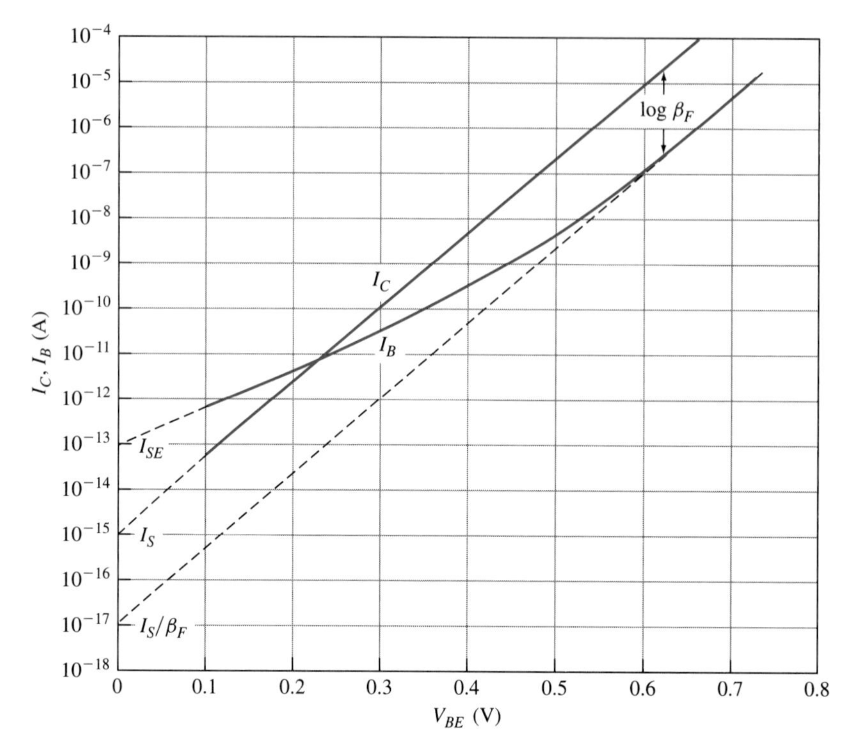

Figure S4.18 Gummel plot for low to medium currents. Values for I_S, I_{SBE}, and n are found from the extrapolated intersections of I_C, I_B, and the slope of the I_B-V_{BE} curve. In this region, $I_B = I_{SBE}e^{qV_{BE}/n_{BE}kT}$ and the slope has the value $d \log I_B/dV_{BE} = q \log e/n_{BE}kT$.

factor called the *B-E emission coefficient* and is on the order of $n_{BE} = 2$, as discussed in Chapter 9.

Values for I_S, β_F, I_{SBE}, and n_{BE} can be obtained from a Gummel plot similar to that of Figure 9.25. Such a plot is shown in Figure S4.18. The value of I_S is found from extrapolating I_C to $V_{BE} = 0$ [Equation (10.13)]. The value of β_F is found from the ratio I_C/I_B in the region indicated. Values for I_{SBE} and n_{BE} are found from the extrapolated intersection of the base current and the slope of the I_B-V_{BE} plot at low currents where the first term in Equation (S4.9) is negligible. Since $\beta_F = I_C/I_B$, from Equations (S4.8) and (S4.9).

$$\beta_F(I_C) = \frac{\beta_F}{1 + \dfrac{\beta_F I_{SBE}}{I_S} e^{-qV_{BE}/kT(1-1/n_{BE})}} \tag{S4.10}$$

From Equation (S4.10) it can be seen that as V_{BE} (and thus I_C) increases, $\beta_F(I_C)$ increases and approaches its maximum value.

S4.7.3 HIGH CURRENTS

At high currents I_C is reduced from its value predicted by Equation (S4.8) because of these effects:

1. The B-E junction voltage V_{BEJ} is reduced from the applied B-E voltage V_{BE}, because of parasitic resistances:

$$V_{BEJ} = V_{BE} - I_B R_B - I_E R_E \tag{S4.11}$$

Since, at high currents, recombination current is negligible compared with injection current, $I_B \approx I_C/\beta_F$ and $I_E \approx I_C$. Note that I_C and I_B as functions of V_{BE} must be solved iteratively:

$$I_C \approx I_S e^{q[V_{BE}-(I_C/\beta_F)R_B - I_C R_E]/kT} \tag{S4.12}$$

$$I_B \approx \frac{I_S}{\beta_F} e^{q[V_{BE}-(I_C/\beta_F)R_B - I_C R_E]/kT} \tag{S4.13}$$

2. The large minority carrier concentration in the base, which in turn increases the majority carrier concentration (for charge neutrality), increases the back injection into the emitter. Thus the base current is increased, as discussed in Section 9.10. This increased base-emitter back-injection reduces the forward current gain.
3. The high current is also responsible for the base push-out (Kirk) effect, which increases the effective base width and thus reduces β, as discussed in Section 9.11.

For $I_C \leq I_{KF}$, Equation (S4.8) is used to determine the I_C-V_{BE} characteristics. For $I_C \geq I_{KF}$, the Gummel-Poon formulation is appreciably more complicated.

The SPICE (Gummel-Poon) treatment of the high collector current can be explained with the aid of Figure S4.19. This figure treats three cases for the I_C-V_{BE} relations. Curve (a) assumes the Ebers-Moll model with no parasitic resistances. Curve (b) uses the Gummel-Poon model with $R_E = R_B = 0$, and the SPICE parameter I_{KF} is taken as 10 mA. The deviation of curve (b) from the dashed line results from the high-injection effects (items 1 and 2 above). For $I_C > I_{KF}$, where for simplification the parasitic effects (e.g., R_B, R_E) are neglected, the SPICE parameter I_{KF} is taken as the current at which the $\log(I_C)$-V_{BE} curve departs from a straight line. Above this current, I_C is assumed to vary as $e^{qV_{BE}/2kT}$ and can be expressed

$$I_C = \sqrt{I_S \cdot I_{KF}}\; e^{qV_{BE}/2kT} \tag{S4.14}$$

In curve (c), an emitter series resistance $R_E = 2\,\Omega$ is assumed (item 1 above), in addition to the high-injection effects of (b).

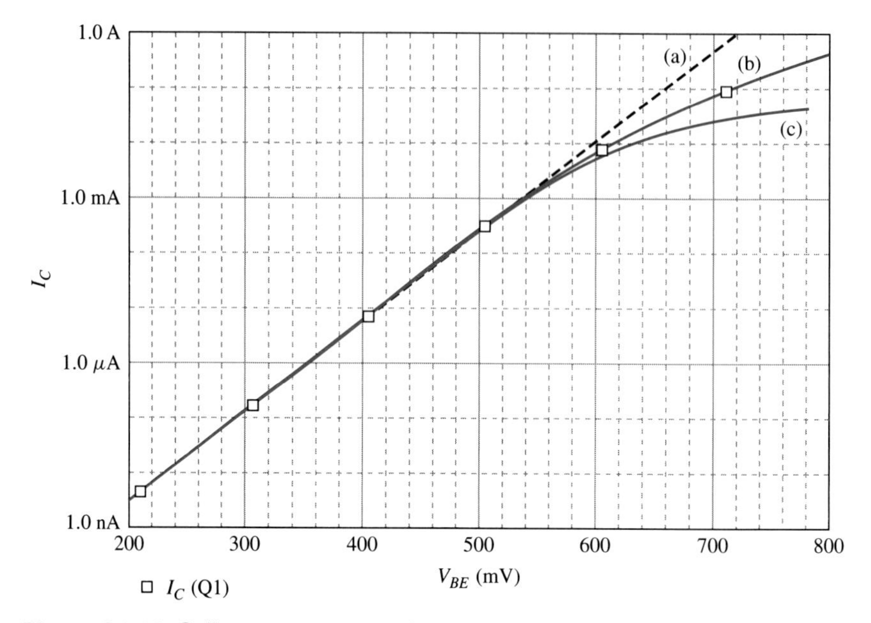

Figure S4.19 Collector current as a function of base-emitter voltage as determined by SPICE for three cases. In (a) the Ebers-Moll model is assumed neglecting parasitic resistances. In (b) the Gummel-Poon model is used with a forward knee current of 10 mA. In (c) in addition to the high-injection effects of (b), an emitter resistance of 2 Ω is assumed.

S4.8 EXAMPLES OF THE APPLICATION OF SPICE TO BJTs

Following are some simple applications using SPICE.

EXAMPLE S4.1 Using SPICE as Curve Tracer

The I_C-V_{CE} characteristics with I_B as a parameter for an npn BJT operating in the normal mode can be simulated by using the circuit of Figure S4.20a. The parameters assumed for this example are

$$I_S = 2 \times 10^{-16}\text{ A}$$
$$\beta_F = 100$$
$$\beta_R = 1$$
$$V_A = 20\text{ V}$$

The collector voltage (V_{CE}) is swept from 0 to 5 V for I_B values of 0, 2, 4, 6, 8, and 10 μA.

The output plots produced by SPICE are indicated in Figure S4.20b. The finite slope in the current saturation region is a result of the finite Early voltage V_A.

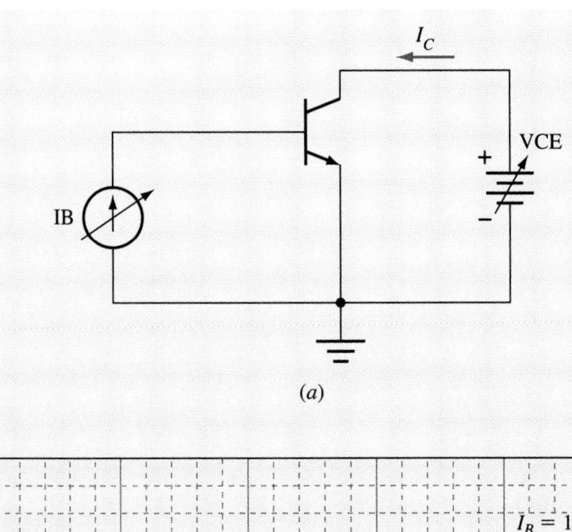

(*a*)

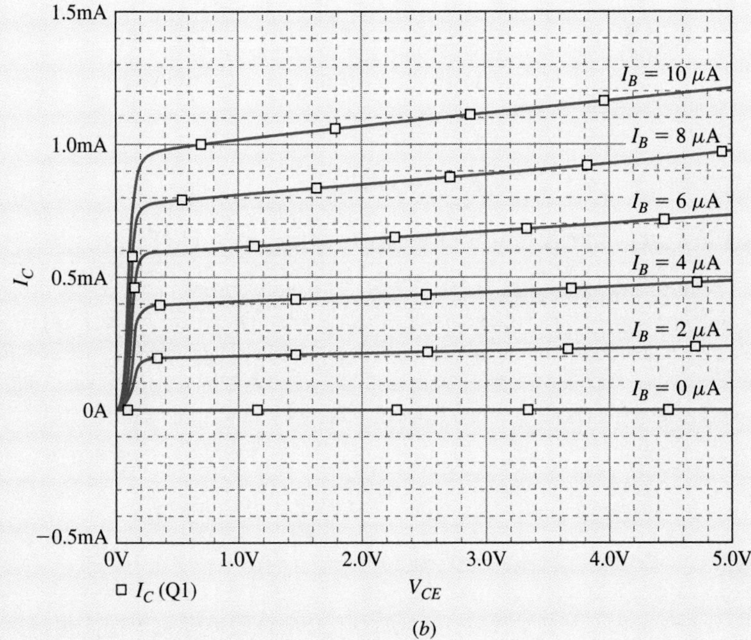

(*b*)

Figure S4.20 Using SPICE as a curve tracer for an npn BJT. (a) Circuit; (b) $I_C = V_{CE}$ results with I_B as a parameter.

Determination of the Operating Points of a Circuit

EXAMPLE S4.2

Given the circuit of Figure S4.21, with the BJT parameters of the previous example, we desire to find the voltages at the emitter and collector. Unless instructed otherwise, SPICE first finds the initial voltages at all nodes of the circuit.

The SPICE printout gives $V_C = 2.286$ V and $V_E = 2.245$ V. Note that $V_C < V_B$ and $V_B > V_E$, so the *B*-*C* and *B*-*E* junctions are both forward biased, or the device is operating in the saturation region.

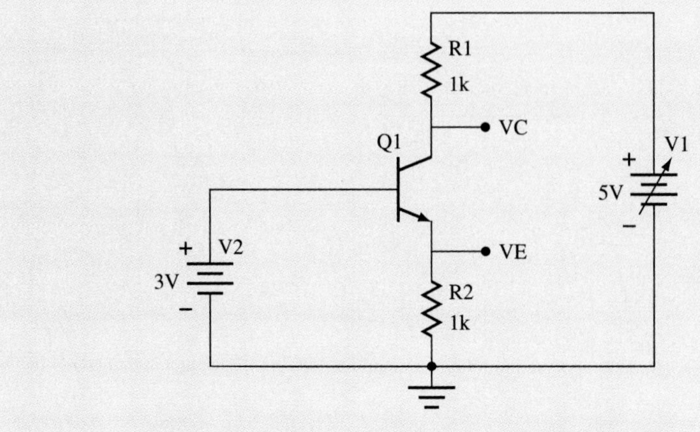

Figure S4.21 Circuit used to find the operating points of an npn BJT with the characteristics of those of Figure S4.18. The results are $V_C = 2.286$ V and $V_E = 2.245$.

EXAMPLE S4.3 Output Voltage for a Given Input Voltage in a BJT Circuit

In the circuit of Figure S4.22a, the BJT has the same parameters as for the previous two examples. The input voltage is $V_{\text{in}} = 2\ \sin(2\pi ft)$, where f is 1 MHz, low enough that parasitic elements need not be considered. The SPICE-generated input and output voltage waveforms are shown in Figure S4.22b. It can be seen that the input voltage is large enough to drive the BJT between cutoff and saturation.

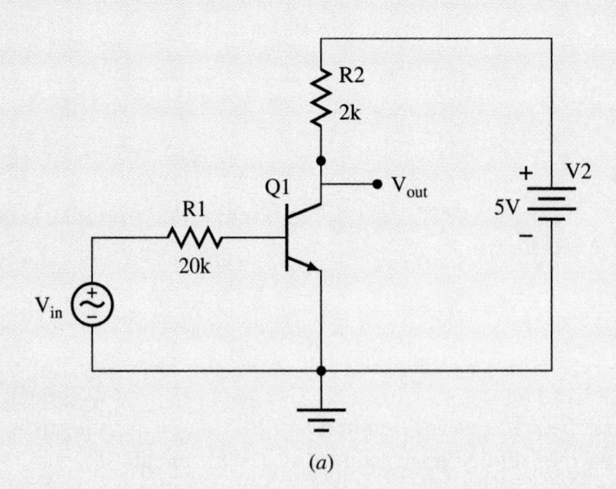

Figure S4.22a Circuit of an npn BJT with sinusoidal input voltage $V_{\text{in}} = 2\ \sin(2\pi ft)$ with $f = 1$ MHz.

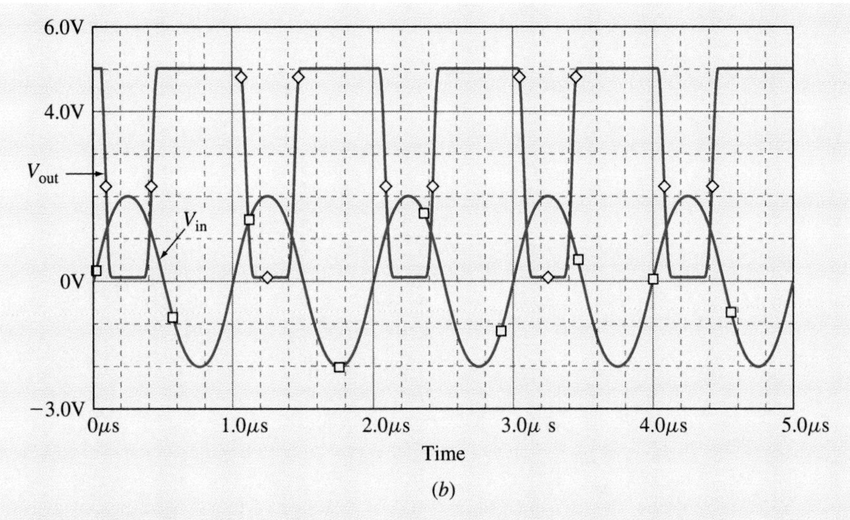

Figure S4.22b Input and output voltages as determined by SPICE.

BJT Transient Response for Square Wave Input

EXAMPLE S4.4

As a final example we consider the transient response of the circuit of Figure S4.23a. The input is a voltage square wave from 0 V to 4 V with rise and fall times of 10 ns and a period of 2 μs. For this case the junction capacitances are considered along with the stored-charge capacitance (as determined in SPICE from the base transit time).

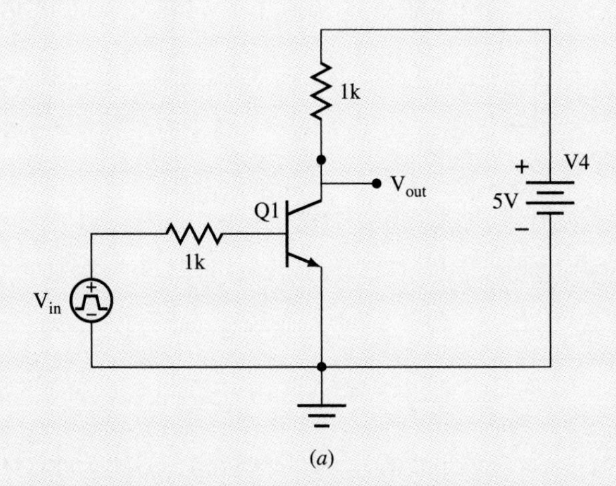

Figure S4.23a Transient response of an npn BJT with a square wave input with voltage from 0 to 4 V, rise and fall times of 10 ns, and period of 2 μs.
(a) The circuit.

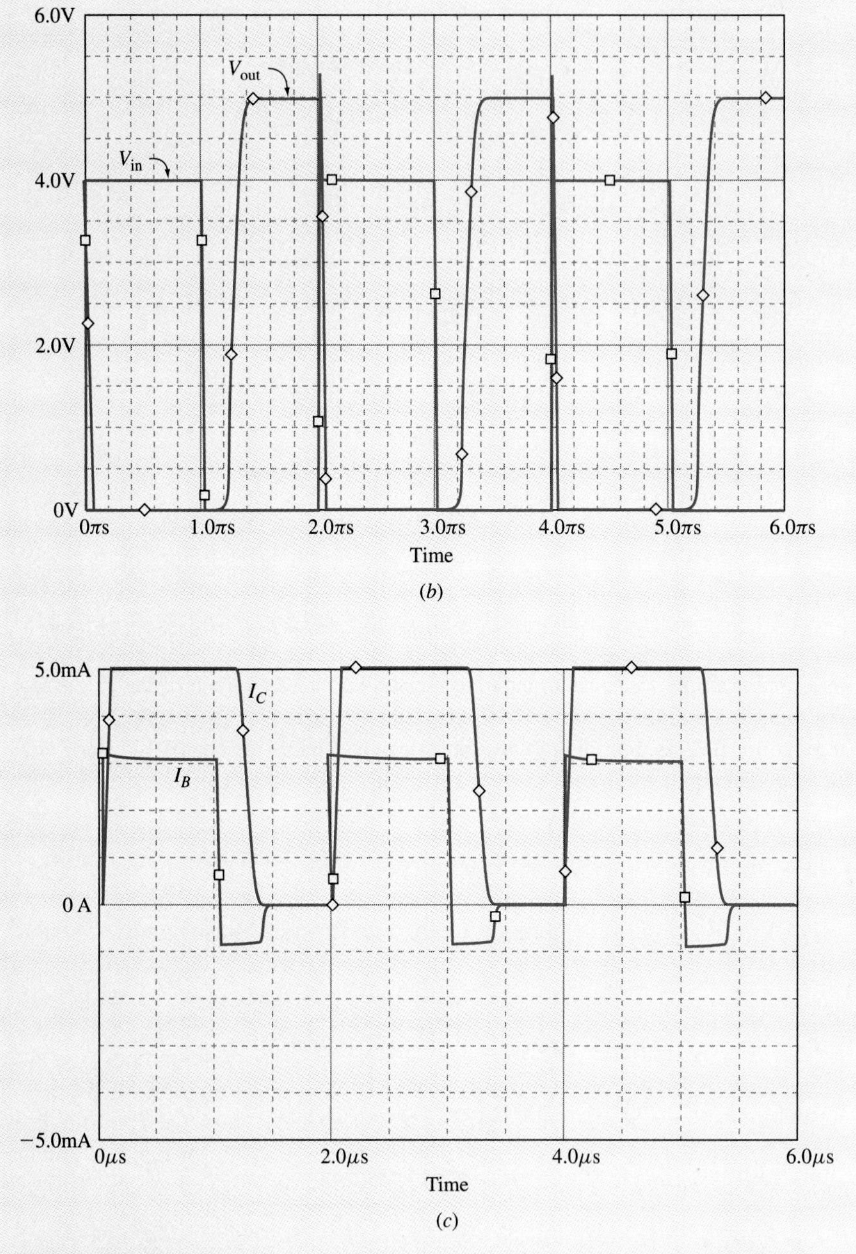

Figure S4.23b–c (b) The input and output voltages; (c) base and collector currents. The finite base transit time causes a delay in the output voltage and current response. Note the reverse base current during the discharge of the stored-charge capacitance.

The BJT parameters are, in SPICE notation,

$$\begin{aligned}
\text{BF} &= 100\\
\text{BR} &= 5\\
\text{IS} &= 2\times 10^{-16}\,\text{A}\\
\text{NF} &= 1\\
\text{NR} &= 1\\
\text{VAF} &= 20\,\text{V}\\
\text{CJE} &= 10^{-11}\,\text{F}\\
\text{VJE} &= 0.85\,\text{V}\\
\text{MJE} &= 0.4\\
\text{CJC} &= 10^{-11}\,\text{F}\\
\text{VJC} &= 0.75\,\text{V}\\
\text{MJC} &= 0.4\\
\text{TF} &= 20\,\text{ns}
\end{aligned}$$

The input and output voltages as functions of time are plotted in Figure S4.23b. There is a time delay of about 0.2 μs between the input going low and the output beginning to rise. The base and collector currents are indicated in Figure S4.23c. The delay in the decay of the collector current is indicated, as is the reversal of the base current as the stored minority carrier charge is removed.

S4.9 SUMMARY

In this chapter we looked at some additional bipolar devices, including heterojunction bipolar transistors and thyristor devices.

Heterojunction bipolar transistors have some inherent advantages over homojunction BJTs. These include increased injection efficiency, thus resulting in higher current gain β and higher operating frequencies. As in a homojunction BJT, the performance can be increased by appropriately grading the base doping. The performance can be further increased by grading the base composition such that the band gap decreases with distance from the emitter edge of the base.

We also examined thyristor npnp, or three-junction, devices. The four-layer diode switch acts as a reverse-biased diode for one polarity of applied voltage, but under the other polarity exhibits a spontaneous change of state, from low current, high voltage to low voltage, high current. A more useful device is the silicon controlled rectifier (SCR), in which an externally applied signal at the gate controls the switching, rather than the input signal itself, and the triac, which increases the power control of the load.

Both the four-layer diode switch and the SCR are essentially back-to-back bipolar transistors. Such structures occur parasitically in the fabrication of CMOS devices, and can be a problem because they can cause latch-up, preventing circuit operation. Latch-up can be controlled by implanting recombination regions between the transistors, or by increasing the distance between transistors.

The use of SPICE to analyze BJT operation was introduced. There are two SPICE levels for use with BJTs. One is based on the Ebers-Moll model in which the current gains are considered to be constant. Only the E-B injection currents are considered. Examples based on the Ebers-Moll and Gummel-Poon models were presented.

S4.10 REFERENCES

1. R. J. Ferro, "Base and Collector Current Mechanisms in Bipolar NPN Gallium Aluminum Arsenide/Gallium Arsenide Heterojunction Transistors," Ph.D. Dissertation, University of Vermont, 1985.
2. A. Marty, G. Rey, and J. P. Bailbe, "Electrical Behavior of an NPN AlGaAs/GaAs Heterojunction Transistor," *Solid State Electronics,* 22, pp. 549–557, 1979.
3. H. Kroemer, "Theory of Wide-Gap Emitter for Transistors," *Proc. IRE,* 45, pp. 1535–1539, 1957.
4. John D. Cressler, "Re-Engineering Silicon: Si-Ge Heterojunction bipolar technology," *IEEE Spectrum,* 32, pp. 49–55, 1995.
5. Guofu Niu, John D. Cressler, Shiming Zhang, William E. Ansley, Charles S. Webster, and David L. Harame, "A unified approach to RF and microwave noise parameter modeling in bipolar transistors," *IEEE Trans. Electron Devices,* ED48, pp. 2568–2574, 2001.
6. David L. Harame and Bernard S. Meyerson, "The Early History of IBM's SiGe Mixed Signal Technology," *IEEE Trans. Electron Devices,* 48, pp. 2555–2567, 2001.
7. Tak H. Ning, "History and Future Perspectives of the Modern Silicon Bipolar Transistor," *IEEE Trans. Electron Devices,* 48, pp. 2485–2491, 2001.

S4.11 REVIEW QUESTIONS

1. What is meant by a heterojunction bipolar transistor? Where is the heterojunction? What is its purpose?
2. In an npn heterojunction transistor, compare the barrier for holes with the barrier for electrons. How does this difference improve the current gain? What effect in realistic HBTs offsets this advantage?
3. How is an energy spike in the energy band diagram of the heterojunction of an HBT avoided in practice?
4. Apart from increased current gain, what is the second advantage of HBTs?
5. How is the current gain temperature dependence of HBTs different from that of homojunction bipolar transistors?

6. Explain in your own words how a four-layer diode switch works.
7. Show how the operating point of a four-layer diode switch moves along the *I-V* characteristics as the voltage is first increased then decreased.
8. How does the addition of a controllable gate impact the operation of a silicon-controlled rectifier?
9. What is meant by latch-up?

S4.12 PROBLEMS

S4.1 Equation (9.28) states that

$$\gamma = \frac{1}{1 + \dfrac{p_E(0^-)}{n_B(0^+)} \dfrac{D_{pE}}{D_{nE}} \dfrac{W_B}{W_E}}$$

Discuss the effect of the heterojunction on the term $p_E(0^-)/n_B(0^+)$, even for equal doping.

S4.2 Consider a nondegenerately doped HBT. Adapt Equation (9.42) to find an expression for the β of this transistor, as a function of the band gaps of the emitter and base.

S4.3 In connection with the npnp device of Figure S4.10, it was stated that at low V_S, the recombination in the forward-biased junction transition regions is significant.

a. Explain why the recombination current is significant.
b. When recombination cannot keep up with the carrier injection, what happens to the energy band diagram? Explain why the device switches to a high-current state.

S4.4 We indicated that in an SCR device, applying a positive voltage pulse V_G to the gate (p_2 region) could turn the device to the on state. Explain how applying a negative current pulse to the n_1 region would have the same effect.

S4.5 Can the SCR be turned off by applying a negative current pulse to the p_2 region?

S4.6 Explain how increasing the spacing between the transistors in a CMOS circuit layout will help prevent latch-up.

S4.7 In the active mode, for currents large enough that the E-B recombination current can be neglected, show that although R_E and R_B influence I_C and I_B, β_F is not affected. *Hint:* Use Equations (S4.12) and (S4.13).

S4.8 Using the circuit of Figure PS4.1, use SPICE to obtain a Gummel plot (i.e., log I_C and log I_B as functions of V_{BE}) from 0.2 V $\le V_{BE} \le$ 0.8 V. The npn BJT has the values BF = 150, IS = 10^{-14} A, ISE = 10^{-13} A.

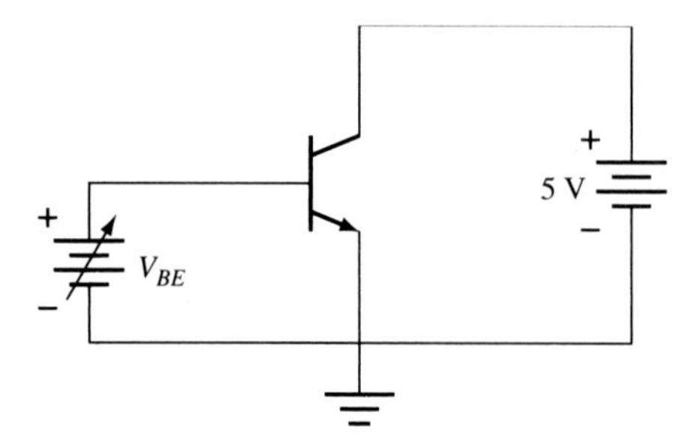

Figure PS4.1

S4.9 Repeat Problem S4.8 with BF = 150, IS = 10^{-14} A, ISE = 10^{-13} A, IKF = 1 mA.

S4.10 Repeat Problem S4.8 with BF = 150, IS = 10^{-14} A, ISE = 10^{-13} A, RE = 3 Ω, RC = 10 Ω.

S4.11 Comment on the differences in the results for the above three problems.

S4.12 For the circuit of Figure PS4.2, using SPICE, plot the output voltage v_{out} for $0 \leq 3$ ms. The BJT has $\beta = 150$ and $I_S = 1$ pA.

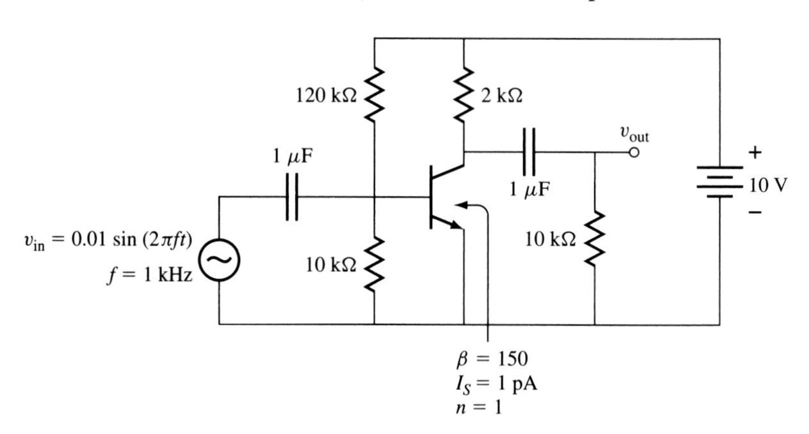

Figure PS4.2

PART 5

Optoelectronic Devices

One could argue that the two areas in which semiconductors have made the most profound changes in our lives are computing and communication. Increasingly, optical communication is becoming the fastest and in many cases the most cost-effective approach to transferring information from one place to another. The bandwidths that have turned the Internet and the World Wide Web into household tools (and playgrounds) are in large part due to optical communications.

In this chapter we will explore the role of semiconductors in photonics, examining some common optical devices such as photodetectors, light-emitting diodes, lasers, and imagers. ■

CHAPTER 11

Optoelectronic Devices

11.1 INTRODUCTION AND PREVIEW

To begin our discussion of optoelectronic devices, let us first consider the electromagnetic spectrum, Figure 11.1, from the subaudio region, $f \approx 10$ Hz to the x-ray region, $\lambda \approx 1$ nm. For semiconductor optoelectronics, the visible and near-infrared regions are of primary importance. The photon wavelengths and energies in the visible region are shown on an expanded scale, and the two most important wavelengths for fiber optics are also indicated.

The success of optical communication is not due just to the advances in optical fibers, but also to the concurrent development of the optical sources (laser diodes and light-emitting diodes) and photodetectors that are fast, cheap, and reliable. These devices are the subjects of this chapter.

11.2 PHOTODETECTORS

The term *photodetector* usually describes a diode (or transistor) that is used to measure the amount of light energy present. When we say "measure" we imply that the output signal should be a function of the light intensity.[1] For example, a fiber-optic receiver contains a photodetector whose output current is proportional to the detected light intensity. The rest of the receiver contains amplification and decision circuitry.

The case of a generic photodetector is considered first. Once the design trade-offs are understood, some specific photodetector types are considered.

11.2.1 GENERIC PHOTODETECTOR

Figure 11.2 shows a generic photodiode structure. It contains a pn junction and is illuminated from the top. Photons with energy greater than the semiconductor

[1]Actually, the term *intensity* is used loosely here. Technically, intensity is in the nonintuitive units of power per solid angle, and is usually used to describe light emitters. For photodetectors, the actual quantity we need is called *irradiance* (measured in W/m^2).

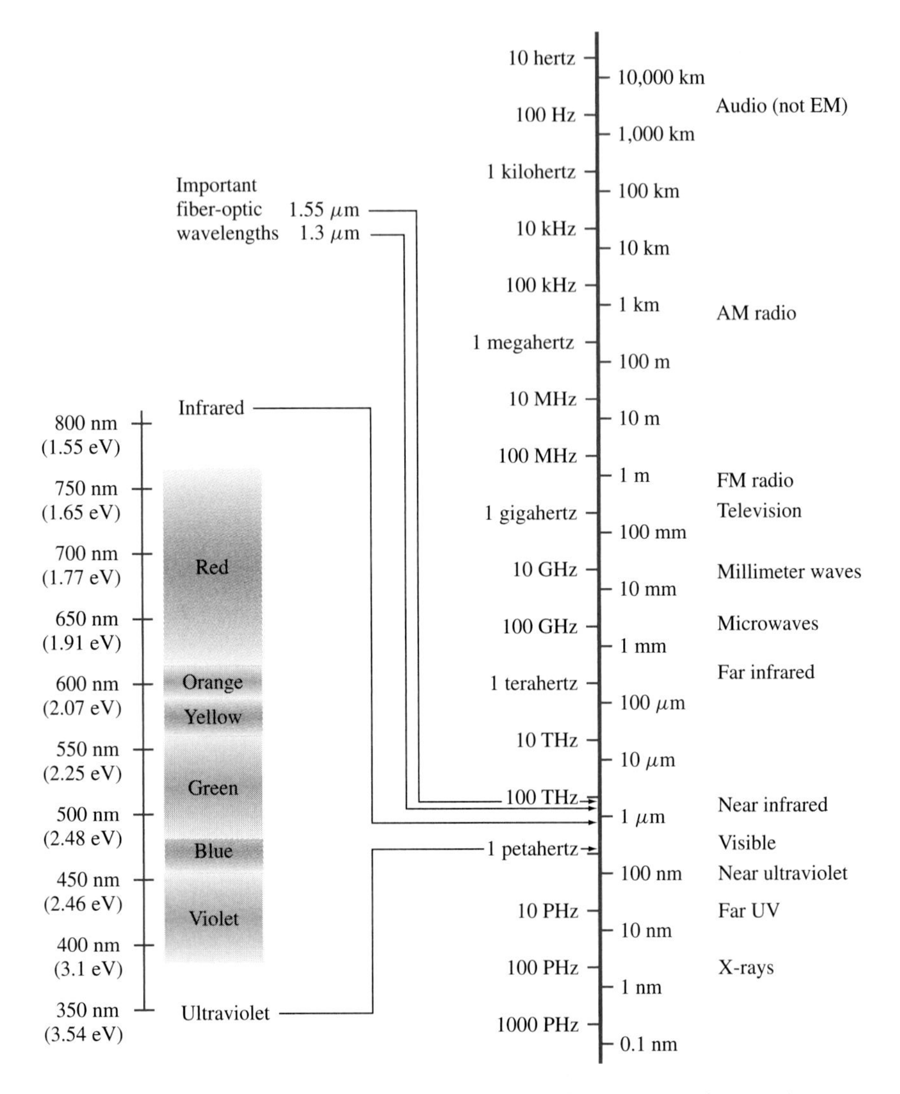

Figure 11.1 The electromagnetic spectrum. (The frequencies of acoustic waves are given for comparison although they are not electromagnetic waves.)

band gap create electron-hole pairs. Electrons and holes are separated by the field in the junction depletion region and flow through top and bottom contacts to an external circuit.

It is important to not block the light with the top contact, so that contact is often in the form of a ring. To get to the contact from the n^+ region, the current flows laterally in the top layer; therefore that layer's sheet resistance should be small. A small sheet resistance can result from a thick layer or from heavy doping, or a combination. In practice, however, it is required that little light be absorbed in this layer, so it is made thin (a fraction of a micrometer) and thus it must be heavily doped.

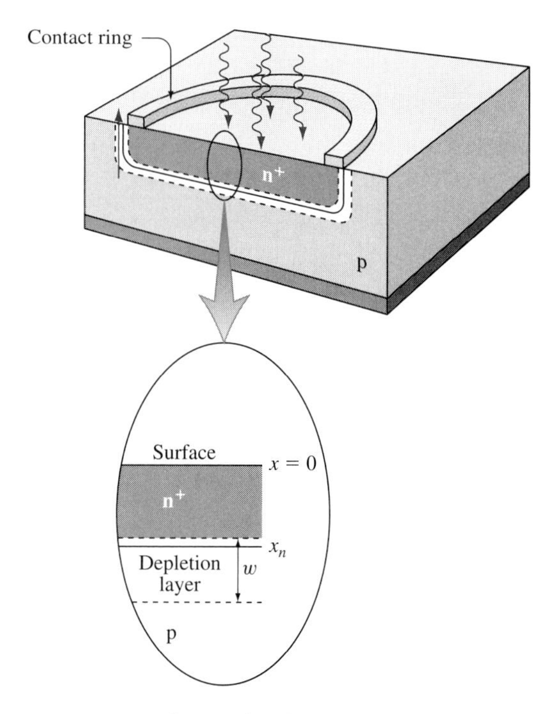

Figure 11.2 A generic photodiode.

As the light penetrates into the semiconductor, it will be absorbed. If the photon flux at a given distance into the material is $F_L(x)$, then the variation of F_L with x within the semiconductor is proportional to $F_L(x)$:

$$\frac{dF_L(x)}{dx} = -\alpha F_L(x) \tag{11.1}$$

where α is the *absorption coefficient*. This can be solved to find

$$F_L(x) = F_L(0)e^{-\alpha x} \tag{11.2}$$

The absorption coefficient α then gives the distance at which the flux is reduced to $1/e$ (37 percent) of its surface value.

When the light is incident on the surface, however, there is a partial reflection. This *Fresnel reflection* occurs when light is incident on an interface between any two materials of different refractive indices. For normal incidence, the reflection coefficient R, or the fraction of incident photons reflected at the interface between material 1 and material 2, is[2]

$$R = \frac{F_{L\text{reflected}}}{F_{Li}} = \left(\frac{n_1 - n_2}{n_1 + n_2}\right)^2 \tag{11.3}$$

[2]Compare this expression, which gives the probability that a given photon is reflected at a step index change between two materials, with Equation (S1A.41), which is the probability that an electron is reflected from a step potential change between two materials. Ultimately the mathematics in deriving these two is the same.

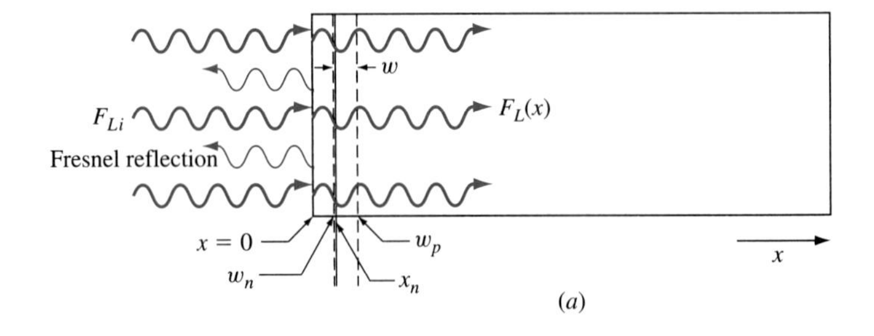

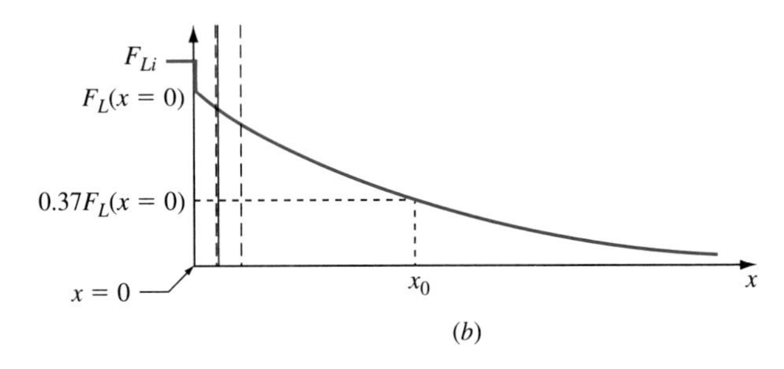

Figure 11.3 Variation of photon flux with distance. (a) A physical diagram showing the sample and the depletion region; (b) a plot of the the flux as a function of distance. There is a loss due to Fresnel reflection at the surface, followed by the decaying exponential loss due to absorption. The photon penetration depth x_0 is defined as the depth at which the photon flux is reduced to e^{-1} of its surface value.

where F_{Li} is the incident photon flux, and the n's are the refractive indices of the materials. The transmission coefficient T at the surface is

$$T = 1 - R = \frac{4n_1n_2}{(n_1 + n_2)^2} \tag{11.4}$$

Figure 11.3 shows the light incident on the semiconductor, and the variation of the optical flux as a function of depth into the material.

EXAMPLE 11.1

Find the variation in light flux with depth in a Si n^+p junction illuminated with monochromatic light of 1.42 eV, such as might be emitted by a GaAs light-emitting diode.

■ **Solution**

If the incident flux is F_{Li}, then the flux at the surface of the semiconductor is

$$F_L(0) = (1 - R)F_{Li} \tag{11.5}$$

The refractive index of Si is a function of photon energy, but in the visible region, it is about $n_{\text{Si}} = 3.6$. The reflection coefficient between air ($n = 1$) and Si then is, from

Equation (11.3),

$$R = \frac{(n_{\text{air}} - n_{\text{Si}})^2}{(n_{\text{air}} + n_{\text{Si}})^2} = \frac{(1 - 3.6)^2}{(1 + 3.6)^2} = 0.32$$

or about 32 percent of the incident photons are reflected. The Fresnel reflection at the surface of the semiconductor can thus result in a considerable loss of photons. To reduce the surface reflection in practice, thin layers of transparent dielectrics having refractive indices intermediate between those of air and the semiconductor are deposited on the semiconductor surface.

For our uncoated example, the flux that penetrates into the silicon is

$$F_L(0) = (1 - 0.32)F_{Li} = 0.68F_{Li}$$

Now we can compute the absorption in the silicon. Figure 11.4 shows absorption coefficients as a function of photon energy for several semiconductors. At this photon

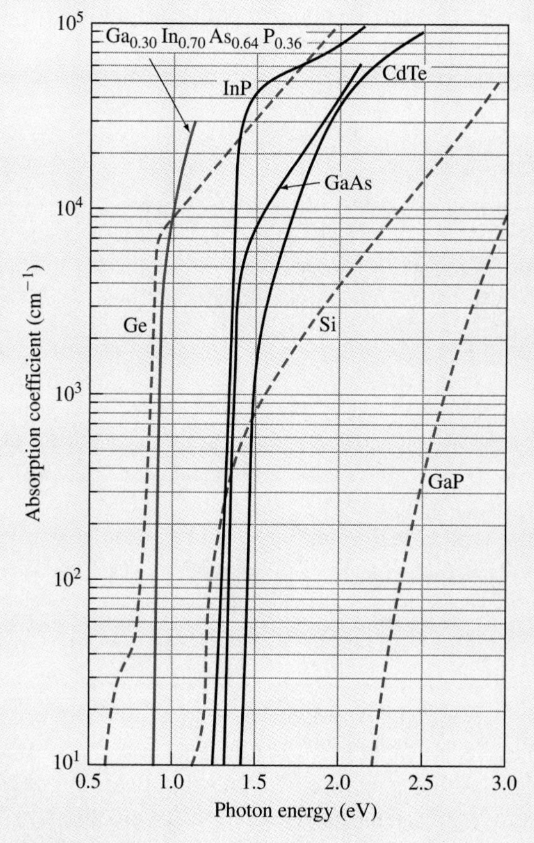

Figure 11.4 Absorption coefficients of some semiconductor materials. The indirect-gap materials are shown with a broken line. Based on data from References 1 and 2.

energy, the absorption coefficient α in Si is on the order of $4 \times 10^2\ \text{cm}^{-1} \approx 0.04\ \mu\text{m}^{-1}$. From Equation (11.2), then, the light is reduced to 37 percent of the surface value at a depth x_0 of

$$\frac{F_L(x)}{F_L(0)} = 0.37 = e^{-\alpha x_0}$$

$$\ln(0.37) = -1 = -\alpha x_0$$

$$x_0 = \frac{1}{0.04\ \mu\text{m}^{-1}} = 25\ \mu\text{m}$$

Let us assume that each photon absorbed creates an electron-hole pair. If the electron and hole are generated deep within a quasi-neutral region where there is no electric field, they will diffuse in random directions and eventually recombine, producing no net current. To obtain a photocurrent in a pn junction, minority carriers created in the quasi-neutral region must diffuse to the junction. Once at the junction, the electric field will accelerate the carriers across the junction, producing current.

EXAMPLE 11.2

Compare the absorption depth with the minority carrier diffusion length for the n^+p device of Example 11.1. Can the carriers diffuse to the junction to create a photocurrent? Assume $V_a = 0$, the n^+ region is 0.3 μm thick, and the p region doping is $10^{17}\ \text{cm}^{-3}$.

■ Solution

The n^+ layer thickness (0.3 μm) is much less than the photon penetration depth (25 μm). The absorption in this region, and thus its contribution to the photocurrent, can be ignored. For $V_a = 0$, from Figure 5.13 the junction width is 0.11 μm, and this region also contributes negligibly to photocurrent.

In the quasi-neutral p region, the minority carrier (electron) diffusion length (Figure 3.23) is 110 μm. This is appreciably greater than the photon penetration depth, thus a large fraction of the optically produced electrons diffuse back to the junction where they are collected and contribute to current.

The rate at which the electron-hole pairs are generated, G_L, is proportional to the photon flux density, $G_L(x) = \alpha F_L(x)$. This is still assuming every photon generates an electron-hole pair. We can find the electron concentration in the steady state by solving the continuity equation for electrons in the p region. In this case we have an optical generation term and a recombination term for the optically produced excess carriers:

$$\frac{1}{q}\frac{dJ_n(x)}{dx} + G_L(x) - \frac{\Delta n_p}{\tau_n} = 0 \tag{11.6}$$

where J_n is the photocurrent and Δn_p is the excess (photogenerated) electron concentration. Since the p region is uniformly doped, there is no field ($\mathscr{E} = 0$) and the total electron current is due to diffusion:

$$J_n = qD_n \frac{dn}{dx} \tag{11.7}$$

Recalling that $n = n_0 + \Delta n$, and that n_0 is a constant, Equation (11.6) becomes

$$D_n \frac{d^2 \Delta n(x)}{dx^2} + \alpha(1 - R)F_{Li}e^{-\alpha x} - \frac{\Delta n(x)}{\tau_n} = 0 \tag{11.8}$$

Solving Equation (11.8) for $\Delta n(x)$, the resultant photocurrent can be obtained by evaluating Equation (11.7) at a convenient location such as $x = 0$. This is treated in more detail in the section on solar cells.

In the above discussion, the contributions to the photocurrent from absorption in the transition region and in the n^+ surface were neglected, since for this example these regions are thin compared with the total penetration depth. For higher absorption coefficients, the penetration depth is reduced and absorption in these regions must be considered. Let us consider the absorption in the junction first.

In the depletion region the electric field is high enough that virtually all of the carriers generated there are accelerated out of the region before they can recombine, so all contribute to photocurrent. We can find this by integrating the absorption over the junction width:

$$J_D = q\alpha \int_{x_n}^{x_n+w} F_L(x)\, dx = qF_L(1 - R)e^{-\alpha x_n}[1 - e^{-\alpha w}] \tag{11.9}$$

where J_D is the photocurrent produced in the depletion region and x_n and w are defined in Figure 11.2.

For high absorption coefficients α, the light is absorbed closer to the surface, and the n^+ surface layer can contribute to photocurrent also. Surface states, however, cause the photogenerated carriers in this thin region to have a high probability of recombining at the surface, and thus the contribution of this region to the photocurrent is low.

The locations at which the carriers are produced also affect the response time. Because of the high field in the depletion region, the response time is much faster for carriers generated there than for the carriers generated in the quasi-neutral regions. The carriers in the quasi-neutral regions must diffuse (a slow process) to the junction to contribute to current. The current contributed by optical generation in the depletion region is sometimes referred to as the *prompt photocurrent*.

Next let us consider the effect of photocurrent on the I-V_a characteristics of the photodiode. The total current is the sum of the dark current and the photocurrent I_L:

$$I = I_{\text{dark}} + I_L \tag{11.10}$$

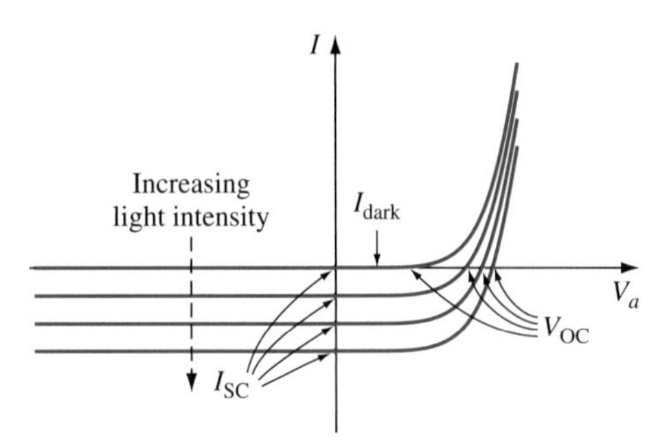

Figure 11.5 The I-V_a characteristics of a solar cell with varying illumination as a parameter. As the intensity increases, the short-circuit current I_{SC} increases linearly, but the open circuit voltage V_{OC} increases sublinearly.

Neglecting the voltage drop across the series resistance, the junction voltage is $V_j = V_{bi} - V_a$. For the cases in which the surface layer and the depletion region contribute a negligible photocurrent, I_L is independent of junction voltage, meaning that the effect of illumination is to translate the dark I-V_a characteristics in the $-I$ direction by the amount I_L. (The photocurrent is in the same direction as the reverse current.) This is illustrated in Figure 11.5. The short-circuit photocurrent I_{SC} (at $V_a = 0$) and the open-current voltage V_{OC} (at $I = 0$) are also shown. While I_{SC} is directly proportional to light intensity, the open-circuit voltage V_{OC} increases sublinearly (logarithmically) with intensity.

The dark current is the diode current discussed in Chapter 5; it just has a different name when discussing photodiodes. It is the sum of injection current and generation-recombination current, and is given by

$$I_{dark} = I_0(e^{qV_a/nkT} - 1) \tag{11.11}$$

[Compare this with Equation (5.81).] Setting $I = 0$ in Equation (11.10), with the aid of Equation (11.11), the open-circuit voltage V_{OC} is

$$V_{OC} = \frac{nkT}{q} \ln\left(1 + \left|\frac{I_L}{I_0}\right|\right) \tag{11.12}$$

As we will see in the next section, for solar cells it is advantageous to make V_{OC} as large as possible. From Equation (11.12) it appears that a large n would be desirable to obtain a large V_{OC}. Just the opposite is the case, however, because a large I_0 is associated with a large n, and in actuality V_{OC} decreases with increasing n. Further, the reverse dark current I_0 reduces the signal-to-noise ratio in the detection process and must be minimized.

Two figures of merit for photodetectors are *quantum efficiency* η_Q and *responsivity* R_{ph}. The quantum efficiency is defined as the photoinduced carrier flux density J_L/q passing the junction per incident photon flux density F_{Li}, or

$$\eta_Q = \frac{J_L/q}{F_{Li}} \tag{11.13}$$

The responsivity is defined as the output current density per watt of incident optical power per unit area. The energy per photon is $h\nu$ and so

$$R_{\text{ph}} = \frac{J_L/q}{h\nu\, F_{Li}} = \frac{q\eta_Q}{h\nu} \tag{11.14}$$

EXAMPLE 11.3

What is the quantum efficiency and responsivity for the prompt response of an InGaAs pn photodiode whose junction is 0.2 μm below the surface, and whose depletion layer is $w = 2\ \mu$m thick at a reverse bias of 10 V? The incident light has a wavelength of 1.55 μm. At this wavelength the absorption coefficient of this material is about $10^4\ \text{cm}^{-1}$, the refractive index is 3.4, and no antireflection coating is used.

■ **Solution**

The Fresnel reflection loss in going from air to the semiconductor is

$$R = \left(\frac{n_{\text{air}} - n_{\text{semi}}}{n_{\text{air}} + n_{\text{semi}}}\right)^2 = \left(\frac{1.0 - 3.4}{1.0 + 3.4}\right)^2 = 0.30$$

Of this, $(1 - R) = 0.7$ of the incident power remains. The photon flux density remaining after absorption in the surface layer is

$$F_n(x_n) = (1 - R)F_{Li}e^{-\alpha x_n} = 0.7F_{Li}e^{-(10^4\ \text{cm}^{-1})(2\times 10^{-5}\ \text{cm})} = 0.573F_{Li}$$

Of this, a fraction,

$$1 - e^{-\alpha w} = 1 - e^{-(10^4\ \text{cm}^{-1})(2\times 10^{-4}\ \text{cm})} = 1 - 0.135 = 0.865$$

is absorbed in the depletion region. Thus the total quantum efficiency for the prompt response is

$$\eta_Q = (1 - R)e^{-\alpha x_n}(1 - e^{-\alpha w}) = 0.573 \times 0.865 = 0.495$$

The corresponding responsivity is

$$R_{\text{ph}} = \frac{q}{h\nu}\eta_Q = \frac{q}{h}\frac{\lambda}{c}\eta_Q = \frac{(1.6 \times 10^{-19}\ \text{C})(1.55 \times 10^{-6}\ \text{m})}{(6.62 \times 10^{-34}\ \text{J}\cdot\text{s})(3 \times 10^8\ \text{m/s})}(0.495) = 0.62\ \text{A/W}$$

*11.2.2 SOLAR CELLS

Solar cells are photodetectors that are used to generate dc power. [1] As such, they are of reasonably large area, on the order of several square centimeters. To reduce the series resistance due to lateral current flow in the thin surface layer, digitated metal contacts are used on the illuminated side as indicated in Figure 11.6. Power is generated in the cell and is dissipated in a load; in this case the load resistance is R_L.

The device's I-V_a characteristics and its load line for a given incident light intensity are shown in Figure 11.7. The load line has a slope of $-1/R_L$, and since the applied voltage is zero (the point is for the solar cell to be a power source, not to require one), the load line goes through the origin. Choosing the load to give maximum power for a given illumination, the operating point is (I_m, V_m) as shown in the figure. Note that the power dissipated by a device is $P = IV$, but

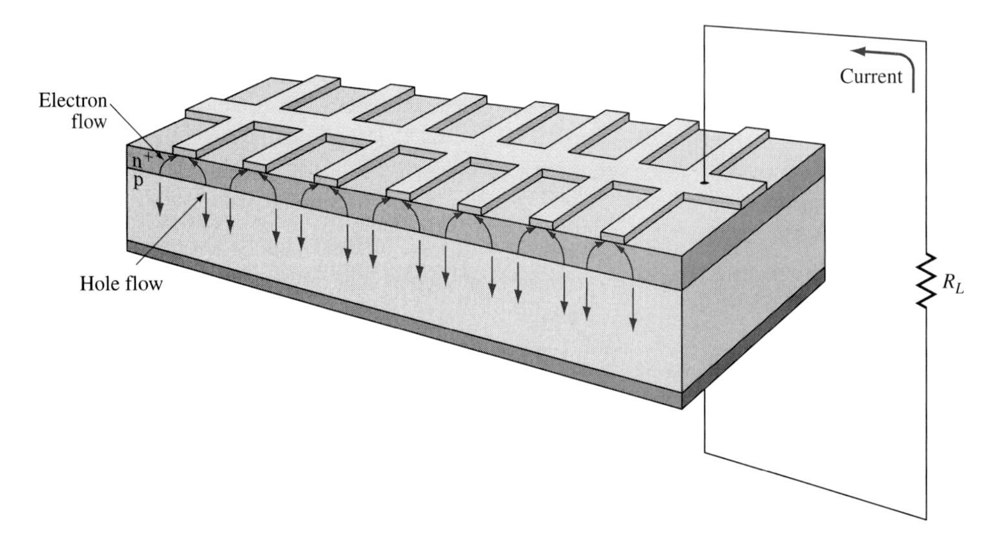

Figure 11.6 A solar cell with interdigitated contacts. The load resistor is R_L.

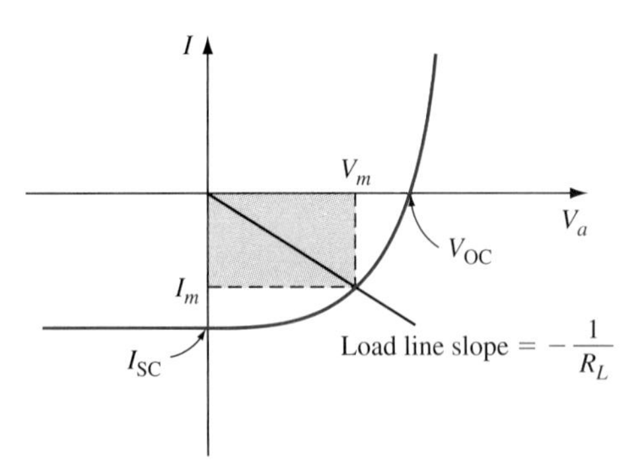

Figure 11.7 The I-V_a characteristic of a solar cell. The maximum power is obtained at $P_m = I_m V_m$.

in this case V is positive and I is negative. Therefore the power *dissipated* by the solar cell is negative, meaning it produces power.[3]

The maximum power output P_m is

$$P_m = I_m V_m \tag{11.15}$$

and is represented by the shaded area of Figure 11.7.

The power conversion efficiency η of a solar cell is defined as the maximum output electrical power divided by the incident optical power P_{Li}:

$$\eta = \frac{P_m}{P_{Li}} \times 100 \qquad \text{percent} \tag{11.16}$$

A parameter called the *fill factor,* FF, measures how well the shaded box in Figure 11.7 fills the quadrant IV portion of the I-V characteristic. The fill factor is defined as

$$\text{FF} = \frac{I_m V_m}{I_{\text{SC}} V_{\text{OC}}} \tag{11.17}$$

A typical value of FF is on the order of 0.7. The efficiency can be expressed in terms of the fill factor by

$$\eta = \text{FF} \frac{I_{\text{SC}} V_{\text{OC}}}{P_{Li}} \tag{11.18}$$

Note that the power conversion efficiency is η and the quantum efficiency η_Q is given by Equation (11.13).

The power conversion efficiency of a solar cell depends on a number of parameters that affect I_{SC} and V_{OC}. One of these is the spectrum of the output from the sun, since the spectrum affects the absorption. The solar spectrum is shown in Figure 11.8 for AM0 (air mass zero) and AM1 (air mass one). Air mass zero is the radiant energy outside the earth's atmosphere, as seen by orbiting satellites. Passing through the atmosphere alters the spectrum, and AM1 refers to the solar spectrum that has passed through one atmosphere at sea level for the sun directly overhead. The difference in the two curves results from scattering of the incident light and absorption in the earth's atmosphere.

For maximum conversion efficiency, the band gap of the material, the absorption coefficient spectrum, and the minority carrier lifetime are all important considerations. The band gap matters because those incident photons with energy less than the band gap are not absorbed and thus cannot contribute to photocurrent. The absorption spectrum affects the probability that a photon will create an electron-hole pair, and the minority carrier lifetime controls the

[3]Recall that all the other I-V characteristics in this book appear in only the first and third quadrants, where the I-V product (dissipated power) is positive.

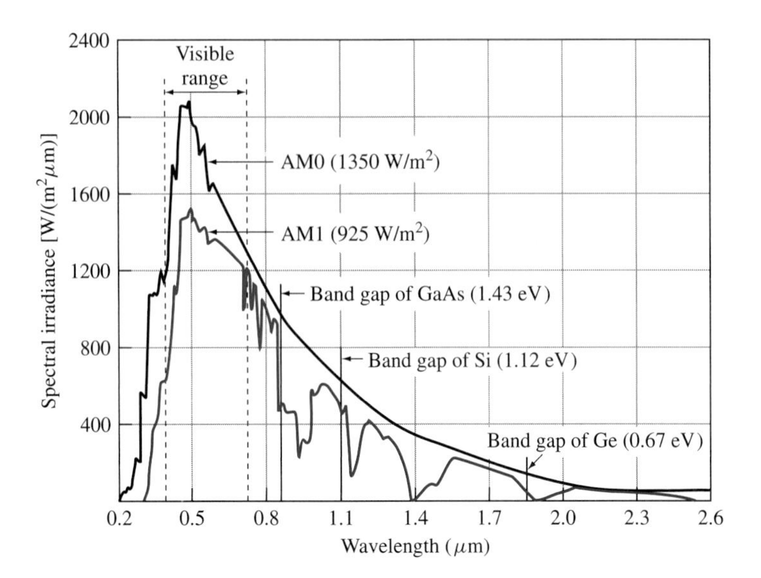

Figure 11.8 The solar spectrum. On earth at sea level, normal incidence, the spectrum is AM1 (one atmosphere). Satellites and other objects outside the atmosphere are exposed to AM0.

diffusion length, and thus the probability of collection of the optically generated minority carriers.

In silicon solar cells, about 20 percent of the incident solar power is lost because that much of the solar spectrum power consists of photons with energy below the band gap. Those photons with energy larger than the band gap are absorbed, but one photon creates only one electron-hole pair. The electron relaxes to the bottom of the conduction band, and the hole relaxes to the top of the valence band. Thus, a fraction of the photon energy greater than the band gap is lost as heat (phonons). This causes another 40 percent of the incident energy to be lost in Si devices. Thus for Si cells the maximum theoretical conversion efficiency $\eta_{\max}$ is on the order of

$$\eta_{\max} \approx 100 - 20 - 40 \approx 40\%$$

Actual conversion efficiencies are approximately half this value.

Consider a Si n^+p solar cell as shown schematically in Figure 11.9. Let $x = 0$ at the junction (we have moved this reference point from the surface, where it was earlier in this chapter). The width of the p region, W_p', is made much larger than L_n, the electron diffusion length. We ignore the photocurrent produced in the thin n^+ region. We wish to calculate the short-circuit current produced by photon absorption in the p region for monochromatic light with absorption coefficient α.

The photon flux density (number of photons entering the p region per unit area per second) is $F_{Li}(1 - R)e^{-\alpha x_n}$. Then at position x within the p region, the

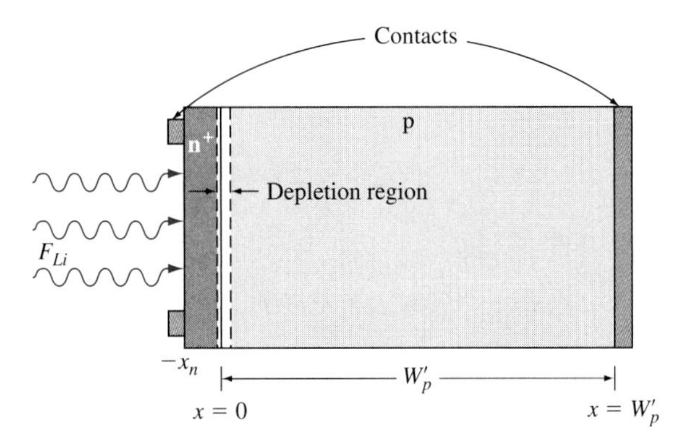

Figure 11.9 A solar cell illuminated from the left.

photon flux density is $F_{Li}(1 - R)e^{-\alpha x_n}e^{-\alpha x}$ and Equation (11.8) becomes

$$D_n \frac{d^2 \Delta n}{dx^2} + \alpha F_{Li}(1 - R)e^{-\alpha x_n}e^{-\alpha x} - \frac{\Delta n}{\tau_n} = 0 \tag{11.19}$$

The solution to this equation is

$$\Delta n(x) = C_1 e^{x/L_n} + C_2 e^{-x/L_n} - \frac{\alpha \tau_n F_{Li}(1 - R)e^{-\alpha x_n}e^{-\alpha x}}{(\alpha^2 L_n^2 - 1)} \tag{11.20}$$

where C_1 and C_2 are constants. To find these, we note that since we are calculating short-circuit current, $V_a = 0$. Also, at $x = 0$, the junction, there is a depletion region, so

$$\Delta n(0) = 0$$

If the silicon is thick enough that $W'_p \gg L_n$, we can approximate $W'_p = \infty$ and

$$\Delta n(W'_p) = 0$$

With these boundary conditions, then,

$$C_1 = 0$$

and

$$C_2 = \frac{\alpha \tau_n F_{Li}(1 - R)e^{-\alpha x_n}}{(\alpha^2 L_n^2 - 1)}$$

Then the excess carrier distribution with depth is

$$\Delta n(x) = \frac{\alpha \tau_n F_{Li}(1 - R)e^{-\alpha x_n}}{(\alpha^2 L_n^2 - 1)}(e^{-x/L_n} - e^{-\alpha x}) \tag{11.21}$$

We can now find the current density crossing the junction from

$$J_n = qD_n \left. \frac{d\Delta n}{dx} \right|_{x=0} \tag{11.22}$$

Taking the derivative of Equation (11.21), we obtain

$$J_n = qD_n\tau_n\alpha F_{Li}(1-R)e^{-\alpha x_n}\frac{\alpha - \dfrac{1}{L_n}}{\alpha^2 L_n^2 - 1} \tag{11.23}$$

Since $D_n\tau_n = L_n^2$ and $(\alpha^2 L_n^2 - 1) = (\alpha L_n + 1)(\alpha L_n - 1)$, Equation (11.23) becomes

$$J_n = \frac{q\alpha L_n(1-R)F_{Li}e^{-\alpha x_n}}{(\alpha L_n + 1)} \tag{11.24}$$

EXAMPLE 11.4

Find the quantum efficiency of a Si solar cell for $\lambda = 1\ \mu\text{m}$ and near the peak solar energy at $\lambda = 0.5\ \mu\text{m}$. Ignore the photocurrent contribution from the n^+ and depletion regions. The cell parameters are:

$$R = 0.2$$
$$x_n = 0.4\ \mu\text{m}$$
$$W'_p = 500\ \mu\text{m}$$
$$N'_A = N_A = 10^{17}\ \text{cm}^{-3}$$

■ **Solution**

We begin with Equation (11.13), using J_n for the photoinduced current J_L:

$$\eta_Q = \frac{J_n}{qF_{Li}}$$

Substituting for J_n from Equation (11.24), we obtain

$$\eta_Q = \frac{\alpha L_n(1-R)e^{-\alpha x_n}}{(\alpha L_n + 1)} \tag{11.25}$$

We must determine α and L_n. We can use Figure 11.4 to find the absorption coefficients, but we need to express our wavelengths in terms of the photon energies. From the golden rule

$$E_{\text{ph}}(\text{eV})\lambda(\mu\text{m}) = 1.24$$

Then

$$E_{\text{ph}}(1) = \frac{1.24}{1} = 1.24\ \text{eV} \qquad \lambda = 1\ \mu\text{m}$$

$$E_{\text{ph}}(0.5) = \frac{1.24}{0.5} = 2.48\ \text{eV} \qquad \lambda = 0.5\ \mu\text{m}$$

From Figure 11.4, the absorption coefficients in Si at these wavelengths are

$$\alpha(1) = 100\ \text{cm}^{-1} = 10^{-2}\ \mu\text{m}^{-1}$$
$$\alpha(0.5) = 10^4\ \text{cm}^{-1} = 1\ \mu\text{m}^{-1}$$

The electron (minority carrier) diffusion length is found from Figure 3.23 to be

$$L_n = 110\ \mu\text{m}$$

Since W_p' is appreciably larger than L_n, Equations (11.24) and (11.25) are reasonable approximations.

Then η_Q becomes, from Equation (11.25),

$$\eta_Q(1) = \frac{10^{-2} \times 110 \times (1 - 0.2)e^{-10^{-2}\times 0.4}}{(10^{-2} \times 110 + 1)} = \frac{1.1 \times 0.8 \times 0.996}{2.1} = 41\%$$

$$\eta_Q(0.5) = \frac{1 \times 110 \times (1 - 0.2)e^{-1\times 0.4}}{(1 \times 110 + 1)} = \frac{110 \times 0.8 \times 0.67}{111} = 53\%$$

We considered two specific wavelengths in the above example, but to obtain the total quantum efficiency of the cell, one would perform a weighted average of η_Q over the solar spectrum.

The power efficiency η of the cell will be less than the quantum efficiency η_Q. This is because the excess photon energy (greater than that of the band gap) does not contribute to the photocurrent.

To design good solar cells, then, we observe that the quantum efficiency is dependent on the absorption coefficient and the minority carrier diffusion length. The diffusion coefficient should be as large as possible to maximize the collection of photogenerated carriers.

We also recall that the optical penetration depth is equal to the reciprocal of the absorption coefficient. A small value of α results in deep penetration and thus requires large diffusion lengths. A large α, on the other hand, results in absorption in the surface n^+ layer, but as we indicated in the previous section, many of the carriers generated here recombine at the surface; they do not contribute to photocurrent. For the two wavelengths considered in the example, the optical penetration distance is 100 μm (large) for $\lambda = 1\ \mu$m and 1 μm (short) for $\lambda = 0.5\ \mu$m. A fraction $(1 - e^{-\alpha x_n})$ of the photons that enter the cell do not reach the p-type region. This amounts to 0.4 percent loss at $\lambda = 1\ \mu$m, but a significant 33 percent at $\lambda = 0.5\ \mu$m in the above example.

11.2.3 THE p-i-n (PIN) PHOTODETECTOR

In a solar cell, most of the optically generated current results from minority carrier diffusion to the pn junction. Since diffusion is a relatively slow process, the speed of response to changes in light is limited. While this is of no concern in solar cells where the illumination and thus the output is dc, it severely limits the frequency response of pn photodetectors used in optical communication applications. One method to increase the speed for such an application is to use a reverse-biased p-i-n (often referred to as PIN) detector.

The PIN diode has a layer of intrinsic (or lightly doped) material between the n- and p-type layers. This structure is designed to extend the physical length of the depletion region to increase the collection of the optically produced electron-hole pairs in this region. Figure 11.10a shows the structure, and Figures 11.10b and (c) show the equilibrium and reverse bias energy band diagrams normal to the plane of the junction respectively.

At equilibrium, it appears as though there are two junctions and two depletion regions. When a reverse bias is applied, however, the energy band diagram looks like Figure 11.10c. We see that there is an electric field throughout the

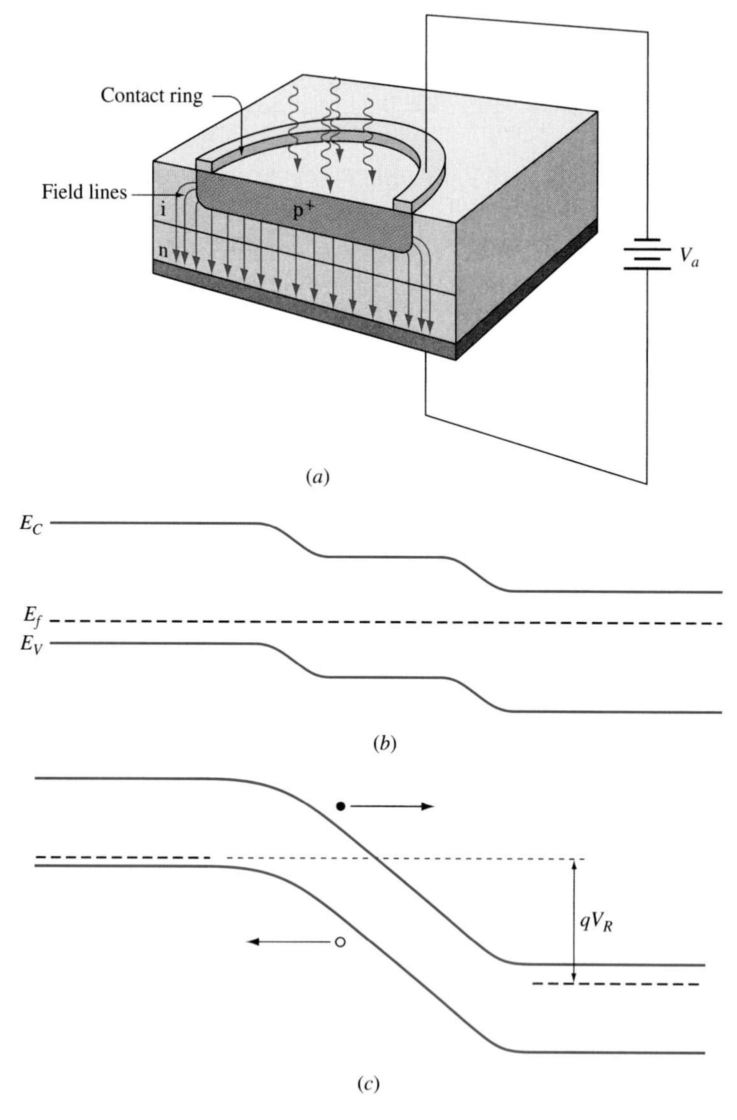

Figure 11.10 The PIN diode. (a) The structure; (b) equilibrium energy band diagram; (c) energy band diagram under reverse bias.

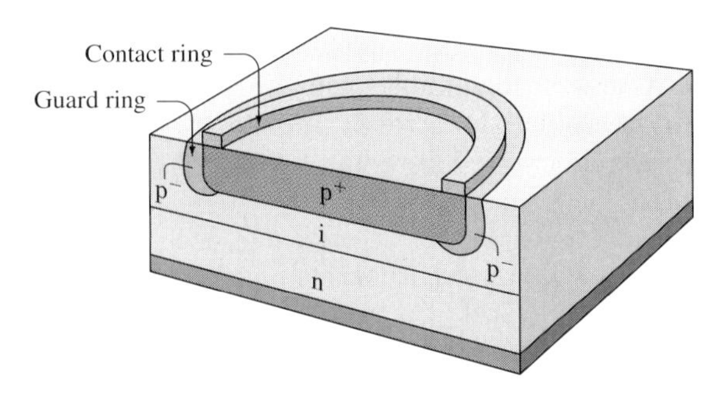

Figure 11.11 A PIN structure will break down at the edge of the p$^+$ region where the electric field lines concentrate. A guard ring prevents premature breakdown.

entire intrinsic region. This is because the intrinsic region is inherently resistive as a result of the small number of carriers available for current transport. This high resistance means a large part of the voltage is dropped across this region. The n- and p-type regions are much more conductive, so field dropped in those regions is negligible.

In a PIN photodetector, the surface layer (in this case the p$^+$ layer) is made thin enough such that little optical absorption occurs there. The intrinsic layer is thick enough (greater than the photon penetration depth) that most of the optical absorption occurs in the depletion region. Therefore most of the photogenerated carriers contribute to the prompt response.

Notice, however, that in Figure 11.10a at the corners of the p region, the electric field lines are close together and thus the field is higher here. The PIN structure of Figure 11.10a has a problem of breakdown in these corners. To permit a higher reverse voltage and increased speed, the field in these regions is reduced by using a lightly doped p-type *guard ring,* as shown in Figure 11.11. Because some of the depletion region exists in this lightly doped guard ring, the maximum field in this region is reduced.

11.2.4 AVALANCHE PHOTODIODES

Another commonly used type of photodiode in communications is the avalanche photodiode (APD). It uses some of the strategies of the PIN diode. The major difference in an APD is that it has internal gain.

The energy band diagram for an APD under high reverse bias was shown in Figure 5.27 and the carrier multiplication process was discussed in "Reverse-Bias Carrier Multiplication and Avalanche," Section 5.3.3. These devices are operated with reverse voltages on the order of 200 V, so there is a very high electric field in the depletion region. An electron excited to the conduction band by an incident photon experiences very rapid acceleration, resulting in impact ionization and carrier multiplication as discussed in Chapter 5. Both electrons and holes are multiplied, producing an amplified response to the photon. Normally the voltage applied is slightly less in magnitude than the avalanche breakdown voltage, where the carrier multiplication factor M is large but the dark current is not excessive. Gains on the order of 50 are common. The responsivity of an avalanche photodiode, then, is

$$R_{\text{ph}} = M\left(\frac{q\eta_Q}{h\nu}\right) \tag{11.26}$$

where M is the multiplication factor [Equation (5.105)].

Avalanche photodiodes are often used in telecommunications, especially in low-light situations, because of their high sensitivity. The avalanche process, however, does take some time, so APDs cannot achieve the high speeds that PINs offer. Another trade-off with APDs is that carriers that are generated in the depletion region by thermal rather than by optical processes are also amplified, creating noise. Also, because the impact ionization process has some randomness to it, it also creates extra noise.

11.3 LIGHT-EMITTING DIODES

In this section we will reverse the optical process and consider emission instead of absorption. Light-emitting diodes (LEDs) are used for a variety of applications from displays to illumination to fiber-optic communication links.

11.3.1 SPONTANEOUS EMISSION IN A FORWARD-BIASED JUNCTION

LEDs operate by spontaneous emission. Electrons in the conduction band and holes in the valence band have finite lifetimes. When they recombine, the excess energy of the electron is released, either as light (a radiative transition), as phonons (nonradiative), or a combination of the two.

To achieve significant emission, a situation is required where there are many electrons at elevated energy states (i.e., in the conduction band) and holes (for them to recombine with) present in the same physical area. A typical approach is to use a double-heterostructure pn junction, as shown in Figure 11.12a. This junction has a wide-band-gap p side and a wide-band-gap n side, with a narrow-band-gap material in between. The result is a potential well for electrons and another for holes. Under forward bias, excess electrons diffuse across the depletion region from the n side, and holes diffuse across in the other direction. The carriers tend to get caught and confined in the wells, increasing the probability of recombination.

In Chapter 3, we discussed optical transitions in semiconductors. We saw that, because not only energy but also K must be conserved, optical devices should be made of direct-gap materials, as shown in Figure 11.12b. These include gallium arsenide, indium phosphide, and many others. Ternary and quaternary compounds (having three and four components respectively) may also be direct gap. Figure 11.13 shows, for example, the band gap of $GaAs_xP_{1-x}$ as a function of the As concentration x. GaAs is a direct gap material, but GaP is

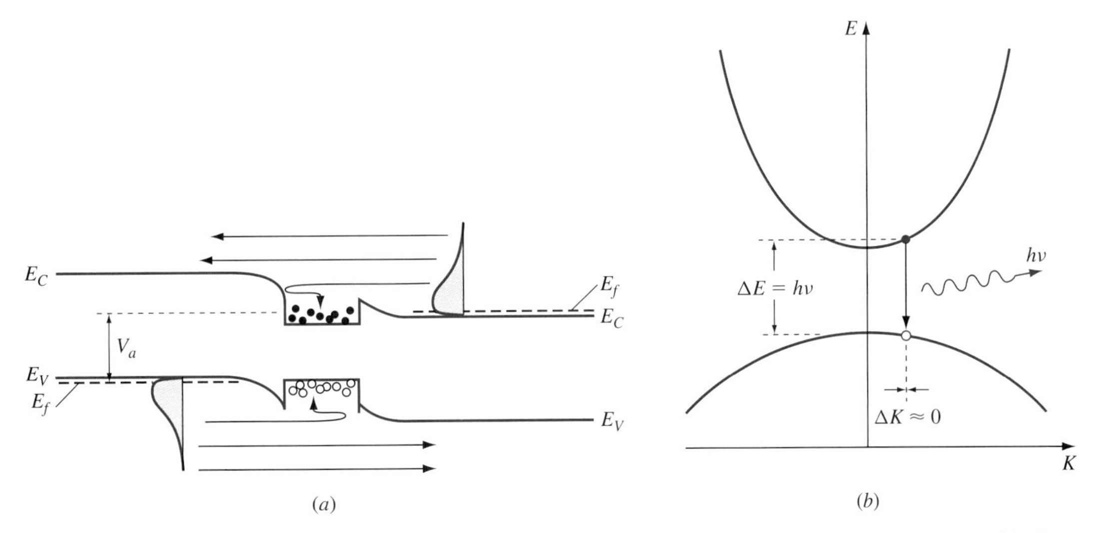

Figure 11.12 (a) A double-heterostructure LED pn diode. The potential wells for electrons and holes capture carriers and increase the probability of recombination. (b) The *E*-*K* diagram reminds us that *K* must also be conserved. Thus LEDs are usually made from direct-gap materials.

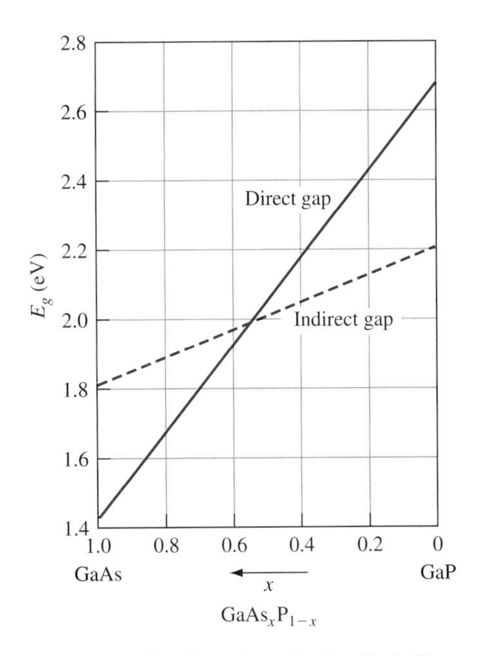

Figure 11.13 The band gap in the GaAsP system. For mole fractions of As less than about 0.55, the material is indirect gap.

indirect. The $GaAs_xP_{1-x}$ alloy makes a transition from direct to indirect gap where $x = 0.55$. GaAsP can be used to make LEDs for wavelengths ranging from about 870 nm corresponding to $x = 1$ (infrared) to about 630 nm (red) for $x = 0.55$.

Figure 11.14 shows a historical perspective of the development of semiconductor LEDs in the visible range. The first display/indicator LEDs were GaAsP ($GaAs_xP_{1-x}$), with efficiency lower than Thomas Edison's first light bulb. In the 1970s and 1980s improvements were made by using GaP doped first with zinc and oxygen, then with nitrogen. These important LEDs exploit isoelectronic traps, as discussed in the next section. Further improvements were obtained by going to quaternary compounds and heterojunctions (e.g., AlInGaP with a thin layer of GaAs or GaP). In the late 1990s semiconductor nitrides extended the available color range to include efficient green and blue.

*11.3.2 ISOELECTRONIC TRAPS

Although usually LEDs are made of direct-gap materials, Figure 11.14 shows that GaP, an indirect material, is frequently used for visible LEDs. This is done by creating *isoelectronic traps*. For example, when nitrogen atoms are introduced into GaP, they tend to replace phosphorus atoms in the lattice (both are in Column 5 of the periodic table). The N atoms, however, have electronegativities (3.0 eV) greater than those of phosphorus (2.1 eV). One state (two including spin) per nitrogen atom that would normally be in the conduction band of GaP resides inside the forbidden band of GaP. These traps are about 10 meV below E_C, and electrons can be trapped temporarily in these states.

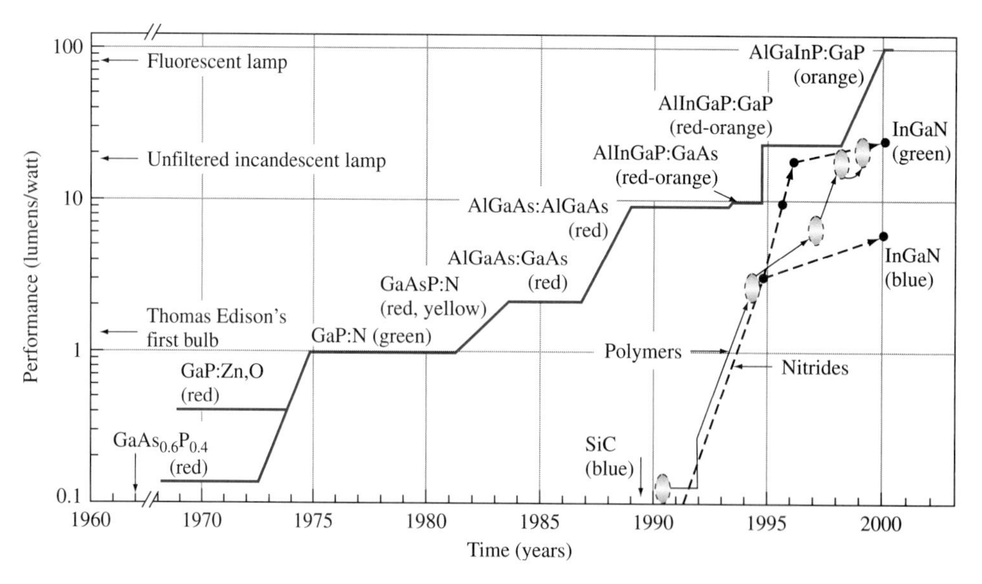

Figure 11.14 A historical view of the development of visible LEDs. (Abstracted with permission from J. R. Sheats, et al., *Science,* 273, p. 5277, 1996. Copyright 1996 The American Association for the Advancement of Science.)

Note, however, that nitrogen is *not* a donor—it does not become ionized by contributing an electron to the conduction band. It is part of the background semiconductor material because it has replaced a P atom of the GaP lattice. A donor, being ionized, can attract electrons from comparatively far away (Coulomb attraction). The nitrogen atom, however, has no long-range forces; the electron has to actually stumble into the state. Figure 11.15a shows schematically the bound state for a donor and for a nitrogen isoelectronic trap in GaP.

An electron in the conduction band may happen to fall into the potential well created by the isoelectronic trap. But the state exists only locally—there are no long-range forces. Therefore the uncertainty in the electron's position is very small; the electron is very close to the nitrogen atom. The uncertainty principle states that if the uncertainty in the position is small, then the uncertainty in the K vector is large. Thus these nitrogen states extend throughout K space, as shown

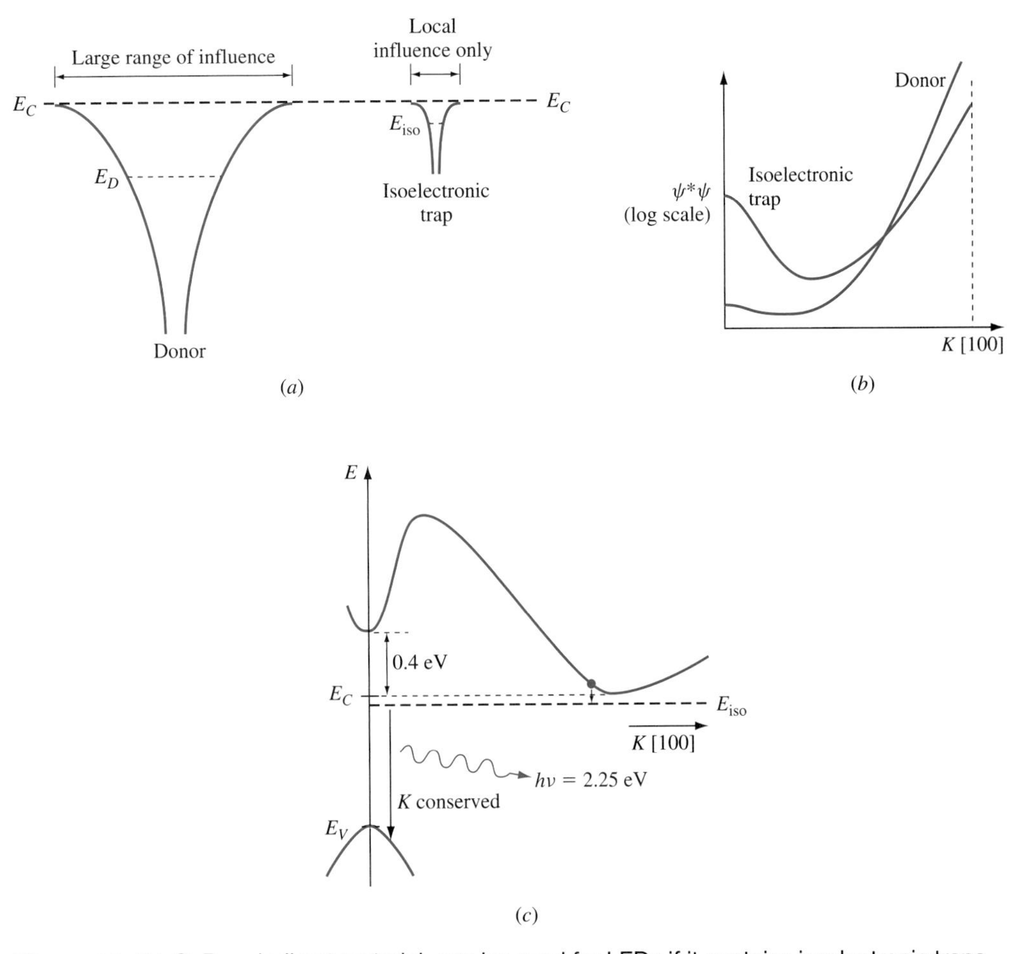

Figure 11.15 GaP, an indirect material, can be used for LEDs if it contains isoelectronic traps (see text). (a) Comparison of donor and isoelectronic states; (b) Fourier transforms of bound donor and isoelectronic trap states in K space (not to scale); (c) an optical transition at $K = 0$.

in Figure 11.15b, which compares the Fourier transform in K space of the donor state and isoelectronic trap in GaP. Because the electron in the isoelectronic trap is so tightly bound, the probability function $\psi^*\psi$ at $K = 0$ is about 3 orders of magnitude larger than that for a bound donor. As indicated in Figure 11.15c, an electron can drop from the conduction band to the trap state, at which point its crystal momentum is highly uncertain. Thus the electron can drop to the valence band while still conserving momentum, and a direct transition is made.[4]

Nitrogen forms an isoelectronic trap in GaP, producing a green LED ($\lambda = 565$ nm). By replacing a Ga atom with Zn and an adjacent phosphorus atom with oxygen, a similar isoelectronic trap can be produced that generates red emission.

By varying the As concentration in the alloy $GaAs_xP_{1-x}$, the band gap can be varied, and thus the energy of the nitrogen state within the gap. The indirect GaAsP:N system can be used to make lamps and displays from green to red. With sufficient As, the transitions become direct and thus band to band.

11.3.3 BLUE LEDs AND WHITE LEDs

For many years, visible LEDs were available only in green, yellow, orange, and red. The blue LED remained an elusive holy grail, because a full-color display requires red, green, and blue light to make all the visible colors. Until blue LEDs existed, LED technology could not be used for color displays.

In the late 1990s, however, gallium nitride of respectable quality became available. GaN is a direct-gap material, emits blue light, and shows considerable promise for commercial applications.

Another approach to making blue LEDs has been to use organic polymers, which were also shown in Figure 11.14. These use electrons and holes in a manner similar to semiconductors, and have the additional advantage of being flexible, so that thin, conformal displays can be envisioned.

All of the LEDs described so far emit only a particular color. A quasi-white LED has been created, however, by using a blue LED and a phosphor. Some of the blue light is absorbed by the phosphor and reradiated in the red. The resulting combination of blue and red appears reasonably white to the human eye.

11.3.4 INFRARED LEDs

LEDs are used not only for displays. They are also frequently used as the optical source in fiber-optic systems. Figure 11.16a shows schematically a fiber optic link coupling a light source (LED or laser) with a photodiode. Figure 11.16b

---

[4]Purists will argue (correctly) that a bound electron cannot be described by a single wave vector, but rather by a Fourier continuum of wave vectors. In materials like GaP, which has conduction band minima at $K = 0$ and at another value of K, the amplitudes of the Fourier waves are large at both $K = 0$ and at the other minimum. Thus the electron entering the state at the other minimum has a high probability of recombining radiatively at $K = 0$. This process does not work in Si because it has a relative maximum in its conduction band at $K = 0$ with a resultant small value of $\psi^*\psi$.

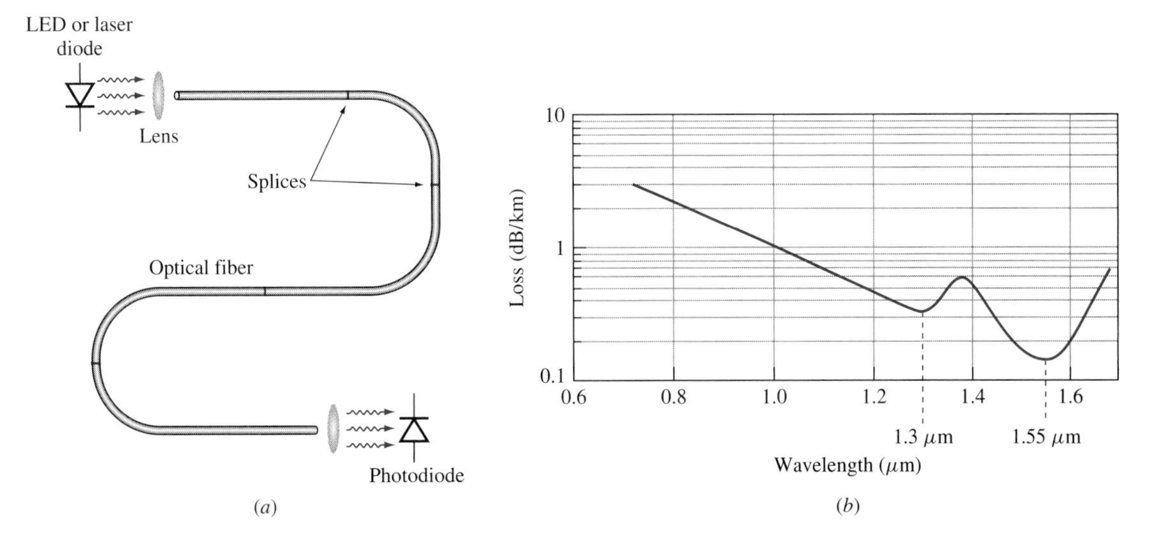

Figure 11.16 (a) A fiber optic link contains a light source, generally either an LED or a laser diode, a fiber that may contain multiple splices, and a photodetector, usually either a PIN diode or an avalanche photodiode. (b) The optical loss of glass used in optical fibers. Loss is least at the favored wavelengths of 1.3 μm and 1.55 μm.

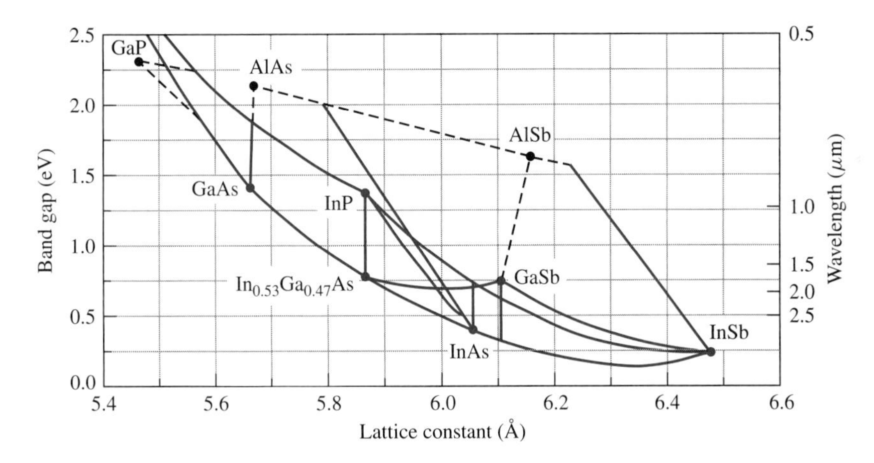

Figure 11.17 The lattice constants of several common semiconductors. The solid lines indicate direct-gap materials; the dashed lines are indirect-gap.

shows the absorption spectrum of the glass used in optical fibers.[5] This shows that the best wavelengths to obtain low loss in fibers are $\lambda = 1.3\ \mu$m and $\lambda = 1.55\ \mu$m, which are in the infrared.

To determine what material systems would be best for fabricating sources at these wavelengths, we consult Figure 11.17, which shows the band gap and

[5]The peak between 1.3 and 1.55 μm is due to water ions in the glass, and has been essentially eliminated in modern fibers.

emission wavelengths for various III-V compounds. It suggests materials such as $In_xGa_{1-x}As$, $InAs_xP_{1-x}$, and some antimonides. The particular materials selected also depend on the lattice constants—the emitting materials must be grown onto a commonly available substrate.

EXAMPLE 11.5

Given that GaAs and InP are commonly available substrates, what material and composition should be used to produce an LED that emits at 1.3 μm?

■ Solution

We require material whose lattice constant is equal to that of a good substrate, and at the same time has a band gap corresponding to 1.3 μm:

$$E_g = \frac{1.24}{1.3\ \mu\text{m}} = 0.95\ \text{eV}$$

All the materials with the same lattice constant (Figure 11.17) as GaAs have band gaps that are larger than this, so GaAs is not an appropriate substrate.

For InP, however, there is a ternary compound of GaAs-InAs that is lattice matched. This occurs at $In_{0.53}Ga_{0.47}As$. It does not, however, emit at 1.3 μm, but emits at $\lambda = 1.5\ \mu$m. By increasing phosphorus content, though, one can move toward the InP point. The movement is not simply vertical, because creating the quaternary compound InGaAsP means some sort of interpolation between the GaAs-InAs curve and the InAs-InP curve. The final result turns out to be $In_{0.76}Ga_{0.24}As_{0.55}P_{0.45}$.

Let us now consider some of the characteristics of spontaneous emission. The emission is a random event, and the light reflects that randomness in the following ways: The direction of propagation of the photons is random, the timing of the emission (the phase of the photon) is random, and the polarization of the light is random.

What is not entirely random is the energy of the light. It is controlled by the energy distributions of the electrons and holes. Consider Figure 11.18. The minimum energy a released photon can have in a band-to-band transition is theoretically equal to the band-gap energy. In fact, this particular transition cannot occur, because it requires an electron at the very bottom edge of the conduction band and a hole at the very top energy of the valence band. The density-of-states functions are zero at those two energies. As discussed in Chapter 2, the peak concentration of electron energies is $\frac{1}{2}kT$ above E_C, and for the holes $\frac{1}{2}kT$ below E_V. Thus the most probable emission energy is slightly higher than the band gap. Transitions above and below this peak value are also possible, just less probable. This means that an LED actually emits over a small range of wavelengths, typically about 50 to 100 nm wide. Figure 11.18b shows a typical spectrum of an LED.

Next, let us look at the physical structure of an LED. Figure 11.19 shows a double-heterostructure surface emitting LED. We have already said that the emission occurs in the narrow-band-gap layer (refer to Figure 11.12), called the

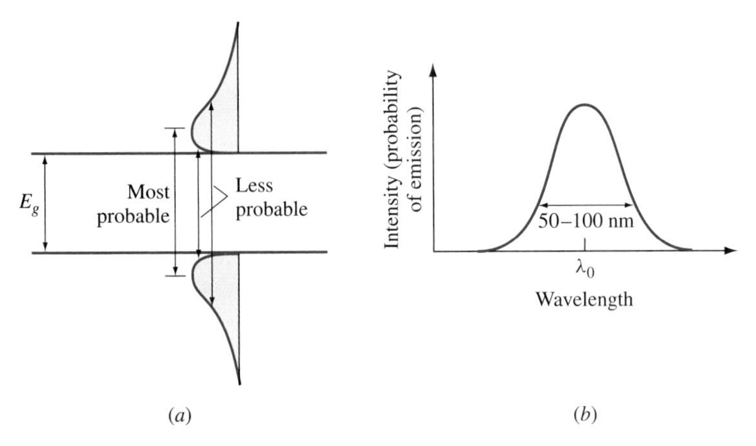

Figure 11.18 (a) The intensity of recombination at a given wavelength depends on the distributions of electrons and holes in energy; (b) the resulting emission spectrum.

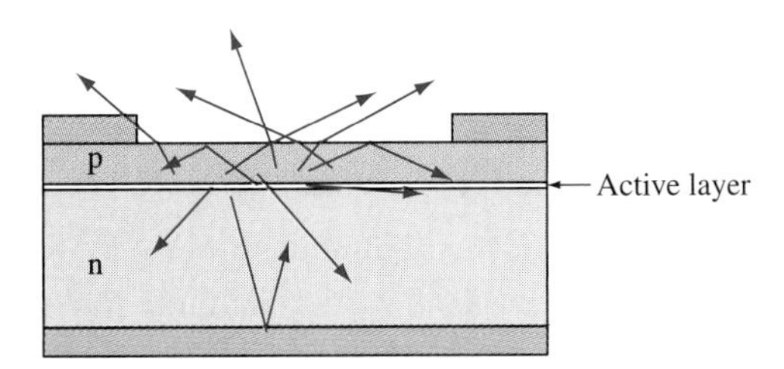

Figure 11.19 A generic surface-emitting LED. Some photons are lost by reabsorption in the bulk, Fresnel reflection from the surface, and total internal reflection.

active layer. Since the direction in which the photon travels is completely random, the emission is uniformly distributed in all directions. Photons that are emitted upward or downward will enter the wider-band-gap material, which cannot absorb the emitted photons (their energy is too small to overcome the band gap). Thus these layers are transparent to the emitted light. Light that is emitted along the active layer will be reabsorbed.

Only light emitted through the surface will be used, but there are other sources of loss. One of these is Fresnel reflection, discussed earlier with respect to photodiodes, and the other is total internal reflection loss. Photons emitted at sufficiently high angle to the surface can be internally reflected according to Snell's law. The critical angle, measured with respect to the semiconductor surface, is

$$\theta_{cr} = \cos^{-1}\left(\frac{n_{air}}{n_{semi}}\right) \tag{11.27}$$

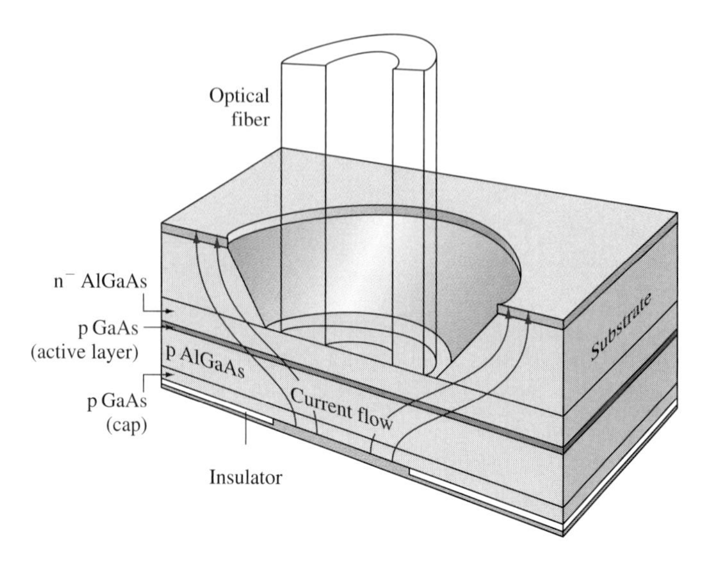

Figure 11.20 A Burrus-type LED. This one uses a double heterostructure to confine the carriers, making recombination more efficient. The etched opening in the LED helps align and couple an optical fiber.

where the n's are the refractive indices. Only photons striking the surface within this angle will be transmitted across the surface and into the air.

Finally, since the light is emitted evenly in all directions, it is essential to bring the light-collection device (a lens or a fiber) as close as possible to the junction to capture as much light as possible. Figure 11.20 shows a *Burrus* LED structure, in which a well is etched in the LED surface and an optical fiber is inserted into the well and epoxied in place.

In general, optical coupling from LEDs to a fiber results in light being lost, but the Burrus structure helps appreciably.

Another solution for coupling the light to an optical fiber is to use an edge-emitting diode. It so happens that the smaller-band-gap materials tend to have higher refractive indices. From Equation (11.27), we see that that when light goes from a higher to lower index, it may be totally internally reflected. Figure 11.21 shows how a thin, high-index layer (the active layer) can also be a waveguide. Those photons that are traveling at shallow enough angles will tend to be reflected at either edge of the active layer.

In this structure, the light leaves the semiconductor from the edge of the chip rather than its surface. This has advantages and disadvantages. The light is emitted from a relatively small region, making it easier to couple the light into a fiber. On the other hand, to get access to the edge, the chip must be accurately sawn or cleaved away from the rest of the wafer, and the tiny devices are difficult to handle.

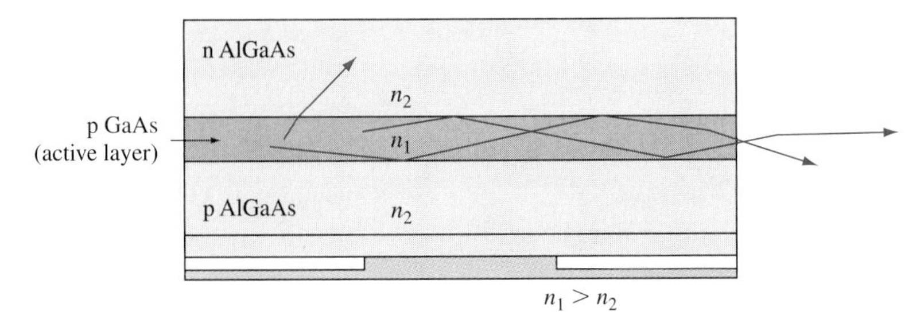

Figure 11.21 In an edge-emitting LED, the higher-index active layer acts as a waveguide for photons traveling at less than the critical angle.

One would expect from Figure 11.21 that most of the light in an edge-emitting LED would be reabsorbed, since it is confined to the emitting layer. That is not, in fact, the case. To understand why we must take a closer look at waveguides.

Consider the optical waveguide of Figure 11.22. It consists of a narrow region of refractive index n_1, called the *core*, and is surrounded on either side by regions of lower index, n_2, called the *cladding*. The structure shown could be a waveguide as is found in edge-emitting LEDs and lasers, or it could be an optical fiber. The optics is the same. We will assume we are discussing an LED, however, and thus every photon originates in the high-index material of the core (which is the same as the active layer).

Although the previous figure showed rays being totally internally reflected, the ray model does not apply to very narrow cores, on the order of 1 μm or less. Instead, we have to go to the full vector electromagnetic wave description, and solve Maxwell's equations for all regions, matching boundary conditions. Such a derivation is left for other courses, but the result is that the waveguide supports certain modes. The figure shows the electric field distribution for the first three modes, along with their directions of propagation. (In practice, however, the active layer is made thin enough that only a single transverse mode is supported.) Note that even though the mode is centered in the small-band-gap layer, some of the mode's energy is actually carried in the transparent cladding layers. Thus the absorption the photons in this mode experience is actually some average of the high absorption of the core layer and the low absorption of the cladding layers.

Edge-emitting LEDs are becoming increasingly common for fiber-optic applications. The big advantage is that while a surface-emitting LED emits in a wide angle, limited basically by the critical angle, the edge-emitting LED emits into a smaller angle, making coupling to fibers much more efficient. As we shall see in the next section, a very similar structure is used to produce laser diodes.

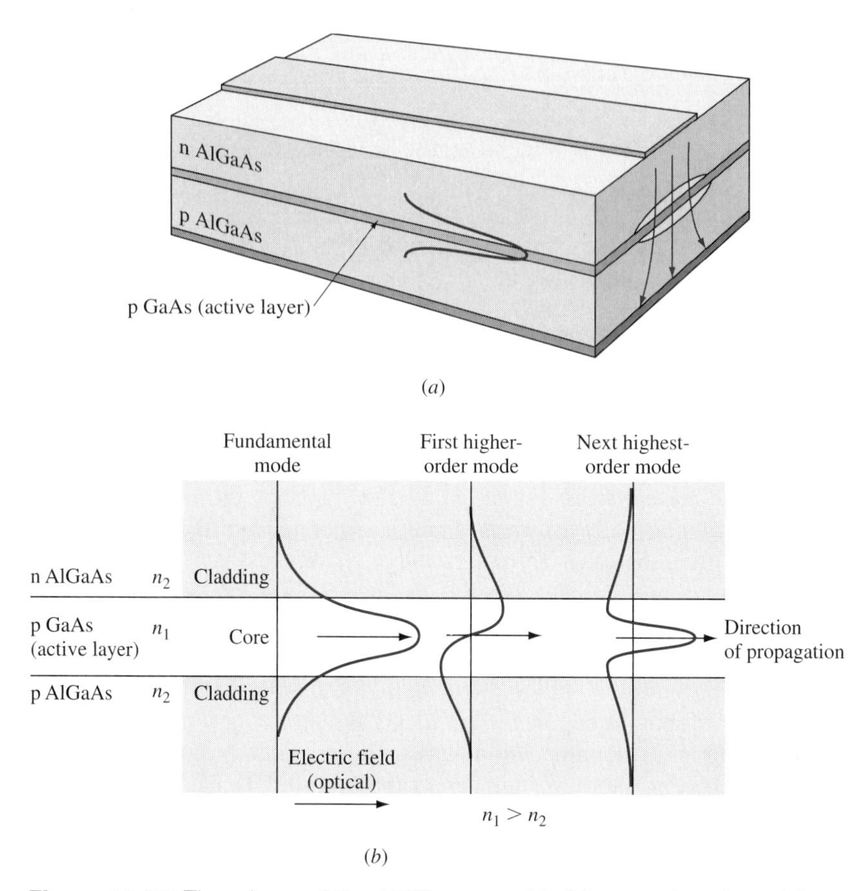

Figure 11.22 The edge-emitting LED's waveguide (a) supports only certain transverse modes, whose field distributions are shown in (b). In practice, only the first mode is allowed. It is not completely confined to the active layer, thus its absorption is reduced.

11.4 LASER DIODES

The key difference between LEDs and laser diodes is that lasers operate by stimulated emission rather than spontaneous emission. For stimulated emission to occur, population inversion[6] must be achieved. That is, there must be a large enough number of electrons at excited energies (in semiconductors, that means in the conduction band), and a large enough population of empty states at lower energies for the electrons to fall to (holes in the valence band), that the probability of stimulated emission exceeds the probability of absorption by an amount

[6]More accurately, as discussed later, a probability inversion. The probability of an electron being in a higher energy state must be greater than that for a lower energy state.

great enough to overcome other losses in the cell. In an optical system, if the number of photons coming out is greater than the number that went in, the system is said to have optical gain.

For lasers to operate, two things are required:

1. Gain
2. Feedback

These two operate together to produce lasing.

11.4.1 OPTICAL GAIN

As mentioned earlier, optical gain exists when the number of photons leaving a system is greater than the number going in (the opposite of optical loss, which occurs in absorption, and the number of photons is reduced). We have said that there are three optical processes: spontaneous emission, absorption, and stimulated emission. Any of these can happen in a semiconductor, and indeed at any given time all are occurring.

Let us consider spontaneous emission first. Consider two electron levels, E_2 and E_1 as shown in Figure 11.23. If the number of electrons in the upper state is N_2, then the upper state is being depleted at the rate

$$\frac{dN_2}{dt} = -A_{21}N_2 \qquad \text{spontaneous emission} \tag{11.28}$$

where A_{21} is the Einstein rate coefficient for spontaneous emission. Note that the actual rate is negative, meaning that if the only process occurring is spontaneous emission, N_2 will decrease. In fact, the rate constant A_{21} is related to the spontaneous lifetime

$$\tau_{\text{radiative,spont}} = \frac{1}{A_{21}} \tag{11.29}$$

In absorption, the number of electrons N_1 in the lower level decreases and N_2 increases. The rate equation is

$$\frac{dN_2}{dt} = B_{21}N_1 g(\nu) \qquad \text{absorption} \tag{11.30}$$

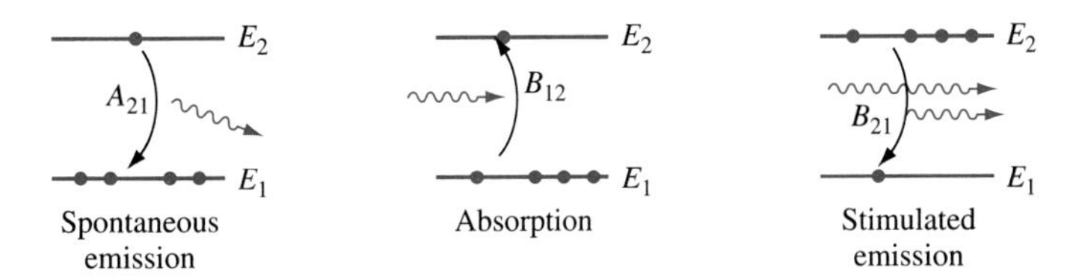

Figure 11.23 Optical processes revisited.

where $g(\nu)$ is called the *lineshape* function. It arises because absorption requires an incident photon to excite the process. That photon must have the correct energy to be absorbed—it must have an energy equal to the energy difference between an occupied lower state and an available upper state. The lineshape describes the probability of absorption in a given material as a function of light frequency ν. For example, in Figure 11.23 there are two discrete levels, and thus only one frequency could be absorbed. In this case $g(\nu)$ would be a delta function. By uncertainty, though, no energy level is infinitely narrow, so every lineshape has some width.

In a semiconductor, instead of discrete states there is a valence band and a conduction band, each with its own density of states. The distribution and occupancy of these governs the probability of absorption as a function of photon energy.

The same probabilities govern spontaneous emission, so the lineshape function is the same as the spectral emission of the spontaneous emission. Thus, the distribution function shown in Figure 11.18 *is* the lineshape function.

Finally, we consider stimulated emission. We expect that the stimulated emission process will depend on the number of carriers at excited energies that are available for recombination, and it will also depend on the frequencies of the incoming photons. The incoming photons in a semiconductor laser actually originate inside the cavity, from spontaneous emission. Thus the rate equation is:

$$\frac{dN_2}{dt} = -B_{21}N_2 g(\nu) \qquad \text{stimulated emission} \tag{11.31}$$

Notice that in the Einstein coefficients A_{21}, B_{21}, and B_{12}, the first subscript indicates the initial state and the second indicates the final state.

At thermal equilibrium, electrons are being excited into the conduction band at the same rate as they recombine. Thus

$$\frac{dN_2}{dt} = -\frac{dN_1}{dt} \qquad \text{equilibrium} \tag{11.32}$$

Furthermore, the states may have some degeneracy. Let them have g_2 and g_1 states respectively. The g's are the degeneracies for each state—the number of ways that state can be occupied without violating the Pauli exclusion principle. If electrons can occupy the lower level in g_1 different ways (e.g., because of different quantum numbers such as angular momentum, i.e., g_1 states), and the upper level similarly has a degeneracy of g_2, then

$$g_2 B_{21} = g_1 B_{12} \tag{11.33}$$

For stimulated emission, if a photon with an appropriate wavelength enters the material, stimulated emission is more probable than absorption when

$$\frac{N_2}{g_2} > \frac{N_1}{g_1} \tag{11.34}$$

This situation is called *probability inversion,* although the term *population inversion* is normally used. It is not a normal situation, since all electrons seek their lowest energies. We expect that usually $N_1 > N_2$. To create population inversion, electrons must somehow be artificially induced to be concentrated at high energies, which for semiconductor lasers means excited into the conduction band. At equilibrium in an n-type material there are many electrons, but few holes for them to recombine with. As we saw earlier when we discussed LEDs, excess electrons and holes can be injected across a double heterojunction, placing large numbers of both in the same physical space. The same technique is used in laser diodes. A double heterostructure is employed and the junction is forward biased. The difference is that the excess carrier concentrations must be much higher than are used in LEDs in order to achieve lasing. In fact, every laser diode is also an LED. When the junction is forward biased by a small amount, a small current flows. Electrons and holes cross the junction where they recombine. Since at low currents, the population is not inverted, the emission is primarily spontaneous. As the current increases, the light production increases linearly as shown in Figure 11.24 (LED region).

As the current continues to increase, however, the population does invert. Photons are already present in the junction from the spontaneous emission, and their energies match the lineshape function. Therefore, under population inversion, these photons can stimulate more emission from the excited electrons in the junction. As the inversion increases, stimulated emission also increases. When the stimulated emission is sufficiently large to overcome not only absorption but other losses in the cavity (such as through the end mirrors discussed in the next section) the light output increases dramatically. There is a distinct current threshold, as reflected in the power-current curve. Below this threshold, the laser behaves as an LED, and above threshold it is a laser.

To maintain a population inversion, current must be continually supplied. Otherwise all the available electrons will be quickly used up. In fact, the lasing

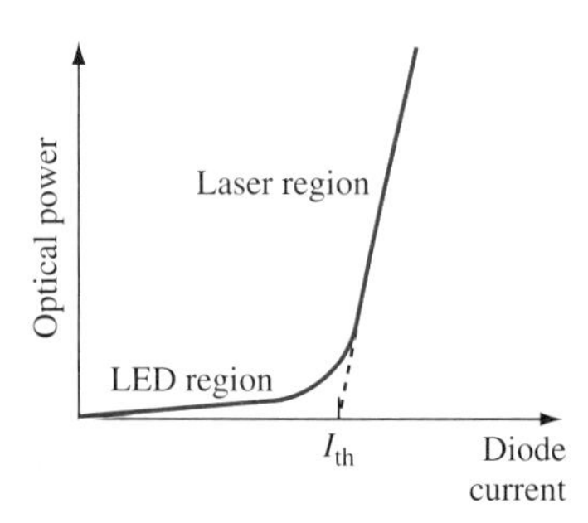

Figure 11.24 The power-current curve of a laser diode. Below threshold, the diode is an LED. Above threshold, the population is inverted and the light output increases rapidly.

process is so fast that above threshold, electrons are demoted back to the valence band by stimulated emission almost as soon as they arrive. Thus the output intensity is limited by the arrival rate of the electrons and holes, or, in other words, is proportional to the current.

11.4.2 FEEDBACK

It would seem, then, that any LED could be made to lase simply by increasing the current. This is not the case, however. Consider the edge-emitting LED of Figures 11.21 and 11.22. Spontaneously emitted photons will have random directions, you recall, so many of the emitted photons will not be traveling along the junction. The gain, however, only exists in or near the junction, so those photons will not be amplified. Spontaneously emitted photons that happen to be traveling along the junction plane, however, will remain in the gain region. They can in fact be amplified, as long as the population is inverted to make the probability of stimulated emission greater than the probability of absorption. Still, the chip is short and the photons don't spend very much time in the gain region before they leave the chip. Furthermore, as we saw before, the optical mode extends into the wide-band-gap layers, so only part of the mode actually overlaps with the gain region.

Optical feedback is used to increase the total optical amplification, by making the photons pass through the gain region multiple times. The optical feedback typically comes from two mirrors, one at each end of the laser. This arrangement is called a *Fabry-Perot cavity*. These mirrors are partially reflecting and partially transmitting, as shown in Figure 11.25. These mirrors are often just the cleaved crystal facets of the semiconductor material itself. The Fresnel reflection is significant, since the refractive index of the semiconductors is appreciably higher than that of air. Thus some percentage of the photons striking the mirror will be reflected back into the laser cavity. These photons will be reflected back and forth inside the laser, and on each reflection the light will be further amplified. After a few passes the optical field will be very strong indeed. The intensity will level off when the rate at which carriers are used up

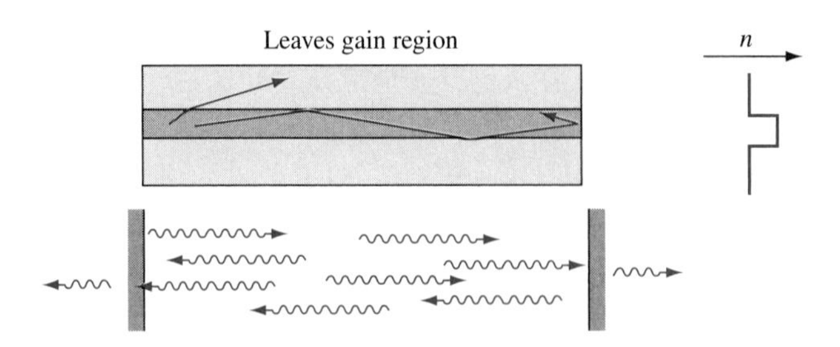

Figure 11.25 The ends of the chip form partially reflective mirrors, which allows the photons to be reflected back and forth and thus be exposed to gain for a longer period of time.

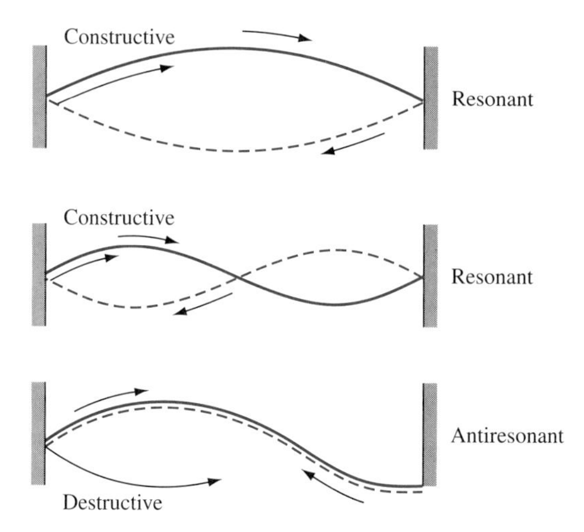

Figure 11.26 Wavelengths that are integer multiples of half the cavity's length can resonate, interfering constructively. Other wavelengths die out eventually.

by amplification just offsets the rate at which carriers are supplied. To increase the intensity, the rates must be increased by increasing the current.

An interesting thing happens in a mirror cavity, however. The electric fields of the light will interfere on successive round trips. Figure 11.26 shows that certain wavelengths will interfere constructively, so that the fields from successive passes add, while others interfere destructively, canceling themselves out. Individual photons emitted spontaneously at the nonresonant wavelengths can still be amplified, as long as $g(\nu)$ is large, but after a few trips through the cavity the interference causes these photons to die out.

On the other hand, photons that originate spontaneously and happen to be at the resonant wavelengths will reinforce themselves after multiple trips through the cavity, and continue to be amplified on every pass. Thus the optical field is very strong at these wavelengths. These resonant wavelengths, shown in Figure 11.27 for several values of mirror reflectivity, are called the *longitudinal modes*—the "longitudinal" part comes from resonating along the length of the cavity.[7] Notice from the figure that the sharpness of the resonance is related to the reflectivity of the mirrors.

The resonant wavelengths of a Fabry-Perot cavity are given by

$$\lambda = \frac{2nd}{q} \tag{11.35}$$

[7]There are transverse modes too, but laser diodes are usually designed to operate in a single transverse mode, so they are not discussed here.

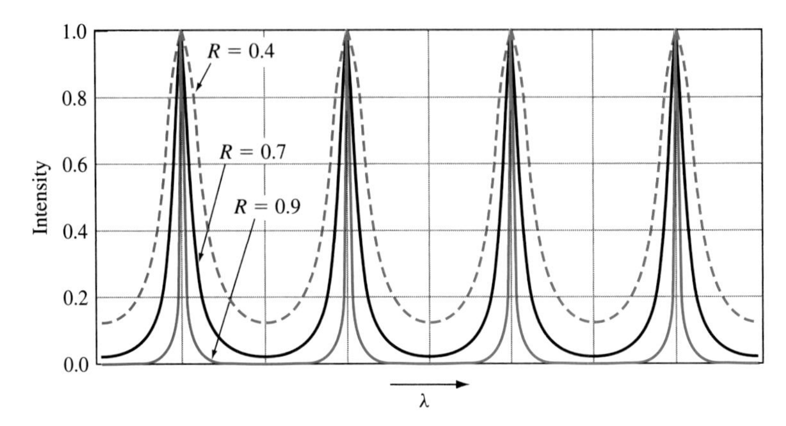

Figure 11.27 The resonances of a Fabry-Perot cavity. The width of the resonances depends on the reflectivity R of the mirrors.

where d is the length of the cavity, n is the refractive index of the material inside the cavity, and q is an integer.[8]

EXAMPLE 11.6

A double heterostructure Fabry-Perot edge-emitting laser in the AlGaAs-GaAs system emits in the neighborhood of 900 nm. The chip is 300 μm long. What is the wavelength difference between two adjacent modes?

■ **Solution**

We start by finding out in what neighborhood of q we are operating. From Equation (11.35) we see that we will need the refractive index n. In a double heterojunction laser, the lowest band gap and highest index material will be the active layer. In the AlGaAs/GaAs system, GaAs has the lowest band gap (Figure 11.17). It also has the highest refractive index. The refractive index of GaAs is about 4.3 and the addition of Al reduces the index somewhat.[9] Solving for q we have

$$q = \frac{2nd}{\lambda} = \frac{2(4.3)(300 \times 10^{-6}\text{ m})}{900 \times 10^{-9}\text{ m}} = 2866.7$$

Since q must be an integer, there must be a cavity mode for $q = 2866$ and another for $q = 2867$. Their wavelengths are

$$\lambda_2 = \frac{2nd}{2866} = 900.2\text{ nm}$$

[8]Note that in this book the symbol q is usually electronic charge. To be consistent with much of the laser literature, we used the same symbol here to indicate which longitudinal mode is being considered. It should be clear from context which q is which.

[9]Since the light mode will actually extend in the lower index cladding regions as well, the index of refraction that the mode experiences is actually some average of the two indices. We will simplify the problem by assuming the index of GaAs.

and

$$\lambda_2 = \frac{2nd}{2867} = 899.9 \text{ nm}$$

The spacing between modes is thus $900.2 - 899.9 = 0.3$ nm.

As we saw in Figure 11.27, the spectral width of the modes is influenced by the mirror reflectivity R. For a GaAs laser in which the mirrors are the uncoated cleaved facets, the reflectivity is low ($R = 0.38$), and the resonances are quite broad. Coating the facets can make more highly reflective mirrors, and in this case the resonance peaks can be quite narrow. The result is that the laser beam is more coherent. The use of higher reflectivity mirrors also reduces the lasing threshold current and thus increases the efficiency.

11.4.3 GAIN + FEEDBACK = LASER

In a laser, then, there are two effects at work. If the probability of stimulated emission exceeds the probability of absorption, there is optical gain. In addition, there is optical feedback from the mirrors to allow the photons to pass through the gain region multiple times. Let us see then, how lasing occurs.

To have stimulated emission, we must have some initial photon to start the process off. Photons are always being produced spontaneously, with a probability determined by the lineshape function (gain curve), shown in Figure 11.28a.

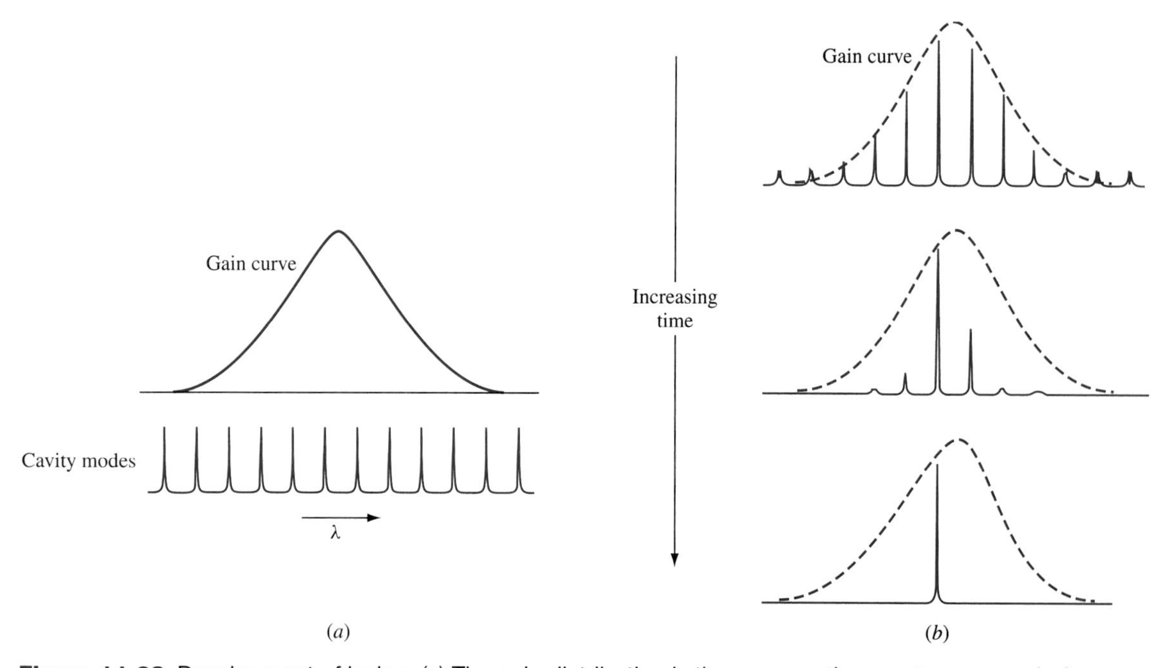

Figure 11.28 Development of lasing. (a) The gain distribution is the same as the spontaneous emission spectrum. (b) Only the photons at the resonance will amplify. The ones near the center of the gain curve will amplify the fastest.

Some of the spontaneously emitted photons will be traveling out of the junction plane, and since they don't remain in the gain region they are lost, eventually reabsorbed or emitted from the surface.

Spontaneously emitted photons that happen to travel along the junction plane can be amplified. Also recall that the double heterostructure acts as a waveguide, so photons traveling at small angles will be reflected back into the core. All of the spontaneously emitted photons traveling along the junction plane are amplified at first. The stimulated photons are identical to the original photons in wavelength, phase, and direction. Initially, the emission is mostly spontaneous and the output spectrum looks like the gain curve. These photons continue to travel along the junction plane and continue to be amplified. When they reach the end of the waveguide, however, they encounter a partially reflecting mirror. Some percentage of the photons is transmitted, and those photons become part of the laser emission beam; the rest are reflected back along the cavity. These are further amplified, and at the other end there is another partial mirror where some more photons are transmitted. Modes near the peak of the lineshape function, however, have more gain and thus are amplified most.

Now the cavity effect comes into play. After several passes, the fields start to add constructively or destructively. The resonant wavelengths will be amplified and the electromagnetic field associated with those wavelengths will grow rapidly, Figure 11.28b. After a few passes, the interference builds up and the modes start to emerge.

Eventually the center mode will be so large in amplitude that it is stimulating new photons just as fast as electrons becomes available, Figure 11.28b. This one mode can, in some circumstances, use up all the electrons as fast as they are delivered, since there are many photons at the required energy. This single mode operation is actually preferred, because it means a more coherent beam, and one that can carry a higher bandwidth of data.

Also notice that if the gain in a laser diode is low (low current level), the photons may be lost through the mirrors at a rate faster than they can be amplified during one pass. If that happens, there is no lasing because there is no net gain. In other words, population inversion does not guarantee lasing; the inversion must be great enough to overcome the cavity loss.

To reduce the loss, the mirrors must be made as reflective as possible. If, for example, the reflectivity is 99 percent, and the laser is emitting 3 mW (enough power to cause eye damage), then the power inside the laser is 99 times larger, or 297 mW.

11.4.4 LASER STRUCTURES

Various structures are used to make laser diodes. For example, the double-heterostructure is often made such that the active layer is thin enough to become a quantum well (the energy states become discrete). Figure 11.29 shows several different double-heterostructure (DH) single-quantum-well (SQW) energy band diagrams and their accompanying refractive index diagrams. The accompanying optical field distributions are also shown. The shaded areas indicate the size of the

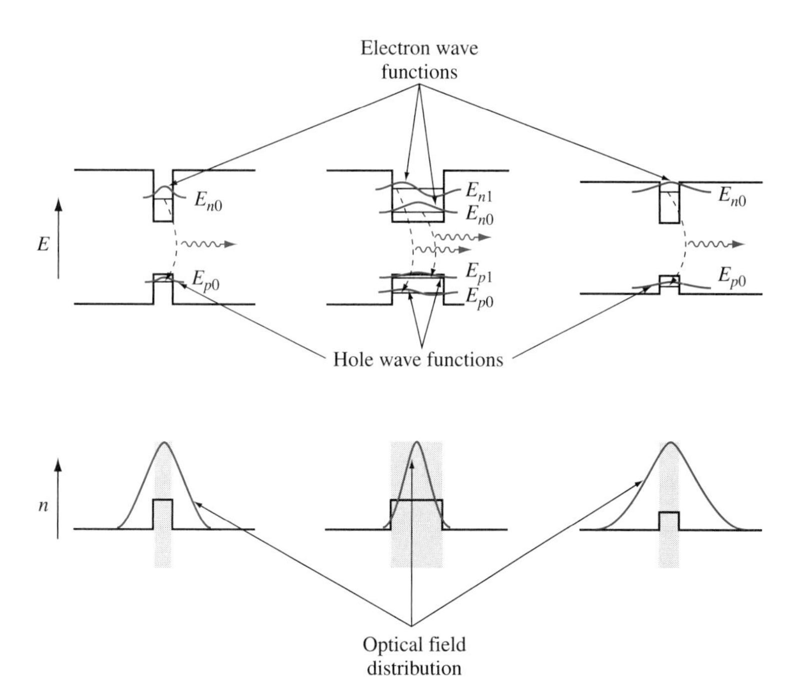

Figure 11.29 Adjusting the depth and width of quantum wells to select the wavelength of emission is one form of band-gap engineering. The shaded areas indicate the width of the well to illustrate the degree of confinement of the mode.

core; the greater the overlap of the mode with this area, the greater the gain that it sees. Notice from Figure 11.29 that the narrower the well, the less confined the optical mode is. Since only the part of the mode that actually overlaps the active layer sees gain, high confinement is desirable (unlike in the LED, where it contributes to excessive re-absorption).

In Supplement A to Part 1, it was shown that the number of states in a quantum well, as well as the energies of those states, depends on the width and depth of the well. The energy difference between the allowed states determines the emission wavelength of the laser. The width of the well is determined by the thickness of the active layer, and the depth of the well is controlled by the energy difference in the conduction band (for electrons) and the valence band (for holes). These, in turn, are determined by the choice of materials. This offers opportunities for *bandgap engineering,* in which optical and electrical properties of materials can be tuned by these parameters.

A narrow well is required to produce discrete states, but it has a low probability of capturing electrons or holes, because electrons and holes can easily pass across the top of the well. A narrow well also means poor overlap between the mode and the gain. Figure 11.30 shows two common structures for improving the performance of lasers over a single quantum well design. One is called the

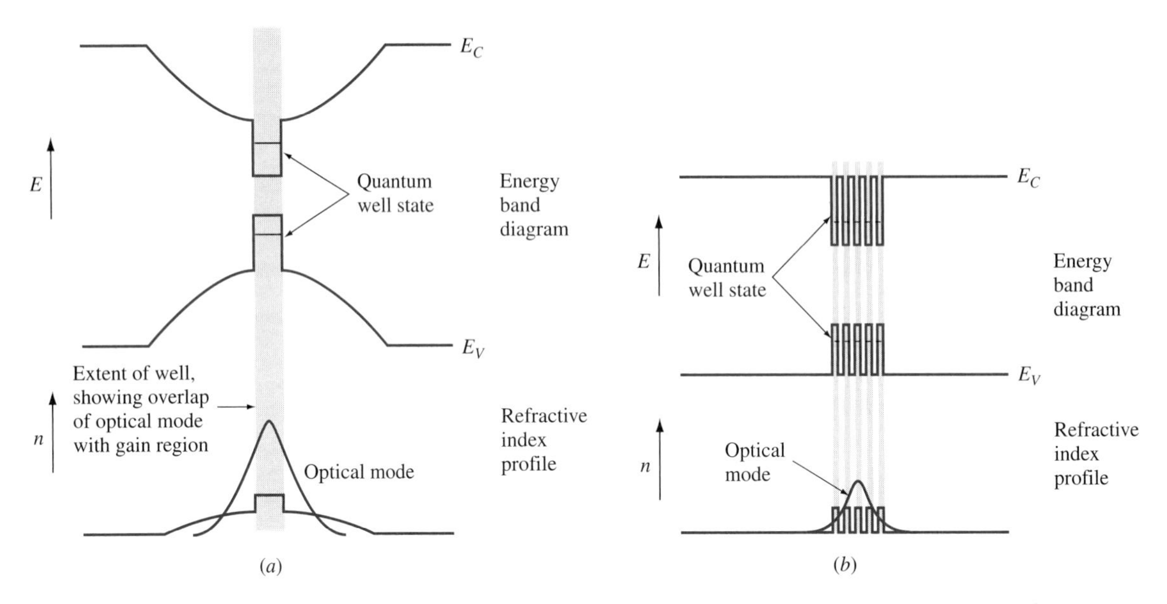

Figure 11.30 (a) A GRINSCH structure helps funnel the carriers into the wells to improve the probability of recombination. (b) A multiple quantum well structure has the advantage of single states, like the SQW, but improves carrier capture.

graded-index separate confinement heterostructure (GRINSCH) laser (a). The active layer is a SQW small-band-gap material as before. In the cladding layers, however, the composition of the alloys is gradually changed. On the energy band diagram, this produces a sort of funnel to direct carriers into the potential well of the active layer. The result is higher gain for a fixed current level. The GRINSCH laser not only helps confine the carriers but it also affects the distribution of the optical field, improving the overlap between the optical mode and the gain region. Another way to use narrow wells while still improving the optical overlap and carrier confinement is to use a multiple-quantum-well (MQW) structure, Figure 11.30b.

Laser diodes need not be edge emitting, however. Vertical-cavity surface-emitting lasers[10] (VCSELs) use two mirrors, one above the active layer and one below (Figure 11.31). The mirrors here are multiple layers of alternating semiconductors of different refractive indices. These are known as *dielectric mirror stacks* or *distributed Bragg reflectors* (DBRs).

These dielectric mirrors use constructive interference between the Fresnel reflections at each dielectric interface. Stacks of 50 to 100 layers are not uncommon, and high values of reflectivity can be made—approaching 100 percent.[11]

[10]Often pronounced "vick-sells."

[11]The constructive interference comes from a careful choice of the thickness of each layer; if different thicknesses are chosen, the Fresnel reflections could interfere destructively, annihilating any reflection and producing an antireflection layer.

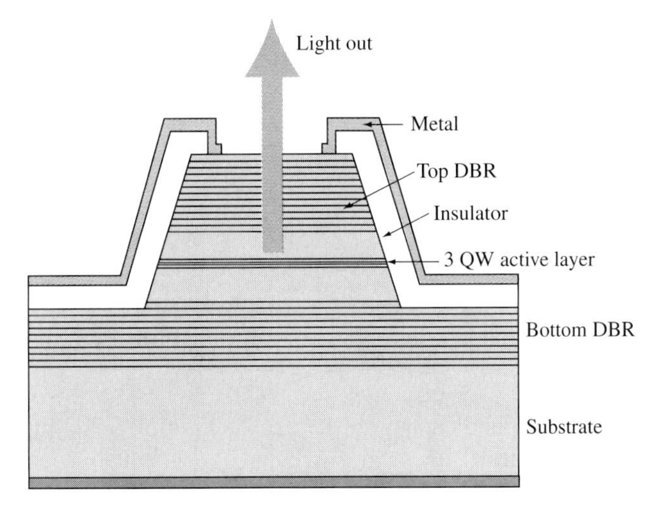

Figure 11.31 A vertical cavity surface-emitting laser. (After Ueki et al., *IEEE Photonics Technology Letters,* 11, no. 12, pp. 1539–1541, 1999, © IEEE.)

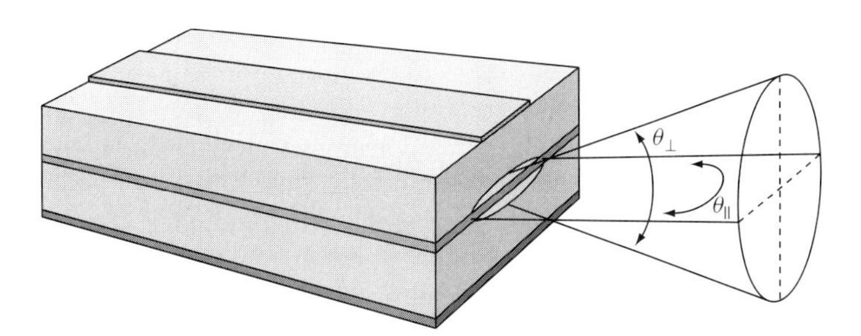

Figure 11.32 The output pattern of an edge-emitting laser is elliptical and widely divergent.

One of the advantages of VCSELs is that the output beam is easier to couple into optical fibers than the beam from edge-emitting lasers. To see why this is so, consider the edge-emitting Fabry-Perot laser diode in Figure 11.32. The active layer is very thin, on the order of 0.1 μm. The width of the lasing spot is usually wider than this; a few micrometers is typical. So the lasing spot, if one looks in the near field (right up against the output facet of the chip), appears elliptical.

In the far field, however, the picture is much different. Recall that when light passes through an aperture, it is diffracted. The smaller the aperture, the larger the diffraction angle. The aperture of the beam at the output facet is much smaller in the direction perpendicular to the junction plane, so the angle of spread, $\theta_\perp$, of the beam is much wider than the angle in the plane parallel to the junction, $\theta_\parallel$. Typical values of $\theta_\perp$ are 20 to 40°, and values of $\theta_\parallel$ are in the range of 5° to 20°. Therefore, in the far field, the beam from an edge-emitting laser (or LED) is

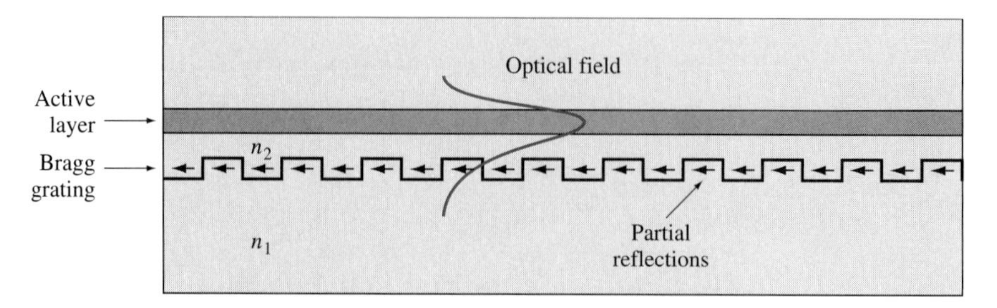

Figure 11.33 The distributed feedback (DFB) laser uses a grating to provide continuous feedback along the laser cavity.

elliptical, oriented perpendicularly to the near-field ellipse. As might be imagined, coupling an elliptical beam to a circular fiber inevitably leads to losses. Vertical cavity lasers have wide beam angles also, since they are also small, but the beams are circular and can be imaged onto a fiber core efficiently with a lens. Also, with VCSELs the laser does not have to be cleaved from the rest of the chip, making it possible to integrate electronics such as logic and drive circuitry, modulators, and photodetectors on the same substrate, producing optoelectronic integrated circuits (OEICs).

Finally, there are other approaches to providing optical feedback apart from using a Fabry-Perot cavity. One common structure is a distributed feedback (DFB) laser, shown in Figure 11.33. A corrugated layer is obtained below the active layer by etching a periodic structure and then filling in the corrugations with a regrowth of a material with a different refractive index. The optical field extends across this periodic variation in refractive index. At each step, there will be a small Fresnel reflection. These repeated reflections accumulate and interfere, producing a laser beam of very narrow spectral width. This type of structure is often used in lasers intended for RF applications.

11.4.5 OTHER SEMICONDUCTOR LASER MATERIALS

Although we have primarily discussed lasers in the AlGaAs and InP systems, many other semiconductors are becoming technologically important for laser diodes. These include III-V compounds such as InGaAsSb, InAsPSb, InGaAsP, and AlGaAs, for the red and near infrared; IV-IV semiconductors including PbSnTe, PbSSe, and HgCdTe for the 5- to 17-μm range; and II-VI compounds like ZnCdTe and ZnTeSe for blue and violet. Some semiconductors and their wavelength ranges are shown in Figure 11.34. III-V nitride compounds have also successfully been used in laser diodes. Difficulties in using some of the materials lie in finding an appropriate lattice-matched substrate on which to grow the layers and in doping. For example, it is hard to dope most of the wide-band-gap II-VI's with acceptors. Without a p-type material, diodes cannot be made. Lasers from some of the material systems shown must be pumped by nondiode means such as optical pumping.

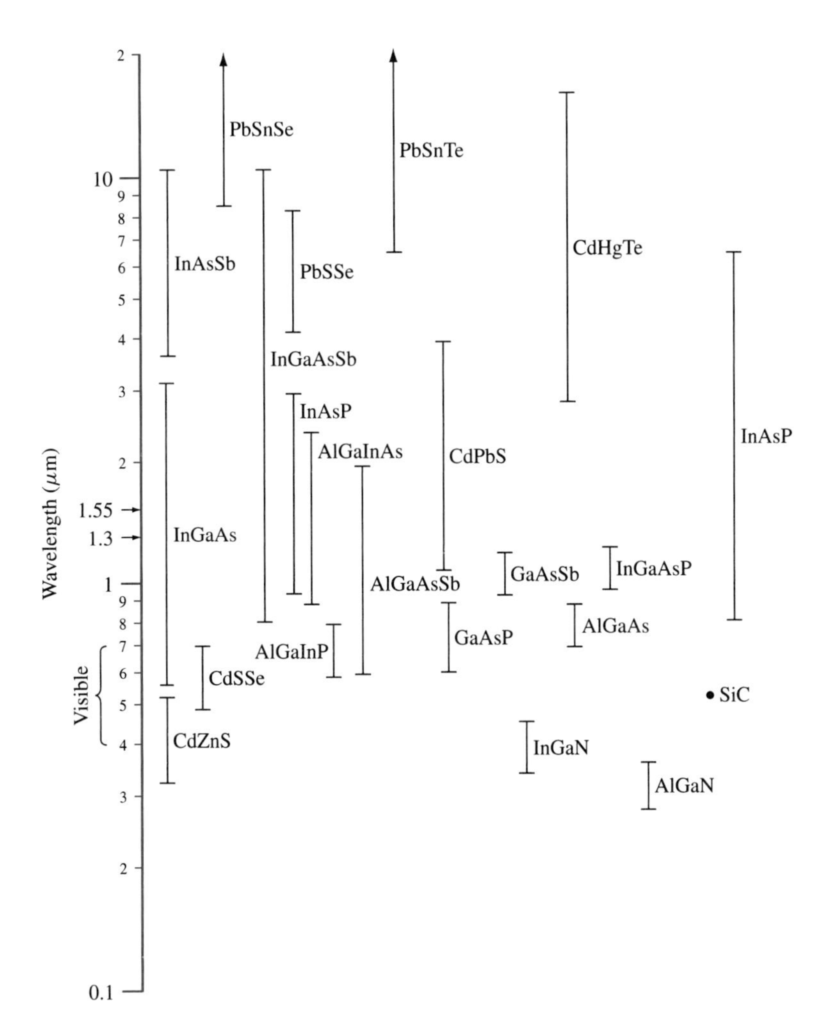

Figure 11.34 Some semiconductor materials and their wavelength ranges. Based on data from References 3 to 5.

11.5 IMAGE SENSORS

Semiconductor-based image sensors are widely used in applications such as digital cameras and camcorders. Here we briefly describe two types of such sensors. These are charge-coupled devices (CCDs) and MOS image sensors using photodiodes for light detection.

11.5.1 CHARGE-COUPLED IMAGE SENSORS

Although a basic CCD structure was described in the Supplement to Part 3, there are a number of variations of the CCD structure. Here we first describe a linear image sensor and then a simple area image sensor such as that used in digital cameras.

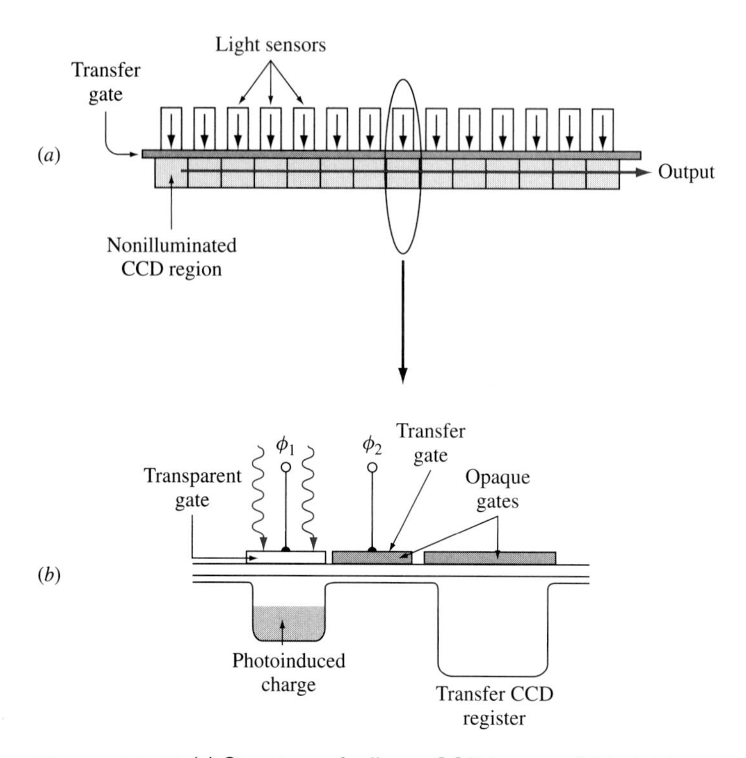

Figure 11.35 (a) Structure of a linear CCD imager; (b) hybrid diagram of one stage. The shaded areas represent opaque gates. The light sensors have optically transparent gates.

Consider the structure of Figure 11.35a, which consists of a number of MOS capacitors with optically transparent gate contacts (labeled "Light sensors"), a transfer gate, a CCD shift register described in the Supplement to Part 3, and a readout device such as a MOSFET (not shown). In operation, the light-sensing capacitor is biased to produce a potential well under it. The transfer gate is biased to form a barrier that confines the photogenerated electrons as indicated in Figure 11.35b. Under illumination, the charge in the well increases with time and is proportional to the light intensity. After a given light integration time, the ϕ_1 contact is pulsed low while ϕ_2 is pulsed high. The accumulated charge is then shifted to a well under the transfer gate, from which it is transferred to the readout device. Phases ϕ_1 and ϕ_2 are then returned to their original values and the cycle continues. The time dependence of the output current is then a measure of the light intensity as a function of position.

A number of the above linear imagers can be incorporated into an area imager with the structure of Figure 11.36. After the light integration time, all the photosensing wells are simultaneously emptied into their respective shift registers. The charge in the wells of these shift registers is transferred to the output shift register and then transferred to the output. The output signal current

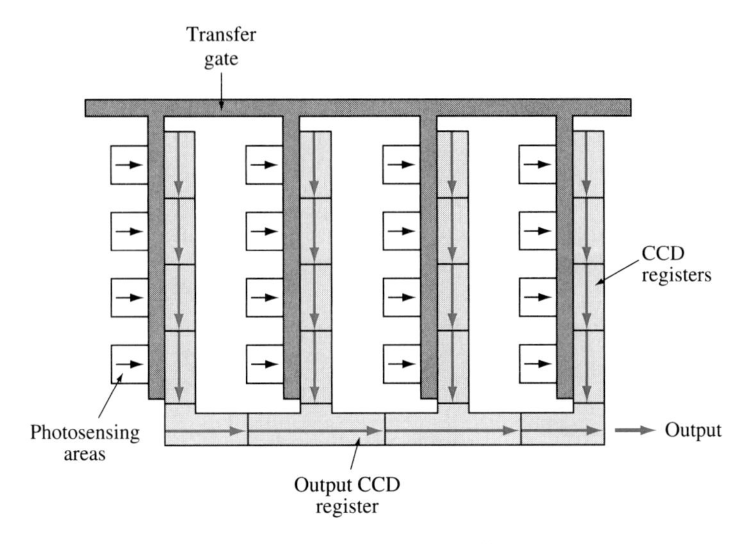

Figure 11.36 Schematic of a 4 × 4 area CCD imager.

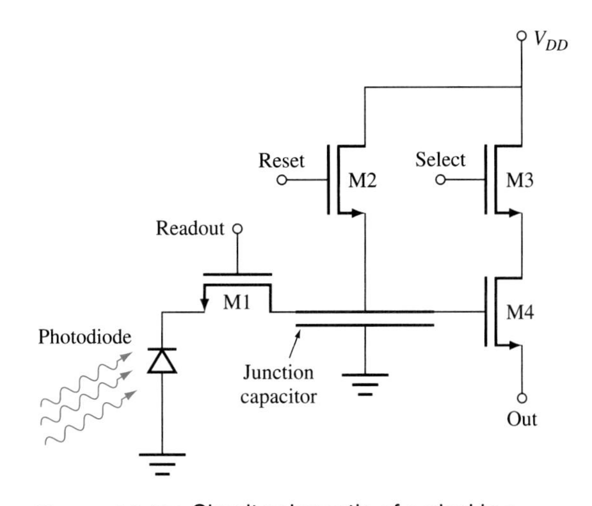

Figure 11.37 Circuit schematic of a pixel in a MOS image sensor.

waveform is then a function of the two-dimensional position dependence of the light intensity.

The figure shows a 4 × 4 (16-pixel) imager, but CCD cameras are routinely made with millions of pixels.

11.5.2 MOS IMAGE SENSORS

Next we consider a CMOS image sensor. A representative single-pixel circuit is shown schematically in Figure 11.37. It consists of a photodiode for detecting light, four MOSFETs, and a junction capacitor (often called a *floating diffusion*), all fabricated in a p-type substrate, which is taken to be at ground potential. [6]

During the period of light detection, M1, M2, and M3 are off and the capacitor is charged to V_{DD} (e.g., 3.3 V). The voltage on the photodiode (open-circuit voltage) depends on the light intensity and the time of integration. To detect the photodiode voltage and thus its charge, M1 is turned on. Electrons then flow from the diode through M1 to the capacitor, thus discharging the capacitor and reducing its voltage to a value dependent on the charge it received. After the charge is transferred (e.g., 1 μs), M1 is turned off and the photodiode again detects the incident light. Meanwhile, the voltage on the capacitor, which represents the light signal, is detected by turning on M3, which puts the drain voltage of M4 at V_{DD}. This activates M4. The output current is then a measure of the capacitor voltage or the detected light. After readout (e.g., 1 μs), M3 is turned off and M2 is turned on to recharge (reset) the capacitor to V_{DD}, and then turned off. After the prescribed light integration time, M1 is turned on and the cycle repeats.

As in a CCD, a two-dimensional image sensor requires a matrix of pixels with appropriate control circuitry.

11.6 SUMMARY

In this chapter we examined some of the optical properties of semiconductor devices, and saw how diodes can be used for both detection and light sources.

For solar cells and photodetectors, only those photons that are absorbed in or within a diffusion length of the junction produce photocurrent. Thus successful diode structures either have their junctions close to the surface where the light is absorbed or the surface layer may be a wide-band-gap material that is transparent to the incident radiation. It is also advantageous to make the depletion region wide to increase photocarrier collection. A very common structure for this purpose is the PIN diode, in which the middle intrinsic layer effectively extends the depletion width. In this case most of the photogenerated current contributes to the prompt response.

High-speed photodetectors are operated under reverse bias. This widens the depletion region, increasing absorption, and the electric field in the junction helps speed up the response time. Under reverse bias, the current is proportional to the light input even under varying loads.

Light-emitting diodes and lasers, on the other hand, are operated under forward bias. By injecting electrons and holes across the junction, both types of carriers are made available in the same physical region for more efficient recombination. The use of a double heterostructure to trap carriers increases the efficiency of both lasers and LEDs.

The double heterostructure also helps to confine the light to the gain region in a laser diode. This is because the refractive index of the narrow-band-gap material is different (usually higher) than that of the surrounding wide-band-gap material. Light encountering a boundary from high to low index can be totally internally reflected.

In a laser, two mirrors at either end of the cavity cause the light to reflect back and forth in the gain medium, so the optical field can be amplified many times. Still, only certain wavelengths will add constructively and actually lase. The narrow spectral width of diode lasers arises from the narrow resonances of the Fabry-Perot cavity rather than from the spectral width of the gain.

III-V semiconductors are the most common for semiconductor sources, but advances in materials technology are creating new possibilities over a wide range of optical wavelengths.

Charge-coupled imagers and MOS-based imagers were briefly discussed. Such devices are used in digital cameras and camcorders.

11.7 READING LIST

Items 35 to 44 in Appendix G are recommended.

11.8 REFERENCES

1. J. J. Loferski, "The first forty years: a brief history of the modern photovoltaic age," *Progress in Photovoltaics: Research and Applications,* Vol. 1, pp. 67–78, 1993.
2. H. Melchior, "Demoduation and photodetection techniques," in F. T. Arecchi and E. O. Schultz-Dubois, eds., *Laser Handbook,* Vol. 1, North-Holland, Amsterdam, pp. 725–835, 1972.
3. E. Kapon, *Semiconductor Lasers II: Materials and Structures,* Academic Press, New York, p. 73, 1999.
4. B. E. A. Saleh and M. C. Teich, *Fundamentals of Photonics,* John Wiley & Sons, New York, p. 633, 1991.
5. J. Singh, *Semiconductor Devices Basic Principles,* John Wiley & Sons, New York, p. 460, 2001.
6. Hideki Mutoh, "3-D Optical and Electrical Simulation for CMOS Image Sensors," *IEEE Trans. Electron Devices,* Vol. 50, pp. 12–16, 2003.

11.9 REVIEW QUESTIONS

1. Why are photodiodes typically reverse biased?
2. Why are solar cells operated in the fourth quadrant?
3. Explain how light energy is converted to electrical current in a photodiode.
4. What is the purpose of the intrinsic region in a PIN diode?
5. What is dark current? From what mechanism(s) does it arise?
6. What is the difference between quantum efficiency and responsivity?
7. What is meant by air mass zero?
8. What is the difference between spontaneous emission and stimulated emission? On which does a light-emitting diode operate?

9. Why are direct-gap materials used for lasers?
10. What is meant by an isoelectronic trap?
11. How is optical feedback achieved in lasers?
12. What governs the spectral width of laser diodes?
13. Why is the gain curve the same as the lineshape function?
14. Explain how the double heterostructure improves the efficiency of lasers. (*Hint:* There is an electrical reason and an optical reason.)
15. Explain in your own words the operation of a CCD imager.

11.10 PROBLEMS

11.1 Consider a photodetector operating in the neighborhood of 60 GHz. How many cycles of green light are there in a single cycle of 60 GHz? Can a photodetector be used to follow the oscillations of the electromagnetic field associated with this light?

11.2 Light with wavelength $\lambda = 700$ nm is incident on a sample of GaAs.

a. Where in the spectrum does this radiation lie?

b. At what depth is the incident flux (neglecting Fresnel loss) reduced to 10 percent of its value at the surface? 1 percent?

c. The color is changed to orange. Now how deep does the light penetrate (to the 10 percent level)?

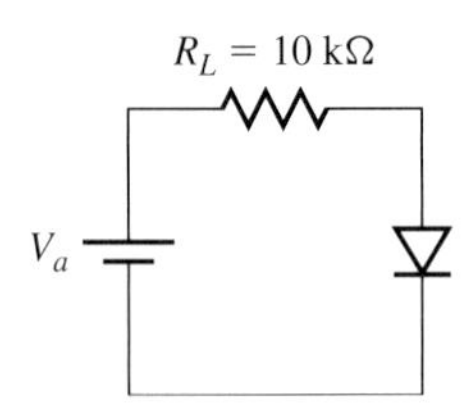

Figure P11.1

11.3 a. Calculate the Fresnel reflection at normal incidence for light going from air to glass ($n = 1.5$).

b. Explain why you can see into a store window and see your reflection at the same time, but at night looking out your window from a lighted room you can see only your reflection.

11.4 Show that Equation (11.8) follows from Equation (11.6).

11.5. For the circuit shown in Figure P11.1,

a. Plot the I-V_a characteristic for the diode with photocurrent $I_L = 100\ \mu$A, 200 μA, and 300 μA. Let $I_{\text{dark}} = I_0 = 10^{-14}$ A and the ideality factor $n = 1$.

b. On your graph, also plot the load lines for $V_a = +5$ V and $V_a = -5$ V.

c. Find the current flowing through the circuit for each load line and plot it against I_L. Recalling that the photocurrent I_L is proportional to the intensity of the light, under which bias regime should one operate photodiodes if one wants the output current to be proportional to intensity?

11.6 A photodiode is made of $Al_{0.1}Ga_{0.9}As$. The refractive index at $\lambda = 800$ nm is 3.65. If the junction depth is 0.3 μm, and the junction width is 1.5 μm, find the quantum efficiency η_Q and the responsivity R_{ph}. Assume the light is incident from air.

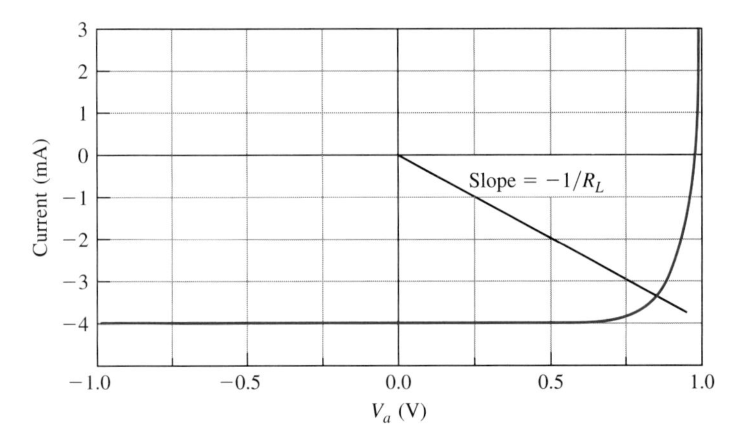

Figure P11.2

11.7 For the solar cell whose I-V characteristics are shown in Figure P11.2, find I_{sc}, V_{oc}, and η. The incident power is 15 mW.

11.8 Which absorbs more of the total solar spectrum, GaAs or Ge?

11.9 If a photon of wavelength at the solar spectrum peak of $\lambda = 0.5\ \mu$m (green) is absorbed by Si, the electron and hole have excess energy as shown on the energy band diagram of Figure P11.3. If both carriers scatter down (or up) to the band edges, what percentage of the absorbed energy is lost as phonons?

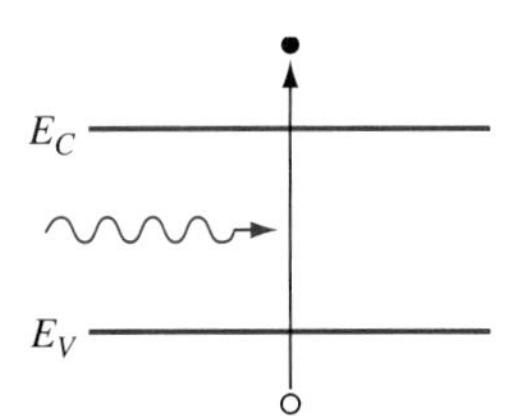

Figure P11.3

11.10 If the doping in the p region of an n$^+$p photodiode is decreased, one would expect the diffusion length in a solar cell to increase. Verify (or contradict) this by repeating Example 11.4 with $N_a' = 10^{16}$ cm^{-3}. For a factor of 10 change in doping, what was the change in η_Q?

11.11 If a GaAs photodiode has a junction depth of 0.2 μm, and if light absorbed in the surface layer is considered lost to surface recombination, what is the total fractional loss in photons in the surface layer? Let the photon energy be 1.4 eV, and repeat for $E_{ph} = 1.8$ eV. Where on the spectrum are these two energies?

11.12 a. Explain why the PIN diode would break down first at the corners if there were no guard ring.

b. Draw the energy band diagram for a PIN diode under high reverse bias and indicate the breakdown mechanism.

c. Draw the energy band diagram for a p^+p^-in junction under the same reverse bias. Explain why this structure will break down at higher voltages.

11.13 a. What should the concentration x of arsenic be in a $GaAs_xP_{1-x}$ LED designed to emit band to band at $\lambda = 670$ nm (Figure 11.13)?

b. If you also take into account the fact that the electrons are concentrated slightly above E_C and the holes are concentrated a little below E_V, how much does that change the band gap you would choose?

11.14 Recall that the peak of the electron distribution with energy is about $\frac{1}{2}kT$ above E_C, and the peak of the hole distribution is about $\frac{1}{2}kT$ below E_V. If each distribution is approximated as having an overall width of kT, estimate the spectral width of the emission. Assume the material emits at 1.3 μm. (*Hint:* To find $\Delta\lambda$, use $E = hc/\lambda$, and take the derivative $dE/d\lambda$ to obtain an expression for $\Delta\lambda$ in terms of ΔE.

11.15 a. Explain why nitrogen is not a donor in GaP.

b. Explain why the N has no long-range forces in GaP.

11.16 Optical fiber manufacturers battled the OH^- ion (resulting from water) for years. These ions, when incorporated into the glass, produce a strong absorption at 1.4 μm (see Figure 11.16). They have finally managed to nearly eliminate it. Is there a similar absorption in the earth's atmosphere?

11.17 What semiconductor materials can be used to produce emission at 1 eV? Of these, are any compatible with readily available substrates (e.g., GaAs or InP)?

11.18 Find the frequency difference between the Fabry-Perot resonances of an edge-emitting laser diode chip in which the effective index that the mode sees is 3.5, the wavelength is 900 nm, and the chip length is 100 μm. If the gain curve is 50 nm wide, how many Fabry-Perot resonances are there in this range for this diode?

11.19 A diode begins to lase when the gain in the cavity exceeds the losses. One source of loss is the partially reflective mirrors at either end of the cavity. Some percentage of the light power is lost each time the light strikes one of the mirrors. How would the power-current curve of a laser be changed if coatings are added to the facets to increase the reflectivity?

11.20 Lasers are often characterized with an L-I-V plot, or one that plots light, current, and voltage, like the one in Figure P11.4 for a VCSEL. The L-I (power, or light, versus current) curve uses the left axis, and

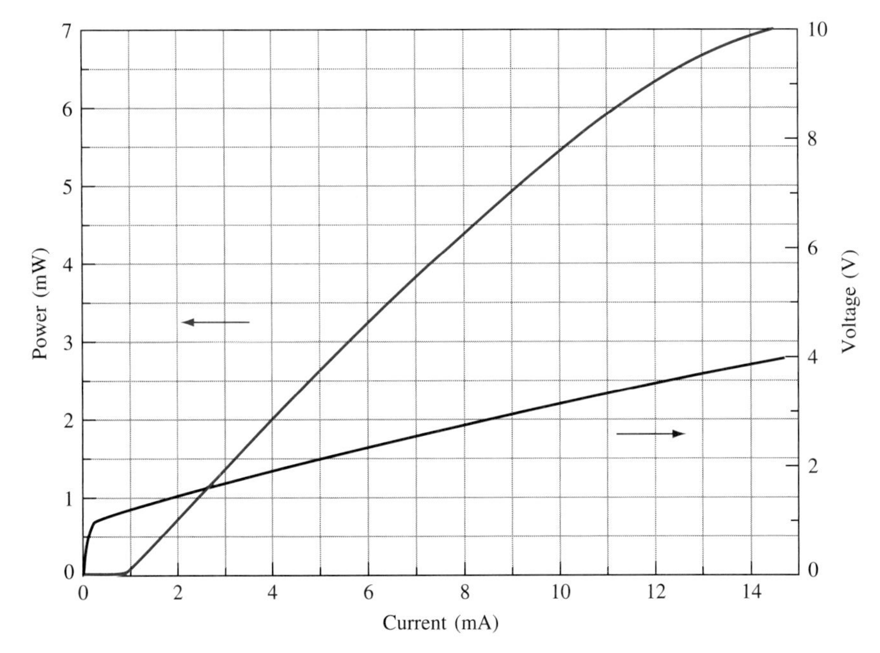

Figure P11.4

the V-I curve uses the right axis. For the laser shown in the figure, what is the ratio of the optical power emitted to the electrical power dissipated in the device at an operating current of 10 mA? What happens to the rest of the power?

11.21 Comment on the probability of absorption (zero, low, medium, high) by a photon of $\lambda = 600$ nm (red) by the following materials:

Si

Ge

GaAs

InAs

SiC

GaN

CdS

APPENDIX

Table A.1 Derived units

Energy	Joule: $\frac{kg \cdot m^2}{s^2}$
Force	Newton: $\frac{kg \cdot m}{s^2}$
Power	Watt: $\frac{kg \cdot m^2}{s^3} = \frac{J}{s}$
Capacitance	Farad: $\frac{C^2 \cdot s^2}{kg \cdot m^2}$
Current	Ampere: $\frac{C}{s}$
Current density	Amperes/area: $\frac{C}{s \cdot m^2}$
Electric potential	Volt: $\frac{kg \cdot m^2}{s^2 \cdot C}$
Inductance	Henry: $\frac{kg \cdot m^2}{C^2}$
Resistance	Ohm: $\frac{kg \cdot m^2}{s \cdot C^2}$
Magnetic induction	Tesla: $\frac{kg}{s \cdot C}$

Table A.2 Some semiconductors and their band gaps

Semiconductor	Band structure	Band gap, eV
Si	Indirect	1.12
GaAs	Direct	1.43
Ge	Indirect	0.67
InP	Direct	1.35
AlAs	Indirect	2.16
AlP	Indirect	2.45
AlSb	Indirect	1.6
SiC	Indirect	2.2
GaN	Direct	3.39
GaP	Indirect	2.34
GaSb	Direct	0.81
InAs	Direct	0.36
InSb	Direct	0.18
CdS	Direct	2.42
CdTe	Direct	1.56
CdSe	Direct	1.70
ZnO	Direct	3.35
ZnS	Direct	3.68
ZnTe	Direct	2.25
ZnSe	Direct	2.7
PbS	Indirect	0.41
PbTe	Indirect	0.31

APPENDIX B

List of Symbols

a	acceleration, lattice constant
A	area
A_{21}	Einstein coefficient for spontaneous optical emission
A_E	area of emitter junction
B	magnetic field
B_{12}	Einstein coefficient for absorption
B_{21}	Einstein coefficient for stimulated emission
BF	forward current gain (SPICE)
BR	inverse mode current gain (SPICE)
BV	reverse breakdown voltage (SPICE)
c	speed of light
C	capacitance
C'_B	substrate (bulk) capacitance per unit area
C_{GD}	gate-to-drain capacitance
C_{GS}	gate-to-source capacitance
C_{in}	input capacitance
C_j	junction capacitance
C_{jBC}	collector-base junction capacitance
C_{jBE}	emitter-base junction capacitance
CJC	zero-bias base-collector junction capacitance (SPICE)
C_{jD}	drain junction capacitance
CJE	zero-bias base-to-emitter junction capacitance (SPICE)
C_{jS}	source junction capacitance
C_L	load capacitance
C'_{ox}	oxide capacitance per unit area
C_{OD}	drain overlap capacitance
C_{OS}	gate to source overlap capacitance
C_{out}	output capacitance
C_{sc}	stored-charge capacitance
$C_{\text{sc}BC}$	collector-base stored-charge capacitance

C_{scBE}	emitter-base stored-charge capacitance
C_w	stray wiring capacitance
C_μ	capacitance between collector and base in hybrid-pi model
C_π	capacitance between base and emitter in hybrid-pi model
d	length of optical cavity
D_n	diffusion coefficient for electrons
D_{nB}	diffusion coefficient for electrons in base (npn)
D_p	diffusion coefficient for holes
D_{pC}	diffusion coefficient for holes in collector (npn)
D_{pE}	diffusion coefficient for holes in emitter (npn)
E	energy
E_a	activation energy
E_0	reference energy; ground state energy
E_A	acceptor energy; acoustic phonon energy
E_A^*	effective acceptor energy
E_B	energy barrier height
E_C	energy at the bottom of the conduction band
E_{Cn}	conduction band edge in n-type material
E_{Cp}	conduction band edge in p-type material
E_{C0}	conduction band edge of non-degenerately doped material
ΔE_C	change in conduction band edge due to degenerate doping
E_D	donor energy
E_D^*	effective donor energy
E_i	intrinsic Fermi level
E_f	Fermi level
E_{fm}	Fermi level in the metal (or polySi)
E_{fn}	quasi-Fermi level for electrons
E_{fp}	quasi-Fermi level for holes
E_{fs}	Fermi level in the semiconductor
E_g	band gap
E_{gn}	band gap in n-type material
E_{gp}	band gap in p-type material
E_{g0}	band gap of nondegenerately doped material
ΔE_g	change in band gap due to degenerate doping
E_g^*	apparent band gap due to degenerate doping
ΔE_g^*	impurity-induced apparent band-gap narrowing
ΔE_g^{**}	alternative form for expressing band-gap narrowing [see Equation (2.113)]
ΔE_{gBE}^*	apparent band-gap narrowing for base-emitter junction
E_K	kinetic energy
E_n	nth energy level
E_P	potential energy
E_{pho}	optical phonon energy
E_{phonon}	phonon energy
E_T	trap energy level
E_V	energy at the top of the valence band

$E_{V\text{bulk}}$	valence band edge in bulk
E_{Vn}	valence band edge in n-type material
E_{Vp}	valence band edge in p-type material
E_{vac}	vacuum energy level
$\mathscr{E}$	electric field; true electric field
$\mathscr{E}_L$	longitudinal electric field
$\mathscr{E}_{Lc}$	critical longitudinal electric field
$\mathscr{E}_e^*$	effective electric field for electrons
$\mathscr{E}_h^*$	effective electric field for holes
$\mathscr{E}_{\max}$	maximum electric field
$\mathscr{E}_T$	transverse electric field
$\mathscr{E}_{T\text{eff}}$	effective transverse electric field
f	frequency
f_{co}	cutoff frequency
$f(E)$	probability of occupancy of a state at energy level E by an electron
$f_p(E)$	probability of occupancy of a state at energy level E by a hole
f_T	unity current gain frequency
F	force
F_e	force on electrons
FF	fill factor
F_h	force on holes
F_{Hn}	Lorentz force on electrons
F_{Hp}	Lorentz force on holes
F_L	photon flux
F_{Li}	incident photon flux
F_n	electron flux
FO	fan-out
F_p	hole flux
g	degeneracy
$g(\nu)$	lineshape function
g_d	output conductance
$g_{d\text{sat}}$	saturation output conductance
g_i	degeneracy of ith state
g_m	transconductance
$g_{m\text{sat}}$	saturation transconductance
G	generation rate
G_L	optical generation rate
G_n	electron generation rate
$G_{n(\text{th})}$	thermal electron generation rate
$G_{n(\text{op})}$	optical electron generation rate
G_p	hole generation rate
G_{op}	optical generation rate
G_P	small-signal conductance of the transition region
G_{th}	thermal generation rate
h	Plank's constant; emitter width

$\hbar$	h-bar ($h/2\pi$)
HOT	higher-order term
$\hat{i}$	unit vector in the x-direction
i_b	small-signal base current
i_c	small-signal collector current
i_d	small-signal drain current
i_e	small-signal emitter current
i_g	small-signal gate current
I	current
I_A	anode current
I_B	base current
IBV	reverse breakdown current (SPICE)
I_C	collector current
I_{C0}	collector-base junction leakage current
I_{CT}	collector current (Ebers-Moll)
I_D	drain current
I_{dark}	dark current
I_{Dn}	drain current in NFET
I_{Dp}	drain current in PFET
$I_{D\text{sat}}$	saturation current
$I_{D\text{satn}}$	saturation current in NFET
$I_{D\text{satp}}$	saturation current in PFET
I_E	emitter current
I_{EE}	current flowing from emitter to ground (or emitter supply)
I_F	forward current
I_{F0}	forward saturation current (Ebers-Moll)
I_{KF}	forward knee current
IKF	forward knee current (SPICE)
I_{KR}	inverse knee current
IKR	inverse knee current (SPICE)
I_L	photocurrent exclusive of dark current
I_m	operating current for a solar cell
I_n	electron current
I_{nC}	electron current that crosses the base from the emitter and reaches the collector (npn)
I_{nE}	electrons injected from emitter to base (npn)
I_0	dark current, leakage current
I_p	hole current
I_{pC}	collector hole current
I'_{pC}	hole current due to electron-hole generation in B-C junction
I''_{pC}	hole current extracted from collector into the base (npn)
I_{pE}	hole current injected from base to emitter (npn)
I_{rec}	recombination current (in base)
I_R	reverse current
I_{R0}	reverse saturation current (Ebers-Moll)

IS	saturation current (SPICE)
ISC	base-to-collector recombination current (SPICE)
I_{SC}	short-circuit current
ISE	base-to-emitter saturation current (SPICE)
j	$\sqrt{-1}$
$\hat{j}$	unit vector in the y direction
J	current density
J_C	collector current density
$J_{C\,\max}$	maximum collector current density to avoid excess base push-out
J_D	photocurrent generated in the depletion region
J_{diff}	diffusion current density
J_{drift}	drift current density
J_E	emitter current density
J_{E0}	emitter saturation current density
J_G	generation current density
J_{GR}	generation-recombination current density
J_{GR0}	generation-recombination leakage current density
J_L	photoinduced current density
J_n	electron current density
J_{nB}	electron diffusion current density in base (npn)
$J_{n\text{diff}}$	electron diffusion current density
$J_{n\text{drift}}$	electron drift current density
J_{np}	current density due to electrons in p-type material
J_0	diffusion dark current, density; diffusion leakage current density
J_{pn}	current density due to holes in n-type material
J_p	hole current density
J_{pB}	hole current density in base
J_{pC}	hole diffusion current density in collector (npn)
J_{pE}	hole diffusion current density in emitter (npn)
$J_{p\text{diff}}$	hole diffusion current density
$J_{p\text{drift}}$	hole drift current density
J_S	a leakage current that is a function of both J_0 and J_{GR}
$\hat{k}$	unit vector in the z direction
k	Boltzmann's constant
K	wave vector for electrons; propagation constant
K_A	wave vector for acoustic phonon
K_i	wave vector for incident particle
K_e	wave vector for an electron
K_{pht}	wave vector for a photon
K_t	wave vector for transmitted particle
K_x	x component of K
K_y	y component of K
K_z	z component of K
$\bar{l}$	mean free path
L	length; channel length

L	channel length (SPICE)
ΔL	channel length shortening
LAMBDA	channel length modulation factor (SPICE)
L_{eff}	effective channel length
L_m	metallurgical channel length
$L_{\min}$	minimum channel length such that short-channel effects need not be accounted for
L_n	diffusion length for electrons, gate length for NMOS
L_{nB}	diffusion length for electron in base (npn)
L_p	diffusion length for holes; gate length for PMOS
L_{pE}	diffusion length for hole in emitter (npn)
m	mass
m_0	mass of a free electron
m^*	effective mass
m_x^*	effective mass for an electron traveling in the x direction
m_y^*	effective mass for an electron traveling in the y direction
m_z^*	effective mass for an electron traveling in the z direction
m_{ce}^*	conductivity effective mass for electrons
m_{ch}^*	conductivity effective mass for holes
m_{dse}^*	density of states effective mass for electrons
m_{dsh}^*	density of states effective mass for holes
m_e^*	effective mass for electrons
m_h^*	effective mass for holes
m_{lh}^*	effective mass for holes in the light-hole band
m_{hh}^*	effective mass for holes in the heavy-hole band
m_{sh}^*	effective mass for holes in the split-off band
$m_{\parallel}^*$	longitudinal effective mass for electrons
$m_{\perp}^*$	transverse effective mass for electrons
M	mass of a particle; multiplication factor; collection efficiency
MJC	collector-base junction grading coefficient (SPICE)
MJE	emitter-base junction grading coefficient (SPICE)
MTTF	mean time to failure
n	electron concentration; quantum number in Bohr model; refractive index; diode quality factor
n_B	electron concentration in base
n_{BC}	base-to-collector recombination current emission coefficient
n_{BE}	base-to-emitter recombination current emission coefficient
n_{B0}	equilibrium electron concentration in base (npn)
$n_{B(\text{norm})}$	normalized electron concentration in base
n^+	heavily doped n-type material
Δn	excess electron concentration
Δn_B	excess electron concentration in base (npn)
Δn_p	excess electron concentration in p-type material
$n(E)$	distribution of electrons with energy
n_{E0}	equilibrium concentration of electrons in emitter

n_i	intrinsic electron concentration
n_{n0}	equilibrium concentration of electrons in n-material
$n_{\min}$	minimum electron concentration in depletion region
n_p	electron concentration in p-type material
n_{p0}	equilibrium concentration of electrons in p-material
n_0	equilibrium electron concentration
n_s	electron concentration at the surface
N	emission coefficient (SPICE)
N_A	concentration of acceptors
N'_A	net concentration of acceptors ($N_A - N_D$)
N'_{AB}	net concentration of acceptors in base (npn)
N_C	effective density of states of electrons in the conduction band
NC	base-to-collector recombination current emission coefficient (SPICE)
N_D	concentration of donors
N'_D	net concentration of donors ($N_D - N_A$)
N'_{DC}	net concentration of donors in collector (npn)
N'_{DE}	net concentration of donors in emitter (npn)
NE	base-to-emitter recombination current emission coefficient (SPICE)
N_{ii}	number of implanted impurity atoms per unit area
NSUB	substrate doping (SPICE)
N_T	concentration of traps
N_V	effective density of states for holes in the valence band
O_{op}	quantum mechanical operator
p	hole concentration; classical momentum
p_{ac}	ac power
p_B	hole concentration in base
p_{B0}	equilibrium hole concentration in base
p_{dc}	dc power
p_E	hole concentration in emitter
p_{E0}	equilibrium hole concentration in emitter (npn)
P_m	maximum power that can be produced by a solar cell at a given illumination condition
p_n	hole concentration in n-type material
p_0	equilibrium concentration of holes
p_{n0}	equilibrium concentration of holes in n-type material
p_{p0}	equilibrium concentration of holes in p-type material
p^+	heavily doped p-type material
p_{C0}	equilibrium hole concentration in collector (npn)
Δp	excess hole concentration
Δp_C	excess hole concentration in collector (npn)
Δp_E	excess hole concentration in emitter (npn)
Δp_n	excess hole concentration in n-type material
$p(E)$	distribution of holes with energy
p_0	equilibrium hole concentration
$P(x, t)$	probability density

P	probability that a carrier generates an electron-hole pair in avalanche breakdown; power
P_{dynamic}	dynamic power dissipation
PHI	surface potential (SPICE)
q	absolute value of the charge of an electron ($q = 1.6 \times 10^{-19}$ C); longitudinal mode number in laser
Q	charge
Q_B	depletion or bulk charge; stored charge in the base
Q_s	stored charge
Q_{ch}	channel charge (per unit area) (mobile channel charges that carry current)
Q_{chlf}	channel charge at low field
Q_{chmin}	minimum channel charge
Q_f	fixed oxide charge
Q_{ii}	implanted charge per unit area
Q_{it}	interface trapped charge
Q_m	mobile ion charge
Q_{ot}	oxide trapped charge
Q_{sR}	reclaimable stored charge
Q_{ss}	surface charge density
Q_V	charge density (charge per unit volume)
Q_{VT}	tunneling charge volume density
r	radius; position; distance
r_b	series base resistance
r_c	collector resistance
r_e	emitter resistance
r_o	output resistance
r_n	radius of *n*th energy level orbit (Bohr model)
r_μ	feedthrough resistance (between base and collector)
r_π	differential input resistance (BJT)
$\vec{r}$	position vector
R	resistance; recombination rate; reflection coefficient
$R_\square$	sheet resistance in the base
R_B	effective resistance under the emitter; base resistor
R'_B	resistance from base contact to emitter edge
RB	base resistance (SPICE)
R_{BD}	drain-to-substrate resistance
R_{BS}	source-to-substrate resistance
R_C	collector resistor
RC	collector resistance (SPICE)
R_D	drain resistance
RD	drain ohmic resistance (SPICE)
R_E	emitter resistor
RE	emitter resistance (SPICE)
R_H	Hall coefficient
R_{Hn}	Hall coefficient for electrons

R_{Hp}	Hall coefficient for holes
R_L	load resistance
$R_{\max}$	maximum recombination rate
R_n	electron recombination rate
R_p	hole recombination rate; parallel resistance; small-signal resistance of the transition region
R_{ph}	responsivity
R_S	series resistance; source resistance
RS	series resistance; source ohmic resistance (SPICE)
S	gate voltage swing
$S(E)$	density of states
$S_a(E)$	number of states at a particular energy E_a
$S_{\min}$	minimum voltage swing
$S_V(E)$	density of states for holes
t	time; thickness
t_{BC}	base-collector transit time
t_d	propagation delay time
t_{dr}	rise time
t_{df}	fall time
t_{lh}	low-to-high transition time
$\bar{t}_n$	mean free time between collisions for electrons
t_{nii}	scattering time for electrons due to ionized impurity scattering
t_{nl}	scattering time for electrons due to lattice scattering
t_{ox}	oxide thickness
$\bar{t}_p$	mean free time between collisions for holes
t_s	storage time
t_T	base transit time
t_{TF}	base transit time in forward mode
t_{TR}	base transit time in reverse mode
T	temperature, transmission coefficient; period, tunneling probability
$T(t)$	time-dependent part of wavefunction
TF	forward base transit time (SPICE)
THETA	mobility modulation factor (SPICE)
TOX	oxide thickness (SPICE)
TR	inverse mode base transit time (SPICE)
U_K	Bloch function
UO	low-field mobility (SPICE)
v	velocity
v_a	applied ac voltage
v_{be}	small-signal base-to-emitter voltage
v'_{be}	small-signal voltage drop across r_π
v_{ce}	small-signal base-to-collector voltage
v_d	small-signal drain voltage
v_{dn}	electron drift velocity
v_{dp}	hole drift velocity

v_g	group velocity; small-signal gate voltage
v_{max}	maximum velocity
v_n	velocity of electron in *n*th energy level
v_p	phase velocity
v_{sat}	saturation velocity
v_{satn}	saturation velocity in n-type material
v_{satp}	saturation velocity in p-type material
V	voltage
V_a	applied voltage
V_A	Early voltage
VA	Early voltage forward mode (SPICE)
V_{AF}	Early voltage forward mode
V_{AK}	anode-to-cathode voltage
V_{AR}	Early voltage reverse mode
V_B	substrate-to-source voltage, base voltage
VB	Early voltage in inverse mode (SPICE)
V_{BE}	base-to-emitter voltage
V_{BF}	forward breakdown voltage
V_{bi}	built-in voltage
$V_{\text{bi}BC}$	built-in voltage of collector-base junction
$V_{\text{bi}EB}$	built-in voltage of emitter-base junction
V_{BR}	reverse breakdown voltage
V_{ch}	channel voltage
V_E	emitter voltage
V_C	collector voltage
V_{CC}	collector supply voltage (in bipolar circuits)
V_{CE}	collector-to-emitter voltage
V_{CB}	collector-to-base voltage
V_D	drain voltage
V_{DD}	drain supply voltage in MOS circuits
V_{Dn}	drain voltage in NFET
V_{Dp}	drain voltage in PFET
V_{DS}	drain-to-source voltage
$V_{D\text{sat}}$	drain saturation voltage
V_F	forward voltage
V_{FB}	flat band voltage
V_G	gate voltage
V_{GS}	gate-to-source voltage
V_H	Hall voltage
V_{Hn}	Hall voltage for electrons
V_{Hp}	Hall voltage for holes
V_{in}	input voltage
V_j	junction voltage
VJC	base-collector junction built-in voltage (SPICE)
VJE	emitter-base junction built-in voltage (SPICE)

V_j^n	junction voltage appearing across n side
V_j^p	junction voltage appearing across p side
V_{jCB}	junction voltage of collector-base junction
V_{jBE}	junction voltage of base-emitter junction
V_m	operating voltage for a solar cell
VMAX	maximum drift velocity (SPICE)
V_{OC}	open-circuit voltage
V_{out}	output voltage
V_ρ	voltage in resistivity measurement
V_R	reverse voltage
V_S	source voltage
V_{SS}	source supply voltage in a MOS circuit
V_{sub}	substrate voltage
V_T	threshold voltage
VTO	threshold voltage (SPICE)
V_{Tn}	threshold voltage for NFET
V_{Tp}	threshold voltage for PFET
w	depletion region width
w_B	depletion width in bulk or substrate
w_{BC}	width of base-collector depletion region
w_D	drain depletion width
w_n	width of depletion region on n side
w_p	width of depletion region on p side
w_{pEB}	width of p side of depletion region of base-emitter junction (npn)
w_{pCB}	width of p side of depletion region of collector-base junction
w_s	source depletion width
w_T	width of depletion region at source at and above threshold
W	width; gate width
W	channel width (SPICE)
W_B	width of quasi-neutral region in base
W_{BM}	width of base region between metallurgical junctions
W_E	width of quasi-neutral region in emitter
W_{EM}	width of the emitter region between metallurgical junctions
W_n	gate width of NFET
W_{nC}	width of quasi-neutral region in n collector (npn)
W_{nCM}	width of n-collector region between metallurgical junctions
W_p	gate width of PFET
W_B	base width
W_T	tunneling width
x	position
x_B^-	base edge of base-collector depletion region
x_B^+	collector edge of base-collector depletion region
x_n	location of edge of depletion region in n-type material
x_0	location of the metallurgical junction
x_p	location of edge of depletion region in p-type material

y	position, position along channel
y_{be}	admittance from base to emitter
z	position
Z_{in}	input impedance
α	common base current gain; absorption coefficient
α_F	common base current gain for forward operation
α_R	common base current gain for reverse operation
α_T	base transport factor
β	probability that an electron will recombine with a hole in a given time; common emitter current gain in a BJT
β_{DC}	low-frequency current gain
β_F	common emitter current gain for forward operation
β_R	common emitter current gain for reverse operation
γ	ionization energy; injection efficiency
γ_n	ionization energy in n-type material
γ_p	ionization energy in p-type material
δ	fraction of reclaimable charge
δ_n	energy difference between conduction band edge and Fermi level $(E_C - E_f)$
δ_p	energy difference between Fermi level and valence band edge $(E_f - E_V)$
ε	permittivity
ε_0	permittivity of free space
ε_{ox}	permittivity of oxide
ε_r	relative permittivity
$\varepsilon_{\text{SiO}_2}$	permittivity of silicon dioxide
η	doping parameter (W_B/λ); power conversion efficiency
η_Q	quantum efficiency
θ	mobility modulation factor
θ_{cr}	critical angle
$\theta_{\perp}$	divergence angle perpendicular to the junction plane
$\theta_{\parallel}$	divergence angle parallel to the junction plane
λ	wavelength; channel length modulation parameter (SPICE); doping characteristic length
μ	mobility
μ_H	Hall mobility
μ_{lf}	low field mobility
μ_n	electron mobility
μ_{nii}	mobility of electrons due to ionized impurity scattering
μ_{nl}	mobility of electrons due to lattice scattering
μ_0	zero-field mobility under effect of transverse field
μ_p	hole mobility
ν	frequency
ρ	resistivity
σ	conductivity
σ_0	conductivity at equilibrium, in the dark

σ_n	conductivity due to electrons
σ_p	conductivity due to holes
σ_{pc}	conductivity due to photocarriers
τ_0	lifetime (generic)
τ_D	dielectric relaxation time
τ_n	electron lifetime
τ_{nB}	electron lifetime in base (npn)
τ_T	minority carrier transit time
τ_p	hole lifetime
τ_{pC}	hole lifetime in n-type collector (npn)
τ_{pE}	hole lifetime in n-type emitter (npn)
$\tau_{\text{radiative,spont}}$	radiative lifetime for spontaneous emission
ϕ_{ox}	voltage drop across oxide
$\phi_{\text{ox}}^{\text{th}}$	oxide voltage at threshold
ϕ_f	potential difference between Fermi level and intrinsic level (in volts)
ϕ_s	surface potential (in volts)
Φ	work function
Φ_M	work function in metal
Φ_n	work function in n-type material
Φ_p	work function in p-type material
Φ_s	work function in semiconductor
χ	electron affinity
χ_n	electron affinity in n-type material
χ_p	electron affinity in p-type material
χ_s	electron affinity in the semiconductor
ψ	time-independent wave function
ψ_n	time-independent wave function for nth state
ψ_r	time-independent wave function for reflected particle
ψ_t	time-independent wave function for transmitted particle
Ψ	wave function
ω	angular frequency
ω_{phn}	phonon angular frequency

C APPENDIX

Fabrication

C.1 INTRODUCTION

Here we briefly examine some of the fabrication techniques used in the semiconductor industry. Integrated circuits, for example, may combine many millions of transistors on a single chip. A process of photolithography is used to define the structures, and various techniques are used to dope the different regions n type or p type. Furthermore, metal or other conductors must be laid down to provide the interconnections, as well as insulating layers to avoid short circuits. The circuits are mass produced on wafers and then cut up into individual chips after being fabricated. The chips are mounted and connections to the outside world are made. Finally the packaging process is completed and the devices are shipped. The emphasis here is on silicon. The entire process starts with the production of ultrapure silicon.

C.2 SUBSTRATE PREPARATION

For semiconductor device fabrication, the industry requires ultrapure and defect-free silicon crystals. Although the semiconductor is intentionally doped with alien elements, very precise control is needed over the concentration of those elements, because other elements may influence the electrical characteristics of the final device. Impurities such as gold and copper can create trap states in the forbidden gap, which can collect and freeze carriers in the lattice.

Similarly, crystalline defects can also create problems. Dangling bonds and other defect states can trap carriers or provide mechanisms by which they can recombine. When carriers recombine, current is lost.

Thus, there is an art as well as a science to fabricating electronic-grade, defect-free, pure silicon. Here some fabrication processes involved in producing high-quality silicon for integrated circuits are discussed.

C.2.1 The Raw Material

One of the reasons integrated circuit (IC) technology is inexpensive is that silicon is highly abundant. The process of producing silicon substrates actually starts with SiO_2 (silica), which is the primary ingredient of sand.

The silica is heated to about 1800°C in the presence of carbon, to get rid of the oxygen. The carbon reacts with the oxygen to create carbon monoxide, leaving silicon behind:

$$SiO_2 + 2C \rightarrow Si + 2CO \tag{C.1}$$

The resultant Si is on the order of 95 percent pure. Although it is suitable for metallurgical applications, this silicon contains too many impurities to be usable for electronics.

Next, the silicon is reacted with hydrochloric acid. The silicon reacts with the chlorine to produce either silicon tetrachloride ($SiCl_4$) or trichlorosilane (SiHCl), both of which are liquids.

$$Si + 4HCl \rightarrow SiCl_4 + 2H_2 \tag{C.2}$$

$$Si + 3HCl \rightarrow SiHCl_3 + H_2 \tag{C.3}$$

Other impurities, particularly iron, also react with the chlorine, and the resulting compounds can be distilled out. After several distillation steps at different temperatures, the silicon-chlorine compound is pure enough (on the order of 99.99999 percent pure) to proceed to the next step.

The ultrapure $SiCl_4$ or $SiHCl_3$ is then reacted with hydrogen to form Si, for example via

$$SiCl_4 + 2H_2 \rightarrow Si + 4HCl \tag{C.4}$$

Although silicon naturally forms a diamond crystal as discussed in Chapter 1, at this stage it is polycrystalline. The next step, then, is to grow a single, defect-free crystal from the high-purity polycrystalline silicon. The process of obtaining a single crystal of Si incorporates more impurities, however, particularly oxygen and carbon.

C.2.2 Crystal Growth

To obtain a large single crystal of silicon, one starts with a single crystal seed. It is carefully oriented so that the surface that will be growing is a particular crystal plane, typically (111) or (100). Then the resulting large crystal also has the desired orientation.

One common method for growing a single crystal ingot of silicon is the *Czochralski method.* Here the seed crystal is dipped into a crucible of molten silicon, Figure C.1. The molten silicon may be doped n type or p type in the melt to produce a doped silicon sample. As the seed crystal is slowly pulled out, the silicon in the melt near the seed cools off and crystallizes onto the seed, extending the crystal downward and outward. In this process, the seed crystal is effectively extended, the new material expanding the edges of the seed crystal. The

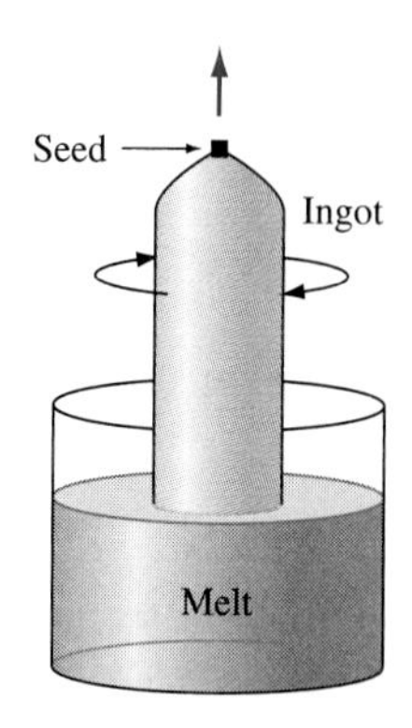

Figure C.1 In the Czochralski method, a seed crystal is pulled slowly from a melt of pure silicon.

diameter of the ingot increases, depending on the rate at which the seed is pulled. When the growing crystal has the desired diameter, the rate of pulling is adjusted so that the diameter remains constant.

The process has to be very carefully controlled, however. If the pull rate slows down, the crystal (boule) gets wider; if it speeds up the boule gets narrower. The crystal is also rotated for better uniformity. Furthermore, the crystal has to be grown slowly (on the order of 50 mm per hour) to avoid defects, and the temperature has to be carefully controlled. This process is difficult, but has developed over the years. In fact, one way to date photographs of silicon fab (fabrication) lines is by the wafer size. In the very early days, the boules could only be grown reliably in 1- or 2-inch diameters. Wafers 3 inches in diameter appeared around the 1970s. At the end of the twentieth century, 200-mm (8-inch) wafers were common and 300-mm (12-inch) wafers were in production, Figure C.2. This figure depicts wafers being automatically stepped through a series of baths for cleaning and etching.

Larger wafers mean greater economy of scale. Consider a very large scale integrated (VLSI) circuit, such as a microprocessor chip with a few million transistors on it. The larger the wafer, Figure C.3a, the more circuits can fit on it, and thus the more economical the process. Figure C.3b shows a 200-mm wafer holding 130 complete circuits (16-Mbyte DRAMS).

Consider what happens if a piece of dust gets on the wafer during the process. A piece of dust can ruin many transistors at a time, but if even one transistor is destroyed, the whole circuit may have to be discarded. In the small

Figure C.2 Wafers are automatically transferred from one bath to the next in a series of cleaning steps. (Used with permission of the Intersil Corporation.)

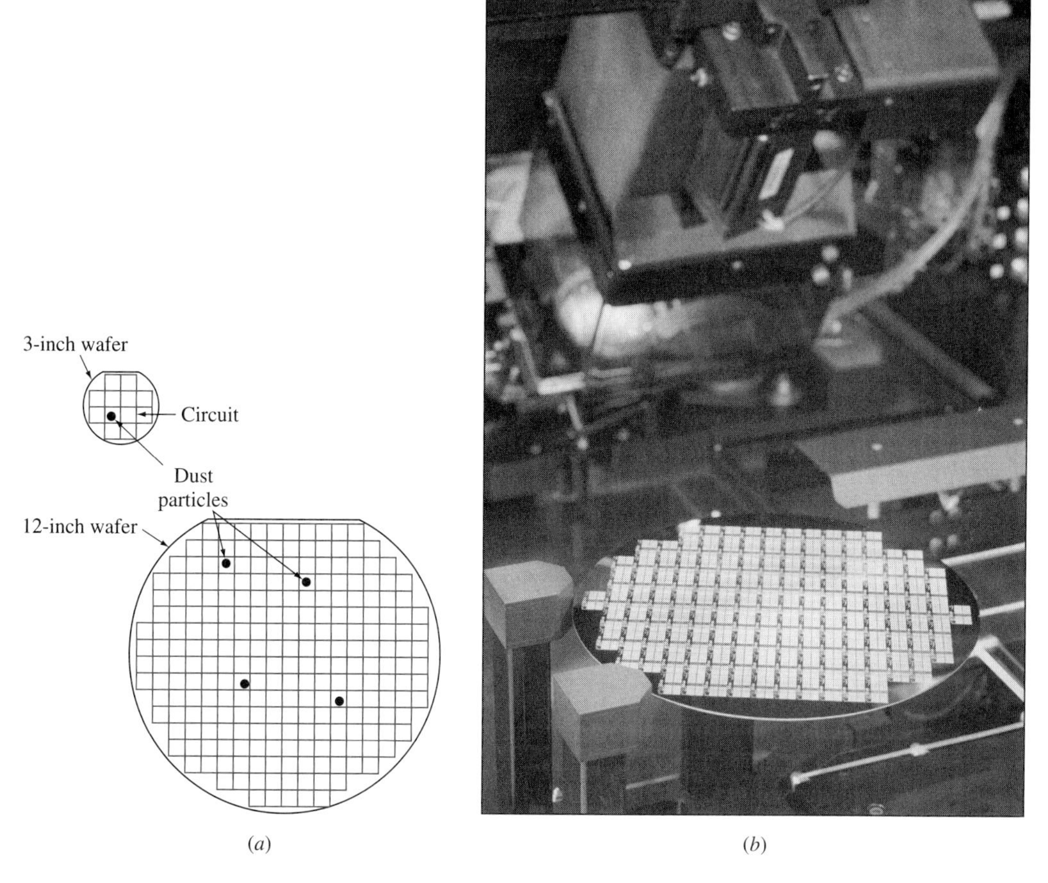

Figure C.3 (a) The larger the wafer, the more circuits it can contain, and thus the more cost-effective the process. (b) A 200-mm wafer of 16 Mbyte dynamic random-access memory (DRAM) chips. [(b) Tom Way/IBM Corporation.]

wafer, if one of the 12 circuits is rejected, the yield of good devices is about 92 percent. In the large wafer case, for the same level of contaminants per unit area, more circuits may be lost, but the percentage yield is much greater. For the example in the figure, the large wafer has four bad circuits but still produces a better than 98 percent yield.

Furthermore, the ovens must be heated up the same number of times and the processes executed the same number of times to produce a 3-inch wafer as a 12-inch wafer, so the economies of scale are enormous. The downside is that the entire fabrication line has to be designed for a particular wafer size. The tubes of the ovens, the boats (wafer holders), and everything must be matched to the size of the wafers, Figure C.4. Therefore, it is expensive to upgrade a given line.

Figure C.4 Everything in a fab line, including these wafer holders, is built for a specific size wafer. Thus upgrading to a larger wafer size incurs significant tooling costs. (Used with the permission of the Intersil Corporation.)

Another method sometimes used to produce single-crystal Si of the appropriate diameter is the *float-zone process.* This method is used to obtain material of greater purity than can be obtained by the Czochralski technique. The starting material is a bar of cast polycrystalline Si. This bar is held in a vertical position as indicated in Figure C.5. Here, an RF coil or other heating device is passed along the ingot, melting the material locally. As the heated zone moves, one end of the melted region is starting to melt and the other end is recrystallizing. As the material solidifies, however, some of the contaminants may preferentially stay in the liquid rather than the solid phase. This is controlled by the *segregation coefficients* of the impurities. The segregation coefficient is the ratio of a particular substance that remains in the liquid form of another substance compared with the fraction that freezes out.

As the liquid region moves along, the impurities are carried along with it. The process is repeated several times to purify the ingot as much as possible. The impurities accumulate at one end, which can be cut off and discarded.

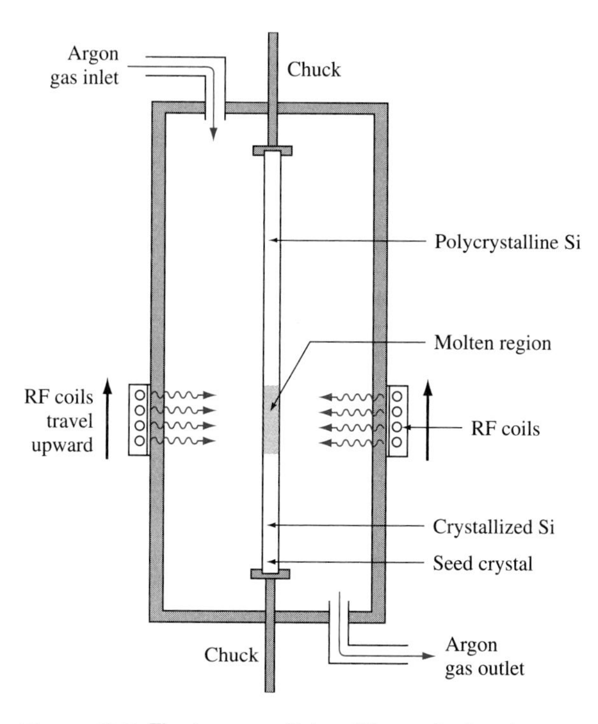

Figure C.5 Float-zone refining. The melted region moves along the boule. As the material on the trailing end of the melted region cools, the impurities preferentially stay in the melt, and travel along in the liquid zone to the end, where they are cut off. The process is generally repeated several times.

This float-zone process is more expensive than the Czochralski process, and, because of the difficulties associated with maintaining a large-diameter molten zone, the resultant crystal diameter is less than for the Czochralski technique. Once the boule is ready, it is sawn into wafers, and then polished.

C.2.3 Defects

The silicon crystal must be of very high purity, but it must also be free of crystal defects. Figure C.6 shows three common crystal defects that can occur; there are many others.

The interstitial defect occurs when an extra atom is inserted into the crystal, but doesn't necessarily bind to the neighboring atoms. It does, however, distort the lattice, disrupting the periodicity locally and altering the energy band structure.

The vacancy defect occurs when a lattice site is empty. This can distort the lattice, and also create dangling bonds. The dangling bonds, in turn, can attract electrons or holes, causing trap states or recombination sites. These traps can collect carriers and reduce the conductivity of the sample.

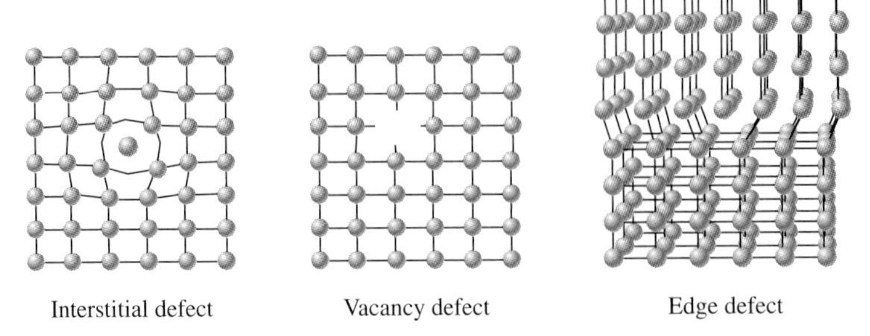

Figure C.6 Some simple crystal defects. The interstitial and vacancy defects are known as *point defects,* and the edge defect is one type of line defect.

The edge defect is one type of line defect. Here an extra plane of atoms exists in the lattice and it ends abruptly. This can also cause dangling bonds and distorts the lattice as well. (There are many excellent references on crystallography for the interested student.[1])

C.2.4 Epitaxy

Often, we wish to grow a uniform layer of one type of semiconductor onto a substrate. For example, one can grow a layer of silicon with one doping level onto a substrate of another. We require that the new layer form a continuous crystal with the substrate, so that there are no defects or interface states that can trap carriers. For this, an epitaxial (epi) growth process is used. In epitaxy, a thin layer of crystal is grown, rather than simply deposited onto the substrate wafer, and the substrate wafer acts as a seed crystal.

Lattice Matching In semiconductors, we often wish to create heterojunctions, meaning that one type of crystal is grown atop another. For example, one may wish to construct a heterostructure in which a layer of InP is sandwiched between two layers of InGaP. One cannot purchase InGaP substrates, so one must start with an available substrate material that is lattice matched.

Lattice matching means that the lattice constants of the two materials must be as nearly equal as possible. Consider what would happen if one attempted to grow the material in Figure C.7 on the substrate shown. The upper material has a slightly larger lattice constant. As the atoms land on the surface and try to fit into the crystal structure, they will bond with the atoms in the layer below. Since the upper material's atoms are normally spaced farther apart, defects will occur occasionally unless the lattice constants of the two materials are very well matched.

In addition, even where defects do not occur, there is strain on the lattice near the junction. The epi layer may have to stretch slightly to match the substrate, or

---

[1]See, for example, S. K. Ghandi, *VLSI Fabrication Principles: Silicon and Gallium Arsenide,* 2nd ed., John Wiley and Sons, New York, 1994.

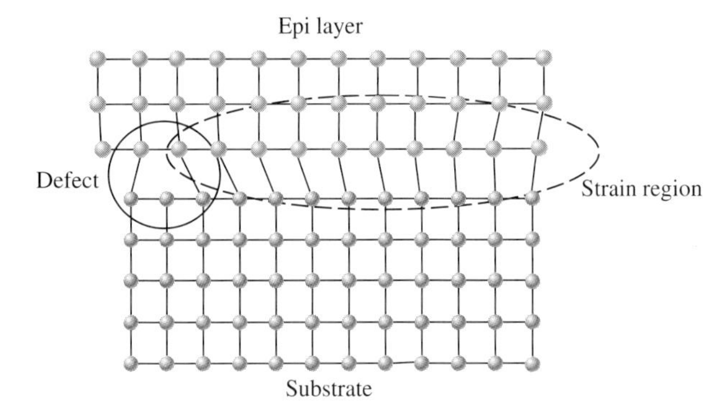

Figure C.7 In heteroepitaxy, the two crystals must be well lattice matched or defects will occur. When the lattice constants are different, one layer may become strained, which changes that material's periodicity and thus the band-gap properties.

it may compress. This strain changes the lattice constant of the epi layer locally, and that in turn will affect the band gap. As the epitaxial layer grows thicker, the atoms will eventually assume their normal spacing, so the layer is strained only near the interface.

If, on the other hand, the epitaxial layer is thin, and another layer of the substrate material is grown on top, then the outer layers can keep the epitaxial layer distorted. Strained-layer epitaxy is being exploited increasingly as engineers get more sophisticated in growth techniques, leading to many band-gap engineering opportunities.

To get back to our example, if one wanted to grow an InGaP-InP-InGaP double heterostructure, one would start with InP, which is a readily available substrate. One would then grow the first layer of InGaP, followed by growth of the InP layer, followed by the InGaP cap layer.

Figure C.8 shows the lattice constants of some common semiconductor crystals. The lines connecting two binary compounds indicate the ternary compounds. For example, tracing the line that connects GaAs to InAs, we can find the lattice constant and band gap for any $In_xGa_{1-x}As$ ternary compound. The x indicates the ratio of indium to gallium in this case. The compound $In_{0.53}Ga_{0.47}As$ has exactly the same lattice constant as InP, making this a popular combination for heteroepitaxy.

Liquid-Phase Epitaxy (LPE) One of the earliest ways used to grow epitaxial layers (primarily for compound semiconductors) was liquid-phase epitaxy. Here the seed crystal, or substrate, is placed in a hollow in a sliding tray, Figure C.9a. The tray can slide to one of several positions, and in each position is a well of molten semiconductor material. For example, if the substrate is GaAs, the molten material in the first well may contain molten aluminum, gallium, and arsenic. The proportions in the melt are chosen such that when they solidify (onto the crystal) the resulting solid composition is the one desired. (The various elements

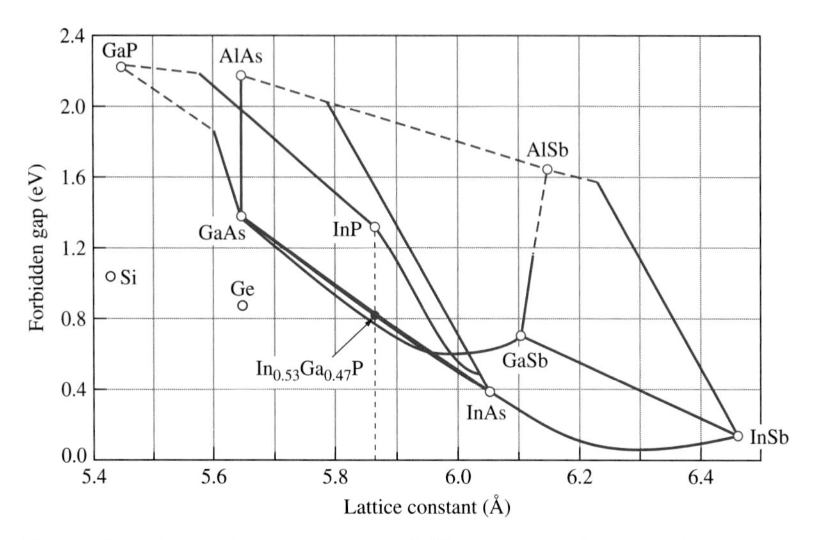

Figure C.8 Energy band-gap versus lattice constant for several common compound semiconductors. We can see that the ternary compound $In_{0.53}Ga_{0.47}As$ is lattice matched to InP.

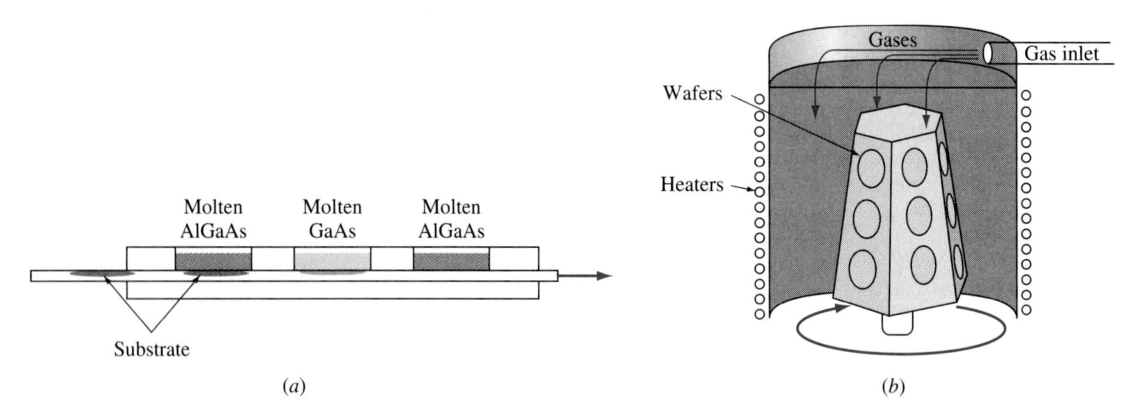

Figure C.9 (a) Liquid-phase epitaxy. The wafers are on a tray that slides through several troughs of liquid semiconductor material. (b) In vapor-phase epitaxy, the materials are introduced in gaseous form.

do not freeze out at the same rates, because of their different segregation coefficients.) Then the tray is slid to the next well, which might contain GaAs again, and the process continues.

Vapor-Phase Epitaxy (VPE) Vapor-phase epitaxy (VPE), or chemical vapor deposition (CVD), is a technique commonly used to grow silicon layers on silicon. The silicon wafers are placed in a chamber, Figure C.9b, and exposed to an atmosphere containing, for example, $SiCl_4$. The wafers are heated to about 1200°C so that the overall reaction

$$SiCl_4 + 2H_2 \rightarrow Si + 4HCl \tag{C.5}$$

can take place. The silicon produced by this reaction can adhere onto the surface of the crystal and the crystal grows.

VPE is also used to grow III-V compounds. For example, a GaAs substrate can be exposed to an atmosphere of arsine (AsH_3), phosphine (PH_3), and gallium chloride (GaCl) gases. The gallium from the GaCl attaches to the existing crystal in the lattice sites where the column III element normally goes. In the alternative sites, either of the two column V elements can attach. The ratio of As to P in this GaAsP film is controlled by the ratio of the arsine to phosphine gasses.

Metal-Organic Vapor-Phase Epitaxy (MOCVD) Vapor-phase epitaxy does not work ideally for all systems, however. Particularly, compounds containing aluminum are difficult to grow in accurately controlled compositions because the

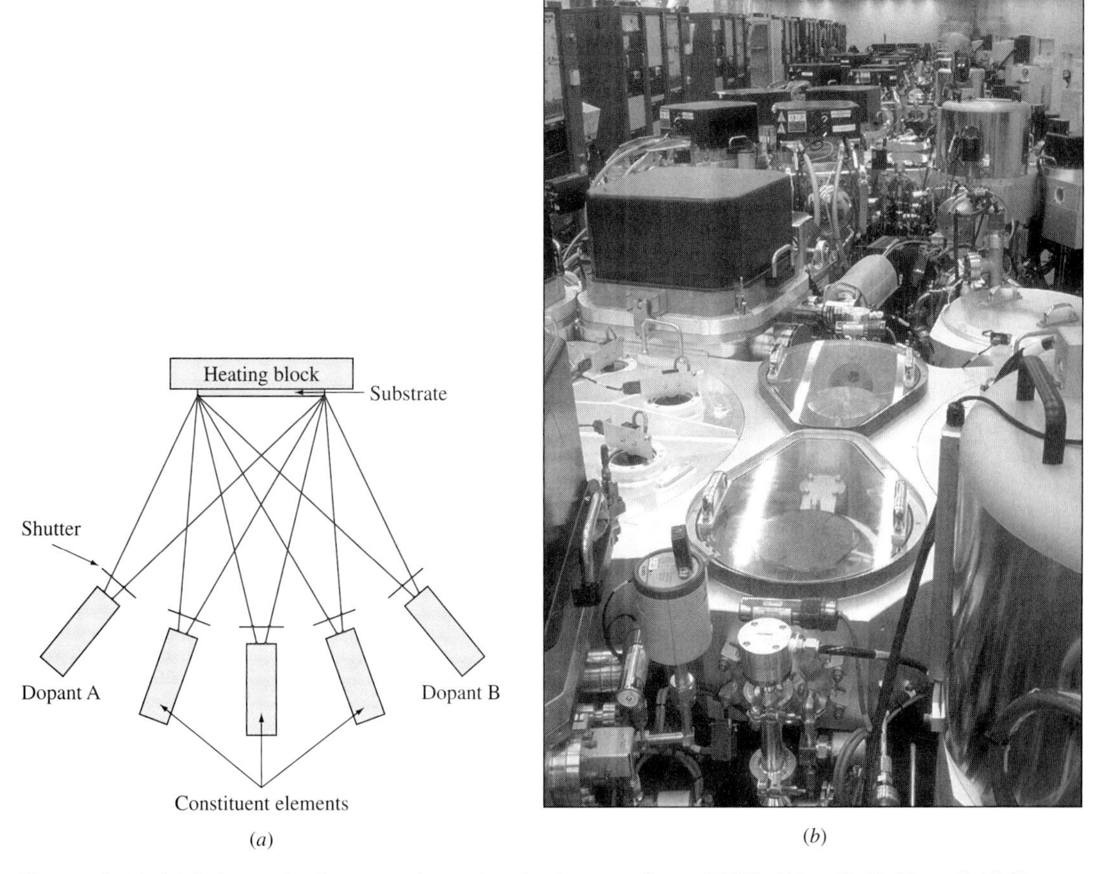

Figure C.10 (a) Schematic diagram of a molecular beam epitaxy (MBE). (After S. K. Ghandi, *VLSI Fabrication Principles: Silicon and Gallium Arsenide,* 2nd ed., p. 294, John Wiley and Sons, 1994. This material is used by permission of John Wiley & Sons, Inc.) (b) Six MBE machines in a production facility.

aluminum does not diffuse well on the surface (to find its correct place in the lattice) and because of its high activity, among other reasons. A variation of the epitaxial growth technique that helps with this is called metal-organic chemical-vapor deposition (MOCVD). Here the Ga and Al metals are introduced in organic compounds such as trimethyl gallium, $Ga(CH_3)_3$, and trimethyl aluminum, $Al(CH_3)_3$. The arsenic and phosphorus are introduced in arsine and phosphine gases as before. MOCVD is capable of growing monolayers (layers one atom thick), which makes possible abrupt changes in composition and highly precise control.

Molecular Beam Epitaxy (MBE) A highly versatile technique for growing epitaxial layers is known as *molecular beam epitaxy* (MBE). It might be considered as sort of a solid-phase epitaxy. The individual elements (and dopants) are heated in their separate crucibles in the MBE machine under high vacuum. As the atoms evaporate, they travel to the substrates and are deposited. The gates to the individual crucibles can be opened and closed to vary the composition of the layers as indicated in Figure C.10a. Figure C.10b shows six MBE machines.

MBE can also put down monolayers for extremely precise control of material growth.

C.3 DOPING

A substrate may be n type or p type (or intrinsic), but integrated circuit engineers need to be able to make certain regions be of the opposite type, to create diodes and transistors. There are two major techniques for doping an existing semiconductor crystal: diffusion and ion implantation.

C.3.1 Diffusion

In diffusion, a substrate at an elevated temperature is exposed to an atmosphere containing the desired dopant. For example, to diffuse an n-type layer into a p-type substrate, the wafer is placed in a diffusion furnace containing a gas of an n-type dopant such as phosphorus. The P atoms are in higher concentration in the atmosphere than in the wafer, so they will diffuse into the surface of the wafer.[2] It requires very high temperatures (800 to 1100°C) for the phosphorus atoms to have enough kinetic energy to work their way into the substrate. The longer the exposure and the higher the temperature, the deeper the diffusion. Diffusion is carried out in a diffusion furnace such as that shown in Figure C.11a.

In diffusion, the distribution of dopants is not uniform, but is instead more concentrated toward the surface. Figure C.11b shows a plot of the concentration of dopants after diffusion. Note that the material is p type until the phosphorus concentration exceeds the background boron concentration in the p-type wafer. Near the surface where $N_D > N_A$, the material is n type.

[2] Recall from Chapter 3 that all things diffuse into regions of lower concentration.

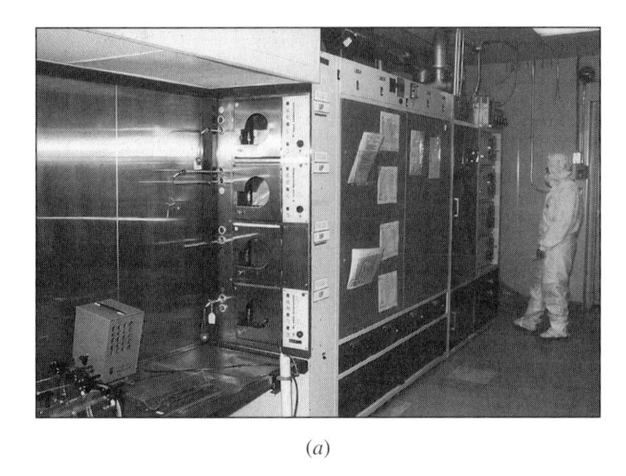

(a)

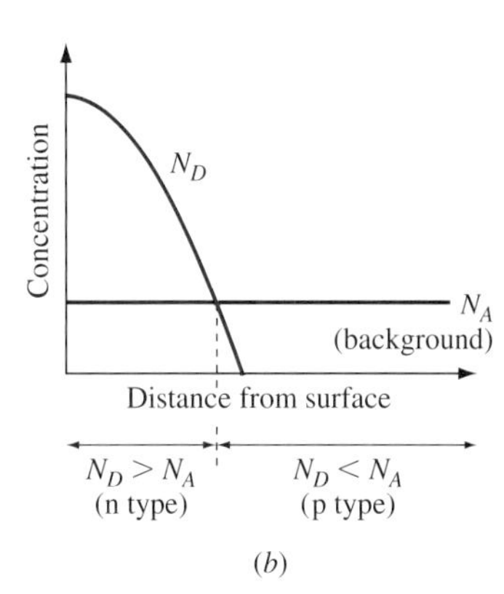

(b)

Figure C.11 (a) Diffusion furnace. (Used with the permission of the Intersil Corporation.) (b) The dopant profiles after a phosphorus diffusion into a p-type substrate.

A second diffusion, this time with acceptors, could also be performed to make the surface p type, producing a pnp structure. The second diffusion would have to be shallower but with an even higher concentration, to overcome the donors deposited earlier. Furthermore, when the wafer is heated for the second diffusion, the donors from the first diffusion will diffuse even further into the semiconductor.

Diffusion is not currently used extensively as a primary method for doping semiconductors, but many other processing steps are done at high temperatures, and every time the temperature is raised, whatever dopants exist in the semiconductor will diffuse, always toward regions of lower concentration.

C.3.2 Ion Implantation

A more common way to dope semiconductors is by ion implantation. In this approach, ions of the required dopants are accelerated toward the substrate. The ions have high kinetic energies, from the keV range up to the MeV range. They arrive with such force at the crystal that they are implanted into the surface. The penetration depth may be on the order of a micrometer—about 1800 lattice constants in silicon. This results in significant damage to the crystal; bonds are broken and atoms are dislodged. Thus, ion implantation is followed by an annealing step. The crystal is heated so that the atoms can move somewhat easily, breaking bonds and shifting positions. Through this process they tend to fall back and regroup into a natural crystal formation.

Figure C.12 shows the bipolar junction transistor from Chapter 9 and its dopant profile. This device is fabricated using BiCMOS technology (an integrated circuit that combines bipolar transistors with complementary metal-oxide-semiconductor field-effect transistors). The plot is measured by a technique known as secondary ion mass spectroscopy (SIMS). It produces a profile of the dopant concentration as a function of depth from the surface. The dopants are added by ion implantation.

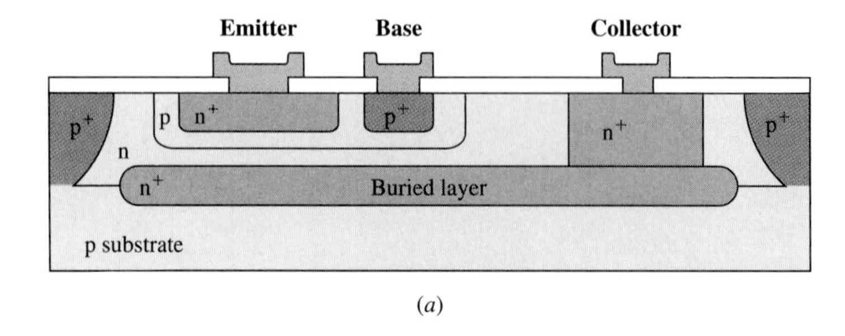

(*a*)

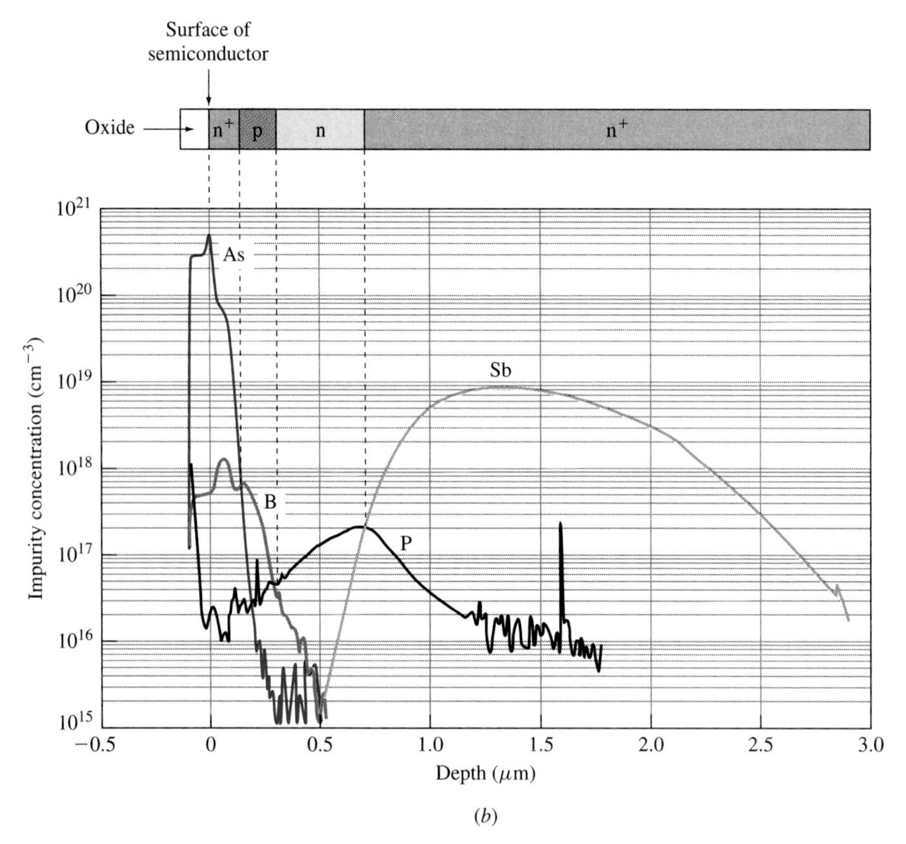

(*b*)

Figure C.12 (a) A cross section of an npn bipolar transistor; (b) a SIMS profile showing the dopant concentration (courtesy IBM). The SIMS was done before metallization, and there is an oxide on the surface at this point.

The deep layer is a heavily doped n-type (n^+) region "buried layer" that is highly conductive. It is doped with antimony, because Sb diffuses slowly in silicon, about an order of magnitude more slowly than the boron and phosphorus used later. Thus, the Sb atoms will not move much during the subsequent processing. The n layer is the *collector* of the transistor, doped with phosphorus.

The p-type base (boron) is then implanted. Finally, the surface layer, the *emitter,* must be doped even more heavily so that the donor concentration again exceeds the acceptor concentration. Here arsenic is used as the emitter dopant because it makes a more uniform emitter-base junction than phosphorus. There is an oxide layer at the surface that also contains dopants, and these also appear on the SIMS plot.

C.4 LITHOGRAPHY

To make transistors and diodes we need to perform the doping only in isolated areas rather than across the entire wafer. In addition, we need to create contacts and connections on the devices. This section indicates how lithography is used to make tiny and precise structures on semiconductor wafers. In lithography, masks with precisely defined features are used to block off parts of the wafer from the various processes.

Let us see how to make a simple diode (Figure C.13). The p-type wafer is covered with a layer of oxide followed by a layer of *photoresist,* Figure C.13a. Photoresist is a photosensitive organic compound, often spun onto the wafer. When it is exposed to light, it undergoes a chemical reaction. For positive photoresist, a developer will remove the photoresist that was exposed to the light. For negative photoresist, the developer removes the unexposed parts. A mask is used, and the photoresist is exposed to light only where the mask is transparent. The photoresist is then developed, leaving holes in the resist.

Next, the wafer is put in the ion implanter. Donor ions embed themselves in the oxide (which is removed later) and the substrate. Where they strike the semiconductor, it becomes n type. In this case an n^+-type well is created in the p-type substrate. Note that the number of donors implanted must exceed the background acceptor concentration to make the material n type where implanted.

Next, a new layer of oxide is grown. This layer is patterned with another mask to allow the ion implantation that creates a p^+ layer for the ohmic p contact.

Another layer of oxide is grown, and another layer of photoresist spun onto the wafer. Now a third mask is used, to form the two holes through which the contact will be made. When the next layer of metal is put down, it will be isolated from the p-type substrate by the oxide left behind.

Finally, the metal is patterned to make electrical contact to the p region and the n region. The contacts are made through the holes in the oxide.

Figure C.14 shows one of the masks used to make a particular 16 Mbyte memory chip. The mask is made large at first and then photographically reduced.

C.5 CONDUCTORS AND INSULATORS

In addition to the semiconductor materials, insulators and connectors are required to interconnect the individual components and to isolate the components and their electrodes from one another. The conductors are typically metal, usually aluminum, copper, or degenerately doped silicon layers. The degenerately

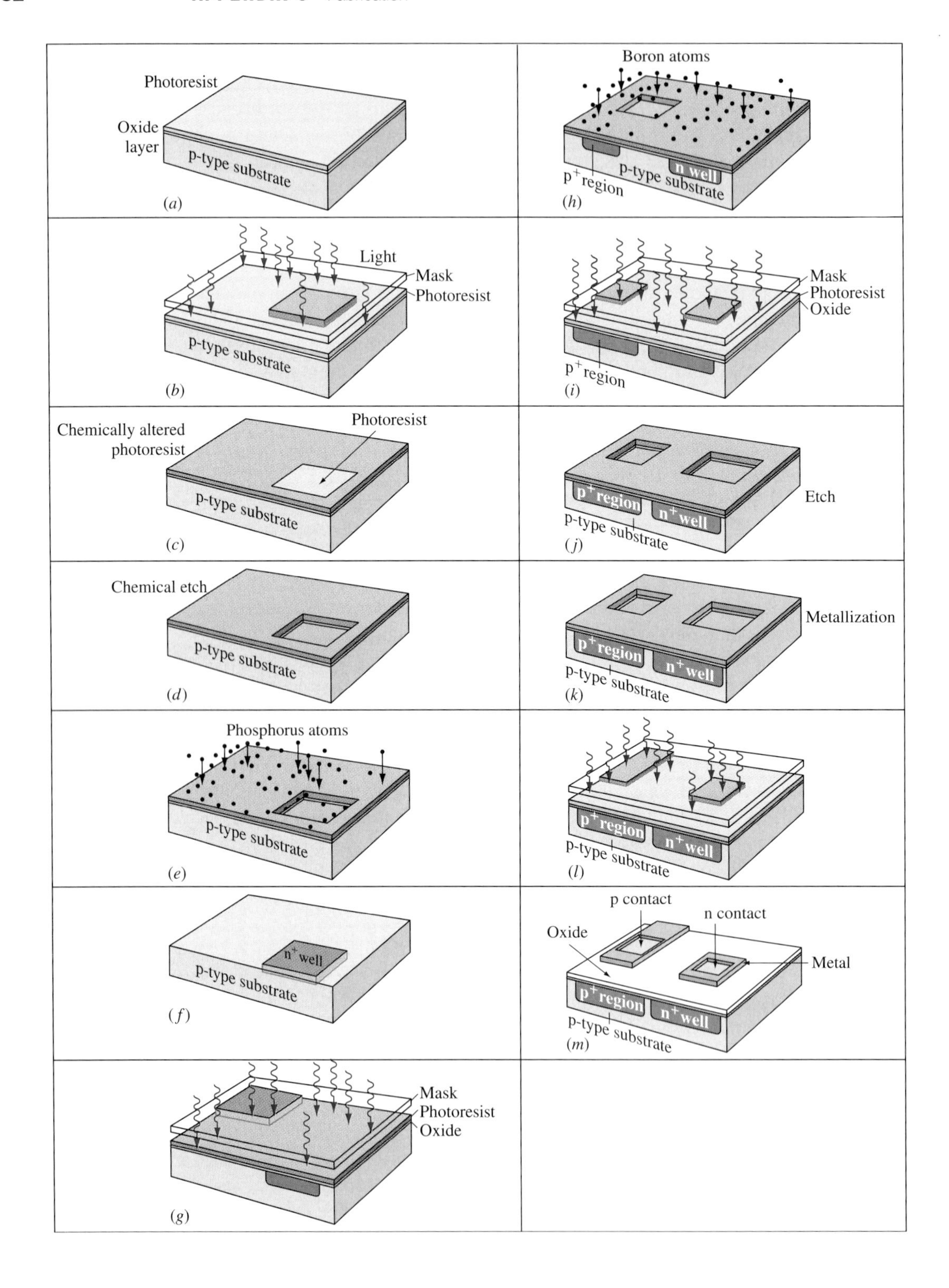
Photoresist
Oxide layer
p-type substrate
(a)
Light
Mask
Photoresist
p-type substrate
(b)
Chemically altered photoresist
Photoresist
p-type substrate
(c)
Chemical etch
p-type substrate
(d)
Phosphorus atoms
p-type substrate
(e)
n+ well
p-type substrate
(f)
Mask
Photoresist
Oxide
(g)
Boron atoms
p+ region
p-type substrate
n well
(h)
Mask
Photoresist
Oxide
p+ region
(i)
Etch
p+ region
n+ well
p-type substrate
(j)
Metallization
p+ region
n+ well
p-type substrate
(k)
p+ region
n+ well
p-type substrate
(l)
p contact
n contact
Oxide
Metal
p+ region
n+ well
p-type substrate
(m)

Figure C.14 A mask for a 16-Mbyte memory chip. The mask is made large and then photographically reduced to the final size. (Tom Way/IBM Corporation.)

doped layers may be part of the semiconductor crystal or they may be polycrystalline silicon.

C.5.1 Metallization

Metal is often used for interconnections on integrated circuits. Aluminum is the principal metal used, because it is inexpensive and highly conductive. Aluminum on silicon can form ohmic junctions, in which the resistance of the contact is the same under both positive and negative voltages. In some cases, however, particularly with n-type silicon, aluminum will form rectifying (Schottky) junctions, which is usually not desired. Schottky barriers were covered in Chapter 6, where we saw that the interface between the metal and the semiconductor forms a potential barrier to carriers. If the barrier is thin enough, substantial tunneling can occur, making the contact conductive in both directions.

Figure C.13 Photolithography is used to make a diode. (a) A layer of silicon dioxide is grown and then covered with photoresist. (b) This is exposed to light through a mask. (c) The light changes the chemistry of the photoresist where it's exposed. (d) The exposed photoresist and the layers under it are etched away. (e) Phosphorus atoms are ion implanted, leaving the n^+ structure shown in (f). In (g), a new layer of oxide is deposited, and a mask is used to open a hole in the oxide. (h) A second implantation is done to create a p^+ contact layer. (i) A third mask is used to pattern the oxide and (j) etched to make contact holes. (k) A layer of metal is deposited over the entire wafer. (l) The metal is patterned using another mask. (m) The final device has two contacts, one to the n region and one to the p substrate.

Figure C.15 A scanning electron micrograph of the metal connections of an integrated circuit. All material except the metal has been removed to show the connections. (This photograph has been reprinted pursuant to a license granted by LSI Logic Corporation, © 2000, LSI Logic Corporation. ALL RIGHTS RESERVED.)

Aluminum is generally deposited on silicon by sputtering. In this process, an aluminum source is placed close to the wafer, and a plasma (gas of highly energetic atoms) is introduced. The plasma atoms, typically argon, collide with the aluminum source and dislodge the Al atoms. The aluminum atoms then deposit on every surface inside the sputtering chamber, including the wafer. Since aluminum has a lower melting temperature than silicon, metallization has to be done after all the high-temperature steps are completed. The patterning is done by photolithography.

In complex circuits, several layers of metallization are needed to provide all the connections, Figure C.15. These layers are separated by silicon dioxide, which is insulating, and metal "vias" (connections from one layer to the next) are formed by making holes in the oxide and filling with metal such as tungsten.

More recently, copper is replacing the aluminum in the top interconnection layer. Aluminum suffers from electromigration, in which the electrons traveling through the conductor have high enough kinetic energy to dislodge the aluminum atoms. Over time, this process moves enough Al atoms to create voids (open circuits) or bridges (short circuits). The copper atoms are much heavier and thus not as susceptible to electromigration. Engineers have only recently learned how to pattern copper, which is enabling the shift to this metal for interconnections.

Gold is also used in metallization, especially in GaAs. Although Au can be used in silicon, it must be combined with layers of other metals. Gold diffuses easily in silicon and creates deep trap states that can ruin the electrical properties of the semiconductor.

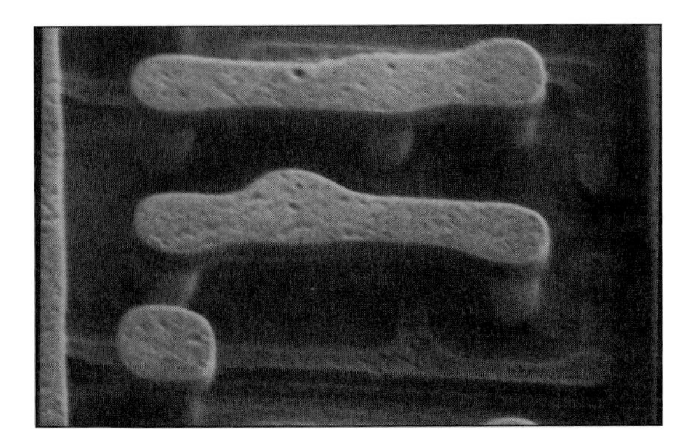

Figure C.16 Polysilicon is often used as a surface conductor, with metal used to make connections on the higher levels. (This photograph has been reprinted pursuant to a license granted by LSI Logic Corporation, © 2000, LSI Logic Corporation. ALL RIGHTS RESERVED.)

C.5.2 Poly Si

Degenerately doped silicon is very conductive, and can be used for interconnections as well. For example, heavily doped layers may be used to transport carriers laterally between different structures on a chip. The buried n layer in the bipolar junction transistor of Figure C.12 conducts electrons from the collector region under the emitter laterally to the collector contact.

Polycrystalline silicon is often used on the surface as a conductor. It can be deposited by a pyrolysis process using silane, the reaction for which is

$$SiH_4 \rightarrow Si + 2H_2 \tag{C.6}$$

Although Si is deposited, it will form not a single-crystal epitaxial layer but rather a polycrystalline layer, because of the conditions prevailing when the deposition is done. While normally not very conductive, poly Si can be doped to high conductivities, appropriate for such VLSI structures as gate electrodes for MOS technology. It is also used as an interface between metal and the silicon substrate in order to ensure a low-resistance electrical contact. Figure C.16 shows a chip whose surface conductor is poly Si and the next connection layer is metal.

C.5.3 Oxidation

Silicon dioxide is an electrical insulator, and is used to electrically isolate various structures on a chip. One of the reasons silicon is the most widely used semiconductor is the ease with which a "native" oxide can be grown on it—silicon dioxide forms whenever the silicon is exposed to oxygen.

Silicon dioxide can be used to protect areas from diffusion and even ion implantation. The SiO_2 is patterned onto the wafer by photolithography, and the diffusion or implant is done through the holes in the oxide. If the oxide is thick enough, and the diffusivities of the dopant atoms are small enough in the oxide, the dopants will not reach the protected areas of the silicon. Fortunately, the diffusivities in SiO_2 of boron, phosphorus, and antimony, to name a few, are much smaller than their diffusivities in silicon.

Oxidation can be carried out through a wet or dry process. In the dry process, the wafer is exposed to dry O_2, producing the reaction

$$\mathrm{Si} + \mathrm{O_2} \rightarrow \mathrm{SiO_2} \tag{C.7}$$

Alternatively, in a wet oxidation process, the oxygen is introduced by water vapor, reacting via

$$\mathrm{Si} + \mathrm{H_2O} \rightarrow \mathrm{SiO_2} + \mathrm{H_2} \tag{C.8}$$

The rate of the reaction is controlled by the temperature. Typical temperatures are in the range of 900 to 1200°C. The wet process occurs considerably faster, while the dry process yields a better-quality interface between the silicon and the oxide. Therefore a common technique is to grow a thin layer of oxide by using the dry process and following it with a thicker wet-oxidation process layer.

When the oxidation process starts, the first oxygen atoms react with the silicon surface. As the oxidation continues, however, the growth of additional SiO_2 occurs at the Si-SiO_2 interface, since that is where more silicon atoms are available for reaction. Thus, as the oxide grows, the silicon is consumed, as shown in Figure C.17. The surface of the wafer moves up (material is being added) but the surface of the silicon moves downward. The rates of these movements are nearly equal.

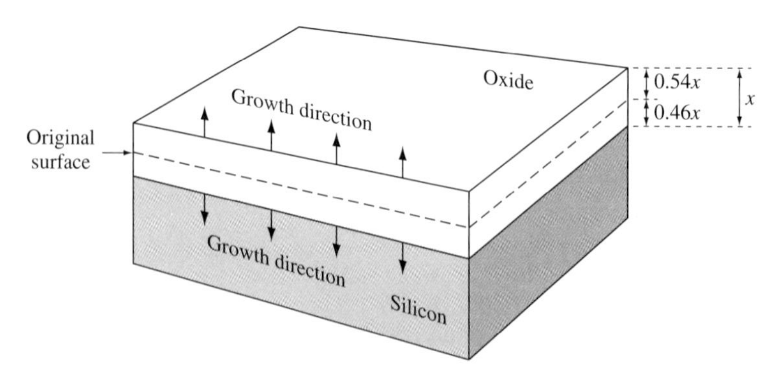

Figure C.17 As an oxide layer grows, the oxide layer expands upward and also downward into the silicon.

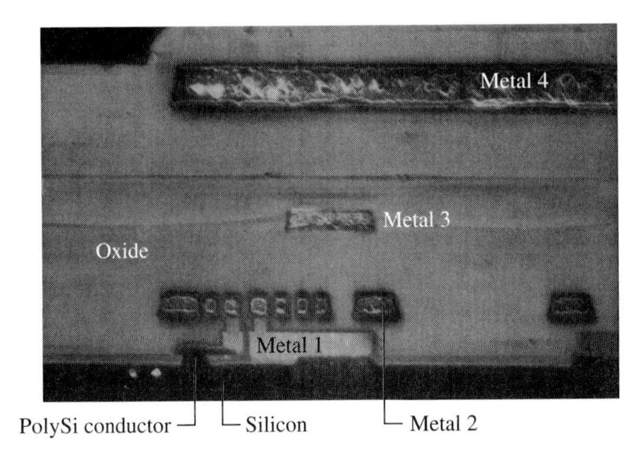

Figure C.18 A cross section of a chip with a polysilicon conductor layer at the surface, followed by four layers of metal, separated by layers of oxide. The oxide layers are planarized by chemical-mechanical polishing (CMP). (This photograph has been reprinted pursuant to a license granted by LSI Logic Corporation, © 2000, LSI Logic Corporation. ALL RIGHTS RESERVED.)

Oxide layers can also be deposited, rather than grown. Deposition techniques such as CVD or sputtering, however, do not produce as high-quality an oxide-silicon interface as thermal oxidation does.

Silicon dioxide is used to insulate the various metal layers from each other, as seen in cross section in Figure C.18. Here, there is a surface conductor layer of polysilicon, to which some metal is also connected (Metal 1). Metal 2 consists of several lines going into the page. Notice that some of the Metal 2 lines overlap the edges of the features in the layers below. To avoid the metal being deposited over a drop-off (like that seen in the surface oxide), oxide is grown over the lower layer, and then planarized by chemical-mechanical polishing (CMP). The ability to planarize layers has been an important development in the production of complex ICs.

Figure C.19a shows the region between two CMOS transistors. The drain of one is on the left, and the source of the other is on the right. A thick layer of oxide (called the *field oxide*) electrically isolates the two. Three polysilicon contacts running into the page can be seen (not connected to the two transistors) and the contact to the source on the right is made via a metal plug connecting the source directly to the Metal 1 layer. The surface of the chip was planarized before depositing the metal. Figure C.19b shows a completed integrated circuit device, including the lightly doped drain discussed in the Supplement to Part 3. The need for planarization can also be seen here as successive layers of oxide build up. This unit has been planarized after four layers of oxide have been deposited.

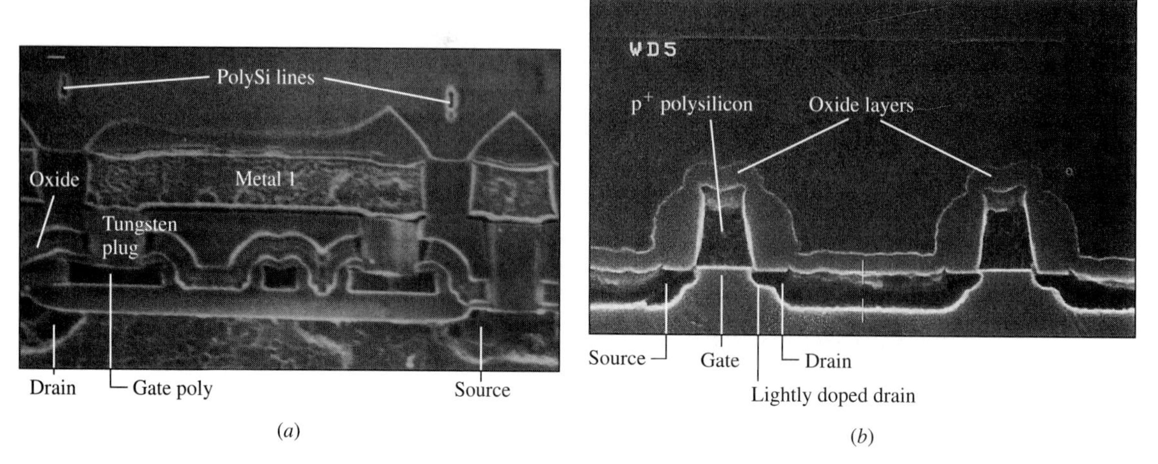

Figure C.19 (a) The region between two CMOS transistors. A field oxide is used to electrically isolate the two devices. The polysilicon lines running into the page are not connected to these two devices. Note the planarization used before depositing the Metal 1 layer. (b) A CMOS device. The gate oxide is too thin to be seen on this scale. The gate electrodes are polysilicon, and the junctions can be seen clearly. (These photographs have been reprinted pursuant to a license granted by LSI Logic Corporation, © 2000, LSI Logic Corporation. ALL RIGHTS RESERVED.)

A final comment about oxidation is that the reaction occurs quickly and easily—meaning that as soon as any silicon wafer is exposed to atmosphere, an oxide layer immediately starts to grow, even at room temperature. Thus, care must be taken when transferring wafers from one process to another, or else the oxide must be etched off before the next step is started.

C.5.4 Silicon Nitride

Another insulating material commonly used in a silicon fabrication process is silicon nitride, Si_3N_4. It is even denser than silicon dioxide, and thus even more impenetrable, which is helpful against rapid diffusers such as water and sodium. This is why Si_3N_4 is often used as a final passivation layer when the chip is complete, to protect it from the environment. Silicon nitride is also used during fabrication as a blocking layer against diffusions and implants.

Silicon nitride can be deposited by using dichlorosilane and ammonia in a liquid-phase chemical-vapor deposition (LPCVD) process:

$$3SiCl_2H_2 + 4NH_3 \rightarrow Si_3N_4 + 6HCl + 6H_2 \tag{C.9}$$

This process is conducted at high temperatures, between 700 and 800°C. When the nitride is to be used as a final passivation and protection layer, however, a high-temperature process can't be used—remember that whenever the temperature is raised, all of the dopants will diffuse. Even more important, the aluminum used in the metal contacts melts at 660°C.

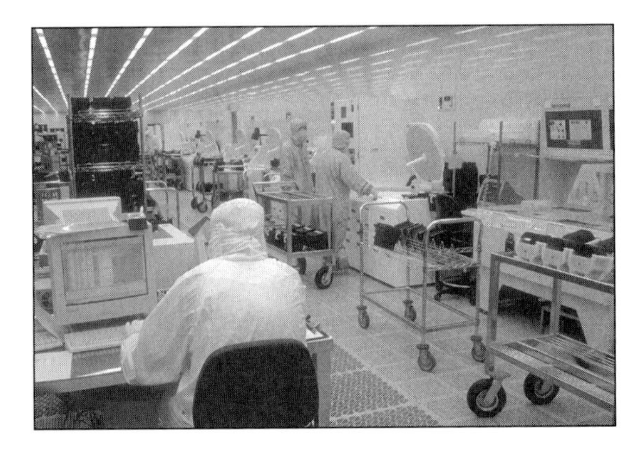

Figure C.20 A production clean room. All personnel must wear special clothing to prevent contaminating the wafers.

For passivation, then, a low-temperature process is needed, and plasma-enhanced CVD (PECVD) can be carried out with silane and ammonia in an argon plasma. In PECVD the RF energy is transferred to the reactants via the Ar ions and assists the reaction:

$$SiH_4 + NH_3 \rightarrow SiNH + 3H_2 \tag{C.10}$$

In the PECVD process, the silicon nitride film is not stoichiometric, and may have varying amounts of hydrogen.

C.6 CLEAN ROOMS

In all of the processing steps above, the need for cleanliness is absolute. The integrated circuits are all made in "clean rooms," which are specially designed to reduce contaminants. Huge filters clean the air continuously. The air pressure in the clean room is higher than that outside, so that when the door is opened, air will flow out and not in. Entry to a clean room is through an anteroom, and the door to the outside hall is closed before the door to the clean room is opened.

The greatest source of contamination is potentially from the workers. Fallen eyelashes, dead skin cells, and contaminants too small to see must be prevented from contacting the wafers. Thus, all clean room personnel wear "bunny suits" (Figure C.20), which cover their clothes, skin, hair, hands, shoes, etc. In this figure, the holes in the floor that help circulate clean air can be seen.

C.7 PACKAGING

Once the chip is complete, it must be packaged so it can be connected to the outside world. The integrated circuit is tested in wafer form, with a series of electrical probes that step and repeat across the wafer. Each circuit is tested, and those that fail are marked with ink.

The wafer is separated into individual chips. It can be sawn or diced. In the dicing process, the wafer is scribed by a diamond tip. The wafer is then placed on a soft backing, and gently rolled to snap apart the dice, cleaving the silicon crystal. The defective, (inked), devices are thrown away.

Next, the functioning parts are attached to a header. Usually, the back of the wafer serves as one of the contacts, so the mounting technique must provide electrical connection to the header. Typical examples are conductive epoxy, solder, or a eutectic bond. In the eutectic bond, gold is deposited onto the back of the wafer. The chip is pressed against the header and heated, and ultrasonic energy is applied. This forms a silicon-gold alloy and bonds the chip to the header.

C.7.1 Wire Bonding

Each chip usually has a series of bonding pads around the outside perimeter. These are large-area contacts (large by chip standards, perhaps 100 to 250 μm square), Figure C.21a). Very thin wires (thin by outside world standards, about 15 to 75 μm in diameter[3]) are bonded to these pads. These wires are used to make connections to the leads of the package.

The two main types of wire bonds are ball bonds and wedge bonds. In a ball bonder, the wire is brought to the bonding pad inside a capillary tube. The tube is heated, and the end of the wire is melted with an electric spark. The melted wire is pressed onto the pad, forming a ball and making good electrical contact with the pad, Figure C.21b. A wedge bond (Figure C.21c) is formed by using pressure combined with ultrasonic vibration (in lieu of melting). The ultrasound causes the metal of the wire to flow, and the pressure causes it to bond with the metallized contact pad. The vibration from the ultrasound also lets the wire *scrub* its way through any residual surface oxide.

Ball bonding machines can make many bonds in parallel, Figure C.21d.

C.7.2 Lead Frame

The other ends of the wire bonds go to the individual leads of the package. There are many types of packages, from TO-style cans used for individual transistors, to dual-in-line packages (DIPs), to ball grid array surface-mount packages for ultrahigh-density circuit boards.

Figure C.22a shows a TO can, typically used for discrete transistors. These come in various sizes and have different numbers of leads. Sometimes there is a window or lens in the can if the package is to be used for an LED, a laser, or a photodetector. For transistors and such, the can is solid metal.

The die is mounted on the header, and wire bonded to one or more leads. The ground contact is often made through the bottom of the chip, and the corresponding lead is in contact with the entire can. The other leads are electrically insulated from the surrounding can.

[3]Human hair is about 60 to 100 μm in diameter.

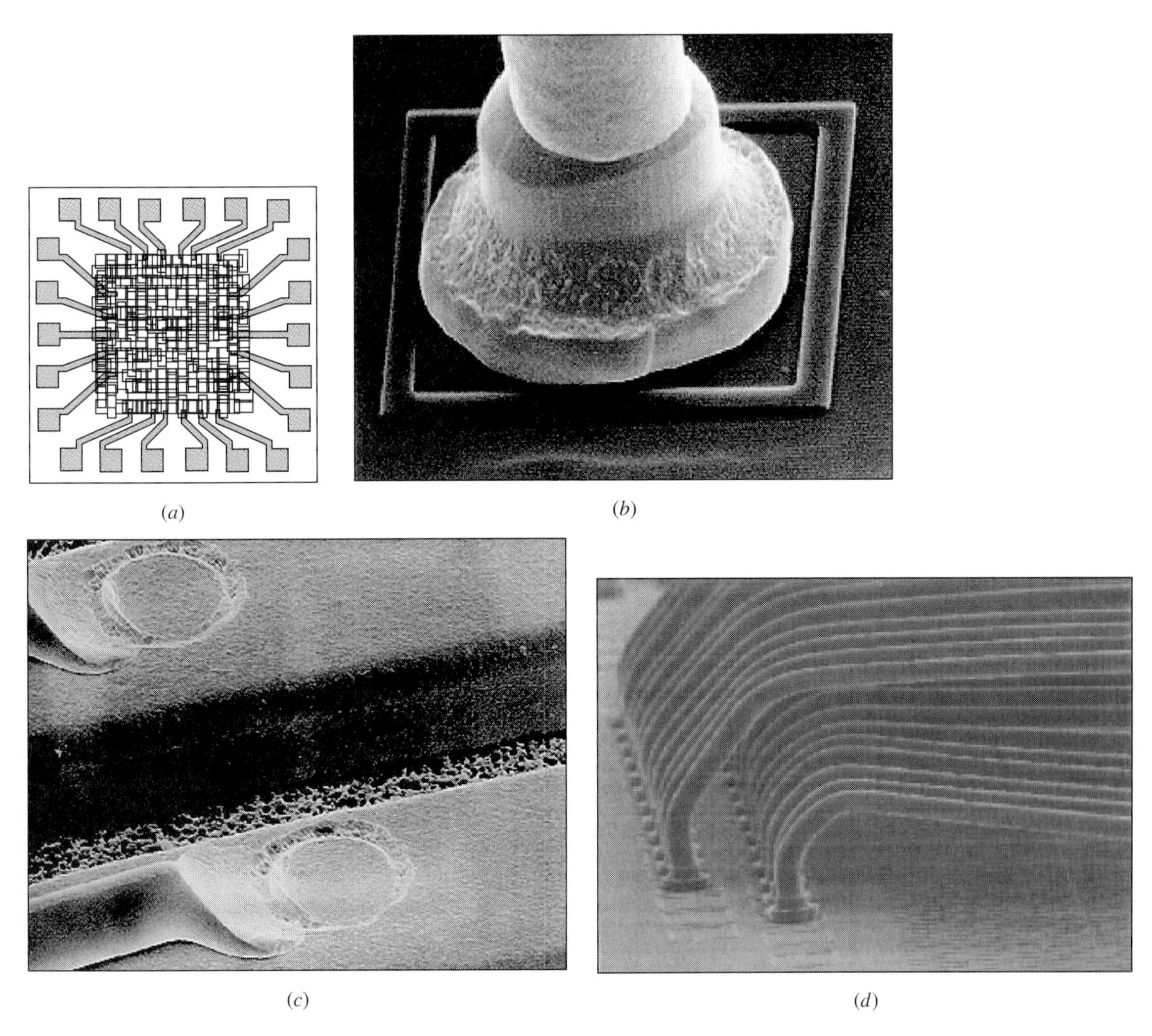

(*a*) (*b*) (*c*) (*d*)

Figure C.21 (a) A chip has bonding pads around its perimeter to which wire bonds can be made; (b) a ball bond; (c) a wedge bond; (d) an array of ball bonds. (Photos used with the permission of the Intersil Corporation.)

The familiar dual-in-line package of Figure C.22b contains a lead frame with various numbers of leads. The die is mounted on a support, and wire bonded to the leads as necessary. Then the package is sealed in epoxy (the common black package) or, for environmentally demanding situations, the package may be hermetically sealed under a metal cap.

C.7.3 Flip Chip

Sometimes instead of wire bonding a chip to a lead frame, a technology known as flip chip is used. Here, the header has a series of contact pads that exactly match the spacing of the contact pads on the chip, Figure C.23. On the chip, the

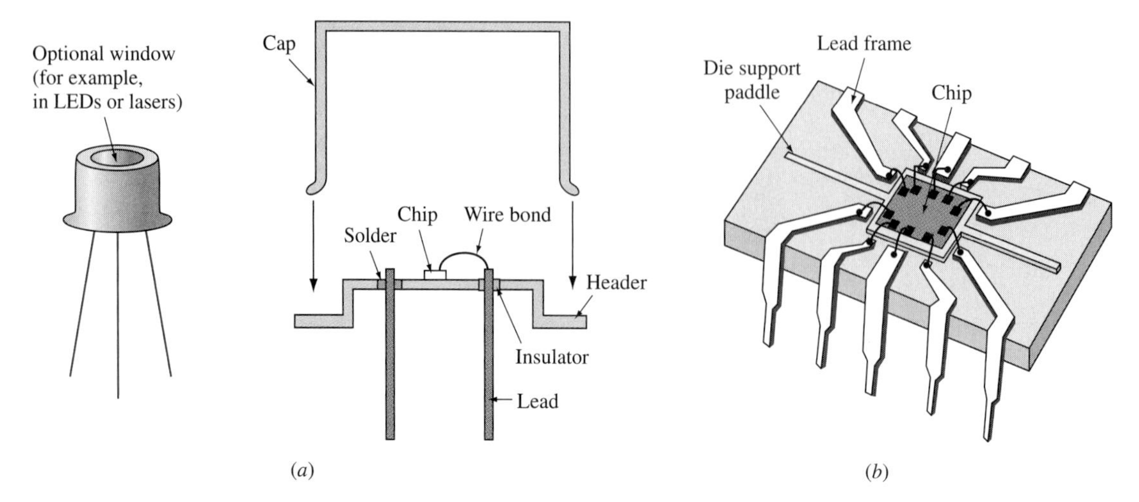

Figure C.22 (a) A TO can used for discrete devices (the window version might be used for LEDs or photodetectors); (b) the lead frame of a dual-in-line package (DIP).

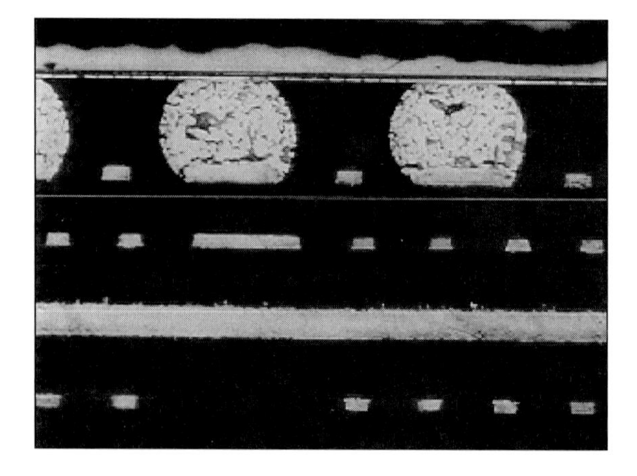

Figure C.23 In flip chip techonology, a header (top) has round contact pads that directly contact the top metal layer of the chip. (Used with the permission of the Intersil Corporation.)

contact pads each have a small ball of solder on them. The chip is then turned upside-down and aligned with the contacts on the packaged header. The advantage of this approach is speed—all the contacts are made in one step instead of requiring many wire bonds.

C.7.4 Surface-Mount Packages

The DIP package and TO can are just two examples of through-hole-mount packages, those having leads that go through holes on a circuit board. The other class

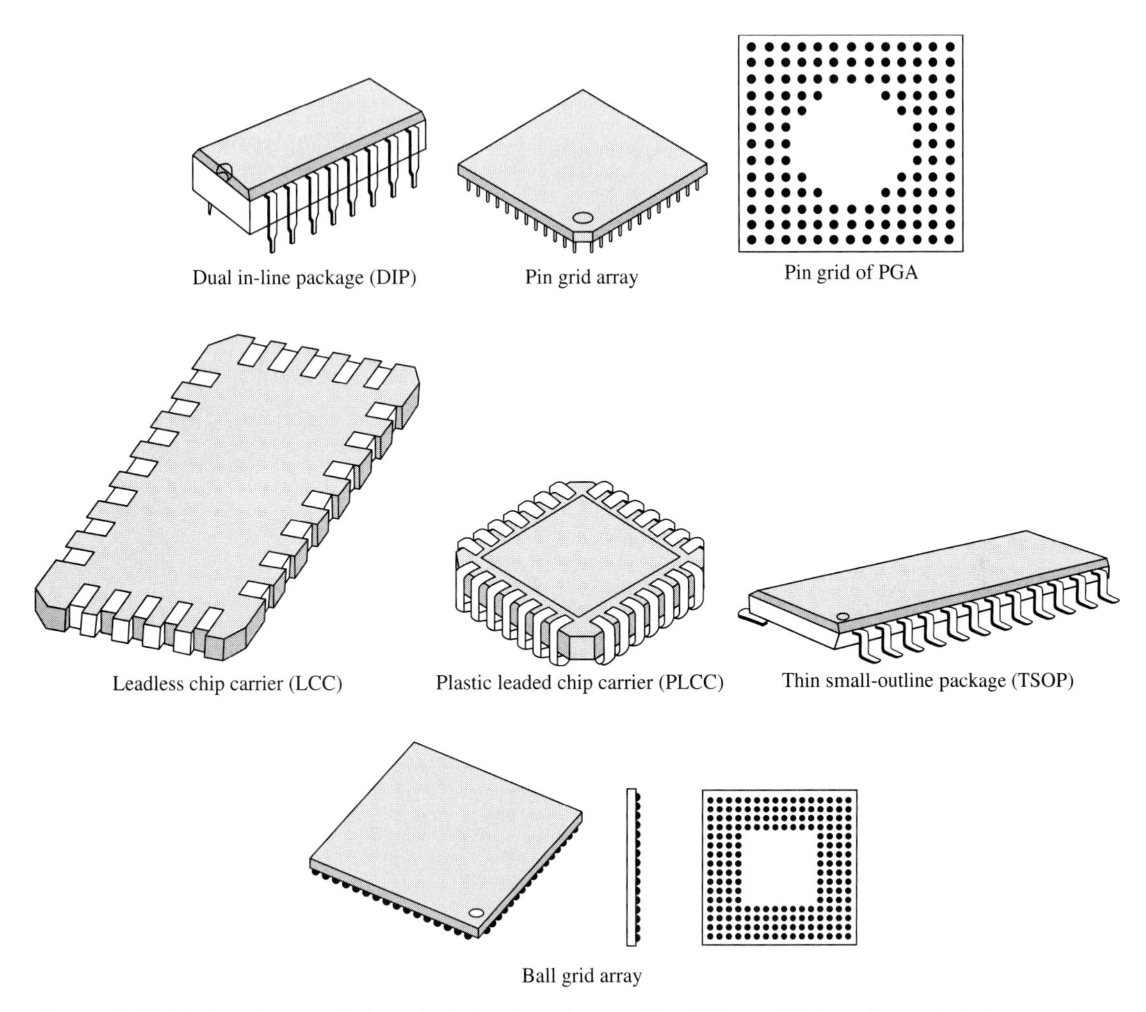

Figure C.24 Various types of integrated circuit packages. The DIP and PGA are through-hole-mount packages; the others mount onto the surface of the circuit board.

of packages is the surface-mount type, the leads of which are soldered or otherwise attached to contacts on the surface of the chip. Examples are the leadless chip carrier (LCC), the thin small-outline package (thin SOP), and the ball grid array (BGA), Figure C.24. The leadless chip carrier has contact pads around the edge of the device that contact directly to the circuit board. A variation is the leaded chip carrier, which has actual leads that bend underneath the package (J leads, shaped liked the letter J) and make contact that way. The thin outline small package (TSOP) has leads that bend outward (gull wings). The ball grid array makes for dense packaging.

C.8 SUMMARY

This has been a very brief outline of the growth, fabrication, and packaging processes used in modern semiconductor manufacturing. We have only scratched the surface, looking at crystal growth and preparation, doping, epitaxial growth, and patterning using photolithography. We saw that conductors on ICs can be metal or degenerately doped silicon, and that silicon dioxide makes for a very useful and convenient insulator. The packaging continues to evolve, and we showed a few of the many package styles that are available.

Semiconductor fabrication involves not just electrical engineers, but physicists, crystallographers, chemists, materials scientists, mechanical and packaging engineers, and many others.

APPENDIX D

Density-of-States Function, Density-of-States Effective Mass, Conductivity Effective Mass

D.1 INTRODUCTION

In this appendix, we derive the density-of-states function for electrons in the conduction band and in the valence band for semiconductors of interest. We then determine the density-of-states effective masses for semiconductors such as GaAs and Si. First, the density-of-states function for the free electron in one dimension is obtained. That is then extended to two and three dimensions. The free electron results in three dimensions are then obtained for the conduction band in a semiconductor with a single minimum of $K = 0$ (e.g., GaAs, InP, InAs), then for valence band electrons, and then for conduction band electrons with multiple equivalent minima.

D.2 FREE ELECTRONS IN ONE DIMENSION

Consider a region of one-dimensional space with length L, Figure D.1a. Let the potential energy E_P be constant everywhere in the region, so that as long as it is in the region, the electron may be considered to be free. We will use the approach of *periodic boundary conditions* in which the region is in the shape of a ring as indicated in Figure D.1b. Here the positions $x = 0$ and $x = L$ are the same.

The wave function for a free electron in one dimension is

$$\psi(x) = Ae^{jKx} \tag{D.1}$$

where A is a constant.

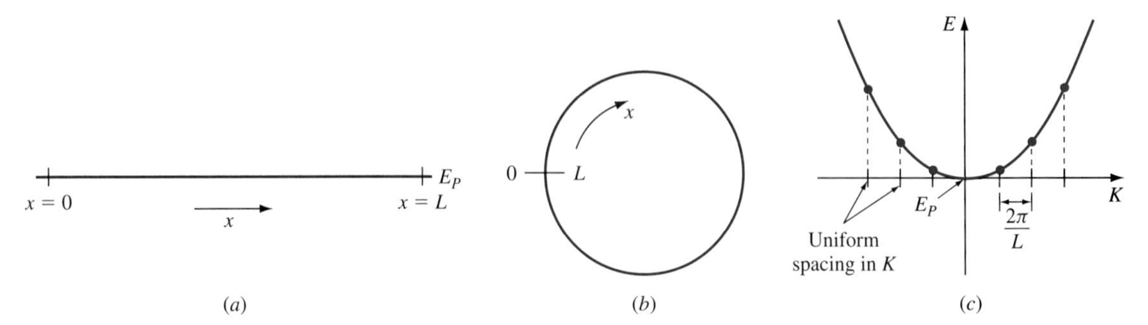

Figure D.1 Electron in a one-dimensional system. (a) The potential energy; (b) the system thought of as a ring, with the point L coinciding with the point $x = 0$ (periodic boundary conditions); (c) the allowed states in K and E.

The periodic boundary condition requires that

$$\psi(0) = \psi(L) \tag{D.2}$$

Then from Equation (D.1),

$$e^0 = e^{jKL} \tag{D.3}$$

which requires that $jKL = j2\pi n$ or

$$K_n = \frac{2\pi n}{L} \qquad (n = \pm 1, \pm 2, \ldots) \tag{D.4}$$

Since, for the free electron, the kinetic energy is

$$E_K = E - E_P = \frac{\hbar^2 K^2}{2m_0} \tag{D.5}$$

and since K is quantized, so are E_K and E.

Note that $n = 0$ is not a solution. Although $n = 0$ satisfies Equation (D.4), it is ruled out on physical grounds, since it implies that K and thus E_K is zero, or the electron is stationary.

The values of K and E for this case are shown in Figure D.1c. From Equation (D.4) the spacing (in K space) between solutions for K, is

$$\Delta K = \frac{2\pi}{L} \tag{D.6}$$

or there are $(L/2\pi) \times 2$ states per unit K. The factor of two arises from considering spin; i.e., at each value of K, there are two states of opposite spin. The density of states $S(K)$ in K space depends on L, the physical length of the space. Therefore, the number of states per unit length of real space (x) per unit K, is $[(L/2\pi) \times 2]/L$,

$$S(K) = \frac{1}{2\pi} \times 2 \tag{D.7}$$

and the states are distributed uniformly in K space.

It is of more interest to see how the states are distributed in energy. Since

$$E - E_P = \frac{\hbar^2 K^2}{2m_0}$$

$$K = \sqrt{\frac{2m_0(E - E_P)}{\hbar^2}} \tag{D.8}$$

and

$$dK = \frac{1}{\hbar}\sqrt{\frac{m_0}{2}} \times \frac{1}{\sqrt{E - E_P}}\, dE \tag{D.9}$$

Letting $S(E)$ be the density of states in energy, we can write

$$S(E)\, dE = S(K)\, dK \tag{D.10}$$

Then from Equations (D.7) and (D.9),

$$S(E)\, dE = \frac{1}{2\pi} \times 2 \times \frac{1}{\hbar}\sqrt{\frac{m_0}{2}} \times \frac{1}{\sqrt{E - E_P}}\, dE$$

or

$$S(E) = \frac{1}{\pi\hbar}\sqrt{\frac{m_0}{2}} \times \frac{1}{\sqrt{E - E_P}} \qquad \text{free electron in 1-D} \tag{D.11}$$

D.3 FREE ELECTRONS IN TWO DIMENSIONS

For the two-dimensional case, we take a two-dimensional region of length L_x and width L_y, again with E_P constant. The wave function for this case is

$$\psi(x, y) = Ae^{j(K_x x + K_y y)} \tag{D.12}$$

Using the periodic boundary conditions

$$\begin{aligned} \psi(0, y) &= \psi(L_x, y) \\ \psi(x, 0) &= \psi(x, L_y) \end{aligned} \tag{D.13}$$

gives

$$e^{j0} = 1 = e^{jK_x L_x}$$

$$e^{j0} = 1 = e^{jK_y L_y}$$

or

$$\begin{aligned} K_x &= \frac{2\pi n_x}{L_x} \qquad (n_x = \pm 1, \pm 2, \ldots) \\ K_y &= \frac{2\pi n_y}{L_y} \qquad (n_y = \pm 1, \pm 2, \ldots) \end{aligned} \tag{D.14}$$

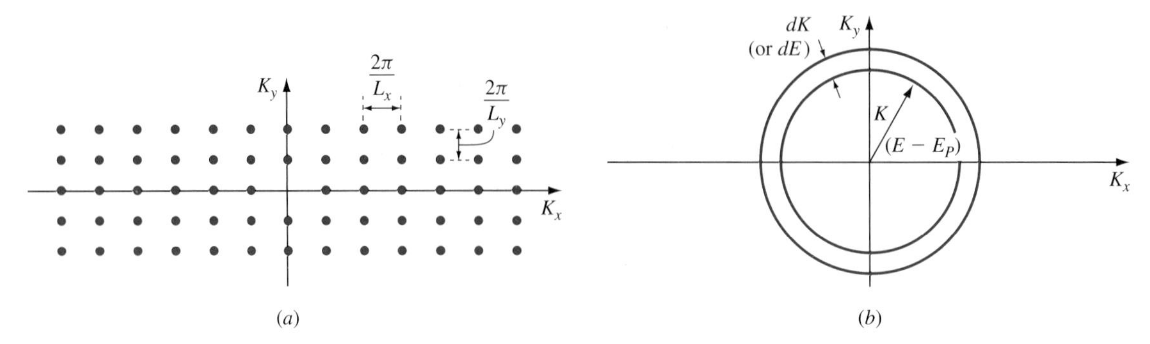

Figure D.2 The free electron model in two dimensions. (a) The states are distributed evenly in K space with separation $\Delta K_x = 2\pi/L_x$ and $\Delta K_y - 2\pi/L_y$. (b) The number of states in the circular ring of radius K and thickness dK are calculated.

In two-dimensional K space, the states are uniformly distributed in both K_x and K_y as shown in Figure D.2a with spacing $\Delta K_x = 2\pi/L_x$ and $\Delta K_y = 2\pi/L_y$. The density of states in K space is the number of states per unit area in K space per unit area of physical space L_x, L_y, or the number of states in the area in K_x, K_y space:

$$S(K_x, K_y) = \left(\frac{1}{2\pi}\right)^2 \times 2 \tag{D.15}$$

The number of states in the area of the region of radius K (or E) and width dK (or dE) as shown in Figure D.2b is

$$S(K)\,dK = S(K_x, K_y) \times 2\pi K\,dK$$

Since regions of constant energy are circles in K space, to express the density of states in terms of energy rather in terms of K, we use Equations (D.10), (D.8), and (D.9):

$$S(E)\,dE = \left(\frac{1}{2\pi}\right)^2 \times 4\pi K \sqrt{\frac{2m_0(E - E_P)}{\hbar^2}} \times \frac{1}{\sqrt{E - E_P}}\,dE$$

$$S(E) = \frac{m_0}{\pi\hbar^2} \qquad \text{free electron in 2-D} \tag{D.16}$$

Note that for the two-dimensional case, $S(E)$ is independent of E.

D.4 FREE ELECTRONS IN THREE DIMENSIONS

Consider now a volume of space of dimensions L_x, L_y, and L_z. The wave function is

$$\psi(x, y, z) = Ae^{j(K_x x + K_y y + K_z z)} \tag{D.17}$$

Again, for periodic boundary conditions,

$$
\begin{aligned}
K_x &= \frac{2\pi n_x}{L_x} \qquad (n_x = \pm 1, \pm 2, \ldots) \\
K_y &= \frac{2\pi n_y}{L_y} \qquad (n_y = \pm 1, \pm 2, \ldots) \\
K_z &= \frac{2\pi n_z}{L_z} \qquad (n_z = \pm 1, \pm 2, \ldots)
\end{aligned}
\tag{D.18}
$$

Then per unit volume in K space there are

$$\frac{L_x}{2\pi} \times \frac{L_y}{2\pi} \times \frac{L_z}{2\pi} \times 2$$

states. The number of states in K space per unit volume is

$$S(K_x, K_y, K_z) = \left(\frac{1}{2\pi}\right)^3 \times 2$$

The number of states per unit volume in a region of dK or energy dE in this case is contained in a spherical shell of radius K (or E) and thickness dK (or dE) as indicated in Figure D.3. Then

$$S(K)\,dK = S(K_x, K_y, K_z) \times 4\pi K^2\,dK$$

and

$$S(E)\,dE = \left(\frac{1}{2\pi^2}\right) \times \left(\frac{2m_0}{\hbar^2}\right)^{3/2} \times (E - E_P)^{1/2}\,dE$$

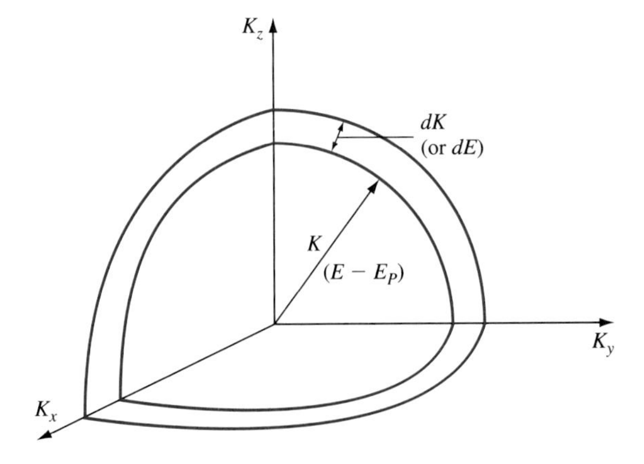

Figure D.3 Spherical surfaces of constant K and constant energy in K space are used to find the three-dimensional density of states.

or

$$S(E) = \frac{1}{2\pi^2}\left(\frac{2m_0}{\hbar^2}\right)^{3/2}(E - E_P)^{1/2} \qquad \text{free electron in 3-D} \tag{D.19}$$

D.5 QUASI-FREE ELECTRONS IN A PERIODIC CRYSTAL

Next we consider an electron in a crystal, where the potential energy is not constant but rather is periodic, with the periodicity of the lattice. The electron wave function in a periodic crystal is a Bloch wave:

$$\psi(x, y, z) = U_K(x, y, z)e^{j(K_x x + K_y y + K_z z)}$$

where the Bloch function U_K has the form

$$U_k(x, y, z) = U_K(x + a, y + b, z + c)$$

where a, b, and c are the lattice constants. Then proceeding as for a free electron, results identical to Equations (D.11), (D.16), and (D.19) are obtained, with the exception that an effective mass m^* is substituted for the free electron mass m_0. However, the effective mass can be a function of direction, and a single number cannot be used. In practice some sort of average is usually employed. In the next section we will derive expressions for one such average, the density-of-states effective mass m^*_{ds} for some cases of interest.

D.6 DENSITY-OF-STATES EFFECTIVE MASS

For the free electron case, the density-of-states function of Equation (D.19) is given in terms of the mass of the free electron. It is convenient to express the density-of-states function for a three-dimensional crystal in the same form as Equation (D.19) but with the free electron mass m_0 replaced by the density-of-states effective mass m^*_{ds}. This, in effect, defines the density-of-states effective mass.

$$S(E) = \frac{1}{2\pi^2}\left(\frac{2m^*_{ds}}{\hbar^2}\right)^{3/2}(E - E_P)^{1/2} \tag{D.20}$$

As mentioned earlier, the density-of-states effective mass is some sort of average of the effective masses for electrons traveling in different directions, for the case of the conduction band, or an average of the effective masses for holes in the various overlapping valence bands.

Since $S(E)$ for the free electron was obtained with the E-K curve being a parabola, Equation (D.20) and thus the expression for m^*_{ds} will be valid only in the parabolic regions of the E-K curves for a given semiconductor.

D.6.1 Case 1: Conduction Band with a Single Minimum at $K = 0$

As discussed in Chapter 2, for a semiconductor whose conduction band has a single minimum at $K = 0$, near the band minimum the E-K curve is parabolic with the same curvature in all directions in the crystal. As indicated then, the effective mass m^* is related to the curvature, so for this case m^* is the same in every direction in K space. Therefore, in this region the procedure for free electrons can be followed with m_0 replaced by m^*.

$$S(E) = \frac{1}{2\pi^2}\left(\frac{2m^*}{\hbar^2}\right)^{3/2}(E - E_C)^{1/2} = \frac{1}{2\pi^2}\left(\frac{2m^*_{dse}}{\hbar^2}\right)^{3/2}(E - E_C)^{1/2} \quad \text{(D.21)}$$

where $m^*_{dse} = m^*$ is scalar and single valued. The E-K plot and $S(E_K)$-E plots are shown in Figure D.4 for this case. Part (a) shows the E-K diagram for the region near the bottom of the band, and (b) gives the density-of-states function. Note that the electron potential energy in this case is E_C.

D.6.2 Case 2: Valence Band with Two Bands Having Maxima at E_V and at $K = 0$

This is the case for most semiconductors of interest for electronic devices. The E-K diagram for this case in the region near the top of the valence band is shown in Figure D.5a. To first approximation the E-K curve is parabolic for each band and with curvature independent of direction of K, and thus in each valence band, m^* is scalar and single valued. However, since the curvatures of the two bands are different, so are their effective masses:

$$m^* = \hbar^2\left(\frac{d^2E}{dK^2}\right)^{-1}$$

Associating m^*_h with the band of smaller curvature and m^*_l with the band of higher curvature as indicated, the density of states for each band can be obtained

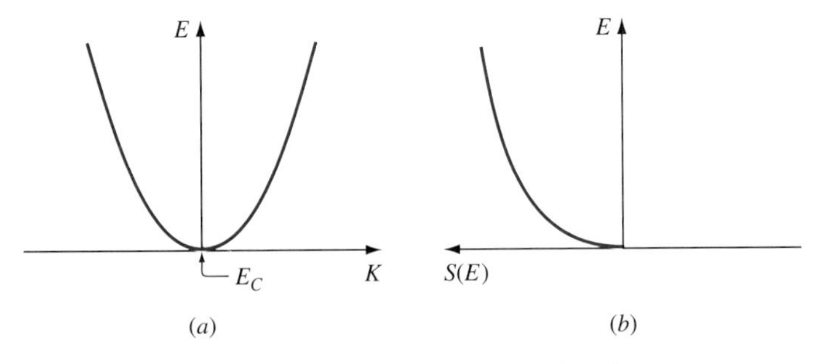

Figure D.4 (a) Energy–wave vector relation for electrons near the bottom of the conduction band; (b) the corresponding density-of-states function.

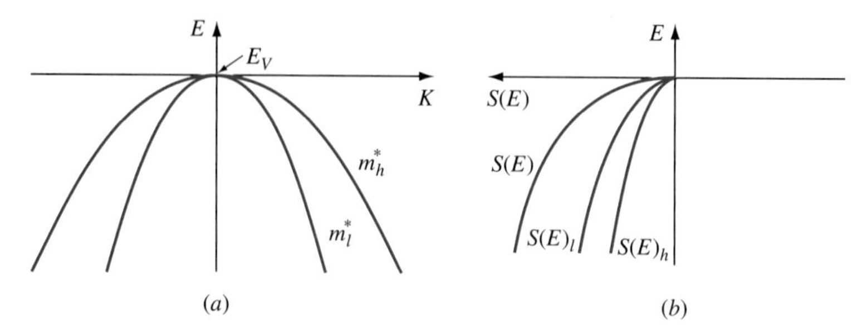

Figure D.5 (a) Energy–wave vector relations for light and heavy holes; (b) corresponding density-of-states functions. Also shown is the total density-of-states function for holes $S(E)$.

by the procedure for the free electron, resulting in:

$$S(E) = \frac{1}{2\pi^2}\left(\frac{2m_h^*}{\hbar^2}\right)^{3/2}(E_V - E)^{1/2} + \frac{1}{2\pi^2}\left(\frac{2m_l^*}{\hbar^2}\right)^{3/2}(E_V - E)^{1/2}$$

$$S(E) = \frac{1}{2\pi^2}\left(\frac{2m_{dsh}^*}{\hbar^2}\right)^{3/2}(E_V - E)^{1/2} \tag{D.22}$$

where

$$(m_{dsh}^*)^{3/2} = (m_h^*)^{3/2} + (m_l^*)^{3/2} \tag{D.23}$$

or

$$m_{dsh}^* = [(m_h^*)^{3/2} + (m_l^*)^{3/2}]^{2/3} \tag{D.24}$$

The symbol m_{dsh}^* indicates that this value is used to calculate the hole concentration in the valence band. In Figure D.5b the density-of-states function for light holes, for heavy holes, and total density-of-states function for holes are indicated. Again this expression for m_{dsh}^* is only valid in the region where both bands can be approximated as parabolic.

D.6.3 Case 3: Conduction Band Has Multiple Equivalent Minima at $K \neq 0$ (e.g., Si, Ge, GaP)

We next consider a case in which the conduction band has equivalent minima, but not at $K = 0$. To be specific, we choose the case of Si, which has equivalent minima in each of the $\langle 100 \rangle$ directions. We find the density-of-states function associated with one minimum and multiply this value by the number of equivalent minima (six in this case).

Let us choose the minimum in the z or $\langle 001 \rangle$ direction as indicated in Figure D.6. For this minimum the energy can be expressed

$$E = E_C + \hbar^2\left[\frac{K_x^2}{2m_\perp^*} + \frac{K_y^2}{2m_\perp^2} + \frac{(K_z - K_{z0})^2}{2m_\parallel^2}\right] \tag{D.25}$$

where the minimum is at $K_x = 0$, $K_y = 0$ and $K_z = K_{z0}$. The quantities $m_\perp$ and

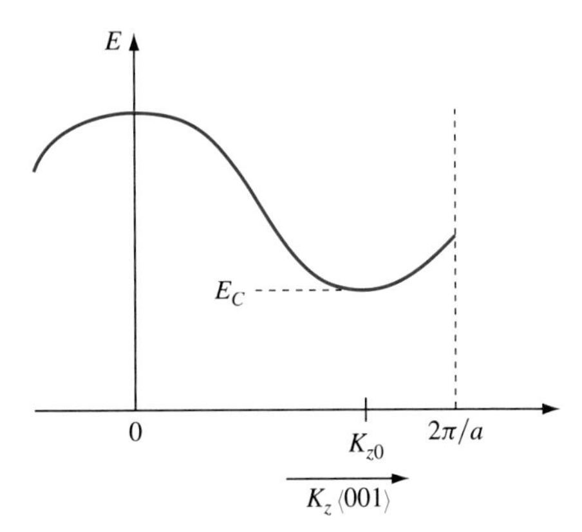

Figure D.6 The *E-K* diagram for the conduction band of silicon.

$m_{\parallel}$ are the effective masses for electrons traveling perpendicular and parallel to the z axis, respectively. Making a change of variables, we let

$$
\begin{aligned}
K_x' &= K_x\sqrt{\frac{m_0}{m_\perp^*}} \\
K_y' &= K_y\sqrt{\frac{m_0}{m_\perp^*}} \\
K_z' &= (K_z - K_{z0})\sqrt{\frac{m_0}{m_\parallel^*}}
\end{aligned}
\tag{D.26}
$$

The energy can be expressed as

$$
E - E_C = \hbar^2\frac{[(K_x')^2 + (K_y')^2 + (K_z')^2]}{2m_0} = \frac{\hbar^2(K')^2}{2m_0}
\tag{D.27}
$$

or the surface of constant energy is a sphere in K' space.

However, since the density of states is uniform in K space,

$$
S(K)\,dK_x\,dK_y\,dK_z = \left(\frac{1}{2\pi}\right)^3 \times 2\,dK_x\,dK_y\,dK_z
$$

From Equation (D.26)

$$
\begin{aligned}
dK_x &= \sqrt{\frac{m_\perp^*}{m_0}}\,dK_x' \\
dK_y &= \sqrt{\frac{m_\perp'}{m_0}}\,dK_y' \\
dK_z &= \sqrt{\frac{m_\parallel^*}{m_0}}\,dK_z'
\end{aligned}
\tag{D.28}
$$

Then following the procedure used for the free electron, but in K' space, yields

$$S(E) = \left(\frac{1}{2\pi}\right)^3 \times 2\frac{(m_\perp^* m_\perp^* m_\parallel^*)}{\hbar^3}(E - E_C) \tag{D.29}$$

This is the density of states for a single minimum. For g equivalent minima,

$$S(E) = g\left(\frac{1}{2\pi}\right)^3 \times 2\frac{(m_\perp^* m_\perp^* m_\parallel^*)}{\hbar^3}(E - E_C) \tag{D.30}$$

which has the form of Figure D.4b.

Putting Equation (D.30) into the form of Equation (D.21), we can find

$$m_{dse}^* = g^{2/3}[(m_\perp^*)^2(m_\parallel^*)]^{1/3} \tag{D.31}$$

D.7 CONDUCTIVITY EFFECTIVE MASS

The density-of-states effective mass represents one method of averaging the effective masses for different directions in the crystal. The conductivity effective mass represents another method.

From the free electron model, the conductivity is

$$\sigma = \frac{nq^2\bar{t}}{m_0} \tag{D.32}$$

We wish to express the conductivity of electrons or holes in the form

$$\sigma = \frac{nq^2\bar{t}}{m_C^*} \tag{D.33}$$

where m_C^* is the conductivity effective mass for an electron in a crystal.

D.7.1 Case 1: Single Minimum in the Conduction Band at $K = 0$

As discussed in Section D.6.1, the effective mass m_e^* for electrons in this minimum is isotropic and single valued, so

$$m_{ce}^* = m_e^* \tag{D.34}$$

D.7.2 Case 2: Holes in the Valence Band

Here the effective mass in each band is isotropic and

$$\sigma_p = q^2\bar{t}\left(\frac{p_l}{m_l^*} + \frac{p_h}{m_h^*}\right) = q^2\bar{t}\left(\frac{p}{m_{ch}^*}\right) \tag{D.35}$$

where it is assumed that the hole scattering time is the same in each band. The total hole concentration is p, while the concentration in the light-hole band is p_l and that in the heavy-hole band is p_h.

For nondegenerate semiconductors, From Equations (2.60) and (2.62),

$$p_l = 2\left(\frac{m_l^* kT}{2\pi\hbar^2}\right)^{3/2} e^{-(E_f - E_V)/kT}$$

$$p_h = 2\left(\frac{m_h^* kT}{2\pi\hbar^2}\right)^{3/2} e^{-(E_f - E_V)/kT} \tag{D.36}$$

$$p = p_l + p_h = 2\left(\frac{m_{dsh}^* kT}{2\pi\hbar^2}\right)^{3/2} e^{-(E_f - E_V)/kT}$$

Solving for m_{ch}^* with the aid of Equation (D.35) yields

$$m_{ch}^* = \frac{m_{dsh}^{*\ 3/2}}{m_l^{*1/2} + m_h^{*1/2}} = \frac{m_l^{*3/2} + m_h^{*3/2}}{m_l^{*1/2} + m_h^{*1/2}} \tag{D.37}$$

D.7.3 Case 3: Electrons in Conduction Band with Multiple Equivalent Minima

For the case of multiple equivalent minima in the conduction band, the effective mass, and thus the conductivity for electrons in any single minimum, is a tensor, i.e., depends on the direction of motion. To determine the total conductivity, the conductivity is summed over all the directions.

To derive an expression for conductivity and thus of conductivity effective mass requires the use of tensor analysis. Here the results are simply stated for semiconductors of interest.[1]

1. Although the conductivity in any one minimum is a tensor, for a cubic structure the sum of the conductivities is a scalar.
2. For a semiconductor whose conduction band minima lie along the coordinate axis in K space (e.g., Si, Ge), the conductivity can be expressed

$$\sigma = \frac{q^2 n \bar{t}}{m_{ce}^*} \tag{D.38}$$

where

$$\frac{1}{m_{ce}^*} = \frac{1}{3}\left(\frac{2}{m_\perp^*} + \frac{1}{m_\parallel^*}\right) \tag{D.39}$$

D.7.4 Case 4: Strained Silicon

The above discussion of conductivity effective mass considered the common semiconductors in which the materials have cubic symmetry. It is found, however,

[1]For a derivation, see for example, John P. McKelvey, *Solid State and Semiconductor Physics,* Harper and Row, New York, 1966, Chapter 9.

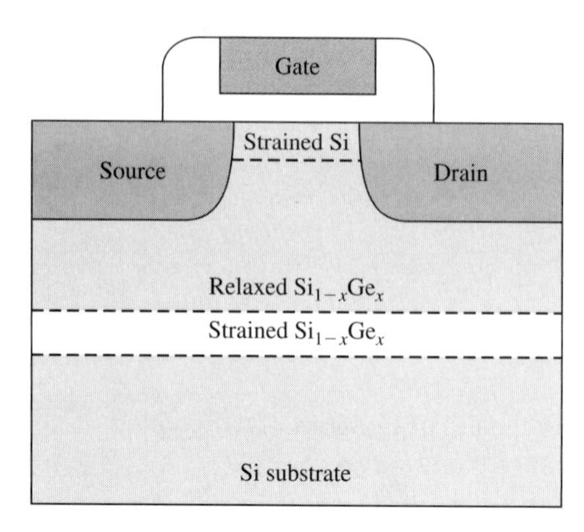

Figure D.7 One technology for producing strained silicon to enhance the low-field mobility for carriers in a MOSFET.

that under mechanical stress the conductivity effective mass of both holes and electrons in Si can be decreased, thus increasing their low-field mobilities. This approach to speeding up device operation appears promising for CMOS applications. There are a number of approaches for producing strained Si for the channels in MOSFETs. We briefly describe one such method, which is compatible with standard CMOS technology.

A $Si_{1-x}Ge_x$ (SiGe) alloy[2] is grown epitaxially onto a Si substrate (See Figure D.7). Epitaxy is discussed in Appendix C, and involves growing layers of one material on top of another atom by atom. The atoms must line up, but if there is a difference in lattice constant, one material is effectively stretched or compressed. The SiGe alloy is thick enough that its lattice constant relaxes gradually with position from that of the Si substrate to the natural atomic spacing of the SiGe alloy. (The SiGe alloy has a slightly larger lattice constant than the Si.) This relieves the strain induced by the difference in the lattice constants of the alloy and the Si substrate. The SiGe alloy then acts as the effective substrate for subsequent growth. A thin ($\sim$20 nm) layer of Si is then grown onto this SiGe layer. Because of the difference in the lattice constants, this Si layer, which will be the channel region in the MOSFETs, is subjected to a tensile stress as indicated in Figure D.8. This strain increases the low-field mobility as discussed below. Note that the Si unit cell in this strained layer is no longer cubic.

Hole Mobility The effect of the strain on the mobility is to split the degeneracy of the two overlapping valence bands of Figure D.5a such that the band for the lower effective mass is at a higher energy than that of the higher effective mass, as indicated in Figure D.9. Since the holes occupy the higher energies, the conductivity effective mass is reduced and the hole mobility is increased.

[2]With x typically in the 15 to 30 percent range.

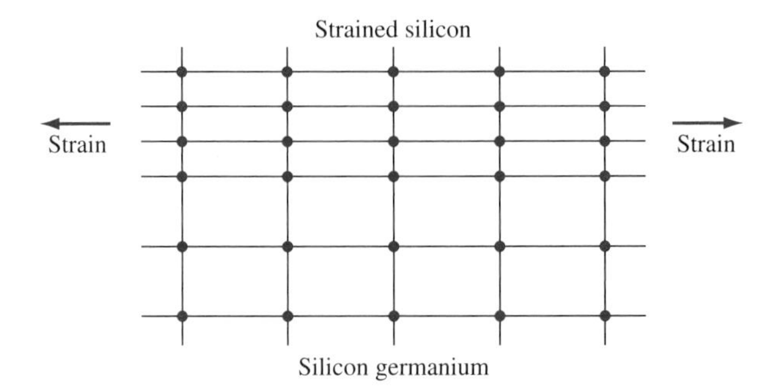

Figure D.8 Illustration (not to scale) of silicon deposited on a SiGe alloy. Because the Si lattice constant is smaller than that of SiGe, tensile strain is induced in the silicon layer.

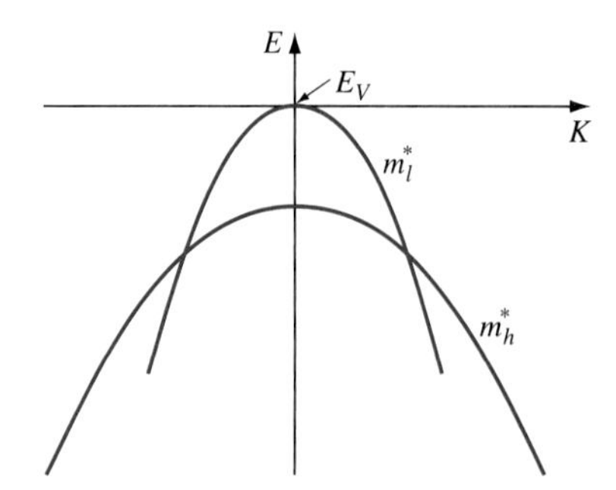

Figure D.9 The two degenerate valence bands in unstrained silicon are split because of strain.

Electrons Recall that the conduction band minima lie in the ⟨001⟩ directions. In each minimum, the electron effective mass is a tensor, but for the cubic structure all minima are at the same energy. Thus the resultant conductivity effective mass, and therefore the mobility, is a scalar. Under tensile stress, however, the minima in different crystalline directions are split in energy. For Si grown on a (110) SiGe surface, the 6-fold degeneracy (12-fold including spin) reduces energy of the conduction band edge and thus reduces the electron mobility in the ⟨001⟩ direction.

Improvements in low-field mobility in excess of 50 percent have been observed for both electrons and holes in strained Si compared with nonstrained Si.

D.8 SUMMARY OF COMMON RESULTS FOR EFFECTIVE MASS

For a single extremum at $K = 0$, there is one scalar effective mass:

$$m^* = m^*_{ds} = m^*_c$$

This usually occurs in the conduction band, so

$$m^*_e = m^*_{dse} = m^*_{ce}$$

For two valence bands degenerate at $K = 0$ there are two scalar effective masses, m_l^* and m_h^*:

$$m^*_{dsh} = [(m^*_h)^{3/2} + (m^*_l)^{3/2}]^{2/3}$$

$$m^*_{ch} = \frac{m_l^{*3/2} + m_h^{*3/2}}{m_l^{*1/2} + m_h^{*1/2}}$$

For g equivalent conduction band minima at $K \neq 0$, but along principal crystallographic directions in K space, the effective mass in a single minimum is a tensor with components $m^*_\perp$ and $m^*_\parallel$. However, for a cubic structure, summing over all equivalent minima, the effective masses are scalars:

$$m^*_{dse} = g^{2/3}[(m^*_\perp)^2(m^*_\parallel)]^{1/3}$$

$$\frac{1}{m^*_{ce}} = \frac{1}{3}\left(\frac{2}{m^*_\perp} + \frac{1}{m^*_\parallel}\right)$$

For extrema in an arbitrary position in K space, the results are more complicated.

For noncubic structures, such as strained Si, the carrier mobility can be increased over that of cubic Si.

E APPENDIX

Some Useful Integrals

$$y_n = \int_0^\infty x^n e^{-ax^2}\, dx$$

for $a > 0$

n even	n odd
$y_0 = \frac{1}{2}\sqrt{\frac{\pi}{a}}$	$y_1 = \frac{1}{2a}$
$y_2 = \frac{1}{4}\sqrt{\frac{\pi}{a^3}}$	$y_3 = \frac{1}{2a^3}$
$y_4 = \frac{3}{8}\sqrt{\frac{\pi}{a^5}}$	$y_5 = \frac{1}{a^3}$

$$y_n = \int_0^\infty x^n e^{-ax}\, dx = \frac{\Gamma(n+1)}{a^{n+1}}$$

for $n > -1$, $a > 0$

$y_0 = \frac{1}{a}$	$y_{-1/2} = \sqrt{\frac{\pi}{a}}$
$y_1 = \frac{1}{a^2}$	$y_{1/2} = \frac{1}{2}\sqrt{\frac{\pi}{a^3}}$
$y_2 = \frac{2}{a^3}$	$y_{3/2} = \frac{3}{4}\sqrt{\frac{\pi}{a^5}}$

$\Gamma_{(n+1)} = n!$ if n is an integer

$\Gamma_{1/2} = \sqrt{\pi}$

$\Gamma_{(n+1)} = n\Gamma_n$ if $n > 0$

$$\int x^m e^{ax}\, dx = \frac{x^m e^{ax}}{a} - \frac{m}{a}\int x^{m-1} e^{ax}\, dx$$

F APPENDIX

Useful Equations

GENERAL PHYSICS

$$\mathscr{E} = -\frac{dV}{dx}$$

$$C = \frac{\varepsilon A}{t}$$

$$\Psi(x, t) = U_K(x)e^{j[Kx-(E/\hbar)t]}$$

$$F = -\nabla E_P = -\frac{dE_P}{dr}$$

SEMICONDUCTOR MATERIALS

$$\gamma = E_{\text{vac}} - E_V \qquad \text{non-degenerate}$$

$$\chi = E_{\text{vac}} - E_C$$

$$E_g = E_C - E_V$$

$$n_0 = N_C e^{-(E_C - E_f)/kT}$$

$$p_0 = N_V e^{-(E_f - E_V)/kT}$$

$$n_i^2 = N_C N_V e^{-E_g/kT}$$

$$n_0 p_0 = n_i^2$$

$$E_g = kT \ln \frac{N_C N_V}{n_i^2}$$

$$\delta_n = E_C - E_f = kT \ln \frac{N_C}{N_D'}$$

$$\delta_p = E_f - E_V = kT \ln \frac{N_V}{N_A'}$$

$$J_n = q\mu_n n\mathscr{E} + qD_n\frac{dn}{dx} = q\mu_n\left(n\mathscr{E} + \frac{kT}{q}\frac{dn}{dx}\right)$$

$$J_p = q\mu_p p\mathscr{E} - qD_p\frac{dp}{dx} = q\mu_p\left(p\mathscr{E} - \frac{kT}{q}\frac{dp}{dx}\right)$$

$$\mathscr{E} = \frac{1}{q}\frac{dE_{\text{vac}}}{dx}$$

$$\Delta E_g^* \approx \Delta E_c^* = E_{c0} - E_c^* \qquad \text{degenerate } n \text{ type}$$

$$n_0 p_0 = \frac{N_D}{N_C}\, e^{\Delta E_g^*/kT} n_i^2 \qquad \text{degenerate } n \text{ type}$$

$$n_0 p_0 = \frac{N_A}{N_V}\, e^{\Delta E_g^*/kT} n_i^2 \qquad \text{degenerate } p \text{ type}$$

JUNCTIONS

$$qV_{\text{bi}} = |\Phi_p - \Phi_n|$$

$$V_{\text{bi}} = \frac{kT}{q}\ln\frac{N_D' N_A'}{n_i^2} \qquad \text{pn junction}$$

$$V_{\text{bi}} = \frac{kT}{q}\ln\frac{N_V N_D'}{n_i^2} \qquad \text{p}^+\text{n junction}$$

$$V_{\text{bi}} = \frac{kT}{q}\ln\frac{N_C N_A'}{n_i^2} \qquad \text{n}^+\text{p junction}$$

$$qV_{\text{bi}} = E_g - (\delta_n + \delta_p)$$

$$V_j^n = \frac{qN_D'}{2\varepsilon} w_n^2$$

$$V_j^p = \frac{qN_A'}{2\varepsilon} w_p^2$$

$$w_n = (x_0 - x_n) = \left(\frac{2\varepsilon V_j^n}{qN_D'}\right)^{1/2} = \left[\frac{2\varepsilon V_j}{qN_D'\left(1 + \frac{N_D'}{N_A'}\right)}\right]^{1/2}$$

$$w_p = (x_p - x_0) = \left(\frac{2\varepsilon V_j^p}{qN_A'}\right)^{1/2} = \left[\frac{2\varepsilon V_j}{qN_A'\left(1 + \frac{N_A'}{N_D'}\right)}\right]^{1/2}$$

$$w = w_n + w_p = \left[\frac{2\varepsilon V_j (N_A' + N_D')}{qN_A' N_D'}\right]^{1/2} \qquad \text{pn junction}$$

$$w = \left(\frac{2\varepsilon V_j}{qN'_A}\right)^{1/2} \qquad \text{n}^+\text{p junction}$$

$$w = \left(\frac{2\varepsilon V_j}{qN'_D}\right)^{1/2} \qquad \text{p}^+\text{n junction}$$

$$V_j = \frac{qN'_D N'_A w^2}{2\varepsilon(N'_D + N'_A)} \qquad \text{pn junction}$$

$$n_{p0} = N'_D e^{-qV_{\text{bi}}/kT} \qquad \text{non-degenerate}$$

$$p_{n0} = N'_A e^{-qV_{\text{bi}}/kT}$$

$$n_p(x_p) = n_{p0} e^{qV_a/kT}$$

$$p_n(x_n) = p_{n0} e^{qV_a/kT}$$

$$\Delta n_p(x_p) = n_{p0}(e^{qV_a/kT} - 1)$$

$$\Delta p_n(x_n) = p_{n0}(e^{qV_a/kT} - 1)$$

$$J = q\left(\frac{D_n n_{p0}}{L_n} + \frac{D_p p_{n0}}{L_p}\right)(e^{qV_a/kT} - 1) = J_0(e^{qV_a/kT} - 1) \qquad \text{pn junction}$$

$$J_0 = q\left(\frac{D_n n_{p0}}{w_B} + \frac{D_p p_{n0}}{L_p}\right) \qquad \text{short p-side diode}$$

$$J_0 = q\left(\frac{D_n n_{p0}}{w_{B(p)}} + \frac{D_p p_{n0}}{w_{B(n)}}\right) \qquad \text{both sides short}$$

$$I = I_0(e^{qV_a/kT} - 1)$$

$$R - G = \frac{np - n_i^2}{\tau_0(n + n_i + p + n_i)}$$

$$C_j = A\left[\frac{q\varepsilon N'_D N'_A}{2(N'_D + N'_A)(V_{\text{bi}} - V_a)}\right]^{1/2} \qquad \text{pn junction}$$

$$C_j = A\left[\frac{q\varepsilon N'}{2(V_{\text{bi}} - V_a)}\right]^{1/2} \qquad \text{one-sided junction}$$

$$C_j = \frac{A\left(\dfrac{\varepsilon_n \varepsilon_p}{w_n w_p}\right)}{\left(\dfrac{\varepsilon_n}{w_n} + \dfrac{\varepsilon_p}{w_p}\right)} \qquad \text{heterojunction}$$

$$C_{\text{sc}} = \frac{dQ_{\text{sr}}}{dV_a} = \delta\frac{dQ_s}{dV_a} = \delta I \tau_n$$

$$t_T = \frac{(w_B)^2}{2D_n}$$

$$\mathscr{E}(x) = \frac{qa}{2\varepsilon}\left[x^2 - \left(\frac{w}{2}\right)^2\right] \qquad -\frac{w}{2} \le x \le \frac{w}{2} \qquad \text{linearly graded junction}$$

$$w = \left(\frac{12\varepsilon V_j}{qa}\right)^{1/3} \qquad \text{linearly graded junction}$$

$$V_{\text{bi}} = \frac{2kT}{q}\ln\left[\frac{a}{2n_i}\left(\frac{12\varepsilon V_{\text{bi}}}{qa}\right)^{1/3}\right] \qquad \text{linearly graded junction}$$

FIELD-EFFECT TRANSISTORS

$$I_D = WQ_{\text{ch}}(y)\mu(y)\mathscr{E}_L(y)$$

$$I_D = 0 \qquad V_{GS} < V_T$$

$$C'_{\text{ox}} = \frac{\varepsilon_{\text{ox}}}{t_{\text{ox}}}$$

$$I_D = \frac{WC'_{\text{ox}}\mu}{L}\left[(V_{GS} - V_T)V_{DS} - \frac{V_{DS}^2}{2}\right] \qquad V_D \le (V_G - V_T) \qquad \text{simple model}$$

$$V_{DS\text{sat}} = (V_{GS} - V_T) \qquad \text{simple model}$$

$$I_{D\text{sat}} = \frac{WC'_{\text{ox}}\mu}{L}\left[\left(V_{GS} - V_T - \frac{V_{DS\text{sat}}}{2}\right)V_{DS\text{sat}}\right] \qquad V_{GS} > V_T,\ V_{DS} > V_{DS\text{sat}}$$

simple model

$$I_{D\text{sat}} = \frac{WC'_{\text{ox}}\mu}{2L}(V_{GS} - V_T)^2 = \frac{WC'_{\text{ox}}\mu}{2L}V_{DS\text{sat}}^2 \qquad V_{GS} > V_T,\ V_{DS} > V_{DS\text{sat}}$$

$$|v| = \frac{\mu_{\text{lf}}|\mathscr{E}_L|}{1 + \dfrac{\mu_{\text{lf}}|\mathscr{E}_L|}{v_{\text{sat}}}} \qquad \mu = \frac{\mu_{\text{lf}}}{1 + \dfrac{\mu_{\text{lf}}|\mathscr{E}_L|}{v_{\text{sat}}}}$$

$$I_D = \frac{WC'_{\text{ox}}\mu_{\text{lf}}}{\left(L + \dfrac{\mu_{\text{lf}}V_{DS}}{v_{\text{sat}}}\right)}\left[(V_{GS} - V_T)V_{DS} - \frac{V_{DS}^2}{2}\right] \qquad V_{DS} \le V_{DS\text{sat}}$$

considering velocity saturation

$$I_{D\text{sat}} = \frac{WC'_{\text{ox}}\mu_{\text{lf}}}{\left(L + \frac{\mu_{\text{lf}}V_{DS\text{sat}}}{v_{\text{sat}}}\right)}\left[(V_{GS} - V_T)V_{DS\text{sat}} - \frac{V_{DS\text{sat}}^2}{2}\right] \qquad V_{DS} \geq V_{DS\text{sat}}$$

considering velocity saturation

$$V_{DS\text{sat}} = \frac{v_{\text{sat}}}{\mu_{\text{lf}}}L\left[\left(1 + \frac{2\mu_{\text{lf}}(V_{GS} - V_T)}{v_{\text{sat}}L}\right)^{1/2} - 1\right]$$

considering velocity saturation

$$I_{D\text{sat}} = C'_{\text{ox}}Wv_{\text{sat}}(V_{GS} - V_T - V_{DS\text{sat}}) \qquad \text{considering velocity saturation}$$

$$\mu_{\text{lf}} = \frac{\mu_0}{1 + \theta(V_G - V_T)}$$

$$w_T = \left[\frac{2\varepsilon_s}{qN'_A}\phi_s(0)\right]^{1/2}$$

$$V_T = \frac{\Phi_{MS}}{q} + 2\phi_f - \frac{Q_f}{C'_{\text{ox}}} - \frac{Q_i(2\phi_f)}{C'_{\text{ox}}} - \frac{Q_B(2\phi_f)}{C'_{\text{ox}}}$$

$$\Delta V_T = -\frac{Q_{ii}}{C'_{\text{ox}}} = +\frac{qN_{ii}}{C'_{\text{ox}}}$$

$$I_D = I_0 e^{q(V_{GS} - V_T)/nkT} \qquad \text{subthreshold}$$

$$S = \frac{2.3kTn}{q}$$

$$P = C_L V_{DD}^2 f$$

$$P = I_{\text{Leakage}}V_{DD}$$

$$t_d = \frac{1}{2}\left(\frac{C_L V_{DD}}{2I_{D\text{satn}}} + \frac{C_L V_{DD}}{2I_{D\text{satp}}}\right)$$

$$g_m \equiv \frac{i_d}{v_g} = \frac{\partial I_D}{\partial V_{GS}}$$

$$g_d \equiv \frac{i_d}{v_d} = \frac{\partial I_D}{\partial V_{DS}}$$

$$g_m = \frac{WC'_{\text{ox}}\mu_{\text{lf}}V_{DS}}{L\left(1 + \frac{\mu_{\text{lf}}V_{DS}}{Lv_{\text{sat}}}\right)}$$

$$g_{m\text{sat}} = W v_{\text{sat}} C'_{\text{ox}} \left\{ 1 - \left[1 + \frac{2\mu_{\text{lf}}(V_{GS} - V_T)}{v_{\text{sat}} L} \right]^{-1/2} \right\}$$

$$g_d = \frac{W C'_{\text{ox}} \mu_{\text{lf}}}{L} \left[\frac{(V_{GS} - V_T - V_{DS}) - \dfrac{\mu_{\text{lf}} V^2{}_{DS}}{2 L v_{\text{sat}}}}{\left(1 + \dfrac{\mu_{\text{lf}} V_{DS}}{L v_{\text{sat}}}\right)^2} \right] \qquad V_{DS} \leq V_{DS\text{sat}}$$

$$g_{d\text{sat}} = 0 \qquad V_D \geq V_{D\text{sat}}$$

$$f_T = \frac{v_{\text{sat}}}{2\pi L} \left\{ 1 - \left[1 + \left(\frac{2\mu_{\text{lf}}(V_{GS} - V_T)}{v_{\text{sat}} L} \right)^{-1/2} \right] \right\} \qquad \text{saturation}$$

$$I_D = \frac{W C'_{\text{ox}} \mu_{\text{lf}} \left(V_{GS} - V_T - \dfrac{V_{DS}}{2} \right) (V_{DS} - 2 I_D R_S)}{L \left[1 + \dfrac{\mu_{\text{lf}} (V_{DS} - 2 I_{DS} R_S)}{L v_{\text{sat}}} \right]} \qquad V_{DS} \leq V_{DS\text{sat}}$$

considering series resistance

$$MTTF \propto e^{E_a/kT}$$

$$\frac{I_{D\text{sat}}(77\text{ K})}{I_{D\text{sat}}(300\text{ K})} = \frac{\mu_{\text{lf}}(77\text{ K})}{\mu_{\text{lf}}(300\text{ K})} \frac{\left(1 + \left.\dfrac{\mu_{\text{lf}} V_{DS\text{sat}}}{L v_{\text{sat}}}\right|_{300\text{ K}}\right)}{\left(1 + \left.\dfrac{\mu_{\text{lf}} V_{DS\text{sat}}}{L v_{\text{sat}}}\right|_{77\text{ K}}\right)}$$

$$I_D = \frac{q W \mu_{\text{lf}} N'_D a}{L} \left\{ V_{DS} - \frac{2}{3}(V_{\text{bi}} - V_T) \left[\left(\frac{V_{DS} + V_{\text{bi}} - V_{GS}}{V_{\text{bi}} - V_T} \right)^{3/2} - \left(\frac{V_{\text{bi}} - V_G}{V_{\text{bi}} - V_T} \right)^{3/2} \right] \right\}$$

bulk channel FETs, below saturation, simple model

$$I_{D\text{sat}} = \frac{q W \mu_{\text{lf}} N'_D a}{L} \left\{ V_{GS} - V_T - \frac{2}{3}(V_{\text{bi}} - V_T) \left[1 - \left(\frac{V_{\text{bi}} - V_{GS}}{V_{\text{bi}} - V_T} \right)^{3/2} \right] \right\}$$

bulk channel FETs, simple model

BIPOLAR JUNCTION TRANSISTORS

$$V_{CE} = V_{CB} + V_{BE}$$

$$I_E = I_C + I_B$$

$$I_E = I_{nE} + I_{pE}$$

$$I_C = I_{nC} + I_{pC}$$

$$I_{nC} = I_{nE} - I_{\text{rec}}$$

$$I_B = I_{pE} + I_{\text{rec}} - I_{pC}$$

$$\alpha = \frac{I_{\text{out}}}{I_{\text{in}}} = \frac{I_C}{I_E}$$

$$\beta = \frac{I_{\text{out}}}{I_{\text{in}}} = \frac{I_C}{I_B}$$

$$\beta = \frac{\alpha}{1-\alpha}$$

$$\alpha = \frac{I_{nE}}{I_E}\frac{I_{nC}}{I_{nE}}\frac{I_C}{I_{nC}} = \gamma\alpha_T M$$

$$\gamma = \frac{I_{nE}}{I_E}$$

$$\alpha_T = \frac{I_{nC}}{I_{nE}} = 1 - \frac{I_{\text{rec}}}{I_{nE}}$$

$$M = \frac{I_C}{I_{nC}} = 1 + \frac{I_{pC}}{I_{nC}}$$

$$\gamma \approx \frac{1}{1 + \dfrac{p_E(0^-)}{n_B(0^+)}\dfrac{D_{pE}}{D_{nB}}\dfrac{W_B}{W_E}}$$ prototype (uniformly doped) npn

$$\gamma \approx \frac{1}{1 + \dfrac{n_E(0^-)}{p_B(0^+)}\dfrac{D_{nE}}{D_{pB}}\dfrac{W_B}{W_E}}$$ prototype (uniformly doped) pnp

$$\alpha_T \approx 1 - \frac{W_B^2}{2L_n^2}$$ prototype (uniformly doped) npn

$$\alpha_T \approx 1 - \frac{W_B^2}{2L_p^2}$$ prototype (uniformly doped) pnp

$$\gamma \approx \frac{1}{1 + \dfrac{N'_{AB}}{N_C}e^{\Delta E_g^*/kT} \times \dfrac{D_{pE}}{D_{nB}} \times \dfrac{W_B}{W_E}}$$

prototype npn with degenerately doped emitter

$$\gamma \approx \frac{1}{1 + \dfrac{N'_{DB}}{N_V}e^{\Delta E_g^*/kT} \times \dfrac{D_{nE}}{D_{pB}} \times \dfrac{W_B}{W_E}}$$

prototype pnp with degenerately doped emitter

$$\beta \approx \frac{N'_{DE}}{N'_{AB}} \frac{D_{nB}}{D_{pE}} \frac{W_E}{W_B} \qquad \text{nondegenerate emitter (npn)}$$

$$\beta \approx \frac{N_C}{N'_{DB}} \frac{D_{nB}}{D_{pE}} \frac{W_E}{W_B} e^{-\Delta E_g^*/kT} \qquad \text{degenerate emitter (npn)}$$

$$\Delta E_g^* = E_g^*(\text{base}) - E_g^*(\text{emitter})$$

$$\eta = \frac{W_B}{\lambda} \qquad \text{graded-doping parameter}$$

$$\eta = \ln \frac{N'_A(0^+)}{N'_A(W_B)} \qquad \text{exponential grading of base doping}$$

$$J_{nB} = \frac{q D_n n_B(0^+)}{W_B} \left(\frac{\eta}{1 - e^{-\eta}} \right) \qquad \text{graded base}$$

$$\frac{\beta(\eta)}{\beta(0)} \approx \left(\frac{\eta}{1 - e^{-\eta}} \right) \qquad \text{graded base}$$

$$\begin{aligned} I_F &= I_{F0}(e^{qV_{BE}/kT} - 1) \\ I_R &= I_{R0}(e^{qV_{BC}/kT} - 1) \\ I_E &= I_F - \alpha_R I_R \\ I_C &= \alpha_F I_F - I_R \end{aligned} \qquad \text{Ebers-Moll model}$$

$$I_E = I_{CT} + \frac{I_F}{\beta_F + 1}$$

$$I_C = I_{CT} + \frac{I_R}{\beta_R + 1}$$

$$I_B = \frac{I_F}{\beta_F + 1} + \frac{I_R}{\beta_R + 1}$$

$$I_{CT} = \frac{\beta_F I_F}{\beta_F + 1} - \frac{\beta_R I_R}{\beta_R + 1}$$

$$R_B = \frac{R_\square L}{4h} \qquad \text{base resistance}$$

$$J_C \leq 0.3|q N_{DC} v_{\text{sat}}| \qquad \text{design rule for base push-out (Kirk) effect}$$

Hybrid-pi models:

$$C_\mu = C_{jBC}$$

$$C_\pi = C_{jBE} + C_{\text{sc}BE}$$

$$r_\pi \approx \frac{kT}{qI_B} \approx \frac{\beta_{DC} kT}{qI_C}$$

$$r_\pi = \frac{\dfrac{kT}{q^2 A_E n_i^2} e^{-qV_{BE}/kT}}{\dfrac{D_{pE}}{W_E N_{DE}}} \qquad \text{npn}$$

$$r_\mu = \frac{1}{\left.\dfrac{\partial I_C}{\partial V_{CB}}\right|_{V_{BE}}}$$

$$r_0 \approx \frac{1}{\left.\dfrac{\partial I_C}{\partial V_{CE}}\right|_{V_{CB}}} = \frac{V_A}{I_C}$$

$$g_m = \left.\frac{\partial I_C}{\partial V_{BE}}\right|_{V_{CE}} = \frac{i_c}{v_{be}} = \frac{qI_C}{kT} = \frac{\beta_{DC}}{r_\pi}$$

$$Q_B = -\frac{qA_E n_B(0^+) W_B}{2}$$

$$dQ_{Br} = \delta dQ_B$$

$$C_{scBE} = \frac{\delta W_B^2 \beta_{DC}}{2D_n r_\pi}$$

$$f_{co} = \frac{1}{2\pi r_\pi (C_\pi + C_\mu)}$$

$$\beta(f) = \frac{g_m v_{be}}{i_b} = \frac{\beta_{DC}}{\sqrt{1 + \left(\dfrac{f}{f_{co}}\right)^2}}$$

$$t_{tB} = \frac{Q_B}{I_{nB}} = \frac{Q_B}{I_C} \qquad t_{tB} = \frac{W_B^2}{2D_n}$$

$$\left.\begin{aligned} I_{B1} &= (1 - \alpha_1) I_{E1} - I_{C01} \\ I_{C2} &= \alpha_2 I_{E2} + I_{C02} \end{aligned}\right\} \qquad \text{npnp two-transistor model}$$

$$I_A = \frac{I_{C01} + I_{C02}}{1 - (\alpha_1 + \alpha_2)} \qquad \text{npnp}$$

OPTOELECTRONIC DEVICES

$$F_L(x) = F_L(0)e^{-\alpha x}$$

$$R = \left(\frac{n_1 - n_2}{n_1 + n_2}\right)^2$$

$$V_{oc} = \frac{nkT}{q} \ln\left(1 + \left|\frac{I_L}{I_0}\right|\right)$$

$$\eta_Q = \frac{J_L/q}{F_{Li}}$$

$$R_{ph} = \frac{J_L/q}{h\nu F_{Li}} = \frac{q\eta_Q}{h\nu}$$

$$\mathrm{FF} = \frac{I_m V_m}{I_{sc} V_{oc}}$$

$$\eta = \mathrm{FF}\frac{I_{sc} V_{oc}}{P_{Li}}$$

$$R_{ph} = M\left(\frac{q\eta_Q}{h\nu}\right)$$

$$\frac{dN_2}{dt} = -A_{21}N_2 \qquad \text{spontaneous emission}$$

$$\tau_{\text{radiative,spont}} = \frac{1}{A_{21}}$$

$$\frac{dN_2}{dt} = B_{12}N_1\rho(\nu)$$

$$\frac{dN_2}{dt} = -B_{21}N_2\rho(\nu) \qquad \text{stimulated emission}$$

$$\lambda = \frac{2nd}{q}$$

G APPENDIX

List of Suggested Readings

1. R. S. Muller and T. I. Kamins, *Device Electronics for Integrated Circuits.* John Wiley & Sons, New York, 1986.
2. D. A. Neamen, *Semiconductor Physics and Devices: Basic Principles,* Irwin, Homewood, IL, 1992.
3. R. E. Pierret, *Advanced Semiconductor Fundamentals,* Addison-Wesley, Reading, MA, 1987.
4. S. Wang, *Fundamentals of Semiconductor Theory and Device Physics,* Prentice Hall, Englewood Cliffs, NJ, 1989.
5. R. Eisberg and R. Resnick, *Quantum Physics of Atoms, Molecules, Solids, Nuclei and Particles,* 2nd ed., John Wiley & Sons, New York, 1985.
6. C. M. Wolfe, G. E. Stillman, and N. Holonyak, Jr., *Physical Properties of Semiconductors,* Prentice Hall, Englewood Cliffs, NJ, 1989.
7. R. E. Hummel, *Electronic Properties of Materials,* 2nd ed., Springer-Verlag, Berlin, 1993.
8. C. T. Sah, *Fundamentals of Solid-State Electronics,* World Scientific, Singapore, 1991.
9. J. Singh, *Semiconductor Devices, an Introduction,* McGraw-Hill, New York, 1994.
10. S. M. Sze, *Semiconductor Devices, Physics and Technology,* John Wiley & Sons, New York, 1985.
11. E. S. Yang, *Microelectronic Devices,* McGraw-Hill, New York, 1988.
12. S. M. Sze, *Physics of Semiconductor Devices,* 2nd ed., John Wiley & Sons, New York, 1981.
13. R. A. Smith, *Wave Mechanics of Crystalline Solids,* Chapman and Hall, London, 1961.
14. W. Shockley, *Electrons and Holes in Semiconductors,* D. Van Nostrand, New York, 1950.

15. John P. McKelvey, *Solid State Physics for Engineers and Materials Science,* Krieger, Malabar, FL, 1993.

16. R. A. Smith, *Semiconductors,* 2nd ed., Cambridge University Press, Cambridge, U.K., 1978.

17. B. G. Streetman and S. Banerjee, *Solid State Electronic Devices,* 5th ed., Prentice Hall, Upper Saddle River, NJ, 2000.

18. R. F. Pierret, *Semiconductor Device Fundamentals,* Addison Wesley, New York, 1996.

19. Y. Taur and T. H. Ning, *Fundamentals of Modern VLSI Devices,* Cambridge University Press, New York, 1998.

20. J. S. Blakemore, *Semiconductor Statistics.* Dover Publications, New York, 1987.

21. J. E. Carroll, *Rate Equations in Semiconductor Electronics,* Cambridge University Press, New York, 1985.

22. D. K. Schroder, "The Concept of Generation and Recombination Lifetimes in Semiconductors," *IEEE Trans. Electron Devices,* ED-29, p. 1336, 1982.

23. S. Dimitrijev, *Understanding Semiconductor Devices,* Oxford University Press, New York, 2000.

24. C. Kittel, *Introduction to Solid State Physics,* 6th ed., John Wiley and Sons, New York, 1986.

25. J. P. McKelvey, *Solid State and Semiconductor Physics,* Harper and Row, New York, 1966.

26. J. Lindmayer and C. Y. Wrigley, *Fundamentals of Semiconductor Devices,* D. Van Nostrand, Princeton, NJ, 1965.

27. D. Kahng, "A Historical Perspective on the Development of MOS Transistors and Related Devices," *IEEE Trans. Electron Devices,* ED-23, pp. 655–657, 1976.

28. C. T. Sah "Characteristics of the Metal-Oxide Semiconductor Transistors," *IEEE Trans. Electron Devices,* ED-11, pp. 324–345, 1964.

29. D. K. Schroder, *Modular Series on Solid State Devices: Advanced MOS Devices,* Addison-Wesley, Reading, MA, 1987.

30. Y. Tsividis, *Operation and Modeling of the MOS Transistor,* 2nd ed., McGraw-Hill, New York, 1999.

31. H. Morkoc, "The HEMT—A Superfast Transistor," *IEEE Spectrum,* 21, pp. 28–35, 1984.

32. C. T. Oak, "Evolution of the MOS Transistor—From Conception to VLSI," *Proceedings of the IEEE,* 76, pp. 1280–1326, 1988.

33. Daniel Foty, *MOSFET modeling with Spice: Principles and Practice,* Prentice Hall, Upper Saddle River, NJ, 1997.

34. "Special Issue on Silicon Bipolar Transistor Technology: Past and Future Trends," *IEEE Trans. on Electron Devices,* ED-48, 2001.

35. F. Bassani and G. P. Parravicini, *Electronic States and Optical Transitions in Solids,* Pergamon Press, New York, 1975.

36. P. Bhattacharya, *Semiconductor Optoelectronic Devices,* Prentice Hall, Englewood Cliffs, NJ, 1994.

37. H. C. Casey, Jr., and M. B. Panish, *Heterostructure Lasers Part A: Fundamental Principles,* Academic Press, New York, 1978.

38. H. C. Casey, Jr., and M. B. Panish, *Heterostructure Lasers Part B: Materials and Operating Characteristics,* Academic Press. New York, 1978.

39. M. Fukuda, *Optical Semiconductor Devices,* John Wiley & Sons, New York, 1999.

40. H. Kressel and J. K. Butler, *Semiconductor Lasers and Heterojunction LEDs,* Academic Press, New York, 1977.

41. B. E. A. Saleh and M. C. Teich, *Fundamentals of Photonics,* John Wiley & Sons, New York, 1991.

42. R. G. Seippel, *Optoelectronics,* Reston Publishing, Reston, VA, 1981.

43. J. Singh, *Optoelectronics: An Introduction to Materials and Devices,* McGraw-Hill, New York, 1996.

44. J. Singh, *Semiconductor Optoelectronics,* McGraw-Hill, New York, 1995.

INDEX